CRC SERIES IN NUTRITION AND FOOD

Editor-in-Chief

Miloslav Rechcigl, Jr.

Handbook of Nutritive Value of Processed Food
Volume I: Food For Human Use
Volume II: Animal Feedstuffs

Handbook of Nutritional Requirements in a Functional Context
Volume I: Development and Conditions of Physiologic Stress
Volume II: Hematopoiesis, Metabolic Function, and Resistance to Physical Stress

Handbook of Agricultural Productivity
Volume I: Plant Productivity
Volume II: Animal Productivity

Handbook of Naturally Occurring Food Toxicants

Handbook of Foodborne Diseases of Biological Origin

Handbook of Nutritional Supplements
Volume I: Human Use
Volume II: Agricultural Use

HANDBOOK SERIES

Nutritional Requirements
Volume I: Comparative and Qualitative Requirements

Nutritional Disorders
Volume I: Effect of Nutrient Excesses and Toxicities in Animal and Man
Volume II: Effect of Nutrient Deficiencies in Animals
Volume III: Effect of Nutrient Deficiencies in Man

Diets, Culture Media, Food Supplements
Volume I: Diets for Mammals
Volume II: Food Habits of, and Diets for Invertebrates and Vertebrates — Zoo Diets
Volume III: Culture Media for Microorganisms and Plants
Volume IV: Culture Media for Cells, Organs and Embryos

CRC Handbook of Foodborne Diseases of Biological Origin

Editor
Miloslav Rechcigl, Jr., Ph.D.
Nutrition Advisor and Director
International Research Staff
Agency for International Development
U.S. Department of State

CRC Series in Nutrition and Food
Series Editor-in-Chief
Miloslav Rechcigl, Jr., Ph.D.

CRC Press, Inc.
Boca Raton, Florida

Library of Congress Cataloging in Publication Data
Main entry under title:

Handbook of foodborne diseases of biological origin.

(CRC series in nutrition and food)
Bibliography:
Includes index.
1. Food poisoning--Handbooks, manuals, etc.
I. Rechcigl, Miloslav. II. Series.
RA1258.H36 1983 615.9'54 82-14678
ISBN 0-8493-3964-2

This book represents information obtained from authentic and highly regarded sources. Reprinted material is quoted with permission, and sources are indicated. A wide variety of references are listed. Every reasonable effort has been made to give reliable data and information, but the author and the publisher cannot assume responsibility for the validity of all materials or for the consequences of their use.

Direct all inquiries to CRC Press, Inc., 2000 Corporate Blvd., N.W., Boca Raton, Florida, 33431.

International Standard Book Number 0-8493-3964-2

Library of Congress Card Number-82-14678
Printed in the United States

PREFACE
CRC SERIES IN NUTRITION AND FOOD

Nutrition means different things to different people, and no other field of endeavor crosses the boundaries of so many different disciplines and abounds with such diverse dimensions. The growth of the field of nutrition, particularly in the last 2 decades, has been phenomenal, the nutritional data being scattered literally in thousands and thousands of not always accessible periodicals and monographs, many of which, furthermore, are not normally identified with nutrition.

To remedy this situation, we have undertaken an ambitious and monumental task of assembling in one publication all the critical data relevant in the field of nutrition.

The *CRC Series in Nutrition and Food* is intended to serve as a ready reference source of current information on experimental and applied human, animal, microbial, and plant nutrition presented in concise tabular, graphical, or narrative form and indexed for ease of use. It is hoped that this projected open-ended multivolume compendium will become for the nutritionist what the *CRC Handbook of Chemistry and Physics* has become for the chemist and physicist.

Apart from supplying specific data, the comprehensive, interdisciplinary, and comparative nature of the *CRC Series in Nutrition and Food* will provide the user with an easy overview of the state of the art, pinpointing the gaps in nutritional knowledge and providing a basis for further research. In addition, the series will enable the researcher to analyze the data in various living systems for commonality or basic differences. On the other hand, an applied scientist or technician will be afforded the opportunity of evaluating a given problem and its solutions from the broadest possible point of view, including the aspects of agronomy, crop science, animal husbandry, aquaculture and fisheries, veterinary medicine, clinical medicine, pathology, parasitology, toxicology, pharmacology, therapeutics, dietetics, food science and technology, physiology, zoology, botany, biochemistry, developmental and cell biology, microbiology, sanitation, pest control, economics, marketing, sociology, anthropology, natural resources, ecology, environmental science, population, law politics, nutritional and food methodology, and others.

To make more facile use of the series, the publication has been organized into separate handbooks of one or more volumes each. In this manner the particular sections of the series can be continuously updated by publishing additional volumes of new data as they become available.

The Editor wishes to thank the numerous contributors many of whom have undertaken their assignment in pioneering spirit, and the Advisory Board members for their continuous counsel and cooperation. Last but not least, he wishes to express his sincere appreciation to the members of the CRC editorial and production staffs, particularly President Bernard J. Starkoff, Earl Starkoff, Sandy Pearlman, Amy G. Skallerup, and Cathy Walker for their encouragement and support.

We invite comments and criticism regarding format and selection of subject matter, as well as specific suggestions for new data which might be included in subsequent editions. We should also appreciate it if the readers would bring to the attention of the Editor any errors or omissions that might appear in the publication.

Miloslav Rechcigl, Jr.
Editor-in-Chief

PREFACE

Acute gastrointestinal disturbances of exogenous origin are usually a result of microbial or parasitic contamination of food or from an ingestion of metallic poisons. The present publication is concerned with the former.

Despite the great advances made in food sanitation, outbreaks of foodborne infections occur from time to time. In more developed countries such as the U.S. such outbreaks may result from malfunctioning of automated food processing equipment or from faulty packaging or storage methods. Increased concentration of food preparation facilities capable of producing large volume of products, coupled with the efficiency with which the products can be distributed over wide geographical areas, can aggravate the potential health hazard even further, with serious consequences.

This handbook provides an overview of different biological agents and important toxins that may cause diseases on ingestion with food or water. Included are agents of bacterial, viral, and fungeal origin as well as foodborne parasitic agents.

This publication should be of interest to all individuals concerned with the safety and quality of the food supply including food scientists, dieticians, nutritionists, and epidemiologists.

THE EDITOR

Miloslav Rechcigl, Jr. is a Nutrition Advisor and Chief of Research and Methodology Division in the Agency for International Development.

He has a B.S. in Biochemistry (1954), a Master of Nutritional Science degree (1955), and a Ph.D. in nutrition, biochemistry, and physiology (1958), all from Cornell University. He was formerly a Research Biochemist in the National Cancer Institute, National Institutes of Health and subsequently served as Special Assistant for Nutrition and Health in the Health Services and Mental Health Administration, U.S. Department of Health, Education and Welfare.

Dr. Rechcigl is a member of some 30 scientific and professional societies, including being a Fellow of the American Association for the Advancement of Science, Fellow of the Washington Academy of Sciences, Fellow of the American Institute of Chemists, and Fellow of the International College of Applied Nutrition. He holds membership in the Cosmos Club, the Honorary Society of Phi Kappa Pi, and the Society of Sigma Xi, and is recipient of numerous honors, including an honorary membership certificate from the International Social Science Honor Society Delta Tau Kappa. In 1969, he was a delegate to the White House Conference on Food, Nutrition, and Health and in 1975 a delegate to the ARPAC Conference on Research to Meet U.S. and World Food Needs. He served as President of the District of Columbia Institute of Chemists and Councillor of the American Institute of Chemists, and currently is a delegate to the Washington Academy of Sciences and a member of the Program Committee of the American Institute of Nutrition.

His bibliography extends over 100 publications including contributions to books, articles in periodicals, and monographs in the fields of nutrition, biochemistry, physiology, pathology, enzymology, molecular biology, agriculture, and international development. Most recently he authored and edited *Nutrition and the World Food Problem* (S. Karger, Basel, 1979), *World Food Problem: a Selective Bibliography of Reviews* (CRC Press, 1975), and *Man, Food and Nutrition: Strategies and Technological Measures for Alleviating the World Food Problem* (CRC Press, 1973) following his earlier pioneering treatise on *Enzyme Synthesis and Degradation in Mammalian Systems* (S. Karger, Basel, 1971), and that on *Microbodies and Related Particles, Morphology, Biochemistry and Physiology* (Academic Press, New York, 1969). Dr. Rechcigl also has initiated a new series on *Comparative Animal Nutrition* and was Associated Editor of *Nutrition Reports International.*

ADVISORY BOARD MEMBERS

ADVISORY BOARD MEMBERS

CONTRIBUTORS

Dhiman Barua, M.D.
Diarrheal Diseases Control Programme
World Health Organization
Geneva, Switzerland

Alex Ciegler, Ph.D.
Food and Feed Safety
Southern Regional Research Center, USDA
New Orleans, Louisiana

Dean O. Cliver, Ph.D.
Food Research Institute
University of Wisconsin
Madison, Wisconsin

Bibhuti R. DasGupta, Ph.D.
Food Research Institute
University of Wisconsin
Madison, Wisconsin

Rudolph DiGirolamo, Ph.D.
Marine Resources Consultant
Sacramento, California

Joost Harwig, Ph.D.
Food Directorate
Health Protection Branch
Health and Welfare Canada
Ottawa, Ontario, Canada

Betty C. Hobbs, Ph.D.
Christian Medical College
Brown Memorial Hospital
Ludhiana, Punjab, India

George J. Jackson, Ph.D.
Food and Cosmetics Microbiology Branch
Division of Microbiology
Bureau of Foods
U.S. Food and Drug Administration
Washington, D.C.

Abraham Z. Joffe, Ph.D.
Laboratory of Mycology and Mycotoxicology
Department of Botany

T. Kuiper-Goodman, Ph.D.
Food Directorate
Health Protection Branch
Health and Welfare Canada
Ottawa, Ontario, Canada

Edward P. Larkin, Ph.D.
Virology Branch
Division of Microbiology
Bureau of Foods
U.S. Food and Drug Administration
Cincinnati, Ohio

Maurice O. Moss, Ph.D., D.I.C.
Department of Microbiology
University of Surrey
Guildford, Surrey
England

Deryck S. P. Patterson, Ph.D., D.Sc., C.Chem., F.R.I.C.*
Department of Biochemistry
Central Veterinary Laboratory
Weybridge, Surrey
England

Mamoru Saito, Ph.D.
Tokyo University
Tokyo, Japan

Edward J. Schantz, Ph.D.
Department of Food Microbiology and Toxicology
University of Wisconsin
Madison, Wisconsin

P. M. Scott, Ph.D.
Food Directorate
Health Protection Branch
Health and Welfare Canada
Ottawa, Ontario, Canada

Nancy H. Shoptaugh, Ph.D.
Department of Food Microbiology and Toxicology
University of Wisconsin
Madison, Wisconsin

* Deceased November, 1981.

CONTRIBUTORS

Yoshio Ueno, Ph.D.
Department of Toxicology
and Microbial Chemistry
Faculty of Pharmaceutical Science
Tokyo University of Science
Tokyo, Japan

Ronald F. Vesonder, Ph.D.
Mycotoxin, Microbiology, and
Biochemistry Research
Northern Regional Research
Laboratory
Peoria, Illinois

Robert I. Wise, M.D., Ph.D., F.A.C.P.
Vetrans Administration Hospital
Togus, Maine

DEDICATION

To my inspiring teachers at Cornell University—Harold H. Williams, John K. Loosli, the late Richard H. Barnes, the late Clive M. McCay, and the late Leonard A. Maynard. And to my supportive and beloved family—Eva, Jack, and Karen.

TABLE OF CONTENTS

Foodborne Toxins

VIRUSES OF VERTEBRATES: THERMAL RESISTANCE

Edward P. Larkin

INTRODUCTION

The thermal resistance of virus in foods is enhanced by the presence of protective agents that reduce the lethal effect of heat on the viruses at temperatures below 60°C. In addition, when solid foods are processed heat transfer is by conduction, and the required internal temperatures must be attained to effectively inactivate viral contaminants.

In Table 1 some of the characteristics of animal viruses are depicted. The various families have been separated by their buoyant densities and are listed alphabetically. The families 11 through 15 have been separated and also listed alphabetically. This same system was used to depict the information published in the literature pertaining to the familial resistance of viruses in laboratory media and in foods.

Examination of the data shown in Tables 2 and 3 indicate that the vast majority of viruses within the various families are inactivated at temperatures of about 60°C or less. This literature review indicates that thermal resistance appears to be restricted to viruses within the last five families in Table 1. Greater resistance to temperature inactivation has been reported for the Parvoviridae, the Papoviridae, and the Picornaviridae. The Adenoviridae and the Reoviridae appear to be only slightly more resistant to thermal inactivation than the first ten families. The greatest number of viruses of public health importance are in the family of Picornaviridae. At the present time the exact position in the classification system of the hepatitis and gastroenteritis viruses is not known but probably they will be included in the Picornaviridae or the Parvoviridae.

The thermal inactivation data published in the literature have been concerned chiefly with inactivation of high titers of viruses in fluid preparations. There was great variation in the techniques and procedures used to evaluate the thermal resistance of viruses. In the majority of studies the viral suspensions were heated in test tubes that were held in constant temperature water baths. In some cases the viral suspensions were mixed in an attempt to equalize the temperature in the suspending media. In addition, a number of containers were used to process the virus suspensions such as capillary tubes, ampules, and varying types of flasks and bottles. It has been our experience that in evaluating the thermal resistance of viruses, even distribution of the heat occurs only when the viral suspension is processed in ampules or capillary tubes that are submerged and agitated in a constant temperature bath. When other types of nonsubmerged containers are utilized, there is uneven distribution of the heat and possible contamination of the heat processed fluids by viruses surviving on the walls and closures of the vessels. Such contamination may be interpreted as apparent survival of virus after the heat treatment process. In some cases these viruses were not detected by cell culture systems but were observed when animal inoculation was utilized. It is possible that free virus nucleic acid may have been the causative agent in some of the reported virus persistence papers, especially when the heat-treated preparation was injected into an animal.

All the reported thermal inactivation data have been incorporated into Figure 1. Normal virus concentrations expected to be encountered in food should be inactivated at the time-temperature points shown in the figure. Only limited virus inactivation data have been reported in solid foods and consideration of this deficit was given when the data were used to plot the solid food slope. This slope is close to that of the U.S. Public Health Service Pasteurization Standard for ice cream mix (68.3°C for 30 min or 79.4°C for 25 sec).

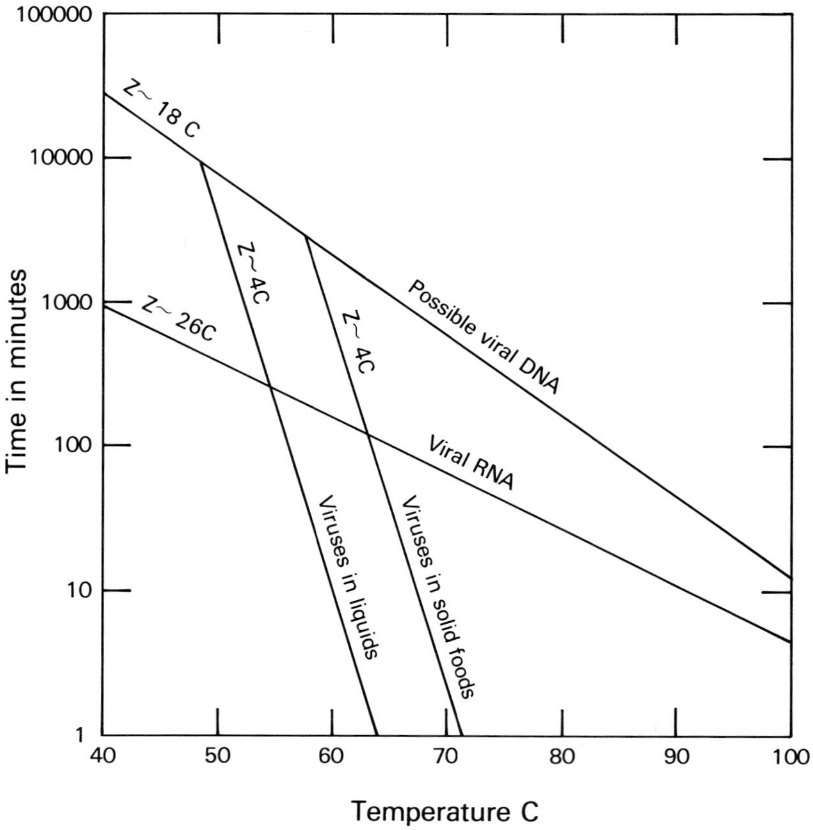

FIGURE 1. Time-temperature requirements for inactivation of natural contamination levels (< 3 logs/g or m*l*) of most viruses or viral nucleic acids in liquids or solid-type foods.

Only a limited number of viruses have been reported to have extremely high thermal resistance. These viruses are foot-and-mouth disease virus, hepatitis "A" (infectious hepatitis), hepatitis "B" (serum hepatitis), and the limited number of viruses in the papova and parvo virus families. In the reported foot-and-mouth disease studies a number of investigations were performed using high titers of viruses, and in the majority of investigations tubes or bottles were used to process the suspension in the water bath.[1] Apparent survival of low levels of viruses is possible with this procedure. Only limited or no viruses were reported to survive the heat processing of foods naturally contaminated by foot-and-mouth disease virus. Expected virus levels in naturally infected animals are shown in Table 5. Research in the Food and Drug Administration laboratories in Cincinnati indicated that virus levels of this magnitude were inactivated by temperatures less than those required for pasteurization of ice cream mix.

The reported thermal inactivation data on the hepatitis viruses are limited. In most cases viruses of unknown titer were heat treated in blood or blood components in an attempt to inactivate viruses, and infectivity was determined by human or primate feeding or inoculation studies. In one series of studies using hepatitis "A" in marmoset serum a temperature of 60°C for 1 hr was not sufficient to prevent infectivity in marmosets injected with the heated product. When the same virus was suspended in water and heat processed, a reduction in infectivity was reported for the same time-temperature process. Because of the limitation in the reported data and the scarcity of information, the thermal resistance of hepatitis viruses is presently unknown.

It is possible that viral persistence may occur in heat processes using long time-temperature procedures such as cooking solid foods where periods of hours are required to reach internal temperatures of 60°C. Some viruses that could be present in animal foods are shown in Table 7. If viral protective agents are present in food, the heat treatment process at levels below 60°C may have only a slight effect on the viruses, and the actual kill will commence only at temperatures above 60°C. This could result in persistence of viruses in the food unless such considerations are made in determining the total food process, and the lethality is calculated for the time-temperature process after reaching the 60°C level.

In Table 4 a number of substances are listed that have a protective effect on viruses when low temperatures (40 to 60°C) are used in thermal inactivation studies. In addition, this protective effect has been observed with other substrates such as serum, ice cream mix, and other high-protein, carbohydrate-containing suspensions. It has been shown that with higher temperatures, in excess of 60°C, the protective effect on the virus decreases significantly. When virus suspensions containing less than 10^4 particles per milliliter are processed, normal inactivation occurs in a short period of time at temperatures of 60 to 70°C and the virus suspensions appear to be inactivated at relatively the same rate.

Thermal inactivation of viruses is directly related to the chemical composition of the particle. The major organic constituents changed by temperature are lipids, proteins, and nucleic acids. The density of the particle is related to the varying compositions of the three organic constituents. Particles containing lipids or lipid complexes will tend to be more buoyant than particles containing carbohydrate and protein.

Low temperature (20 to 35°C) inactivation of viral inactivity is dependent upon disruption of the nucleic acids. At higher temperatures (>50°C), inactivation is due to denaturation or disruption of the proteins of the virus particles.

The nucleic acids present in viruses are of two types: ribonucleic acid and deoxyribonucleic acid. In a minority of the virus families the nucleic acid is double stranded. The nucleic acids are resistant to thermal inactivation at high temperatures for short periods of time. However, the nucleic acid molecule is disrupted gradually at low temperatures over relatively long time periods.

Some of the thermal characteristics of nucleic acids are shown in Table 6. In the case of the RNA-containing viruses, nucleic acid inactivation is probably due to breakage of the phosphate bonds, whereas, in the DNA-containing viruses, inactivation is due to cleavage of the purine or pyrimidine bases of the nucleic acid complex. At low temperatures, DNA is reported to be about 30-fold more resistant than the RNA. In Figure 1 inactivation of the RNA at 40°C occurred in a period of about 18 hr, whereas the limited data on DNA inactivation indicated times as long as 21 days. Complete inactivation is dependent on the viral titer; the higher the nucleic acid content, the longer the period of time necessary for inactivation. At temperatures of 30 to 45°C it is possible to inactivate nucleic acid with little apparent denaturation of the protein.

Virus particles whose outer surface contain envelope structures are susceptible to mechanical injury. Such particles appear to be very sensitive to thermal inactivation. In many cases the envelope is composed of lipid complexes. Whether loss of infectivity is due to mechanical injury of the envelope or to a change in protein or lipid structures is at present unknown.

Viruses with protein outer surfaces are thermally inactivated by denaturation of the protein that occurs at higher temperatures. In some cases the coat is ruptured and the nucleic acid is liberated into the medium. A certain percentage of particles entrap the nucleic acid in the core as the protein components on the surface are denatured. This particle has lost its ability to attach to the cell due to disruption of the structural integrity of the attachment site. However, these particles still containing nucleic acid may enter into the cell system by some mechanical process such as pinocytosis or engulfment and the nucleic acid may be liberated within the cell and produce infection.

It has been demonstrated that viral nucleic acid is liberated into the medium during a thermal inactivation process. If nucleases are present, the nucleic acids are rapidly inactivated. However, in the absence of nucleases, the liberated nucleic acid may be infectious. Laboratory studies have shown that demonstrated infectivity by ribonucleic acid is 2 to 4 logs less sensitive than that of the intact viroid. Thus, 10^5 virus particles may be inactivated but infectivity is still demonstrated in the cell or animal system because of the infectivity of the free nucleic acid.[2] Infectivity of nucleic acid via the oral route has not been demonstrated.[3]

The thermal process used to inactivate enzymes in foods, not less than 73°C or more than 77°C, probably will be greater than that required to inactivate viruses. In Figure 1, a virus inactivation process is depicted that should be sufficient to inactivate most viruses in naturally contaminated foods.

ACKNOWLEDGMENTS

This work was supported in part by the U.S. Army Natick Research and Development Command. I thank Roger Dickerson for his helpful advice and Ruth Dixon for assistance in the preparation of this manuscript.

Table 1
SOME CHARACTERISTICS OF ANIMAL VIRUSES

Family	Density (g/cm³) (cesium chloride)	Nucleic acid (type-stranded)	Lipid	Envelope
1. Arenaviridae	1.18[a]	RNA, SS	+	+
2. Bunyaviridae	1.20—1.23	RNA, SS	−	+
3. Coronaviridae	1.19—1.23	RNA, SS, + +	−	+
4. Herpetoviridae	1.27—1.29	DNA, DS,+ +	+	+
5. Orthomyxoviridae	1.17—1.20	RNA, SS	−	+
6. Paramyxoviridae	1.21—1.24	RNA, SS	+	+
7. Poxviridae		DNA, DS	+	−
8. Retroviridae	1.16—1.18[a]	RNA, SS	+	+
9. Rhabdoviridae	1.20	RNA, SS	−	+
10. Togaviridae	1.25	RNA, SS, + +	+	+
11. Adenoviridae	1.33—1.35	DNA, DS, + +	−	−
12. Papovaviridae	1.34	DNA, DS, + +	−	−
13. Parvoviridae	1.38—1.46	DNA, SS, + +	−	−
14. Picornaviridae	1.32—1.41	RNA, SS, + +	−	−
15. Reoviridae	1.31—1.38	RNA, DS	−	−

Note: +, present; −, absent; + +, nucleic acid is infectious; RNA, ribonucleic acid; DNA, deoxyribonucleic acid; SS, single stranded; DS, double stranded.

[a] In sucrose gradient.

Table 2
VIRUS THERMAL INACTIVATION

Family	Virus	Titer	Thermal Process: Temp (°C)	Thermal Process: Time (min)	Remarks	Method of inactivation[a]	Ref.
Arenaviridae	Lymphocytic choriomeningitis				Extremely thermolabile		Andrewes
	American hemorrhagic fever viruses				Readily inactivated		
Bunyaviridae					Thermal characteristics similar to Togaviridae		Andrewes
Coronaviridae	Mouse hepatitis		56	30	Inactivated		Andrewes
	Infectious bronchitis (birds)		56	30	Inactivated		
	Transmissible gastroenteritis (pigs)		56	30	Inactivated		
	Hemagglutinating encephalomyelitis		56	30	Inactivated		
Herpesviridae	Herpes simplex	$\sim 10^5$	50	20	∼5 log loss in infectivity in tris buffer		Plummer
	Cytomegalovirus	$\sim 10^4$	50	40	∼ 2.5 log loss in infectivity in tris buffer		
	Cytomegalovirus	$\sim 10^6$	44	19.9 Half-life	Tissue culture fluid and 2% calf serum	WB, T	Krugman
	Laryngotracheitis		37	>3 yr	Persisted, lyophilized allentoic fluid		Hofstad
	Herpes		50—52	30	Inactivated		Andrewes
	Cytomegalovirus		50	40	Inactivated		Dupré
			56	30	Inactivated		
	Herpes simplex		52	30	Inactivated		
Orthomyxoviridae	NM strains	10^5 to 10^8	62	30	Virus survived, mouse brain emulsions or infected fluids		Burnet
	Influenza A		61		Inactivated		Francis
	Influenza B		56	15—30	Inactivated, allentoic fluid		

Table 2 (continued)
VIRUS THERMAL INACTIVATION

Family	Virus	Titer	Thermal Process: Temp (°C)	Thermal Process: Time (min)	Remarks	Method of inactivation[a]	Ref.
	Influenza A2/Aichi/2/68		56	22	Four log loss, allentoic fluid		De Flora
	Influenza A PR-8		56	25	Inactivated, phosphate buffered saline		Siegert
	Influenza A	~ 10^4 to 10^5	45	10	Inactivated, allentoic fluids	WB-T	Lauffer
	Orthomyxoviruses		50	30	Usually inactivated		Andrewes
Paramyxoviridae	Respiratory cyncytial	10^3 to $10^{6.9}$	50	min	Pellet suspended in distilled water or sodium chloride solution (0.3 *M*) < 3 log drop in infectivity	WB-T	Rechsteiner
	Newcastle disease		37	~ 10 months	Inactivated, lyophilized allentoic fluid		Hofstad
	Newcastle disease	10^{12}	79—81	0.033 sec	Inactivated, allentoic fluid	HT-ST tubular apparatus	Dutcher
	Measles	>10^4	56	30	Inactivated, tissue culture fluid		Black
	Paramyxoviruses				Heat labile		Andrewes
	Measles		56	25	Inactivated,		Dupré
	Mumps		55	20	buffered		
	Mumps		56	20	saline		
Poxviridae	Fowl pox		37	>2 yr	Persisted, lyophilized allentoic fluid		Hofstad
	Variola		55	Half-life 2 min	In saline	WB, a	Nahon
	Vaccinia	10^8	60	10	>7 log loss in infectivity, last fraction persisted for more than 60 min, citrate buffer	WB, T	Kaplan
	Vaccinia	~10^6	56	30	~ 5 log loss in infectivity		Sharp

	Vaccinia	5×10^8	56		~6 log drop in titer		Galasso
			60		Inactivated, tissue culture medium		
	Goat poxvirus	$\sim 10^5$	50	60	Extracted from scabs,		Pandey
	Sheep poxvirus	$\sim 10^5$	50	60	~ 4 log drop in infectivity of both viruses		
Retroviridae	Avian erythroblastosis	High	56	6	99% drop in infectivity, plasma	WB, T	Bonar
		High	37	min	In plasma, rapid	WB, T	Eckert
			56	sec	inactivation, tailing		
	Rous sarcoma	High	60	6	Phosphate buffer saline + 2% horse serum — half-life at 60° is 0.7 min		
	Moloney	$\sim 10^3$	56		Inactivated, sodium citrate buffer	WB, T	Moloney
	Murine sarcoma (Moloney)	$\sim 10^4$	60	~6.5	Inactivated, tissue culture fluid	WB, T	Nakata
	AK	$\sim 10^3$	50—55	30	Inactivated		Simkovics
	Graffi	$\sim 10^3$	56—65		Inactivated		
	Friend	$\sim 10^{3-4}$	56	30	Inactivated		
	Moloney leukemia	$\sim 10^{3-4}$	56	30	Inactivated		
	Mouse lymphocytic leukemia	$\sim 10^3$	56	30	Inactivated, cell free filtrate		Buffet
	Rauscher leukemia	$\sim 10^4$	56	30	Inactivated, sodium citrate buffer	WB, T	Zeigel
	Bovine leukemia		60	30 sec	Inactivated, in tissue culture fluid	WB, T	Baumgartener
Rhabdoviridae	Rabies		60	5	Inactivated		Schultz
	Rhabdoviruses				Relatively heat sensitive		Andrewes
Togaviridae	EEE	$10^{7.2}$	37	80 hr	Tissue	WB, T	Nir
	WEE	$10^{5.6}$	37	48 hr			
	Sindbis	$10^{5.6}$	37	36 hr	culture		
	SLE	$10^{6.0}$	37	48 hr			
	West Nile	$10^{5.6}$	37	72 hr	fluid,		
	Ntaya	$10^{4.0}$	37	30 hr	inactivated		
	Semliki Forest		50	20	>2 log loss in infectivity	WB, T	Fleming

Table 2 (continued)
VIRUS THERMAL INACTIVATION

Family	Virus	Titer	Thermal Process: Temp (°C)	Thermal Process: Time (min)	Remarks	Method of inactivation[a]	Ref.
	Modoc	$10^{6.0}$	37	72 hr	∼2.5 log loss, phosphate buffer saline + 0.5% calf serum		Davis
	Japanese B encephalitis	$10^{6.0}$	37	90 hr	>6 log loss in infectivity, tissue culture fluid	WB, a	Darwish
	Tickborne encephalitis	10^7 to 10^9	50	12	Tissue culture fluid, inactivated	WB, T	Mayer
	WEE	$10^{7.6}$	56	30	Extract from mouse brain, ∼ 4 log drop		Fastier
	EEE	∼ 10^7	50	>4 hr	Survived, ∼ log drop	WB, a	Mika
	VEE	∼ 10^7	50	>7 hr	PBS, survived, ∼ 5 log drop		
	EEE	10^8	56	2 hr	Inactivated		Mahdy
Adenoviridae	Adenovirus		56	30	Inactivated		Andrewes
	Adenovirus	>10^2	56	30	Inactivated		Huebner
	Infectious bronchitis		37	6 months	Inactivated, lyophilized allentoic fluid		Hofstad
	Ovine adenovirus		60	30	Inactivated		Andrewes
	Adenoviruses		56	2.5—5	Inactivated		Dupré
Papovaviridae	S. E. polyoma		70	30	Tissue culture fluid survived,	WB, a	Eddy
			80	30	inactivated		
	Mouse polyoma		60	30	Tissue culture fluids no effect on infectivity	WB-vials screw cap	Brodsky
			65	30	4 log drop in infectivity		
			70	30	>6 log drop in infectivity		
	Rabbit papilloma		70	30	Inactivated		Andrewes
			67	30	Survived		
	Oral papillomatosis		65	30	Survived		
	Human wart		50	30	Survived		Andrewes
	Canine oral papillomatosis		58	60	Inactivated		Andrewes

	Mouse polyoma		70	30	Usually inactivated	WB, T	Andrewes
	K		70	4.5 hr	Inactivated		Andrewes
			70	3 hr	Survived		
Parvoviridae	Feline panleucopenia of cats		50		Fairly stable		Andrewes
			80—85	30	Inactivated		Andrewes
	Porcine parvo		70	2 hr	Survived		Andrewes
	Adeno-associated		60	30			Andrewes
				Half-life			
Picornaviridae	SV-2 enterovirus		45	120 sec	Tissue culture fluid, rapidly inactivated	Screw cap vial, WB	Heberling
	Foot-and-mouth disease (FMDV)	10^7	55	60	∼ 6 log inactivation, resistant fraction, tissue culture fluids	WB, T	Bachrach (1960)
	FMDV	∼ 10^7	61	3 sec	∼ 5 log reduction tissue culture fluids, diluted in veronal-acetate buffer	WB, a	Bachrach (1957)
	Coxsackie A-21		50	30	>4.2 logs Loss		Dimmock
	Poliovirus 1		50	30	>6.0 logs		
	ECHO-4		50	30	>3.0 logs		
	ECHO-11		50	60	>3.7 logs in		
	Rhinoviruses HGP		50	60	>2.8 logs		
	B.632		50	30	>2.4 logs		
	FEB		50	60	>2.5 logs ineffectivity		
	No.		50	60	3.6 logs Loss		
	DC		50	60	>3.3 logs		
	1098		50	60	2.7 logs in		
	C.V. 11		50	60	2.7 logs		
	S.D. 1		50	60	>3.9 logs infectivity		
	Duck hepatitis	10^2 to 10^6	56	30	2 to 5 log loss in infectivity	WB, T	Hwang
	Poliovirus		60	30	Inactivated		Andrewes
	ECHO		65	30	Inactivated		
	Coxsackie A		55	30	Inactivated		Dupré
			53—55	30	Inactivated		
	Murine polio		60	30	Water, inactivated	WB sealed tubes	Lawson
	SK	10^3	60	30	Water, inactivated		
	Modified Lansing	1300/mℓ	55	30	Water, inactivated		
	TO_4	400/mℓ	65	30	Water, inactivated		
	Coxsackievirus		60	30	Water, inactivated		

Table 2 (continued)
VIRUS THERMAL INACTIVATION

Family	Virus	Titer	Thermal Process: Temp (°C)	Thermal Process: Time (min)	Remarks	Method of inactivation[a]	Ref.
	Coxsackievirus		55	30	Brain suspension, inactivated		Dalldorf
	Poliovirus 1		50		Most sensitive to heat		Papaevangelou
	2		50		Least sensitive to heat		
	3		50		Intermediately sensitive to heat		
	Poliovirus		44—55	30	Inactivated		Schultz
	Poliovirus		55	5	Inactivated, extracts of cord and brain		Shaughnessy
			52.5	30	Inactivated		
		Low level	42.5	30	Inactivated		
	Poliovirus 2	$\sim 10^5$	56	30	Inactivated	WB, capillary tubes	Stanley
	Poliovirus	$10^{6.5}$—$10^{7.3}$	75	60	Few survivors	WB, a	Medearis
	Poliovirus 1-2	$>10^8$	50	24 hr	$\sim$ 8 log drop	WB, vials, stoppers	Younger
	Poliovirus 3	$>10^6$	50	24 hr	$\sim$ 6 log drop		
Reoviridae	Reovirus 3		56	Half-life of 1.6 min	Tissue culture fluid + 5% calf serum	WB, T	Gomatos
	Reovirus		55	15	Inactivated		Dupre
	African horse sickness		60	60	Persisted		Andrewes
Other	Hepatitis B		98	1	1/10 dilution of serum, inactivated	WB, T	Krugman
	Hepatitis B		60	10 hr	Inactivated, serum		Soulier
	Hepatitis A		100	5	Inactivated, water	WB, large, a	Provost
			60	60	Titer reduced, water		
			60	60	In serum, no apparent reduction in titer		
	Hepatitis A		60	60	Inactivated		Dupré
	Hepatitis A		60	19	Human feces, oysters, titer reduced	WB, a	Peterson

[a] WB, water bath; T, test tubes; a, ampules.

Table 3
THERMAL INACTIVATION OF VIRUSES IN FOODS

Family	Virus	Titer	Thermal Process: Temp. (°C)	Thermal Process: Time (min)	Remarks	Method of inactivation[a]	Ref.
Herpesviridae	Herpes simplex	$\sim 10^4$	65	2 sec	Inactivated in ice cream mix	WB, a	Sullivan
	Porcine herpes		68		Inactivated in sausage and ham		Kounev
Paramyxoviridae	Newcastle disease (NCDV)	10^9	64.4	160 sec	Suspended in whole egg $\sim$ 8 log drop in infectivity in 160 sec. Survivors after 200 sec	WB	Gough
	NCDV	Allentoic fluid diluted 1/100	60	5—7	Dependent on virus titer, inactivated	WB, a	Foster
Retroviridae	Rauscher leukemia	$\sim 10^3$	50	5	Inactivated	WB, a	Sullivan
			55	5 sec	ice		
	Moloney sarcoma	$\sim 10^3$	55		cream		
			55	15 sec	mix		
	Rous sarcoma	$\sim 10^5$	55	25	Inactivated,		
			65	3 sec	in ice cream mix		
	Bovine leukemia		73.5	42 sec	Inactivated, in milk	Stainless steel tank	Baumgartener
Togaviridae	Swine fever	$10^{4.7}$—$10^{6.0}$	65	30	Inactivated, in ham	Canned	Terpestra
	Hog cholera		65	90	Inactivated, in ham		Stewart
	Rift Valley fever		56	40	Inactivated, in blood		Dupré
	Tickborne encephalitis	$10^{5.7}$	70	20	Inactivated, in milk	WB, a	Gresikova
			55	20	5 log loss, in serum		
			62	20	Inactivated, in serum		
		$10^{3.7}$	72	10 sec	Inactivated, in milk	Laboratory pasteurizer	Gresikova
Adenoviridae	Adenovirus 14	$\sim 10^4$	65	2 sec	Inactivated, in ice cream mix	WB, a	Sullivan
Picornaviridae	Swine vesicular disease	$\sim 10^{6.5}$	56	30	Inactivated, in milk		Herniman
			60	2	Inactivated, in milk		

Table 3 (continued)
THERMAL INACTIVATION OF VIRUSES IN FOODS

Family	Virus	Titer	Thermal Process: Temp. (°C)	Thermal Process: Time (min)	Remarks	Method of inactivation[a]	Ref.
			60	10	Inactivated, in fecal slurry		
			64	2	Inactivated, in fecal slurry		
	Murine poliomyelitis		70	30	Inactivated, in milk	WB, a	Lawson
	Poliovirus 1		75—80	4	Inactivated, in meat balls fried in oil		Aizen
	Coxsackievirus B-3		80—85	4			
	Poliovirus	2.5×10^7	60	30 sec	>6 log inactivation, in milk	Cheese process	Cliver
	Poliovirus 1	$\sim 10^4$			∼2 Log	Stewed oysters	Di Girolamo
				10	∼2.5 drop	Fried oysters	
				20	∼2.5 in	Baked oysters, oven	
				30	∼2 virus	Steamed oysters	
	Poliovirus 1	10^5—10^8	80	5	Survived, 47% fat most protective, ground beef	WB, T, screw cap	Filppi
	Coxsackievirus A-9	7.5×10^5	49	6 hr	In Thuringer sausage ∼3 log drop in virus		Herrman
	Poliovirus 1	1×10^8	60	30	In sausage, internal temperature, ∼1 log drop		Kantor
	Echovirus 6	6×10^7	60	30			
	Poliovirus	$10^{2.5} \times 10^{3.5}$	61.7	30	Inactivated, in milk, cream and ice cream; 1 sample in stainless steel tubes indicated survivors	WB, a, C and stainless steel tubes	Kaplan (1952)
	Poliovirus Lansing	$10^{2.3} \times 10^{4.0}$	79.5	15 sec	Survived in ice cream	WB, C	Kaplan (1954)
	MEF-1 Y-S-K		82.2	25 sec	(10% ground brain)		
	Coxsackievirus	$10^{7.5}$	61.7	15	Inactivated,		Kaplan (1954)
			71.6	30 sec			
			72.1	30	in		
			67.7	30			
			71.1	15 sec	milk		

Coxsackievirus A-9		85	20 sec	Inactivated, in milk		Kostenko
		75	15 sec	Survivors, in milk		
Poliovirus I	3×10^5	71		Inactivated, in beef patties	Broiled	Sullivan
Coxsackievirus B-2	$1\text{-}3 \times 10^4$	71				
FMDV	$10^{6.7}$—$10^{7.5}$	65	64.5	Survived,	WB, tubes	Hyde
		72	15—17 sec	in		
		80	15—17 sec	milk		
		65		Survived, evaporation to 50% of liquid volume		
FMDV	Natural contamination	90		Inactivated		
		90	35 sec	Inactivated, dried virus		Felkai
		80	70 sec	In milk		
FMDV	$1 \times 10^{7.8}$	80	6 hr	Inactivated, tissue suspension		Demopoullos
		80	4 hr	Not inactivated, tissue suspension		
FMDV	$\sim 1 \times 10^6$	72	0.25	Survived, pelleted cell debris	WB, T	Blackwell
		72	5	Survived, whole milk		
		72	3	Survived, evaporated milk		
		93	0.25	Survived, cream		
FMDV				Cheddar cheese and Camenbert cheese process, survived inactivated in Mozzarella cheese process		Blackwell
		63	6 sec, 28	Survived in milk		
FMDV		55	20	Inactivated, in blood		Dupré
		60	5	Inactivated, in vesicular fluid		
		85	360	Cattle tongue suspension		
Porcine enterovirus F-7		65	2	Inactivated		Kelly
Echovirus 6	9.3×10^3	56.7	3.5	Inactivated,		Strock
and Poliovirus 1	9.3×10^6	57.8	2			
		58.9	1.1	in egg white		
Echovirus 6	2.4×10^4	56.7	20	Inactivated,		
Poliovirus 1	2.4×10^6	57.8	11	in		
		58.9	7			
		60.0	3	egg yolk		

Table 3 (continued)
THERMAL INACTIVATION OF VIRUSES IN FOODS

Family	Virus	Titer	Thermal Process: Temp. (°C)	Thermal Process: Time (min)	Remarks	Method of inactivation[a]	Ref.
	Foot-and-mouth disease	~ $10^{4.8}$			Boiling water bath 65 min, cooled in 45 min, peak internal temperature 74°C, inactivated	In cans	Heidelbaugh
	FMDV	$10^{4.5}$—$10^{7.1}$	56	6	~ 5 logs inactivated,	In bottles	Sellers
			63	1	pH		
			72	17 sec			
			80—85	>5 sec	6.7		
			56	30	~ 5 logs inactivated,		
			63	2			
			72	55 sec			
			80—85	> 5 sec	pH 7.6		
	FMDV		65	30 sec	Inactivated in milk, skim milk, cream and powdered milk	WB, T	Moosbregger
	FMDV		65	30 sec	Inactivated, in milk		
Reoviridae	Reovirus 1	~ 10^{4}	65	12 sec	Inactivated, ice cream mix	WB, a	Sullivan

[a] WB, water bath; T, test tubes; a, ampules; C, capillary tubes.

Table 4
AGENTS MODIFYING VIRUS THERMAL INACTIVATION

Virus	Remarks	Ref.
Poliovirus	Mg^{++} stabilized virus at 45°C	Ackerman
Poliovirus 1	Cations stabilized virus at 45°C	Fujioka
Polioviruses 1, 2, 3	$MgCl_2$ stabilized virus during storage at 4°C	Melnick
Polioviruses 1 + 2 Echo 6 and 7 Coxsackie B-5	Reduced sulfhydryl groups stabilized viruses to inactivation at less than 50°C	Halsted
Poliovirus 1	L-cystine stabilized five different stocks of virus to temperatures below 50°C	Pohjanpelto
Poliovirus 1	Elemental sulfides (tetrasulfide) stabilized virus against inactivation at 50°C	Pons
Poliovirus 1, 2, 9, 3	50 μg/mℓ of L-cystine stabilized viruses	Pohjanpelto
Coxsackie A-9, B-3 and Echo-1	500—2500 μg/mℓ stabilized the viruses	
Echo-3, 6 and 19	No stabilization	
Echo-32	$MgCl_2$ stabilized virus at 50°C	Branche
Poliovirus 2	0.1 *M* NaCl at 56°C for 1 hr ∿ 6 log drop in infectivity 2.0 *M* NaCl at 56°C for 1 hr ∿ 2 log drop in infectivity	Speir
Rhinoviruses and Enteroviruses	$MgCl_2$ less effectively stabilized Rhinoviruses at 50°C	Dimmock
Simian virus SV-2	$MgCl_2$ stabilized virus at 50°C	Heberling
Herpesvirus (JES)	Na_2SO_4 and Na_2HPO_4 stabilized at 50°C $MgCl_2$, $MgSO_4$, KH_2PO_4, 2 M KCl, or NaCl did not stabilize virus; very thermosensitive in isotonic salt solution	Wallis
Rhabdoviruses	EDTA and serum demonstrated protective effect on virus at 37 and 56°C	Michalski
Newcastle disease	Casein protected virus at 45°C, $MgSO_4$ did not protect virus	Ballesteros
Tickborne encephalitis	Monovalent metallic cations stabilized virus at 50°C	Mayer
DNA viruses	Not stabilized by divalent cations	Wallis
Reovirus	Stabilized by divalent cations	Wallis
Vesicular exanthema of swine	No cationic stabilization at 50°C, a characteristic of human enteroviruses not animal viruses	Zee

Table 5
NATURALLY OCCURRING VIRUS CONCENTRATIONS IN FOOD ANIMAL PRODUCTS

Virus	Titer	Remarks	Ref.
Foot-and-mouth Disease virus (FMDV)	10^{4}	In serum, bovine	Burrows
	$10^{4.5-5.3}$	Pharynx	
	$10^{3.3-5.2}$	Milk	
	10^{3}	Saliva	
FMDV	$10^{4.5}$	Bone marrow, bovine	Cox
FMDV	$10^{2.0}$	Muscle, brain	Cottral
	$10^{4.0-5.0}$	Blood	
FMDV	$10^{0.7}$—$10^{2.1}$	Beef, bovine	Henderson
	$10^{1.5}$—$10^{2.2}$	Liver	
	$10^{1.7}$	Kidney	
	$10^{2.4}$	Blood	
	$10^{0.7}$—$10^{2.8}$	Rumen	
	$10^{1.8}$—$10^{2.3}$	Lymph node	
Tickborne encephalitis	$\sim 10^{3.7}$	Goat milk	Gresikova
Enteroviruses	$<10^{1}$	Shellfish	Denis
Swine fever virus	$10^{4.7}$—$10^{6.0}$	Blood	Terpstra

Table 6
THE EFFECT OF TEMPERATURE ON NUCLEIC ACIDS

Virus	Remarks	Ref.
Foot-and-mouth disease (FMDV)	Virus heated to 61° and 85°C in RNase free medium produced infectious nucleic equivalent to that produced by phenol extraction	Bachrach
FMDV	Prolonged incubation at 25°C for 24 hr or 37°C for 8 hr resulted in a loss of ~ 3 logs of infectious nucleic acid	Brown
Tobacco mosaic virus (TMV)	Loss of RNA biological activity occurs at moderate heat in marked contrast to DNA which is inactivated with a high-temperature coefficient	Ginoza
Poliovirus 1 and TMV	~ 90% loss of RNA infectivity in 80—120 min at 65°C	Gordon
	~ 99% loss of RNA infectivity in 40—45 min at 80°C	
	~ 99% loss of RNA infectivity in 3—5 min at 100°C	
Poliovirus 1	Virus rapidly degraded at 56°C liberating viral RNA	Jordan
Polioviruses	Protein denaturation precedes dissociation of viral RNA at 50°C	Meitens
EEE and VEE	RNA inactivated ~ 90% in 4 hr at 50°C	Mika
Poliovirus	Heating RNA for 5 min at 60°C results in loss of 1 log of infectivity; virus heated same procedure results in 7 log loss of infectivity	
Poliovirus mutants	Variation in sensitivity to heat	Papaevangelou
Bacteriophage	Heat (50—60°C) inactivation is accompanied by release of native DNA. The molecules are intact; 90% released from virus in 60 min at 60°C	Ritchie

Table 7
VIRUSES PRODUCING SYSTEMIC INFECTIONS IN FOOD ANIMALS (NORTH AMERICA)

Virus group	Type of animal: Bovine	Porcine	Avian
RNA viruses			
Picornaviruses	Enteroviruses (+?), Encephalomyocarditis virus +	Enteroviruses (+?)	Enteroviruses (+?)
Myxoviruses	Parainfluenza 3 +	Swine influenza (type A) +	Influenza virus (type A) +, Newcastle disease virus +
Reoviruses	Types 1, 2, and 3 +	Probably types 1, 2, and 3 +	Probably types 1, 2 and 3 +
Togaviruses	Encephalitis viruses +		Eastern encephalitis + Western encephalitis +
Retroviruses	Bovine leukemia virus		Rous sarcoma +, Avian Leukosis complex
Other RNA viruses	Bovine diarrhea virus, BVD mucosal disease, vesicular stomatitis virus +, rabies virus +, bovine syncytial virus (?)	Hog cholera virus, vesicular stomatitis virus +, lymphocytic choriomeningitis +	Infectious bronchitis virus
DNA viruses			
Herpes viruses	Pseudorabies virus (+?) malignant catarrhal fever virus	Pseudorabies virus (+?)	Marek's disease virus, Herpes turkey virus
Adenoviruses	3 serotypes (?)	At least 3 serotypes (?)	At least 10 serotypes(?)
Other DNA viruses		Hemagglutinating encephalomyelitis	Avian encephalomyelitis, viral arthritis agent

Note: (−?), Systemic infection and infectious for man possible; +, infectious for man; ?, systemic infection probable.

REFERENCES

1. **Tierney, J. T. and Larkin, E. P.**, Potential sources of error during virus thermal inactivation, *J. Appl. Environ. Microbiol.*, 36, 432—437, 1978.
2. **Larkin, E. P. and Fassolitis, A.**, Viral heat resistance and infectious RNA, *J. Appl. Environ. Microbiol.*, 38, 650—655, 1979.
3. **Israel, M. A., Chan, H. W., Hourihan, S. L., Rowe, W. P., and Martin, M. A.**, Biological activity of polyoma viral DNA in mice and hamsters, *J. Virol.*, 29, 990—996, 1979.
4. **Larkin, E. P.**, Thermal Inactivation of Viruses, U.S. Army Natick Research and Development Command, Natick, Mass., Technical Report, Natick TR-78-002, 1—44, 1978.

BIBLIOGRAPHY

Some Characteristics of Animal Viruses

Andrewes, S. C. and Pereira, H. G., *Viruses of Vertebrates*, 3rd Ed., Williams and Wilkins, Baltimore, 1972.

Matthews, R. E. F., Classification and nomenclature of viruses. Third report of the International Committee on Taxonomy of Viruses, *Intervirology* 12(3-5), 1—296, 1979.

Nucleic Acids

Alexander, H. E., Koch, G., Mountain, I. M., and Van Damme, O., Infectivity of ribonucleic acid from poliovirus in human cell monolayers, *J. Exp. Med.,* 103, 493—506, 1958.

Colter, J. S., Bird, H. H., Moyer, A. W., and Brown, R. A., Infectivity of ribonucleic acid isolated from virus-infected tissues, *Virology,* 4, 522—532, 1957.

Dimmock, N. J., Biophysical studies of a rhinovirus: extraction and assay of infectious ribonucleic acid, *Nature (London),* 209, 792—794, 1966.

Gierer, A. and Schramm, G., Infectivity of ribonucleic acid from tobacco mosaic virus, *Nature (London),* 177, 702—703, 1956.

Herriott, R. M., Infectious nucleic acids, a new dimension in virology, *Science,* 134, 256—260, 1961.

Holland, J. J., McLaren, L. C., and Syverton, J. T., Mammalian cell-virus relationship. III. Poliovirus production by non-primate cells exposed to poliovirus ribonucleic acid, *Proc. Soc. Exp. Biol. Med.,* 100, 843—847, 1959.

Holland, J. J., McLaren, L. C., and Syverton, J. T., The mammalian cell virus relationship. IV. Infection of naturally insusceptible cells with enterovirus ribonucleic acid, *Exp. Med.,* 110, 65—80, 1959.

Iglewski, W. J. and Franklin, R. M., Purification and properties of reovirus ribonucleic acid, *J. Virol.,* 1(2), 302—307, 1967.

Love, D. N., The effect of DEAE-Dextran on the infectivity of a feline calicivirus and its RNA, *Arch. Gesamte Virusforsch.,* 41, 52—58, 1973.

Weiss, R., Persistent infections without proviral DNA, *Nature (London),* 265, 295—296, 1977.

Yukhananova, S. A., Nikolayeva, O. V., Fadeyeva, L. L., and Parfanovich, M. I., Isolation of infectious RNA from arboviruses of the California and C groups, *Acta Virol.,* 18, 88, 1974.

Viral Thermal Inactivation

Andrewes, S. C. and Pereira, H. G., *Viruses of Vertebrates,* 3rd ed., Williams and Wilkins, Baltimore, 1972.

Andrewes, C. H., The viruses of the common cold, *Sci. Am.,* 203(6), 88—102, 1960.

Bachrach, H. L., Breese, S. S., Jr., Callis, J. J., Hess, W. R., and Patty, R. E., Inactivation of foot-and-mouth disease virus by pH and temperature changes and by formaldehyde, *Proc. Soc. Exp. Biol. Med.,* 95, 147—152, 1957.

Bachrach, H. L., Patty, R. E., and Pledger, R. A., Thermal resistant populations of foot-and-mouth disease virus, *Proc. Soc. Exp. Biol. Med.,* 103, 540—542, 1960.

Baumgartener, L., Olson, C., and Onuma, M., Effect of pasteurization and heat treatment on bovine leukemia virus, *J. Am. Vet. Med. Assoc.,* 169(11), 1189—1191, 1976.

Black, F. L., Growth and stability of measles virus, *Virology,* 7, 184—192, 1959.

Bonar, R. A., Beard, D., Beaudreau, G. S., Sharp, D. G., and Beard, J. W., Virus of avian erythroblastosis. IV. pH and thermal stability, *J. Natl. Cancer Inst.,* 18(6), 831—842, 1957.

Brodsky, I., Rowe, W. P., Hartley, J. W., and Lane, W. T., Studies of mouse polyoma virus infection. II. Virus stability, *J. Exp. Med.,* 109(5), 439—447, 1959.

Buffett, R. F., Grace, J. T., Jr., and Mirand, E. A., Properties of a lymphocytic leukemia agent isolated from Ha/ICR Swiss mice, *Proc. Soc. Exp. Biol. Med.,* 116, 293—297, 1964.

Burnet, F. M. and Lind, P. E., A genetic approach to variation in influenza viruses, *J. Gen. Microbiol.,* 5, 59—66, 1951.

Dalldorf, G., The coxsackie viruses, *Ann. Rev. Microbiol.,* 9, 277—296, 1955.

Darwish, M. A. and Hammon, W. McD., Studies on Japanese B encephalitis virus vaccines from tissue culture. VI. Development of a hamster kidney tissue culture inactivated vaccine for man. 2. The characteristics of inactivation of an attenuated strain of OCT-541, *J. Immunol.,* 96(5), 806—813, 1966.

Davis, J. W. and Hardy, J. I., In vitro studies with Modoc virus in Vero cells: plaque assay and kinetics of growth, neutralization, and thermal inactivation, *Appl. Microbiol.,* 26(3), 344—348, 1973.

De Flora, S. and Badolati, G., Thermal inactivation of untreated and gamma-irradiated A2/Aichi/2/68 influenza virus, *J. Gen. Virol.,* 20, 261—265, 1973.

Dimmock, N. J., Differences between the thermal inactivation of picornaviruses at "high" and "low" temperatures, *Virology,* 31, 338—353, 1966.

Dimopoullos, G. T., Effects of physical environment on the virus of foot-and-mouth disease, *Ann. N.Y. Acad. Sci.,* 83, 706—726, 1960.

Dougherty, R. M., Heat inactivation of Rous sarcoma virus, *Virology,* 14, 371—372, 1961.

Dupré, M. V. and Frobisher, M., Thermal inactivation: animal viruses, in environmental biology, Altman, P. L. and Dittmer, D. S., Eds., FASEB, Bethesda, Md., 1966.

Dutcher, R. M., Read, R. B., Jr., and Litsky, W., The immunological antigenicity of rapid heat inactivated viruses. I. Newcastle disease virus, *Avian Dis.,* 4, 205—217, 1960.

Eckert, E. A., Green, I., Sharp, D. G., Beard, D., and Beard, J. W., Virus of erythromyeloblastic leukosis. VII. Thermal stability of virus infectivity; of the virus particle; and the enzyme dephosphorylating adenosine-triphosphate, *J. Natl. Cancer Inst.*, 16, 153—161, 1955.

Eddy, B. E., Stewart, S. E., and Grubbs, G. E., Influence of tissue culture passage, storage, temperature and drying on viability of SE polyoma virus, *Proc. Soc. Exp. Biol. Med.*, 99, 289—292, 1958.

Fastier, L. B., Toxic manifestations in rabbits and mice associated with the virus of western equine encephalomyelitis, *J. Immunol.*, 68, 531—541, 1952.

Fleming, P., Thermal inactivation of Semliki Forest virus, *J. Gen. Virol.*, 13(3), 385—391, 1971.

Francis, T. J., Jr., Respiratory viruses, *Ann. Rev. Microbiol.*, 1, 351—384, 1947.

Galasso, G. J. and Sharp, D. G., Effects of heat on the infecting, antibody-absorbing and interfering powers of vaccinia virus, *J. Bacteriol.*, 89(3), 611—616, 1965.

Gomatos, P. J., Tamm, I., Dales, S., and Franklin, R. M., Reovirus type 3: physical characteristics and interaction with L cells, *Virology*, 17, 441—454, 1962.

Hahon, N. and Kozikowski, E., Thermal inactivation studies with variola virus, *J. Bacteriol.*, 81(4), 609—613, 1961.

Heberling, R. L. and Cheever, F. S., Simian enterovirus SV2 hemaagglutination studies. I. Relationship of infectivity to hemagglutination, *Proc. Soc. Exp. Biol. Med.*, 151—154, 1965.

Hofstad, M. S. and Yoder, H. W., Jr., Inactivation rates of some lyophilized poultry viruses at 37 and 3 degrees C, *Avian Dis.*, 7(2), 170—177, 1963.

Huebner, R. S., Rowe, W. P., and Chanock, R. M., Newly recognized respiratory tract viruses, *Ann. Rev. Microbiol.*, 12, 49—75, 1958.

Hwang, J., Thermostability of duck hepatitis virus, *Am. J. Vet. Res.*, 36(11), 1683—1684, 1975.

Kaplan, C., The heat inactivation of vaccinia virus, *J. Gen. Microbiol.*, 18, 58—63, 1958.

Krugman, R. D. and Goodheart, C. R., Human cytomegalovirus. Thermal inactivation, *Virology*, 23, 290—291, 1964.

Krugman, S., Giles, J. P., and Hammond, J., Viral hepatitis type B (MS-2 strain). Studies on active immunization, *JAMA*, 217, 41—45, 1971.

Lauffer, M. A., Carnelly, H. L., and MacDonald, E., Thermal destruction of influenza A virus infectivity, *Arch. Biochem.*, 16, 321—328, 1948.

Lawson, R. B. and Melnick, J. L., Inactivation of murine poliomyelitis viruses by heat, *J. Infect. Dis.*, 80, 201—205, 1947.

Mahdy, M. S. and Ho, M., Potentiation effect of fractions of eastern equine encephalomyelitis virus on interferon production, *Proc. Soc. Exp. Biol. Med.*, 116, 174—177, 1964.

Mayer, V., Study of the virulence of tick-borne encephalitis virus. IV. Thermosensitivity of virions and its relationship to other genetic markers, *Acta Virol.*, 9, 397—408, 1965.

Medearis, D. N., Jr., Arnold, J. H., and Enders, J. F., Survival of poliovirus at elevated temperatures (60-75 C), *Proc. Soc. Exp. Biol. Med.*, 104, 419—423, 1960.

Melnick, J. L., Studies on the coxsackie viruses: properties, immunological aspects and distribution in nature, *Bull. N.Y. Acad. Med.*, 26, 342—356, 1950.

Mika, L. A., Officer, J. E., and Brown, A., Inactivation of two arboviruses and their associated infectious nucleic acids, *J. Infect. Dis.*, 113, 195—203, 1963.

Moloney, J. B., The murine leukemias, *Fed. Proc.*, 21, 19—31, 1962.

Nakata, Y., Nakata, K., and Sakamoto, Y., Heat sensitivity of murine sarcoma virus, *Gann*, 66(2), 193—195, 1975.

Nir, Y. D. and Goldwasser, R., Biological characteristics of some of the arbor viruses *in vivo* and *in vitro*. I. Thermal inactivation in cell-culture medium, *Am. J. Hyg.*, 73, 294—296, 1961.

Pandey, R. and Singh, I. P., Heat, chloroform, and ether sensitivity of sheep pox and goat pox viruses, *Acta Virol. (Prague)*, 14(4), 318—319, 1970.

Papaevangelou, G. J. and Youngner, J. S., Correlation between heat-resistance of polioviruses and other genetic markers, *Proc. Soc. Exp. Biol. Med.*, 108, 505—507, 1961.

Peterson, D. A., Wolfe, L. G., Larkin, E. P., and Deinhardt, F. W., Thermal treatment and infectivity of hepatitis A virus in human feces, *J. Med. Virol.*, 2, 201—206, 1978.

Plummer, G. and Lewis, B., Thermoinactivation of *Herpes simplex* virus and cytomegalovirus, *J. Bacteriol.*, 89(3), 671—674, 1965.

Provost, P. J., Wolanski, B. S., Miller, W. J., Ittensohn, O. L., McAleer, W. J., and Hilleman, M. R., Physical, chemical and morphologic dimensions of human hepatitis A virus strain CR326 (28578), *Proc. Soc. Exp. Biol. Med.*, 148, 532—539, 1975.

Rechsteiner, J., Thermal inactivation of respiratory syncytial virus in water and hypertonic solutions, *J. Gen. Virol.*, 5, 397—403, 1969.

Schultz, E. W., The neutrotropic viruses, *Ann. Rev. Microbiol.*, 2, 335—377, 1948.

Sharp, D. G., Sadhukhan, P., and Galasso, G. J., The slow decline in quality of vaccinia virus at low temperatures (37°C to −62°C), *Proc. Soc. Exp. Biol. Med.*, 115, 811—814, 1964.

Shaughnessy, H. J., Harman, P. H., and Gordon, F. B., The heat resistance of the virus of poliomyelitis, *J. Prev. Med.*, 4, 149—155, 1930.
Siegert, R., and Braune, P., The pyrogens of myxoviruses, *Virology*, 24, 218—223, 1964.
Sinkovics, J. G., Viral leukemias in mice, *Ann. Rev. Microbiol.*, 16, 75—100, 1962.
Soulier, J. P., Blatix, C., Courouce, A. M., Benamon, D., Amouch, P., and Drouet, J., Prevention of virus B hepatitis (SH hepatitis), *Am. J. Dis. Child.*, 123, 429—434, 1972.
Stanley, N. F., Lorman, D. C., Ponsford, J., and Larkin, M., Variants of poliovirus hominis. IV. Inactivation of heat, ultraviolet light and formaldehyde, *Aust. J. Exp. Med. Sci.*, 34, 297—300, 1956.
Youngner, J. S., Thermal inactivation studies with different strains of poliovirus, *J. Immunol.*, 78, 282—290, 1957.
Zeigel, R. F. and Rauscher, F. J., Electron microscope and bioassay studies on a murine leukemia virus (Rauscher). I. Effects of physiochemical treatments on the morphology and biological activity of the virus, *J. Natl. Cancer Inst.*, 32, 1277—1307, 1964.

Thermal Inactivation of Viruses in Foods

Aizen, M. S. and Pille, E. R., An experimental study of the survival of certain enteroviruses in meat products (in Russian; brief English summary), *Gig. Sanit.*, 3, 28—30, 1972.
Baumgartener, L., Olson, C., and Onuma, M., Effect of pasteurization and heat treatment on bovine leukemia virus, *J. Am. Vet. Med. Assoc.*, 169(11), 1189—1191, 1976.
Blackwell, J. H., Survival of foot-and-mouth disease virus in cheese, *J. Dairy Sci.*, 59(9), 1574—1579, 1976.
Blackwell, J. H. and Hyde, J. L., Effect of heat on foot-and-mouth disease virus (FMDV) in the components of milk from FMDV-infected cows, *J. Hyg. Camb.*, 77:77—83, 1976.
Cliver, D. O., Cheddar cheese as a vehicle for viruses, *J. Dairy Sci.*, 56, 1329—1331, 1973.
DiGirolamo, R., Liston, J., and Matches, J. R., Survival of virus in chilled, frozen and processed oysters, *Appl. Microbiol.*, 20, 58—63, 1970.
Demopoullos, G. T., Fellowes, O. N., Collis, J. J., Poppensiek, G. C., Edwards, A. G., and Graves, J. H., Thermal inactivation and antigenicity studies of heated tissue suspensions containing foot-and-mouth disease virus, *Am. J. Vet. Res.*, 20, 510—521, 1959.
Dupré, M. V. and Frobisher, M., Thermal inactivation: animal viruses, in Environmental Biology, Altman, P. L. and Dittmer, D. S., Eds., FASEB, Bethesda, Md., 1966.
Felkai, V. T., Solyom, F., Szent-Ivanyi, M., and Wagner, A., The heat tolerance of foot-and-mouth disease virus in milk, *Magyar Allatorv. Lapja*, 25, 378—384, 1970.
Filppi, J. A. and Banwart, G. J., Effect of the fat content of ground beef on the heat inactivation of poliovirus, *J. Food Sci.*, 39, 865—866, 1974.
Foster, B. S. and Thompson, C. H., Report of the 60th Annual Meeting, U.S. Livestock Sanitary Association, Chicago, Illinois, *Vet. Med.*, 52, 118—121, 1957.
Gough, R. E., Thermostability of Newcastle disease virus in liquid whole egg, *Vet. Rec.*, 93(24), 632—633, 1973.
Grěsiková-Kohútová, M., The effect of heat on infectivity of the tick-borne encephalitis virus, *Acta Virol.*, 3, 215—221, 1959.
Heidelbaugh, N. D. and Graves, J. H., Effects of some techniques applicable in food processing on the infectivity of foot-and-mouth disease virus, *Food Technol.*, 22(2), 120—124, 1968.
Herniman, K. A. J., Medhurst, P. M., Wilson, J. N., and Sellers, R. F., The action of heat, chemicals and disinfectants on swine vesicular disease virus, *Vet. Rec.*, 93, 620—624, 1973.
Hermann, J. E. and Cliver, D. O., Enterovirus persistence in sausage and ground beef, *J. Milk Food Technol.*, 36(8), 426—428, 1973.
Hyde, J. L., Blackwell, J. H., and Callis, J. J., Effect of pasteurization and evaporation on foot-and-mouth disease virus in whole milk from infected cows, *Can. J. Comp. Med.*, 39, 305—309, 1975.
Kantor, M. A. and Potter, N. N., Persistence of echovirus and poliovirus in fermented sausages. Effects of sodium nitrite and processing variables, *J. Food Sci.*, 40, 968—972, 1975.
Kästli, P. and Moosbrugger, G. A., Destruction of foot-and-mouth disease virus by heat in milk products (in French; brief summary in English), *Schweizer Arch. Tierheilkd.*, 110, 89—94, 1968.
Kaplan, A. S. and Melnick, J. L., Effect of milk and other daily products on the thermal inactivation of coxsackievirus, *Am. J. Pub. Health*, 44, 1174—1184, 1954.
Kaplan, A. S. and Melnick, J. L., Differences in thermostability of antigenically related strains of poliomyelitis virus, *Proc. Soc. Exp. Biol. Med.*, 86, 381—384, 1954.
Kaplan, A. S. and Melnick, J. L., Effect of milk and cream on the thermal inactivation of human poliomyelitis virus, *Am. J. Pub. Health*, 42, 525—534, 1952.
Kelly, D. F., Studies on enteroviruses of the pig. VII. Some properties of a porcine enterovirus (F_7), *Res. Vet. Sci.*, 5, 56—69, 1964.

Kostenko, An. N. and Botsman, N. E., Transmission of enteroviruses by milk and milk products (in Russian), *Materialakh U ShC'' yezda gigiyenistov Ukrainskoy SSR v 1971 g. v g. Kieve*, 165—168, 1971.
Kounev, Z., Studies on the Stability of Aujesky's Disease Virus in Meat Processing, Ph.D. thesis, Central Veterinary Research Institute, Sofia 1606, Bulgaria, 1978.
Lawson, R. B. and Melnick, J. L., Inactivation of murine poliomyelitis viruses by heat, *J. Infect. Dis.*, 80, 201—205, 1947.
Moosbrugger, P. K., La destruction du virus aphteux par la chaleur dans les produits laitiers, *Extr. Schweizer Arch. Tierheilleundex Fasc.*, 2(110), 89—94, 1968.
Sellers, R. F., Inactivation of foot-and-mouth disease virus in milk, *Br. Vet. J.*, 125, 163—167, 1969.
Stewart, W. C., Downing, D. R., Carbsey, E. A., Knesse, J. I., and Synder, M. L., Thermal inactivation of hog cholera virus in ham, *Am. J. Vet. Res.*, 40, 739—741, 1979.
Strock, N. R. and Potter, N. N., Survival of poliovirus and echovirus during simulated commercial egg pasteurization treatments, *J. Milk Food Technol.*, 35, 247—251, 1972.
Sullivan, R., Marnell, R. M., Larkin, E. P., and Read, R. B., Jr., Inactivation of poliovirus 1 and coxsackievirus B-2 in broiled hamburgers, *J. Milk Food Technol.*, 38(8), 473—475, 1975.
Sullivan, R., Tierney, J. T., Larkin, E. P., and Read, R. B., Jr., Thermal resistance of certain oncogenic viruses suspended in milk and milk products, *Appl. Microbiol.*, 22(3), 315—320, 1971.
Terpstra, C. and Krol, B., Effect of heating on the survival of swine fever virus in pasteurised canned ham from experimentally infected animals, *Tijdschr. Diergeneesk.*, deel 101, afl. 22, 1237—1241, 1976.

Agents Modifying Virus Thermal Inactivation

Ackerman, W. W., Fujioka, R. S., and Kurtz, H. B., Cationic modulation of the inactivation of poliovirus by heat, *Arch. Environ. Health*, 21(3), 377—381, 1970.
Ballesteros, S. and Lucio, B., Effect of magnesium sulfate and casein hydrolysate on thermoinactivation of Newcastle disease virus vaccine, *Avian Dis.*, 16(4), 724—728, 1972.
Branche, W. C., Jr., Young, V. M., Houston, F. M., and Koontz, L. W., Characterization of prototype virus Echo-32, *Proc. Soc. Exp. Biol. Med.*, 118(1), 186—190, 1965.
Dimmock, N. J. and Tyrrell, D. A. J., Some physiochemical properties of rhinoviruses, *Br. J. Exp. Pathol.*, 45, 271—280, 1964.
Fujioka, R., Kurtz, H., and Ackermann, W. W., Effects of cations and organic compounds on inactivation of poliovirus with urea, guanidine, and heat, *Proc. Soc. Exp. Biol. Med.*, 132(3), 825—829, 1969.
Halsted, C. C., Seto, D. S. Y., Simkins, J., and Carver, D. H., Protection of enteroviruses against heat inactivation by sulfhydryl-reducing substances, *Virology*, 40(3), 751—754, 1970.
Heberling, R. L. and Cheever, F. S., Simian enterovirus SV-2 hemagglutination studies. I. Relationship of infectivity to hemagglutination, *Proc. Soc. Exp. Biol. Med.*, 151—154, 1965.
Mayer, V. and Slavik, I., Thermostabilization of the tick-borne encephalitis virus, *Virology*, 29, 492—493, 1966.
Melnick, J. L. and Wallis, C., Effect of pH on thermal stabilization oral poliovirus vaccine by magnesium chloride, *Proc. Soc. Exp. Biol. Med.*, 112, 894—897, 1963.
Michalsik, F., Parks, N. F., Sokol, F., and Clark, H. F., Thermal inactivation of rabies and other rhabdoviruses: stabilization of the chelating agent Ethylenediaminetetraacetic acid at physiological temperatures, *Infect. Immun.*, 14(1), 135—143, 1976.
Pohjanpelto, P., Response of enteroviruses to cystine, *Virology*, 15, 225—230, 1961.
Pohjanpelto, P., Stabilization of poliovirus by cystine, *Virology*, 6, 472—487, 1958.
Pons, M., Stabilization of poliovirus by sodium tetrasulfide, *Virology*, 22(2), 253—261, 1964.
Speir, R. W., Thermal stability of mengo and poliovirus in hypertonic salt, *Virology*, 14, 382—383, 1961.
Wallis, C. and Melnick, J. L., Thermostabilization and thermosensitization of herpesvirus, *J. Bacteriol.*, 90, 1632—1637, 1965.
Wallis, C. and Melnick, J. L., Reovirus activation by heating and inactivation by cooling in $MgCl_2$ solutions, *Virology*, 22, 608—619, 1964.
Wallis, C., Yang, C. S., and Melnick, J. L., Effect of cations on thermal inactivation of vaccinia, herpes simplex, and adenoviruses, *J. Immunol.*, 89, 41—46, 1962.
Zee, Y. C. and Hackett, A. J., The influence of cations on the thermal inactivation of vesicular exanthema of swine virus, *Arch. Gesamte Virusforsch*, 20(4), 473—475, 1967.

Naturally Occurring Virus Concentrations in Food Animal Products

Burrows, R., Mann, J. A., Greig, A., Chapman, W. G., and Goodridge, D., The growth and persistence of foot-and-mouth disease virus in the bovine mammary gland, *J. Hyg. Camb.*, 69, 307—321, 1971.
Cottral, G. E., Con, B. F., and Baldwin, D. E., The survival of foot-and-mouth disease virus in cured and uncured meat, *Am. J. Vet. Res.*, 21, 288—297, 1960.

Cox, B. F., Cottral, G. E., and Baldwin, D. E., Further studies on the survival of foot-and-mouth disease virus in meat, *Am. J. Vet. Res.*, 22, 224—226, 1961.
Denis, F. and Brisou, J. F., Contamination virale des fruits de mer: etude portant sur Panalyse de 15,000 coquillages, *Bull. Acad. Natl. Med. (Tome)*, 160(1), 18—22, 1976.
Gresikova, M., Havranek, I., and Gorner, F., The effect of pasteurization on the infectivity of tick-borne encephalitis virus, *Acta Virol. (Prague)*, 5, 31—36, 1961.
Henderson, W. M. and Brooksby, J. B., The survival of foot-and-mouth disease virus in meat and offal, *J. Hyg.*, 46, 394—402, 1948.
Terpstra, C. and Krol, B., Effect of heating on the survival of swine fever virus in pasteurised canned ham from experimentally infected animals, *Tijdschr. Diergeneesk.*, deel 101, afl. 22, 1237—1241, 1976.

The Effect of Temperature on Nucleic Acids

Bachrach, H. L., Thermal degradation of foot-and-mouth disease virus into infectious ribonucleic acid, *Proc. Soc. Exp. Biol. Med.*, 107, 610—613, 1961.
Brown, F., Cartwright, B., and Stewart, D. L., The effect of various inactivating agents on the viral and ribonucleic acid infectivities of foot-and-mouth disease and on its attachment to susceptible cells, *J. Gen. Microbiol.*, 31, 179—186, 1963.
Ginoza, W., Kinetics of heat inactivation of ribonucleic acid of tobacco mosaic virus, *Nature (London)*, 181, 958—961, 1958.
Gordon, M. P., Huff, J. W., and Holland, J. J., Heat inactivation of the infectious ribonucleic acids of polio and tobacco mosaic viruses, *Virology*, 19(3), 416—418, 1963.
Jordan, L. and Mayor, H. D., Studies on the degradation of poliovirus by heat, *Microbios*, 9, 51—60, 1974.
Larkin, E. P. and Fassolitis, A., Viral heat resistance and infectious RNA, *Appl. Environ. Microbiol.*, 38, 650—655, 1974.
Mietens, C. and Koschel, K., RNA content and antigenicity of poliovirus before and after inactivation by heating, *Z. Naturforsch.*, 26b, 945—950, 1971.
Mika, L. A., Officer, J. E., and Brown, A., Inactivation of two arboviruses and their associated infectious nucleic acids, *J. Infect. Dis.*, 113, 195—203, 1963.
Norman, A. and Veomett, R. C., Heat inactivation of poliovirus ribonucleic acid, *Virology*, 12(1), 136—139, 1960.
Papaevangelou, G. J. and Youngner, J. S., Thermal stability of ribonucleic acid from poliovirus mutants, *Virology*, 15, 509—511, 1961.
Ritchie, D. A., Physical characterization of the DNA released from phage particles by heat inactivation, *Fed. Eur. Biochem. Soc. Lett.*, 11(1), 257—260, 1970.

Viral Thermal Inactivation Theory

Apostolov, K. and Damjanovic, V., Selective inactivation of the infectivity of freeze-dried Sendai virus by heat, *Cryobiology*, 10, 255 and 259, 1973.
Banks, B. E. C., Damjanovic, V., and Vernon, C. A., So-called thermodynamic compensation law and thermal death, *Nature (London)*, 240(5377), 147—148, 1972.
DiGioia, G. A., Licciardello, J. J., Nickerson, J. T. R., and Goldblith, S. A., Thermal inactivation of Newcastle disease virus, *Appl. Microbiol.*, 19(3), 451—454, 1970.
Dimmock, N. J., Differences between the thermal inactivation of picornaviruses at "high" and "low" temperatures, *Virology*, 31, 338—353, 1966.
Eylar, O. R. and Wisseman, C. L., Jr., Thermal inactivation of type 1 Dengue virus strains, *Acta Virol.*, 19, 167—168, 1975.
Hiatt, C. W., Kinetics of the inactivation of viruses, *Bacteriol. Rev.*, 28(2), 150—163, 1964.
Milo, G. E., Jr., Thermal inactivation of poliovirus in the presence of selective organic molecules (cholesterol, lecithin, collagen, and B-carotene), *Appl. Microbiol.*, 21(2), 198—202, 1971.
Pollard, E. C., The Physics of Viruses. Thermal Inactivation of Viruses, Academic Press, New York, 1953, 103—104.
Rosenberg, B., Kemeny, G., Switzer, R. C., and Hamilton, T. C., Quantitative evidence for protein denaturation as the cause of thermal death, *Nature (London)*, 232, 471—473, 1971.
Wallis, C. and Melnick, J. L., Thermosensitivity of poliovirus, *J. Bacteriol.*, 86(3), 499—504, 1963.
Woese, C., Thermal inactivation of animal viruses, *Ann. N.Y. Acad. Sci.*, 83, 741—751, 1960.

MICROBIAL FOOD TOXICANTS: *CLOSTRIDIUM BOTULINUM* TOXINS

Bibhuti R. DasGupta

INTRODUCTION

Clostridium botulinum has been a subject of awe and alarm for about 100 years. The reason for awe, alarm, and vigilance on food sanitation is a protein-botulinum toxin — produced by the Gram-positive, spore forming, anaerobic bacteria; as far as is known, the organism without this particular protein is as benign as a bacterium can be.[1-8] This article presents a critical evaluation of what is known about botulinum toxin. Certain aspects of the toxin are discussed in a wide perspective.

The pharmacological action of this toxin is on the cholinergic neuromuscular junction and peripheral autonomic synapses. The toxin, according to the contemporary view,[9-12] reacts with a specific receptor on the presynaptic membrane and the consequent chemical events alter the membrane structure in such a way that efflux of acetylcholine is blocked. Interruption of electrochemical conduction leads to flaccid muscle paralysis. The toxin is regarded therefore as a neurotoxin. An alternative to this chemical view of poisoning is the pipe and valve hypothesis.[13,14] The escape routes of acetylcholine from the interior of the nerve cell are pipes in the membranes. The valve of a pipe, opening in response to a nerve impulse, permits one way flow of the transmitter substance. When the valve is open the neurotoxin, acting as a molecular plug, enters the pipe and blocks it.

Botulism is a type of food poisoning that is caused by ingestion of the toxin. In improperly preserved food[5,7,8,15,16] germination of spore(s) occurs with subsequent multiplication of the bacteria and neurotoxin production. A few cases of botulism resulting from wounds have been documented.[7,17,18] Paralytic toxico-infection in humans of ages 3 to 35 weeks, termed infant botulism,[19] is believed to be due to the ingestion of spores and not the preformed toxin (see below for further discussion).

CHARACTERISTICS OF THE NEUROTOXINS

The eight immunologically distinguishable, but pharmacologically similar, neurotoxins are called types A, B, C_1, C_2, D, E, F, and G.[7,9,20] In general a pure culture produces one immunological type of neurotoxin, e.g., type A neurotoxin is produced by type A cultures. Exceptions to this general rule are (1) strain 84 produces types A and F;[21,22] (2) certain strains of types C and D produce mixtures of C_1, C_2, and D;[23,24] and (3) one nontoxigenic type C strain can be converted to produce type D neurotoxin by infection with a phage obtained from a type D toxigenic strain.[25] Conversely, three nontoxigenic type D strains can be converted to produce type C neurotoxin by phage recovered from type C toxigenic strains.[26-28] See Reference 9 for further discussion on phage and toxigenicity.

The eight neurotoxin types are neutralized by the corresponding homologous antiserum but a low level of cross neutralization between types E and F is known.[29,30] Type B from the proteolytic strain Okra and nonproteolytic strain QC appear to be slightly different antigenically; the difference is so small that "...a high degree of cross reaction takes place...."[31-33]

Botulinum toxin types have been given various names.[7,9,34,35] Of these, the crystalline-, progenitor-, M-, L-, 19S-, 16S-, 12S-, and 10S-toxins are various complexes of the neurotoxin and some other macromolecules(s). The terms α-fraction, proto-E_α, neurotoxin, 7S-, derivative-, nicked-, unnicked-, single chain-, and dichain-toxin, refer only to the neurotoxin protein. The term progenitor toxin was used at least in one publication[36] to refer to unnicked type E neurotoxin. In this article the following convention will be used:

Toxic crystals — crystalline preparation of a protein or proteins, the solution of which is neurotoxic. The terms crystalline type A and C toxin will not be used because crystals of type A or C neurotoxin protein have not yet been prepared. The known crystallized products are actually complexes of the neurotoxin and other protein(s).[37,38]

Nontoxin protein (or nontoxin) — protein(s) other than neurotoxin (commonly associated with botulinum neurotoxin) but does not include detoxified neurotoxin.

Serologically reactive protein (SRP) — detoxified neurotoxin and also proteins that are not intrinsically toxic but immunologically and serologically are active and similar to the neurotoxin.

The neurotoxin molecules produced by *C. botulinum* in a culture media exist as aggregates of neurotoxin and nontoxin proteins.[39-44] These nontoxin proteins are also produced by the bacteria. A small amount of the neurotoxin can be isolated as molecules of 1.5×10^6 to 2.0×10^6 mol wt from the culture fluids by Sephadex® gel filtration.[39,45] Thus only a small percentage of the neurotoxin molecules present in the culture medium remain unaggregated with other proteins or in the form of loose complexes that dissociate during gel filtration. Some of these complexes, referred to as progenitor toxin,[34,46] contain 50 to 80% w/w nontoxin protein and are stable enough for isolation and characterization. The nature of the nontoxin proteins in these complexes within a type[30,40,41,47-49] or between types[30,48] is not always identical. The S values of these complexes also vary: type A yields 19-, 16-, and 12S;[40] type B produces 16- and 12S;[41] type D produces 16- and 12S;[42] and types E[43] and F[44] produce 12- and 10S, respectively. The toxic crystals of types A[37,50] and C[38] are preparations of such complexes. Smith[7] has pointed out that "the term 'crystalline toxin' is as misleading as it is true — the substance to which it refers could easily be called 'crystalline hemagglutinin'."

The toxic complexes share an interesting property; for a given i.p. LD_{50}, the oral toxicity of a neurotoxin type is markedly less than the corresponding complex,[51-53] and in cases of types A and B, the larger the size of the complex the greater is the oral toxicity to mice.[51] A probable explanation of these findings comes from in vitro studies as types A and B neurotoxins were more sensitive to detoxification, when exposed to acid, pepsin, pancreatin, and rat gastric and intestinal juices, than the corresponding toxic complexes. Additionally, the larger the complex, the greater was the resistance to detoxification by pepsin and gastric juice.[53] The undissociated toxic complex was absorbed when placed in the rat duodenum but absorption was insignificant when placed in the stomach. The neurotoxin and nontoxin components were absorbed at the same rate. Most of the absorbed neurotoxin was transported to the lymphatic stream. The toxin titer in lymph depended mostly on the stability of the toxin in the duodenum and little on the rate of absorption.[54,55]

The labile nature of the neurotoxin indicates that without the protective effect of a stabilized form it may not act as an oral poison. The fact that the neurotoxin produced in food exists as a toxic complex suggests the significant role of these complexes in food poisoning.[43,56]

The neurotoxins purified from cultures (or toxic crystals) are single chain (unnicked) or dichain (nicked) molecules or a mixture of single and dichain molecules (Figure 1). The two chains are held together by at least one disulfide bond. The molecular weight of one chain (fragment I, heavy- or H-chain) $\sim$ 100,000 is about double the weight of the other (fragment II, light- or L-chain). Conversion of the single-chain molecule to the dichain form, referred to as nicking (Figure 2), depends on the physiology of the

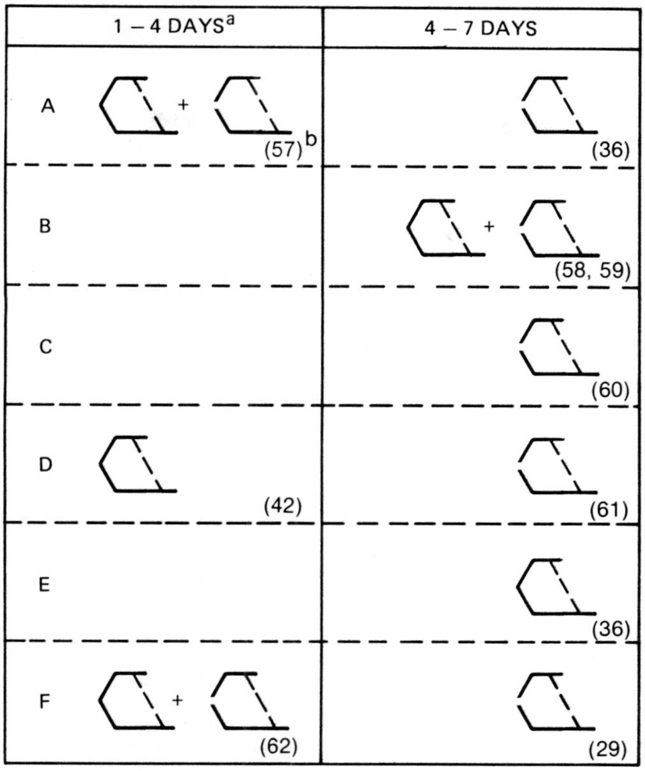

FIGURE 1. Days of incubation of a culture before the neurotoxin was harvested; reference in parentheses; dashed line represents –S–S– bond.

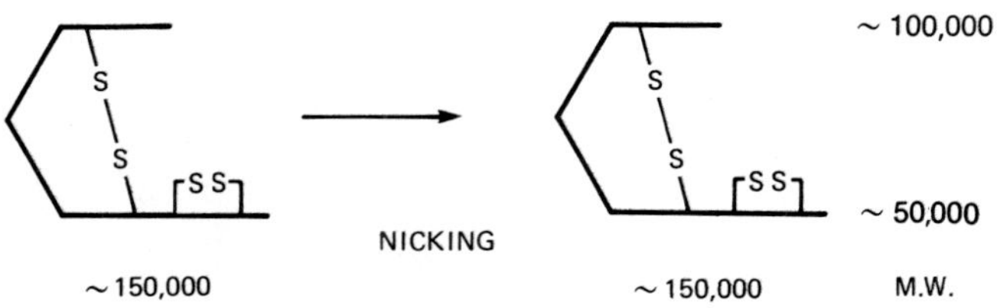

FIGURE 2.

organism and age of the culture. In one study, the ratios of unnicked to nicked neurotoxin in a type F culture after 48, 72, 96, and 120 hr incubation were 1:2.5, 1:3.5, 1:11, and no unnicked form, respectively.[62] Although nicking occurs naturally in some, but not in all, cultures, mild trypsinization of the single-chain molecules of any type produces the nicked form. It appears that the nicked neurotoxins produced naturally or by trypsinization are similar molecules, being products of cleavage of the same or very closely situated bonds.[9,57,59,62,63] A difference in molecular weight between the nicked and unnicked neurotoxins has not been detected. An interchain –S–S– bond was recently located near one end of the H chain of types A and B neurotoxin.[64] The size of the disulfide loop created by this –S–S– bond (Figure 2) is not known.

Some of the properties of the neurotoxin types (Table 1) and their molecular weights (Table 2) are presented. The two molecular weight values for each type cited in Table 1 are those that agree within ±5% and were obtained by different methods. The

Table 1
SOME OF THE PROPERTIES OF BOTULINUM NEUROTOXINS

	A	B	C	D	E	F
Neurotoxin (mol wt)	145,000(36)[d] 150,000(37) 153,000(57)	152,000(65) 150,000(41) 163,000(57)	141,000(60) 141,000(60)	170,000(42)	147,000(36) 150,000(43)	155,000(29) 150,000(29) 157,000(57)
H-chain (mol wt)	97,000(36) 99,000(57)	104,000(65) 104,000(66)	98,000(60)	90,000(61) 110,000(42)	102,000(36)	105,000(29) 108,000(57)
L-chain (mol wt)	53,000(36) 51,000(57)	51,000(65) 50,000(66)	53,000(60)	56,000(61) 60,000(42)	50,000(36)	56,000(29) 51,000(57)
E^{1}_{278}mg/mℓ	1.67(67)	1.85(66)	1.42(60)	?	?	?
S[a]	7.2—7.3(68)	?			7.3(43)	?
IEP[b]	6.1(69)	5.25(66)			?	5.7(73)
MLD/1.0 A_{278}[c]	(9—10) × 10^7(69,70)	(5.3—5.9) × 10^7(66,70)	3.1 × 10^7(60)	?	?	?
MLD/mg protein[c]	?	9.8 × 10^7(66)	4.4 × 10^7(60)	?	?	?
LD_{50}/1.0 A_{278}[c]	8.97 × 10^7(65)	(4.7—6.2) × 10^7(59)[e]	?	?	(1.7—2.9) × 10^7(72)[e]	9.6 × 10^6(29)
LD_{50}/mg protein[c]	(1.05—1.86) × 10^8(67,71)	1.14 × 10^8(59)[e]	?	?	?	?
Nitrogen content of dry toxin	?	?	15.73%(60)	?	?	?

[a] S, sedimentation constant.
[b] IEP, isoelectric point.
[c] Specific toxicity (mice).
[d] Numbers within parentheses are references.
[e] Trypsin activated.

Table 2
MOLECULAR WEIGHTS OF BOTULINUM NEUROTOXINS AND THEIR SUBUNIT CHAINS

Type	A	B	C	D	E	F	Method[b]
Neurotoxin	128,000(68)[a]	167,000(66)	141,000(60)		151,000(72)		U
	150,000(37)	165,000 (70)			140,000(72)	128,000(77)	G
	140,000(74)	150,000(41)			150,000(43)	150,000(29)	G
		155,000(65)			86,000(76)		G
		140—150,000(75)					G
	145,000(36)	170,000(58)	141,000(60)	170,000(42)	147,000(36)	155,000(29)	P
	153,000(57)	152,000(65)				157,000(57)	P
		163,000(57)					P
H-chain		110,000(66)					U
	97,000(36)	104,000(66)	98,000(60)	90,000(61)	102,000(36)	105,000(29)	P
	99,000(57)	110,000(58)		110,000(42)		108,000(57)	P
		104,000(65)					P
		112,000(57)					P
L-chain		50,000(66)					U
	53,000(36)	59,000(66)	53,000(60)	56,000(61)	50,000(36)	56,000(29)	P
		60,000(58)					P
		51,000(65)					P
	51,000(57)	57,000(57)		60,000(42)		51,000(57)	P

[a] Reference.
[b] U, ultracentrifuge; G, gel filtration; P, PAGE-SDS.

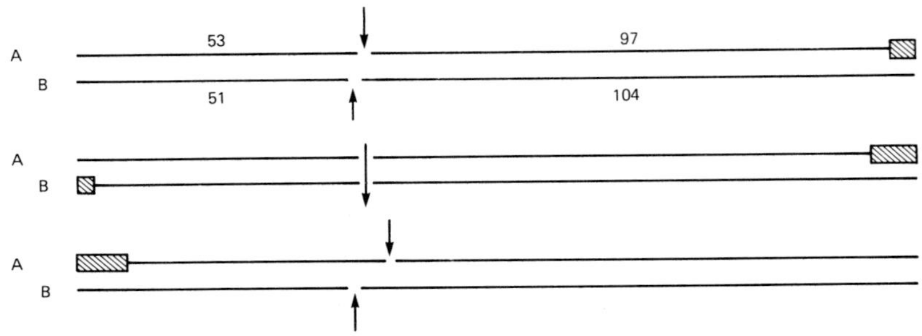

FIGURE 3. 53 and 97 are molecular weights, in thousands, of the light and heavy chains of type A neurotoxin. For the type B neurotoxin the corresponding numbers are 51 and 104. Molecular weight is proportional to the length of a line. Hatched areas are the hypothesized missing portions of the polypeptides.

170,000 mol wt reported for type D neurotoxin[42] is yet to be corroborated. Molecular weights of the H- and L-chains of type D neurotoxin[61] suggest that the molecular weight of the neurotoxin is 156,000. The reasons for not including here a number of other molecular weight values were discussed elsewhere.[9] The nitrogen content and specific toxicity values cited were determined by direct analysis of the neurotoxin and not based on reference proteins (e.g., bovine serum albumin).

The partial specific volume of type B neurotoxin estimated by two different methods is 0.73 and 0.74 mℓ/g[66] and for type C the value is 0.72.[60] Stokes radius of type A neurotoxin, determined by gel filtration, is 4.8 μm[37] and 4.8 to 5.8 μm.[74] Type E neurotoxin appeared, in electron micrograph, to be particles of 4.5 to 5.0 μm in radius.[78] Certain structural parameters of a protein can be calculated from its amino acid composition.[79] Based on this calculation type B neurotoxin is estimated to have 26% helix, 32% β-sheet, and 44% turns. The Q_{10} of the toxic action of type A neurotoxin is 2.36.[80] The activation energy for toxin induced paralysis is 1.55×10^4 cal/mol.[80] Isolated neuromuscular preparations (rat phrenic nerve hemidiaphragm) were used in these in vitro studies. Reference temperature points were 27° and 37°C.

Although all neurotoxin types studied so far seem to have a molecular weight of approximately 150,000, is there a difference in molecular weight between any two types? If so, is the difference due to the use of different techniques and/or variables within a method? To eliminate these uncertainties, types A and B neurotoxins were compared by cochromatography (gel filtration) and coelectrophoresis in polyacrylamide gels in the presence of sodium dodecyl sulfate.[65] The molecular weights of types A and B were found to be 145,000 to 150,000 and 152,000 to 155,000, respectively; based on coelectrophoresis the H-chain of type B was slightly larger than type A, 104,000 vs. 97,000; and the L-chain of type B was slightly smaller than type A, 51,000 vs. 53,000. These are the only known structural differences between the two neurotoxin types besides the difference in their amino acid compositions.[9,66,81]

Depiction of type A and B neurotoxin polypeptides in Figure 3 suggests at least three implications: (1) When the left end of the pair is aligned (the top pair), a small portion of type A, indicated by the hatched area, appears to be missing and the sites of the nicking on the two types appear at different positions; (2) when the nicking sites of the two neurotoxins are aligned (middle pair), it becomes apparent that portions of types A and B are missing; and (3) alignment of the right end of the pair shows that a portion of type A is missing and that the sites of nicking on the two types are different. The missing portions in these proteins could originate from the translational step and/or result from posttranslational modification due to proteolytic degradation. These differences in structure are derived from conjecture and model building, hence their validity in reality is yet to be seen.

Table 3
DISTRIBUTION OF AMINO ACID RESIDUES OF BOTULINUM NEUROTOXIN CORRESPONDING TO THE CODONS

	Code level	Summarized from 68 proteins	Summarized from 207 proteins	Botulinum neurotoxin	Tetanus neurotoxin
Column 1	2	3	4	5	6
Ala	4	5.3	5.2	2.3	2.6
Arg	6	2.6	2.7	1.9	1.7
Asn	2	3.0 }	6.5		
Asp	2	3.6 }		9.8	10.2
Cys	2	1.3	1.4	0.5	0.5
Gln	2	2.4 }	6.5	6.3	5.6
Glu	2	3.3 }			
Gly	4	4.8	4.9	3.0	3.2
His	2	1.4	1.3	0.4	0.6
Ile	3	3.1	3.0	6.3	5.2
Leu	6	4.7	3.9	4.2	5.6
Lys	2	4.1	3.9	5.2	5.2
Met	1	1.1	1.1	1.1	1.0
Phe	2	2.3	2.3	3.3	2.6
Pro	4	2.5	2.9	1.9	2.6
Ser	6	4.5	3.8	4.4	4.6
Thr	4	3.7	3.5	2.7	3.3
Tyr	2	2.3	2.0	3.5	3.6
Try	1	0.8	0.8	1.5	0.4
Val	4	4.2	4.1	2.7	2.6

Note: Column 3 and 4: summarized amino acid composition of 68 (Reference 82) and 207 (Reference 82) proteins expressed as per 61 residues. Column 5 and 6: amino acid composition of type B botulinum (Reference 66) and tetanus (Reference 85) neurotoxins, expressed as per 61 residues. Amino acid composition of botulinum neurotoxin was normalized to 150,000 mol wt. Last three columns do not list separate values of Asp and Asn; Glu and Gln.

Frequency of occurrence of an amino acid in a protein is dictated by the codon(s) of which there are 61. Nevertheless, the amino acid composition of proteins, independent of their length, function, and origin, show an overall deviation from the distribution of amino acids corresponding to the 61 codons. Molecular evolution is responsible for this code randomness.[82-84] Viewed in this light type B botulinum neurotoxin is examined here (Table 3). The amino acid composition, per 61 residues, of a hypothetical protein based on one amino acid per codon is shown in Column 2. Disproportionate occurrence of many amino acids with respect to code level is evident in proteins (Columns 3 and 4); exceptions are Ile, Met, Phe, Try, Tyr, and Val. Botulinum neurotoxin deviates from the code oriented distribution of amino acids more than the family of proteins. Ten amino acids of the neurotoxin are below the code level and the six amino acids, Asp (plus Asn), Glu (plus Gln), Ile, Lys, Phe, and Tyr, that are above the code level belong to codes with low degeneracy. Distribution of four amino acids, Glu (plus Gln), Leu, Met, and Ser, are similar between the neurotoxin and the family of proteins. These deviations become substantially narrowed when botulinum neurotoxin is compared with tetanus neurotoxin. Distributions of nine amino acids among the two proteins are very close; these are Ala, Arg, Cys, Gly, Lys, Met, Ser, Tyr, and Val.

STRUCTURE-ACTIVITY OF THE NEUROTOXINS

The two chains of types A,[63] B,[86] and C[87,88] neurotoxins were separated; the H- and L-chains of types B, C, and the H-chain of type A were purified. The purified polypeptide chains of types A and B were nontoxic. The H- and L-chains of both type B and C were antigenically distinct. Antigenicity of the H-chain of type A and the two chains of type B were partially identical to the homologous dichain neurotoxins. The molecules reconstructed by combining the two complementary chains of type B[86,89] or C[87,88] were neurotoxic. Antigenicity and molecular weight of the reconstructed and native type B neurotoxin were indistinguishable.[86,89]

The H-chains of types A[64] and B[64,89] neurotoxins were susceptible to cleavage near their midpoints by chymotrypsin[64] and trypsin.[64,89] The L-chain of type B was also cleaved at about its midpoint by trypsin.[89] Trypsinization of H- and L-chains of type B neurotoxin produced subfragment I (mol wt 57,000) and subfragment II (mol wt 28,000), respectively. Subfragment II and fragment II (i.e., L-chain) were antigenically identical; subfragment I and fragment I (i.e., H-chain) were partially identical.[89] The type B neurotoxin therefore appears to have at least three antigenic determinants.[89]

Two different approaches were used to locate the structure on the neurotoxin with which it binds to the cellular target site. Rat brain synaptosomes preincubated with fragment I or subfragment I failed to bind the homologous type B neurotoxin; fragment II did not compete.[89] Relatively large amounts of the H-chain of type A neurotoxin that were injected into mice to saturate all receptor sites failed to protect the animals from the subsequent injection of neurotoxin.[63] Mice preinjected with the H-chain died as quickly as those without the H-chain. The researchers actually might have been investigating somewhat different phenomena as they were using different systems.

The role of certain amino acid residues in the structure and biological activity of botulinum toxins was discussed elsewhere.[9] Selective modification of tryptophan detoxified type A toxin.[90,91] Types A and E neurotoxins, both in the nicked form, were detoxified when treated with 1,2-cyclohexanedione.[92] This reagent selectively modifies arginine residues.

The two chains of reduced type B and C neurotoxins were separated by gel filtration[86] and anion exchange chromatography,[87] respectively. Insolubility of reduced type A neurotoxin prevented the separation of its two chains by standard techniques;[63] when the neurotoxin was reduced in the presence of 1% NADH enough of the H-chain protein remained in solution while essentially all of the L- and some of the H-chain precipitated. This unconventional approach was based on some work done to develop a system of affinity chromatography for botulinum neurotoxins on bluedextran-Sepharose 4B column (unpublished). For further discussion see the purification section.

ACTIVATION AND NICKING OF THE NEUROTOXINS

The similar pharmacological activity of the eight neurotoxin types and their distinctive immunological characters suggest that the toxigenic sites (the structure(s) responsible for toxicity) of these molecules are well conserved while the structure(s) responsible for their antigenicity are not as well conserved. Differences in the specific toxicities among the toxin types indicate some structural differences at the toxigenic or at a distal site(s). These may arise from differences in the amino acid sequences and/or posttranslation modification of these proteins. The only two known effects of posttranslational modification of botulinum neurotoxin are the activation phenomenon[9] and nicking, but are they related? The answer discussed below is complex.

Trypsinization enhances toxicity of 8 to 12-hr-old culture fluids of types A, B, E, and F and purified type E neurotoxin as well as certain purified preparations of type B and F neurotoxins. Rooted to these observations (for further discussion see Reference 9) are the two currently held views:

1. Types A, B, E, and F botulinum neurotoxins are synthesized as molecules possessing low toxicity and certain enzymatic modification(s), endogenous and/or-artificial, enhances (i.e., activates) their toxicity.
2. A fully activated molecule is one whose toxicity cannot be enhanced further with trypsin, or in other words, the maximum potential toxicity is that toxicity which is achieved after trypsinization.

These views now warrant additional considerations:

1. A nascent protein molecule may have the maximum attainable toxicity, i.e., its intrinsic toxicity may not be enhanced by any modification.
2. A molecule may not be fully activated by trypsin.
3. Modifications by means other than trypsin alone may activate it further.

Toxicity of a trypsinized neurotoxin therefore need not be regarded as its maximum potential toxicity. Examples for the above cases appear from recent studies.

Toxicity of supernatant fluids of type D (strain 1873) cultures, incubated for 24 to 168 hr, were not enhanced by trypsinization,[93-95a] except in one study samples from 48- and 96-hr-old cooked meat cultures were activated approximately tenfold.[93] The toxicity of the supernatant fluids of type D (strain 4947 and South African) cultures, incubated for 120 hr, was activated sixfold by trypsin.[95a] The enhanced toxicity was not due to C_2 neurotoxin which is also produced by type D organisms.[95a] The neurotoxin partially purified from 72-hr-old cultures of strains 1873 and CB-16 was in the single-chain form. Trypsinization nicked the neurotoxin from both sources and activated the toxic material from CB-16 by twofold, but not the one from 1873.[42] Nonactivation of partially purified type D neurotoxin was noted earlier.[23] The action of trypsin on the toxic material from strain 1873 led the authors [42] to conclude that "fully active botulinum toxin is not necessarily in the nicked form." Was the neurotoxin from strain 1873 fully active? Specific toxicities of the toxic preparations isolated from strain 1873 and CB-16 were not compared. Specific toxicity of the CB-16 preparation, trypsinized or untrypsinized, could have been higher than, equal to, or less than the 1873 preparation. A higher value would suggest that the neurotoxin from strain 1873 is not the fully active type D. A lower or equal value would suggest what is the maximum toxicity of fully active type D neurotoxin.

Toxicity of type B progenitor toxin (M toxin) enhanced by treating it first with a protease and then with trypsin was more than when the M toxin material was treated with trypsin alone.[96] An opposite effect is also known: toxicity of type E neurotoxin activated in two steps, first with a protease and then with trypsin, was less than attained by trypsin alone.[97]

It is not known if the primary cause for activation is nicking.[9] The possible observations one may make in investigating this question are

1. Activation is accompanied by nicking.
2. Activation is not accompanied by nicking.
3. Nicking is not accompanied by activation.

Each of these situations has been documented. Type E[36] and certain preparations of type B[57,59] neurotoxins upon trypsinization undergo activation and nicking. When treated with a trypsin-like enzyme, type E neurotoxin becomes activated without nicking.[59] Certain toxic preparations of types A[63] and D[42] are nicked with trypsin without activation. What does this mean?

Situation 1 neither proves nor disproves that nicking is responsible for activation. A careful kinetic study of nicking and activation may reveal that the rate of nicking is faster, equal, or slower than the rate of activation. Whereas the slower rate of nicking would indicate nicking is not responsible for activation, the faster rate of nicking would leave the question open; an equal rate would indicate a causal relationship. Situations 2 and 3 may result from a difference in detection limits of activation ratio (toxicity after activation/toxicity before activation) and nicking.

One can argue from the calculations described below that a mixture of activable and activated toxin in a 1:1 ratio when completely converted to the activated form will yield an activation ratio less than 2 (or less than twofold activation).

Let x = specific toxicity of an activable molecule
Let y = specific toxicity of an activated molecule
Let N = total number of molecules
Let m = total number of activable molecules
Total toxicity before activation = $m \cdot x + (N-m)y$
Total toxicity after activation = $N \cdot y$

$$\text{Activation ratio} = \frac{N \cdot y}{m \cdot x + (N - m)y} \tag{1}$$

When the number of activable molecules = number of activated molecules (i.e., $m = N/2$ or $N = 2m$), Equation 1 becomes

$$\frac{2my}{m \cdot x + (2m - m)y} \quad \text{or} \quad \frac{2}{x/y + 1} \tag{2}$$

When $y = x$, Equation 2 attains a value of 1, i.e., no activation.
When $y >> x$, Equation 2 approaches 2; thus limit value of Equation 2 is $\leqslant 2$.

The two parameters followed in activation studies are the increase in toxicity and the activation ratio. Their usefulness or sensitivity in the detection of differences in toxicity are not the same. An example presented in Table 4 shows the range of the different proportions of activable and active molecules (total number of molecules 10; toxicity of each activable and active molecule is 1 and 100 units, respectively). The potential activation ratio at the two ends of this range are 100 and 1. Mixtures of activable and active molecules in the ratios of 6:4 and 1:9 upon activation will yield activation ratios of 2.46 to 1.11. The *change* in activation ratio becomes progressively smaller as the mixtures become enriched with active molecules. Toxicity, in contrast to activation ratio, increases by an *identical amount* as each activable molecule is replaced by an active molecule. An activation ratio $\leqslant 2$, derived from mouse toxicity assay, is not generally regarded as a significant change in toxicity. These low values of activation ratios therefore may suggest deceptively the absence of activable molecules; yet the population of activable molecules in the sample may be as high as 50%.

These considerations warrant reexamination of the interpretations of the following two sets of data: (1) Types A and F neurotoxins, each a mixture of unnicked and

Table 4
CHANGES IN TOXICITY AND ACTIVATION RATIO DURING AN HYPOTHETICAL CASE OF TOXIN ACTIVATION

Number of molecules in the mixture			
Activable	Active	Toxicity	Potential activation ratio
10	0	10	100
9	1	109	9.17
8	2	208	4.81
7	3	307	3.25
6	4	406	2.46
5	5	505	1.98
4	6	604	1.65
3	7	703	1.42
2	8	802	1.24
1	9	901	1.11
0	10	1000	1.0

nicked molecules, were not activated by trypsinization.[57] This led to the conclusion that the neurotoxins were endogenously fully activated with only partial nicking, an observation somewhat similar to Situation 2 (see above). Electrophoretic analysis of the two neurotoxin types shows (Figure 1 in Reference 57) that the ratio of unnicked:nicked (sum of the H- and L-chain) forms, in each sample prior to trypsinization, was probably close to 1:2 or at least <1:1. A mixture of activable and activated molecules of such a ratio yields too little an activation to be regarded as valid. Hence the author's[57] conclusion may not be warranted. (2) When type B neurotoxin, a mixture of unnicked and nicked forms, was trypsinized at pH 6.0 the increase in toxicity appeared to be parallel to the extent of nicking; but after 60 min reaction at pH 4.5 maximum toxicity was attained prior to the completion of nicking.[57] The conclusion that the increment in toxicity was completed while nicking was incomplete is not warranted. Toxicity of the neurotoxin-trypsin digest at pH 4.5, 35°C increased linearly up to 60 min and then remained constant for the next 30 min. After incubation for 90 min the digest contained substantial amounts of unnicked material. The horizontal portion of the toxicity vs. digestion-time curve between 60 and 90 min, suggested attainment of maximum activation; this was very likely the net effect of activation and inactivation because type B neurotoxin is 50% inactivated in 30 min at pH 4.5, 35°C.[53] Unless the toxicity was increasing during the 60 to 90 min period, the observed toxicity would have declined due to acid inactivation. In other words, activation was occurring beyond 60 min and the increasing toxicity was compensated for by the declining toxicity. The effect of completion of nicking at pH 4.5 on the level of toxicity was not described.

What causes endogenous nicking of the neurotoxin and what change(s) in the molecule is responsible for its activation are not yet known. The speculations about the activation mechanism discussed earlier[9] remain operative because no contradictory information has appeared yet.

Substrate specificity of trypsin indicates that the site of nicking is either an arginyl or a lysyl bond. When the single-chain type E neurotoxin was treated with 1,2-cyclohexanedione, which specifically modifies arginine residues, the neurotoxin was not nicked by trypsin.[92] This suggests that an arginyl bond is either the nicking site or the modification of arginine residues somehow renders the protein resistant to nicking by trypsin.

Activation of the botulinum neurotoxins with a number of bacterial enzymes was attempted (Table 5) but only one enzyme (see No. 9 in Table 5) was tested for its ability to nick the neurotoxin. The single-chain type E neurotoxin remained unnicked following activation by a trypsin-like enzyme isolated from type B cultures; the neurotoxin was nicked when it was activated further by trypsin or by treatment with trypsin alone.[59,97] The degree of activation attained with the clostridial enzymes, or the enzyme and trypsin, was substantially less than that obtained by trypsin. A similar enzyme isolated from commercial pronase[104] and type F culture[103] were found to activate toxic preparations of type E as much as trypsin but the reports did not mention whether the neurotoxin was nicked.

Crude preparations of proteases from clostridial species activated the type B and E toxic complexes but activation was less than by trypsin.[96] Subsequent trypsinization further activated the type B toxic complexes; total gain in toxicity in some cases was below and in four cases higher than the level attained by trypsin treatment alone.[96,103] The type B and E toxic complexes that could be activated 933- and 173-fold, respectively, by trypsin,[96] behaved differently when exposed to the highly purified protease from *C. botulinum* type F or a crude preparation of protease from other sources; the toxicity of type E was activated more than type B.[96,103]

The following results derived from toxicity activation experiments await clearer explanation(s).

1. The specific toxicity of single-chain type A neurotoxin was ¼ to ½ that of the dichain form. Toxicities of the supernatant fluids of the 8-hr-old culture, from which the single-chain neurotoxin was isolated, and the 96-hr-old culture, from which dichain neurotoxin was isolated, were activated by trypsin 30- and 1-fold, respectively. Trypsinization nicked the single-chain molecule but did not significantly increase toxicity of the single- or dichain neurotoxins.[63] Does this suggest that the molecules which were activable 30-fold in the 8-hr-old culture became partly activated during purification and/or purification procedures modified the neurotoxin in such a way that it could not be activated by trypsin?
2. Specific toxicity of the type E neurotoxin (7S material) first isolated from the toxic complex (12S material) and then trypsinized, in two studies, was 1.38×10^8 (43) and 1.55×10^8 LD_{50}/mg N.[57] Specific toxicity of activated neurotoxin isolated from trypsinized toxic complex was 1.38×10^8 LD_{50}/mg N.[43] Trypsinization of the neurotoxin and its source the toxic complex produced 228- and 522-fold activation, respectively;[43] in another study these values were 57 and 367, respectively.[57] Does this suggest that the type E neurotoxin during its isolation from the 12S complex had undergone 2.3- (43) and 6-fold (57) activation without trypsin?
3. Activations of the type E and B neurotoxins as well as a type B toxic complex with SHP (a highly purified sulfhydryl dependent protease, isolated from *C. botulinum* type F) were less than that by trypsin. In contrast, activation of the toxic complex of type E by SHP was slightly more than that attained by trypsin.[103] The authors suggest that the nontoxic component of the type E complex protected the neurotoxin from degradation during activation. Or did the conformation of the neurotoxin resulting from the complex allow SHP to modify the substrate just like trypsin?
4. Trypsinization of fluids from cultures of various type E strains, grown in similar medium, resulted in different activation ratios.[105,106] Changes in the composition of the media and incubation temperature also affected the activation ratio.[106]

Table 5
EFFECTS OF BACTERIAL ENZYMES ON THE ACTIVATION OF BOTULINUM NEUROTOXIN TYPES

Source of enzyme	Purity	Optimum pH	SH requirement	Hydrolyzed				Activates toxic activity of type	Substrate specificity confined to	Inhibited by	Ref.
				Pr	Am	TAME	LME				
1. *C. botulinum* type A	P	6.0	+	+	+	+	?	B,E	?	NEM,MIA,EDTA	96
2. *C. botulinum* type A (a)	P	8.5	?	+	?	?	?	A(−)	?	EDTA	98
3. *C. botulinum* type A (b)	P	7.5	?	+	?	?	?	A(−)	?	EDTA	98
4. *C. botulinum* type A (c)	P	6.0	?	+	?	?	?	A	?	EDTA	98
5. *C. botulinum* type A I	P	?	?	+	?	?	?	A	?	?	99
6. *C. botulinum* type A II	P	?	?	+	?	?	?	A	?	?	99
7. *C. botulinum* type A	P	7.0	−	+	?	?	?	A(−)	?	DFP(−),STI(−),EDTA	100
8. *C. botulinum* type B	P	6.0	+	+	+	+	?	B,E	?	NEM,MIA,EDTA	96
9. *C. botulinum* type B	NH	6.0	+	+	+	+	?	B,E	Arg,Lys	STI(−),EDTA	97
10. *C. botulinum* type B I	P	?	?	+	?	?	?	B	?	?	99
11. *C. botulinum* type B II	P	?	?	+	?	?	?	B	?	?	99
12. *C. botulinum* type B	P	7.4—7.6	?	+	?	?	?	B	?	?	101
13. *C. botulinum* type E I	P	≥7	−	+	−	−	−	E(−),F(−)	?	EDTA(±),PMSF(−)	102
14. *C. botulinum* type E II	P	6—6.3	+	−	+	+	±	B,E,F,C_2(−)	?	EDTA,PMSF(−)	102
15. *C. botulinum* type E III	P	≥7	−	+	−	−	−	E(−),F(−)	?	EDTA(−),PMSF(±)	102
16. *C. botulinum* type EIV	P	≥7	−	+	−	−	−	E(−), F(−)	?	EDTA(−), PMSF (±)	102
17. *C. botulinum* type F	H	6.0	+	+	+	+	+	B,E	?	NEM,MIA,EDTA	103
18. *C. botulinum* type F	P	6.5	+	+	+	+	?	B,E	?	NEM,MIA,EDTA	96
19. *C. sporogens*	P	7.0	+	+	+	+	?	B,E	?	NEM,MIA,EDTA	96
20. *C. histolyticum*	P	7.5	+	+	+	+	?	B,E	Arg,Lys	NEM,MIA,EDTA	96
21. *C. perfringens* (1)	P	6.0	+	+	+	+	?	E	?	NEM,MIA,EDTA(−)	96
22. *C. perfringens* (2)	P	6.0	+	+	+	+	?	E	?	NEM,MIA,EDTA(−)	96
23. *Bacillus cereus*	P	?	?	+	?	?	?	A,B(−)	?	?	99
24. Pronase	H	?	?	+	+	+	+	E	Arg,Lys	STI	104

Note: *C. botulinum, Clostridium botulinum;* P, partially purified; NH, nearly homogeneous; H, homogeneous; Pr, protein (in case of type E enzymes, azocasein); Am, amide; TAME, tosyl arginine methyl ester (in case of type E enzymes, benzoyl-arginine ethyl ester); LME, lysyl methyl ester; NEM, *N*-ethylmaleimide; MIA, monoiodoacetate; STI, soybean trypsin inhibitor; PMSF, phenyl methyl sulphonyl fluoride; (−), toxocity not increased, and inhibitor not effective.

Does this suggest that the neurotoxin produced by these strains differs in its intrinsic or potential toxicity or both of these properties; also does it suggest the conditions of culturing alter these properties?

NEUROTOXIN PRODUCTION

Various media and culture conditions designed to produce types A, B, C, D, E, and F neurotoxins in high titer and in moderate to large volume are listed in Table 6. The quantity of toxin harvested from each culture was large enough to undertake its purification. Some of the media and culture conditions that produced lower toxin yield are not included.

In a culture the neurotoxin exists in the cell-bound (perhaps also cell fragment bound) and/or excreted (or released) form. Efficient recovery of the neurotoxin, for its purification, depends on an understanding of this distribution and the factors involved in the distribution; e.g., 80 to 90% of the total amount of type E neurotoxin, even after 4 days of incubation, was contained in the cells.[76,119] Type A[120,121] and certain B[120] cultures autolyzed very rapidly in the presence of glucose or maltose in the medium although neither of these sugars was needed for support of growth. Lysis of cells in a 24-hr type A culture was greatly increased by adding filtrates of 1-week-old homologous culture; the filtrate boiled for 5 min did not enhance lysis.[122] Washed cells of type A, harvested at logarithmic growth phase, and the cell walls prepared therefrom autolyzed rapidly in 0.05 *M* phosphate buffer, pH 6.8 to 7 at 37°C.[123] An autolytic enzyme, partially purified from the autolyzed cell walls,[124] cleaved the amide bond between *m*-acetylmuramic acid and L-alanine, and the glycosidic bond between *N*-acetylglucosamine and acetylmuramic acid.[125] Lytic action of the enzyme was specific for the types and strains of the organism; whereas sodium dodecylsulfate-treated cell walls of type A and B (proteolytic type B, and *C. perfringens* were resistant).[124] The enzyme was active between pH 6 to 8 with maximum activity at pH 6.8. Ca^{++}, Mg^{++}, and Co^{++} at 10^{-4} *M* and mercaptoethanol at 10^{-4} to 10^{-3} *M* activated the enzyme.[124]

A number of monovalent cations induced lysis of growing clostridial cells.[126] Lysis was most rapid in presence of 0.3 to 0.5 *M* Na^{+} and when the cells were from a 3- to 4-hr culture. The cells became increasingly resistant to lysis as they aged. Divalent metal cations inhibited Na^{+} induced lysis. The inhibitory effect, observed at or above 5×10^{-3} *M* Ca^{++}, increased with increasing concentration of the cation and decreased in the order of Ni, Zn>Co>Ca>Mg>Ba>Sr, Mn. Chelating agents (e.g., 0.05 *M* EDTA or citrate) enhanced Na^{+} induced lysis. Although *C. botulinum* was not tested the eight other species of *Clostridium* tested were susceptible to the lytic action of Na^{+}. Bacteria belonging to nine other genera were resistant.

One method of isolation of type E neurotoxin from the whole culture is to extract the cell-bound toxic material. The harvested cells are washed with water[76] or 0.05 *M* Na-acetate buffer[49] and then suspended in 1.0 *M* Na-acetate[76] or 0.2 *M* KH_2PO_4-Na_2HPO_4 buffer, pH 6.0.[49] The toxic material appears in the buffer containing higher amounts of Na. The authors did not report whether the cells lysed during extraction of the toxin.

PURITY OF THE NEUROTOXINS

Critical examination of purified types A, B, E, and F neurotoxins showed that these preparations appear to be homogeneous or nearly so.[9] The type A[57] and C[60] neurotoxins purified recently pass the criteria of purity.[9] The available preparations of type D are the 16- and 12S complexes[42] but not pure neurotoxin; its molecular weight was determined from the 12S complex. Purification of type G has not been reported.

Table 6
COMPOSITIONS AND DESCRIPTIONS OF CULTURE MEDIA USED TO PRODUCE *C. BOTULINUM* NEUROTOXIN TYPES

	Type A								Type B					
Year reported (Ref.)	1946 (107)	1946 (108)	1957 (109)	1965 (110)	1970 (69)	1975 (40)	1977 (7)	1977 (47)	1947 (111)	1957 (112)	1966 (113)	1969 (66)	1974 (41)	1976 (59)
Casein (or bactocasitone, BC; lactoalbumin hydrolyzed, LA; casitone, C)	0.3								1		(In beef infusion)	2		
Corn steep liquor	1.0	0.75							1					
N-Z amine			2		2			2						1
Pepticase		2												
Proteose peptone (or polypeptone, P)				1	4						1			2
Peptone						2							2	
Trypticase							2			1.5				
Autolyzed yeast			0.5					0.5						
Yeast extract					2	0.5	2	0.5		0.5		0.5	0.5	1
Glucose (or glycerol, G; dextrin, D)	0.5	0.5	0.5	0.5	1	0.5	0.5	0.5	0.5	0.5	0.5	0.5	0.5	1
NaCl				0.5	0.85						0.5			
Cysteine·HCl										0.075		0.075		
Na-thioglycolate				0.1		0.025					0.1	0.1	0.025	0.05
$CaCl_2$ or $[Ca(CO_3)_2]$ Na_2HPO_4 or $[K_2HPO_4]$				0.2							0.2			
pH	7.2	7.5	7.2	?	7.3	7.0	7.2—7.4	7.2	7.2	7.1	?	7.3	7.0	7.3
Incubation temperature (°C)	34	33—34	35	30	37	30	30	37	34	35	30	37	30	37
Incubation period (hr)	80	96	96	168	96	120	120	96	336	120	120	72	96	96
Strain of organism	Hall	Hall	Hall	Corn T	Hall	Hall	?	Hall	Okra	Bean	Lamanna	Lamanna	Okra	Okra
Saline in dialysis sac (mℓ)				1000	400		75				1000			
Culture volume harvested (liter)	16	16	3			5		16	16	3		6	5	8
Toxin titer: × 10^6/mℓ (T, toxicity after trypsinization)	0.8—1.5	0.8	1.5	0.1—0.3	4—6	2.5	20	1.5	1	2	0.3—1	1—2	7	0.8
LD_{50} or MLD (mice)	MLD	MLD	LD_{50}	MLD	MLD	LD_{50}	LD_{50}	LD_{50}	MLD	LD_{50}	MLD	MLD	LD_{50}	MLD

Table 6 (continued)
COMPOSITIONS AND DESCRIPTIONS OF CULTURE MEDIA USED TO PRODUCE *C. BOTULINUM* NEUROTOXIN TYPES

	Type C						Type D				Type E			Type F			Type G
Year reported (Ref.)	1950 (114)	1958 (115)	1965 (116)	1972 (38)	1972 (23)	1977 (60)	1950 (114)	1960 (117)	1977 (42)	1977 (42)	1956 (105)	1968 (49)	1974 (76)	1974 (44)	1975 (29)	1976 (118)	1977 (20)
Casein (or bactocasitone, BC; lactoalbumin hydrolyzed, LA; casitone, C)	(In corn steep)		2 BC			2 LA	(In corn steep)						1.5 C				
Corn steep liquor			2	2				25								17	
N-Z amine		2							1	1					1		
Pepticase		4									2				2		4
Proteose peptone (or polypeptone, P)				2 P		2 P											
Peptone			4	4		2			2	2		2		2			
Trypticase					3												
Autolyzed yeast		2	1	2	2	1.5			0.5	0.5	2	0.5	0.25	1	1		1
Yeast extract																	
Glucose (or glycerol, G; dextrin, D)		1	1	1	0.2	40	0.5 G	1.0 G	0.5	0.5	1 D	1	1	1	1	1 G	1
NaCl			0.5	0.5		10											
Cysteine·HCl					0.1								0.37				
Na-thioglycolate			0.05	0.05		2.4			0.025	0.025		0.025		0.025	0.05		
$CaCl_2$ or [$Ca(CO_3)_2$]	0.1						0.1	[0.1]					0.05			0.12	
Na_2HPO_4 or [K_2HPO_4]													[0.1]				
pH	7.2	7.6	7.8—8.0	7.9	?	?	7.2	7.2	7.0	7.0	7.2	6.3	?	7.0	7.4	7.4	7.2
Incubation temperature (°C)	?	33	33	33	33	33	?	36	30	30	30	30	30	30	30	35	30—31
Incubation period (hr)	144—192	120	168	144	120	120	240—288	456	72	72	120	96	96	96	120	408—480	168
Strain of organism	?	S. African	Stockholm	Stockholm	468C	Stockholm	?	D6F	CB-16	1873	VH	35396	VH	Langeland	Langeland	8GF	89
Saline in dialysis sac (mℓ)	3500						3500	500								1000	50(?)
Culture volume harvested (liter)	2.5	3	?	3.5	3	0.033	2.5		5	5	3	5	2—8	5	15		
Toxin titer: × 10^6/mℓ (T, toxicity after trypsinization)	2	0.8	1	0.08—0.13	0.06	2.75	100	1.42	0.06 T	0.04	0.18 T	0.05 T	?	0.2	0.02	0.79	0.03
LD_{50} or MLD (mice)	MLD	LD_{50}	MLD	MLD	LD_{50}	MLD	MLD	LD_{50}	LD_{50}	LD_{50}	LD_{50}	LD_{50}	?	LD_{50}	LD_{50}	LD_{50}	LD_{50}

PURIFICATION OF THE NEUROTOXINS

Some new techniques deployed in purification procedures are as follows:

1. Type A,[40,57,109] B,[41,112] and F[44] neurotoxins are easily recovered as precipitates after lowering the pH of the cultures to 3.4 to 4.2 with sulfuric acid. Although ammonium sulfate can be used to precipitate type A,[69] B,[59,66] C[60], and F,[29] sulfuric acid is inexpensive and simpler to use. Recovery of the neurotoxins from type C,[42,115] D,[42,117] E[127], and nonproteolytic B[33] cultures by acid precipitation is poor. An elegant approach has solved this problem for types C and D. Iwasaki and Sakaguchi[128] found that the RNA/protein ratio, w/w, in cultures of types C, D, and E were substantially less than those of types A, B, and F, 0.55 to 0.1 vs. 1.36 to 0.77. This prompted these investigators to presume that more neurotoxin would be precipitated if the RNA/protein ratio was increased to 0.77 or more, by adding RNA from an external source prior to acidification. Indeed, in the presence of added yeast RNA, final concentration of 0.4 mg/mℓ, the recoveries of types C and D neurotoxins from cultures of various strains were satisfactory (73 to 94%). The technique did not significantly improve recovery of type E neurotoxin.
2. The toxic materials extracted from type E cells and precipitated from culture fluids of types A, B, C, D, and F are rich in nucleic acids. Two techniques, introduced in recent years to remove nucleic acids are protamine sulfate and chromatography on DEAE-Sephadex® A-50. Protamine sulfate quickly precipitates the nucleic acids from types A,[40] B,[41] C,[128] D,[42,128] and F[44] preparations. The optimum amount of protamine seems to be 1 mℓ of a 2% solution per 1000 A_{260} units of the crude toxic solution.[103] Excess protamine is removed by passing the precipitate-free fluid through a SP-Sephadex® C-50 column. For the second procedure the crude toxic materials of type A[47] and B[59] are dissolved in 0.05 *M* Na citrate-citric acid buffer, pH 5.5, and loaded on a DEAE-Sephadex® A-50 column. The same buffer is used to equilibrate and wash the column. Essentially all of the toxin applied is recovered in the first protein peak comprised of fractions of A_{260}/A_{280} ratio of 0.5 to 0.55. Use of this buffer and column is a modification of earlier attempts in the purification of crude toxic material in a single step using DEAE-cellulose and 0.067 *M* phosphate-citrate buffer pH 5.6.[110,113] An independent study[129] of the effect of buffer composition, pH, ionic strength, and DEAE-Sephadex® A-25 on the partitioning of nucleic acid from protein points to the efficacy and optimization of the buffer and column selected for purification of the neurotoxin.[59]
3. The neurotoxin and hemagglutinin proteins of type A toxic crystals are separated by allowing the complex to bind, via the hemagglutinin, to a ligand coupled to CH-Sepharose 4B at pH 6.3 and then dissociating the two proteins with a buffer of pH 7.9 containing a higher amount of NaCl. This buffer elutes the neurotoxin leaving the hemagglutinin bound to the ligand, *p*-aminophenyl-β-D-thiogalactopyranoside.[130] Purity of the recovered neurotoxin is comparable to that obtained by ion-exchange chromatography.[9] This technique was developed from the studies of inhibition of hemagglutination by D-galactose, some of its derivatives, and isopropyl-β-D-thiogalactoside, and the observations that the hemagglutinins of *C. botulinum* types A and B are strongly bound by *p*-aminophenyl-β-D-thiogalactopyranoside coupled to CH-Sepharose 4B.[48]

Another approach has been only partially successful. The currently accepted molecular weight of type A neurotoxin was first established by gel filtration.[9,37,71] During these studies when a mixture of blue dextran and the neurotoxin were cochromatographed the neurotoxin was eluted unexpectedly with the blue dextran at the void volume of the gel filtration column (unpublished). When chromatographed separately the protein and blue dextran were eluted at widely different positions. The two substances, when mixed, had apparently formed a stable complex. This association also was noted in another laboratory.[45] These observations were recently reinvestigated by this author (unpublished) with the knowledge that (1) the dinucleotide folds in the β-sheet core of a number of proteins have binding sites for the blue dye (of blue dextran) and NAD; (2) the blue dye can assume a conformation similar to parts of NAD; and (3) nucleotide ligands may be used to elute protein bound to a blue dextran-Sepharose 4B affinity column.[131] Blue dextran-Sepharose 4B, prepared according to Ryan and Vestling,[132] equilibrated with 0.02 *M* Na phosphate or TRIS-HCl buffer, pH 7.5, bound type A neurotoxin. A small amount of the bound protein was eluted with NAD, NADH, or ADP; 1.0 *M* NaCl eluted more protein but not all. The protein eluted was highly toxic. The neurotoxin incubated with blue dextran (in 0.01 *M* phosphate buffer containing 0.005 *M* NaCl, pH 6.2, 23°C for 1 hr) did not lose toxicity (final concentration of neurotoxin A_{278} = 0.038, blue dextran A_{625} = 3.26, i.v. assay). Gel filtration (Sephadex® G-50, 0.04 *M* Na-phosphate buffer, pH 7.5) resolved the complex of the neurotoxin and β-NAD. The UV spectral characteristics of the eluted protein did not reveal the presence of any NAD bound to the protein.

ASSAY

The current standard procedure for detection of botulinum neurotoxin is the mouse assay.[133] Survival of mice protected by specific antitoxin serum and death of unprotected mice following injection of a toxic sample is the positive test. Quantitation of the neurotoxin involves i.p. injection of a serially diluted sample and determination of the highest dilution that kills 50% of the mice over a 96-hr period. A much faster mouse assay, not useful with low doses ($<10^2$ LD_{50}/mℓ), is the i.v. technique. Boroff and Fitzgerald[134] first noted the correlation between the amount of botulinum toxin injected i.v. into mice and the time elapse between injection and death. Careful studies showed that a plot of log of toxin titer vs. log of survival time is linear over the range 4×10^5 to 4×10^3 MLD/mℓ and 24.7 to 71.7 min for type A,[135] and 1.7×10^6 to $\sim 5 \times 10^2$ LD_{50}/mℓ and 60 to 160 min for type E.[136] The method also has been successfully applied to the assay of types B,[41,58,66] C,[38,60] D,[42] E,[136] and F[29,44,73] toxic preparations. In the currently used method, three to four mice (20 to 25 g) are injected in the tail vein with a 0.1-mℓ sample for each and the time (in minutes) between challenge and death are noted. The average time to death is converted into i.p. LD_{50} value from a standard curve. The curve is obtained by determining time to death of mice injected i.v. with toxic samples of known i.p. LD_{50} doses. The measure of toxicity obtained is therefore an expression relative to that of the reference neurotoxin.

One study showed that for a given dose, type A neurotoxin killed mice faster than the toxic crystals:[137]

1. A dose of 4×10^5 to 4×10^3 LD_{50} toxin (mol wt 150,000) by i.v. route killed in 23.1 to 50 min.
2. A dose of 4×10^5 to 4×10^3 LD_{50} toxin (mol wt 150,000) by i.p. route killed in 35.3 to 69.2 min.
3. A dose of 4×10^5 to 4×10^3 LD_{50} toxic crystals (mol wt 900,000) by i.v. route killed in 30.8 to 89 min.
4. A dose of 4×10^5 to 4×10^3 LD_{50} toxic crystals (mol wt 900,000) by i.p. route killed in 51.9 to 126 min.

The time toxin molecules take to migrate from the site of injection to the cellular target receptor sites was thought to be influenced by the dimensions of the toxic complex because the large neurotoxin-nontoxin protein complex (toxic crystals) was slower to act. Contrary to this observation, Sugii and Sakaguchi[40] found that dose-response (time to death after i.v. injection) of different type A toxic preparations, whether 7S (neurotoxin free of other proteins) or 12-, 16- and 19S were the same. The 19S complex and toxic crystals are similar in their composition.[40]

A possibility of error in the assay, due to incorrect placement of the inoculum during i.p. injection, warrants consideration. Studies with substances other than botulinum toxin have revealed a 14%,[138] 10 to 20%,[139] and 12%[140] error in the placement of i.p. injection of mice with one-person procedure of injection. All or part of the misplaced inoculum was at the lumen of the stomach, small bowel, subcutaneous, retroperitoneal, intravascular and uterine horn. The incidence of error was consistently reduced to 1.2% level with a two-person procedure of injection.[140]

Many techniques were developed to improve upon or replace the mouse assay but none has yet surpassed or equaled the sensitivity (= lower limit of detection) and specificity of the mouse assay.

The fluorescent-antibody techniques that detect cellular bound toxin[141,142] are of little value in detecting toxin in solution. Some of the reported serological assays were not specific for the toxin because the antisera used in the assays were developed with impure preparations of the toxin; the contaminants were often multiple antigens and/or cross-reactive. Consequently antigens other than the toxin also were detected and the lower limit of detection was camouflaged. These problems have been pointed out before.[5,30,143,144]

A reverse passive hemagglutination (RPHA) procedure was claimed to be more sensitive than the mouse bioassay.[145] In these studies culture fluids of types A and B, toxin titer 1×10^3 and 2×10^2 MLD/mℓ, respectively, at 400× dilution produced no indirect hemagglutination reaction but did not kill mice. Although the quality of antigens used to develop antiserum was not reported, the likelihood is that they were very impure; the best preparations of types A[109] and B[112] toxins available at that time were mixtures of neurotoxin and other proteins in about 1:4 w/w proportion.[37,70] Use of antiserums prepared against such antigens and culture fluids of very low toxin titer seems to have produced the misleading sensitivity observed.

The RPHA procedure reported to be sensitive enough to detect 0.25 to 0.38 ng toxin N of type A, B, and E was nonspecific; the antiserum cross reacted between the three types as well as with the culture filtrates of *C. butyricum, C. sporogenes,* and *C. perfringens* type C.[146]

An RPHA proposed to be a standardized technique for detection of type A toxin with a sensitivity of 27 mouse LD_{50}/mℓ,[147] as well as passive hemagglutination and bentonite floculation procedures developed for assay of types A, B, and E[144] utilized antiserum preparations that were developed against impure neurotoxins. It has been shown[143] that (1) the proposed standardized technique[147] was based on the reaction of a nontoxin protein rather than the neurotoxin; (2) the reported correlation between RPHA titers and toxicity is not evidence that the test measure neurotoxin; and (3) the titer should change proportionately with the nontoxin protein (the hemagglutinin). The assay[147] was developed to detect in food the toxin as it existed as a complex.

A microcapillary agar-gel diffusion[148] technique was used to detect 1 mouse MLD of type E toxin in a 0.01-mℓ sample; a gel diffusion (Ouchterlony) method was used to detect 37 to 55 mouse LD_{50} of types A, B, and E toxin in a 0.1-mℓ sample.[149] These studies did not eliminate the possibility that the precipitin line observed might be due to other substance(s) present in the sample that could serologically react with the antiserum. The antisera used were not developed against pure neurotoxin types.

As little as 0.7 mouse LD_{50} of type A toxin, in a 5-$\mu\ell$ sample, was detected by the electroimmunodiffusion technique.[150] This amounted to detection of about 4 pg of neurotoxin protein — a feat very difficult to conceive outside of the radioimmunoassay (RIA) system. The antiserum was developed with a toxin preparation that was purified by a novel procedure.[150] DasGupta at the Food Research Institute (unpublished data) was unable to obtain pure neurotoxin with this procedure. Gasper (unpublished work, Food Research Institute) was unable to confirm the detection limit claimed and found that more than 10 ng of neurotoxin (equivalent to 1000 mouse LD_{50}) was required to produce a visible precipitin cone in the electroimmunodiffusion plates. The antiserum used by Gasper was of much higher titer than the titer of 45 mouse $LD_{50}/m\ell$ used by Miller and Anderson.[150]

In contrast to the above examples, the following assays are specific for the neurotoxin type and the neurotoxin protein. Reportedly one can detect with the enzyme-linked immunosorbent assay (ELISA) 50 to 100 mouse $LD_{50}/m\ell$ of type A toxin,[151] about 80 mouse $LD_{50}/m\ell$ of type E,[151a] and about 400 mouse $LD_{50}/m\ell$ of type B[151b] neurotoxin. Because pure type A neurotoxin[40] (150,000 mol wt) was used to develop the antiserum, the concentration of neurotoxin in the test sample is reliably related to the amount of enzyme fixed in the tubes and therefore to the amount of spectrophotometrically determined product of the enzymatic reaction. In the case of type E assay low levels of cross-reaction with culture filtrates of *C. botulinum* type A and B were reported. This suggests presence of contaminants in the type E neurotoxin preparation that was used to prepare the toxoid. Also in the case of type B assay a slight cross-reaction occurred with type A culture filtrate. This indicates once again the impurity, though slight, of the type B neurotoxin used as antigen.

In an RPHA[30] the antitoxin serum was made monospecific (i.e., free of the anti-nontoxin antibodies) by affinity chromatography. The lower limit of detection of types A and B toxin was 8 to 59 mouse $LD_{50}/m\ell$.

Antisera developed with purified type A neurotoxin (150,000 mol wt) was used in an RIA.[152] The authors claimed a lower limit of detection of 400 mouse MLD (equivalent to 4 ng protein), but this was not experimentally established. The value was derived from extrapolation of a curve with the last experimental point at 1000 MLD. Furthermore, it is not clear if the amount of toxin equivalent to 400 MLD must be present in an assay tube or in 1 mℓ of sample.

Quantitation of 8 ng of type A neurotoxin per milliliter was accomplished by an RIA recently developed.[153] The purest neurotoxin available (150,000 mol wt) was the source of toxoid for production of the antitoxin serum. Better preparations of ^{125}I-toxin were obtained by using the chloramine-T method, a modification of Haberman's technique,[154] than with the lactoperoxidase method.[155] A radioiodinated sample could be used for 49 days after its preparation.

The detection limits of RIA for insulin,[156] adenocorticoprotein,[157] and glucagon[158] are 50, 10, and 50 pg/mℓ, respectively, and for staphylococcal enterotoxin A,[159] and choleratoxin[160] are 0.3 and 1 ng/mℓ, respectively. In view of this, is the least amount of botulinum neurotoxin that can be quantitated, 8 ng/mℓ, too high?[153] The answer in terms of the number of molecules is no, because the apparent wide differences narrow when molecular weight of these substances, ranging from 6000 to 150,000 is considered. Thus, the detection limits are 8.3×10^{-12} *M* for insulin, 2.2×10^{-12} *M* for adenocorticoprotein, 1.4×10^{-11} *M* for glucagon, 1.1×10^{-11} *M* for staphylococcal enterotoxin A, 1.2×10^{-11} *M* for cholera toxin, and 5.3×10^{-11} *M* for botulinum toxin.

The published serological assays express serological activity in terms of toxicity,[145,148-152] or as amount of protein.[30,144,147] Equating the two properties is misleading in terms of what the serological assays detect. Betley[153] cultured several type A strains in liquid medium[109] and also in canned carrots and peas. After an appropriate

Table 7
TOXIN CONCENTRATIONS IN FRESH AND AGED CULTURE FILTRATES OF *C. BOTULINUM* TYPE A STRAINS AS DETERMINED BY RIA AND I.P. ASSAY

	Day 0[a]			Day 30[b]		
	Toxin (ng/mℓ)			Toxin (ng/mℓ)		
Strain	i.p.	RIA	% (i.p./RIA)	i.p.	RIA	% (i.p./RIA)
Hall	4,000	3,052	131	3,200	1,992	161
56A	2,430	1,372	177	960	912	105
62A	3,030	1,764	172	450	816	55
69A	3,200	1,876	171	54	528	10
Standard toxin	194,000	183,232	106	—	—	—

[a] Culture filtrates were assayed on day 0 of preparation.
[b] Culture filtrates were reassayed after holding for 30 days at room temperature.

period of incubation, toxic culture filtrates from the liquid medium and toxic fluid from the canned vegetables were obtained. These samples were allowed to age and were assayed in two ways: neurotoxin protein on the basis of toxicity (1 mg ≡ 1 × 10^8 LD_{50}) and serological activity (RIA being specific for the neurotoxin and SRP). On day 0 as much or more protein was detected in terms of toxicity than serological activity (see Tables 7, 8). The ratio of pharmacologically active toxin protein and SRP changed as the samples aged. Toxicity decreased faster than the serological activity. The only exception was the Hall strain in the liquid medium. These studies point out that:

1. The unit of serological activity should be mass/vol and not toxicity/vol because serologically active molecules need not be toxic.
2. If a serological assay is used to estimate the amount of pharmacologically active toxin, one is measuring both nontoxic-serologically reactive (i.e., SRP, see earlier) and serologically active toxin molecules. The relative proportion of these two species may vary from sample to sample. Therefore the amount of neurotoxin detected by serological methods may be erroneous due to the presence of SRP.

Prior to the assay of the toxin its extraction from a test specimen and concentration is a desirable step; 15- to 112-fold concentrations of types A, B, C, D, and E toxin from canned beans have been reported.[161]

INFANT BOTULISM

C. botulinum apparently behaves as an enteric infectious agent in human infant botulism. It appears that germination of ingested spores and subsequent multiplication of the bacteria leads to toxin production in the intestine. Constipation is almost always associated with infant botulism and precedes signs of neuromuscular paralysis by a few days or weeks. Some infants show only mild weakness, lethargy, and reduced feeding. Many have shown more severe and acute signs such as weakened sucking, swallowing, and crying; generalized muscle weakness; and diminished gag reflex with a pooling of oral secretions. Generalized muscle weakness and loss of head control in some reaches a degree of severity so that the patient appears "floppy". In a number

Table 8
TOXIN CONCENTRATIONS IN FRESH AND AGED TOXIC VEGETABLE SAMPLES OF *C. BOTULINUM* TYPE A STRAINS AS DETERMINED BY RIA AND I.P. ASSAY

		Day 0[a]			Day 47[b]		
		Toxin (ng/mℓ)			Toxin (ng/mℓ)		
Vegetable	Strain	i.p.	RIA	% (i.p./RIA)	i.p.	RIA	% (i.p./RIA)
Carrot	Hall	453	435	104	—	—[c]	—
Carrot	56A	269	290	93	45	93	48
Carrot	62A	1210	1305	93	90	140	64
Pea	Hall	1130	870	130	540	570	95
Pea	56A	1420	960	148	4.5	21	21
Pea	62A	2420	900	269	—	—[c]	—

[a] All samples were assayed on day 0 of preparation.
[b] All samples were assayed after holding for 47 days at room temperature.
[c] Protein concentration was too low to be assayed by RIA; therefore values derived from RIA and i.p. assay could not be compared.

of documented hospitalized cases, respiratory arrest has occurred, but with vigilant monitoring procedures most have been resuscitated and have ultimately recovered.[19,162,163] Very acute onset of the disease leading to rapid respiratory arrest and death is thought to be one possible cause of the Sudden Infant Death Syndrome (SIDS) or crib death.[164,165] In California, autopsy studies of 280 infants who died suddenly and unexpectedly revealed *C. botulinum* in fecal specimens of nine cases and spleen of one case; botulinum toxin was found in two of the ten cases.[164] The serum of 1 case[166] out of more than 100 studied contained the toxin. The organism and toxin has been detected in fecal excretion for 6 and 5 months, respectively, after onset of the disease, and also long after discharge of the patient from the hospital.[162,167] This raises the question: what prevents enough of the toxin from being continuously absorbed from the intestine to cause paralysis? The infants hospitalized with clinically recognized botulism were 3 to 35 weeks old with no bias of sex. Prior to the onset of symptoms of the disease these infants had essentially normal history. Extensive investigation of numerous recognized cases of infant botulism failed to identify an infant who had ingested food contaminated with botulinum toxin;[19,162,163,168] the infant was the only family member infected. The disease has been reported in the U.S. (from 16 states), England, and Australia. The high case report in California is perhaps due to the intensive surveillance.

The potential vehicle of *C. botulinum* for affected infants was identified only in a few cases; they were vacuum cleaner dust, yard soil, and honey. The isolated organisms and the toxin type, detected in the sick infants, were A and B. Among the various food and drink fed to the infants who developed botulism only honey was contaminated with *C. botulinum*.[162,168,169] Studies in four laboratories showed presence of the organism in 7.5 to 13% of the honey samples tested.[170-172] Spores were found in samples of honey fed to five infants who had developed botulism. Of the first 43 documented infant botulism cases in California 13 had ingested honey prior to the onset of illness. Honey is not the only vehicle for *C. botulinum* because only about ⅓ of the number of infants infected with botulism in California had ingested honey prior to illness.[171] It would be wise to follow the cautionary comment "...honey is now an identified and avoidable source of *C. botulinum* spores, and it therefore should not be fed to infants."[168]

The mouse is being used as a model for the study of infant botulism.[173,174] Experimental infection of infant mice with a few hundred *C. botulinum* spores produced toxin in the large intestine of 50% of the mice. Adults of the same inbred strain could not be infected even with ten million spores of the same isolate. The mice were susceptible to infection between 7 to 13 days of age; intraintestinal toxin was detectable up to 7 days postinfection.[175] Germfree mice older than 5 days were intraintestinally infected with very few spores but younger mice were not.[173] This suggests that the development of normal microbial flora of conventional infant mice prohibits outgrowth of *C. botulinum* and its toxin production in the lower intestinal tract, but other factors must account for the refractory character of mice younger than 5 days. Spores of several laboratory strains and isolates from human infant showed a marked variation in their virulence for conventional infant mice; depending on the strain, an inoculum of a few hundred to several million spores was required for successful infection. Most strains, when given to mice in sufficient numbers to establish intraintestinal infection and toxin production, did not produce any sign of illness. The reason for the difference in virulence is unclear although difference in potency of toxin is a possible factor. Compared to the toxins of most strains tested, the toxins of three type B strains, two of which were isolated from human cases of infant botulism, were 500 to 5000 times more potent for infant than adult mice.[174] Some isolates of *C. botulinum* from human infant cases have a greater virulence for infant mice in terms of the potency of the toxin as compared to that of most laboratory strains. (D. C. Mills, personal communication).

NEUROTOXIN ANTAGONISM

Tris buffer in the pH range of 7 to 9 and concentration of 5 to 10 m*M* attenuates acetylcholine response in Aplysia neurons.[176] The mechanism of attenuation is perhaps related to the structural similarities between tris and acetylcholine molecules.[176,177] This antagonizing effect of tris therefore warrants consideration in designing biochemical and physiological studies of botulinum neurotoxin on synaptic action.[176]

Exposure of muscle preparations or test animals to the following agents delays onset of botulinum toxin-induced paralysis and time to death: β-bungarotoxin,[178] serotonin,[179] calcium,[180] X-ray irradiation,[181] theophylline,[182] tetraethylammonium, guanidine, and 4-aminopyridine.[183-185] None of the substances have been shown to compete for the target sites in the nerve endings or to interact with the toxin molecules.

β-bungarotoxin — Time to paralysis of rat phrenic nerve-diaphragm treated (1) first with β-bungarotoxin and then with botulinum toxin, (2) in a reverse order, or (3) with the two together was longer than when treated with either (4) botulinum toxin or (5) β-bungarotoxin. Times to paralysis were (1) = 277 min, (2) = 174 min, (3) = 253 min, (4) = 147 min, and (5) = 126 min.

Serotonin — Serotonin injected into mice at least 30 min prior to i.v. injection of botulinum toxin increased the survival time. Similar apparent antagonism was observed also with sodium pentabarbitone and formaldehyde.[186] The two compounds and serotonin produce hypothermia in mice.[186] Because the Q_{10} of botulinum toxin is high,[80,187] Simpson[186] proposed that lowered body temperature slows the binding and/or activity of the toxin. The effect of incubation temperature on binding of type neurotoxin to the large synaptosomes (prepared from rat cerebral and cerebellar cortices) is dramatic: of a given amount of neurotoxin >99% was bound at 37°C, only a small percentage at 4°C, and not at all at 0°C.[188]

Calcium — The paralytic action of botulinum toxin in in vitro studies was delayed by increasing the ambient concentration of Ca^{++}; delay in paralysis was a direct linear function of Ca^{++}/Mg^{++} ratio.[180]

X-ray — Irradiated mice (whole body exposed, 250 kV, 10 mA, filter 0.5 Cu + Al, dose rate 76 rad/min, dose 760 rad) injected (i.p.) with toxin survived longer than nonirradiated-toxin injected mice; with type C toxin survival times were 118.0 vs. 99.3 min, and with type A toxin these values were 129 vs. 109 min. When diluted toxin (which increases survival time) was used the difference between survival time, 620 ± 30 vs. 620 ± 23 min, disappeared.

Theophylline — The interval between addition of toxin and complete inhibition of muscle (mice phrenic nerve-diaphragm) response to a single nerve pulse was 97 ± 15 min. Exposure to theophylline 15 min prior to toxin increased the interval to 207 ± 23 min. The drug did not restore neuromuscular transmission if added at the onset of toxin induced paralysis. Mice injected with toxin and theophylline survived longer than those injected with toxin alone. The drug did not prevent death.

Tetraethylammonium and guanidine — These substances, individually, restored neuromuscular transmission in vitro from complete paralysis to about the normal level. The compounds were ineffective on completely paralyzed muscle in vivo but improved neuromuscular transmission in muscles that were not completely paralyzed (as demonstrated by single fiber EMG). The combined effect of guanidine and germine in bringing about reversal of neuromuscular block of botulism in rabbits was more than the effect of individual drugs;[189] these two drugs also produced clinical improvement of botulism in rabbits.[189] Therapy of human botulism with guanidine has shown improvement in some but not in all cases,[190,191] while germine had no beneficial effect.[191]

4-aminopyridine — This was 20 to 30 times more potent than tetraethylammonium and guanidine in restoring neuromuscular transmission following paralysis; it also was active in vivo. Rats, on the second day after injection of a lethal amount of toxin, were paralyzed, unable to move, and showed signs of respiratory failure. "At that time 4-aminopyridine 2-3 mg/kg body weight was given as an i.p. injection. About 10 min later the animals were able to lift their heads and move around. At higher doses 4-aminopyridine produced generalized convulsions. The recovery lasted 1-2 hr after which paralysis reappeared. A second injection of 4-aminopyridine relieved the symptoms."[183] The authors suggest that the mode of action of 4-aminopyridine, somewhat similar to tetraethylammonium and guanidine, is related to the intracellular and ambient concentration of Ca^{++}. 4-aminopyridine seems to be a promising drug for treating humans in botulinum toxin poisoning. It has been used as an anticurare agent. In treating Eaton Lambert disease, continuous oral administration of the drug for 3 weeks restores neuromuscular transmission with no serious side effects.[185]

ACKNOWLEDGMENTS

This work was supported in part by the College of Agricultural and Life Sciences, the University of Wisconsin, Madison; by funds (DAMD 17-80-C-0100) from the U.S. Army Medical Research and Development Command, Fort Detrick, Frederick, Md., and by the University of Wisconsin Food Research Institute.

REFERENCES

1. Lewis, K. H. and Cassel, K., Jr., Eds., *Botulism,* Publ. No. 999-FP-1, U.S. Public Health Service, Washington, D.C., 1964.
2. Ingram, M. and Roberts, T. A., Eds., *Botulism 1966,* Chapman and Hall, London, 1967.
3. Boroff, D. A. and DasGupta, B. R., Botulinum toxin, in *Microbial Toxins,* Vol. 2A, Kadis, S., Montie, T. C., and Ajl, S. J., Eds., Academic Press, New York, 1971, chap. 1.

4. **Hobbs, G. C.**, Botulinum and its importance in fishery products, in *Advances in Food Research*, Vol. 22, Chichester, C. O., Mrak, E. M., and Stewart, G. F., Eds., Academic Press, New York, 1976, 135.
5. **Crowther, J. S. and Holbrook, R.**, Trends in methods for detecting food-poisoning toxins produced by C. botulinum and S. aureus, in *Microbiology in Agriculture, Fisheries and Food*, Skinner, F. A. and Carr, J. G., Eds., Academic Press, New York, 1976, 215.
6. **Kimble, C. E.**, C. botulinum and C. perfringens, in *Immunological Aspects of Foods*, Catsimpoolas, N., Ed., AVI Publishing, Westport, Conn., 1977, chap. 11.
7. **Smith, L. DS.**, *Botulism, The Organism, Its Toxins, The Disease*, Charles C Thomas, Springfield, Ill., 1977.
8. **Foster, E. M.**, Foodborne hazards of microbial origin, *Fed. Proc.*, 37, 2577, 1978.
9. **DasGupta, B. R. and Sugiyama, H.**, Biochemistry and pharmacology of botulinum and tetanus neurotoxins, in *Perspectives in Toxinology*, Bernheimer, A. W., Ed., John Wiley & Sons, New York, 1977, chap. 5.
10. **Simpson, L. L.**, The neuroparalytic and hemagglutinating activities of botulinum toxin, in *Neuropoisons, their Pathophysiological Actions*, Simpson, L. L., Ed., Plenum Press, New York, 1971, chap. 14.
11. **Koenig, M. G.**, The clinical aspects of botulism, in *Neuropoisons, their Pathophysiological Actions*, Simpson, L. L., Ed., Plenum Press, New York, 1971, chap. 13.
12. **Drachman, D. B.**, Botulinum toxin as a tool for research on the nervous system, in *Neuropoisons, their Pathophysiological Actions*, Simpson, L. L., Ed., Plenum Press, New York, 1971, chap. 15.
13. **Lamanna, C.**, The pipe and valve hypothesis of the mechanism of action of botulinal toxin, *Ann. Acad. Sci. Arts Bosnia Herzegovinia Special Publ. XXIX*, 4, 213, 1976.
14. **Hanig, J. P. and Lamanna, C.**, Toxicity of botulinum toxin: a stoichiometric model for the locus of its extraordinary potency and persistence at the neuromuscular junction, *J. Theor. Biol.*, 77, 107, 1979.
15. **Speck, M. L., Ed.**, *Compendium of Methods for the Microbiological Examination of Foods*, American Public Health Association, Washington, D.C., 1976, chap. 34.
16. Microorganisms in Foods. I. Their Significance and Methods of Enumeration, 2nd ed., Publ. Int. Comm. Microbiol. Specifications for Foods of the Int. Assoc. Microbiol. Soc., University of Toronto Press, 1978, 257.
17. **Cherington, M. and Ginsburg, S.**, Wound botulism, *Arch. Surg.*, 110, 436, 1975.
18. **de Jesus, P. V., Slater, R., Spitz, L. K., and Penn, A. S.**, Neuromuscular physiology of wound botulism, *Arch. Neurol.*, 29, 425, 1974.
19. **Dowell, V. R., Jr.**, Infant botulism: New guise for an old disease, *Hosp. Pract.*, 13, 67, 1978.
20. **Ciccarelli, A. S., Whaley, D. N., McCroskey, L. M., Gimenez, D. F., Dowell, V. R., Jr., and Hatheway, C. L.**, Cultural and physiological characteristics of *C. botulinum* type G and the susceptibility of certain animals to its toxin, *Appl. Environ. Microbiol.*, 34, 843, 1977.
21. **Sugiyama, H., Mizutani, K., and Yang, K. H.**, Basis of type A and F toxicities of *C. botulinum* strain 84, *Proc. Soc. Exp. Biol. Med.*, 141, 1063, 1972.
22. **Giménez, D. F. and Ciccarelli, A. S.**, New strains of *Clostridium botulinum* subtype Af, *Zbl. Bakt. Hyg., I. Abt. Orig. A*, 240, 215, 1978.
23. **Eklund, M. W. and Poysky, F. T.**, Activation of a toxic component of *C. botulinum* types C and D by trypsin, *Appl. Microbiol.*, 24, 108, 1972.
24. **Jensen, B. C.**, The toxic antigenic factors produced by *C. botulinum* types C and D, *Onderstpoort J. Vet. Res.*, 38, 93, 1971.
25. **Eklund, M. W. and Poysky, F. T.**, Interconversion of type C and D strains of *C. botulinum* by specific bacteriophages, *Appl. Microbiol.*, 27, 251, 1974.
26. **Inoue, K. and Iida, H.**, Phage-conversion of toxigenicity in *C. botulinum*, types C and D, *Jpn. J. Med. Sci. Biol.*, 24, 53, 1971.
27. **Oguma, K., Iida, H., and Inoue, K.**, Bacteriophage and toxigenicity in *C. botulinum*: an additional evidence for phage conversion, *Jpn. J. Microbiol.*, 17, 425, 1973.
28. **Oguma, K., Iida, H., Shiozaki, M., and Inoue, K.**, Antigenicity of converting phages obtained from *C. botulinum* types C and D, *Infect. Immun.*, 13, 855, 1976.
29. **Yang, K. H. and Sugiyama, H.**, Purification and properties of *C. botulinum* type F toxin, *Appl. Microbiol.*, 29, 598, 1975.
30. **Sakaguchi, G., Sakaguchi, S., Kozaki, S., Sugii, S., and Ohishi, I.**, Cross reaction in reversed passive hemagglutination between *C. botulinum* type A and B toxins and its avoidance by the use of antitoxic component immunoglobulin isolated by affinity chromatography, *Jpn. J. Med. Sci. Biol.*, 27, 161, 1974.
31. **Shimizu, T. and Kondo, H.**, Immunological difference between the toxin of a proteolytic strain and that of a nonproteolytic strain of *C. botulinum* type B, *Jpn. J. Med. Sci. Biol.*, 26, 269, 1973.

32. **Shimizu, T., Kondo, H., and Sakaguchi, G.,** Immunological heterogeneity of *C. botulinum* type B toxins, *Jpn. J. Med. Sci. Biol.*, 27, 99, 1974.
33. **Miyazaki, S., Kozaki, S., Sakaguchi, S. and Sakaguchi, G.,** Comparison of progenitor toxins of nonproteolytic with those of proteolytic *C. botulinum* type B, *Infect. Immun.*, 13, 987, 1976.
34. **Lamanna, C. and Sakaguchi, G.,** Botulinal toxins and the problem of nomenclature of simple toxins, *Bacteriol. Rev.*, 32, 242, 1971.
35. **Kozaki, S., Sugi, S., Ohishi, I., Sakaguchi, S., and Sakaguchi, G.,** Molecular structure of *C. botulinum* toxins, in *Animal, Plant and Microbial Toxins,* Vol. 1, Ohsaka, A., Hayashi, K., and Sawai, Y., Eds., Plenum Press, New York, 1976, 375.
36. **DasGupta, B. R. and Sugiyama, H.,** A common subunit structure in *C. botulinum* type A, B and E toxins, *Biochem. Biophys. Res. Commun.*, 48, 108, 1972.
37. **DasGupta, B. R. and Boroff, D. A.,** Separation of toxin and hemagglutinin from crystalline toxin of *C. botulinum* type A by anion exchange chromatography and determination of their dimensions by gel filtration, *J. Biol. Chem.*, 243, 1065, 1968.
38. **Syuto, B. and Kubo, S.,** Purification and crystallization of *C. botulinum* type C toxin, *Jpn. J. Vet. Res.*, 20, 19, 1972.
39. **Schantz, E. J.,** Some chemical and physical properties of *C. botulinum* toxins in culture, *Jpn. J. Microbiol.*, 11, 380, 1967.
40. **Sugii, S. and Sakaguchi, G.,** Molecular construction of *C. botulinum* type A toxins, *Infect. Immun.*, 12, 1262, 1975.
41. **Kozaki, S., Sakaguchi, S., and Sakaguchi, G.,** Purification and some properties of progenitor toxins of *C. botulinum* type B, *Infect. Immun.*, 10, 750, 1974.
42. **Miyazaki, S., Iwasaki, M., and Sakaguchi, G.,** *C. botulinum* type D toxin: purification, molecular structure and some immunological properties, *Infect. Immun.*, 17, 395, 1977.
43. **Kitamura, M., Sakaguchi, S., and Sakaguchi, G.,** Significance of 12S toxin of *C. botulinum* type E, *J. Bacteriol.*, 98, 1173, 1969.
44. **Ohishi, I. and Sakaguchi, G.,** Purification of *C. botulinum* type F progenitor toxin, *Appl. Microbiol.*, 28, 923, 1974.
45. **Hauschild, A. H. W. and Hilsheimer, R.,** Heterogeneity of *C. botulinum* type A toxin, *Can. J. Microbiol.*, 14, 805, 1968.
46. **Kozaki, S., Sugii, S., Ohishi, I., Sakaguchi, S., and Sakaguchi, G.,** *C. botulinum* type A, B, E and F 12S toxins, *Jpn. J. Med. Sci. Biol.*, 28, 70, 1975.
47. **Sugiyama, H., Moberg, L. J., and Messer, S. L.,** Improved procedure for crystallization of *C. botulinum* type A toxic complexes, *Appl. Environ. Microbiol.*, 33, 963, 1977.
48. **DasGupta, B. R. and Sugiyama, H.** Inhibition of *C. botulinum* types A and B hemagglutinins by sugars, *Can. J. Microbiol.*, 23, 1257, 1977.
49. **Kitamura, M., Sakaguchi, S., and Sakaguchi, G.,** Purification and some properties of *C. botulinum* type-E toxin, *Biochem. Biophys. Acta*, 168, 207, 1968.
50. **Lamanna, C. and Lowenthal, J. P.,** The lack of identity between hemagglutinin and the toxin of type A botulinal organism, *J. Bacteriol.*, 61, 751, 1951.
51. **Ohishi, I., Sugii, S., and Sakaguchi, G.,** Oral toxicities of *C. botulinum* toxins in response to molecular size, *Infect. Immun.*, 16, 107, 1977.
52. **Sugiyama, H., DasGupta, B. R., and Yang, K. H.,** Toxicity of purified botulinal toxin fed to mice, *Proc. Soc. Exp. Biol. Med.*, 147, 589, 1974.
53. **Sugii, S., Ohishi, I., and Sakaguchi, G.,** Correlation between oral toxicity and in vitro stability of *C. botulinum* type A and B toxins of different molecular sizes, *Infect. Immun.*, 16, 910, 1977.
54. **Sugii, S., Ohishi, I., and Sakaguchi, G.,** Intestinal absorption of botulinum toxins of different molecular sizes in rats, *Infect. Immun.*, 17, 491, 1977.
55. **Ohishi, I., Sugii, S., and Sakaguchi, G.,** Absorption of botulinum type B progenitor and derivative toxins through the intestine, *Jpn. J. Med. Sci. Biol.*, 31, 161, 1978.
56. **Sugii, S., Ohishi, I., and Sakaguchi, G.,** Molecular sizes of botulinum toxins in foods, *Jpn. J. Med. Sci. Biol.*, 31, 159, 1978.
57. **Ohishi, I. and Sakaguchi, G.,** Activation of botulinum toxins in the absence of nicking, *Infect. Immun.*, 17, 402, 1977.
58. **Kozaki, S. and Sakaguchi, G.,** Antigenicities of fragments of *C. botulinum* type B derivative toxin, *Infect. Immun.*, 11, 932, 1975.
59. **DasGupta, B. R. and Sugiyama, H.,** Molecular forms of neurotoxins in proteolytic *C. botulinum* type B cultures, *Infect. Immun.*, 14, 680, 1976.
60. **Syuto, B. and Kubo, S.,** Isolation and molecular size of *C. botulinum* type C toxin, *Appl. Environ. Microbiol.*, 33, 400, 1977.
61. **Boroff, D. A. and Shu-Chen, G.,** Chemistry and biological activity of the toxin of *C. botulinum* type D, *Am. Soc. Microbiol. Abstr.*, 1973, 19.

62. **DasGupta, B. R. and Sugiyama, H.**, Single chain and dichain forms of neurotoxin in type F *C. botulinum* culture, *Toxicon*, 15, 466, 1977.
63. **Krysinski, E. P.**, Studies on Structure-Activity Relationships of *C. botulinum* type A Toxin, Ph.D. thesis, University of Wisconsin, Madison, 1978.
64. **DasGupta, B. R. and Sugiyama, H.**, Limited p roteolysis of *C. botulinum* types A and B neurotoxins, *Am. Soc. Microbiol. Abstr.*, 1978, 25.
65. **DasGupta, B. R. and Sugiyama, H.**, Comparative sizes of type A and B botulinum neurotoxins, *Toxicon*, 15, 357, 1977.
66. **Beers, W. H. and Reich, E.**, Isolation and characterization of *C. botulinum* Type B toxin, *J. Biol. Chem.*, 244, 4473, 1969.
67. **Knox, J. H., Brown, W. P., and Spero, L.**, The role of sulfhydryl groups in the activity of type A botulinum toxin, *Biochim. Biophys. Acta*, 214, 350, 1970.
68. **Boroff, D. A., Townend, R., Fleck, U., and DasGupta, B. R.**, Ultracentrifugal analysis of the crystalline toxin and isolated fractions of *C. botulinum* type A, *J. Biol. Chem.*, 241, 5165, 1966.
69. **DasGupta, B. R., Berry, L. J., and Boroff, D. A.**, Purification of *C. botulinum* type A toxin, *Biochim. Biophys. Acta*, 214, 343, 1970.
70. **DasGupta, B. R., Boroff, D. A., and Cheong, K.**, Isolation of chromatographically pure toxin of *C. botulinum* type B, *Biochem. Biophys. Res. Commun.*, 32, 1057, 1968.
71. **DasGupta, B. R., Boroff, D. A., and Rothstein, E.**, Chromatographic fractionation of the crystalline toxin of *C. botulinum* type A, *Biochem. Biophys. Res. Commun.*, 22, 750, 1966.
72. **Heimsch, R.**, Purification and Properties of *C. botulinum* Type E Toxin, Ph.D. thesis, University of Wisconsin, Madison, 1973.
73. **Yang, K. H.**, Purification and Characterization of *C. botulinum* Type F Toxin, Ph.D. thesis, University of Wisconsin, Madison, 1974.
74. **Hauschild, A. H. W. and Hilscheimer, R.**, Antigenic and chromatographic identity of two apparently distinct toxins of *C. botulinum* type A, *Can. J. Microbiol.*, 14, 1129, 1969.
75. **Remmers, V. J.**, Results of chromatographic purification of *C. botulinum* serotype B, *Arch. Libensmittlhyg.*, 28, 86, 1977.
76. **Sacks, H. S. and Covert, S. V.**, *C. botulinum* type E toxin: effect of pH and method of purification on molecular weight, *Appl. Microbiol.*, 28, 374, 1974.
77. **Ohishi, I. and Sakaguchi, G.**, Molecular construction of *C. botulinum* type F progenitor toxin, *Appl. Microbiol.*, 29, 444, 1975.
78. **Kitamura, M. and Sakaguchi, G.**, Dissociation and reconstitution of 12-S toxin of *C. botulinum* type E, *Biochem. Biophys. Acta*, 194, 564, 1969.
79. **Singleton, R., Jr., Middaugh, C. R., and MacElroy, R. D.**, Comparison of proteins from theromophilic and nonthermophilic sources in terms of structural parameters inferred from amino acid composition, *Int. J. Peptide Protein Res.*, 10, 39, 1977.
80. **Simpson, L. L.**, Ionic requirements for the neuromuscular blocking action of botulinum toxin: implications with regard to synaptic transmission, *Neuropharmacology*, 10, 673, 1971.
81. **Boroff, D. A., Meloche, H. P., and DasGupta, B. R.**, Amino acid analysis of the isolated and purified components from crystalline toxin of *C. botulinum* type A, *Infect. Immun.*, 2, 679, 1970.
82. **Jukes, T. H., Holmquist, R., and Moise, H.**, Amino acid composition of proteins: selection against the genetic code, *Science*, 189, 50, 1975.
83. **Holmquist, R.**, Deviations from compositional randomness in eukaryotic and prokaryotic proteins: the hypothesis of selective-stochastic stability and a principle of charge conversion, *J. Mol. Evol.*, 4, 277, 1975.
84. **Holmquist, R. and Moise, H.**, Compositional nonrandomness: a quantitatively conserved evolutionary invariant, *J. Mol. Evol.*, 6, 1, 1975.
85. **Robinson, J. P., Picklesimer, J. B., and Puett, D.**, Tetanus toxin. The effect of chemical modifications on toxicity, immunogenicity, and conformation, *J. Biol. Chem.*, 250, 7435, 1975.
86. **Kozaki, S., Miyazaki, S., and Sakaguchi, G.**, Development of antitoxin with each of two complementary fragments of *C. botulinum* type B derivative toxin, *Infect. Immun.*, 18, 761, 1977.
87. **Syuto, B. and Kubo, S.**, Purification and characterization of *C. botulinum* type C toxin, *Jpn. J. Med. Sci. Biol.*, 31, 169, 1978.
88. **Syuto, B. and Kubo, S.**, Structure and toxicity of *C. botulinum* type C toxin, *Jpn. J. Med. Sci. Biol.*, 32, 132, 1979.
89. **Kozaki, S., Miazaki, S., and Sakaguchi, G.**, Structure of *C. botulinum* type B derivative toxin: inhibition with a fragment of toxin from binding to synaptosomal fraction, *Jpn. J. Med. Sci. Biol.*, 31, 163, 1978.
90. **Boroff, D. A. and DasGupta, B. R.**, Study of the toxin of *C. botulinum*. VII. Relation of tryptophan to the biological activity of the toxin, *J. Biol. Chem.*, 239, 3694, 1964.

91. **Boroff, D. A. and DasGupta, B. R.**, Study of the toxin of *C. botulinum.* Effects of 2-hydroxy-5-nitrobenzyl bromide on the biological activity of botulinum toxin, *Biochim. Biophys. Acta*, 117, 289, 1966.
92. **DasGupta, B. R. and Sugiyama, H.**, Role of arginine residue in the structure and biological activity of botulinum toxin, *Am. Soc. Microbiol. Abstr.*, 1979, 213.
93. **Iida, H.**, Experimental studies on toxin production of *C. botulinum.* II, *Jpn. J. Bacteriol.*, 19, 463, 1964.
94. **Iida, H.**, Activation of *C. botulinum* toxin by trypsin, in Toxic Microorganisms, Herzberg, M., Ed., U.S. Department of the Interior, Washington, D.C., 1970, 336.
95. **Nakane, A., Oguma, K., Shiozaki, M., and Iida, H.**, Production of trypsin-activable toxic components by *C. botulinum* types C and D, *Jpn. J. Med. Sci. Biol.*, 31, 166, 1978.
95a. **Nakane, A., Oguma, K., and Iida, H.**, Activation of some *C. botulinum* type D toxin by trypsin, *J. Gen. Microbiol.*, 111, 429, 1979.
96. **Ohishi, I., Okada, T., and Sakaguchi, G.**, Responses of *C. botulinum* type B and E progenitor toxins to some clostridial sulfhydryl-dependent proteases, *Jpn. J. Med. Sci. Biol.*, 28, 157, 1975.
97. **DasGupta, B. R. and Sugiyama, H.**, Role of a protease in natural activation of *C. botulinum* neurotoxin, *Infect. Immun.*, 6, 587, 1972.
98. **Inukai, Y.**, Role of proteolytic enzyme in toxin production by *C. botulinum* type A, *Jpn. J. Vet. Res.*, 11, 143, 1963.
99. **Skulberg, A.**, *Studies on the Formation of Toxin by C. botulinum*, Orkanger, Norway, A/S Kaare Grytting, 1964.
100. **Kodama, H.**, Studies on the proteolytic enzyme of *C. botulinum*, *Jpn. J. Vet. Res.*, 9, 127, 1961.
101. **Ivanova, L. G., Blagoveschensky, V. A., and Sergeeva, T. I.**, The action of homologous proteinase on *C. botulinum* type B prototoxin, *Zn. Mikrobiol. Epidemiol. Immunobiol.*, No. 2, 115, 1974.
102. **Nakane, A.**, Proteases produced by a proteolytic mutant of *C. botulinum* type E, *J. Gen. Microbiol.*, 107, 85, 1978.
103. **Ohishi, I. and Sakaguchi, G.**, Response of type B and E botulinum toxins to purified sulfhydryl-dependent protease produced by *C. botulinum* type F, *Jpn. J. Med. Sci. Biol.*, 30, 179, 1977.
104. **Miura, T.**, Isolation and characterization of an esterase-active enzyme from pronase with special reference to activation of *C. botulinum* type E progenitor toxin, *Jpn. J. Med. Sci. Biol.*, 27, 285, 1974.
105. **Duff, J. T., Wright, G. G., and Yarinsky, A.**, Activation of *C. botulinum* type E toxin by trypsin, *J. Bacteriol.*, 72, 455, 1956.
106. **Sakaguchi, G. and Sakaguchi, S.**, Studies on toxin production of *C. botulinum* type E. III. Characterization of toxin precursor, *J. Bacteriol.*, 78, 1, 1959.
107. **Lamanna, C., Eklund, H. W., and McElroy, O. E.**, Botulinum toxin (type A); including a study of shaking with chloroform as a step in the isolation procedure, *J. Bacteriol.*, 52, 1, 1946.
108. **Abrams, A., Kegeles, G., and Hottle, G. A.**, The purification of toxin from *C. botulinum* type A, *J. Biol. Chem.*, 164, 63, 1946.
109. **Duff, J. T., Wright, G. G., Klerer, J., Moore, D. E., and Bibler, R. H.**, Studies on immunity to toxins of *C. botulinum.* I. A simplified procedure for isolation of type A toxin, *J. Bacteriol.*, 73, 42, 1957.
110. **Gerwing, J., Dolman, C. E., and Bains, H. S.**, Isolation and characterization of a toxic moiety of low molecular weight from *C. botulinum* type A, *J. Bacteriol.*, 89, 1383, 1965.
111. **Lamanna, C. and Glassman, H. N.**, The isolation of type B botulinum toxin, *J. Bacteriol.*, 54, 575, 1947.
112. **Duff, J. T., Klerer, J., Bibler, R. H., Moore, D. E., Gottfried, C., and Wright, G. G.**, Studies on immunity to toxins of *C. botulinum.* II. Production and purification of type B toxin for toxoid, *J. Bacteriol.*, 73, 597, 1957.
113. **Gerwing, J., Dolman, C. E., Kason, D. V., and Tremaine, J. H.**, Purification and characterization of *C. botulinum* type B toxin, *J. Bacteriol.*, 91, 484, 1966.
114. **Sterne, M. and Wentzel, L. M.**, A new method for the large-scale production of high-titer botulinum formol-toxoid types C and D, *J. Immunol.*, 65, 175, 1950.
115. **Cardella, M. A., Duff, J. T., Gottfried, C., and Begel, J. S.**, Studies on immunity to toxins of *C. botulinum.* IV. Production and purification of type C toxin for conversion to toxid, *J. Bacteriol.*, 75, 360, 1958.
116. **Syuto, B.**, Studies on purification of *C. botulinum* type C toxin, *Jpn. J. Vet. Res.*, 13, 63, 1965.
117. **Cardella, M. A., Duff, J. T., Wingfield, B. H., and Gottfried, C.**, Studies on immunity to toxins of *C. botulinum.* VI. Purification and detoxification of type D toxin and the immunological response to toxoid, *J. Bacteriol.*, 79, 372, 1960.
118. **Hatheway, C. L.**, Toxoid of *C. botulinum* type F: purification and immunogenicity studies, *Appl. Environ. Microbiol.*, 31, 234, 1976.

119. **Sakaguchi, G. and Sakaguchi, S.**, A simple method for purification of type E botulinal toxin from the precursor extract of the bacterial cells, *Jpn. J. Med. Sci. Biol.*, 14, 243, 1961.
120. **Bonventre, P. F. and Kempe, L. L.**, Physiology of toxin production by *C. botulinum* types A and B. II. Effect of carbohydrate source on growth, autolysis, and toxin production, *Appl. Microbiol.*, 7, 372, 1959.
121. **Inukai, Y.**, Effect of carbohydrate on toxin production by *C. botulinum* type A, *Jpn. J. Vet. Res.*, 10, 64, 1962.
122. **Bonventre, P. F. and Kempe, L. L.**, Physiology of toxin production by *C. botulinum* types A and B. I. Growth, autolysis and toxin production, *J. Bacteriol.*, 79, 18, 1960.
123. **Kawata, T., Takumi, K., Sato, S., and Yamashita, H.**, Autolytic formation of spheroplasts and autolysis of cell walls in *C. botulinum* type A, *Jpn. J. Microbiol.*, 12, 445, 1968.
124. **Kawata, T. and Takumi, K.**, Autolytic enzyme system of *C. botulinum*. I. Partial purification and characterization of an autolysin of *C. botulinum* type A, *Jpn. J. Microbiol.*, 15, 1, 1971.
125. **Takumi, K., Kawata, T., and Hisatsune, K.**, Autolytic enzyme system of *C. botulinsm*. II. Mode of action of autolytic enzymes in *C. botulinum* type A, *Jpn. J. Microbiol.*, 15, 131, 1971.
126. **Ogata, S. and Hongo, M.**, Bacterial lysis of *Clostridium* species. I. Lysis of *Clostridium* species by univalent cation, *J. Gen. Appl. Microbiol.*, 19, 251, 1973.
127. **Gordon, M., Fiock, M. A., Yarinsky, A., and Duff, J. T.**, Studies on Immunity to toxins of *C. botulinum*. III. Preparation, purification, and detoxification of type E toxin, *J. Bacteriol.*, 74, 533, 1957.
128. **Iwasaki, M. and Sakaguchi, G.**, Acid precipitation of *C. botulinum* type C and D toxins from whole culture by addition of ribonucleic acid as a precipitation aid, *Infect. Immun.*, 19, 749, 1978.
129. **Johnson, T. J. A. and Bock, R. M.**, Enzyme fractionation and simultaneous nucleic acid removal from crude cellular extracts by preformed gradient ion exchange gel filtration, *Anal. Biochem.*, 59, 375, 1974.
130. **Moberg, L. J. and Sugiyama, H.**, Affinity chromatography purification of type A botulinum neurotoxin from crystalline toxic complex, *Appl. Environ. Microbiol.*, 35, 878, 1978.
131. **Thompson, T. S., Cass, K. H., and Stellwagen, E.**, Blue dextran-Sepharose: an affinity column for the dinucleotide fold in proteins, *Proc. Natl. Acad. Sci. U.S.A.*, 72, 669, 1975.
132. **Ryan, L. D. and Vestling, C. S.**, Rapid purification of lactate dehydrogenase from rat liver and hepatoma: a new approach, *Arch. Biochem. Biophys.*, 160, 279, 1974.
133. **Schantz, E. J. and Kautter, D. A.**, Standardized assay for *C. botulinum* toxins, *J. Assoc. Off. Anal. Chem.*, 61, 96, 1978.
134. **Boroff, D. A. and Fitzgerald, J. E.**, Fluorescence of the toxin of *C. botulinum* and its relation to toxicity, *Nature (London)*, 181, 751, 1958.
135. **Boroff, D. A. and Fleck, U.**, Statistical analysis of a rapid in vivo method for the titration of the toxin of *C. botulinum*, *J. Bacteriol.*, 92, 1580, 1966.
136. **Sakaguchi, G., Sakaguchi, S., and Kondo, H.**, Rapid bioassay for *C. botulinum* type E-toxins by intravenous injection into mice, *Jpn. J. Med. Sci. Biol.*, 21, 369, 1968.
137. **Lamanna, C., Spero, L., and Schantz, E. J.**, Dependence of time to death on molecular size of botulinal toxins, *Infect. Immun.*, 1, 423, 1970.
138. **Steward, J. P., Ornellas, E. P., Beernink, K. D., and Northway, W. H.**, Errors related to different techniques of intraperitoneal injection in mice, *Appl. Microbiol.*, 16, 1418, 1968.
139. **Miner, N. A., Koehler, J., and Greenaway, L.**, Intraperitoneal injection of mice, *Appl. Microbiol.*, 17, 250, 1969.
140. **Arioli, V. and Rossi, E.**, Errors related to different techniques of intraperitoneal injection in mice, *Appl. Microbiol.*, 19, 704, 1970.
141. **Hunter, B. F. and Rosen, M. N.**, Detection of *C. botulinum* type C cells and toxin by the fluorescent antibody technique, *Avian Dis.*, 9, 345, 1967.
142. **Aalvik, B., Sakaguchi, G., and Riemann, H.**, Detection of type E botulinal toxin in cultures by fluorescent-antibody microscopy, *Appl. Microbiol.*, 25, 153, 1973.
143. **Sugiyama, H., Ohishi, I., and DasGupta, B. R.**, Evaluation of type A botulinal toxin assays that use antitoxin to crystalline toxin, *Appl. Microbiol.*, 27, 333, 1974.
144. **Johnson, H. M., Brenner, K., Angelotti, R., and Hall, H. E.**, Serological studies of types A, B, and E botulinal toxins by passive hemagglutination and bentonite flocculation, *J. Bacteriol.*, 91, 967, 1966.
145. **Sinitsyn, V. A.**, Use of the indirect hemagglutination reaction in detection of botulinic toxins. I. Detection of botulinic toxins types A and B by means of the indirect hemagglutination reaction as modified by Rytsai, *Zh. Microbiol. Epidemiol. Immunobiol.*, 31, 408, 1960.
146. **Sonnenschein, B.**, Use of the reversed passive hemagglutination in detection of *C. botulinum* type A, B, and E toxins, *Zbl. Bakt. Hyg., I. Abt. Orig.*, 240, 221, 1978.
147. **Evancho, G. M., Ashton, D. H., Briskey, E. J., and Schantz, E. J.**, A standardized reversed passive hemagglutination technique for the determination of botulinum toxin, *J. Food Sci.*, 38, 764, 1973.

148. **Mestrandrea, L. W.**, Rapid detection of *C. botulinum* toxin by capillary tube diffusion, *Appl. Microbiol.*, 27, 1017, 1974.
149. **Vermilyea, B. L., Walker, H. W., and Ayres, J. C.**, Detection of botulinal toxins by immunodiffusion, *Appl. Microbiol.*, 16, 21, 1968.
150. **Miller, C. A. and Anderson, A. W.**, Rapid detection and quantitative estimation of type A botulinum toxin by electroimmunodiffusion, *Infect. Immun.*, 4, 126, 1971.
151. **Notermans, S., Dufrenne, J., and Schothorst, M. V.**, Enzyme-linked immunosorbent assay for detection of *C. botulinum* toxin type A, *Jpn. J. Med. Sci. Biol.*, 31, 81, 1978.
151a. **Notermans, S., Dufrenne, J., and Kozaki, S.**, Enzyme linked immunosorbent assay for detection of *C. botulinum* type E toxin, *Appl. Environ. Microbiol.*, 37, 1173, 1979.
151b. **Kozaki, S., Dufrenne, J., Hagenaars, A. M., and Notermans, S.**, Enzyme linked immunosorbent assay (ELISA) for detection of *C. botulinum* type B toxin, *Jpn. J. Med. Sci. Biol.*, 32, 199, 1979.
152. **Boroff, D. A. and Shu-Chen, G.**, Radioimmunoassay for type A toxin of *C. botulinum*, *Appl. Microbiol.*, 25, 545, 1973.
153. **Betley, M. J.**, Radioimmunoassay for *C. botulinum* Type A Toxin, M.S. thesis, University of Wisconsin, Madison, 1978.
154. **Haberman, E.**, ^{125}I-Labelled neurotoxin from *C. botulinum* A: to synaptosomes and ascent to the spinal cord, *Naunyn-Schmiedeberg's Arch. Pharmacol.*, 281, 47, 1974.
155. **Robern, H., Dighton, M., Yano, Y., and Dickie, N.**, Double-antibody radioimmunoassay for staphylococcal enterotoxin C_2, *Appl. Microbiol.*, 30, 525, 1975.
156. **Yallow, R. S. and Berson, S. A.**, Immunoassay of endogenous plasma insulin in man, *J. Clin. Invest.*, 39, 1157, 1960.
157. **Felber, J. P.**, ACTH antibodies and their use for a radioimmunoassay for ACTH, *Experientia*, 19, 227, 1963.
158. **Unger, R. H., McCall, M. S., and Madison, L. L.**, Glucagon antibodies and an immunoassay for glucagon, *J. Clin. Invest.*, 40, 1280, 1961.
159. **Miller, B. A., Reiser, R. F., and Bergdoll, M. S.**, Detection of staphylococcal enterotoxins A, B, C, D and E in foods by radioimmunoassay using staphylococcal cells containing protein A as immunoadsorbent, *Appl. Environ. Microbiol.*, 36, 421, 1978.
160. **Ceska, M., Effenberger, F., and Grossmüler, F.**, Highly sensitive solid-phase radioimmunoassay suitable for determination of low amounts of cholera toxin and cholera toxin antibodies, *J. Clin. Microbiol.*, 7, 209, 1978.
161. **Sonnenschein, B. and Bispinǵ, W.**, Extraction and concentration of *C. botulinum* toxins from specimens, *Zbl. Bakt. Hyg., I. Abt. Orig. A*, 234, 247, 1976.
162. **Arnon, S. S., Midura, T. F., Clay, S. A., Wood, R. M., and Chin, J.**, Infant botulism: epidemiological clinical, and laboratory aspects, *JAMA*, 237, 1946, 1977.
163. **Arnon, S. S. and Chin, J.**, The clinical spectrum of infant botulism, *Rev. Infect. Dis.*, 1, 614, 1979.
164. **Arnon, S. S., Midura, T. F., Damus, K., Wood, R. M., and Chin, J.**, Intestinal infection and toxin production by *C. botulinum* as one cause of sudden infant death syndrome, *Lancet*, 1, 1273, 1978.
165. **Marx, J. L.**, Botulism in infants: a cause of sudden death?, *Science*, 201, 799, 1978.
166. Infant Botulism — Arizona, Morbidity Mortality Weekly Rep. 27, Center for Disease Control, Atlanta, Ga., 1978, 411.
167. **Arnon, S. S., Midura, T. F., Dumas, K., Snowden, S., Thompson, B., and Chin, J.**, Persistent fecal excretion of *C. botulinum* toxin and organisms in patients recovering from infant botulism, *Am. Soc. Microbiol. Abstr.*, 1979, 357.
168. **Arnon, S. S., Midura, T. F., Dumas, K., Thompson, B., Wood, R. M., and Chin, J.**, Honey and other environmental risk factors for infant botulism, *J. Pediatr.*, 94, 331, 1979.
169. **Chin, J., Arnon, S. S., and Midura, T. F.**, Food and environmental aspects of infant botulism in California, *Rev. Infect. Dis.*, 1, 693, 1979.
170. **Sugiyama, H., Mills, D. C., and Kuo, L-J. C.**, Number of *C. botulinum* spores in honey, *J. Food Prot.*, 41, 848, 1978.
171. Honey Exposure and Infant Botulism — United States, Morbidity Mortality Weekly Rep. 27, Center for Disease Control, Atlanta, Ga., 1978, 249.
172. **Midura, T. F., Snowden, S., Wood, R. M., and Arnon, S. S.**, Isolation of *C. botulinum* from honey, *J. Clin. Microbiol.*, 9, 282, 1979.
173. **Moberg, L. J. and Sugiyama, H.**, Intraintestinal germination and growth of *C. botulinum* spores fed to germ free mice, *Am. Soc. Microbiol. Abstr.*, 1979, 213.
174. **Mills, D. C. and Sugiyama, H.**, The comparative sensitivities of infant and adult mice to botulinum toxin, *Am. Soc. Microbiol. Abstr.*, 1979, 212.
175. **Sugiyama, H. and Mills, D. C.**, Intraintestinal toxin in infant mice challenged with *C. botulinum* spores, *Infect. Immun.*, 21, 59, 1978.
176. **Wilson, W. A., Clark, M. T., and Pellmar, T. C.**, Tris buffer attenuates acetylcholine responses in Aplysia neurons, *Science*, 196, 440, 1977.

177. **Rudman, R., Eilerman, D., and LaPlaca, S. J.**, The structure of crystalline Tris: a plastic crystal precursor, buffer, and acetylcholine attenuator, *Science,* 200, 531, 1978.
178. **Chang, C. C., Huang, M. C., and Lee, C. Y.**, Mutual antagonism between botulinum toxin and β-bungarotoxin, *Nature (London),* 243, 166, 1973.
179. **Boroff, D. A. and Fleck, U.**, Effects of serotonin on the toxin of *C. botulinum, J. Pharmacol. Exp. Ther.,* 157, 427, 1967.
180. **Simpson, L. L. and Tapp, J. T.**, Actions of calcium and magnesium on the rate of onset of botulinum toxin paralysis of the rat diaphragm, *Int. J. Neuropharmacol.,* 6, 485, 1967.
181. **Forssberg, A., Sundius, G., Tribukait, B., and Nyman, CL.**, Observations on the combined effect of X-rays and botulinum toxin on the survival of mice, *Int. J. Rad. Biol.,* 6, 351, 1963.
182. **Howard, B. D., Wu, W. C-S., and Gunderson, C. B., Jr.**, Antagonism of botulinum toxin by theophylline, *Biochem. Biophys. Res. Commun.,* 71, 413, 1976.
183. **Lund, H., Leander, S., and Thesleff, S.**, Antagonism of paralysis produced by botulinum toxin in the rat. The effects of tetraethylammonium, guanidine and 4-aminopyridine, *J. Neurol. Sci.,* 32, 29, 1977.
184. **Lund, H.** Effects of 4-aminopyridine on neuromuscular transmission, *Brain Res.,* 153, 307, 1978.
185. **Lund, H.**, Paralysis in Botulinum Toxin Poisoning. An Experimental Study on Pathophysiology and Drug Treatment, Ph.D. thesis, University of Lund, Lund, 1978.
186. **Simpson, L. L.**, Mechanism of the antagonism by 5-hydroxytryptamine of the toxicity due to certain cholinergic blocking agents, *Neuropharmacology,* 10, 335, 1971.
187. **Burgen, A. S. V., Dickens, F., and Zatman, L. J.**, The action of botulinum toxin on the neuromuscular junction, *J. Physiol. (London),* 109, 10, 1949.
188. **Hirokawa, N. and Kitamura, M.**, Binding of *C. botulinum* neurotoxin to the presynaptic membrane in the central nervous system, *J. Cell. Biol.,* 81, 43, 1979.
189. **Cherington, M., Greenberg, H., and Soyer, A.**, Guanidine and germine in botulism, *Clin. Toxicology,* 6, 83, 1973.
190. **Cherington, M.**, Botulism: Ten-year experience, *Arch. Neurol.,* 30, 432, 1974.
191. **Cherington, M. and Schultz, D.**, Effect of guanidine, germine and steroids in a case of botulism, *Clin. Toxicol.,* 11, 19, 1977.

MICROBIAL FOOD AND FEED TOXICANTS: FUNGAL TOXINS

Alex Ciegler and Ronald F. Vesonder

Mycotoxins are toxic and/or carcinogenic secondary metabolites produced by fungi on a wide array of agricultural commodities that can cause diseases referred to as mycotoxicoses; the fungi per se are not involved in the disease process. The presence of mycotoxins in agricultural commodities and ensuing mycotoxicoses represent an extremely complex series of interactions between the causative fungi, the contaminated product, physicochemical environmental factors, and the intoxicated host. This complexity is reflected in the difficulty in achieving some measure of control over the problem.

Certain useful diagnostic features characterize outbreaks of mycotoxicoses: (1) the diseases are not transmissible; (2) drug and antibiotic treatment have little or no effect on the disease; (3) in field outbreaks, the trouble is often seasonal; (4) the outbreak is often associated with a specific food or feedstuff; and (5) examination of the suspected food or feed reveals signs of fungal activity.

Mycotoxicoses constitute a worldwide problem affecting both man and domesticated animals. These diseases occasionally are manifested in an acute manner, but probably insidious effects constitute the current major portion of the problem, e.g., carcinogenesis in man and loss of weight gains and feed efficiency in farm animals.

An interrelationship between host nutrition and mycotoxicoses has been demonstrated primarily for only one of the mycotoxins, aflatoxin. Protein deficiencies in monkeys and rats, pyridoxine-deficient baboons, and marginal lipotrope (methionine, choline, vitamin B_{12}) in the diet of rats appear to increase the susceptibility of these animals to liver damage and hepatomas by aflatoxin. Similar deficiencies have been noted in native populations suffering from an unusually high incidence of hepatomas.

Additional data on this and other aspects of mycotoxins and mycotoxicoses may be found in the following reviews:

1. Goldblatt, L. A., *Aflatoxin,* Academic Press, New York, 1969.
2. Ciegler, A., Kadis, S., and Ajl, S., *Microbial Toxins,* Vols. 6, 7, 8, Academic Press, New York, 1971.
3. Mycotoxins, in *Microbiology,* Schlessinger, D., Ed., American Society for Microbiology, Washington, D.C., 343, 1975.
4. Rodricks, J. V., *Mycotoxins and Other Fungal Related Food Problems,* Advances in Chemistry Series 149, American Chemical Society, Washington, D.C., 1976.
5. Wyllie, T. D. and Morehouse, L. G., *Mycotoxic Fungi, Mycotoxins, Mycotoxicoses,* Vols. 1, 2, 3, Marcel Dekker, New York, 1977.
6. Rodricks, J. V., Hesseltine, C. W., and Mehlman, M. A., *Mycotoxins in Human and Animal Health,* Pathotox Publishers, Park Forest South, Ill., 1977.

It should be noted that some of the compounds listed in the following tables may or may not be true mycotoxins; e.g., some of the compounds may represent intermediates in the pathway of biosynthesis to a mycotoxin or may be degradative products of a mycotoxin. However, the authors believe that these substances warrant inclusion in the present tables.

Table 1
CHEMICAL STRUCTURES OF FUNGAL TOXINS

Name — **Structure**

4-Acetoxyscirpendiol

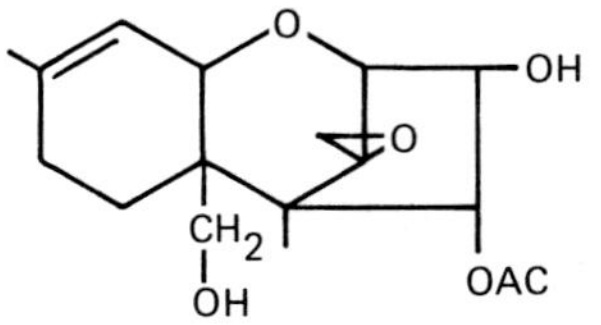

4 β-Acetoxy-3α, 15 diOH-12,13-epoxy-trichotec-9-ene

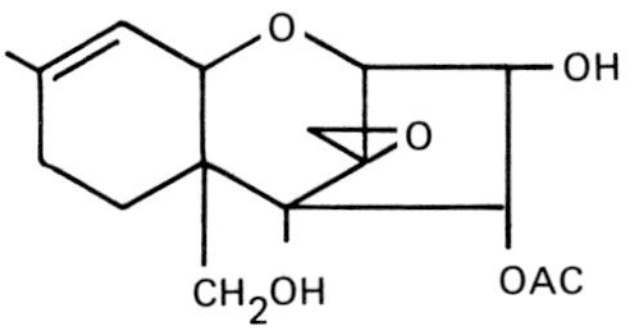

8-Acetylneosolaniol

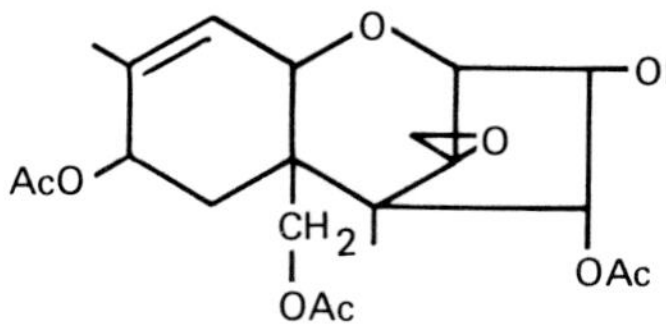

Aflatoxin B_1

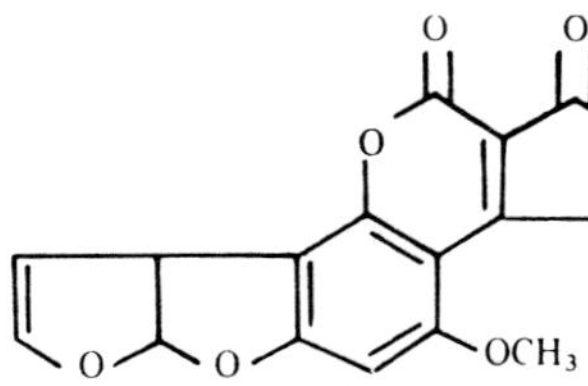

Aflatoxin B_1 aldehyde

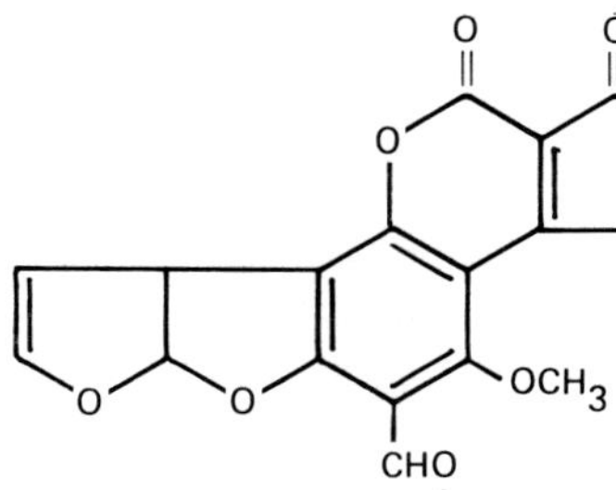

3-Hydroxyaflatoxin B_1

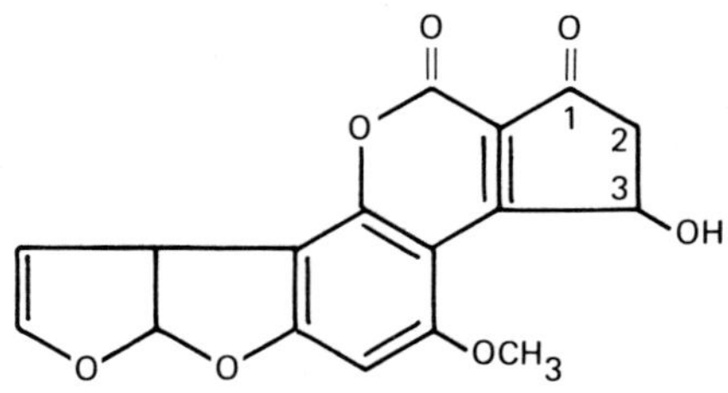

Table 1 (continued)
CHEMICAL STRUCTURES OF FUNGAL TOXINS

Name	Structure
Aflatoxin B_2	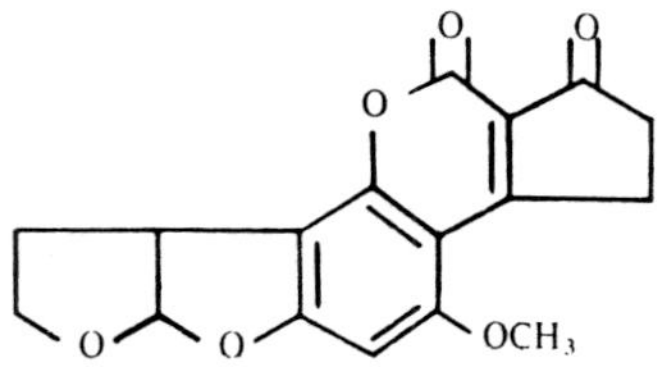
1-Acetoxy-aflatoxin B_2	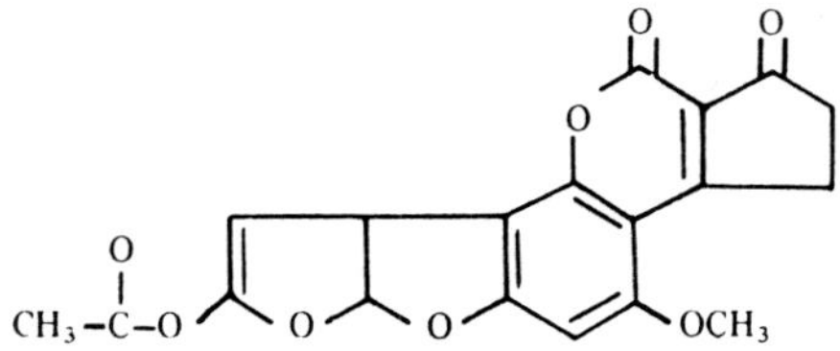
1-Ethoxy-aflatoxin B_2	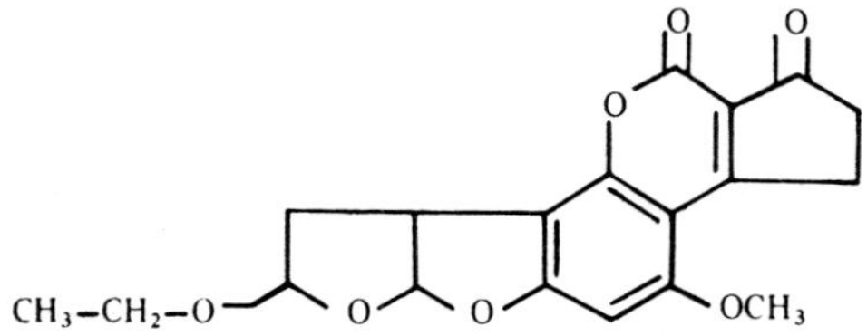
1-Methoxy-aflatoxin B_2	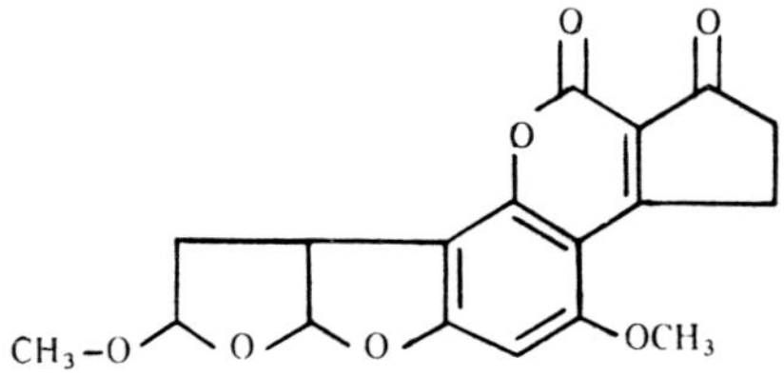
2-Methoxy-aflatoxin B_2	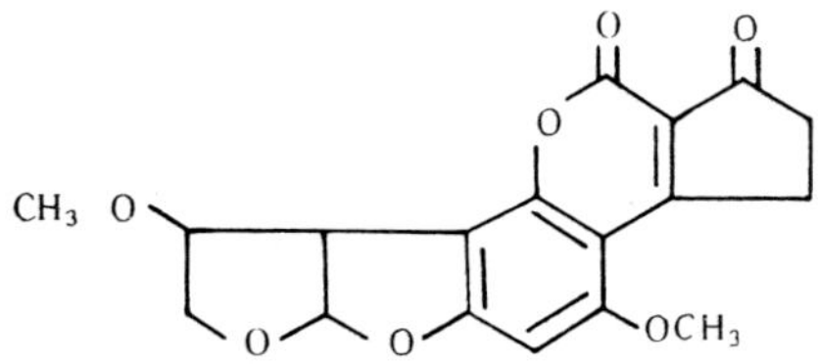
Aflatoxin B_{2a}	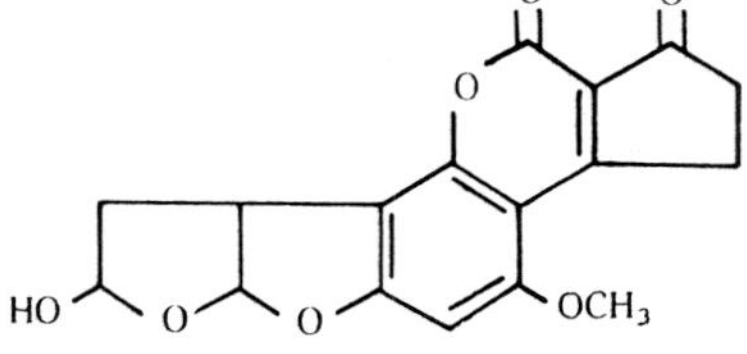
Aflatoxin B_3 (Parasiticol)	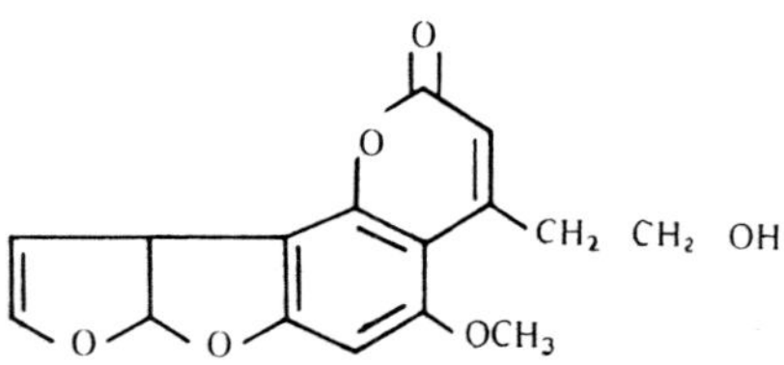

Table 1 (continued)
CHEMICAL STRUCTURES OF FUNGAL TOXINS

Name — **Structure**

Aflatoxin D_1

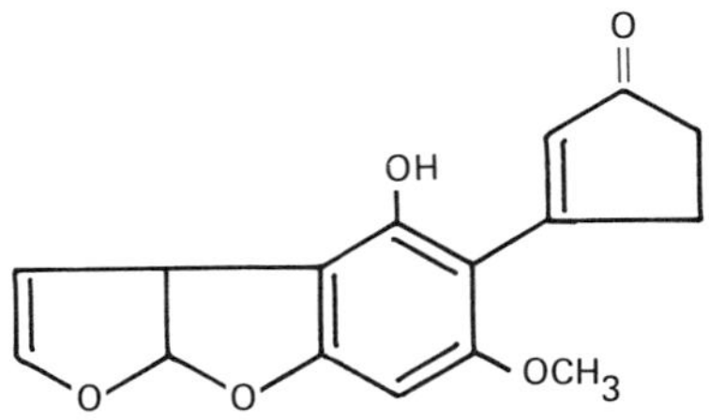

Aflatoxin deshydro D_1

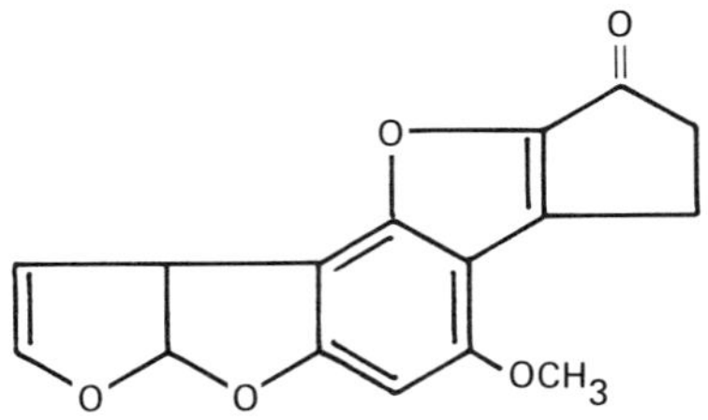

Aflatoxin G_1

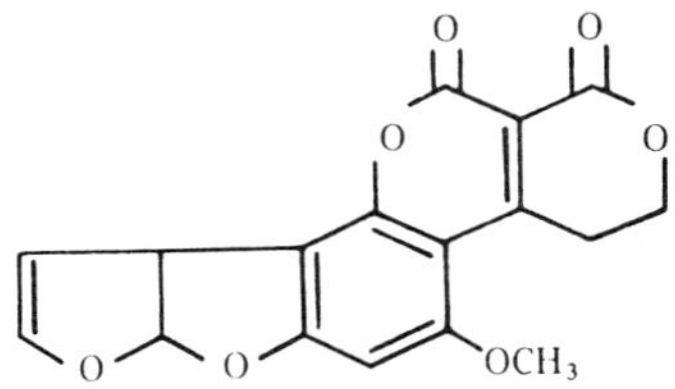

Aflatoxin G_2

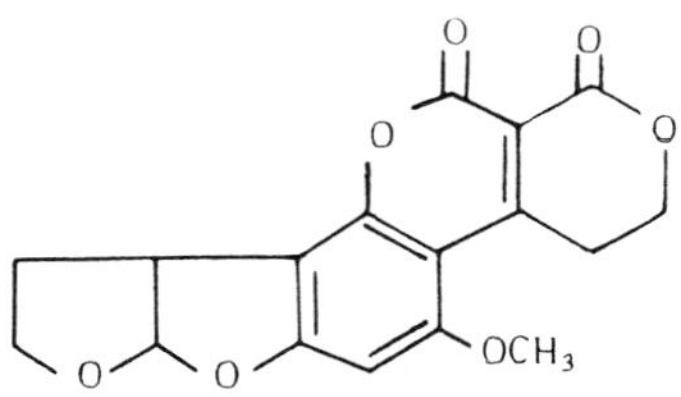

Alternaric acid

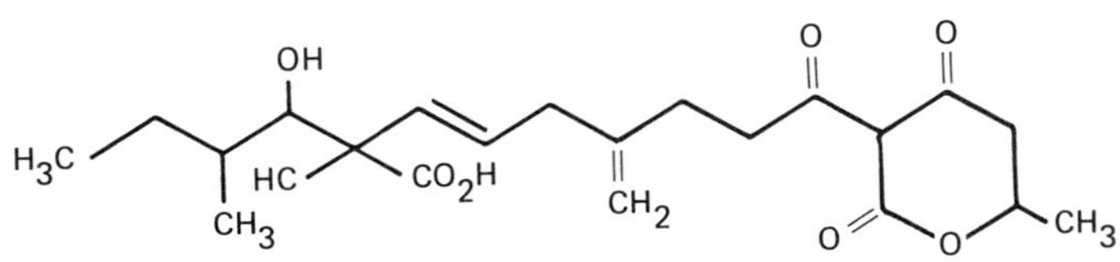

Alternariol

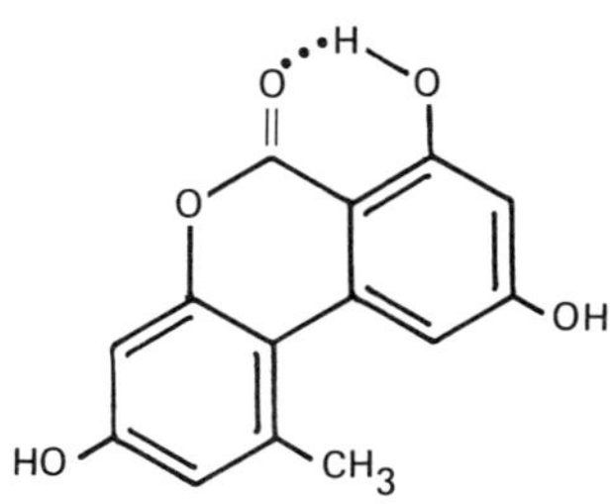

Table 1 (continued)
CHEMICAL STRUCTURES OF FUNGAL TOXINS

Name	Structure
Alternariol methylether	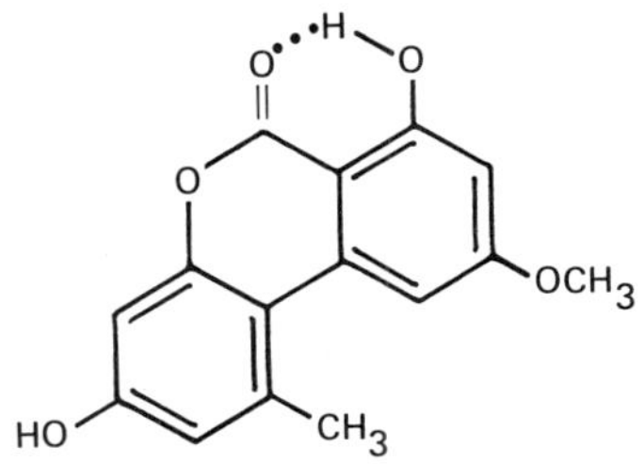
β-Amantine	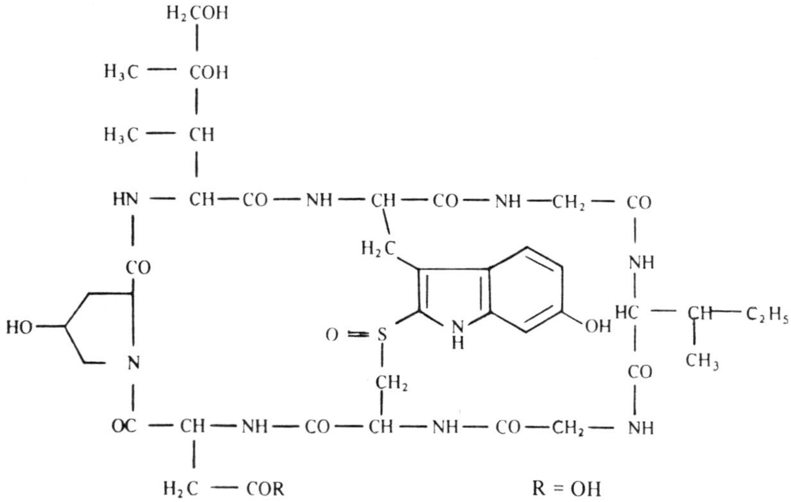
γ-Amantine	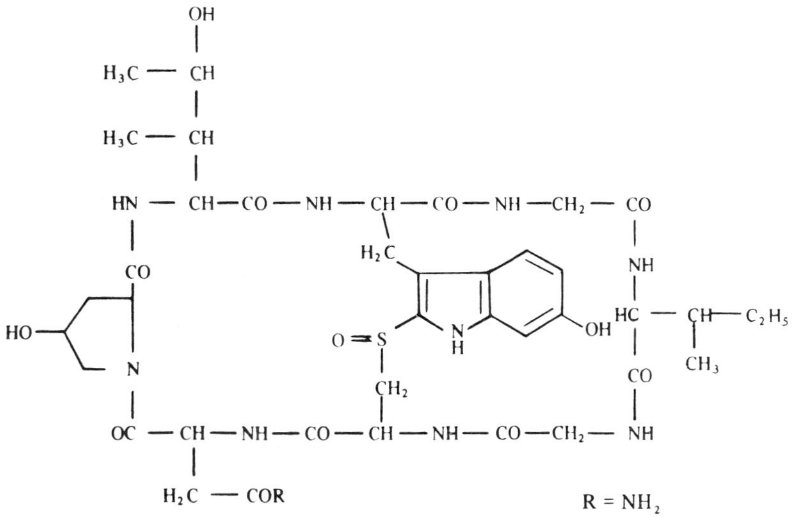

Table 1 (continued)
CHEMICAL STRUCTURES OF FUNGAL TOXINS

Name	Structure
ε-Amantine	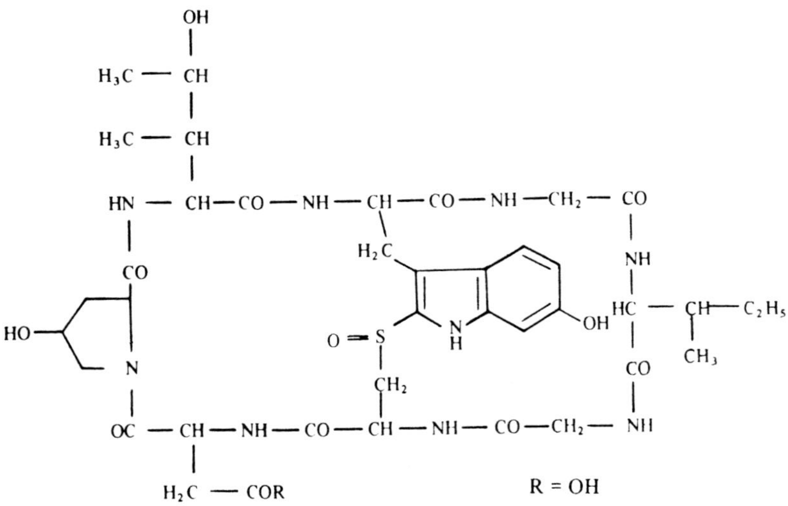
Amanullin	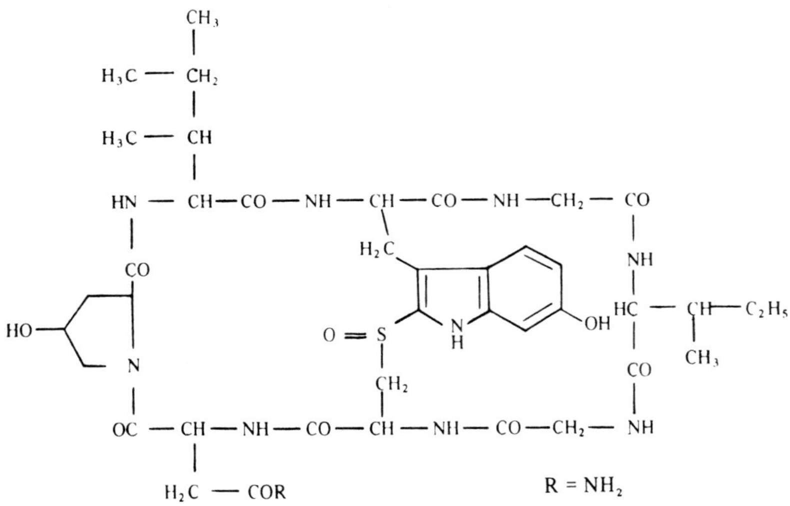
Amatoxins	See α-, β-, γ-, and ε-Amanitine
Antimycin	See citrinin
Ascladiol	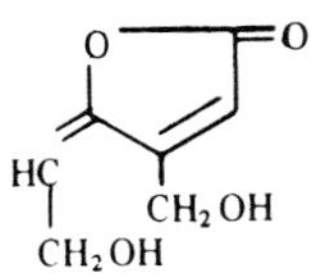
Ascotoxin	See decumbin

Table 1 (continued)
CHEMICAL STRUCTURES OF FUNGAL TOXINS

Name	Structure
Aspercolorin	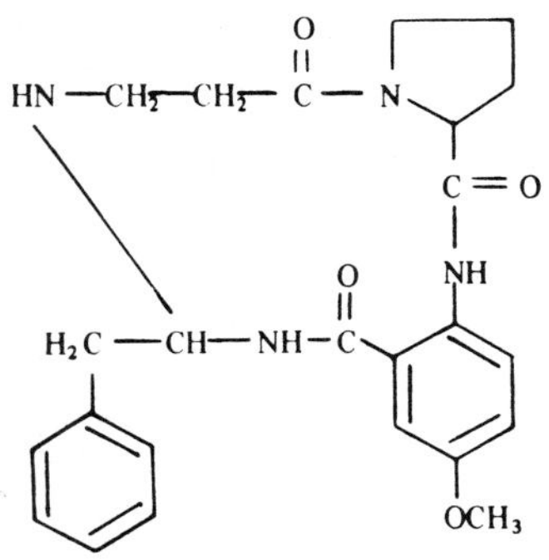
Aspergillic acid	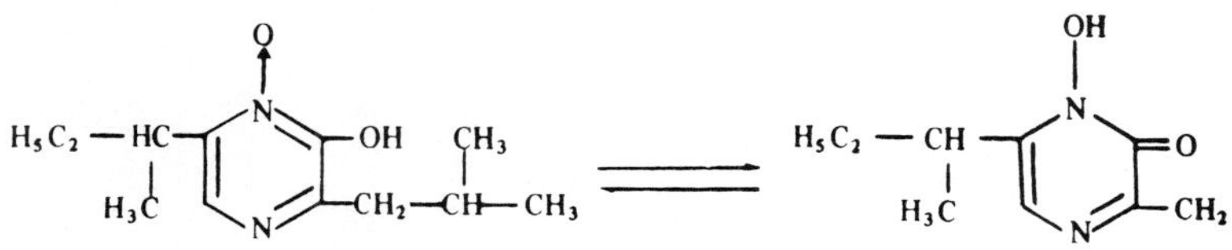
Aspertoxin	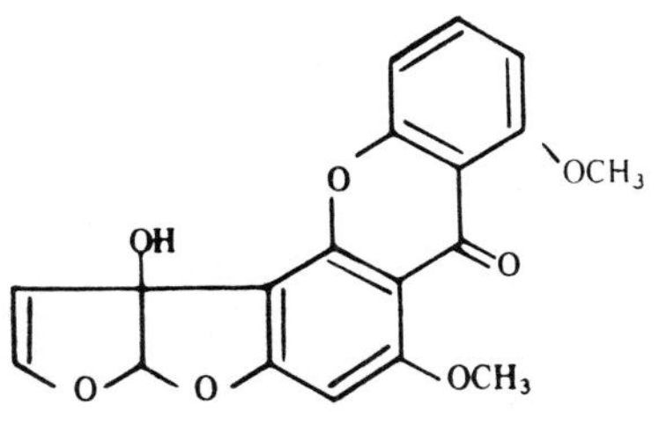
Averufin	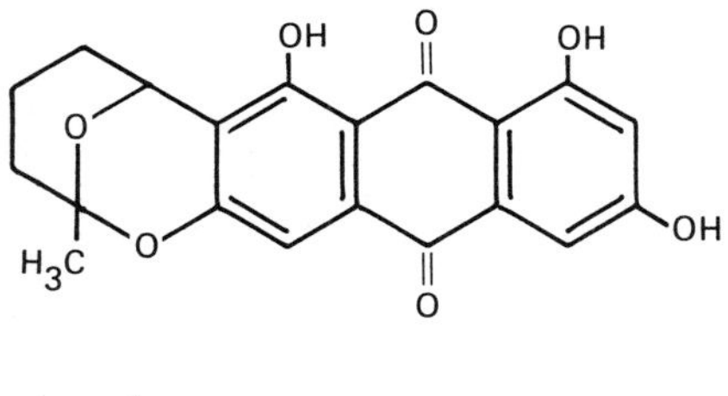
Bergapten	See 5-methoxypsoralen
B-24 toxin	See diacetoxyscirpenol
Beta-toxin	See muscimol
Bufotenine	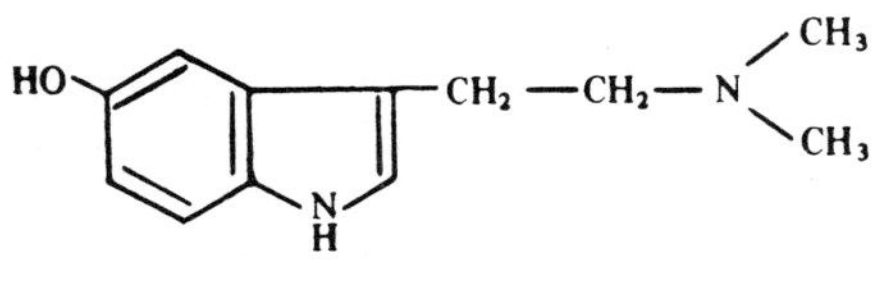
Butenolide	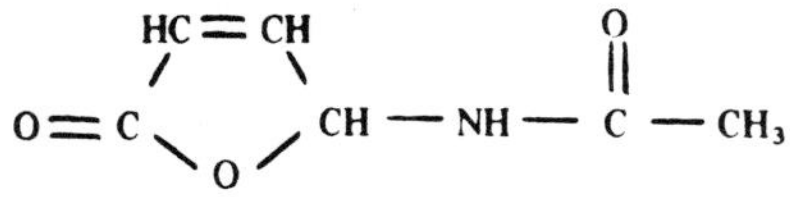

Table 1 (continued)
CHEMICAL STRUCTURES OF FUNGAL TOXINS

Name	Structure
Chaetoglobosin A	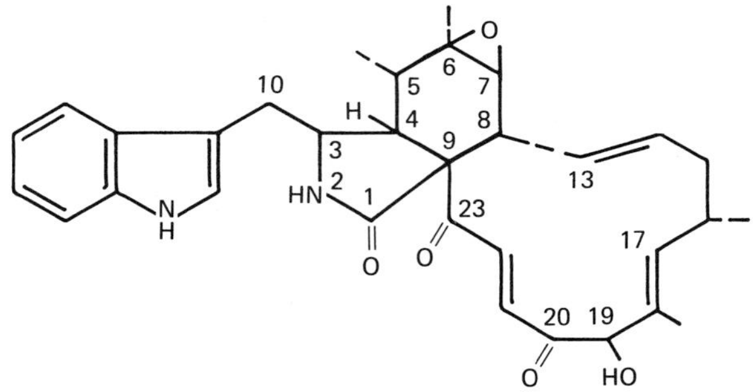
Chaetoglobosin B	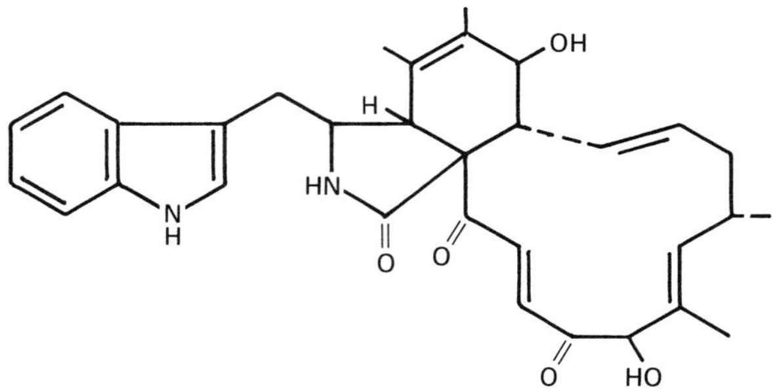
Chaetoglobosin C	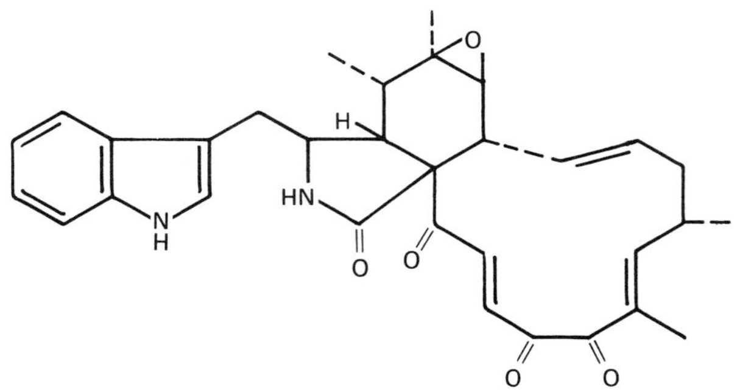
Chaetoglobosin D	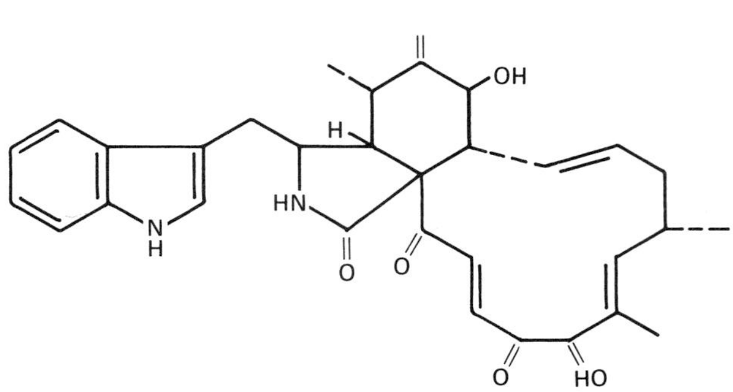

Table 1 (continued)
CHEMICAL STRUCTURES OF FUNGAL TOXINS

Name	Structure
Chaetoglobosin E	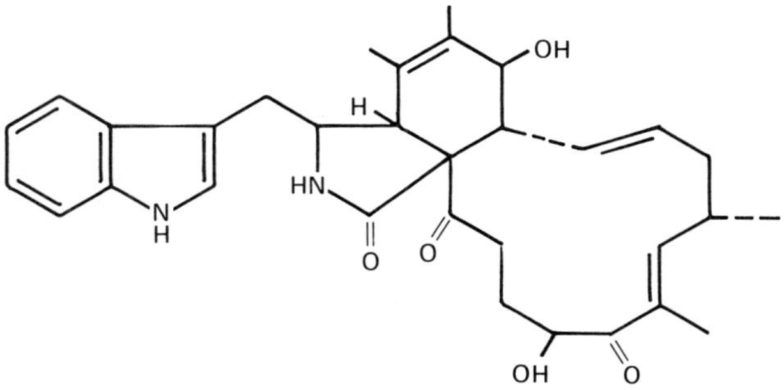
Chaetoglobosin F	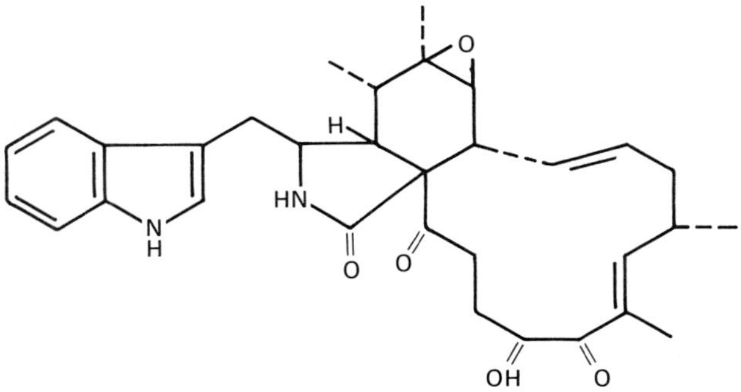
Chaetoglobosin G	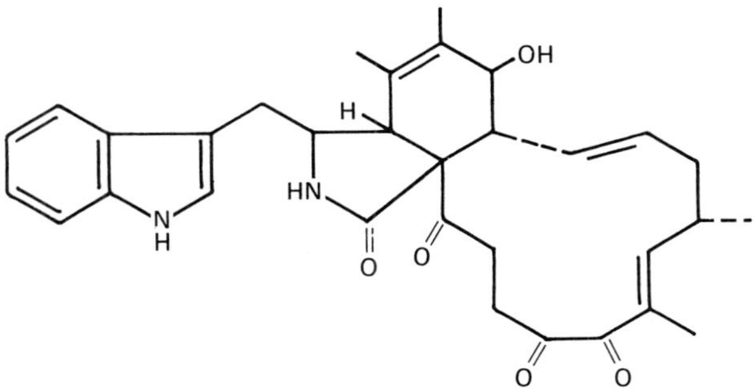
Chaetoglobosin J	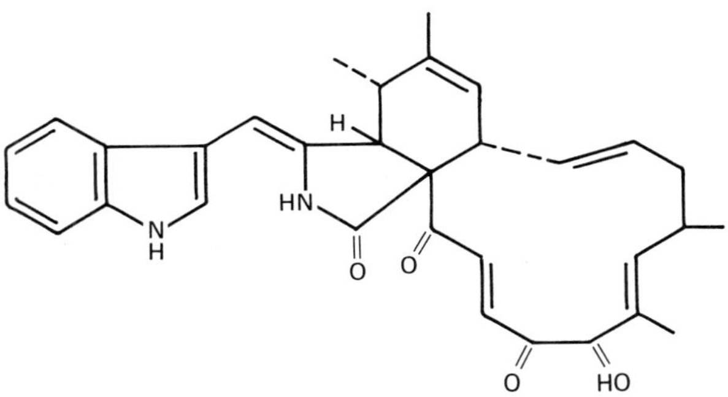

Table 1 (continued)
CHEMICAL STRUCTURES OF FUNGAL TOXINS

Name	Structure
Citreoviridin	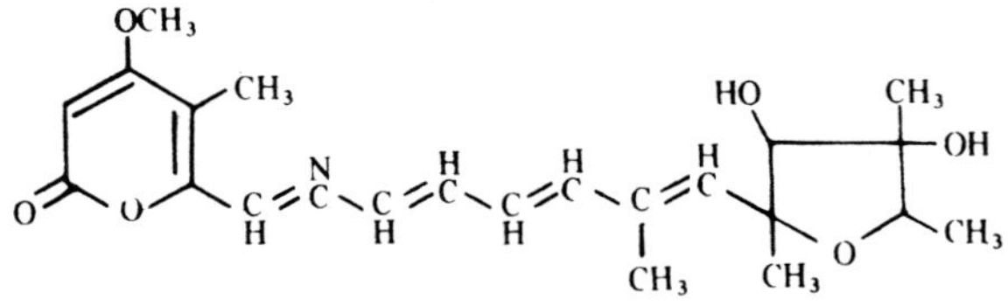
Citrinin	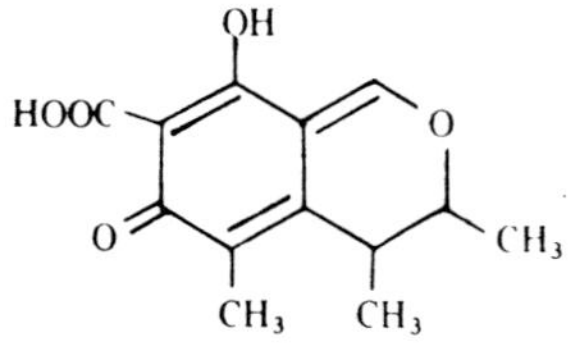
Clavicin	See patulin
Clavitin	See patulin
Claviformin	See patulin
Cochliobolus toxin	Not known
Cyclochlorotine	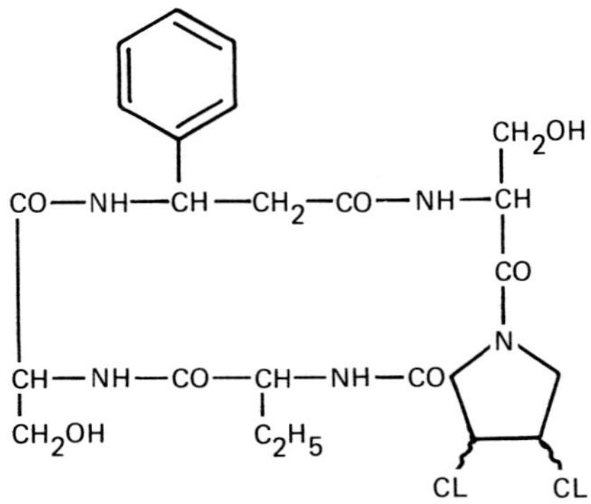
Cyclopiazonic acid	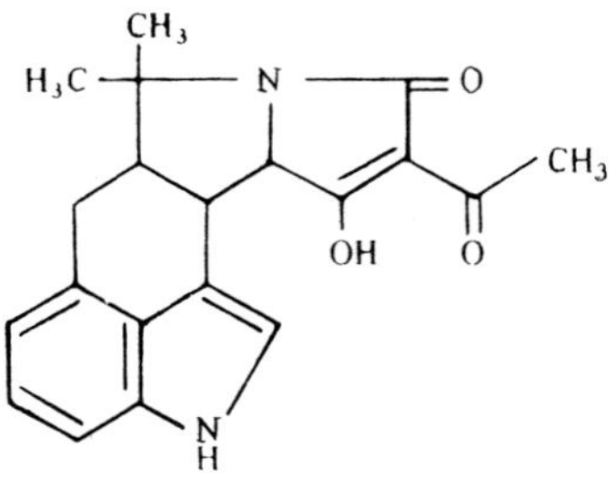
Cytochalasins	See chaetoglobosins

Table 1 (continued)
CHEMICAL STRUCTURES OF FUNGAL TOXINS

Name	Structure
Cytochalasin E	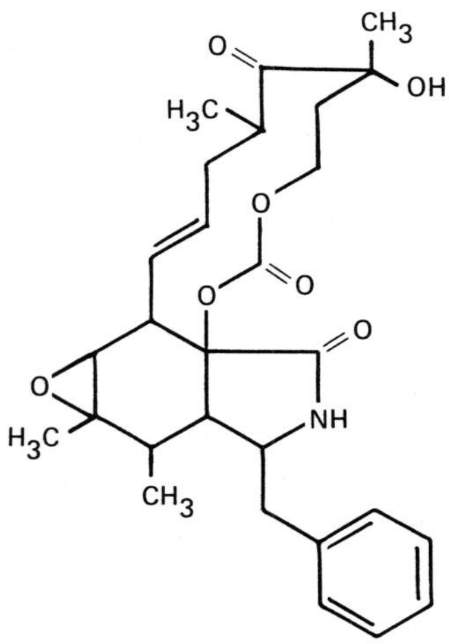
Cytochalasin H	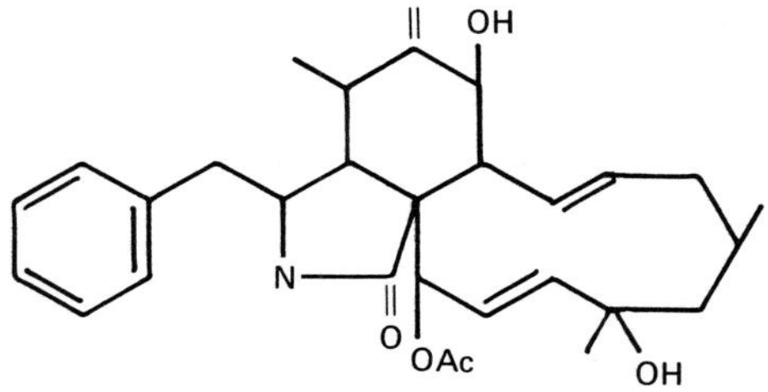
Decumbin	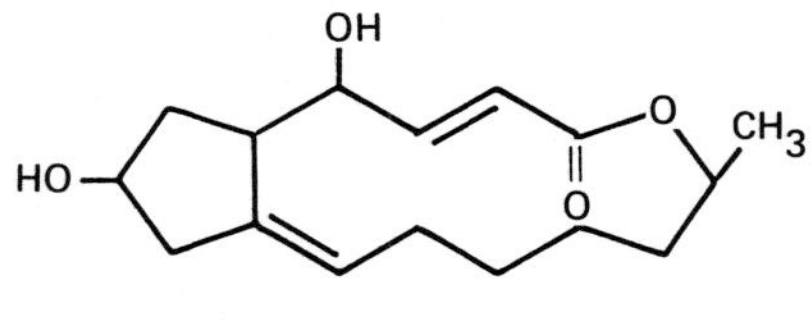
Dendrochine	Not known
Deoxytryptoquivaline	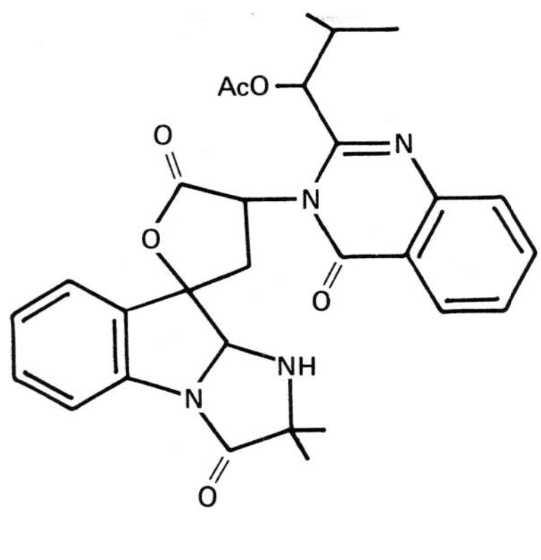

Table 1 (continued)
CHEMICAL STRUCTURES OF FUNGAL TOXINS

Name	Structure

Deoxynortryptoquivaline

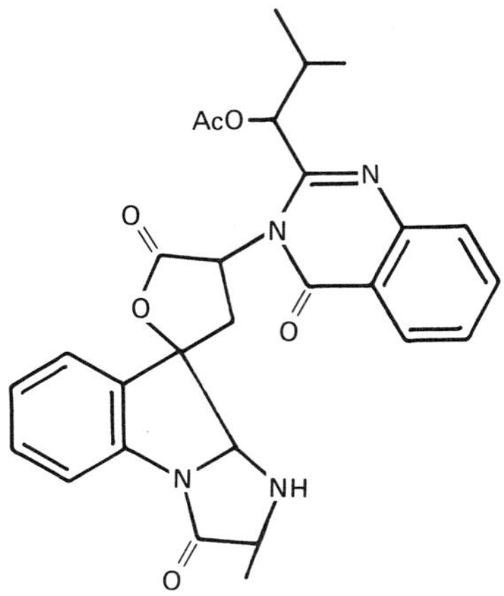

Deoxynortryptoquivalone

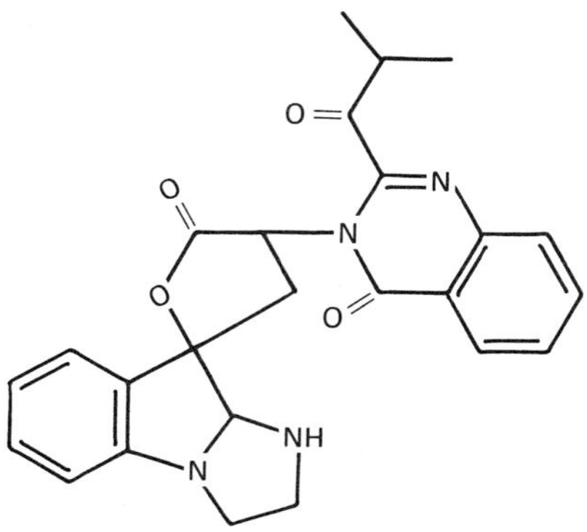

Desmethoxyviridiol

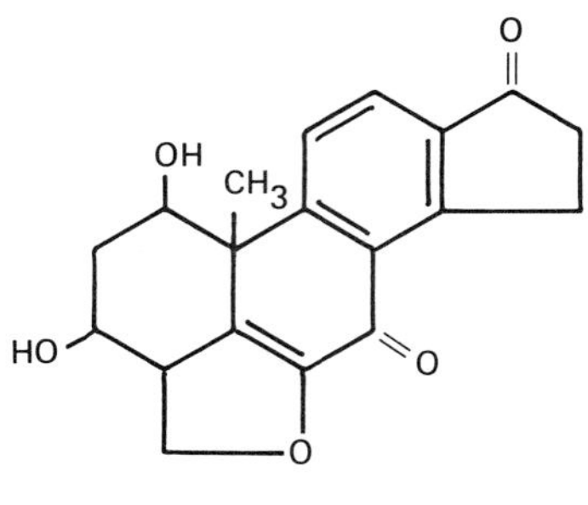

Diacetoxyscirpenol

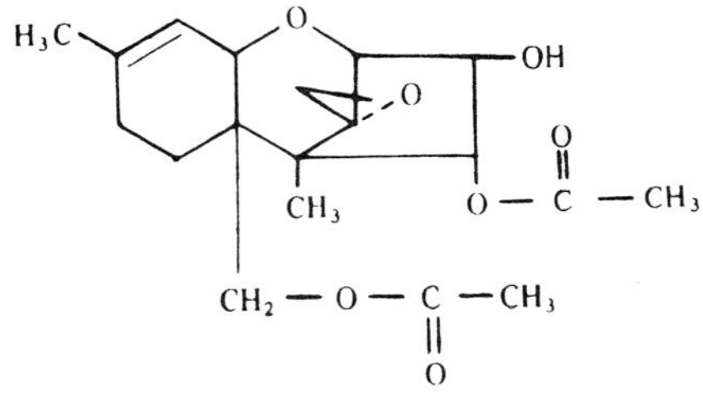

Table 1 (continued)
CHEMICAL STRUCTURES OF FUNGAL TOXINS

Name — Structure

4,β8α-Diacetoxy-12,13-epoxytrichotec-9-ene-3α,15-diol

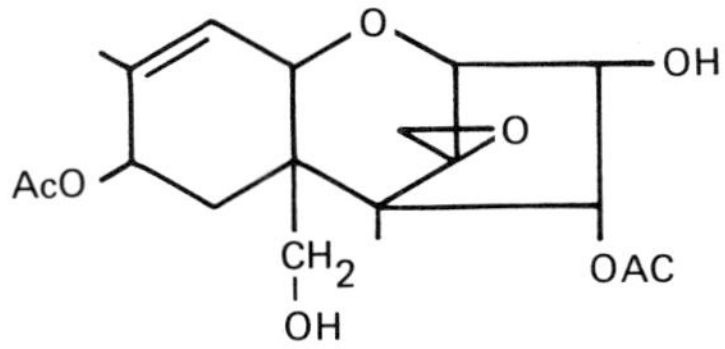

Dihydrodemethylsterigmatocystin

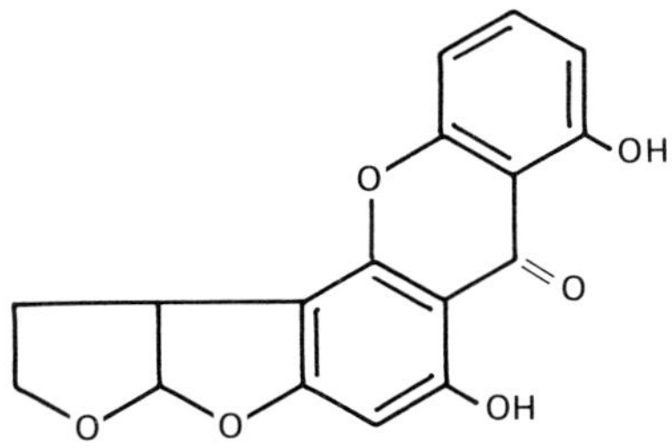

Dihydrosterigmatocystin

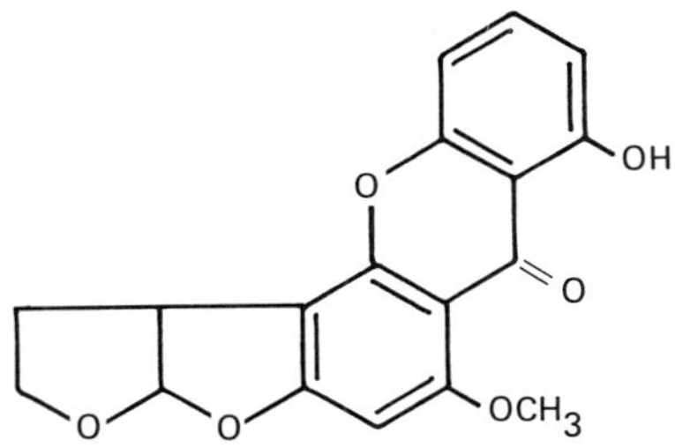

5,6-Dimethoxysterigmatocystin

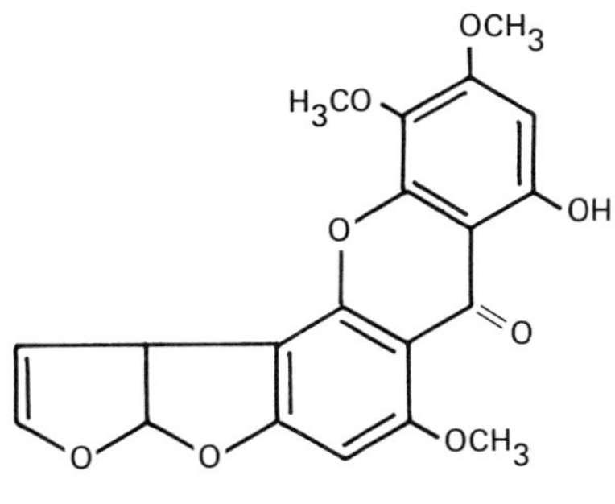

Emodin

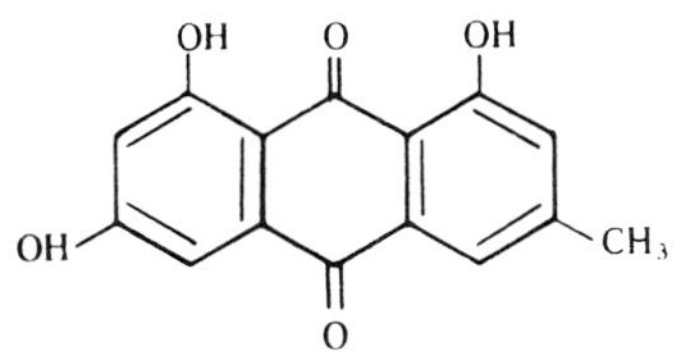

Table 1 (continued)
CHEMICAL STRUCTURES OF FUNGAL TOXINS

Name	Structure
Ergobasine	See ergometrine
Ergoclinine	See ergometrine
Ergocornine	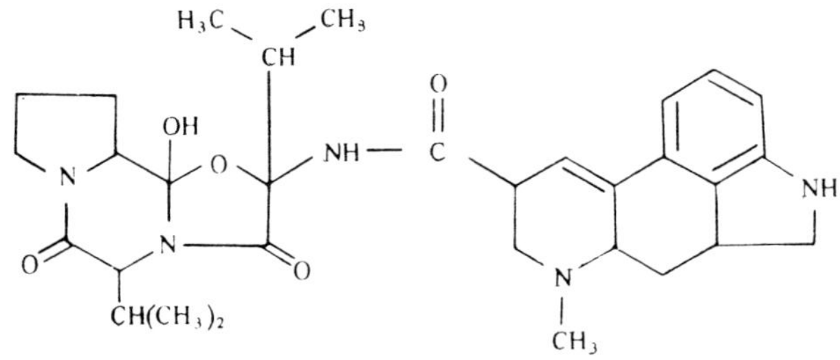
Ergocristine	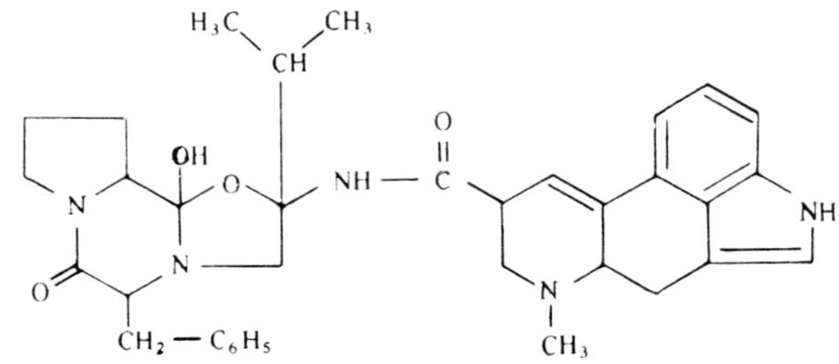
Ergocryptine	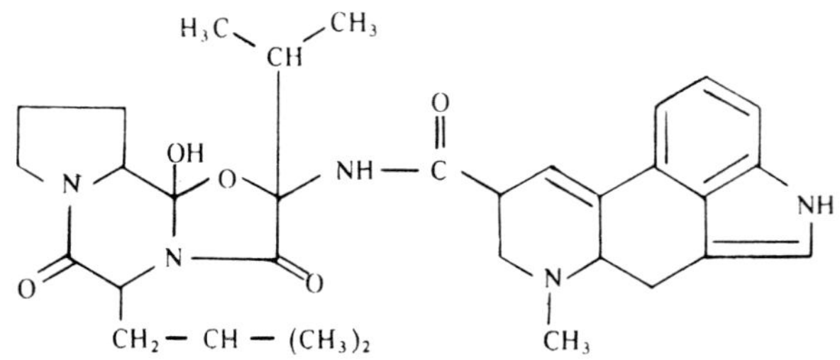
Ergometrine	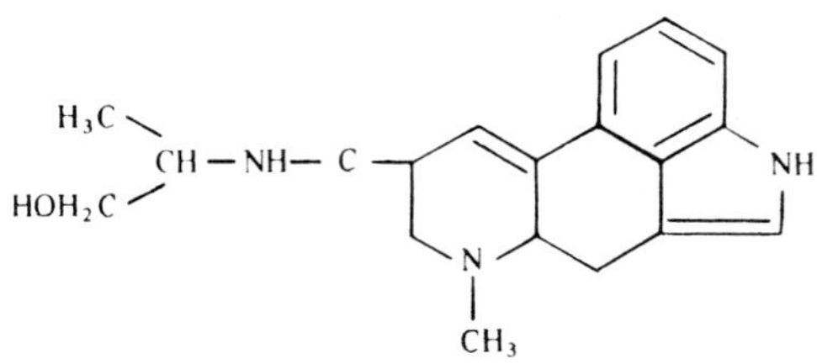
Ergonovine	See ergometrine
Ergostetrine	See ergometrine
Ergot alkaloids	See argoclavine, ergocornine, ergocristine, ergocryptine, ergormetrine, and ergotamine

Table 1 (continued)
CHEMICAL STRUCTURES OF FUNGAL TOXINS

Name — **Structure**

Ergotamine

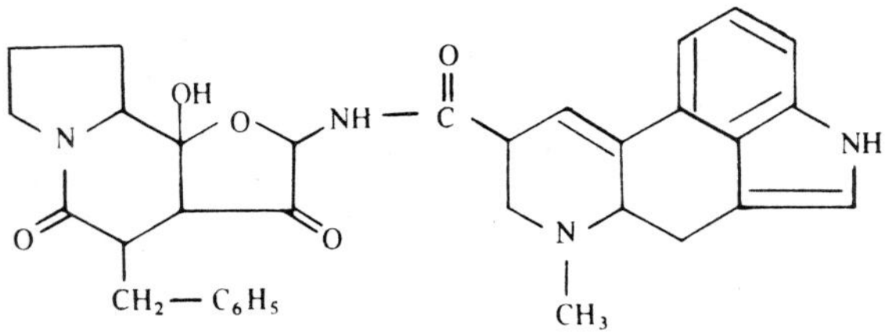

Name	Structure
Ergotocine	See ergometrine
Ergotrate	See ergometrine
Expansin	See patulin
F-2 toxin	See zearalenone
Fescue toxin	See butenolide
Flavutoxin	Not known

Fumagillin

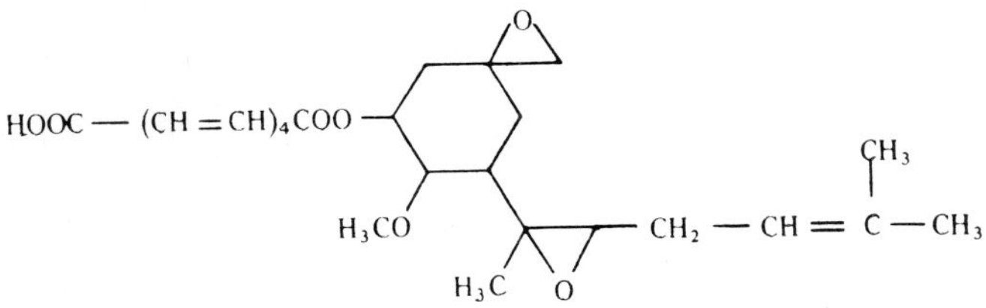

Name	Structure
Fumigacin	See helvolic acid

Fumigaclavine A

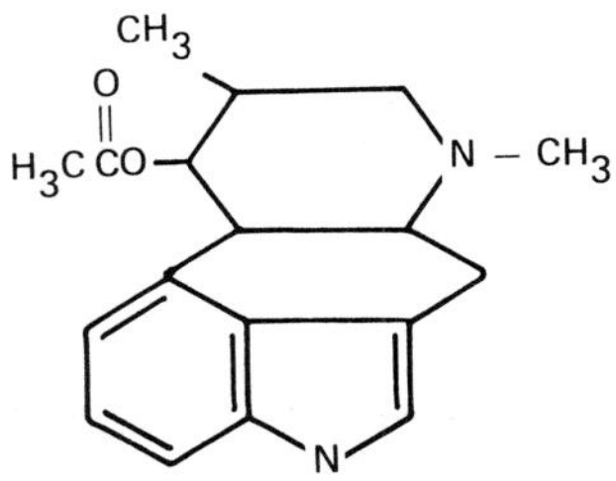

Fumigaclavine B

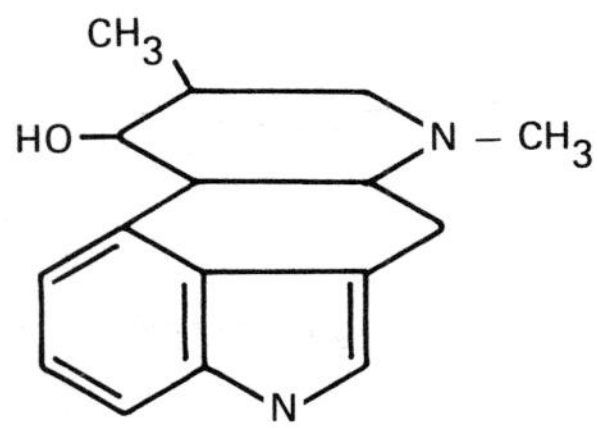

Fumigaclavine C

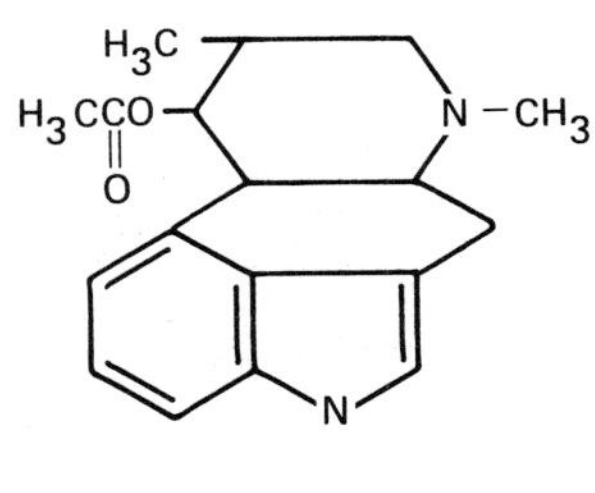

Table 1 (continued)
CHEMICAL STRUCTURES OF FUNGAL TOXINS

Name	Structure
Fumigatoxin	Not known
Fumitremorgin A	Not known
Fumitremorgin B	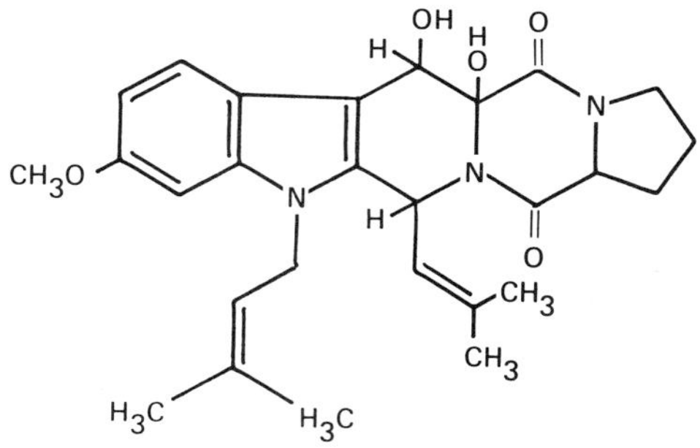
Fusarenon-X	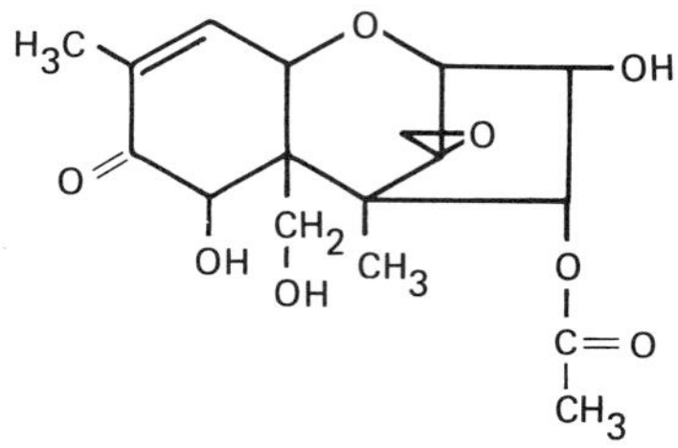
Fusarium culmorum toxin	Not known
Gliotoxin	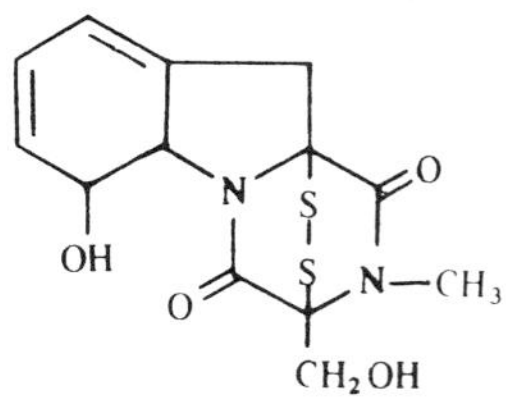
Gyromitrin	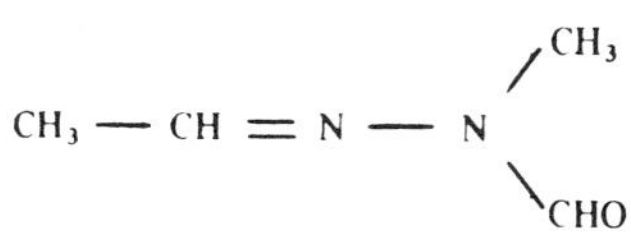
Helvolic acid	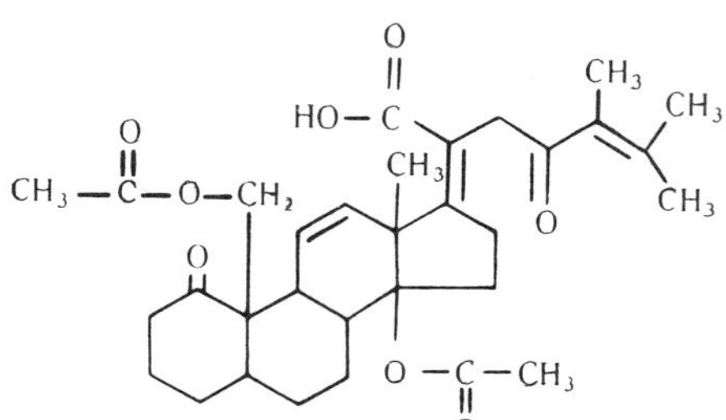

Table 1 (continued)
CHEMICAL STRUCTURES OF FUNGAL TOXINS

Name	Structure
Hiptagenic acid	See β-Nitropropionic acid
HT-2 toxin	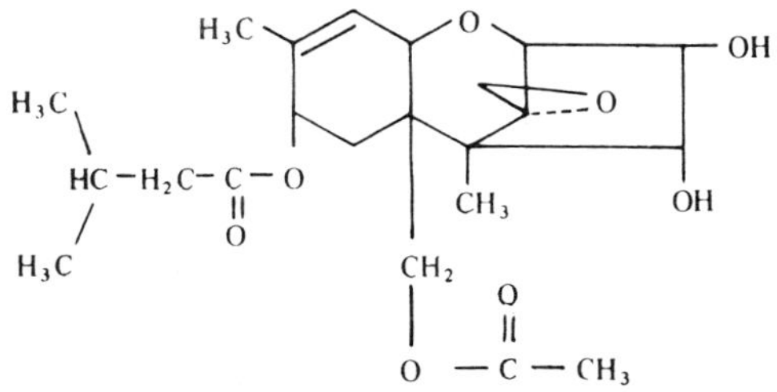
4-Hydroxyochratoxin A	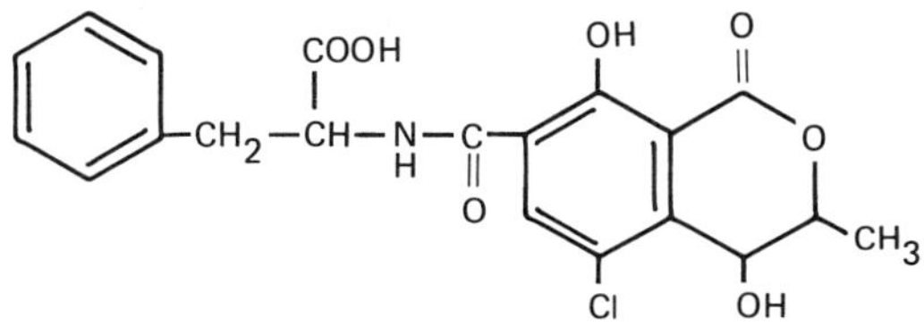
Ibotenic acid	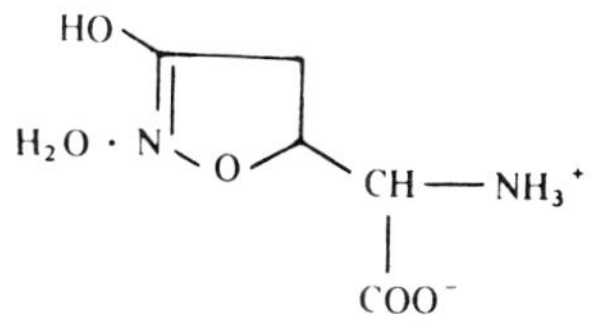
Islanditoxin	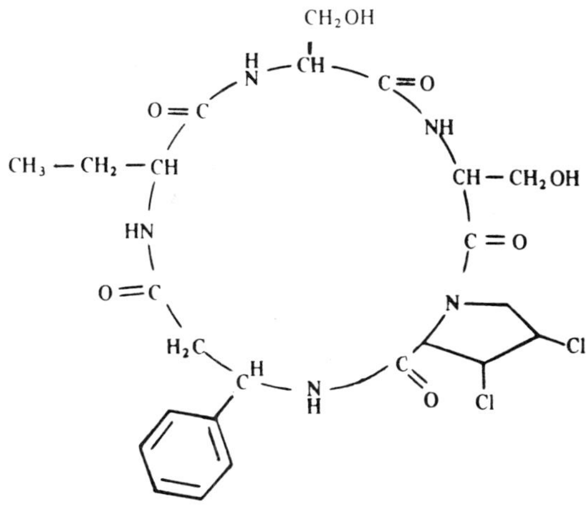
Kojic acid	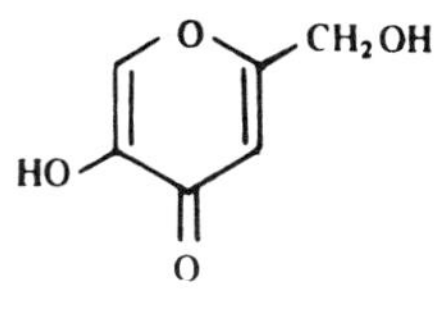

Table 1 (continued)
CHEMICAL STRUCTURES OF FUNGAL TOXINS

Name	Structure
Luteoskyrin	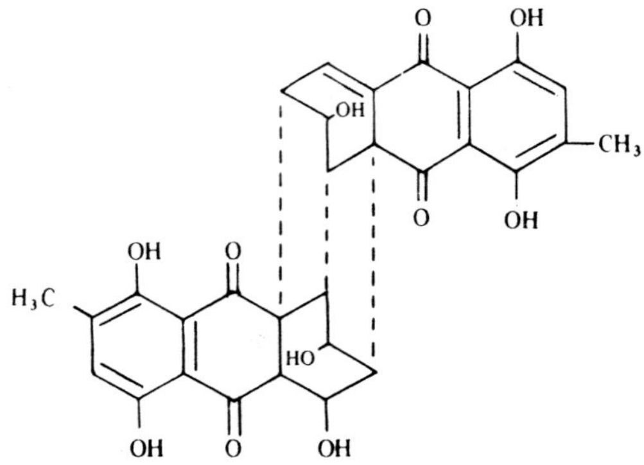
Lysergic acid	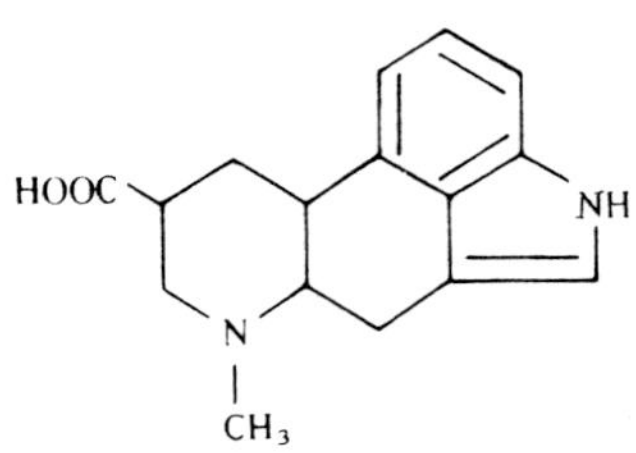
Malformin C	D-Cys-D-Cys-L-Val-L-Leu-D-Leu
Maltoryzine	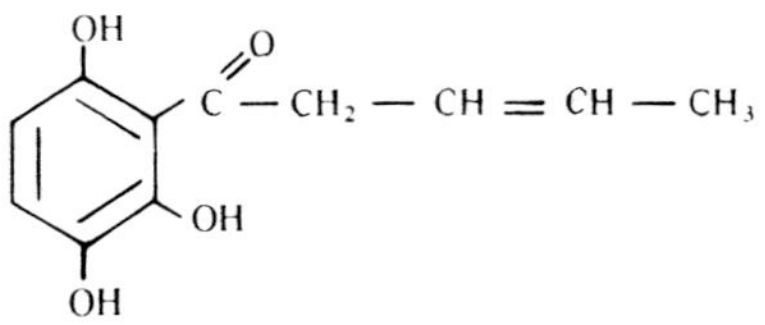
Mappine	See bufotenine
Mescaline	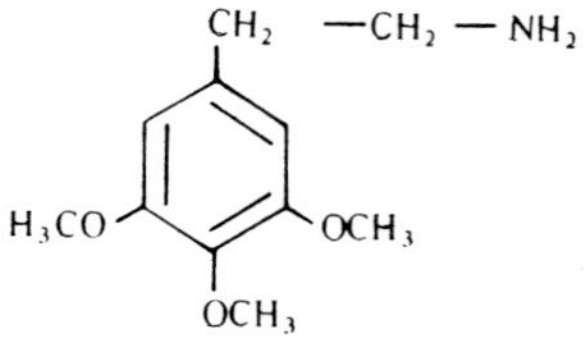
8-Methoxypsoralen	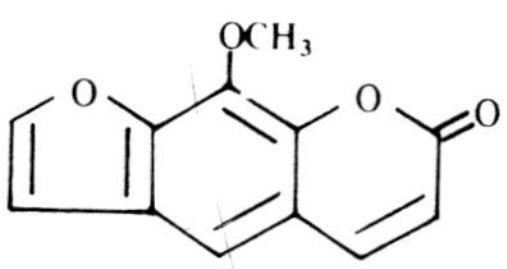

Table 1 (continued)
CHEMICAL STRUCTURES OF FUNGAL TOXINS

Name	Structure
O-Methylsterigmatocystin	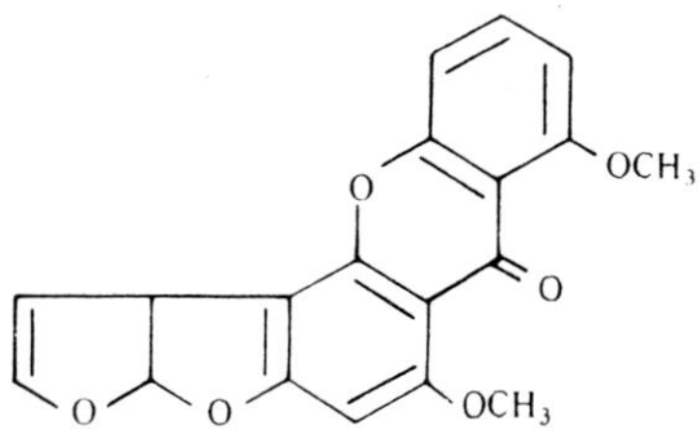
3-Methyl-2-butenylpaspalinine	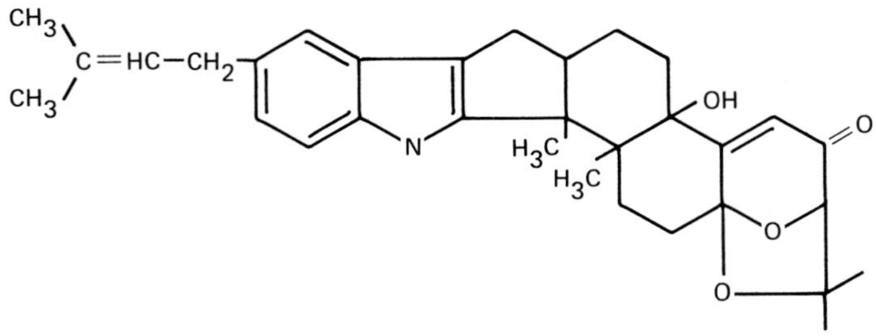
3-Methyl-3-hydroxy-1-butenylpaspalalinine	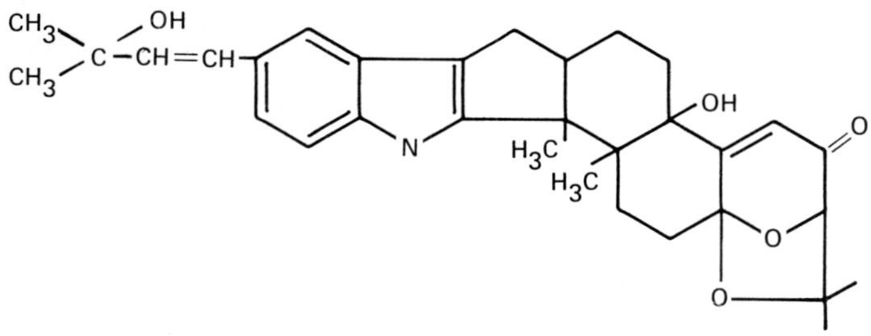
Moniliformin	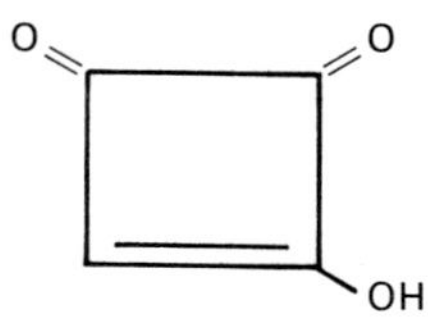
Monoacetoxyscirpenol	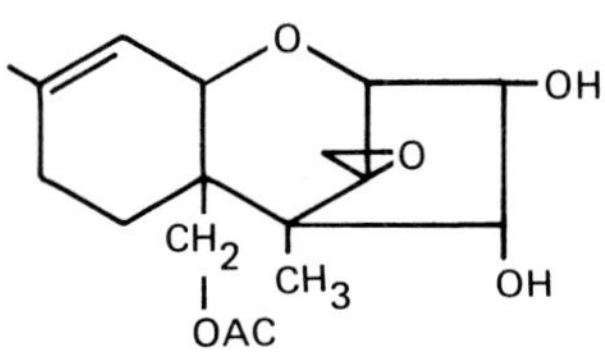
Muscarine	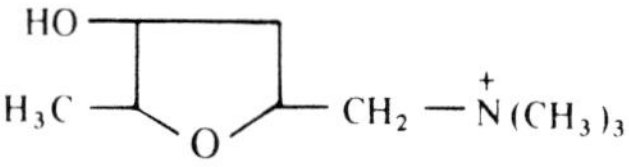

Table 1 (continued)
CHEMICAL STRUCTURES OF FUNGAL TOXINS

Name	Structure
Muscazone	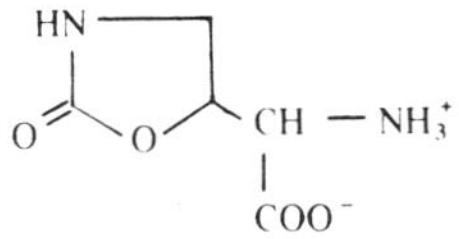
Muscimol	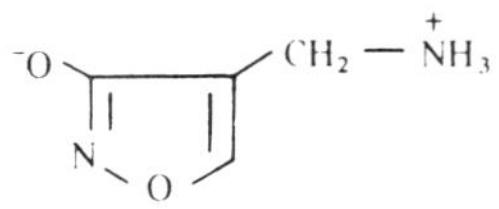
Mycoin	See patulin
Mycophenolic acid	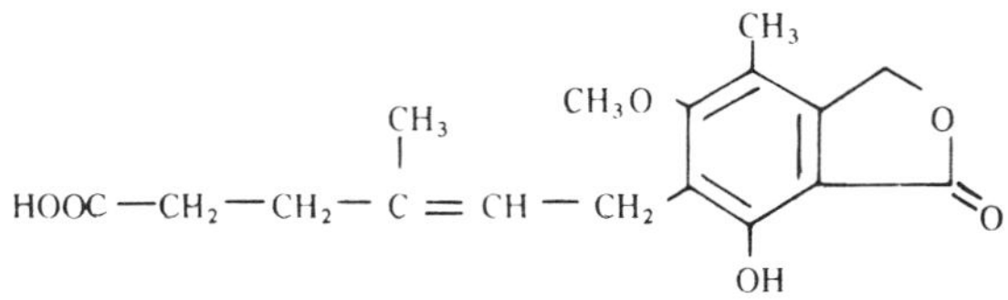
Neosolaniol monoacetate	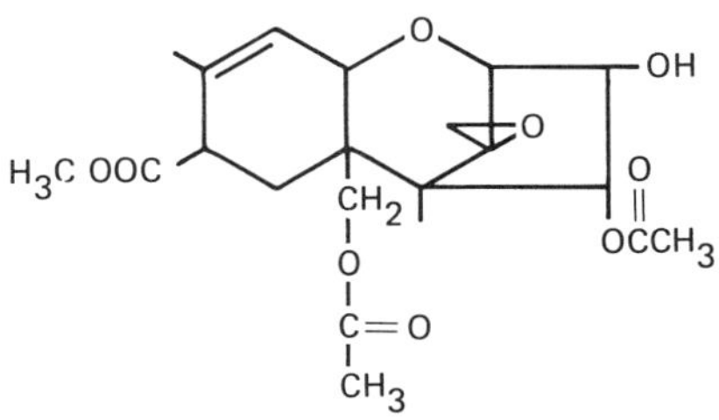
Nidulotoxin	Not known
β-Nitropropionic acid	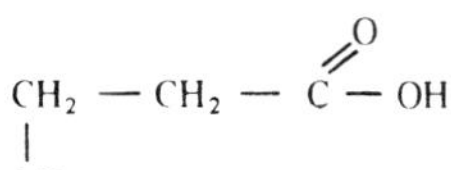
Nivalenol	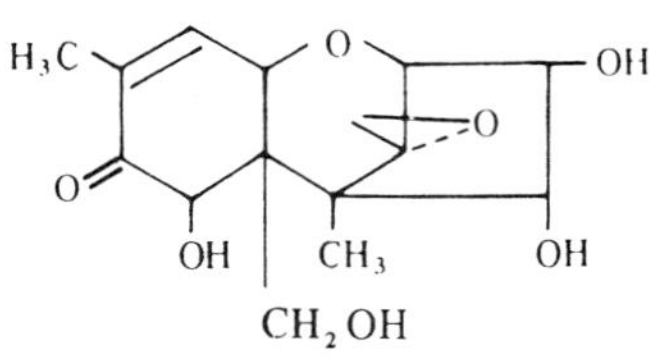
Norsolorinic acid	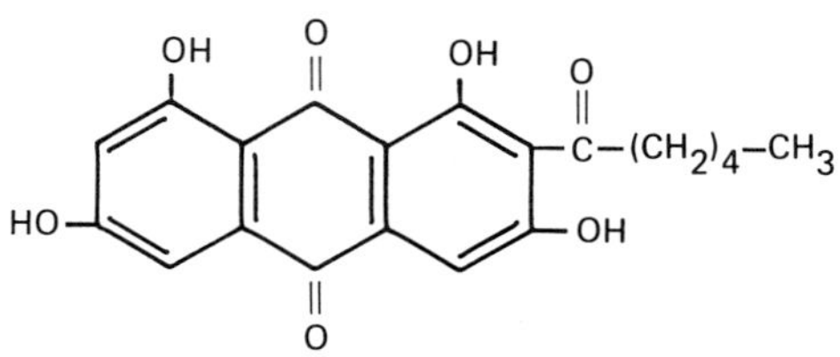

Table 1 (continued)
CHEMICAL STRUCTURES OF FUNGAL TOXINS

Name	Structure
Nortryptoquivaline	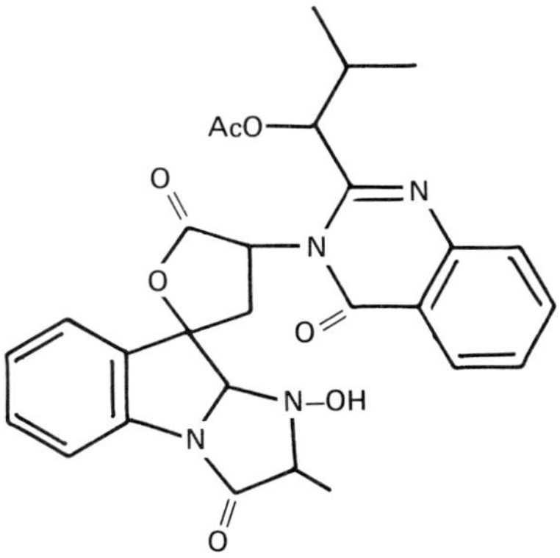
Ochratoxin A	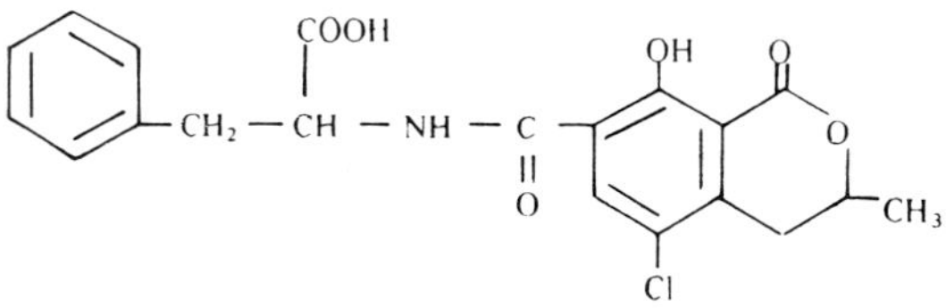
Ochratoxin A ethyl ester	See ochratoxin C
Ochratoxin A methyl ester	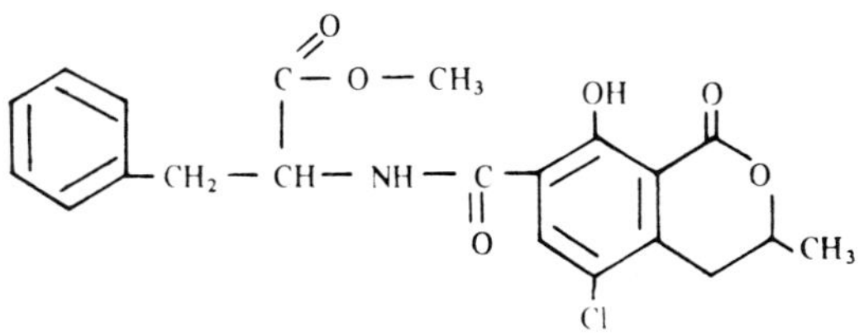
Ochratoxin B	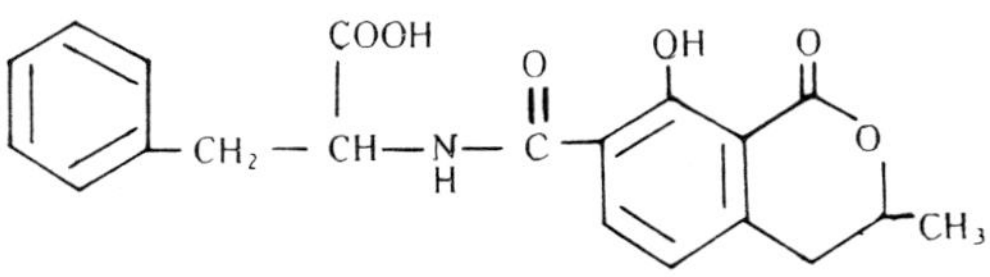
Ochratoxin B ethyl ester	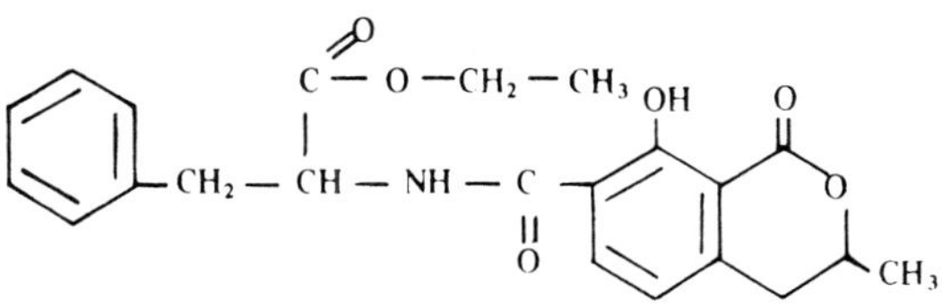
Orchratoxin B methyl ester	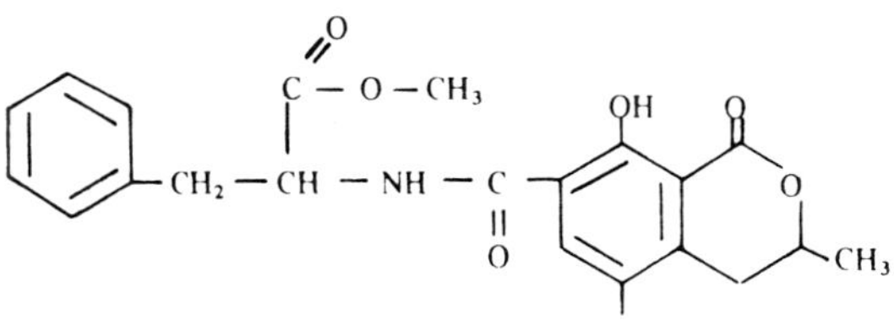

Table 1 (continued)
CHEMICAL STRUCTURES OF FUNGAL TOXINS

Name	Structure
Orchratoxin C	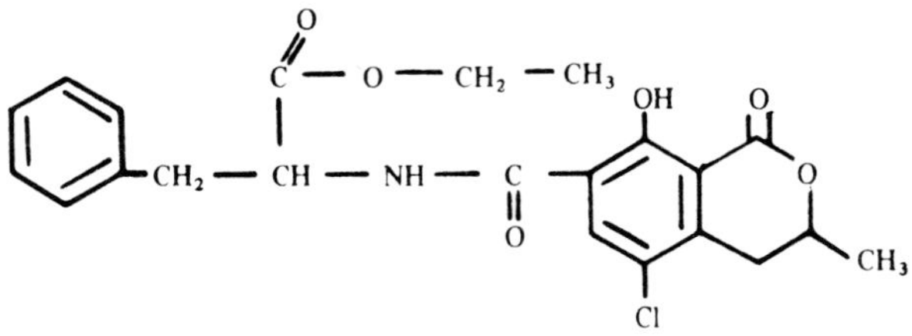
Oxalic acid	$(COOH)_2 2H_2O$
Pantherine	See Muscimol
Parasiticol	See aflatoxin B, aflatoxin B_3
Paspalinine (see also: 3-methyl-3-hydroxy-1-butenylpaspalinine; 3-methyl-2-butenylpaspalinine	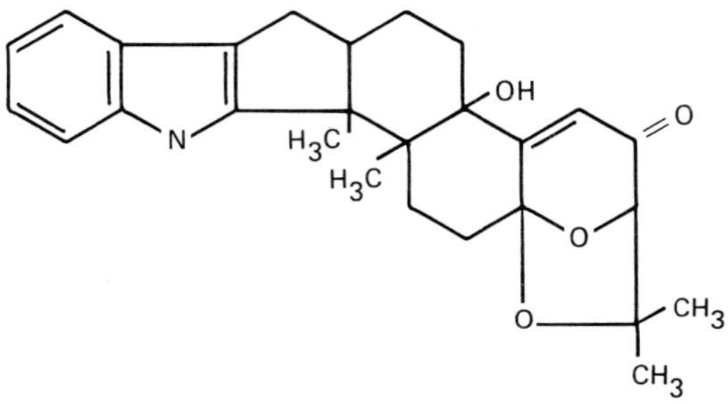
Patulin	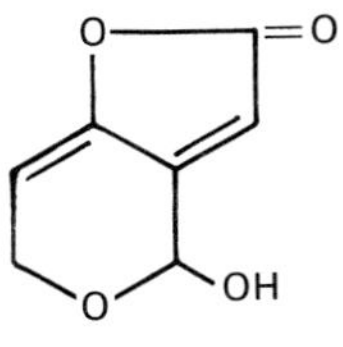
PAXILLINE	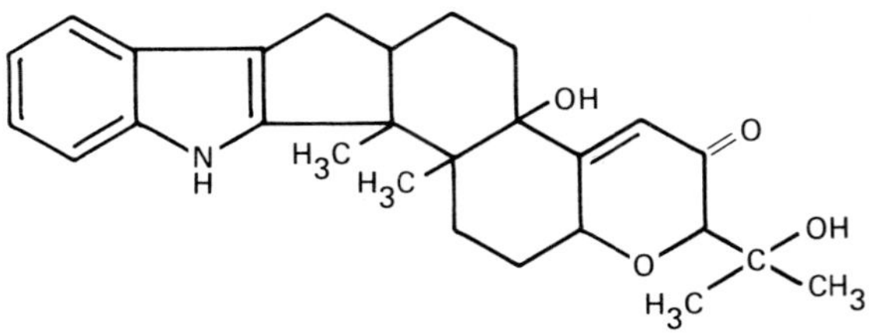
Penicidin	See patulin
Penicillic acid	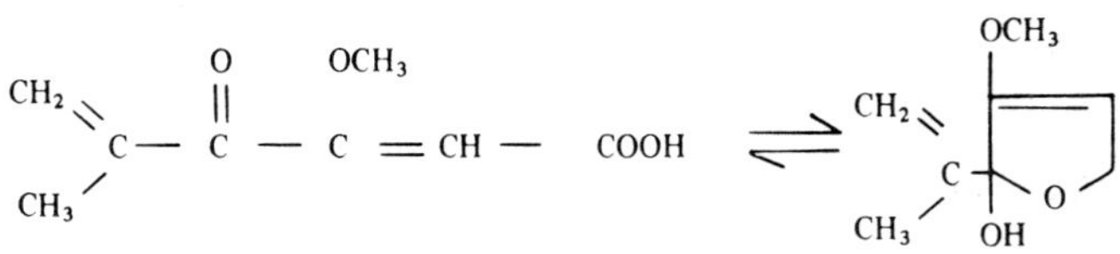

Table 1 (continued)
CHEMICAL STRUCTURES OF FUNGAL TOXINS

Name	Structure
Phallicidine	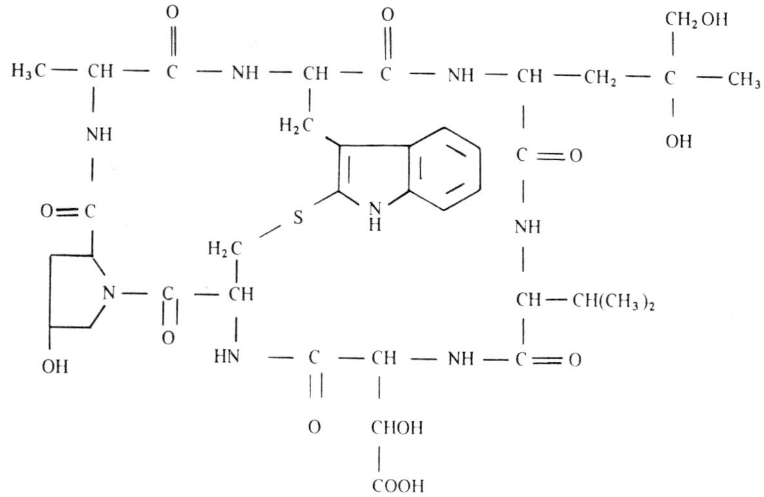
Phallin B*	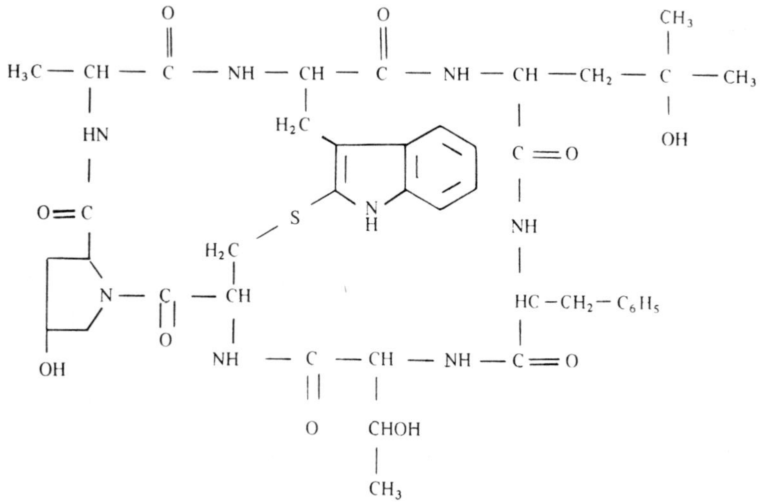
Phalloidin	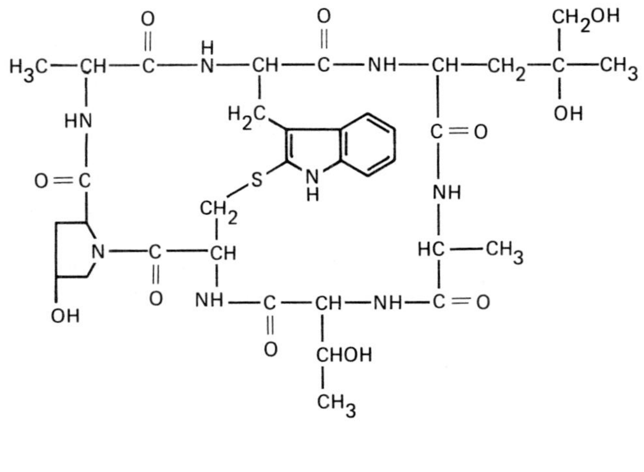

Table 1 (continued)
CHEMICAL STRUCTURES OF FUNGAL TOXINS

Name	Structure
Phalloin	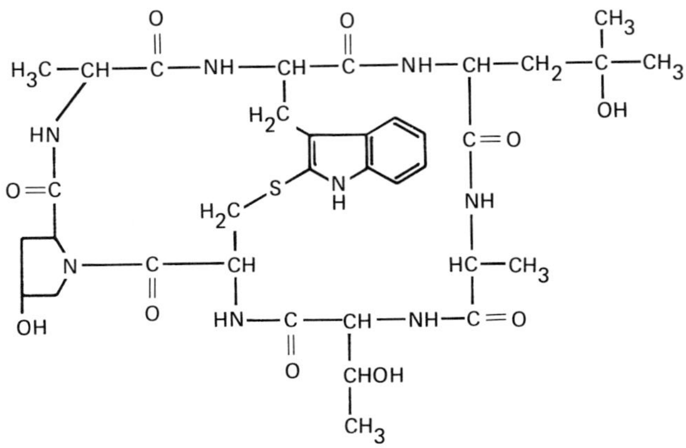
Phallotoxins	See phallacidine, phallin B, phalloidin, and phalloin
Phomenone	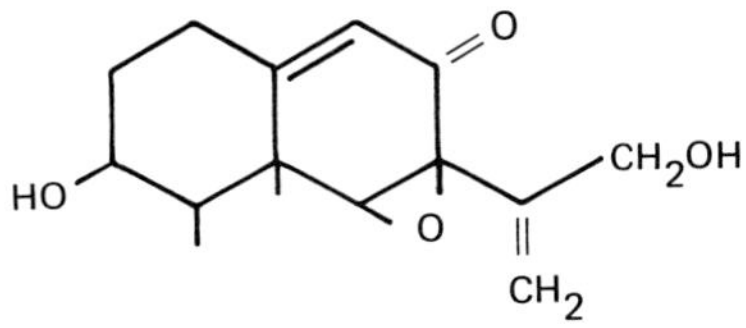
Pigment A (see secalonic acid D)	
PR toxin	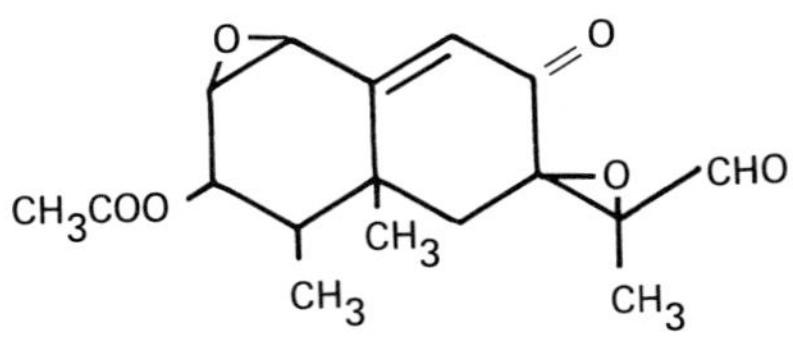
Pramuscimol	See ibotenic acid
Psilocybin	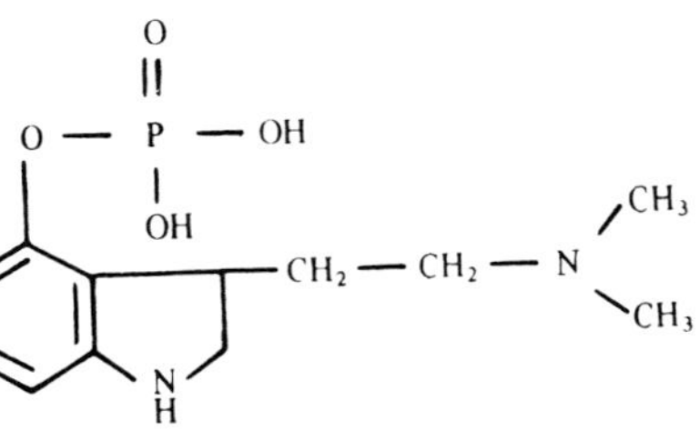
Psoralens	See 8-methoxypsoralen and 4,5′,8-trimethylpsoralen

Table 1 (continued)
CHEMICAL STRUCTURES OF FUNGAL TOXINS

Name	Structure
Roridin E (satratoxin D, stachybotrys toxin)	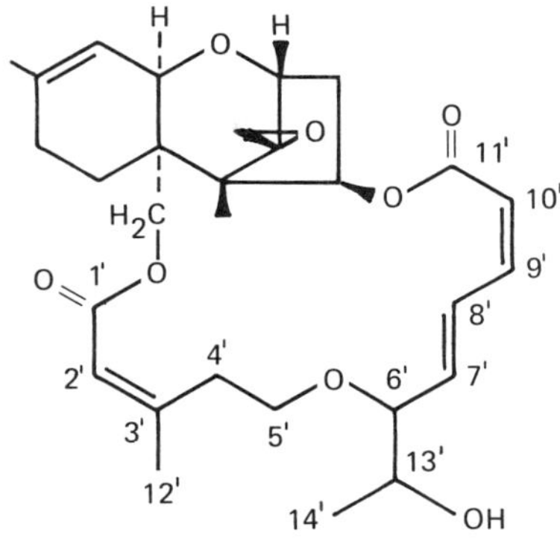
Roseotoxin B	Not known
Rubratoxin A	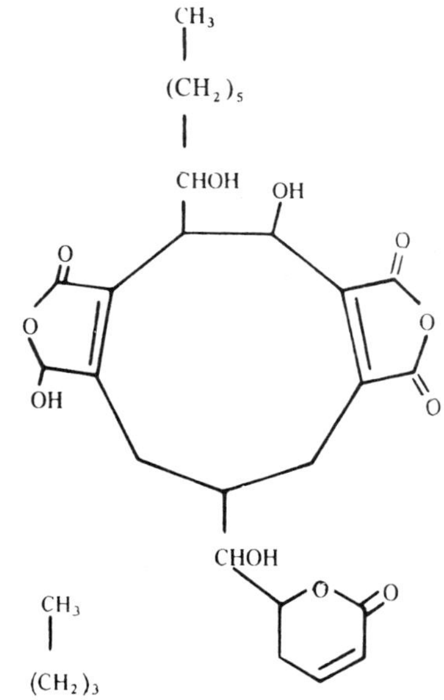
Rubratoxin B	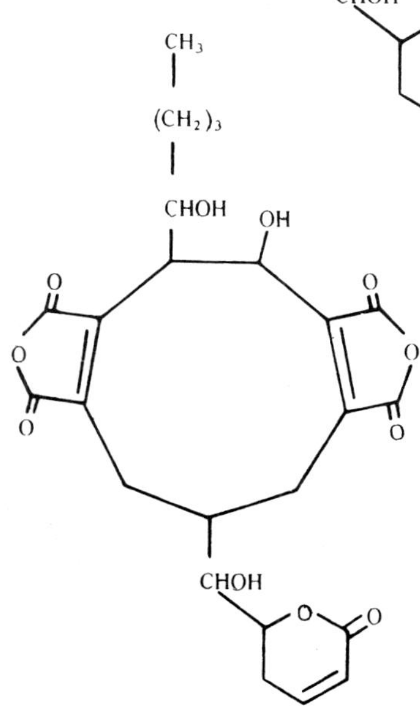

Table 1 (continued)
CHEMICAL STRUCTURES OF FUNGAL TOXINS

Name	Structure
Rugulosin	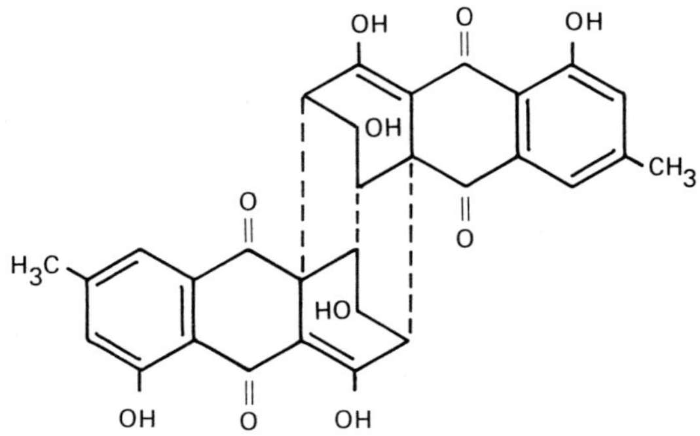
Satratoxin D	See roridin E
Satratoxin H	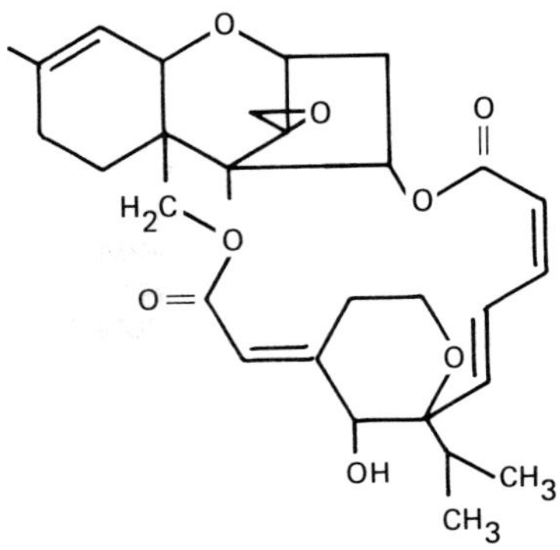
Secalonic acids (A-F differ stereochemically)	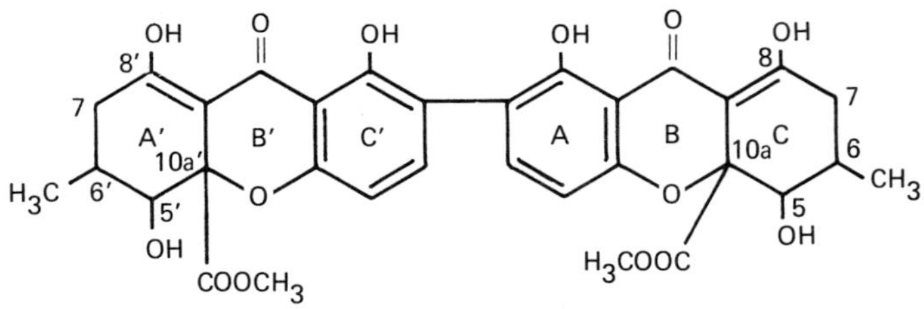
Slaframine	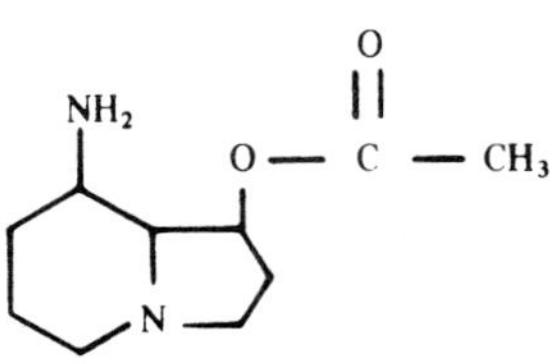

	5-OH	5′-OH	6-Me	6′-Me	10a-CO_2Me	10a′-CO_2Me
A:	α	β	β	α	α	β
C:	α	α	β	α	α	β
D:	β	α	α	β	β	α
E:	β	α	β	α	α	β

Table 1 (continued)
CHEMICAL STRUCTURES OF FUNGAL TOXINS

Name	Structure
Slobber factor	See slaframine
Sporidesmin A	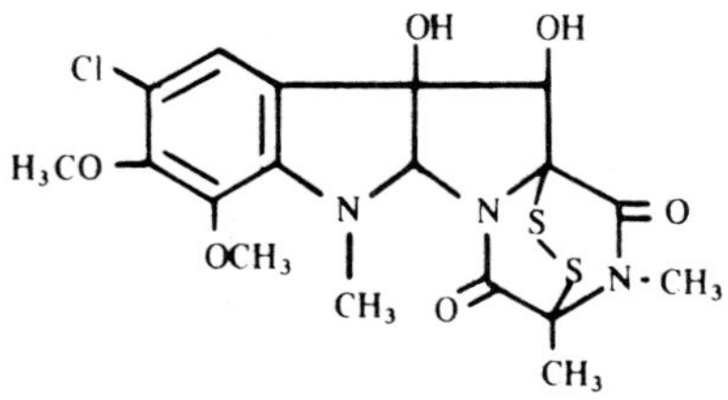
Sporidesmin B	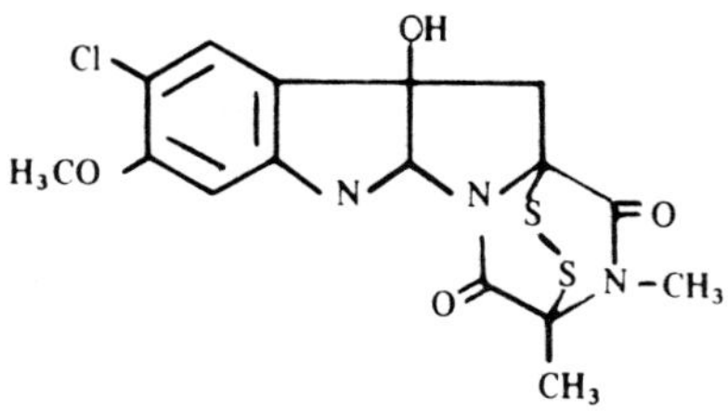
Sporidesmin C	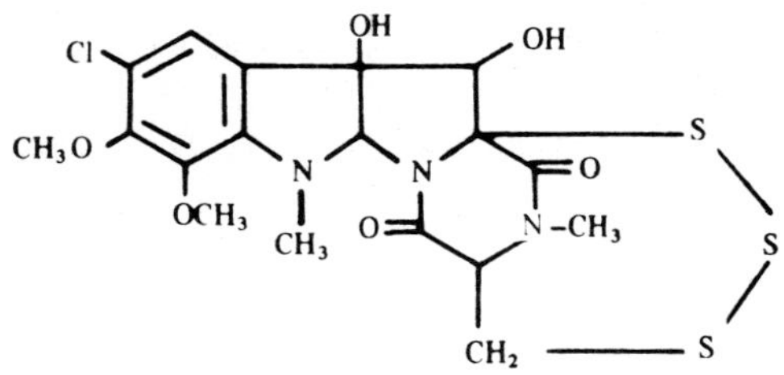
Sporidesmin D	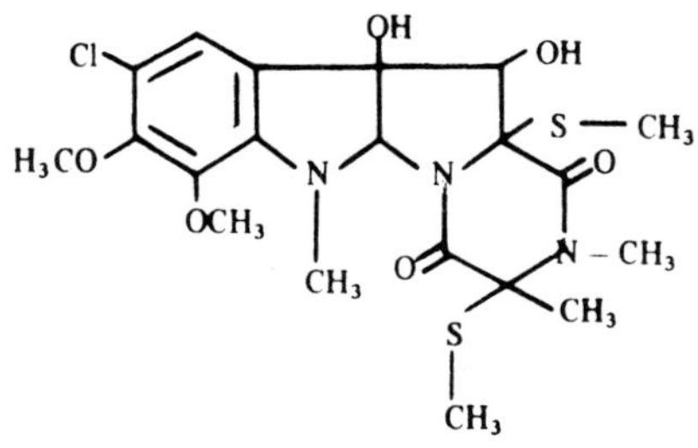
Sporidesmin E	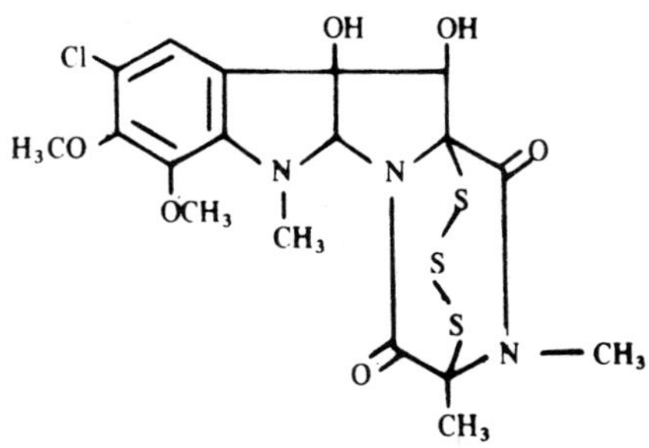

Table 1 (continued)
CHEMICAL STRUCTURES OF FUNGAL TOXINS

Name	Structure
Sporidesmin F	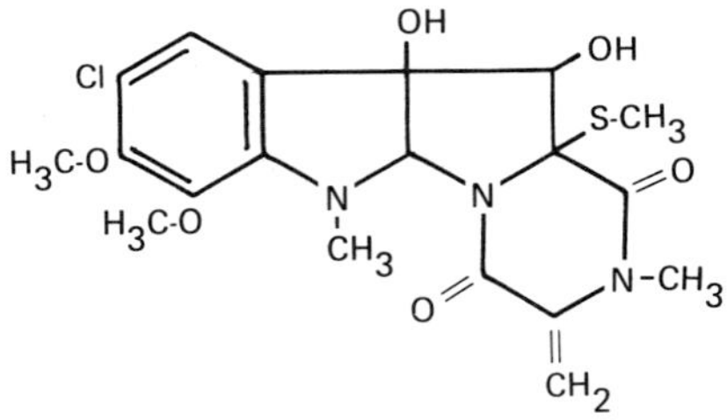
Sporidesmin G	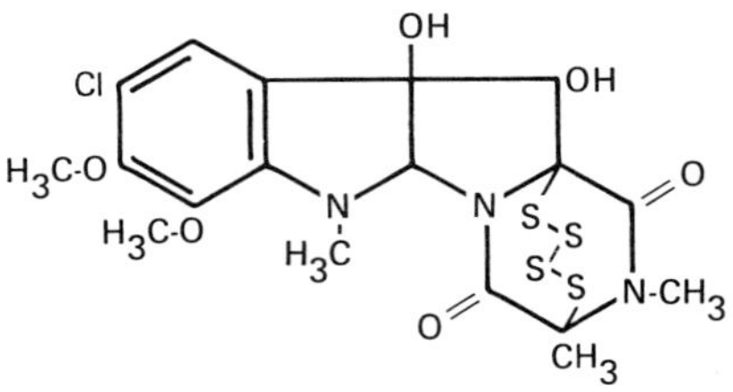
Stachybotrys toxin	(See roridin E)
Stemphone	Not known
Sterigmatocystin	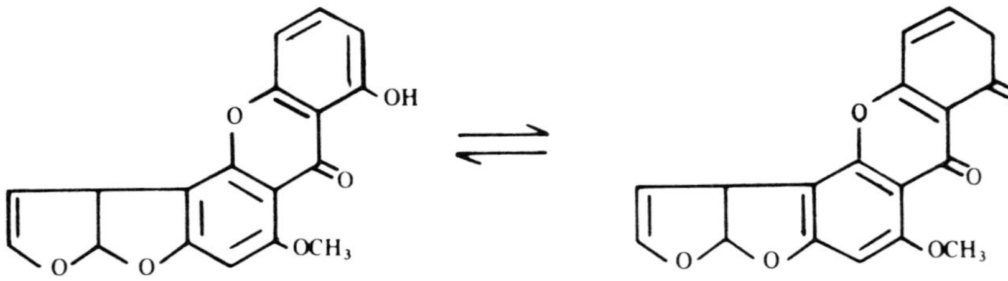
T-2 toxin	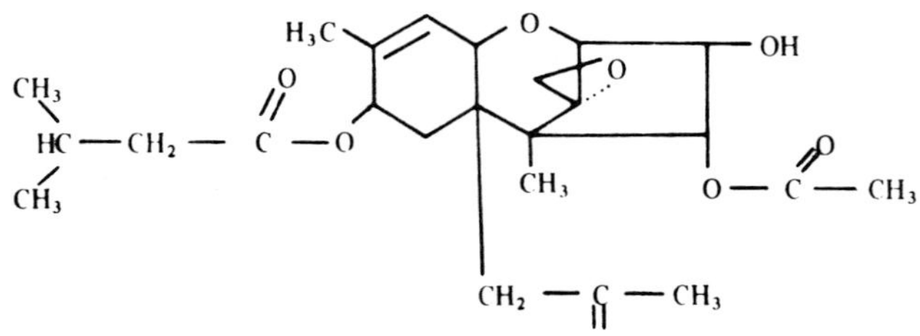
Terreic acid	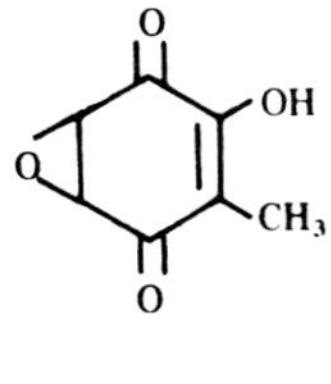
TR-1	(See verruculogen)

Table 1 (continued)
CHEMICAL STRUCTURES OF FUNGAL TOXINS

Name	Structure
TR-2	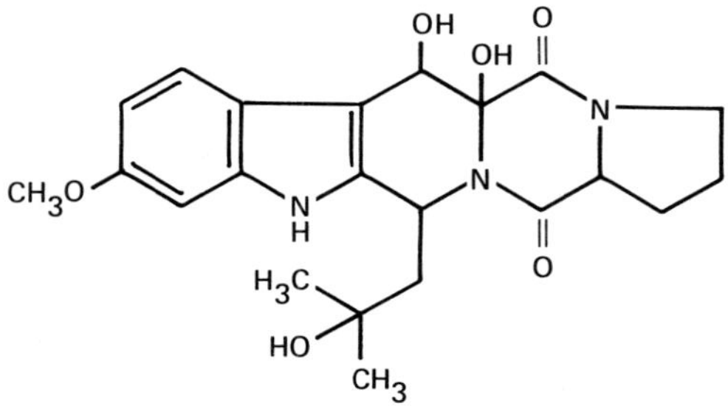
Tremorgen	Not known
Tremortin A (penitrem A)	Not known
Tremortin B (penitrem B)	Not known
Trichodermin	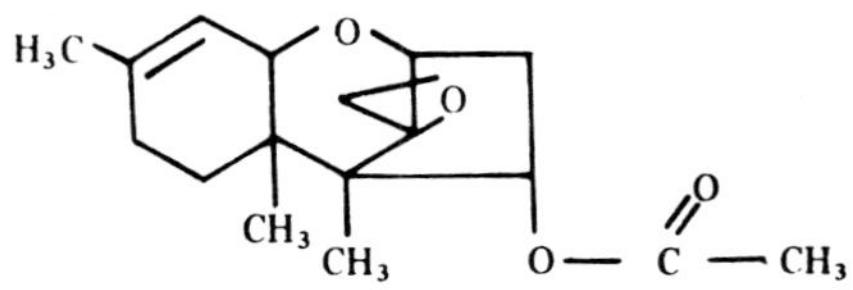
Trichothecin	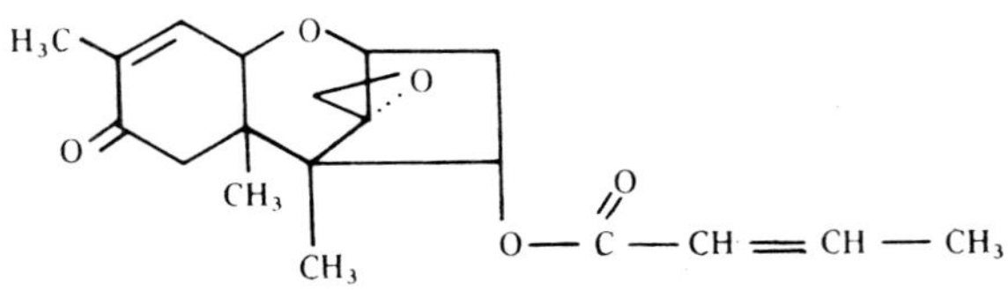
Tryptoquivaline	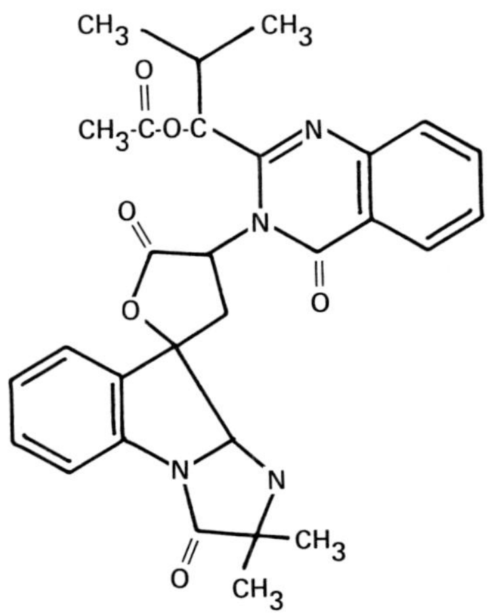

Table 1 (continued)
CHEMICAL STRUCTURES OF FUNGAL TOXINS

Name	Structure
Tryptoquivalone	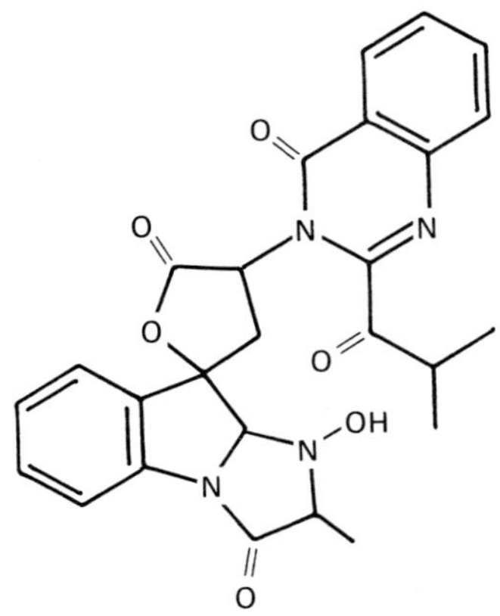
4,5′,8-Trimethylpsoralin	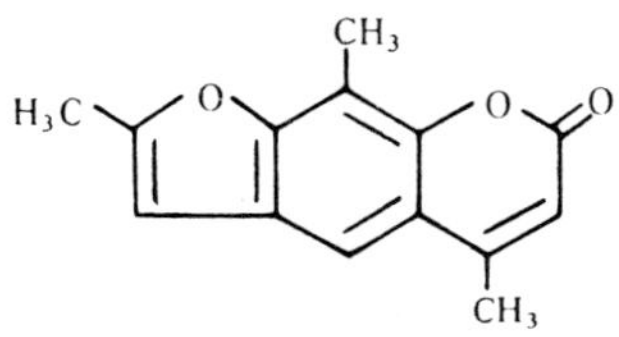

Verrucarin A

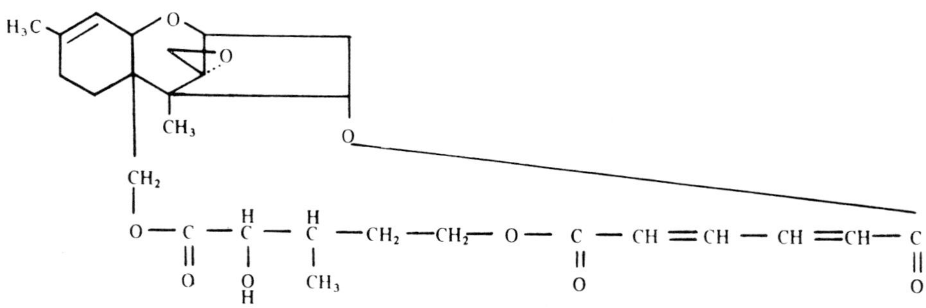

Verrucarin B

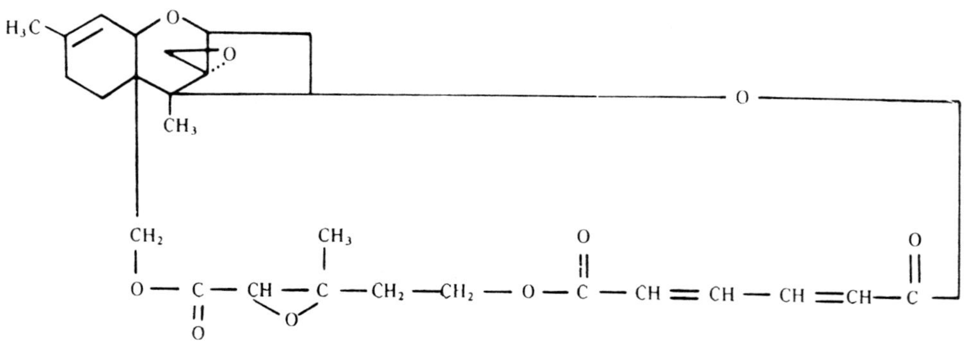

Table 1 (continued)
CHEMICAL STRUCTURES OF FUNGAL TOXINS

Name	Structure
Verruculogen (TR-1)	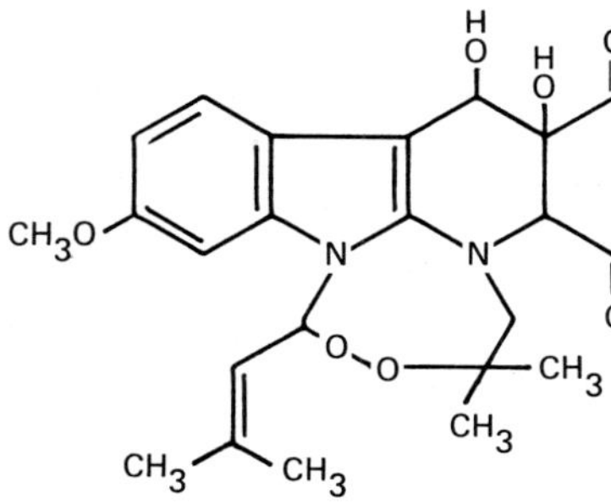
Verruculotoxin	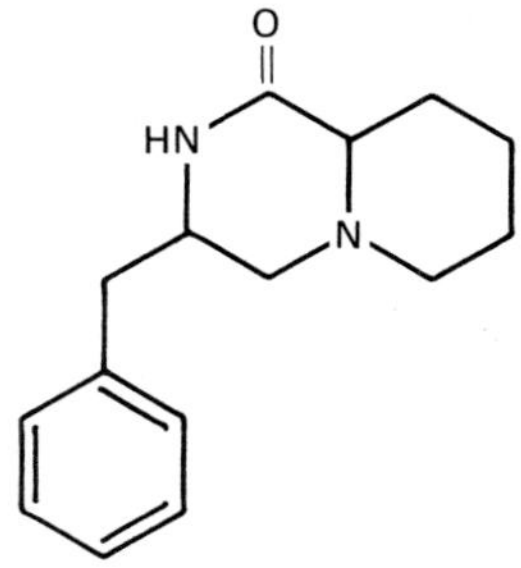
Versicolorin A	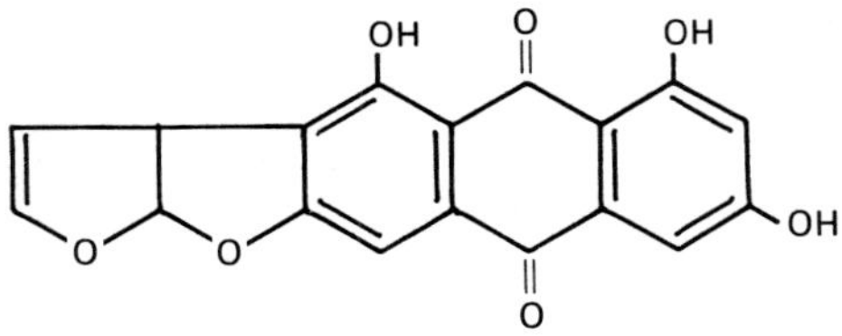
Versiconal acetate	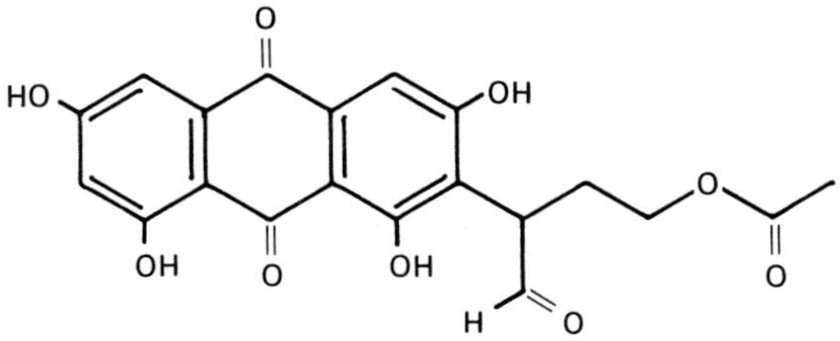
Viomellein	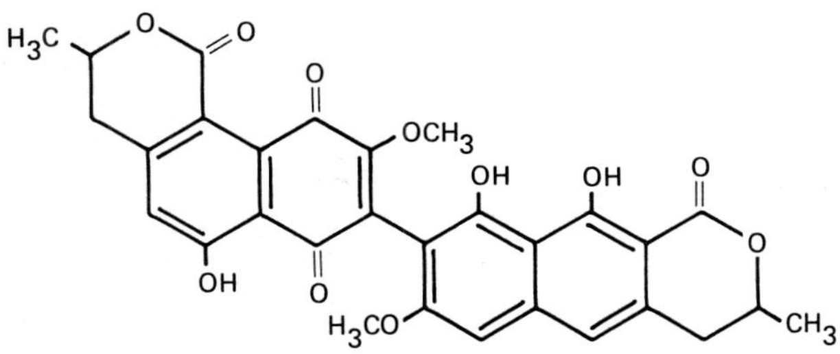

Table 1 (continued)
CHEMICAL STRUCTURES OF FUNGAL TOXINS

Name	Structure
Viriditoxin	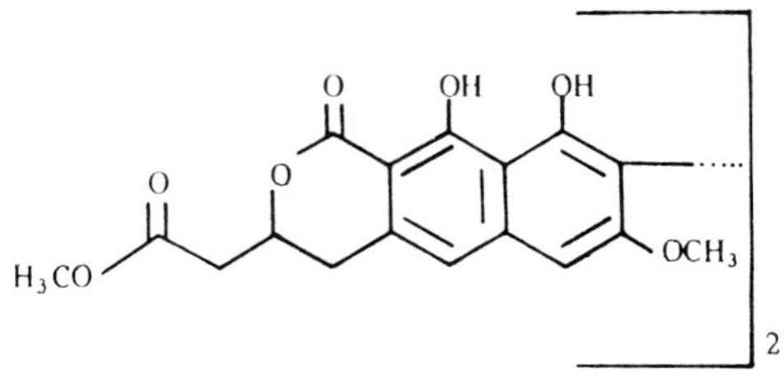
Vomitoxin (deoxynivalenol)	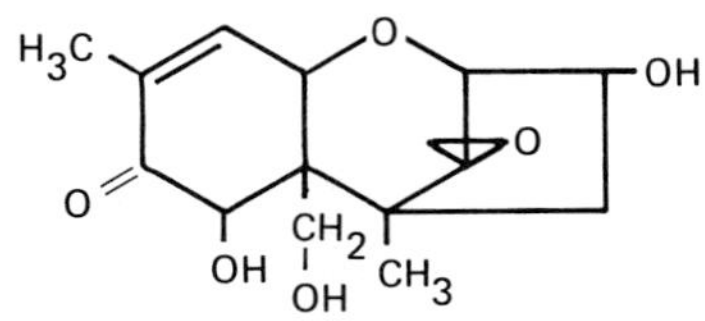
XANTHOMEGNIN	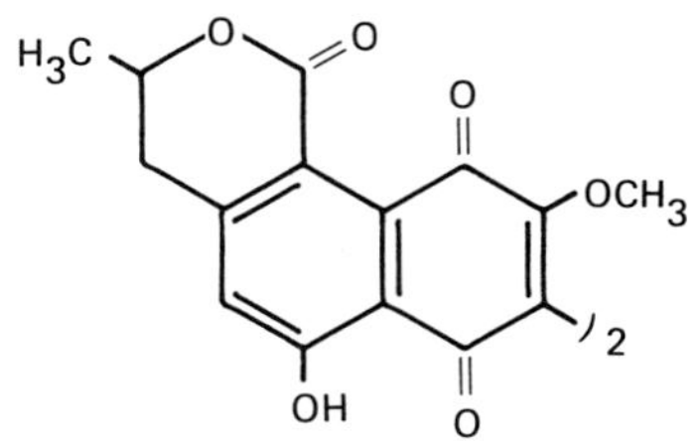
Xanthotoxin	(See 8-methoxypsoralen)
Zearalenone	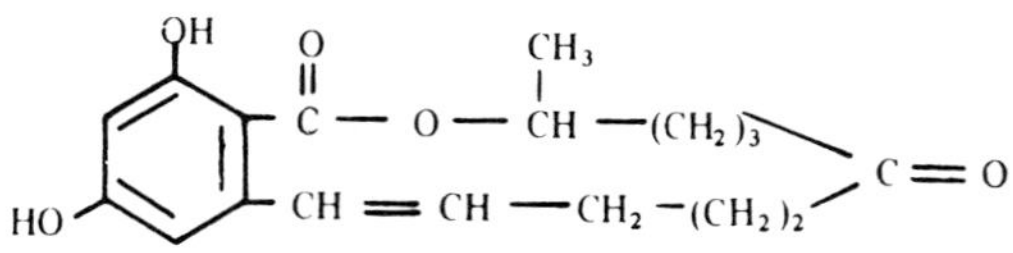

Table 2
CHEMICOPHYSICAL CHARACTERISTICS OF FUNGAL TOXINS

Name	Formula	Mol wt	Melting point (°C)	UV spectrum (Mμ) extinction coefficient (ε)	IR spectrum (cm^{-1})	NMR spectrum	Fluorescence emission	Specific rotation
Acetylscirpediol	$C_{17}H_{24}O_6$	324, m/e⁺306, 278, 234, 219, 43	100—110	No data	3450, 2950, 1720, 1435, 1375, 1240, 1165, 1110, 1080, 1050, 960,	3.66 (d, J = 5), 4.22 (dd), 5.51 (d, J = 3), 3.66 (q, J = 12)	No data	No data
Acetylneosolaniol	$C_{21}H_{18}O_7$	424, m/e⁺424, 352, 364, 304, 43	189—190	No data	No data	3.86 (d, J = 5), 4.18 (m), 5.26 (d, J = 3), 2.30 (d), 2.45 (d), 5.28 (d), 4.08 (q, J = 12)	No data	No data
β-Acetoxy-3α,15-diOH-12,13-epoxytrichotec-9-ene		324	No data	End absorption	($CHCl_3$) 3480, 1720,	0.84 (s), 1.74 (s), 2.15 (s), 2.78 (d, J = 4 Hz), 3.06 (d, J = 4 Hz), 3.64 (d, J = 12 Hz), 3.8 (d, J = 12 Hz), 3.67 (d, $J_{2,3}$ = 5 Hz), ca. 4.25, 5.48, 5.53 (d, $J_{3,4}$ = 3.5)	No data	No data
Aflatoxin B_1	$C_{17}H_{12}O_6$	312	268—269	223 (25,600), 265 (13,400), 362 (21,800)	1760, 1684, 1632, 1598, 1562	($CDCl_3$) 6.89 (d, J = 7 Hz), 6.52 (t, J = 2.5 Hz), 5.53 (t, J = 2.5), 4.81 (t of d, J = 2.5, 7), 6.51 (s), 4.02 (s), A_2B_2 type absorption 3.42, 2.61	Excitation 352 nm, emission 425 nm	$(\alpha)_D$ −558
Aflatoxin B_1 aldehyde	$C_{18}H_{13}O_7$	341.28	273—276 (dec)	(EtOH) 275 (24,900), 310 (15,000), 390 (11,500)	As aflatoxin B_1	As Aflatoxin B_1	No data	$(\alpha)^{23}$ −511
3-Hydroxyaflatoxin B_1	$C_{17}H_{12}O_7$	328	280 (dec)	223 (17,000), 266 (9850), 367 (15,800)	(KBr) 1758, 1690, 1628, 1595	See Ref. 335	No data	No data
Aflatoxin B_2	$C_{17}H_{14}O_6$	314	286—289	220 (17,900), 240 (infl.) (12,800), 266 (12,000), 363 (23,400)	1765, 1695, 1635, 1600, 1565	No data	Excitation 365 nm, emission 425 nm	$(\alpha)_D$ −492
1-Acetoxyaflatoxin B_2	$C_{19}H_{16}O_8$	372	225	221 (19,000), 266 (14,870), 362 (23,400)	No data reported	($CDCl_3$) 3.45 (d), 5.80 (m), 7.55 (m), 362 (m), 3.68 (s), 6.10 (s), 6.62 (t), 7.55 (m), 8.25 (s)	No data	No data

Table 2 (continued)
CHEMICOPHYSICAL CHARACTERISTICS OF FUNGAL TOXINS

Name	Formula	Mol wt	Melting point (°C)	UV spectrum (Mμ) extinction coefficient (ε)	IR spectrum (cm^{-1})	NMR spectrum	Fluorescence emission	Specific rotation
1-Ethoxyaflatoxin B_2	$C_{19}H_{18}O_7$	358	247	226 (14,640), 266 (12,140), 364 (12,580)	No data reported	4.65 (m), 9.06 (t), 6.49 (m), 7.63 (m), 5.89 (m), 3.46 (d), 3.68 (s), 6.04 (s), 6.57 (t), 7.40 (t)	No data reported	No data reported
1-Methoxyaflatoxin B_2	$C_{18}H_{16}O_7$	360	220—223	As aflatoxin B_1	Similar to aflatoxin B_1	2.36 (m), 2.37 (m), 3.25	Not reported	$(\alpha)_D$ −550
2-Methoxyaflatoxin B_2	$C_{18}H_{16}O_7$	360	245—247	No data reported	No data reported	No data reported	No data	$(\alpha)_D$ −360
Aflatoxin B_{2a}	$C_{17}H_{14}O_7$	330	240	(MeOH) 228 (17,600), 256 (10,300), 363 (20,400)	As aflatoxin B_1 with additional strong band at 3440	(Deutero pyridine) 3.21 (d, J = 6), 5.89 (m), 7.43 (m), 4.0 (m), 3.49 (s), 6.25 (s), 6.82 (t, J = 6), 7.43 (m)	No data	No data
Aflatoxin B_3	$C_{14}H_{14}O_6$	302	217 (233, 234)	(MeOH) 217 (17,300), 225 (sh) (12,600), 253 (6800), 262 (7400), 325 (17,300)	3350, 1728, 1625, 1605	(Deutero pyridine) 3.40 (t), 4.58 (t, J = 1.5), 5.32 (m), 3.14 (d, J = 3.0), 6.40 (s), 3.54 (s), coupled triplets at 6.75 and 5.97 (J = 3.0), 3.73 (s)	No data	No data
Aflatoxin D_1	$C_{16}H_{14}O_5$	286	255—258 (dec)	(MeOH) 227 (15,920), 324 (12,440)	3436, 3155, 1665, 1703	Inconclusive	No fluorescence	No data
Aflatoxin deshydro D_1	$C_{16}H_{10}O_4$	284	No data	No data	No data	No data	No data	No data
Aflatoxin G_1	$C_{17}H_{12}O_7$	328	244—246	243 (11,500), 257 (9900), 264 (10,000), 362 (16,100)	($CHCl_3$) 1760, 1695, 1630, 1595	A_2X_2 pattern (4.47, t, J = 6; 3.48, t, J = 6); remaining spectrum same as aflatoxin B_1	Excitation 365 nm, emission 450 nm	$(\alpha)_D$ −556
Aflatoxin G_2	$C_{17}H_{14}O_7$	330	237—240	362 (21,000)	As G_1	7.75 (t), 6.53 (t), 6.05 (s), 5.80 (broad), 5.56 (t), 3.65 (s), 3.52 (d)	Excitation 365 nm, emission 450 nm	$(\alpha)^{23}_D$ −473
2-Ethoxyaflatoxin G_2	$C_{19}H_{18}O_8$	374	203	(MeOH) 223 (17,140), 244 (12,840), 266 (11,700), 366 (19,500)	No data reported	4.70 (m), 9.04 (t), 6.48 (m), 7.62 (m), 5.86 (m), 3.42 (d), 3.69 (s), 6.04 (s), 6.84 (t), 5.73 (t)	No data	No data
Aflatoxin G_{2a}	$C_{17}H_{14}O_8$	346	190	(MeOH) 223 (18,600), 242 (10,100), 262 (8700), 365 (18,000)	Similar to G_2 except for additional band at 3620	(Deuteropyridine) 3.22 (d, J = 6), 5.64 (m), 7.52 (m), 3.95 (m), 3.47 (s), 6.24 (s), 6.78 (t, J = 6), 5.64 (m)	No data	No data

Aflatoxin GM_1	$C_{17}H_{12}O_8$	344	276					
Aflatoxicol H_1	$C_{17}H_{14}O_7$	330	No data	(MeOH) 327 (15,530), 259 (16,010), 252 (14,560)	3480, 1760, 1720, 1685	No data	No data	No data
Aflatoxin M_1	$C_{17}H_{12}O_7$	328	299	(MeOH) 357 (21,250), 265 (14,150), 227 (27,650); ($CHCl_3$) 357 (19,950), 267 (12,950), 244 (9100); (acetonitrile) 350 (19,850), 265 (13,750), 227 (27,250)	As aflatoxin B_1 with peak at 3425 in addition	(Deuterodimethylsulfoxide) 602, 4.36 (d, J = 3), 3.17 (d, J = 3.0), 3.54 (s), 3.22 (s)	Excitation 365 nm, emission 425 nm	No data
Aflatoxin M_2	$C_{17}H_{14}O_7$	330	293	(MeOH) 357 (22,900), 264 (12,100), 221 (21,400); ($CHCl_3$) 357 (21,250), 264 (11,650), 244 (10,100); (acetonitrile) 350 (21,400), 264 (12,050), 225 (20,950)	As aflatoxin M_1 but missing vinyl ether bands at 3100, 1067 and 722	No spectrum published	No data	No data
Aflatoxin P_1	$C_{16}H_{10}O_6$	298	No data	(EtOH) 226 (20,400), 267 (11,200), 342 (14,900), 362 (15,400), 425 (2500); (EtOH/NaOH) 233 (25,000), 298 (12,900), 338 (13,400), 422 (14,400); (EtOH/HCl) 227 (24,300), 256 (12,400), 345 (21,600), 356 (23,200)	3450, 1745, 1660 sh, 1630, 1600, 1550, 1480	No data	No data	$(\alpha)^{20}{}_D$ −574
Aflatoxin Q_1	$C_{17}H_{12}O_7$	328.06	266 (dec)	(EtOH) 224 (20,500), 242 (10,000), 266 (11,700), 265 (18,800)	($CHCl_3$) 3610, 3040, 2980, 1780, 1710, 1640, 1615, 1580	($CHCl_3$) 2.67 A-part of ABX J_{AB} = 19 Hz, J_{AX} = 3 Hz, 1H; 2.97 B-part of ABX J_{AB} = 19 Hz, J_{BX} = 6 Hz, 1 H; 3.6 broad OH (?), 1 H; 4.08 s, 3 H; 4.80 d, J = 7 Hz of t J - 2 Hz, 1 H; 5.48 t (J = 2 Hz), 1 H; 5.55 and 5.60, two singlet-like signals due to X-part of ABX; 6.48 t J = 2 Hz, 1 H; 6.52, s, 1 H; 6.83 d, J = 7 Hz, 1 H	No data	No data
Aflatoxin R_o	$C_{17}H_{16}O_6$	314	230—234	(MeOH) 325 (14,100), 261 (10,800), 254 (6790)	1590, 1600, 1130, 2850, 1485, 1440, 1360	6.45 (t), 5.48 (t), 4.77 (m), 6.77 (d), 6.37 (s), 3.85 (s), 3.27 (m), 2.23 (m), 2.40 (m), 5.28 (m)	Excitation 365 nm emission 425 nm	No data

Table 2 (continued)
CHEMICOPHYSICAL CHARACTERISTICS OF FUNGAL TOXINS

Name	Formula	Mol wt	Melting point (°C)	UV spectrum (Mμ) extinction coefficient (ε)	IR spectrum (cm^{-1})	NMR spectrum	Fluorescence emission	Specific rotation
Aflatoxin RB_1	$C_{17}H_{18}O_6$	318.14	No data	(MeOH) 330 (1750), 274 (2670), 260 (3200)	3500, 3450, 3000, 1630, 1600, 1480	6.50 (triplet), 5.40 (triplet), 6.82 (doublet), 4.81 (multiplet), 6.45 (singlet), 3.89 (singlet), 2.51 (multiplet), 3.40 (multiplet), 3.79 (singlet), 6.22 (singlet), 4.59 (multiplet)	Activation 335 nm emission 420 nm	No data
Aflatoxin RB_2	$C_{17}H_{20}O_6$	320.16	No data	330 (1950), 280 (1030), 260 (1735)	3500, 3450, 2950, 1620, 1485; minor bands 1710	6.50 (triplet), 5.40 (triplet), 6.82 (doublet), 4.81 (multiplet), 6.45 (singlet), 3.89 (singlet), 2.51 (multiplet), 3.40 (multiplet)	Activation 335 nm Emission 420 nm	No data
Agarin (see Muscimol)								
Agroclavine	$C_{16}H_{18}N_2$	238.32	198—203 (dec)	225, 284, 293, 29,512, 7585, 6456			Water, excitation (280), fluorescence (350); Ethanol, excitation (285), fluorescence (345)	$[\alpha]^{20}_D$ −184 (Pyridine)
Agrocybin	$C_8H_5O_2N$	147.13	140 (conflagrates)	215 (67,608), 224 (87,096), 269 (1737), 286 (2398), 304 (3019), 325 (1949)	No data	No data	No data	No data
Alternaric acid	$C_{21}H_{30}O_8$	408	138	273	—	—	—	—
Alternariol	$C_{14}H_{10}O_5$	258	350 (dec)	(MeOH) 330—340, 201, 290, 256, 230	3460, 3100, 1670, 1590, 1430, 1365, 1200, 850	—	Blue fluorescence	—
Alternariol methyl ether	$C_{15}H_{12}O_5$	272	266—270 (dec)	—	—	—	Blue fluorescence	—
Amanin (see β-amanitine)								
Amanita toxins (see α, β, γ and ε-amanitine)								
α-amanitine	$C_{40}H_{56}O_{13}N_{10}S$	917	254—55	302 (11,500)	No data	No data	No data	$[\alpha]^{20}_D$ + 193°

β-amanitine	$C_{40}H_{55}O_{14}N_9S$	918	300	302 (12,600)	No data	No data	No data	No data
γ-amanitine	$C_{39}H_{54}O_{12}N_{10}S$	890	No data	302 (10,000)	No data	No data	No data	No data
ε-amanitine	$C_{39}H_{53}O_{13}N_9S$	888	No data	302 (12,600)	No data	No data	No data	No data
Amanullin	$C_{39}H_{54}O_{11}N_{10}S$	814	No data	No data	No data	No data	No data	No data
Ascladiol	$C_7H_8O_4$	156	65—66	(EtOH) 271, no extinction coeff. reported	3300, 1750, 1735	6.29, 5.87, 4.74, 4.30	No data	No data
Ascotoxin (see decumbin)								
Aspercolorin	$C_{23}H_{27}O_4N_4$	423.48	No data	210 (28,080), 226 (18,920), 260 (13,620), 315 (4170)	No data	No data	No data	No data
Aspergillic acid	$C_{12}H_{20}O_2N_2$	224.3	97—99	328 (8500), 235 (10,500) EtOH	3120, 2940, 2850, 2800—2250 (broad), 2040, 1640, 1585, 1150, 710	8.98, 8.85, 8.55, 8.43, 8.21—7.48, 6.90, 6.79, 6.62, 6.50, 6.38, 6.27, 2.17	No data	$[\alpha]^{18}_D$ +13.3 (EtOH)
Aspertoxin	$C_{19}H_{14}O_7$	354.07	325—327					
Averufin	$C_{20}H_{16}O_7$	368	283—289	(EtOH) 223 (33,000), 256 (s), (16,500), 265 (18,500), 286 (24,800), 294 (30,800), 319 (12,500), 453 (10,500)	(Nujol) 3400, 1677, 1622, 1569, 1270, 1210, 1025	(^{13}C NMR) 16.3 (t), 67.1 (d), 102.1 (s), 133.9 (s), 135.7 (s), 158.9 (s), 160.7 (s), 165.0 (s), 166 (s), 189.7 (s)	No data	No data
Bufotenine	$C_{12}H_{16}ON_2$	204.26	146—147	No data	No data	No data	No data	No data
Butenolide	$C_6H_7O_3N$	141	116.5—118.5	202 (11,300)	3440, 3340, 1705, 1760, 1790	See Ref. 229, 230	No data	No data
Chaetoglobosin A_1	$C_{32}H_{36}O_5N_2$	528.29	168—170	(EtOH) 223, 245 (sh), 274, 282, 292 (log E: 4.61, 3.96, 3.82, 3.82, 3.73)	(KBr) 3438, 3259, 2960, 1689 (br.), 1615, 1432, 1248, 1052, 983, 969, 760	See Ref. 329	No data	$[\alpha]_D$ −270 (MeOH)
Chaetoglobosin B	$C_{32}H_{36}O_5N_2$	528.29	186—187	(EtOH) 222, 245 (sh), 274, 281, 290 (log E: 4.64, 3.99, 3.90, 3.90, 3.83)	(KBr) 3440, 2950, 2900, 1690 (br.), 1620, 1245, 1040, 972, 746	See Ref. 329	No data	$[\alpha]_D$ −176 (MeOH)

Table 2 (continued)
CHEMICOPHYSICAL CHARACTERISTICS OF FUNGAL TOXINS

Name	Formula	Mol wt	Melting point (°C)	UV spectrum (Mμ) extinction coefficient (ε)	IR spectrum (cm^{-1})	NMR spectrum	Fluorescence emission	Specific rotation
Chaetoglobosin C	$C_{32}H_{36}O_5N_2$	528.26	260—263	(EtOH) 222, 273, 281, 291 (log E: 4.56, 3.83, 2.91)	(KBr) 3445, 3305, 2915, 1697 (s), 1642, 1441, 1327, 1099, 1056, 986, 851, 832, 745	See Ref. 330	No data	$[\alpha]_D$ −30 (MeOH)
Chaetoglobosin D	$C_{32}H_{36}O_5N_2$	528.26	216	(EtOH) 221, 273, 281, 290 (log E: 4.64, 3.96, 3.96, 3.88)	(KBr) 3421, 3280, 2945, 2918, 2888, 1686 (s), 1606, 1430, 1250, 1052, 972, 908, 750	See Ref. 330	No data	$[\alpha]_D$ −267 (MeOH)
Chaetoglobosin E	$C_{32}H_{38}O_5N_2$	530.27	(?)	(EtOH) 221, 275, 281, 291 (log E: 4.75, 3.85, 3.85, 3.80)	(KBr) 3410, 2885, 1704 (s), 1676, 1455, 1048, 746	See Ref. 330	No data	$[\alpha]_D$ +158 (MeOH)
Chaetoglobosin F	$C_{32}H_{30}O_5N_2$	530.27	177—178	(EtOH) 222, 276, 283, 292 (log E: 4.68, 3.84, 3.83, 3.78)	(KBr) 3346, 2955, 2920, 1676 (s), 1618, 1430, 1385, 1260, 1110, 979, 879, 740	See Ref. 330	No data	$[\alpha]_D^{CHCl_3}$ −69
Chaetoglobosin G	$C_{32}H_{36}O_5N_2$	528.27	251—253	(EtOH) 222, 275, 282, 291 (log E: 4.51, 3.79, 3.79, 3.73)	(KBr) 3455, 3300, 1713, 1693, 1646, 1623, 987, 948, 741	See Ref. 331	No data	$[\alpha]_D^{MeOH}$ +89

Chaetoglobosin J	$C_{32}H_{36}O_4N_2$	512.25	149—151	(EtOH) 224, 270, 280, 290 (log E: 4.68, 3.86, 3.86, 3.78)	(KBr) 3412, 3273, 1683, 1639, 1612, 980, 975, 925, 750	See Ref. 331	No data	No data
Citreoviridin	$C_{23}H_{30}O_6$	402	107—111	(EtOH) 388 (48,000), 294 (27,100), 286 sh (24,600), 234 (10,200), 204 (17,000)	3500, 1702, 1689, 1654, 1626, 1562, 1531, 1452, 1405, 1249, 1094, 1069, 999, 821, 811		Fluoresces yellow (no spectral data)	No data
Citrinin	$C_{13}O_{14}O_5$	250.24	175	220 (log = 4.32), 255 (log = 3.92), 325 (log = 3.79)	3484, 2985, 1675, 1639, 1108	8.75 (d, J = 6.5), 8.62 (d, J = 6.5), 7.97 (s), 6.96 (q, J = 6.5), 5.16 (q, J = 6.5), 1.7 (s), −3.7 (s), −5.2 (s)	Yellow fluorescence, excitation 395 nm, emission 504 nm	$[\alpha]_D^{22}$ −39
Clavicin (see patulin)								
Clavatin (see patulin)								
Claviformin (see patulin)								
Cyclochlorotine	$C_{25}H_{36}N_5O_8Cl_2$	605	251 (dec)	No data	No data	No data	No data	No data
Cyclopiazonic acid	$C_{20}H_{20}O_3N_2$	336.15	245—246	225 (log 4.60) 253 (log 4.22), 275 (sh) (log 4.28), 284 (log 4.31), 292 (log 4.24)	3478, 3200—2600, 1708, 1618	1.82 (s), 2.85—3.2, 2.95 (d, J = 2), 2.5—3.2 (m), 2.71, 2.91 (m), 3.20 (m)	No data	No data
Cytochalasin E	$C_{28}H_{33}NO_7$	495.22	206—208 (dec)	—	3475, 1765, 1720, 1660	—	—	$[\alpha]_D^{25}$ −25.6
Cytochalasin H	$C_{30}H_{39}NO_5$	493.29	255—257	218 (13,400), 248 (5957), 254 (7659), 259 (7446), 264 (7446)	(KBr) 3315, 1752, 1693, 1375, 1235, 965, 705	See Ref. 336 for H and ^{13}C NMR	Fluoresces rust color after spray with H_2SO_4	No data
Decumbin	$C_{16}H_{24}O_4$	280.35	203	216 (8500)	3330, 1260, 1710, 1120, 990, 975, 965	No data	No data	$[\alpha]_D^{24}$ +94
Dendrochine	Not known	No data	No data					
Deoxytryptoquivaline	$C_{29}H_{30}N_4O_6$	530	150—152	(EtOH) 227 (44,500), 232 (sh, 41,900), 252 (sh, 18,500), 267 (sh, 12,000), 278 (sh, 10,300), 304 (3300), 318 (sh, 2700)	($CHCl_3$) 3360, 3310, 2980, 2935, 2875, 1790, 1720, 1676, 1604, 1483, 1439	($CDCl_3$) 1.04 (d, 3, J = 7), 1.20 (d, 3, J = 7), 1.53 (s, 6), 2.16 (s, 3), 2.54 (m, 1), 3.06 (d, 2, J = 10), 5.24 (s, 1), 5.52 (d, 1, J = 9), 5.65 (t, 1, J = 10), 7.11—7.74 (m, 7), 8.20 (m, 1)	No data	$[\alpha]^{25}_D$ +130

Table 2 (continued)
CHEMICOPHYSICAL CHARACTERISTICS OF FUNGAL TOXINS

Name	Formula	Mol wt	Melting point (°C)	UV spectrum (Mμ) extinction coefficient (ε)	IR spectrum (cm⁻¹)	NMR spectrum	Fluorescence emission	Specific rotation
Deoxynortryptoquivaline	$C_{28}H_{28}N_4O_6$	516	158—160	(EtOH) 228 (43,900), 233 (sh, 40,100), 259 (sh, 15,600), 268 (11,700), 278 (sh, 10,500), 305 (4100), 317 (3300)	($CHCl_3$) 3360, 2975, 2935, 2880, 1790, 1724, 1676, 1607, 1483	($CDCl_3$) 1.00 (d, 3, J = 7), 1.15 (d, 3, J = 7), 1.55 (d, 3, J = 7), 2.16 (s, 3), 2.58 (m, 1), 2.87 (d of d, 1, J = 10, 13), 3.07 (d of d, 1, J = 10, 13), 4.12 (q, 1, J = 7), 5.22 (s, 1), 5.55 (d, 1, J = 9), 5.65 (t, 1, J = 10), 7.02—7.74 (m, 7), 8.14 (m, 1)	No data	$[\alpha]_D^{25}$ +69.5
Deoxynortryptoquivalone	$C_{26}H_{24}N_4O_5$	472	192—193	(EtOH) 232 (32,400), 288 (9250), 320 (sh, 6250)	($CHCl_3$) 3360, 2950, 2935, 2880, 1790, 1705, 1680, 1610, 1588, 1484, 1469	($CDCl_3$) 1.26 (d, 3, J = 7), 1.30 (d, 3, J = 7), 1.56 (d, 3, J = 7), 3.02 (d of d, J = 10, 13), 3.32 (d of d, 1, J = 10, 13), 4.08 (m, 1, J = 7), 4.12 (q, 1, J = 7), 5.36 (s, 1), 5.48 (t, 1, J = 10), 7.04—7.54 (m, 7), 8.24 (m, 1)	No data	$[\alpha]_D^{25}$ +171
Desmethoxyviridiol	$C_{19}H_{16}O_5$	324	155—157	(EtOH) 251 (30,000), 322 (16,000)	3390, 1670, 760	7.77 (s), 7.81 (d), 8.61 (d), 3.65 (m), 2.65 (m), 1.59 (s), 2.55 (m), 4.07 (t), 4.82 (m)	No data	No data
Diacetoxyscirpenol	$C_{19}H_{26}O_7$	366.40	161—162	End absorption only	3415, 3350, 1739, 1681, 1241, 820	($CDCl_3$) 3.58 (d), 4.31 (dd), 5.02 (d), 2.0 (m), 5.38 (d), 4.01 (m), 3.06 (d), 2.78 (d), 0.82 (s), 4.05 (q), 1.72 (s), 1.99 (s), 2.08 (s)	No fluorescence	No data
4β,8α-Diacetoxy-12,13-epoxy-trichotec-9-ene-3α-diol	$C_{19}H_{26}O_8$	382 m/e 322, 292, 232	No data	No data	3450 (s), 2940 (s), 1720 (s), 1230 (s)	(MHz; $CDCl_3$): δ0.84 (3H, s; 14-Me), 1.75 (3H, br, s; 16-Me), 1.96 (1H, d, J = 15.4 Hz; αH-7), 2.09 (3H, sh), 2.16 (3H), 2.34 (1H, dd, J = 15.4 and 5.8 Hz; βH-7), 2.80 (1H, dd = 3.9 Hz, H-13), 3.04 (1H, d, J = 3.9 Hz; H-13), 3.60 (1H, d, J = 12.8 Hz, H-15), 3.66 (1H, d, J = 4.8 Hz; H-2), 3.91 (1H, d, J	Fluoresces blue at 254 nm after spraying with *p*-anisaldihyde and heating	No data

						= 12.8 Hz; H-15), 4.24 (1H, dd, J = 4.8 and 3.2 Hz; βH-3), 4.27 (1H, d, J = 5.7 Hz; H-11), 5.36 (1H, br, d, J = 5.8 Hz; βH-8), 5.4 (1H, d, J = 3.2 Hz; αH-4), 5.82 (1H, br, d, J = 5.7 Hz; H-10)		
Emodin	$C_{15}H_{10}O_5$	270.23	256—257	(EtOH) 222, 252, 265, 289, 437	3400, 1635, 1625	2.26 (s), 6.49 (d, J = 2.0 Hz), 7.05 (d, J = 1.5 Hz), 7.00 (d, J = 2.0 Hz), 7.40 (d, J = 1.5 Hz). Solvent: dimethyl-d_6-sulfoxide	No data	No data
Ergobasine (see ergometrine)								
Ergoclinine (see ergometrine)								
Ergocornine	$C_{31}H_{39}O_5N_5$	561.66	182—184	—	—	—	Water, excitation (320), fluorescence (435); Ethanol, excitation (330), fluorescence (420)	$[\alpha]^{20}_D$ −188° ($CHCl_3$)
Ergocristine	$C_{35}H_{39}O_5N_5$	609.74	155—157	—	—	—	Water, excitation (320), fluorescence (435); Ethanol, excitation (325), fluorescence (415)	$[\alpha]_D^{20}$−183° ($CHCl_3$)
Ergocryptine	$C_{32}H_{41}O_5N_5$	375.89	212 (dec)	312 (8511)	—	—	Water, excitation (320), fluorescence (435); Ethanol, excitation (325), fluorescence (420)	$[\alpha]_D^{20}$ −187° ($CHCl_3$)

Table 2 (continued)
CHEMICOPHYSICAL CHARACTERISTICS OF FUNGAL TOXINS

Name	Formula	Mol wt	Melting point (°C)	UV spectrum (Mμ) extinction coefficient (ε)	IR spectrum (cm^{-1})	NMR spectrum	Fluorescence emission	Specific rotation
Ergometrine	$C_{19}H_{23}O_2N_3$	325	162—163 (dec)	—	—	—	Water, excitation (320), fluorescence (435); Ethanol, excitation (325), fluorescence (415)	$[\alpha]_D^{20}$ +90° (H_2O)
Ergonovine (see ergometrine)								
Ergostetrine (see ergometrine)								
Ergotalkaloids (see agroclavine, ergocornine, ergocristine, ergocryptine, ergometrine, ergotamine)								
Fumigaclavine A	$C_{18}H_{22}N_2O_2$		84—85	No data	No data	1.35 (d), 1.90 (s), 2.47 (s), 2.68 (m), 3.4 (m), 5.7 (d), 6.76, 6.86, 7.12, 7.16, 7.94; see Ref. 342 for ^{13}C NMR	No data	No data
Fumigaclavine B	$C_{16}H_{20}N_2O$		244—245	No data	No data	1.38, 2.43, 2.62 (m), 2.8, 3.14, 4.54 (d), 6.87, 6.96, 7.18, 7.21, 8.04; see Ref. 342 for ^{13}C NMR	No data	—
Fumigaclavine C	$C_{23}H_{30}N_2O_2$	366.2	194	225, 277, 283, 292 (log E: 4.54, 4.01, 4.04, 3.98)	3395, 1715, 1380, 875, 755	1.35 (d, J = 7 Hz), 1.56 (s), 1.90 (s), 2.46 (s), 2.68 (m), 3.32 (m), 3.54 (m), 5.05, 5.00, 6.10, 5.67, 6.72, 7.04, 7.08, 7.70; see Ref. 342 for ^{13}C NMR	No fluorescence	No data
Ergotamine	$C_{33}H_{35}O_5N_5$	581.65	212—214 (dec)	318 (7244)	—	—	Water, excitation (320), fluorescence (425); Ethanol, excitation (325), fluorescence (415)	$[\alpha]_D^{20}$ −160° ($CHCl_3$)
	Ergotocine (see ergometrine)							

Ergotoxine (a 1:1:1 mixture of ergocornine, ergocristine, and ergocryptine)								
Ergotrate (see ergometrine)								
Expansine (see patulin)								
F-2 toxin (see zearalenone)								
Fescue toxin (see butenolide)								
Flavutoxin	$C_{10}H_{13}O_4N$	211	350 (dec)	(MeOH) 225 (20,000), 255 (8800), 282 (11,000), 292 (9500)	1670, 1555, 1410, 1360, 1112, 749	No data	No fluorescence	No data
Fumagellin	$C_{26}H_{34}O_7$	438.33	194—195	(EtOH) 239, 304 (sh), 322 (sh), 336, 351	1710, 1632, 1377	No data	No data	$[\alpha]_D^{25}$ −26.6
Fumitremorgen A	$C_{33}H_{45}O_6N_3$	579	202.5—203.5	(EtOH) 225.5 (66,900), 275.5 (12,000), 295 (8500)	3420, 2940, 1670, 1565, 1440, 1370, 1300, 1160, 1070, 1035	7.68 (1H, doublet, J = 2.0 Hz), 6.82 (1H, doublet of doublet, J = 9 and 2 Hz), 6.67 (1H, doublet, J = 9 Hz), 3.84 (3H, singlet)	No data	Not reported
Fumitremorgen B	$C_{27}H_{33}N_3O_5$	479	211—212	(MeOH) 225.5 (66,900), 275.5 (12,000), 295 (8500)	3420, 2940, 1670, 1565, 1440, 1370, 1300, 1160, 1070, 1035	Not reported		$[\alpha]_D^{25}$ +24
Fusarenon-x (Fusarenone)	$C_{15}H_{20}O_7$	312.31	91—92	End absorption only	3400, 1710, 1690, 1380, 1250, 1170, 1080, 1030, 955	See nivalenol except that the C-4 hydrogen resonates at 1.2 ppm downfield compared to nivalenol	No fluorescence	$[\alpha]^{20}_D$ +58.0°
Gliotoxin	$C_{13}H_{14}O_4N_2S_2$	326.39	221 (dec)	270 (4500), 245 (3500)	No data	No data	No data	$[\alpha]^{22}_D$ −245
Gyromitrin	$C_4H_8ON_2$	100.12	No data					
Helvolic Acid	$C_{33}H_{44}O_8$	568.68	215	332 (98), 231 (17,300)	3—4 μm, 8.05 μm, 7.92 μm, 8.20 μm, 9.64 μm	See: Allinger, N. L., *J. Org. Chem.*, 21, 1180, 1956	No data	$[\alpha]_D^{22}$ −245
Hiptagenic acid (see β-nitropropionic acid)								
HT-2 toxin	$C_{21}H_{32}O_8$	412.47	No data	End absorption	Similar to T-2 toxin	3.60 (d), 4.23 (dd), 4.38 (d), 2.0 (m), 2.3 (dd), 5.30 (dd), 5.75 (dd), 4.14 (d), 3.04 (d), 2.75 (d), 0.80 (s), 4.15 (q), 1.75 (d), 2.05 (s), 0.96 (d), 2.0 (m)	No fluorescence	No data

Table 2 (continued)
CHEMICOPHYSICAL CHARACTERISTICS OF FUNGAL TOXINS

Name	Formula	Mol wt	Melting point (°C)	UV spectrum (Mμ) extinction coefficient (ε)	IR spectrum (cm^{-1})	NMR spectrum	Fluorescence emission	Specific rotation
4-Hydroxyochratoxin A	$C_{20}H_{18}ClNO_7$	419.64	216—218	(EtOH) 213 (32,500), 334 (6400)	1723, 2500—3000 (broad), 1655, 1535, 3380, 1678	See Hutchison et al., *Tetrahedron Lett.*, p. 4033, 1971	Green fluorescence	No data
Ibotenic Acid	$C_5H_8O_5N_2$	176.13	145 (dec)	421 (6150), pH 6—7 208, 240, 6260, 2420	No data	D_2O 5.03 (1 H), 2.23 (1 H)	No data	No data
Islanditoxin	$C_{25}H_{36}O_8N_5Cl_2$	605.49	No data	No data	No data	No data	No data	No data
Kojic Acid	$C_6H_4O_4$	142.11	153—154	270 (7820) in methanolic HCl, 268 (8080) in water, 315 (6560) in 0.1 *N*NaOH	3020, 1660, 1580, 1350, 1280, 1220, 1120, 1070, 935, 875	No data	None	None
Luteoskyrin	$C_{30}H_{22}O_{12}$	574.48	278	No data	3378, 1623	(d_6 DMSO) 2.96 (d, J = 5.5 Hz), 3.36 (br, s), 4.53 (br, d, J = 5.5 Hz), 2.28 (s), 7.28 (s), 11.28 (s), 12.38 (s), 14.53 (s), 5.48	No data	$[\alpha]_D^{25}$ −880 (acetone)
Lysergic acid	$C_{16}H_{16}O_2N_2$	268.32	240 (dec)	310 (10,000)	No data	No data	No data	$[\alpha]_D^{20}$ +40° (pyridine)
Malformin C	$C_{23}H_{39}N_5O_5S_2$	529.24	300 (dec)	No data	3280, 2930, 1630, 1500	(Me_2SO-d_6) δ8.72 (d, 1 H, J = 5 Hz, exchangeable with D_2O), 8.42 (d, 1 H, J = 6 Hz, exchangeable with D_2O), 7.73 (d, 1 H, J = 8 Hz, exchangeable with D_2O), 7.28 (d, 1 H, J = 9 Hz, exchangeable with D_2O), 7.06 (d, 1 H, J = 11 Hz, exchangeable with D_2O), 4.80—3.69 (m, 5 H), 3.46 (d, 1 H, J = 15 Hz), 3.17 (m, 3 H), 1.98 (m, 1 H), 1.78—1.00 (m, 6 H), 0.86 (m, 18 H)	No data	$[\alpha]_D^{25}$ −37.4

Maltorhyzine	$C_{11}H_{14}O_4$	210.22	69 (dec)	220 (log E, 4.1), 280 (s) (log E, 3.1), 320 (log E, 2.1)	3300, 1700, 1600, 1500	No data	No data	No rotation
Mappine (see bufotenine)								
Mescaline	$C_{11}H_{17}O_3N$	211.25	35—36	225, 270 (15, 484), (1584)	No data	No data	No data	No data
8-Methoxypsoralen	$C_{12}H_8O_4$	206.18	145—148	244 (sh), 250 (30,300), 263 (sh), 300 (broad)	See Scheel et al., *Biochemistry,* 2, 1127, 1963	No data	No data	No data
3-Methyl-2-butenyl-paspalinine	$C_{32}H_{39}O_4N$	501.29	No data	(MeOH) 234, 278	1665—1675, 3310—3500, 1355—1380	1.18 (s), 1.35 (s), 1.42 (s), 1.77 (s), 1.8—2.9 (m), 3.62 (m), 4.29 (s), 5.38 (m), 5.73 (s), 6.8—8.0 (m)	Fluoresces on spraying, heating with 50% EtOH:H_2SO_4	No data
3-Methyl-3-hydroxy-1-butenyl-paspalanine	$C_{32}H_{37}O_5N$	517.28	No data	(MeOH) 227, 248, 305, 336	1665—1675, 3310—3500, 1355—1380	1.21 (s), 1.25 (s), 1.41 (s), 1.47 (s), 1.50 (s), 1.90—2.77 (m), 4.32 (s), 5.81 (s), 6.42 (d, J = 16 Hz), 7.00—7.30 (m), 7.76 (s)	Fluoresces on spraying, heating with 50% EtOH:H_2SO_4	No data
O-Methylsterigmatocystin	$C_{19}H_{14}O_6$	328.30	265 (dec)	236 (log 4.614), 310 (log 4.224)	See Ref. 77	3.92 (s), 3.96 (s), 6.83 (d, J = 7), 6.50 (t, J = 2.5), 5.46 (t, J = 2.5), 4.81 (triplets of doublet, J = 2.5, 7), 6.38 (s), 7.48 (t, J = 8.1)	Yellow fluorescence	No data
Moniliformin	$C_4H_2O_3$	98	158	260, 229	3570, 3370, 3300, 3130, 1775, 1705, 1675, 1600, 1100, 840	No data	No data	No data
Monoacetoxyscirpenol	$C_{17}H_{24}O_6$	324	172—173	End absorption	(KBr) 3400, 1715, 1250	($CHCl_3$) 2.7 (d, J = 4 Hz), 3.04 (d, J = 4 Hz), 2.16 (s), 1.73 (br. s), 0.83 (s)	No data	No data
Muconomycin A (see verrucarin A)				No data				
Muscarine	$C_9H_{20}O_2N^+$	174.26	No data	No data		d, 1-Muscarine HCl (KBr) see Matsumoto, T. and Maekawa, H., *Angew. Chem.*, 70, 507, 1958	No data	$[\alpha]_D$ +6.7 (water) muscarine hydrochloride
Muscazone	$C_5H_6O_4N_2$	158.11	190 (dec)	212 (8700) pH 2—7, 220 (7500) pH 12	No data	No data	No data	$[\alpha]_D^{17}$ −13.01°

Table 2 (continued)
CHEMICOPHYSICAL CHARACTERISTICS OF FUNGAL TOXINS

Name	Formula	Mol wt	Melting point (°C)	UV spectrum (Mμ) extinction coefficient (ε)	IR spectrum (cm^{-1})	NMR spectrum	Fluorescence emission	Specific rotation
Muscimol	$C_4H_6O_2N_2$	114.10	172—174 (dec)	pH 6—7 200 (6360), 230 (2810). pH 2 203 (6650)	No data	D_2O (acid) 5.85 ppm (1 H), 4.17 ppm (2 H)	No data	No optical activity
Mycoin (see patulin)								
Mycophenolic acid	$C_{17}H_{20}O_6$	320.35	141	No data	No data	No data	No fluorescence	No rotation
Neosolaniol monoacetate	$C_{21}H_{28}O_9$	424.17	190	End absorption	3370, 3010, 1730, 1230, 1370, 1460, 1070, 3040, 1265, 920, 799	See Ref. 340	No data	No data
Nidulotoxin	Not known	No data	No data					
β-Nitropropionic acid	$C_3H_5O_4N_2$	124	68	No data	No data	No data	No data	
Nivalenol	$C_{15}H_{20}O_7$	312	222—223	(MeOH) 218 (6300)	3500—3200, 1680, 1610	(DMSO + D_2O) 0.95, 1.17 (d, J = 1), 2.87 and 2.97 (d, J = 5), 6.58 (dd, J = 5, 1)	No fluorescence	No data
Norsolorinic acid	$C_{20}H_{18}O_7$	370	256—257	(EtOH) 234 (23,667), 265 (16,650), 283 (173,527,297 sh (19,872), 313 (23,763), 465 (7336)	No data	2.46, 2.65 (d, J = 2.2 Hz), 3.21 (d, J = 2.2 Hz), 2.45, 2.66 (d, J = 2.5 Hz), 3.21 (d, J = 2.5 Hz)	No data	No data
Nortryptoquivaline	$C_{25}H_{28}N_4O_7$	532	256—258	(EtOH) 228 (43,600), 233 (sh 42,000), 254 (sh 18,700), 267 (sh 11,900), 279 (10,200), 306 (4500), 319 (3500)	($CHCl_3$) 3490, 2980, 2940, 2880, 1790, 1728, 1670, 1610, 1455, 1471, 1410	($CDCl_3$) 1.02 (d, 3, J = 7), 1.16 (d, 3, J = 7), 1.58 (d, 3, J = 7), 2.16 (s, 3), 2.57 (m, 1), 2.99, (d of d, 1, J = 10, 13), 3.18 (d of d, J = 10,13), 4.28 (q, 1, J = 7), 5.10 (s, 1), 5.54 (d, 1, J = 9), 5.65 (t, 1, J = 10), 7.01—7.79 (m, 7), 8.12 (m, 1)	No data	$[\alpha]_D^{25}$ + 170
Ochratoxin A	$C_{20}H_{18}O_6NCl$	403.83	169	213 (36,800), 332 (6400)	1723, 2500—3000 (broad), 1655, 1535, 3380, 1678, 1132	See Ref. 383	Green fluorescence	$[\alpha]_D$ −118

Ochratoxin A methyl ester	$C_{21}H_{20}O_6NCl$	417	No data	213 (32,700), 331 (4100), 378 (2050)	1743, 1677, 1655	8.46 (d, J = 7.0 Hz), ABX system at 5.54, complex pattern 6.8, 1.61 (s), 2.74 (s), complex pattern 6.8, complex pattern 4.98, 1.63 (d), 6.21 (s), 6.28 (s)	No data	$[\alpha]_D$ −160
Ochratoxin B	$C_{20}H_{19}O_5N$	369.45	221	(EtOH) 218 (36,200), 318 (6700)	1730, 1680, 1535	See Merwe et al., *J. Chem. Soc.*, p. 7083, 1965	No data	$[\alpha]_D$ −35
Ochratoxin B ethyl ester	$C_{22}H_{23}O_6N$	397	102—103 (from ether)	218 (32,000), 318 (5200), 364 (1250)	3385, 1655, 1530, 1677, 1140, 1743	No data	Light blue	$[\alpha]_D$ −49
Ochratoxin B methyl ester	$C_{21}H_{21}O_6N$	383.48	134—135 (from benzene)	Similar to Ochratoxin B ethyl ester	Similar to ochratoxin B ethyl ester	No data	No data	$[\alpha]_D$ −62
Ochratoxin C	$C_{22}H_{22}O_6NCl$	431	No data	214 (30,000), 333 (7000)	1730, 1680	See Merwe et al., *J. Chem. Soc.*, p. 7083, 1965	No data	$[\alpha]_D$ −100
Oxalic acid	$C_2H_2O_4$	126.07	101—102					
Paspalinine	$C_{27}H_{30}O_4N$	433	No data	(MeOH) 232, 275	1665, 3310—3500, 1355—1380	1.21 (s), 1.25 (s), 1.41 (s), 1.47 (s), 1.50 (s), 1.90—2.77 (m), 4.32 (s), 5.81 (s), 6.42 (d, J = 16.0 Hz), 7.00—7.30 (m), 7.76 (s)	Fluor. on spraying with 50% EtOH-H_2SO_4 and heating	No data
Patulin	$C_7H_6O_4$	154.12	111.0	276 (14,600)	1765, 1740, 1200, 1160, 1030	($CDCl_3$) 5.97 (complex), 4.73 (dd, A part of ABX system, J = 17 Hz), 3.46 (d, J = 5 Hz)	No fluorescence	No rotation
Paxilline	$C_{27}H_{33}NO_4$	435.24	252	(MeOH) 230 (41,500), 281 (8000)	3420, 1650, 1355 (doublet), 1365 (doublet), 740	Acetone d_6, 1.04 (s), 1.22 (s), 1.40 (s), 1.7—2.95 (bm), 3.70 (d, J = 2 Hz), 4.05 (s), 4.08 (s), 4.94 (m), 5.85 (d, J = 2 Hz), 6.95 (m), 7.30 (m), 9.82 (s)	None	No data
Penicidin (see patulin)								
Penicillic acid	$C_8H_{10}O_4$	170.16	83—84	(MeOH) 221 nm (H_2O) 227 nm	3270, 1728, 1643, 1352, 1223, 909, 811	(Deut. benzene) 3 proton signal at 1.72, 3.14; 1 proton signal at 4.19, 4.91, 5.02, 5.62	No fluorescence, amoniated derivative: excitation, 350 nm; emission, 440 nm	No data
Phallicidine	$C_{36}H_{50}O_{12}N_8S$	786	No data	292 (14,125)	No data	No data	No data	No data
Phallin B	$C_{41}H_{52}O_9N_8S$	832.95	No data	292 (14,125)	No data	No data	No data	No data

Table 2 (continued)
CHEMICOPHYSICAL CHARACTERISTICS OF FUNGAL TOXINS

Name	Formula	Mol wt	Melting point (°C)	UV spectrum (Mμ) extinction coefficient (ε)	IR spectrum (cm^{-1})	NMR spectrum	Fluorescence emission	Specific rotation
Phalloidin	$C_{35}H_{48}O_{11}N_8S$	780	280—282	292 (14,125)	No data	Patel, D. J., Tonelli, A. E., Pafaender, P., Faulatich, H., and Wieland, T., *J. Mol. Biol.*, 79, 185—196, 1973	No data	No data
Phalloin	$C_{35}H_{48}O_{10}N_8S$	772	250—280 (dec)	292 (14,125)	No data	No data	No data	No data
Phomenone	$C_{15}H_{20}O_4$	264	148—149	240 (17,000)	KBr (1675, 1110, 1640, 880, 3300—3380)	(CD_3OD) 2.67 (m), 3.58 (d), 3.40 (s), 5.70 (d), 5.25 (m), 4.20 (s), 1.25 (s), 1.20 (d); see Ref. 332		$[\alpha]_D$ +225
PR toxin	$C_{17}H_{20}O_6$	320.16	155—157	(EtOH) 247 (8800)	2945, 1735, 1720, 1680, 1620, 1460, 1435, 1380, 1245, 1035	Three proton doublet 1.03 (J = 7 Hz), three proton doublet 1.45 (J = 0.8 Hz), three proton singlet 1.49, three proton singlet 2.16, one proton doublet 3.65 (J = 3.5 Hz), one proton doublet of doublets 3.96 (J = 3.5, 5 Hz), one proton doublet of doublets 5.16 (J = 5, 5 Hz), one proton singlet 6.43, one proton singlet 9.75, three protons between 1.6 to 2.3	Excitation 300 nm, Emission 360 nm	
Psilocybin	$C_{12}H_{17}O_4N_2P$	254.27	220—228 (crystals from water)					
Roridin E	$C_{29}H_{38}O_8$	514.3	220—221	(EtOH) 195 (log E, 4.2), 223 (4.40), 263 (4.30)	(KBr) 3500, 2970, 1715, 1645, 1598, 1218, 1176, 1080, 1030, 1000, 965, 810	6.69 (s), 4.0 (m), 2.46 (dd, J = 15.5, 11 cps), 3.39 (t, J = 11), 4.20 (d, J = 11), 8.38 (s), 8.81 (d, J = 6)	No fluorescence	$[\alpha]_D^{25°}$ −24°

Roseotoxin B	$C_{30}H_{49}O_7N_5$	591	199—200	No data	1730, 1680—1635, 1525, 3380, 1180	See Ref. 104	—	$[\alpha]_D^{25}$ −23.9
Rubratoxin A	$C_{26}H_{32}O_{11}$	520	No data	(EtOH) 204 (31,900), shoulder 252 (4430), inflection 225	1852, 1815, 1770, 1728, 1712, 1700	5.70, 6.75, 4.46, 7.91, 6.30, 5.26, 7.4, 2.91, 4.08, 4.3	No data	$[\alpha]_D^{18}$ +86.6°
Rubratoxin B	$C_{26}H_{30}O_{11}$	518	185—186 (dec)	250 (9700)	(Nujol mull) 3550, 1860, 1820, 1790, 1760, 1710, 1695	(Deuterioacetone) 5.60, 6.57, 4.38, 7.90, 6.23, 5.35, 7.55, 3.00, 4.12	No data	$[\alpha]_D^{20}$ +67
Rugulosin	$C_{30}H_{22}O_{10}$		290 (dec)	No data	3450, 1690, 1620	d_6-DMSO 2.78 (d, J = 5.5 Hz), 3.38 (br. s), 4.38 (br. d, J = 5.5 Hz), 2.42 (s), 7.16 (d, J = 1 Hz), 7.43 (d, J = 1 Hz), 11.37 (s), 14.54 (s), 5.38	No data	No data
Satratoxin D (see Roridine)								
Satratoxin H	$C_{29}H_{36}O_9$	—	No data	No data	No data	See Ref. 339	No data	No data
Secalonic acid A	$C_{32}H_{30}O_{14}$	638	243 (acetone), 248 ($CHCl_3$), 208—209 (acetic acid)	(EtOH) 252 (17,200), 346 (3100)	(KBr) 3490, 2950, 1726, 1608, 1587, 1560, 1435, 1325, 1235, 1160, 1130, 1063, 1005, 987, 905, 825	TMS (in d-pyridine), 1.26 (d, J = 6.0), 2.2—3.0 (m), 3.56 (s), 4.16 (d, J = 10.5), 6.70 (d, J = 8.5), 7.54 (d, J = 8.5)	No data	—
Secalonic acid D	$C_{32}H_{30}O_{14}$	638	255—259	(EtOH) 236 (17,800), 265 (15,100), 338 (37,800)	(KBr) 3505, 1735, 1610, 1585, 1432, 1232, 1061	(DMSO-d_6) 1.05 (d, J = 4 Hz), 2.0—3.0 (m), 3.63 (s), 3.80 (dd, J = 6 and 10 Hz), 6.00 (d, J = 6 Hz), 6.64 (d, J = 8 Hz), 7.47 (d, J = 8 Hz), 11.70 (s), 13.72 (bs)	No data	—
Secalonic acid E	$C_{32}H_{30}O_{14}$	638	206—208	(Dioxane) 265, 338 (log E, 4.37, 4:64)	(KBr) 3600, 3200, 1740, 1600	[$(CD_3)_2SO$] 1.64 (d, J = 6 Hz), 3.65 (s), 3.96 (d, J = 4 Hz), 5.78 (d, J = 4 Hz), 6.56 (d, J = 8 Hz), 7.44 (d, J = 8 Hz), 11.60 (s), 13.85 (s), Spectrum for half molecule only	No data	—

Table 2 (continued)
CHEMICOPHYSICAL CHARACTERISTICS OF FUNGAL TOXINS

Name	Formula	Mol wt	Melting point (°C)	UV spectrum (Mμ) extinction coefficient (ε)	IR spectrum (cm^{-1})	NMR spectrum	Fluorescence emission	Specific rotation
Secalonic acid F	$C_{32}H_{30}O_{14}$	630	218—221	(EtOH) 236 (19,250), 263 (17,300), 388 (37,000)	(KBr) 3520, 1748, 1610, 1590, 1442, 1238, 1068, 1045	($CDCl_3$) 1.14 (d, J = 7 Hz), 2.0—3.0 (m), 2.67 (b), 286 (b), 3.67 (s), 3.87 (d, J = 10 Hz), 4.09 (b), 6.52 (d, J = 9 Hz), 6.58 (d, J = 9 Hz), 7.35 (d, J = 9 Hz), 7.39 (d, J = 9 Hz), 11.65 (s), 11.80 (s), 13.70 (s), 13.88 (s)	No data	—
Slaframine	$C_{10}H_{18}O_2N_2$	198.26	183—184	No data	*N*-acetylslaframine: 3420, 1665, 1510	Slaframine hydrochloride: (D_2O): 7.85 (s), 4.45 (m)	No data	(*N*-acetylslaframine) $[\alpha]_D^{25}$ −15.9
Slobber factor (see slaframine)								
Sporidesmin A	$C_{18}H_{20}O_6N_3Cl$ S_2	473.95	179	218.5, 254, 302. Sporidesmin A benzene solvate 39,810, 13,182, 2818	Sporidesmin benzene solvate (in paraffin) 3558, 3355, 1715, 1664	Sporidesmin A. ($CDCl_3$) τ 2.92, 4.70, 5.16, 5.42, 6.13, 6.18, 6.6, 6.70, 6.93, 6.9, 7.97, 8.0	No data	$[\alpha]_D^{20}$ −45° (CH_3OH) Sporidesmin A
Sporidesmin B	$C_{18}H_{20}O_5N_3Cl$ S_2	457.95	183	218, 256, 307. 31,622, 12,022, 2570	(In paraffin) 1697, 1666	($CDCl_3$) τ 2.93, 4.6, 6.14, 6.19, 6.47, 6.76, 6.97, 7.97, and 6.8, 7.2 with JAB = 16.1 c/sec	No data	$[\alpha]_D^{21}$ −27° (MeOH); $[\alpha]_D^{21}$ +12° ($CHCl_3$)
Sporidesmin C	$C_{18}H_{20}O_6N_3Cl$ S_3	506.05	No data	222, 255, 309. Diacetate of Sporidesmin C. 22,300, 9900, 2900	—	—	No data	$[\alpha]_D$ −215° ($CHCl_3$)
Sporidesmin D	$C_{19}H_{23}O_6N_3Cl$ S_2	478.99	No data	216, 252, 300. Sporidesmin D ethanolate. 28,183, 10,000, 1905	Sporidesmin D etherate (KBr) 3450, 33330, 1680, 1665, 1605	Sporidesmin D ethanolate ($CDCl_3$) τ 2.90, 4.70, 5.35, 6.95 (J = 3 Hz), 6.05, 6.13, 6.19, 6.29 [q, J = 7 Hz (EtOH)], 6.63, 6.93, 7.60, 7.68, 8.13, 8.43, 8.79 [t, J =	No data	$[\alpha]_D^{23}$ +58° ($CHCl_3$) Sporidesmin D ethan-

						7 Hz, (EtOH)]. Sporidesmin D etherate ($CDCl_3$) τ 2.90, 4.70, 5.33, 6.12, 6.19, 6.50 [q, J = 7 Hz (EtO_2)], 6.62, 6.92, 7.60, 7.68, 8.13, 8.79 [t, J = 7 Hz, (EtO_2)]		olate
Sporidesmin E	$C_{18}H_{20}O_6N_3ClS_3$	506.05	180—185	217, 252, 295. Sporidesmin E etherate. 33,113, 16,595, 3162	Sporidesmin E etherate (KBr) 3325, 1690, 1655	Sporidesmin E etherate ($CDCl_3$) τ 2.92, 2.93, 4.59, 4.71, 5.38, 5.48, 6.10, 6.13, 6.17, 6.14—6.50 (q, J_{AB} = 7 Hz), 6.50, 6.67, 6.86, 6.98, 8.00, 8.05, 8.67—8.92 (t, J_{AB} = 7 Hz)	No data	$[\alpha]_D^{20}$ −131° ($CHCl_3$) Sporidesmin E ethanolate
Sporidesmin F	$C_{19}H_{22}ClN_3O_6S$	455.5	65—75 (Amorphous solid)	216, 250, 298. 28,840, 13,803, 1584	(KBr) 3430, 1690, 1615, 1605	($CDCl_3$) 2.91, 4.00, 4.04, 4.90 and 4.94 (J = 2 Hz), 4.64, 5.37, 6.12, 6.18, 6.60, 6.70, 7.69	No data	No data
Sporidesmin G	$C_{18}H_{20}ClN_3O_6S_4$	537.5	148—153 (Sporidesmin G etherate)	216, 250, 298. Sporidesmin G etherate. 44,668, 11,481, 2951	Sporidesmin G etherate ($CHCl_3$) 3570, 3520, 1690, 1660. Sporidesmin G etherate (KBr) 1210, 825, 725, 585, 510, 505, 480, 410, 395	Sporidesmin G from Sporidesmin and Sporidesmin E. ($CDCl_3$) 2.82, 4.90, 5.35, 6.10, 6.16, 6.60, 6.93, 8.03	No data	$[\alpha]_D^{20}$ −217° ($CHCl_3$) Sporidesmin G etherate
Stachybotrys toxin (see Roridin E)	Not known	No data	No data					
Stemphone	$C_{30}H_{42}O_8$	530	160.5—161.5	(MeOH) 267 (9890), 390 (1030)	(CCl_4) 3578, 3510, 1735, 1673, 1645, 1620, 1596	Not reported		$[\alpha]_D^{29}$ +146°

Table 2 (continued)
CHEMICOPHYSICAL CHARACTERISTICS OF FUNGAL TOXINS

Name	Formula	Mol wt	Melting point (°C)	UV spectrum (Mμ) extinction coefficient (ε)	IR spectrum (cm^{-1})	NMR spectrum	Fluorescence emission	Specific rotation
Sterigmatocystin	$C_{19}H_{14}O_6$	338	265 (dec)	205 (log E, 4.40), 233 (4.49), 246 (4.53), 325 (4.21)	3450, 3099, 2995, 2975, 2920, 1650, 1627, 1610, 1590, 1496, 1482, 1447, 1415, 1400, 1362, 1347, 1322, 1300, 1272, 1239, 1197, 1179, 1122, 1098, 1067, 1044, 1019, 993, 978, 952, 932, 904, 895, 846, 823, 774, 756, 735, 722, 702, 668	No data	No data	$[\alpha]_D^{20}$ −387 $CHCl_3$
T-2 toxin	$C_{24}H_{34}O_9$	466	151—152	End absorption only	3400, 2940, 1720, 1635, 1365, 1240	3.48 (d), 4.10 (dd, J = 2.5), 5.47 (d), 2.0 (m), 2.28 (dd, J = 5.0), 5.20 (dd), 5.72 (dd, J = 5.5, 1.5), 4.30 (d), 2.97 (d), 2.70 (d), 0.71 (s), 4.1 (q), 1.74 (d), 1.99 (s), 2.09 (s), 0.97 (d), 2.0 (m)	No fluorescence	$[\alpha]_D^{26°}$ +15°
Terreic acid	$C_7H_6O_4$	154	127—127.5	213 (10,150), 316 (6230)	(Tetrachloroethane) 2.88 u, 5.86, 5.94, 6.00, 6.88, 7.14, 7.28, 7.56, 8.82, 11.44, 15.58	No data	No data	$[\alpha]_D^{22}$ −16.6 $CHCl_3$

TR-1 (see verruculogen)								
TR-2	$C_{22}H_{27}N_3O_6$	429.19	No data	(EtOH) 224 (37,400), 268 (6830), 294 (7540)	3450, 1660, 1380	($CHCl_3$) 7.71 (doublet, J = 9.0 Hz), 6.87 (doublet, J = 3.0 Hz), 6.60 (doublet of doublets, J = 3.0 and 9.0), 5.37 (multiplet)	—	No data
Tremortin A	$C_{37}H_{44}O_6NCl$	633	210—230 (dec)					
Tremortin B	$C_{37}H_{45}O_5N$	583	185—195 (dec)					
Trichodermin	$C_{17}H_{24}O_4$	292.36	46	(EtOH) 200 (2400)	(KBr) 1730, 1682, 1245, 1225, 1085	3.82 (m), 2.6 (m), 5.58 (dd), 2.0 (m), 2.5 (m), 5.42 (dd), 3.62 (d), 3.13 (d), 2.83 (d), 0.93 (s), 0.72 (s), 1.72 (d), 2.07 (s)	No data	$[\alpha]_D^{20}$ −30°
Tricholomic acid	$C_5H_6O_4$	132.11	207					
Trichothecin	$C_{19}H_{24}O_5$	332.38	118	217 (18,000)	2900, 1698, 1677, 1645, 1645—1700, 1460, 1367, 1290, 1180, 1080, 970, 847, 815	No data	No data	No data
4,5,8-Trimethylpsoralen	$C_{14}H_{12}O_3$	228.24	226—228	250 (21,800)	See Scheel et al., *Biochemistry*, 2, 1127, 1963	No data	No data	No data
Tryptoquivaline	$C_{29}H_{30}N_4O_7$	546.21	153—155	(EtOH) 228 (37,000), 275 (8550), 305 (3800), 317 (3040)	($CHCl_3$) 3520, 1790, 1735, 1680, 1615	($CHCl_3$) δ 1.03 (d, 3, J = 7 Hz), 1.17 (d, 3, J = 7 Hz), 1.50 (s, 3), 1.52 (s, 3), 2.19 (s, 3), 2.63 (m, 1), 3.10 (d, 1, J = 10 Hz), 3.15 (d, 1, J = 10 Hz), 3.63 (b, 1), 4.04 (b, 1), 5.00 (s, 1), 5.61 (d, 1, J = 9 Hz), 5.70 (t, 1, J = 10 Hz), 7.12—7.90 (m, 7), 8.22 (d, 1, J = 8 Hz)	—	$[\alpha]_D^{25}$ 142°

Table 2 (continued)
CHEMICOPHYSICAL CHARACTERISTICS OF FUNGAL TOXINS

Name	Formula	Mol wt	Melting point (°C)	UV spectrum (Mμ) extinction coefficient (ε)	IR spectrum (cm^{-1})	NMR spectrum	Fluorescence emission	Specific rotation
Tryptoquivalone	$C_{26}H_{24}N_4O_6$	488.17	202—204	(EtOH) 234 (34,950), 292 (9550), 320 (6300)	($CHCl_3$) 3525, 1790, 1735, 1715, 1680	($CDCl_3$) δ 1.24 (d, 3, J = 7 Hz), 1.31 (d, 3, J = 7 Hz), 1.59 (d, 3, J = 7 Hz), 3.09 (d of d, 1, J = 10, 14 Hz), 3.47 (d of d, 1, J = 11, 14 Hz), 4.12 (quintet, 1, J = 7 Hz), 4.36 (quartet, 1, J = 7 Hz), 5.24 (s, 1), 5.51 (t, 1, J = 10 Hz), 7.12—7.94 (m, 7), 8.28 (d, 1, J = 8 Hz)	—	$[\alpha]_D^{25}$ 254
Verrucarin A	$C_{27}H_{34}O_9$	501.53	212—215	(EtOH) 260 (log E, 4.25)	2.84 μm, 5.83, 6.12, 6.29	5.83 (dd, J = 7.5, 5.5), 5.46 (d, J = 5), 2.97 (q), 0.87 (s), 4.3, 1.79 (s), 4.2, 0.89 (d, J = 7), 6.06 (d, J = 16), 8.08 (dd, J = 11), 6.70 (t, J = 11), 6.17 (d)	No data	$[\alpha]^D$ 120 $CHCl_3$
Verrucarin B	$C_{27}H_{32}O_9$	500	176—179 (hexahydro derivative)	258.5 (log E, 4.37) (EtOH)	5.76 μm, 5.88, 5.97 (?), 6.16, 6.30	5.90 (dd, J = 7.5, 5.5), 5.47 (d, J = 5), 3.0 (q), 0.88 (s), 4.3, 1.74 (s), 3.41 (s), 1.56 (s), 6.10 (d, J = 16), 7.98 (dd, J = 11), 6.69 (t, J = 11), 6.19 (d)	No data	$[\alpha]_D$ +94
Verruculogen (TR-1)	$C_{27}H_{33}N_3O_7$	511.23	233—235 (dec)	(EtOH) 224 (47,900), 268 (8760), 294 (9710)	3520, 3460, 1655, doublet 1355 and 1365	($CDCl_3$) 1.01 (s, 3H), 1.72 (s, 6H), 1.99 (s, 3H), 1.8—2.6 (bm, 6H), 3.61 (t, 2H), 3.82 (s, 3H), 4.13 (s, 1H), 4.48 (m, 1H), 4.79 (d, J = 3 Hz, 1H), 5.05 (d, J = 8 Hz, 1H), 6.05 (d, J = 10 Hz, 1H), 6.58 (d, J = 2 Hz, 1H), 6.63 (d, J = 8 Hz), 6.81 (d of d, J = 9 Hz, 2 Hz), 7.89 (d, J = 9 Hz, 1H)	—	$[\alpha]_D$ −27
Verruculotoxin	$C_{15}H_{20}N_2O$	244.16	152	Strong end absorption, weak bands 240—260	3320, 3030, 1688, 1615,			

					1640, 1200—1300, 750, 705			
Versicolorin A	$C_{18}H_{10}O_7$	338	287—288	222 (31,488), 254 (15,227), 265 (17,756), 290 (26,547), 321 (12,118), 453 (8166)	3340, 1679, 1625	(Trimethyl ether derivative) 2.57 (s), 2.70 (d, J = 2.5 Hz), 3.25 (d, J = 2.5 Hz), 3.18 (d, J = 7 Hz), 3.52 (t, J = 2.5 Hz), 4.66 (t, J = 2.5 Hz), 5.13 (t of d, J = 2.5, 7.0 Hz), 5.98, 6.06, 6.08 (s)	No data	No data
Versiconal acetate	$C_{18}H_{12}O_7$	340	216—220	(EtOH) 225 (23,800), 267 (14,000), 298 (23,000), 323 (11,300), 480 (7260)	3540, 3250, 1670, 1630, 1619, 1715, 1725	7.04 (J = 4.0 Hz), 6.48 (J = 4.0 Hz), overlap singlet at 7.04, 4.08 (t), 1.98 (4.5 proton signal)	No data	No data
Viomellein	$C_{30}H_{24}O_{11}$	560	260 (dec)	(MeOH) 225 (16,800), 264 (20,200), 395 (8200), 266 (82,000), 380 (22,500)	(KBr) 3395, 2980, 2965, 1730, 1680, 1640, 1590, 3400 (wk), 1740, 1635	1.56 (d), 4.63 (m), 3.02 (d), 6.66 (s), 6.96 (s), 3.90 (s), 9.80 (s), 13.88 (s), 3.84 (s), 13.44 (s), 7.49 (s), 9.72, 13.70, 3.66, 3.74, 6.24 (s), 6.78 (s), 4.96 (m), 2.76, 2.81 (d)	No data	No data
Viriditoxin	$C_{34}H_{30}O_{14}$	662	245				Excitation, 280, 385 nm; emission, 450, 550 nm	$[\alpha]_D^{28°}$ −202°
Vomitoxin (deoxynivalenol)	$C_{15}H_{20}O_6$	296	151—153	(EtOH) 218 (4500)	(KBr) 3470, 3430, 3350, 1680	($CDCl_3$) 1.15 (s), 1.92 (d, J = 1.5 Hz), 3.14 and 3.25 (d, J = 4 Hz), 3.94 (d, J = 4 Hz), 4.95 (d, J = 6 Hz), 6.79 (dd, J = 6 Hz, 1.5 Hz)	No data	$[\alpha]_D^{25°}$ +6.3°
Xanthomegnin	$C_{30}H_{22}O_{12}$	574	260	(MeOH) 222 (26,000), 264 (19,300), 380 (7,900)	(KBr) 3430, 1721, 1675, 1616, 1590, 840	1.51, 3.02, 4.67, 7.50, 4.11 ($CDCl_3$) 12.12 (s), 7.26 (s), 6.47 and 6.39 (AB-system, J = 2.5 Hz), 7.02 (d), 5.69 (m), 5.01 (m), 1.40 (d)	No data	No data
Zearalenone	$C_{18}H_{22}O_5$	318	164—165	(MeOH) 236 (29,925), 274 (13,373), 314 (6240)	No data		No data	$[\alpha]_D^{25°}$ −170°

Table 3
SOURCES AND BIOLOGICAL EFFECTS OF MYCOTOXINS

Producing fungi or other sources	Substrates (opt temp °C)	Animals affected	Toxin effects	LD_{50}	Ref.
4-Acetoxyscirpendiol *Fusarium roseum*	Czapek-peptone (25)	Skin test rats	Vesicant	No data	145
8-Acetylneosolaniol *F. roseum*	Czapek-peptone (25)	Skin test rats	Vesicant	No data	145
4β-Acetoxy-3α,15-diOH-12,13-epoxytrichotec-9-ene *F. Sulphureum*	Corn (25)	Duckling	Lethal	No data	377
Aflatoxin B_1					
Aspergillus flavus, parasiticus	Peanuts, corn, wheat, rice, nuts, coconut, cottonseed meal, various foods, milk (25, 30)	Most mammals, man, birds, fish	Hepatotoxic, carcinogenic, teratogenic, acute and chronic toxicity	Dog, 1.0 mg/kg; duckling, 0.3—0.4 mg/kg; guinea pig, 1.4 mg/kg; monkey (i.p.), 2.5—5.0 mg/kg; rabbit, weanling (i.p.), 0.5 mg/kg; rat, adult female, 180 mg/kg; rat, adult male, 72 mg/kg	1, 2, 3, 237, 250, 251, 252, 253, 254, 255, 256, 289, 380. Also see references in introduction
Aflatoxin B_1 aldehyde synthesis	—	Chicks	Liver lesions, degenerative changes in all organs of chick embryo	35 μg/chick p.o.	4
3-Hydroxyaflatoxin B_1 From aflatoxin B_1 by in vitro metabolism with Verret monkey liver	—	No data	No data	No data	335

Aflatoxin B_2					
Aspergillus flavus, parasiticus	Same as aflatoxin B_1 (25—30)	Same as aflatoxin B_1	Same as aflatoxin B_1	Duckling, 1.7 mg/kg	1, 2, 3, 5
1-acetoxy-aflatoxin B_2					
Synthesis		No data	No data	No data	6
1-ethoxy-aflatoxin B_2					
Synthesis					7
1-methoxy-aflatoxin B_2					
Synthesis		No data	No data	No data	8
2-methoxy-aflatoxin B_2					
Synthesis		No data	No data	No data	8
Aflatoxin B_{2a}					
Acid catalysis of aflatoxin B_1		No data	Almost nontoxic (in vitro); may be highly toxic when formed in vivo	No data	6, 9, 248, 249, 237 (p. 205)
Aflatoxin B_3 (parasiticol)					
Aspergillus flavus, parasiticus	Rice, wheat (no data)	No data	No data	Chick embryo, 5—10 μg/egg	10, 11
Aflatoxin D_1					
Ammoniation reaction product	—	No toxicity	—	—	12, 247
Aflatoxin deshydro D_1					
Produced in gastric intestinal mucus membrane from aflatoxin B_1	—	No data	No data	No data	334
Aflatoxin G_1					
Aspergillus flavus, parasiticus	Same as aflatoxin B_1 (15—20)	Same as aflatoxin B_1	Same as aflatoxin B_1	Duckling, 0.78 mg/kg	1, 2, 3
Aflatoxin G_2					
Aspergillus flavus, parasiticus	Same as aflatoxin B_1 (no data; probably 15—20)	Duckling; no other data available	Probably the same as aflatoxin G_1	Duckling, 3.45 mg/kg	1, 2, 3

Table 3 (continued)
SOURCES AND BIOLOGICAL EFFECTS OF MYCOTOXINS

Producing fungi or other sources	Substrates (opt temp °C)	Animals affected	Toxin effects	LD_{50}	Ref.
1-ethoxy-aflatoxin G_2					
Synthesis		No data	No data	No data	7
Aflatoxin G_{2a}					
Acid catalysis of aflatoxin G_1		No data	No data	No data	6, 9
Aflatoxin GM_1					
Aspergillus flavus	Aflatoxin G_1 is converted in the animal body to aflatoxin GM_1	No data	No data	No data	13
Aflatoxicol H_1					
Aflatoxin B_1 converted to H_2 by human and monkey liver	—	No toxicity to chick embryo and *Salmonella typhimurium*	—	—	14
Aflatoxin M_1					
Aspergillus flavus, parasiticus	Same as aflatoxin B_1; found in milk and urine of animals fed aflatoxin B_1 (probably the same as aflatoxin B_1)	Duckling; no other data available	Carcinogenic; probably the same as aflatoxin B_1	Duckling, 0.32 mg/kg	15, 16, 238, 237 (p. 200), 239, 240
Aflatoxin M_2					
Aspergillus flavus, parasiticus	No data (no data)	Duckling; no other data available	No data	Duckling, 1.23 mg/kg	17
Aflatoxin P_1					
Isolated from the urine of rhesus monkeys		No data	No data	No data	18, 235, 236, 237

Aflatoxin Q_1					
Liver conversion of aflatoxin B_1	—	Chick embryo	Lethality	207 ng/embryo (99—2442 ng)	19, 20, 21, 232, 233, 234
Aflatoxin R_0					
Liver conversion of aflatoxin B_1; transformation of aflatoxin B_1 by *Dactylium dendroides*		Duckling; no other data available	Bile duct hyperplasia	Duckling, 0.48—0.64 mg/kg	22, 251, 257
Aflatoxin RB_1					
Synthesis	—	Chick embryo	Mortality	No data	23
Aflatoxin RB_2					
Synthesis	—	Chick embryo	Lethality	No data	23
Agarin (see Muscimol)					
Agroclavine					
Claviceps purpurea	Grasses, grains	Pig, mouse	Agalactia	No data	24, 368
Agrocybin					
Agrocybe dura		Man	Extremely toxic		25
Alpha toxin (see Ibotenic acid)					
Alternaria longipes toxin					
Alternaria longipes	Corn (25—28)	Chick	Anorexia, diarrhea, loss of muscular control, gizzard erosion, proventriculous hemorrhage, coma, death	No data	26
Alternaria mali toxin					
Alternaria mali	Mycological media (26)	Mouse	Hyperkeratosis of the forestomach, pulmonary hemorrhage, vacuolated liver cord cells	No data	27

Table 3 (continued)
SOURCES AND BIOLOGICAL EFFECTS OF MYCOTOXINS

Producing fungi or other sources	Substrates (opt temp °C)	Animals affected	Toxin effects	LD_{50}	Ref.
Alternaric acid *Alternaria solani*	Corn, grains (25—30)	No data, antifungal	No data	No data	160
Alternariol *Alternaria alternata*	Cz Dox, sorghum (25)	Mice, heLa cells	Teratogen	Very little data	161, 162
Alternariol methyl ether *Alternaria alternata*	Cz Dox, sorghum	No data	No data	No data	161, 162
Amanin (see β-Amanitine)					
Amanita toxins (see α-, β-, γ-, and ε-Amanitine)					
α-Amanitine					
Amanita bisporigera, phalloides (Fr.) Seer., *tenuifolia, verna (virosa) Galerina* spp.		Man, monkey, guinea pig, rabbit, rat, mouse, horse, goat, sheep, pigeon	Man: salivation, vomiting, bloody stools, cyanosis, muscular twitching, convulsions: can be fatal; also fatty degeneration of the liver, liver necrosis, kidney necrosis	Mouse, 0.3 mg/kg	28, 29, 30, 31
β-Amanitine					
Same as α-amanitine		Same as α-amanitine	Same as α-amanitine	Mouse, 0.4 mg/kg	28, 29, 30, 31
γ-Amanitine					
Same as α-amanitine		Same as α-amanitine	Same as α-amanitine	Mouse, 0.15 mg/kg	28, 29, 30, 31

ε-Amanitine					
Same as α-amanitine		Same as α-amanitine	Same as α-amanitine	Mouse, 0.5 mg/kg	28, 29, 30, 31
Amanullin					
Same as α-amanitime			Nontoxic		28, 29, 30, 31
Amatoxins (see α-, β-, γ-, and ε-Amanitine)					
Antimycin (see Citrinin)					
Ascladiol					
Aspergillus clavatus	Wheat flour (no data)	No data	No data	No data	32
Ascotoxin (see Decumbin)					
Aspercolorin					
Aspergillus versicolor	Corn (no data)	Duckling	No data	No data	33
Aspergillic acid					
Aspergillus flavus	No data (no data)	Mouse	Lethal	Mouse (LD_{100}), 150 mg/kg, i.p., 250 mg/kg, oral	34, 231
Aspertoxin					
Aspergillus flavus	Shredded wheat (no data)	Chick embryo	Beak malformation, general edema, loss of muscle tone, hemorrhage from umbilical vessels	Chick embryo, 0.7 μg/egg	35, 36
Averufin					
Aspergillus flavus, parasiticus, versicolor	Yeast extract — sucrose (30)	No data	No data	No data	150, 151
B-24 toxin (see Diacetoxyscirpenol)					
Beta-toxin (see Muscimol)					

Table 3 (continued)
SOURCES AND BIOLOGICAL EFFECTS OF MYCOTOXINS

Producing fungi or other sources	Substrates (opt temp °C)	Animals affected	Toxin effects	LD_{50}	Ref.
Bufotenine					
Amanita citrina, mappa, muscoria, pantherina		Man	Hallucinogenic, respiratory arrest, ataxia, cardiovascular effects		37
Butenolide					
Fusarium nivale *Aspergillus terreus*	Tall fescue (*Festuca arundinacea*) (15)	Cattle (?), mouse	Gangrene of the extremities	Mouse, 43.6 mg/kg	38, 229, 230
Chaetoglobosin A					
Chaetomium globosum	Rice (no data)	HeLa cells	Acute toxicity in mice; polynucleation and multipolar division of HeLa cells	6.5 mg/kg male mice s.c.; 17.8 mg/kg female mice s.c.	40, 329
Chaetoglobosin B					
C. globosum	Rice (no data)	Mice, HeLa cells	Acute toxicity in mice; polynucleation and multipolar division of HeLa cells	No data	40, 329
Chaetoglobosins C, D, E, F, G, J (Data as for A and B)					40, 330, 331
Chaetomium globosum	Corn, commercial feed (20—25)	Rat	CNS damage, hemoglobinuria, hemorrhagic enteritis, subdural hemorrhaging	No data	39, 40

Citreoviridin					
Penicillium citreo-viride, ochrosalmoneum, toxicarium	Rice (no data)	Rat	Neurotoxic, damage to CNS, adrenal cortex and liver, kidney paralysis, respiratory failure	No data; Rat (LD_{100}), 8—30 mg/kg	41, 42, 227, 228
Citrinin					
Aspergillus candidus, niveus, terreus *Penicillium citreo-viride, citrinum, fellutonum, implicatum, jenseni, lividum, viridicatum*	Rice, corn, wheat, barley (30)	Cattle (?), swine, poultry, laboratory animals	Nephrotoxic	Mouse, 35 mg/kg	43, 44, 222, 223, 224, 225
Clavacin (see Patulin)					
Clavatin (see Patulin)					
Claviformin (see Patulin)					
Cochliobolus toxin					
Cochliobolus carbonum	Cereals (25)	Mouse	Respiratory difficulty, ataxia, pile erection, death	No data	45
Cyclochlorotine					
Penicillium islandicum	Rice (25)	Rat, mouse	In mice, liver fibrosis, cirrhosis, no necrosis; hepatomas	0.335 mg/kg mice i.v.; 0.475 mg/kg mice s.c.; 6.55 mg/kg mice p.o.	46, 220, 221

Table 3 (continued)
SOURCES AND BIOLOGICAL EFFECTS OF MYCOTOXINS

Producing fungi or other sources	Substrates (opt temp °C)	Animals affected	Toxin effects	LD_{50}	Ref.
Cyclopiazonic acid					
Penicillium cyclopium	Corn (26)	Duckling, rat	Lethal	No data	47, 217, 218, 219
Cytochalasin E					
Aspergillus clavatus, Rosellinia necatrix	Rice (30)	Rat, human (?)	Circulatory collapse	2.6 mg/kg rat i.p.; 9.1 mg/kg rat p.o.	48, 215, 216
Cytochalasin H					
Phomopsis sp.	Wheat (supplemented with nutrients) (21)	Cockerel	Lethal	12.5 mg/kg cockerel p.o.	337, 338
Decumbin					
Ascochyta imperfecya *Penicillium decumbens*	Corn, potato medium (23—26)	Rat, goldfish	Anorexia, diarrhea, cyanosis, respiratory difficulty, nasal bleeding	Rat (oral), 275 mg/kg	49, 50
Deoxytryptoquivaline					
Aspergillus clavatus	Barley (30)	No data	No data	No data	144
Deoxynortryptoquivaline					
A. clavatus	Barley (30)	No data	No data	No data	144
Deoxynortryptoquivalone					
A. clavatus	Barley (30)	No data	No data	No data	144
Desmethoxyviridiol					
Nodulisporium hinnuleum	Not reported	Cockerel; inhibits monocotyledon growth	Ataxia, coma, convulsions, death	4.2 mg/kg cockerel p.o.	51
Dendrochine					
Dendrodochium toxicum	Wheat, straw, oats, Sudan	Man, domestic animals	Skin necrosis, damage to the	No data	52

	grass (25)		hemopoietic system and alimentary tract		
Diacetoxyscirpenol					
Fusarium diversisporum, equiseti, sembucinum, scirpi, tricinctum	Corn (8)	Cattle, laboratory animals	Internal hemorrhage, skin necrosis	Rats, 0.75 mg/kg (i.p.); 7.3 mg/kg (oral)	53, 213, 214
Gibberella intricans					
4β,8α-diacetoxy-12,13-epoxytrichotec-4-ene-3α, 15-diol	Not reported	Not reported	Not reported	Not reported	147
Fusarium tricinctum					
Emetic toxin (see vomitoxin)					
Emodin					
Cladosporium fulvum	Rice, wheat (28)	Cockerels	Severe diarrhea, death	See Ref. 130	35, 129, 130
Chaetomium affine					
Cortinarius sanguineus					
Penicillium brunneum					
Aspergillus wentii					
A. ochraceus					
Ergobasine (see ergot alkaloids)					
Ergoclinine (see ergot alkaloids)					
Ergocornine (see ergot alkaloids)					
Ergocristine (see ergot alkaloids)					
Ergocryptine (see ergot alkaloids)					
Ergometrine (see ergot alkaloids)					
Ergonovine (see ergot alkaloids)					
Ergostetrine (see ergot alkaloids)					

Table 3 (continued)
SOURCES AND BIOLOGICAL EFFECTS OF MYCOTOXINS

Producing fungi or other sources	Substrates (opt temp °C)	Animals affected	Toxin effects	LD_{50}	Ref.
Ergot alkaloids					
Aspergillus fumigatus *Claviceps purpurea* *Rhizopus arrhizus*	Rye, wild grasses (25)	Man, sheep, cattle	Vasoconstriction, uterus contraction, adrenalin and serotonin antagonism, damage to CNS (reduced activity of vasomotor center)		24, 54
Ergotamine (see Ergot alkaloids)					
Ergotocine (see Ergot alkaloids)					
Ergotrate (see Ergot alkaloids)					
Expansine (see Patulin)					
F-2 toxin (see Zearalenone)					
Fescue toxin (see Butenolide)					
Flavutoxin					
Aspergillus flavus	Wheat supplemented with nutrients (27)	Cockerel	Lethality	19 mg/kg day-old cockerel p.o.	55
Frangulic acid (see Emodin)					
Frequentic acid (see Citromycetin)					
Fumigaclavine A, B					
Aspergillus fumigatus	No data	No data	No data	No data	341, 342
Fumigaclavine C					
Aspergillus fumigatus *Rhizopus arrhizus*	Cereal grains, mycological broth (35—40)	Cockerels, cattle	Lethal, diarrhea, anorexia	Cockerel: 150 mg/kg	341, 342

Fumagillin					
Aspergillus fumigatus	Cereal grains (no data)	Man, mouse	Abdominal pain, skin rash	Mouse (s.c.), 800 mg/kg	56
Fumigacin (see Helvolic acid)					
Fumigatoxin					
Aspergillus fumigatus	Laboratory media (27)	Mouse	Paralysis of the extremities, cyanosis	Mouse (LD_{100}), 150 μg/kg	57
Fumitremorgin A					
Aspergillus fumigatus	Synthetic medium, not reported	Mice	Tremorgen	Not reported	58, 59, 60, 212
Fumitremorgin B					
Aspergillus fumigatus *Penicillium piscarium*	Oats, synthetic medium (see Ref. 58), not reported	Mice	Tremorgenic	Not reported but less than 5 mg/mouse i.p.	58, 59, 60, 143, 211
Fusarenon-X					
Fusarium nivale	Wheat, rice (27)	Mouse	Damage to hematopoietic tissue, proliferating cells of the intestinal epithelium, testes	Mouse, 4.2 mg/kg	61, 62
Fusarium culmorum toxin					
Fusarium culmorum	Barley, corn (No data)	Pig, dairy cattle	Vomiting, anorexia, loss of milk production	No data	63
Gliotoxin					
Aspergillus chevaleri, fumigatus *Gliocladium fimbriatum*	Grain (25)	No data on occurrence as a mycotoxin, but closely related to toxic spori-		Mouse (s.c.), 45—65 mg/kg	64

Table 3 (continued)
SOURCES AND BIOLOGICAL EFFECTS OF MYCOTOXINS

Producing fungi or other sources	Substrates (opt temp °C)	Animals affected	Toxin effects	LD_{50}	Ref.
Penicillium cinerescens, jenseni, obscurum, restrictum, terlikowski		desmins			
Trichoderma viride					
Gyromitrin					
Helvella esculenta, gigas, infula, underwoodii		Man, guinea pig, rabbit	Vomiting, diarrhea, jaundice, liver damage, convulsions, delirium, heart failure	No data	65
Helvolic acid					
Aspergillus fumigatus	Cereal grains (no data)	Mouse	Fatty livers	Mouse (i.p.), 400 mg/kg	66
Hiptagenic acid (see β-Nitropropionic acid)					
HT-2 toxin					
Fusarium tricinctum	Corn (24)	Rat	Dermatitis	No data	67
4-Hydroxyochratoxin A *Penicillium viridicatum*	Yeast extract sucrose (25)	No noted toxicity	None reported at 40 mg/kg rat	—	68
Ibotenic acid					
Amanita muscaria, pantherina, pantherina-gemmata, strobiliformis		Man, fly	Man: psychotomimetic; fly: lethal		69

Islanditoxin					
Penicillium islandicum	Rice (room temperature)	Laboratory animals, man (?)	Hepatotoxic, hemorrhage	Mouse, 0.475 mg/kg (s.c.); 6.55 mg/kg (oral); 0.3 mg/kg (i.v.)	70, 210
Kojic acid					
Aspergillus flavus, glaucus, oryzae, tamarii	Corn (35)	Mouse, dog	Convulsant	Mouse (i.p.), 250 mg/kg	34
Luteoskyrin					
Penicillium islandicum	Rice (no data)	Mouse, rat	Hepatotoxic, carcinogen	Mouse, 6.65 mg/kg (i.v.); 147.0 mg/kg (s.c.); 221.0 mg/kg (oral)	71, 72, 207, 208, 209
Lysergic acid					
Claviceps paspali	Mycological media (25)	Man	Psychotomimetic (principal moiety of D-lysergic acid diethylamide)		73
Malformin C					
Aspergillus niger	Wheat (30)	Rats	Death (no other data)	Rats, 0.90 mg/kg (i.p.) newborn; 0.87 mg/kg (i.p.) 28-days old	146
Maltoryzine					
Aspergillus oryzoe var., *microspora*	Malt sprouts (30)	Mouse, dairy cattle	Mouse, muscular narcotism	Mouse (i.p.), 3 mg/kg	74
Mappine (see Bufotenine)					
Mescaline					

Table 3 (continued)
SOURCES AND BIOLOGICAL EFFECTS OF MYCOTOXINS

Producing fungi or other sources	Substrates (opt temp °C)	Animals affected	Toxin effects	LD_{50}	Ref.
Lophophora williamsii		Man, guinea pig, rat, mouse	Psychotomimetic for man, convulsions, respiratory arrest	Rat (i.p.), 370 mg/kg	75
8-Methoxypsoralen					
Sclerotinia sclerotiorum	Celery (no data)	Man	Dermatitis	No data	76, 203
3-Methyl-2-butenyl-paspalinine					
Claviceps paspali	Isolated from sclerotia on *Paspalum dilatatum*	Cattle, mice	Tremorgen	Mice, 14 mg/kg i.p.	159
3-Methyl-3-hydroxy-1-butenylpaspalinine					
Claviceps paspali	Isolated from sclerotia on *Paspalum dilatatum*	Cattle, mice	Tremorgen	Mice, 14 mg/kg i.p.	159
0-Methylsterigmatocystin					
Aspergillus flavus	Laboratory media (room temperature)	May be nontoxic	—	—	77
Moniliformin					
Fusarium moniliforme, fusariodes	Corn	Cockerels, duckling	Lethal, muscular weakness, cyanosis, respiratory distress, coma	Cockerel, 4.0 mg/kg p.o.; duckling, 3.68 mg/kg p.o.; rat, 50.0 mg/kg male p.o.; 41.6 mg/kg female p.o.; mice, 24 mg/kg i.p.	352, 353, 354, 355, 356, 357, 367
Monoacetoxyscirpenol					
Fusarium roseum	Rice (25—12)	Turkey, chicken, rats, swine	Bilateral inflammation of beak; GI hem-	Rat s.c. 0.752 ± 0.029 mg/kg	374

			orrhage; loss weight gain, egg production; death		
Muconomycin A (see Verrucarin A)					
Muscarine					
Amanita muscaria *Clitocybe cerussala, dealbata, virulosa, truncicola*		Man, cat, frog	Sialorrhea, lacrimation, diaphoresis, nausea, vomiting, diarrhea, bradycardia, circulatory collapse		78, 79
Muscazone					
Amanita muscaria		Man	Sialorrhea, lacrimation, diaphoresis, nausea, vomiting, diarrhea, bradycardia, circulatory collapse		69
Muscimol					
Amanita muscaria, pantherina, pantherina-gemmata		Man	Psychotomimetic		69
Mycoin (see patulin)					
Mycophenolic acid					
Penicillium brevi-compactum, stoloniferum, urticae	Corn (no data)	Mouse, monkey, rat	Affects lekocytes	Mouse (LD_{100}), 500 mg/kg (i.v.), 2,000 mg/kg (oral); rat LD_{50}, 700 mg/kg p.o., 450 mg/kg i.v.	80, 81

Table 3 (continued)
SOURCES AND BIOLOGICAL EFFECTS OF MYCOTOXINS

Producing fungi or other sources	Substrates (opt temp °C)	Animals affected	Toxin effects	LD_{50}	Ref.
Neosolaniol monoacetate					
Fusarium tricinctum	Mycological broth (27)	Cockerels	Lethal	0.789 mg/kg	340
Nidulotoxin					
Aspergillus nidulans	Laboratory media (30)	Chick, duck embryo	Growth retardation	Chick embryo: 1 μg/embryo	82
Ochratoxin A					
Aspergillus alliaceus, melleus, ochraccus, ostianus, sclerotiorum, sulfureus *Penicillium viridicatum*	Cereals (25)	Rat, duckling, chick, pig, poultry, man (?)	Tubular necrosis of the kidney, mild liver degeneration, enteritis	Duckling (oral), 3.0 mg/kg; rat (oral), 28 mg/kg	85, 192, 199, 200, 201, 202, 379
Ochratoxin A ethyl ester (see Ochratoxin C)					
Ochratoxin A methyl ester					
Aspergillus ochraceus	Corn, wheat (no data)	Same as ochratoxin A	Same as ochratoxin A	Duckling, 135—170 μg/bird	86
β-Nitropropionic acid					
Aspergillus flavus, oryzae *Penicillium atrovenetum*	Laboratory media (24)	Dairy cattle	Vasodilation	No data	34, 83
Nivalenol					
Fusarium nivale	Wheat, rice (27)	Mouse	Cell degeneration of bone marrow, lymph nodes,	Mouse (i.p.), 4 mg/kg	84

			intestines, testes, thymus		
Norsolorinic acid *Aspergillus flavus*	Synthetic, rice	No data	No data	No data	156, 157
Nortryptoquivaline *Aspergillus clavatus*	Barley (30)	No data	No data	No data	144
Ochratoxin B Same as ochratoxin A	Corn, other cereals (no data)	Trout (?)	Comparatively nontoxic	—	87
Ochratoxin B ethyl ester *Aspergillus ochraceus*	Corn (no data)	Trout	Acute toxicity	13.0 mg/kg	86, 187
Ochratoxin B methyl ester *A. ochraceus*	Corn (no data)	—	Nontoxic	—	86
Ochratoxin C *A. ochraceus*	Corn (no data)	Same as ochratoxin A	Same as ochratoxin A	Duckling, 135—170 μg/bird	86
Oxalic acid *A. niger* *Penicillium oxalicum*	Mycological and natural products (no data)	No data as mycotoxin	Calcium depletion in rats, neurological coma, gastric irritation	Mouse (i.p.), 150 mg/kg	34, 88
Pantherine (see Muscimol)					
Parasiticol (see Aflatoxin B_3)					
Paspalinine *Claviceps paspali*	Isolated from sclerotia on *Paspalum dilatatum*	Cattle, mice	Tremorgen	Mice, 14 mg/kg i.p.	159

Table 3 (continued)
SOURCES AND BIOLOGICAL EFFECTS OF MYCOTOXINS

Producing fungi or other sources	Substrates (opt temp °C)	Animals affected	Toxin effects	LD_{50}	Ref.
Patulin					
Aspergillus clavatus, giganteus, terreus *Byssochlamys nivea* *Penicillium claviforme, expansum, urticae*	Moldy feed (25)	Cattle, laboratory animals	Edema, carcinogenic	Mouse, rat (i.v.), 15—35 mg/kg	89, 187
Paxilline					
Penicillium paxilli	Wheat with supplements (25)	Cockerel, mice	Tremoring	Not reported	90, 91
Penicidine (see Patulin)					
Penicillic acid					
Aspergillus melleus, ochraceus, quercinus, sulfureus *Penicillium baarnense, cyclopium, madriti, martensii, palitans, puberulum, roqueforti, thomii*	Corn, cereals (5—10)	Laboratory animals	Digitalis-like action on heart, dilator action on systemic blood vessels, antidiuretic action	Mouse, 110 mg/kg (s.c.), 250 mg/kg (i.v.)	92, 93, 168

Penicillium puberulum toxin					
Penicillium puberulum	Corn, wheat, millet, oats, peanuts (28)	Mouse, duckling	Incoordination, ataxia, convulsions, apnea	No data	94
Phallicidine					
Amanita bisporigera, phalloides (Fr.) Secr., *tenuifolia, verna (virosa)* *Galerina* spp.		Man, monkey, guinea pig, rabbit, rat, mouse, horse, goat, sheep, pigeon	Man: salivation, vomiting, bloody stools, cyanosis, muscular twitching, convulsions; can be fatal; also fatty degeneration of the liver, liver necrosis, kidney necrosis	Mouse, 2.5 mg/kg	28, 29, 30, 31
Phallin B					
Same as phallicidine		Same as phallicidine	Same as phallicidine	Mouse, 15 mg/kg	28, 29, 30
Phalloidin					
Same as phallicidine		Same as phallicidine	Same as phallicidine	Guinea pig, 3.5 mg/kg; mouse, 2.0 mg/kg	28, 29, 30, 31
Phalloin					
Same as phallicidine		Same as phallicidine	Same as phallicidine	Mouse, 1.8 mg/kg	28, 29, 30, 31
Phallotoxins (see Phallicidine, Phallin B, Phalloidin, and Phalloin)					
Phomenone *Phoma exigua*	Mycological media	Mice	Inhibits transcription	No toxicity	332, 333

Table 3 (continued)
SOURCES AND BIOLOGICAL EFFECTS OF MYCOTOXINS

Producing fungi or other sources	Substrates (opt temp °C)	Animals affected	Toxin effects	LD_{50}	Ref.
Secalonic acid A *Phoma terrestris, Claviceps purpuvea, Aspergillus ochraceus*	Rice (27)	Rat	Paw edema, increased vascular permeability, no toxicity to chick embryo	No data	96, 344, 345, 346, 347, 348
Secalonic acids B, C No data available					96, 346
Secalonic acid D *Penicillium oxalicum, Aspergillus aculcatus, A. japonicum*	Corn, rice (25, 28)	Laboratory animals	Pulmonary toxin, lethal	42 mg/kg mice i.p.	95, 96, 346, 350
Secalonic acid E *Phoma terrestris*	Mycological broth (see Ref.) (25)	No data	No data	No data	96, 344, 349
Secalonic acid F *Aspergillus aculeatus, japonicum*	Rice (25)	No data	No data	No data	350
PR toxin *Penicillium roqueforti*	Yeast extract-sucrose (24)	Weanling rats	Congestion and edema of lung, brain and kidney, degenerative changes in liver and kidney with hemorrhage in latter	11 mg/kg weanling rat i.p.; 115 mg/kg weanling rat p.o.	97, 98

Pramuscimol (see Ibotenic acid)					
Psilocybin					
Conocybe siligenoides, Panacolus companulatus, Psilocybe aztecorum, caerulescens, cordispora, cubensis, hoogshageni, mexicana, mixaeensis, semperviva, wassonii, yungensis, zapotecorum, Strophana cubensis		Man	Hallucinogenic	Mouse, 280 mg/kg	75
Roridin E (Satratoxin D, Stachybotrys toxin)	Straw, feed (25)	Man, horse	Hemorrhage necrosis of mucus membrane	No data	100, 101
Stachybotrys atra					
Roseotoxin B	Rice	Ducklings, mice, rabbit, pig	Lethality, dermal necrosis	No data	102, 103, 104
Trichothecium roseum					
Rubratoxin A					
Penicillium purpurogenum, rubrum	Grains, laboratory media (25)	Swine, cattle, laboratory animals	Hepatotoxic, hemorrhage	Mouse (i.p.), 6.3 mg/kg	99
Rubratoxin B					
Same as rubratoxin A	Same as rubratoxin A (25)	Same as rubratoxin A	Same as rubratoxin A, HeLa cells	Mouse (i.p.), 3.0 mg/kg	99, 105, 106

Table 3 (continued)
SOURCES AND BIOLOGICAL EFFECTS OF MYCOTOXINS

Producing fungi or other sources	Substrates (opt temp °C)	Animals affected	Toxin effects	LD_{50}	Ref.
Rugulosin					
Myrothecium verrucaria, Penicillium rugulosum, P. wortmanni, P. tardum, P. brunneum, P. variabila, Endothia parasitica	Rice (25)	Mice	Hepatotoxic; enlarged liver, centrolobular necrosis, carcinogenic	83.0 mg/kg mice (toxin in NaCl) 44.0 mg/kg mice (toxin in olive oil)	
Satratoxin D (see Roridin E)					
Secalonic acid (see Pigment A)					
Slaframine					
Rhizoctonia leguminicola	Red clover, hay (25)	Ruminants, laboratory animals	Excessive salivation, bloat, anorexia, lacrimation, diarrhea	No data	107
Slobber factor (see Slaframine)					
Sporidesmin A					
Pithomyces chartarum	Pasture grasses, cereals (24)	Sheep, cattle	Hepatotoxic, facial eczema, edema	Sheep (oral), 0.5—1.0 mg/kg	108
Sporidesmin B					
Same as sporidesmin	Same as sporidesmin (no data)	Same as sporidesmin	Same as sporidesmin	No data	108
Sporidesmin C					
Same as sporidesmin	Same as sporidesmin (no data)	No data	No data	No data	109
Sporidesmin D					
Same as sporidesmin	Rye (no data)	No data	No data	No data	110
Sporidesmin E					
Same as sporidesmin	Pasture grasses	Same as spori-	Same as spori-	No data	111, 112

	(no data)	desmin	desmin; also toxic to HeLa cells		
Sporidesmin F *Pithomyces chartarum*	Rye (24)	No data	Sporidesmin F showed little or no biological activity; (no data on toxicity)	No data	113
Sporidesmin G *Pithomyces chartarum*	Rye(24)	No data	No data	No data	114
Stachybotrys toxin *Stachybotrys atra*	Straw, feed (25)	Man, horse	Hemorrhage, necrosis of mucous membranes	No data	115, 168, 169
Stemphone *Stemphylium sarcinaeforme*	Red clover (25)	Chick embryo, fish larvae	Lethal	Chick embryo (LD_{33}), 100 μg/embryo	116
Sterigmatocystin *Aspergillus nidulans, versicolor, Bipolaris* spp.	Corn (no data)	Rat, mouse	Carcinogenic, hepatotoxic, nephrotoxic, diarrhea	Mouse (oral), 800 mg/kg; rat, 65 mg/kg (i.p.), 120 mg/kg (oral)	117, 118, 148, 149, 167, 351
T-2 toxin *Fusarium tricinctum,* strain T-2	Corn (8)	Cattle, trout	Hemorrhage, skin necrosis, shedding of intestinal mucosa in trout	Mouse (LD_{100}), 100—150 mg/kg	119, 369, 370, 378
Terreic acid *Aspergillus terreus*	Cereal grains (no data)	Man, mouse	Irritation of human mucous membranes	Mouse (i.v.), 71—119 mg/kg	120

Table 3 (continued)
SOURCES AND BIOLOGICAL EFFECTS OF MYCOTOXINS

Producing fungi or other sources	Substrates (opt temp °C)	Animals affected	Toxin effects	LD_{50}	Ref.
TR-1 (see Verruculogen)					
TR-2					
Synthesis from Verruculogen (TR-1) and possibly produced by *Penicillium verruculosum*	—	Cockerel	Tremorgen	No data	121
Tremorgen					
Aspergillus flavus	Cereals (no data)	Mouse	Tremors (sustained)	No data	122
Tremortin A (Penitrem A)					
Penicillium cyclopium, crustosum, granulatum, martensii, palitans, puberulum	Feedstuffs (25)	Cattle, sheep, horse, mouse, rat	Tremors, convulsions, ataxia, anorexia, pilo erection	Mouse (i.p.), 1.2 mg/kg	123, 124, 125
Tremortin B (Penitrem B)					
Same as tremortin A	Same as tremortin A (25)	Same as tremortin A	Same as tremortin A	Mouse (i.p.), 5 mg/kg	124
Trichodermin					
Fusarium spp., *Trichoderma viride*	Corn (low temp ?)	Mouse, cattle (?)	Hemorrhage (?)	Mouse, 500—1,000 mg/kg e (s.c.), > 1,000 mg/kg (oral)	126
Trichothecin					
Fusarium spp., *Trichothecium roseum*	Corn (8)	Cattle (?)	Hemorrhage (?)	No data	126, 127
Tryptoquivaline					
Aspergillus clavatus	Rice (30)	Humans, rats	Tremorgen	No data	128

Tryptoquivalone *A. clavatus*	Rice (30)	Rats, humans	Tremorgen	No data	128
4,5′,8-Trimethylpsoralen *Sclerotinia sclerotiorum*	Celery	Man	Contact dermatitis	No data	76
Verrucarin A *Myrothecium roridum, verrucaria*	Laboratory media (25 or lower)	Man	Dermatitis	Mouse (i.v.), 1.5 mg/kg	131, 132
Verrucarin B Same as verrucarrin A	Same as Verrucarin A	Man	Dermatitis	Mouse (i.v.), 7.0 mg/kg	131, 132
Verruculogen (TR-1) *Penicillium verruculosum*	Wheat supplemented with nutrients (28)	Mice, chickens	Tremorgen	Mice, 2.4 mg/kg i.p., 127 mg/kg p.o.; chickens, 15.2 mg/kg i.p., 366 mg/kg p.o.	133, 134
Verruculotoxin *P. verruculosum*	Wheat plus supplements (28)	Cockerel	Loss muscular coordination, coma, death	Cockerel 20 mg/kg p.o.	135, 158
Versicolorin A *Aspergillus versicolor, flavus, parasiticus*	Rice (25)	No data	No data	No data	152, 153
Versiconal acetate Same as versicolorin A	Rice (25)	No data	No data	No data	154, 155
Viriditoxin *A. viridenutens*	Laboratory media, corn (28)	Mouse	Lethal	Mouse, 2.8 mg/kg i.p.	136, 166
Viomellein *Penicillium viridicatum, cyclopium, Aspergillus ochraceus, melleus, sulphureus*	Rice (25)	Laboratory animals	Hepatotoxin, nephrotoxin	No data	141, 142

Table 3 (continued)
SOURCES AND BIOLOGICAL EFFECTS OF MYCOTOXINS

Producing fungi or other sources	Substrates (opt temp °C)	Animals affected	Toxin effects	LD_{50}	Ref.
Vomitoxin (deoxynivalenol) *Fusarium roseum*	Corn, barley (10—20), rice	Pigs, laboratory animals, humans (?)	Feed refusal by pigs, vomit ng; in mice, dilation with hemorrhage of the gastroentestinal tract and engorgement of the testes, hepatotoxin, nephrotoxin	70 mg/kg mice i.p. (males), 46 mg/kg mice p.o. (males)	137, 138, 139
Xanthomegnin *P. viridicatum* *A. sulphureus* *A. melleus* *Trichophyton* spp. *P. cyclopium* *Microsporum cookei*	Rice	Laboratory animals		No data	141, 142
Xanthotoxin (see 8-Methoxypsoralen)					
Zearalenone *Fusarium graminearum*	Corn, commercial feed (12—27)	Swine, rat, sheep	Enlarged vulvae and mammary glands, vaginal prolapse, abortion	No data	140

Table 4
MYCOTOXIN-PRODUCING FUNGI

Fungus	Toxin	Ref.
Agrocybe dura	Agrocybin	25
Alternaria alternata	Alternariol	101, 102
	Alternariolmonomethyl ether	101, 102
Alternaria solani	Altenaric acid	160
A. mali	Unknown toxin	27
A. longipes	Unidentified toxin	26
Amanita bisporigera	α-Amanitine	28—31
	β-Amanitine	28—31
	γ-Amanitine	28—31
	ε-Amanitine	28—31
	Amanullin	28—31
	Phallicidine	28—31
A. citrina	Bufotenine	37
A. mappa	Bufotenine	37
A. muscaria	Bufotenine	37
	Ibotenic acid	69
	Muscarine	78, 79
	Muscazone	69
	Mucimol	69
A. pantherina	Bufotenine	37
	Ibotenic acid	69
	Muscimol	69
A. phalloides	α-Amanitine	28—31
	β-Amanitine	28—31
	γ-Amanitine	28—31
	ε-Amanitine	28—31
	Amanullin	28—31
	Phallicidine	28—31
A. strobiliformis	Ibotenic acid	69
A. tenuifolia	α-Amanitine	28—31
	β-Amanitine	28—31
	γ-Amanitine	28—31
	ε-Amanitine	28—31
	Amanullin	28—31
	Phallicidine	28—31
A. verna (virosa)	α-Amanitine	28—31
	β-Amanitine	28—31
	γ-Amanitine	28—31
	ε-Amanitine	28—31
Ascochyta imperfecta	Decumbin	49, 50
Aspergillus aculeatus	Secalonic acid D	95, 96, 346, 350
	Secalonic acid F	350
A. alliaceus	Ochratoxin A	85, 192, 199—202
A. auranteo-brunneus	Sterigmatocystin	117, 118, 149, 167
A. candidus	Citrinin	43, 44, 222—225
A. chevaleri	Gliotoxin	64
A. clavatus	Ascladiol	32
	Cytochalasin E	48, 215, 216
	Deoxytryptoquivaline	
	Deoxynortryptoquivaline	144
	Deoxynortryptoquivalone	144
	Nortryptoquivaline	144
	Patulin	144

Table 4 (continued)
MYCOTOXIN-PRODUCING FUNGI

Fungus	Toxin	Ref.
	Tryptoquivaline	89
	Tryptoquivalone	128
		128
A. flavus	Aflatoxins B_1, B_2, M_1	1—3, 237,
	Aspergillic acid	250—256,
	Aspertoxin	5, 16,
	Averufin	237—240
	Flavutoxin	24, 231
	Kojic acid	35, 36
	O-methylsterigmatocystin	150, 151
	β-nitropropionic acid	55
	Norsolorinic acid	34
		77
		34, 53
		156, 157
A. fumigatus	Ergot alkaloids	24, 54
	Fumigaclavine A, B, C	341, 342
	Fumigillin	56
	Fumitremorgin A, B	50, 59, 60,
	Fumigatoxin	212
	Gliotoxin	57
	Helvolic acid	64
		66
A. giganteus	Patulin	89
A. glaucus	Kojic acid	34
A. melleus	Ochratoxin A	85,
	Penicillic acid	192—202
	Viomellein	92, 93, 168
	Xanthomegnin	141, 142
		141, 142
A. nidulans	Nidulotoxin	82
	Sterigmatocystin	117, 118,
		149, 167
A. niger	Malformin C	146
	Oxalic acid	34
A. niveus	Citrinin	43, 44,
		222—225
A. ochraceus	Emodin	35, 129,
	Ochratoxin A, B, C	130
	Ochratoxin A methyl ester	85, 192,
	Ochratoxin B ethyl ester	199—202
	Ochratoxin B methyl ester	86
	Penicillic acid	86, 187
		86
		92, 93, 168
A. oryzae	Kojic acid	34
	β-nitropropionic acid	34, 83
A. oryzae var. *microspora*	Maltoryzine	74
A. ostianus	Ochratoxin A	85, 192,
		199—202
A. parasiticus	Aflatoxins B_1, B_2, G_1, G_2,	1—3, 237,
	M_1, GM_1	250—256,
	Aspercolorin	15, 16,
	Averufin	237—240
	Versicolorin A	33

Table 4 (continued)
MYCOTOXIN-PRODUCING FUNGI

Fungus	Toxin	Ref.
	Versiconal acetate	152, 151
		152, 153
		154, 155
A. quadrilineatus	Sterigmatocystin	117, 118,
		149, 167
A. quercinus	Penicillic acid	92, 93, 168
A. sclerotiorum	Ochratoxin A	85, 192,
		199—202
A. sulfureus	Ochratoxin A	85, 192,
	Penicillic acid	199—202
	Viomellein	92, 93, 168
	Xanthomegnin	141, 142
		141, 142
A. tamarii	Kojic acid	34
A. terreus	Butenolide	38, 229,
	Citrinin	230
	Patulin	43, 44,
	Terreic acid	222—225
		89
		120
A. ustus	Sterigmatocystin	117, 118,
		149, 167
A. versicolor	Aspercolorin	33
	Averufin	150, 151
	Dihydrodemethylsterigmato-	351
	cystin	351
	Dihydrosterigmatocystin	148
	5,6-Dimethoxysterigmatocys-	117, 118,
	tin	149
	Sterigmatocystin	152, 153
	Versicolorin A	154, 155
	Versiconal acetate	
A. viridi-nutens	Viriditoxin	136, 166
A. wentii	Emodin	35, 129,
		130
Bipolaris sp.	Sterigmatocystin	117, 118,
		149, 167
Byssochlamys nivea	Patulin	89
Chaetomium affine	Emodin	35, 129,
		130
C. globosum	Chaetoglobosin A, B, C, D,	40, 329
	E, F, G	
Claviceps paspali	Lysergic acid	73
	3-Methyl-2-butenyl-paspali-	159
	nine	159
	3-Methyl-3-hydroxy-1-bu-	159
	tenyl-	
	paspalinine	
	Paspalinine	
C. purpurea	Agroclavine	24
	Ergot alkaloids	54
	Secalonic acid A	96,
		344—348
Clitocybe cerussata	Muscarine	78, 79
C. dealbata	Muscarine	78, 79
C. rivulosa	Muscarine	78, 79
C. truncicola	Muscarine	78, 79

Table 4 (continued)
MYCOTOXIN-PRODUCING FUNGI

Fungus	Toxin	Ref.
Cochliobolus carbonum	Unknown toxin	45
Conocybe siligenoides	Psilocybin	75
Cortinarius orileanus	Orcelanine	30
C. sanguineus	Emodin	35, 129, 130
Dendrodochium toxicum	Dendrochine	52
Endothia fluens	Rugulosin	106
E.gyrosa	Rugulosin	106
E. parasitica	Rugulosin	106
Fusarium avenaceum	T-2 toxin	370
	Neosolaniol	370
	Diacetoxyscirpenol	370
F. culmorum	Unknown toxin	63
	Neosolaneol	370
	Diacetoxyscirpenol	370
F. diversisporum	Diacetoxyscirpenol	53, 213, 214
F. episphaeria	Nivalenol	370
	Fusarinone	370
F. equiseti	Diacetoxyscirpenol	53, 213, 214
	Nivalenol diacetate	375
F. fusariodes	Moniliformin	352—357, 367
F. graminearum	Vomitoxin	137—139
	Zearalenone (F-2)	140
F. moniliforme	Moniliformin	352—357, 367
F. nivale	Butenolide	38, 229, 230
	Fusarenon-x	61, 62
	Nivalenol	84
	Nivalenol diacetate	376
F. niveum	Nivalenol	370
	Fusarenone	370
F. poae	T-2 toxin	369
	HT-2 toxin	370, 371
	Neosolaniol	369
F. rigidiusculum	Neosolaniol	370
	Diacetoxyscirpenol	370
F. roseum gibbosum	4-Acetoxyscirpendiol	145
	8-Acetylneosolaniol	145
	Vomitoxin	137—139
	Scirpentriol	373
	Monacetoxyscirpenol	374
	Nivalenol	370
	Fusarenone	370
	Monoacetoxydeoxynivalenol	138, 139
F. sambucinum	Diacetoxyscirpenol	52, 213, 214
F. scirpi	Diacetoxyscirpenol	53, 213, 214
	T-2	370
	Neosolaniol	370

Table 4 (continued)
MYCOTOXIN-PRODUCING FUNGI

Fungus	Toxin	Ref.
F. solani	T-2 toxin	370
	HT-2 toxin	370
	Neosolaniol	372
	Diacetoxyscirpenol	370
F. sporotrichiodes	T-2 tetraol	369
	T-2 toxin	369
F. sulphureum	4β-Acetoxy-3α,15-diOH-12,13-epoxytrichothecene-9-ene	377
F. tricinctum	Diacetoxyscirpenol	53, 213, 214
	HT-2 toxin	67
	Neosolaniol monoacetate	340
	T-2 toxin	119
	Neosolaniol	370
Galerina spp.	α-Amanitine	28—31
	β-Amanitine	28—31
	γ-Amanitine	28—31
	ε-Amanitine	28—31
	Amanullin	28—31
	Phallicidine	28—31
Gibberella intricans	Diacetoxyscirpenol	53, 213, 214
Gliocladium fimbriatum	Gliotoxin	64
Helvella esculenta	Gyrometrin	65
H. gigas	Gyrometrin	65
H. infula	Gyrometrin	65
H. underwoodii	Gyrometrin	65
Inocybe lacera	Muscarine	78—79
I. napipes	Muscarine	78—79
I. patouillardii	Muscarine	78—79
I. picrosma	Muscarine	78—79
Myrothecium roridum	Verrucarin A	131, 132
M. verrucaria	Rugulosin	106
	Verrucarin A	131, 132
	Verrucarin B	131, 132
Nodulisporium hinnuleum	Desmethoxyviridiol	51
Oinphalotus olcarius	Muscarine	78, 79
Penicillium atrovenetum	β-Nitropropionic acid	34, 83
P. baarnense	Penicillic acid	92, 93, 168
P. brevi-compactum	Mycophenolic acid	80, 81
P. brunneum	Emodin	35, 129, 130
	Rugulosin	106
P.cinerascens	Gliotoxin	64
P. citreo-viride	Citreoviridin	41, 42, 227, 228
	Citrinin	43, 44, 222—225
P. citrinum	Citrinin	43, 44, 222—225
P. claviforme	Patulin	89

Table 4 (continued)
MYCOTOXIN-PRODUCING FUNGI

Fungus	Toxin	Ref.
P. cyclopium	Cyclopiazonic acid	47, 217—219
	Penicillic acid	92, 93, 168
	Penitrem A	123, 124, 125
	Viomellein	141, 142
P. crustosum	Penitrem A	123—125
P. decumbens	Decumbin	49, 50
P. expansum	Patulin	89
P. fellutanum	Citrinin	43, 44, 222—225
P. granulatum	Penitrem A	123—125
P. implicatum	Citrinin	43, 44, 222—225
P. islandicum	Cyclochlorotine	46, 220, 221
	Islanditoxin	70, 210
	Luteoskyrin	71, 72, 207—209
P. jenseni	Citrinin	43, 44, 222—225
	Gliotoxin	64
P. lividum	Citrinin	43, 44, 222—225
P. madriti	Penicillic acid	92, 93, 168
P. martensii	Penicillic acid	92, 93, 168
	Penitrem A	123—125
P. obscurum	Gliotoxin	64
P. ochrasalmoneum	Citreoviridin	41, 42, 227, 228
P. oxalicum	Oxalic acid	34
	Secalonic acid D	95, 96, 346, 350
P. palitans	Penicillic acid	92, 93, 168
	Penitrim A	123—125
P. paxilli	Paxilline	90, 91
P. puberulum	Penicillic acid	92, 93, 168
	Unknown toxin	94
	Penitrem A, B, C	123—125
P. purpurogenum	Rubratoxin A, B	99, 105
P. restrictum	Gliotoxin	64
P. roqueforti	Penicillic acid	92, 93, 168
	PR toxin	
P. rubrum	Rubratoxin A, B	99, 105
P. rugulosum	Rugulosin	106
P. stoloniferum	Mycophenolic acid	80, 81
P. tardum	Rugulosin	106
P. terlikowski	Gliotoxin	64
P. thomii	Penicillic acid	92, 93, 168
P. toxicarium	Citreoviridin	41, 42, 227, 228

Table 4 (continued)
MYCOTOXIN-PRODUCING FUNGI

Fungus	Toxin	Ref.
P. urticae	Mycophenolic acid	80, 81
	Patulin	89
P. variabile	Rugulosin	106
P. verruculosum	Verruculotoxin	135, 158
P. viridicatum	Citrinin	43, 44, 222—225
	4-Hydroxyochratoxin A	68
	Ochratoxin A	85, 199—202
	Viomellein	141, 142
	Xanthomegnin	141, 142
P. wortmanni	Rugulosin	106
Phoma exigua var. *nonoxydabilis*	Phomenone	332, 333
P. terrestris	Secalonic acid A	96, 344—348
	Secalonic acid E	96, 344, 349
Phomopsis sp.	Cyctochalasin H	337, 338
Pithomyces chartarum	Sporidesmin A	108
	Sporidesmin B	108
	Sporidesmin C	109
	Sporidesmin D	110
	Sporidesmin E	111, 112
	Sporidesmin F	113
	Sporidesmin G	114
Psilocybe aztecorum	Psilocybin	75
P. caerulescens	Psilobycin	75
P. cordispora	Psilocybin	75
P. cubensis	Psilocybin	75
P. hoogshageni	Psilocybin	75
P. mexicana	Psilocybin	75
P. mixacensis	Psilocybin	75
P. semperviva	Psilocybin	75
P. wassonii	Psilocybin	75
P. yungensis	Psilocybin	75
P. zapotecorum	Psilocybin	75
Rhizopus arrhizus	Ergot alkaloids	24, 54
	Fumigaclavine C	341, 342
Rhizoctonia leguminicola	Slaframine	107
Rosellinia necatrix	Cytochalasin E	48, 215, 216
Sclerotinia sclerotiorum	8-Methoxypsoralen	76, 203
	4,5,8-Trimethylpsoralen	76
Stachybotrys atra	Roridin E (Satratoxin D)	100, 101
	Satratoxin H	339
Stemphylium sarcinaeforme	Stemphone	116
Strophana cubensis	Psilocybin	75
Trichoderma viride	Gliotoxin	64
	Trichodermin	126
Trichothecium roseum	Roseototoxin B	102—104
	Trichothecin	126, 127

Table 5
NATURAL OCCURRENCE OF MYCOTOXINS IN FOODS AND FEEDS: REPRESENTATIVE FINDINGS

Toxin Commodity contaminated	Country	Ref.
Alternariol	U.S.	163
Alternariol monomethylether		
Pecans		
Sorghum	U.S.	164, 165
Emodin	U.S.	130
Chestnuts		
Sterigmatocystin	U.S.	167
Pecans		
Coffee	S.W. Africa	174
Sporidesmins	S. Africa	170
Grass litter		
Grass	New Zealand	171
Grass	Australia	172, 173
Patulin	England	177
Apples		
Apple cider	France	178
Apple juice, pears	Germany	179, 183
Apples	Canada	180
Grape juice	Canada	184
Apple juice	Canada	185
Bananas, pineapples	Germany	186
grapes, peaches,		
apricots		
Apple juice	Sweden	187
Penicillic acid	U.S.	328
Corn, dried beans		
Ochratoxin	Canada, Denmark	87, 190, 191
Corn, barley, wheat,		
oats, rye, green coffee beans	Finland, France,	
beans, peanuts, hay,	Norway, Poland,	
animal products (pork	Sweden, U.K.,	
and poultry)	U.S., Yugoslavia	
Corn	Bulgaria	188
Poultry feed, barley, oats	Denmark	189, 317
Corn	U.S.	193, 318
Barley, oats	Sweden	194
Wheat	Canada	195
Rice	Japan	196
White beans, wheat, oats, rye,	Canada	197
peanuts, mixed feed	U.S.	198
Wheat		
Psoralens	U.S.	204, 206
Celery		
Celery, parsley	Italy	205
Citrinin	India	226
Peanuts		
Wheat, oats, barley, rye	Canada	197
Aflatoxin M_1 (milk toxin)	France	241
Cheese		
Dried milk products	Germany	242
Milk	Germany	243, 261
Buffalo milk	India	244
Milk, milk products	India	245
(barfi, khoa, paneer),	S. Africa	246, 259
cheese		

Table 5 (continued)
NATURAL OCCURRENCE OF MYCOTOXINS IN FOODS AND FEEDS: REPRESENTATIVE FINDINGS

Toxin Commodity contaminated	Country	Ref.
Milk		
Milk	Iran	258
Milk, milk powder, yogurt, fresh cheese, hard cheese, camembert, processed cheese	Germany	260

Aflatoxin B_1

(Aflatoxin B_1 has been analyzed for extensively throughout most of the world and the number of publications on the subject are too numerous to list. Only representative data are shown to illustrate the magnitude of the problem. Toxin levels are not given in the table to avoid misinterpretation of data.)

Corn (maize)	U.S.	262, 271
Corn	Uganda	272, 273
Corn	India	274, 275
Corn	Thailand, Hong Kong	276
Corn	Philippines	277, 278
Corn	Dominican Republic	279
Corn	Australia	280
Corn	France	281
Peanuts (groundnuts)	Thailand, Vietnam Mainland China	282
Peanut oil	Malaya	283
Peanuts	Nigeria	284
Peanuts	Uganda	285, 272
Peanuts, expeller cake, peanut meal	Brazil, Nigeria, Sudan Senegal, Indonesia, Kenya, Uganda, Argentina, Ghana, Congo	286
Peanuts	India	287
Peanuts	Thailand	276
Peanut butter	Philippines	277, 278
Peanuts	Swaziland	288
Peanuts	Kenya	289
Peanut meal	Indonesia, U.S., Brazil, India	290
Peanut plant parts	France	291
Farmer's stock peanuts	U.S.	292
Virginia peanuts	U.S.	293
Spanish peanuts	U.S.	294
Peanuts	Angola, Gambia, Ghana, Madagascar, Malawi, Mali, Mozambique, Nigeria, Bangladesh, Burma, India, Indonesia, Philippines, Taiwan, France, Poland, Spain	256
Peanut oil	India	295, 310
Peanut oil, peanut products	Taiwan	296
Cottonseed meal	U.S.	297

Table 5 (continued)
NATURAL OCCURRENCE OF MYCOTOXINS IN FOODS AND FEEDS: REPRESENTATIVE FINDINGS

Toxin Commodity contaminated	Country	Ref.
Cottonseed and products	U.S.	298, 299, 300, 301, 303, 381
Cottonseed meal, crush, cake	Brazil, Colombia, Guatemala, Nicaragua, El Salvador, Syria, Turkey, U.S.S.R.	302
Cottonseed	India	304
Nuts and fruit		
Pecans	U.S.	306
Pistachio	Turkey, Iran	307, 311, 382
Figs	Near East	308
Aleppo pine nuts	Tunisia	309
Hazelnut, walnut	Germany	310
Almonds	U.S.	312
Coconut and products	Ceylon	313
Walnut, almond	U.S.	314
Hazelnut, almonds, mixed nuts	Not stated	315
Haricot bean and other pulses (legumes)		316
	Trichothecenes	
T-2		
Barley	Canada	324
Corn	U.S.	319
Mixed feed	U.S.	320
Grain (?)	Russia	321
Sorghum	India	325
Vomitoxin		320, 322, 323, 327
Corn	U.S.	
Barley	Japan	138
Corn	S. Africa	358
Diacetoxyscirpenol		
Feed corn	Germany	326
Mixed feed	U.S.	320
Zearalenone (F-2)		
Corn	S. Africa	358
Corn flakes	Canada	359
Sorghum	U.S.	360
Feedstuffs, corn, sorghum, sesame meal	U.S.	320
Hay	U.S.	361
Corn	Yugoslavia	362
Barley	Scotland	363
Corn	France	364
Feed	Finland	365
Corn, barley	U.S.	366

Table 6
MYCOTOXIN MELTING POINTS: CROSS INDEX

Melting point	Compound number
35—36	101
46	165
65—66	42
65—75	155
68	116
69 (dec)	100
83—84	131
84—85	81
91—92	89
97—99	45
100—110	1
101—102	126
102—103	123
107—111	58
111	128
116—118	49
118	167
127	161
134—135	124
138	32
140	31
141	113
145 (dec)	95
145—148	102
146—147	48
148—149	136
148—153	156
151—152	160
151—153	179
152	174
153—154	97
153—155	169
155—156	75
155—157	69, 137
158	106
158—160	67
160—161	158
161—162	70
162—163	77
164—165	181
168—170	50
169	120
172—173	107
172—174 (dec)	111
175	59
176—199	172
177—178	55
179	150
180—185	154
182—184	74
183	151
183—184	149
185—186	142
185—195	164

Table 6 (continued) MYCOTOXIN MELTING POINTS: CROSS INDEX

Melting point	Compound number
186—187	51
189—190	2
194	83—86
198—203	30
199—200	140
202—204	87, 170
203	18, 64
296—208	62, 147, 166
210—230 (dec)	163
211—212	88
212	76, 84, 171
215	92
216	53
216—218	94, 176
217 (233, 234)	13
218—221	148
220—223	10, 138, 139
221	90, 122
222—223	117
225	8
226—228	168
230—234	26
233—235	173
237—240	17
240	12, 99
244—245	82
244—246	16
245	178
245—246	11, 61
247	9
248	145
251	56, 60
252	129
254—255	37
255—257	63
255—258	14, 146
256—257	72, 118, 119
260 (dec.)	52, 177, 180
265 (dec.)	105, 159
266	25
266—270	34
268—269	4
273—276	5
276	20
278	98
280—282	6, 134
283—289	47
286—289	7, 175
290 (dec)	143
293	23
299	22
300	38, 99a
325—327	46
350	33, 85

Table 7
MYCOTOXINS BY CHEMICAL FORMULAE: CROSS INDEX

Chemical formula	Compound number
$C_2H_2O_4$	126
$C_3H_5O_4N_2$	116
$C_4H_2O_3$	106
$C_4H_6O_2N_2$	111
$C_4H_8N_2O$	91
$C_5H_6O_4$	166
$C_5H_6O_4N_2$	110
$C_5H_8N_2O_5$	95
$C_6H_4O_4$	97
$C_6H_7NO_3$	49
$C_7H_6O_4$	128
$C_7H_6O_4$	161
$C_7H_8O_4$	42
$C_8H_5O_2N$	31
$C_8H_{10}O_4$	131
$C_9H_{20}O_2N$	109
$C_{10}H_{13}NO_4$	85
$C_{10}H_{18}O_2N_2$	149
$C_{11}H_{14}O_4$	100
$C_{11}H_{17}NO_3$	101
$C_{12}H_8O_4$	102
$C_{12}H_{16}N_2O$	48
$C_{12}H_{20}O_2N_2$	45
$C_{13}H_{14}O_4N_2S_2$	90
$C_{13}H_{14}O_5$	59
$C_{14}H_{10}O_5$	33
$C_{14}H_{12}O_3$	168
$C_{14}H_{14}O_6$	13
$C_{15}H_{10}O_5$	72
$C_{15}H_{12}O_5$	34
$C_{15}H_{20}N_2O$	174
$C_{15}H_{20}O_4$	136
$C_{15}H_{20}O_6$	179
$C_{15}H_{20}O_7$	89, 117
$C_{16}H_{10}O_4$	15
$C_{16}H_{10}O_6$	24
$C_{16}H_{14}O_5$	14
$C_{16}H_{16}O_2N_2$	99
$C_{16}H_{18}N_2$	30
$C_{16}H_{20}N_2O$	82
$C_{16}H_{24}O_4$	64
$C_{17}H_{12}O_6$	4
$C_{17}H_{12}O_7$	6, 22, 27, 16
$C_{17}H_{12}O_8$	20
$C_{17}H_{14}O_6$	7, 12
$C_{17}H_{14}O_7$	17, 21, 23
$C_{17}H_{14}O_8$	19
$C_{17}H_{16}O_6$	26
$C_{17}H_{18}O_6$	27
$C_{17}H_{20}O_6$	28, 113, 137
$C_{17}H_{24}O_4$	1, 107, 165
$C_{18}H_{10}O_7$	175
$C_{18}H_{12}O_7$	176
$C_{18}H_{13}O_7$	5
$C_{18}H_{16}O_7$	10, 11
$C_{18}H_{20}N_3O_5Cl\,S_2$	150, 151

Table 7 (continued)
MYCOTOXINS BY CHEMICAL FORMULAE: CROSS INDEX

Chemical formula	Compound number
$C_{18}H_{20}O_6N_3Cl\,S_3$	152, 154
$C_{18}H_{20}Cl\,N_3O_6S_4$	156
$C_{18}H_{22}N_2O_2$	81
$C_{18}H_{22}O_5$	181
$C_{19}H_{14}O_6$	105
$C_{19}H_{14}O_6$	159
$C_{19}H_{14}O_7$	46
$C_{19}H_{16}O_5$	69
$C_{19}H_{16}O_8$	8
$C_{19}H_{18}O_7$	9
$C_{19}H_{18}O_8$	18
$C_{19}H_{22}ClN_3O_6S$	155
$C_{19}H_{23}O_6N_3ClS_2$	153
$C_{19}H_{23}O_2N_3$	77
$C_{19}H_{24}O_5$	167
$C_{19}H_{26}O_7$	70
$C_{19}H_{26}O_8$	71
$C_{20}H_{16}O_7$	47
$C_{20}H_{18}O_6NCl$	120
$C_{20}H_{18}ClNO_7$	94
$C_{20}H_{18}O_7$	118
$C_{20}H_{19}NO_5$	122
$C_{20}H_{20}O_3N_2$	61
$C_{21}H_{13}O_7$	2
$C_{21}H_{20}O_6NCl$	121
$C_{21}H_{21}O_6N$	124
$C_{21}H_{28}O_9$	114
$C_{21}H_{30}O_8$	32
$C_{21}H_{32}O_8$	93
$C_{22}H_{22}O_6NCl$	125
$C_{22}H_{23}O_6N$	123
$C_{22}H_{27}N_3O_6$	162
$C_{23}H_{27}O_4N_4$	44
$C_{23}H_{30}N_2O_2$	83
$C_{23}H_{30}O_6$	58
$C_{23}H_{39}N_5O_5S_2$	99a
$C_{24}H_{34}O_9$	160
$C_{25}H_{28}N_4O_7$	119
$C_{25}H_{36}N_5O_8Cl_2$	60, 96
$C_{26}H_{24}N_4O_5$	68
$C_{26}H_{24}N_4O_6$	170
$C_{26}H_{30}O_{11}$	142
$C_{26}H_{32}O_{11}$	141
$C_{26}H_{34}O_7$	86
$C_{27}H_{33}NO_4$	127
$C_{27}H_{33}NO_4$	129
$C_{27}H_{32}O_9$	172
$C_{27}H_{33}N_3O_5$	88
$C_{27}H_{33}N_3O_7$	173
$C_{27}H_{34}O_9$	171
$C_{28}H_{28}N_4O_6$	67
$C_{29}H_{30}N_4O_6$	66
$C_{28}H_{33}NO_7$	62
$C_{29}H_{30}N_4O_7$	169
$C_{29}H_{36}O_9$	144

Table 7 (continued)
MYCOTOXINS BY CHEMICAL FORMULAE: CROSS INDEX

Chemical formula	Compound number
$C_{29}H_{38}O_8$	139
$C_{30}H_{22}O_{10}$	143
$C_{30}H_{22}O_{12}$	98, 180
$C_{30}H_{24}O_{11}$	177
$C_{30}H_{39}NO_5$	63
$C_{30}H_{42}O_8$	158
$C_{30}H_{49}N_5O_7$	140
$C_{31}H_{39}O_5N_5$	74
$C_{32}H_{30}O_{14}$	145, 146, 147, 148
$C_{32}H_{30}O_5N_2$	55
$C_{32}H_{36}O_4N_2$	57
$C_{32}H_{36}O_5N_2$	50, 52, 53, 56
$C_{32}H_{37}NO_5$	104
$C_{32}H_{38}O_5N_2$	54
$C_{32}H_{39}NO_4$	103
$C_{32}H_{41}O_5N_5$	76
$C_{33}H_{35}O_5N_5$	84
$C_{33}H_{44}O_8$	92
$C_{33}H_{45}N_3O_6$	87
$C_{34}H_{30}O_{14}$	178
$C_{35}H_{39}O_5N_5$	75
$C_{35}H_{48}N_8O_{10}S$	135
$C_{35}H_{48}N_8O_{11}S$	134
$C_{36}H_{50}N_8O_{12}S$	132
$C_{37}H_{44}O_6NCl$	163
$C_{37}H_{45}O_5N$	164
$C_{39}H_{53}O_{13}N_9S$	40
$C_{39}H_{54}O_{11}N_{10}S$	41
$C_{39}H_{54}O_{12}N_{10}S$	39
$C_{40}H_{55}O_{14}N_9S$	38

REFERENCES

1. **Detroy, R. W., Lillehoj, E. B., and Ciegler, A.,** Aflatoxin and related compounds, in *Microbial Toxins,* Vol. 6, Ciegler, A., Kadis, S., and Ajl, S., Eds., Academic Press, New York, 1971, chap. 1.
2. **Goldblatt, L. A.,** *Aflatoxin,* Academic Press, New York, 1969.
3. **Lillehoj, E. B., Ciegler, A., and Detroy, R. W.,** Fungal toxins, in *Essays in Toxicology,* Blood F., Ed., Academic Press, New York, 1970, chap. 1.
4. **Osiyemi, F. O. and Adekunle, A. A.,** Synthesis, structure and biological properties of aflatoxin B_1 aldehyde, *Acta Pharm. Suecica,* 10, 526, 1973.
5. **Hartley, R. D., Nesbitt, B. F., and O'Kelly, J.,** Toxic metabolites of *Aspergillus flavus, Nature (London),* 198, 1056, 1963.
6. **Dutton, M. F. and Heathcote, J. G.,** The structure, biochemical properties and origin of the aflatoxins B_{2a} and G_{2a}, *Chem. Ind.,* p. 418, 1968.
7. **Dutton, M. F. and Heathcote, J. G.,** Derivatives of aflatoxins B_{2a} and G_{2a}, *Chem. Ind. (London),* p. 983, 1969.
8. **Waiss, A. C. and Wiley, M.,** Anomalous photochemical addition of methanol to 6-methoxydifurocoumarone, *Chem. Commun.,* p. 512, 1969.
9. **Lillehoj, E. B. and Ciegler, A.,** Biological activity of aflatoxin B_{2a}, *Appl. Microbiol.,* 17, 516, 1969.
10. **Heathcote, J. G. and Dutton, M. F.,** New metabolites of *Aspergillus flavus, Tetrahedron,* 25, 1497, 1969.

11. **Stubblefield, R. D., Shotwell, O. L., Shannon, G. M., Weisleder, D., and Rohwedder, W. K.**, Parasiticol: a new metabolite from *Aspergillus parasiticus, J. Agric. Food Chem.*, 18, 391, 1970.
12. **Cucullu, A. F., Lee, L. S., Pons, W. A., and Stanley, J. B.**, Ammoniation of aflatoxin B_1: isolation and characterization of a product with molecular weight 206, *J. Agric. Food Chem.*, 24, 408, 1976.
13. **Heathcote, J. G. and Hibbert, J. F.**, New aflatoxins from cultures of *Aspergillus flavus, Biochem. Soc. Trans.*, p. 301, 1974.
14. **Salhab, A. S. and Hsieh, D. P. H.**, Aflatoxin H_1: a major metabolite of aflatoxin B_1 produced by human and Rhesus monkey livers in vitro, *Res. Commun. Chem. Pathol. Pharmacol.*, 10, 419, 1975.
15. **Büchi, G. and Weinreb, S. M.**, The total synthesis of racemic aflatoxin M_1 (milk toxin), *J. Am. Chem. Soc.*, 91, 5408, 1969.
16. **Sinnhuber, R. O., Lee, D. J., Wales, J. H., Landers, M. K., and Beyl, A. C.**, Aflatoxin M_1, a potent liver carcinogen for rainbow trout, *Fed. Proc.*, 29, 568, 1978.
17. **Purchase, I. F. H.**, Acute toxicity of aflatoxins M_1 and M_2 in one-day-old ducklings, *Food Cosmet. Toxicol.*, 5, 339, 1967.
18. **Dalezios, J., Wogan, G. N., and Weinreb, S. M.**, Aflatoxin P_1: a new aflatoxin metabolite in monkeys, *Science*, 171, 584, 1971.
19. **Masri, M. S., Haddon, W. F., Lundin, R. E., and Hsieh, D. P. H.**, Aflatoxin, Q_1: a newly identified major metabolite of aflatoxin B_1 in monkey liver, *J. Agric. Food Chem.*, 22, 512, 1974.
20. **Büchi, G. H., Muller, P. M., Roebuck, B. D., and Wogan, G. N.**, Aflatoxin Q_1: a major metabolite of aflatoxin B_1 produced by human liver, *Res. Commun. Chem. Pathol. Pharmacol.*, 8, 585, 1974.
21. **Hsieh, D. P. H., Salhal, A. S., Wong, J. J., and Yang, S. L.**, Toxicity of aflatoxin Q_1 as evaluated with the chicken embryo and bacterial auxotrophs, *Toxicol. Appl. Pharmacol.*, 30, 237, 1974.
22. **Detroy, R. W. and Hesseltine, C. W. H.**, Isolation and biological activity of a microbial conversion product of aflatoxin B_1, *Nature (London)*, 219, 967, 1968.
23. **Ashoor, S. H. and Chu, F. S.**, Reduction of aflatoxins B_1 and B_2 with sodium borohydride, *J. Assoc. Off. Anal. Chem.*, 58, 492, 1975.
24. **Hofmann, A., von Bruner, R., Kobel, H., and Brock, A.**, Neue Alkaloide aus der saprophytishen Kultur des Mutterkornpilzes von *Pennisetum typhocdeum* Rich., *Helv. Chim. Acta*, 40, 1358, 1957.
25. **Bu'dock, J. D., Jones, E. P. H., Mansfield, G. H., Thompson, J. W., and Whiting, M. C.**, Structure of two polyacetylenic antibiotics, *Chem. Ind. (London)*, p. 990, 1954.
26. **Doupnik, B., Jr. and Sobers, E. K.**, Mycotoxicosis: toxicity to chicks of *Alternaria longipes* isolated from tobacco, *Appl. Microbiol.*, 16, 1596, 1968.
27. **Slifkin, M. K. and Spalding, J.**, Studies of the toxicity of *Alternaria mali, Toxicol. Appl. Pharmacol.*, 17, 375, 1970.
28. **Dessey, G. and Francioli, M.**, *Amanita phalloides* and *verna* and their toxins, *Boll. Inst. Fur Milan*, 17, 779, 1938.
29. **Tyler, V. E., Brady, L. R., Benedict, R. G., Khanna, J. M., and Malone, M. H.**, Chromatographic and pharmacologic evaluation of some toxic *Galerina* species, *Lloydia*, 26, 154, 1963.
30. **Wieland, O.**, Changes in liver metabolism induced by the poisons of *Amanita, Clin. Chem.*, 11, 323, 1965.
31. **Wieland, T.**, The toxic peptides of *Amanita phalloides, Fortschr. Chem. Org. Naturst.*, 25, 214, 1967.
32. **Suzuki, T., Takeda, M., and Tanabe, H.**, A new mycotoxin produced by *Aspergillus clavatus, Chem. Pharm. Bull. (Japan)*, 19, 1786, 1971.
33. **Aucamp, P. J. and Holzapfel, C. W.**, Toxic substances from *Aspergillus versicolor, J. S. Afr. Chem. Inst.*, 22, S35, 1969.
34. **Wilson, B. J.**, Toxins other than aflatoxins produced by *Aspergillus flavus, Tetrahedron*, 30, 478, 1966.
35. **Rodricks, J. V., Lustig, E., Campbell, A. D., and Stoloff, L.**, Aspertoxin, a hydroxy derivative of o-methyl-sterigmatocystin from aflatoxin-producing cultures of *Aspergillus flavus, Tetrahedron Lett.*, p. 2975, 1968.
36. **Rodricks, J. V., Lustig, E., Campbell, A. D., and Stoloff, L.**, Aspertoxin, a hydroxy derivative of o-methylsterigmatocystin from aflatoxin-producing cultures of *Aspergillus flavus, Tetrahedron Lett.*, p. 2975, 1968.
37. **Fischer, R.**, Chemistry of the brain, *Nature (London)*, 220, 411, 1968.
38. **Yates, S. G., Tookey, H. L., Ellis, J. J., and Burkhardt**, Mycotoxins produced by *Fusasium nivale* isolated from tall fescue, *Phytochemistry*, 1, 139, 1968.
39. **Christensen, C. M., Nelson, G. H., Mirocha, C. J., Bates, F., and Dorworth, C. E.**, Toxicity to rats of corn invaded by *Chaetomium globosum, Appl. Microbiol.*, 14, 774, 1966.
40. **Umeda, M., Ohtsubo, K., Saito, M., Sekita, S., Yoshihira, K., Natori, S., Udagawa, S., Sakabe, F., and Kurata, H.**, Cytotoxicity of new cytochalasins from *Chaetomium globosum, Experientia*, 31, 435, 1975.

41. **Sakabe, N., Goto, T., and Hirata, Y.**, Structure of citreoviridin, a mycotoxin produced by *Penicillium citreo-viride* molded on rice, *Tetrahedron*, 33, 3077, 1977.
42. **Uraguchi, K.**, Mycotoxic origin of cardiac beri beri, *J. Stored Prod. Res.*, 5, 227, 1969.
43. **Friis, P., Hasselager, E., and Krogh, P.**, Isolation of citrinin and oxalic acid from Penicillium viridicatum westling and their nephrotoxicity in rats and pigs, *Acta Pathol. Microbiol. Scand.*, 77, 559, 1969.
44. **Carlton, W. W. and Tuite, X.**, Metabolites of *P. viridicatum* toxicology, in *Mycotoxins in Human and Animal Health*, Rodricks, J. V., Hesseltine, C. W., and Mehlman, M. A., Eds., Pathotox Publishers, Park Forest South, Ill., 1977, 532.
45. **Hamilton, P. B., Nelson, R. R., and Harris, B. S. H.**, Murine toxicity of *Cochliobolus carbonum*, *Appl. Microbiol.*, 16, 1719, 1968.
46. **Uraguchi, K., Saito, M., Noguchi, Y., Takahashi, K., Enomoto, M., and Tatsuno, T.**, Chronic toxicity and carcinogenicity in mice of the purified mycotoxins luteoskyrin and cyclochlorotine, *Food Cosmet. Toxicol.*, 10, 193, 1972.
47. **Holzapfel, C. W.**, The isolation and structure of cyclopiazonic acid, a toxic metabolite of *Penicillium cyclopium* Westling, *Tetrahedron*, 24, 2101, 1968.
48. **Büchi, G., Kitaura, Y., Yuan, S.-S., Wright, H. E., Clardy, J., Demain, A. L., Glinsukon, T., Hunt, N., and Wogan, G. N.**, Structure of cytochalasin E, a toxic metabolite of *Aspergillus clavatus*, *J. Am. Chem. Soc.*, 95, 5423, 1973.
49. **Singleton, V. L., Bohonos, N., and Ullstrup, A. J.**, Decumbin, a new compound from a species of *Penicillium*, *Nature (London)*, 181, 1072, 1958.
50. **Suzuki, Y., Tanaka, H., Aoki, H., and Tamura, T.**, Ascotoxin (decumbin), a metabolite of *Ascochyta imperfecta* Peck, *Agric. Biol. Chem. (Japan)*, 34, 395, 1970.
51. **Cole, R. J., Kirksey, J. W., Springer, J. P., Clardy, J., Cutler, H. G., and Garren, K. H.**, Desmethoxyviridiol, a new toxin from *Nodulisporium hinnuleum*, *Phytochemistry*, 14, 1429, 1975.
52. **Tullock, M., C. M. I.**, Mycological Papers No. 130, Commonwealth Mycology Institute, Kew, Surrey, England, 42.
53. **Bamburg, J. R., Marasas, W. F., Riggs, N. V., Smalley, F. H., and Strong, F. M.**, Toxic spiroepoxy compounds from *Fusaria* and other hyphomycetis, *Biotechnol. Bioeng.*, 10, 445, 1968.
54. **Stoll, A. and Hofmann, A.**, The ergot alkaloids, in *The Alkaloids*, Vol. 8, Manske, R. H. F., Ed., Academic Press, New York, 1965, 725.
55. **Kirksey, J. W. and Cole, R. J.**, New toxin from *Aspergillus flavus*, *Appl. Microbiol.*, 26, 827, 1973.
56. **Tarbell, D. S., Carman, R. M., Chapman, D. D., Huffman, K. R., and McCorkindale, N. J.**, The structure of fumigillin, *J. Am. Chem. Soc.*, 82, 1005, 1960.
57. **Iwata, K., Nagai, T., and Okuaaira, M.**, Fumigatoxin, a new toxin from a strain of Aspergillus fumigatus, *J. S. Afr. Chem. Inst.*, 22, S131, 1969.
58. **Yamazaki, M., Suzuki, S., and Miyaki, K.**, Tremorgenic toxins from *Aspergillus fumigatus*, *Fres. Chem. Pharm. Bull.*, 19, 1739, 1971.
59. **Yamazaki, M., Sasago, K., and Miyaki, K.**, The structure of fumitremorgen B (FTB), a tremorgenic toxin from *Aspergillus fumigatus*, *Chem. Commun.*, p. 408, 1974.
60. **Dix, D. T., Martin, J., and Moppett, C. E.**, Molecular structure of the metabolite lanulosin, *Chem. Commun.*, p. 1168, 1972.
61. **Ueno, Y., Ueno, I., Iitoi, Y., Tsunoda, H., Enomoto, M., and Ohtsubo, K.**, Toxicological approaches to the metabolites of *Fusaria*. III. Acute toxicity of fusarenon-x, *Jpn. J. Exp. Med.*, 41, 521, 1971.
62. **Ueno, Y., Ueno, I., Amakai, K., Ishikawa, Y., Tsunoda, H., Okubo, K., Saito, M., and Enomoto, M.**, Toxicological approaches to the metabolites of *Fusaria*. II. Isolation of Fusarenon-x from the culture filtrate of *Fusarium nivale* Fn2B, *Jpn. J. Exp. Med.*, 41, 507, 1971.
63. **Fisher, E. E., Kellock, A. W., and Wellington, N. A. M.**, Toxic strains of *Fusarium submorum* (W. G. Sm.) Sacc. from *Zea mays* L. associated with sickness in dairy cattle, *Nature (London)*, 215, 322, 1967.
64. **Bell, M. R., Johnson, J. R., Wildi, B. S., and Woodward, R. B.**, The structure of gliotoxin, *J. Am. Chem. Soc.*, 80, 1001, 1958.
65. **List, P. H. and Luft, P.**, Gyromitrin, das Gift der Fruehjahrslorchel, *Gyromitra (Helvella) esculenta* Fr., *Tetrahedron Lett.*, p. 1893, 1967.
66. **Okuda, S., Iwasaki, S., Sair, M. I., Machida, Y., Inoue, A., Tsuda, K., and Nakayama, Y.**, Stereochemistry of helvolic acid, *Tetrahedron Lett.*, 2295, 1967.
67. **Bamburg, J. R. and Strong, F. M.**, Mycotoxins of the trichothecane family produced by *Fusarium tricinctum* and *Trichoderma lignorum*, *Photochemistry*, 8, 2405, 1969.
68. **Hutchison, R. D., Steyn, P. S., and Thompson, D. L.**, The isolation and structure of 4-hydroxyochratoxin A and 7-carboxy-3,4-dehydro-8-hydroxy-3-methylisocoumarin from *Penicillium viridicatum*, *Tetrahedron Lett.*, p. 4033, 1971.

69. **Benedict, R. G., Tyler, V. E., and Brady, L. R.**, Chemotaxonomic significance of isoxazole derivatives, *Lloydia,* 29, 333, 1966.
70. **Marumo, S.**, Islanditoxin, a toxic metabolite produced by *Penicillium islandicum* Sopp, *Bull. Agric. Chem. Soc. Jpn.,* 19, 258, 1955.
71. **Pham, P., Chuong, V., Bouhet, J. C., Thiery, J., and Fromageot, P.**, Conformations of luteoskysin and rugulosin in solution, *Tetrahedron,* 29, 3533, 1973.
72. **Uraguchi, K., Saito, M., Noguchi, Y., Takahashi, K., Enomoto, M., and Tatsuno, T.**, Chronic toxicity and carcinogenicity in mice of the purified mycotoxins luteoskyrin and cyclochlorotine, *Food Cosmet. Toxicol.,* 10, 193, 1972.
73. **Arcamone, F., Bonnino, C., Chain, E. B., Ferretti, A., Minghetti, A., Pennella, P., Tonola, A., and Vero, L.**, Production of a new lysergic acid derivative in submerged culture by a strain of *Claviceps paspali* Stevens and Hall, *Proc. Royal Soc. (London) Ser. B,* 15, 26, 1961.
74. **Izuka, H. and Iida, M.**, Maltoryzine, a new toxic metabolite produced by a strain of *Aspergillus oryzae* var *microsporus* isolated from the poisonous malt sprout, *Nature (London),* 190, 681, 1962.
75. **Hoffer, A. and Osmond, H.**, *The Hallucinogens,* Academic Press, New York, 1967.
76. **Scheel, L. D., Perone, V. B., Larkin, R. L., and Kupel R. E.**, The isolation and characterization of two phototoxic furanocoumarins (psoralens) from diseased celery, *Biochemistry,* 2, 1127, 1963.
77. **Burkhardt, H. J. and Forgacs, J.**, O-Methylsterigmatocystin, a new metabolite from *Aspergillus flavus,* Link ex Fries, *Tetrahedron,* 24, 717, 1968.
78. **Wieland, T.**, Poisonous principles of mushrooms of the genus *Amanita, Science,* 159, 946, 1968.
79. **Wilkinson, S.**, The history and chemistry of muscarine, *Q. Rev. Chem. Soc. London,* 15, 153, 1961.
80. **Carter, S. B., Franklin, T. J., Jones, D. F., Leonard, B. J., Mills, S. D., Turner, R. W., and Turner, W. B.**, Mycophenolic acid: an anticancer compound with unusual properties, *Nature (London),* 223, 848, 1969.
81. **Logan, W. R. and Newbold, G. T.**, Lactones. V. Experiments relating to mycophenolic acid, *J. Chem. Soc.,* p. 1946, 1957.
82. **Lafont, P., Lafont, J., and Frayssinet, C.**, La nidulotoxine: toxine di *Aspergillus nidulans* Went., *Experientia,* 15, 61, 1970.
83. **Raistrick, H. and Stössl, X.**, Studies in the biochemistry of microorganisms. 104. Metabolites of *Penicillium atrovenetum,* O. Smith: β-nitropropionic acid, a major metabolite, *Biochem. J.,* 68, 647, 1958.
84. **Tatsuno, T., Fujimoto, Y., and Morita, Y.**, Toxicological research on substances from *Fusarium nivale.* III. The structure of nivalenol and its monoacetate, *Tetrahedron Lett.,* p. 2823, 1969.
85. **Purchase, I. F. H. and van du Watt, J. J.**, The long-term toxicity of ochratoxin A to rats, *Food Cosmet. Toxicol.,* 9, 681, 1971.
86. **Steyn, P. S. and Holzaplel, C. W.**, The isolation of methyl and ethyl esters of ochratoxin A and B, metabolites of *Aspergillus ochraceus* Wilh., *J. S. Afr. Chem. Inst.,* 20, 186, 1967.
87. **Krogh, P.**, Ochratoxins, in *Mycotoxins in Human and Animal Health,* Rodricks, J. V., Hesseltine, C. W., and Mehlman, M. A., Eds., Pathotox Publishers, Forest Park South, Ill., 1977, 489.
88. **Krogh, P., Hasselager, E., and Friis, P.**, Studies on fungal nephrotoxicity. II. Isolation of 2 nephrotoxic compounds from *Penicillium viridicatum* West., Citrinin and oxalic acid, *Acta Pathol. Microbiol. Scand. Sect. B,* 78, 401, 1978.
89. **Stott, W. T. and Bullerman, X.**, Patulin: a mycotoxin of potential concern in foods, *J. Milk Food Technol.,* 38, 695, 1975.
90. **Cole, R. J., Kirksey, J. W., and Wells, J. M.**, a new tremorgenic metabolite from *Penicillium paxilli, Can. J. Microbiol.,* 20, 1159, 1974.
91. **Springer, J. P., Clardy, J., Wells, J. M., Cole, R. J., and Kirksey, J. W.**, The structure of paxilline, a tremorgenic metabolite of *Penicillium paxilli* Bainier, *Tetrahedron Lett.,* p. 2531, 1975.
92. **Kurtzman, C. P. and Ciegler, A.**, Mycotoxin from a blue-eye mold of corn, *Appl. Microbiol.,* 20, 204, 1970.
93. **Ciegler, A. and Kurtzman, C. P.**, Penicillic acid production by blue-eye fungi on various agricultural commodities, *Appl. Microbiol.,* 20, 761, 1970.
94. **Wilson, B. J., Harris, T. M., and Hayes, A. W.**, Mycotoxin from *Penicillium puberulum, J. Bacteriol.,* 93, 1737, 1967.
95. **Steyn, P. S.**, The isolation, structure and absolute configuration of secalonic acid D, the toxic metabolite of *Penicillium oxalicum, Tetrahedron,* 26, 51, 1970.
96. **Howard, C. C. and Johnstone, R. A. W.**, Fungal metabolites. I. Steriochemical features and mass spectrometry of secalonic acids, *J. Chem. Soc., Perkins I,* 2033, 1973.
97. **Wei, R.-D., Stell, P. E., Smalley, E. B., Schnoes, H. K., and Strong, F. M.**, Isolation and partial characterization of a mycotoxin from *Penicillium roqueforti, Appl. Microbiol.,* 25, 111, 1973.
98. **Wei, R.-D., Schnoes, H. K., Hart, P. A., and Strong, F. M.**, The structure of PR toxin, a mycotoxin from *Penicillium roqueforti, Tetrahedron,* 31, 109, 1975.

99. **Moss, M. O., Wood, A. B., and Robinson, F. V.**, The structure of rubratoxin A, a toxic metabolite of *Penicillium rubrum, Tetrahedron Lett.*, p. 367, 1969.
100. **Rodricks, J. V. and Eppley, R. M.**, Stachybotrys and stachybotryotoxicosis, in *Mycotoxins,* Purchase, I. F. H., Ed., Elsevier, New York, 1974, chap. 9.
101. **Eppley, R. M. and Bailey, W. J.**, 12,13-Epoxy-Δ^9-trichothecenes as the probable mycotoxins responsible for stachybotryotoxicosis, *Science,* 181, 758, 1973.
102. **Richard, J. L., Pier, A. C., and Tiffany, L. H.**, Biological effects of toxic products from *Trichothecium roseum* Link, *Mycopathol. Mycol. Appl.*, 40, 161, 1970.
103. **Richard, J. L., Tiffany, L. H., and Pier, A. C.**, Toxigenic fungi associated with stored corn, *Mycopathol. Mycol. Appl.*, 38, 313, 1969.
104. **Engstrom, G. W., DeLance, J. V., Richard, J. L., and Baetz, A. L.**, Purification and characterization of roseotoxin B, a toxic cyclodepsipeptide from *Trichothecium roseum, J. Agric. Food Chem.*, 23, 244, 1975.
105. **Hayes, A. W. and Wilson, B. J.**, Bioproduction and purification of rubratoxin B, *Appl. Microbiol.*, 16, 1163, 1968.
106. **Takida, N., Seo, S., Ogihara, Y., Sankawa, U., Iitara, I., Kitagawa, I., and Shibata, S.**, Anthraquinoid coloring matters of *Penicillium islandicum* Sopp and some other fungi, Luteoskyrin, Rubroskyrin, rugulosin and their related compounds, *Tetrahedron,* 29, 3703, 1973.
107. **Aust, S. D., Broquist, H. P., and Rinehart, K. L., Jr.**, Slaframine: a parasympathomimetric from *Rhizoctonia leguminicola, Biotechnol. Bioeng.*, 10, 403, 1968.
108. **Taylor, A.**, The chemistry and biochemistry of sporidesmins and other 2,5-epidithia-3,6-dioxopiperazines, in *Biochemistry of Some Food-Borne Microbial Toxins,* Taylor, A., Ed., MIT Press, Cambridge, Mass., 1968, 69.
109. **Hodges, R. and Shannon, J. S.**, The isolation and structure of sporidesmin C, *Aust. J. Chem.*, 19, 1059, 1966.
110. **Rahman, R. and Taylor, A.**, A new metabolite of *Pithomyces chartarum* related to the sporidesmins, *Chem. Commun.*, p. 1032, 1967.
111. **Brewer, D., Rahman, R., Safe, S., and Taylor, A.**, A new toxic metabolite of *Pethomyces chartarum* related to the sporidesmins, *Chem. Commun.*, p. 1571, 1968.
112. **Rahman, R., Safe, S., and Taylor, A.**, Sporidesmins. IX. Isolation and structure of sporidesmin E, *J. Chem. Soc., Sect. C,* p. 1665, 1969.
113. **Jamieson, W. D., Rahman, R., and Taylor, A.**, Sporidesmins. VIII. Isolation and structure of sporidesmin D and sporidesmin F, *J. Chem. Soc. Sect. C,* p. 1564, 1969.
114. **Rahman, F. R., Safe, S., and Taylor, A.**, Sporidesmins. XII. Isolation and structure of sporidesmin G, a naturally occurring 3,6-epitetrathiopiperazine-2,5-dione, *J. Chem. Soc. Perkin I,* p. 470, 1972.
115. **Forjacs, J.**, Stachybotryotoxicosis and moldy corn toxicosis, in *Mycotoxins in Foodstuffs,* Wogan, G. N., Ed., MIT Press, Cambridge, Mass., 1965, 87.
116. **Scott, P. M. and Lawrence, J. W.**, Stemphone, a biologically active yellow pigment produced by *Stemphylium sarcinaeforme* (Cev.) Wiltshire, *Can. J. Microbiol.*, 14, 1015, 1968.
117. **Holzapfel, C. W., Purchase, I, F. H., Steyn, P. S., and Gouws, L.**, The toxicity and chemical assay of sterigmatocystin, a carcinogenic mycotoxin and its isolation from two new fungal sources, *S. Afr. Med. J.*, 40, 1100, 1966.
118. **Lillehoj, E. B. and Ciegler, A.**, Biological activity of sterigmatocystin, *Mycopathol. Mycol. Appl.*, 35, 373, 1968.
119. **Bamburg, J. R., Riggs, N. V., and Strong, F. M.**, The structures of toxins from two strains of *Fusarium tricinctum, Tetrahedron,* 24, 3329, 1968.
120. **Sheenan, J., Lawson, W., and Gaul, R.**, The structure of terreic acid, *J. Am. Chem. Soc.*, 80, 5536, 1958.
121. **Cole, R. J., Kirksey, J. W., Cox, R. H., and Clardy, J.**, Structure of the tremor-producing idole, TR-2, *J. Agric. Food Chem.*, 23, 1015, 1975.
122. **Wilson, B. J. and Wilson, C. H.**, Toxin from *Aspergillus flavus*: production on food materials of a substance causing tremors in mice, *Science,* 144, 177, 1964.
123. **Ciegler, A.**, Tremorgenic toxin from *Penicillium palitans, Appl. Microbiol.*, 18, 128, 1969.
124. **Hou, C. T., Ciegler, A., and Hesseltine, C. W. H.**, Tremorgenic toxins from Penicillia. II. A new tremorgenic toxin, Tremortin B, from *Penicillium palitans, Can. J. Microbiol.*, 17, 599, 1971.
125. **Wilson, B. J., Wilson, C. H., and Hayes, A. W.**, Tremorgenic toxin from *Penicillium cyclopium* grown on food materials, *Nature (London),* 220, 77, 1968.
126. **Bamburg, J. R., Marasas, W. F. O., Riggs, N. V., Smalley, E. B., and Strong, F. M.**, Toxic spiroepoxy compounds from *Fusaria* and other hyphomycetes, *Biotechnol. Bioeng.*, 10, 445, 1968.
127. Evans, R., Hanson, J. R., and Marten, T., Formation of the sesquiterpenoid trichothecin, *J. Chem. Soc. Perkin I,* p. 1212, 1976.
128. **Clardy, J., Springer, J. P., Büchi, G., Matsuo, K., and Wightman, R.**, Tryptoquivaline and tryptoquivalone, two tremorgenic toxins of *Aspergillus clavatus, J. Am. Chem. Soc.*, 97, 663, 1975.

129. Wu, M. T., Ayers, J. C., Koehler, P. E., and Chassis, G., Toxic metabolite produced by *Aspergillus wentii, Appl. Microbiol.*, 27, 337, 1974.
130. Wells, J. M., Cole, R. J., and Kirksey, J. W., Emodin, a toxic metabolite of *Aspergillus wentii* isolated from weevil-damaged chestnuts, *Appl. Microbiol.*, 30, 26, 1975.
131. Härri, E., Loeffler, W., Sigg, H. P., Stähelin, H., Stoll, C. H., Tamm, Ch., and Wiesing, D., Uber die Verrucarin und Roriden, eine Gruppe von cytostatisch hochwirksamen Antibiotika aus *Myrothecium*- arten, *Helv. Chim. Acta*, 45, 839, 1962.
132. Böhner, B., Fetz, E., Härri, E., Sigg, H. P., Stoll, C. H., and Tamm, Ch., Uber die Isolierung von Verrucarin H, Verrucarin J., Roridin D and Roridin E aus *Myrothecium*- arten, *Helv. Chim. Acta*, 48, 1079, 1965.
133. Cole, R. J., Kirksey, J. W., Moore, J. H., Blankenship, B. R., Diener, U. L., and Davis, N. D., Tremorgenic toxin from *Penicillium verruculosum, Appl. Microbiol.*, 24, 248, 1972.
134. Fayos, J., Lokensgard, D., Clardy, J., Cole, R. J., and Kirksey, J. W., Structure of verruculogen, a tremor producing peroxide from *Penicillium verruculosum, J. Am. Chem. Soc.*, 96, 6785, 1974.
135. Cole, R. J., Kirksey, J. W., and Morgan-Jones, G., Verruculotoxin, a new mycotoxin from *Penicillium verruculosum, Toxicol. Appl. Pharmacol.*, 31, 465, 1975.
136. Lillehoj, E. B. and Ciegler, A., A toxic substance from *Aspergillus* viridinutans, *Can. J. Microbiol.*, 18, 193, 1972.
137. Vesonder, R. F., Ciegler, A., and Jensen, A. H., Isolation of the emetic principle from *Fusarium*-infected corn, *Appl. Microbiol.*, 26, 1008, 1973.
138. Yoshizawa, T. and Marooka, N., Deoxynivalenol and its monoacetate: new mycotoxins from Fusarium roseum and moldy barley, *Agric. Biol. Chem.*, 37, 2933, 1973.
139. Yoshizawa, T. and Morooka, N., Acute toxicities of the new trichothecene mycotoxins: Deoxynivalenol and its monoacetate, *J. Food. Hyg. Soc. Jpn.*, 15, 261, 1974.
140. Mirocha, C. J., Christiansen, C. M., and Nelson, G. H., Estrogenic metabolite produced by *Fusarium graminearum* in stored corn, *Appl. Microbiol.*, 15, 497, 1967.
141. Stack, M. E., Eppley, R. M., Dreifuss, P. A., and Pohland, A. E., Isolation and identification of xanthomegnin, viomellein rubrosulphin and viopurpurin as metabolites of *Penicillium veridicatum, Appl. Environ. Microbiol.*, 33, 351, 1977.
142. Carlton, W. W., Stack, M. E., and Eppley, R. M., Hepatic alterations produced in mice by xanthomegnin and viomellein metabolites of *Penicillium viridicatum, Toxicol. Appl. Pharmacol.*, 38, 455, 1976.
143. Gallagher, R. T. and Latch, G. C. M., Production of the tremorgenic mycotoxins verruculogen and fumitremorgen B by *Penicillium piscarium* Westling, *Appl. Environ. Microbiol.*, 33, 730, 1977.
144. Buchi, G., Luk, K. C., Kobbe, B., and Townsend, J. M., Four new mycotoxins of *Aspergillus clavatus* related to tryptoquivaline, *J. Org. Chem.*, 42, 244, 1977.
145. Ishii, K., Pathre, S. V., and Mirocha, C. J., Two new trichothecenes from *Fusarium roseum, J. Agric. Food Chem.*, 26, 649, 1978.
146. Anderegg, R. J., Biemann, K., Büchi, G., and Cushman, M., Malformin C, a new metabolite of *Aspergillus niger, J. Am. Chem. Soc.*, 98, 3365, 1976.
147. Ilus, T., Ward, P. J., Nummi, M., Adlercreutz, H., and Gripenberg, J., A new mycotoxin from *Fusarium, Phytochemistry*, 16, 1839, 1977.
148. Hamasaki, T., Nakagomi, T., Hatsuda, Y., Fukuijama, K., and Katsube, Y., 5,6-Dimethoxysterigmatocystin, a new metabolite from *Aspergillus multicolor, Tetrahedron. Lett.*, No. 32, 2765, 1977.
149. Rabie, C. J., Steyn, M., and van Schalkwyk, G. C., New species of *Aspergillus* producing sterigmatocystin, *Appl. Environ. Microbiol.*, 33, 1023, 1977.
150. Lin, M. T. and Hsich, D. P. H., Averufin in the biosynthesis of aflatoxin B *J. Am. Chem. Soc.*, 95, 1668, 1973.
151. Donkersloot, J. A., Mateles, R. I., and Yang, S. S., Isolation of averufin from a mutant of *Aspergillus parasiticus* impaired in aflatoxin biosynthesis, *Biochem. Biophys. Res. Commun.*, 47, 1051, 1972.
152. Lee, L. S., Bennett, J. W., Cucullu, A. F., and Staley, J. B., Synthesis of versicolorin A by a mutant strain of *Aspergillus parasiticus* deficient in aflatoxin production, *J. Agric. Food Chem.*, 23, 1132, 1975.
153. Lee, L. S., Bennett, J. W., Cucullu, A. F., and Ory, R. L., Biosynthesis of aflatoxin B_1. Conversion of versicolorin A to aflatoxin B_1 by *Aspergillus parasiticus, J. Agric. Food Chem.*, 24, 1167, 1967.
154. Schroeder, H. W., Cole, R. J., Grigsby, R. D., and Hein, H., Jr., Inhibition of aflatoxin production and tentative identification of an aflatoxin intermediate "versional acetate" from treatment with dichlorvos, *Appl. Microbiol.*, 27, 394, 1974.
155. Cox, R. H., Churchill, F., Cole, R. J., and Dorner, J. W., Carbon-13 nuclear magnetic resonance studies of the structure and biosynthesis of versiconal acetate, *J. Am. Chem. Soc.*, 99, 3158, 1977.
156. Lee, L. S., Bennett, J. W., Goldblatt, L. A., and Lundin, R. E., Norsolorinic acid from a mutant strain of Aspergillus parasiticus, *J. Am. Oil Chem. Soc.*, 48, 93, 1971.

157. **Hsieh, D. P., Lin, M. T., Yao, R. C., and Singh, R.**, Biosynthesis of aflatoxin. Conversion of norsolorinic acid and other hypothetical intermediates into aflatoxin B_1, *J. Agric. Food Chem.*, 24, 1171, 1976.
158. **Macmillan, J. G., Springer, J. P., Clardy, J., Cole, R. J., and Kerksey, J. W.**, Structure and synthesis of verruculotoxin, a new mycotoxin from *Penicillium verruculosum* Peyronel, *J. Am. Chem. Soc.*, 98, 246, 1976.
159. **Cole, R. J., Dorner, J. W., Lansden, J. A., Cox, R. H., Pape, C., Cunfer, B., Nicholson, S. S., and Bedell, D. M.**, Paspalum staggers: isolation and identification of tremorgenic metabolites from sclerotia of *Claviceps paspali*, *J. Agric. Food Chem.*, 25, 1197, 1977.
160. **Harvan, D. J. and Pero, R. W.**, The structure and toxicity of the Alternaria metabolites, in *Mycotoxins and Other Fungal Related Food Problems*, Advances in Chemistry Series 149, Rodricks, J. V., Ed., American Chemical Society, Washington, D.C., 1976, chap. 15.
161. **Raistrick, H., Stickings, C. E., and Thomas, R.**, Alternariol and alternariol monomethyl ether, metabolic products of *Alternaria tenuis*, *Biochemistry*, 55, 421, 1953.
162. **Burroughs, R., Seitz, L. M., Sauer, D. B., and Mohr, H. E.**, Effect of substrate on metabolite production by *Alternaria alternata*, *Appl. Environ. Microbiol.*, 31, 685, 1976.
163. **Schroeder, H. W. and Cole, R. J.**, Natural occurrence of alternariols in discolored pecans, *J. Agric. Food Chem.*, 25, 204, 1977.
164. **Seitz, L. M., Sauer, D. B., Mohr, H. E., Burroughs, R., and Paukstelis, J. V.**, Metabolites of *Alternaria* in grain sorgham. Compounds which could be mistaken for zearalenone and aflatoxin, *J. Agric. Food Chem.*, 23, 1, 1975.
165. **Seitz, L. M., Sauer, D. B., Mohr, H. E., and Burroughs, R.**, Weathered grain sorghum: natural occurrence of alternariols and storability of the grain, *Phytopathology*, 65, 1259, 1975.
166. **Weisleder, D. and Lillehoj, E. B.**, Structure of viriditoxin, a toxic metabolite of *Aspergillus viridinutans*, *Tetrahedron Lett.*, No. 48, 4705, 1971.
167. **Schroeder, H. W. and Hein, H., Jr.**, Natural occurrence of sterigmatocystin in in-shell pecans, *Can. J. Microbiol.*, 23, 639, 1977.
168. **Ciegler, A., Mintzlaff, H.-J., Weisleder, D., and Leistner, L.**, Potential production and and detoxification of penicillic acid in mold-fermented sausage (salami), *Appl. Microbiol.*, 24, 114, 1972.
169. **Eppley, R. M., Mazzola, E. P., Highet, R. J., and Bailey, W. J.**, Structure of satratoxin H, a metabolite of *Stachybotrys atra*. Application of proton and carbon-13 nuclear magnetic resonance, *J. Org. Chem.*, 42, 240, 1977.
170. **Marasas, W. F. O., Adelaar, T. F., Kellerman, T. S., Minni, J. A., van Rensburg, I. B. J., and Burroughs, G. W.**, First report of facial eczema in sheep in South Africa, *Onders. J. Vet. Res.*, 39, 107, 1972.
171. **Synge, R. L. M. and White, E. P.**, Photosensitivity diseases in New Zealand. XXIII. Isolation of sporidesmin, a substance causing lesions characteristic of facial eczema from *Sporidesmium bakeri*, Syd., New Zealand, *J. Agric. Res.*, 3, 907, 1960.
172. **Janes, B. S.**, *Spordesmium bakeri* recorded from Victoria, Australia, *Nature (London)*, 184, 1327, 1959.
173. **Flynn, D. M., Hore, D., Leaver, D. D., and Fischer, E. E.**, Facial eczema, *J. Agric. Vict. Dep. Agric.*, 60, 49, 1962.
174. **Purchase, I. F. H. and Pretorius, M. E.**, Sterigmatocystin in coffee beans, *J. Am. Oil Chem. Soc.*, 57, 225, 1973.
175. **Scott, P. M.**, Patulin, in *Mycotoxins*, Purchase, T. F. H., Ed., Elsevier Scientific Press, New York, 1974, chap. 18.
176. **Ciegler, A.**, Patulin, in *Mycotoxins in Human and Animal Health*, Rodricks, J. V., Hesseltine, C. W., and Mehlman, M. A., Eds., Pathotox Publishers, Park Forest South, Ill., 1977, 609.
177. **Brian, P. W., Elson, G. W., and Lowe, D.**, Production of patulin in apple fruits by *Penicillium expansum*, *Nature (London)*, 178, 263, 1956.
178. **Drilleau, J.-F. and Bohuon, G.**, La patuline dans les products cidricoles, *C. R. Seances Acad. Agric. Fr.*, 59, 1031, 1973.
179. **Eyrich, W.**, Zum nachweis von Patulin in Apfelsaft, *Chem. Mikrobiol. Technol. Lebensm.*, 4, 17, 1975.
180. **Harwig, J., Chen, Y.-K., Kennedy, B. P. O., and Scott, P. M.**, Occurrence of patulin-producing strains of *Penicillium expansum* in natural rots of apple in Canada, *Can. Inst. Food Sci. Technol.*, 6, 22, 1973.
181. **Wilson, D. M. and Nuovo, G. J.**, Patulin production in apples decayed by *Penicillium expansum*, *Appl. Microbiol.*, 26, 124, 1973.
182. **Ware, G. M., Thorpe, C. W., and Pohland, A. E.**, Liquid chromatographic method for determination of patulin in apple juice, *J. Assoc. Off. Anal. Chem.*, 57, 1111, 1974.
183. **Frank, H. K., Orth, R., and Hermann, R.**, Patulin in Lebensmitteln pflanzlicher Herkunft. I. Kernobst und daraus hergestellte Produkte, *Z. Lebensm. Unters. Forsch.*, 162, 149, 1976.

184. **Scott, P. M., Fulcki, T., and Harwig, J.**, Patulin content of juice and wine produced from moldy grapes, *J. Agric. Food Chem.*, 25, 434, 1977.
185. **Scott, P. M., Miles, W. F., Toft, P., and Dube, J. G.**, Occurrence of patulin in apple juice, *J. Agric. Food Chem.*, 20, 450, 1972.
186. **Frank, H. K., Orth, R., and Figge, A.**, Patulin in Lebensmitteln pflanzlicher Herkunft. II. Verschiedene Obstarten, Gemüse und daraus hergestellte Produkte, *Z. Lebensm. Unters. Forsch.*, 163, 111, 1977.
187. **Josefsson, E. and Andersson, A.**, Patulin i äppledrycker, Särtryck ur Vår föda, 28, 189, 1977.
188. **Barnes, J. M., Carter, R. L., Peristianes, Austwick, P. K. C., Flynn, F. V., and Aldridge, W. N.**, Balkan (endemic) nephropathy and a toxin-producing strain of *Penicillium verrucosum* var. *cyclopium:* an experimental model in rats, *Lancet*, p. 671, 1977.
189. **Elling, F., Hald, B., Jacobsen, C., and Krogh, P.**, Spontaneous toxic nephropathy in poultry associated with ochratoxin A, *Acta Pathol. Microbiol. Scand.*, Sect. A, 83, 739, 1975.
190. **Krogh, P., Hald, B., Plestina, R., and Ceovic, S.**, Balkan (endemic) nephropathy and foodborn ochratoxin A: preliminary results of a survey of foodstuffs, *Acta Pathol. Microbiol. Scand. Sect. B*, 85, 238, 1977.
191. **Krogh, P.**, Ochratoxin A residues in tissues of slaughter pigs with nephropathy, *Nord. Vet. Med.*, 29, 402, 1977.
192. **Krogh, P. and Elling, F.**, Mycotoxic nephropathy, *Vet. Sci. Commun.*, 1, 51, 1977.
193. **Shotwell, O. L., Hesseltine, C. W., and Goulden, M. L.**, Ochratoxin A: occurrence as a natural contaminant of a corn sample, *Appl. Microbiol.*, 17, 765, 1969.
194. **Krogh, P., Hald, B., and Pedersen, E. J.**, Occurrence of ochratoxin A and citrinin in cereals associated with mycotoxic porcine nephropathy, *Acta Pathol. Microbiol. Scand. Sect. B*, 81, 689, 1973.
195. **Prior, M. G.**, Mycotoxin determinations on animal feedstuffs and tissues in Western Canada, *Can. J. Comp. Med.*, 40, 75, 1976.
196. **Uchiyama, M., Isohata, E., and Takeda, Y.**, A case report on the detection of ochratoxin A from rice, *J. Food Hyg. Soc.*, 17, 103, 1976.
197. **Scott, P. M., Walbeck, W., van, Kennedy, B., and Anyetti, D.**, Mycotoxins (ochratoxin A, citrinin, sterigmatocystin) and toxigenic fungi in grains and other agricultural products, *J. Agric. Food Chem.*, 20, 1103, 1972.
198. **Shotwell, O. L., Goulden, M. L., and Hesseltine, C. W.**, Survey of U.S. wheat for ochratoxin and aflatoxin, *J. Assoc. Off. Anal. Chem.*, 59, 122, 1976.
199. **Suzuki, S., Kozuka, Y., Satoh, T., and Yamaki, M.**, Studies on the nephrotoxicity of ochratoxin A in rats, *Toxicol. Appl. Pharmacol.*, 34, 479, 1975.
200. **Brown, M. H., Szczech, G. M., and Purmalis, B. P.**, Teratogenic and toxic effects of ochratoxin A in rats, *Toxicol. Appl. Pharmacol.*, 37, 331, 1976.
201. **Kanisawa, M., Suzuki, S., Kozuka, Y., and Yamazaki, M.**, Histopathological studies on the toxicity of ochratoxin A in rats. I. Acute oral toxicity, *Toxicol. Appl. Pharmacol.*, 42, 55, 1977.
202. **Chang, F. C. and Chu, F. S.**, The fats of ochratoxin A in rats, *Food Cosmet. Toxicol.*, 15, 199, 1977.
203. **Wu, C. M., Koehler, P. E., and Ayres, J. C.**, Isolation and identification of xanthotoxin (8-methoxypsoralen) from celery infected with *Sclerotinia sclerotiorum*, *Appl. Microbiol.*, 23, 852, 1972.
204. **Birmingham, D. J., Key, M. M., Tubich, G. E., and Perone, V. B.**, Phototoxic bullae among harvesters, *Arch. Dermatol.*, 83, 73, 1962.
205. **Musazo, L., Caporale, G., and Rodighiero, G.**, The isolation of bergapten from celery and parsley, *Gaz. Chem. Ital.*, 84, 870, 1954.
206. **Floss, H., Guenther, H., and Hadwiger, L. A.**, Biosynthesis of furanocoumarins in disease celery, *Phytochemistry*, 8, 585, 1969.
207. **Ueno, I., Hayashi, T., and Ueno, Y.**, Pharmacokinetic studies on the hepatotoxicity of luteoskyrin. I. Intracellular distribution of radioactivity on the liver of mice administered ^{3}H-luteoskyrin, *Jpn. J. Pharmacol.*, 24, 535, 1974.
208. **Ueno, Y. and Ishikawa, I.**, Production of luteoskyrin, a hepatotoxic pigment by *Penicillium islandicum* Sopp, *Appl. Microbiol.*, 18, 406, 1969.
209. **Vouhet, J.-C., van Chuong, P. P., Toma, F., Kirszenbaum, M., and Fromegeot, P.**, Isolation and characterization of luteoskyrin and rugulosin, two hepatotoxic anthraquinoids from *Penicillium islandicum* Sopp and *Penicillium rugulosum* Thom., *J. Agric. Food Chem.*, 24, 964, 1976.
210. **Marumo, S.**, Islanditoxin, a toxic metabolite produced by *Penicillium islandicum* Sopp. III. Structure of islanditoxin, *Bull. Agric. Chem. Soc. Jpn.*, 23, 428, 1959.
211. **Schroeder, H. W., Cole, R. J., Hein, H., Jr., and Kirksey, J. W.**, Tremorgenic mycotoxins from *Aspergillus crespitosus*, *Appl. Microbiol.*, 29, 857, 1975.
212. **Eickman, N., Clardy, J., Cole, R. J., and Kirksey, J. W.**, The structure of fumitremorgen A, *Tetrahedron Lett.*, p. 1051, 1975.

213. Grove, D. F. and Mortimer, P. H., The cytotoxicity of some transformation products of diacetoxyscirpenol, *Biochem. Pharmacol.*, 18, 1473, 1969.
214. Cole, M. and Robinson, G. N., Microbial metabolites with insecticidal properties, *Appl. Microbiol.*, 24, 660, 1972.
215. Glinsukon, T., Yuan, S. S., Wightman, R., Kitaura, Y., Buchi, G., Shank, R. C., Wogan, G. N., and Christensen, Isolation and purification of cytochalasin E and two tremorgens from *Aspergillus clavatus*, *Plant Foods for Man*, 1, 113, 1974.
216. Demain, A. L., Hunt, N. A., Melik, V., Kobbe, B., Hawkins, H., Matsuo, K., and Wogan, G. N., Improved procedure for production of cytochalasin E and tremorgenic mycotoxins by *Aspergillus clavatus*, *Appl. Environ. Microbiol.*, 31, 138, 1976.
217. Ohmono, S., Sugita, M., and Abe, M., Isolation of cyclopiazonic acid, cyclopiazonic acid imine and bissecodehydrocyclopiazonic acid from the cultures of *Aspergillus vers color* (Vuill.) Tiraboshi, *J. Agric. Chem. Soc. (Japan)*, 47, 57, 1973.
218. Purchase, I. F. H., The acute toxicity of the mycotoxin cyclopiazonic acid to rats, *Toxicol. Appl. Pharmacol.*, 18, 114, 1971.
219. Luk, K. C., Koble, B., and Townsend, J. M., Production of cyclopiazonic acid by *Aspergillus flavus* Link, *Appl. Environ. Microbiol.*, 33, 211, 1977.
220. Ihu, K. and Ueno, Y., Production of the hepatotoxic chlorine-containing peptide by *Penicillium islandicum* Sopp, *Appl. Microbiol.*, 26, 359, 1973.
221. Ghosh, A. C., Manmade, A., Townsend, J. M., Bousquet, A., Howes, J. F., and Demain, A. L., Production of cyclochlorotine and a new metabolite Simatoxin by *Penicillium islandicum* Sopp, *Appl. Environ. Microbiol.*, 35, 1074, 1978.
222. Jordan, W. H. and Carlton, W. W., Citrinin mycotoxicosis in the mouse, *Food Cosmet. Toxicol.*, 15, 29, 1977.
223. Washburn, K. W., Ames, D., Wyatt, R. D., Effect of dietary citrinin and aflatoxin on the problems of broilers growing on wire floors, *Poultry Sci.*, 55, 1977, 1976.
224. Ames, D. D., Wyatt, R. D., Marks, H. L., and Washburn, K. W., Effect of citrinin, a mycotoxin produced by *Penicillium citrinum* on laying hens and young broiler chicks, *Poultry Sci.*, 55, 1294, 1976.
225. Shinohara, Y., Arai, M., Hirao, K., Sugihara, S., Nakanishi, K., Tsunoda, H., and Ito, N., Combination effect of citrinin and other chemicals on rat kidney tumorigenesis, *Gann*, 67, 147, 1976.
226. Subrahmanyam, P. and Rao, A. S., Occurrency of aflatoxins and citrinin in groundnut (*Arachis hypogaea* L.) at harvest in relation to pod condition and kernel moisture content, *Curr. Sci.*, 43, 717, 1977.
227. Ueno, Y. and Ueno, I., Isolation and acute toxicity of citreoviriden, a neurotoxic mycotoxin of *Penicillium citreo-viride* Biourge, *Jpn. J. Exp. Med.*, 42, 91, 1972.
228. Nagel, D. W., Steyn, P. S., and Scott, D. B., Production of citreoviriden by *Penicillium pulvillorum*, *Phytochemistry*, 11, 627, 1972.
229. Tookey, H. L., Yates, S. G., Ellis, J. J., Grove, M. D., and Nichols, R. E., Toxic effects of a butenolide mycotoxin and of *Fusarium tricinctum* cultures in cattle, *J. Am. Vet. Med. Assoc.*, 160, 1522, 1972.
230. Ojima, N., Takenaka, S., and Seto, S., New butenolides from *Aspergillus terreus*, *Phytochemistry*, 12, 2527, 1973.
231. MacDonald, J. C., Toxicity, analysis and production of aspergillic acid and its analogues, *Can. J. Biochem.*, 51, 1311, 1973.
232. Hsieh, D. P. H., Daleqios, J. I., Krieger, R. I., Masri, M. S., and Haddon, W. F., Use of monkey liver microsomes in production of aflatoxin Q_1, *J. Agric. Food Chem.*, 22, 515, 1974.
233. Masri, S. M., Booth, A. N., and Hsieh, D. P. H., Comparative metabolic conversion of aflatoxin B_1 to M_1 and Q_1 by monkey, rat and chicken liver, *Life Sci.*, 15, 203, 1974.
234. Büchi, G., Luk, K.-C., and Muller, P. M., Synthesis of aflatoxin Q_1, *J. Org. Chem.*, 40, 3458, 1975.
235. Dalezios, J. I. and Wogan G. N., Metabolism of aflatoxin B_1 in Rhesus monkeys, *Cancer Res.*, 32, 2297, 1972.
236. Dalezios, J. I., Hsieh, D. P. H., and Wogan, G. N., Excretion and metabolism of orally administered aflatoxin B_1 in Rhesus monkeys, *Food Cosmet. Toxicol.*, 11, 606, 1973.
237. Campbell, T. C. and Hayes, J. R., The role of aflatoxin metabolism in its toxic lesion, *Toxicol. Appl. Pharmacol.*, 35, 199 and 209, 1976.
238. Canton, J. H., Kroes, R., van Logten, M. J., van Schothorst, M., Stavenuiter, J. F. C., and Verhülsdonk, The carcinogenicity of aflatoxin M_1 in rainbow trout, *Food Cosmet. Toxicol.*, 13, 441, 1975.
239. Holzapfel, C. W. and Steyn, P. S., Isolation and structure of aflatoxins M_1 and M_2, *Tetrahedron Lett.*, p. 2799, 1966.
240. Masri, M. S., Lundin, R. E., Page, J. R., and Garcia, V. C., Crystalline aflatoxin M_1 from urine and milk, *Nature (London)*, 215, 753, 1967.

241. Neumann-Kleinpaul, A. and Terplan, G., Zum Vorkommen von Aflatoxin M_1 in Trockenmilch-produkten, *Arch. Lebensm. Hyg.*, p. 128, 1972.
242. **Jacquet, J., Boutibonnes, P., and Teherani, A.**, Sur la présence des flavatoxines dans les abonents des animaux et dans les aliments d'origine animals distinés à l'homme, *Bull. Acad. Vet.*, 43, 36, 1970.
243. **Kiermeir, F. and Mücke, W.**, Uber den Nachweis von Aflatoxin M in Milch, *Z. Lebensm. Unters. Forsch.*, 150, 137, 1972.
244. **Yadagiri, B. and Tulpule, P. G.**, Aflatoxin in buffalo milk, *Indian J. Dairy Sci.*, 27, 293, 1974.
245. **Paul, R., Kalra, M. S., and Singh, A.**, Incidence of aflatoxins in milk and milk products, *Indian J. Dairy Sci.*, 29, 318, 1976.
246. **Purchase, I. F. H. and Vorster, L. J.**, Aflatoxin in commercial milk samples, *S. Afr. Med. J.*, 42, 219, 1968.
247. **Lee, L. S., Stanley, J. B., Cucullu, A. F., Pons, W. A., Jr., and Goldblatt, L. A.**, Ammoniation of aflatoxin B_1: isolation and identification of the major reaction product, *J. Assoc. Off. Anal. Chem.*, 57, 626, 1974.
248. **Gurtoo, H. L. and Dahms, R.**, On the nature of the binding of aflatoxin B_{2a} to rat hepatic microsomes, *Res. Commun. Chem. Pathol. Pharmacol.*, 9, 107, 1974.
249. **Patterson, D. S. P. and Roberts, B. A.**, The formation of aflatoxins B_{2a} and G_{2a} and other degradation products during the in vitro detoxification of aflatoxin by livers of certain avian and mammalian species, *Food Cosmet. Toxicol.*, 8, 527, 1970.
250. **Roebuck, B. D. and Wogan, G. N.**, Species comparison of in vitro metabolism of aflatoxin B_1, *Cancer Res.*, 37, 1649, 1977.
251. **Salhab, A. S. and Edwards, G. S.**, Comparative in vitro metabolism of aflatoxicol by liver preparations from animals and humans, *Cancer Res.*, 37, 1016, 1977.
252. **Ayres, J. C.**, Aflatoxins as contaminants of feeds, fish, and foods, in *Microbial Safety of Fishery Products*, Chichester, C. O. and Graham, H. E., Eds., Academic Press, 1973, 261.
253. **Edds, G. T.**, Acute aflatoxicosis, a review, *J. Am. Vet. Med. Assoc.*, 162, 304, 1973.
254. **Wogan, G. N.**, Biochemical effects of aflatoxins, *Israel J. Med. Sci.*, 10, 441, 1974.
255. **Patterson, D. S. P.**, Structure, metabolism, and toxicity of the aflatoxins: a review, in *Mycotoxins*, Soc. Nutr. Diet. Langue Francaise, Paris, 1976, 71.
256. **Jones, E. D.**, Aflatoxin in Feeding Stuffs, Its Incidence, Significance and Control, in Animal Feeds of Tropical and Subtropical Origin, Tropical Products Institute Conference, 1974.
257. **Schoenhard, G. L., Lee, D. J., Howell, S. E., Pawlowski, Libbey, L. M., and Sinnhuber, R. O.**, Aflatoxin B_1 metabolism to aflatoxicol and derivatives lethal to *Bacillus subtilis* GSY 1057 by rainbow trout (*Salmogardneri*) liver, *Cancer Res.*, 36, 1048, 1976.
258. **Suzangar, M., Emami, A., and Barnett, R.**, Aflatoxin contamination of village milk in Isfahan, Iran, *Trop. Sci.*, 18, 155, 1976.
259. **Luck, H., Steyn, M., and Wehner, F. C.**, A survey of milk powder for aflatoxin content, *S. Afr. J. Dairy Technol.*, 8, 85, 1976.
260. **Polzholer, K.**, Aflatoxinbestimmung in Milch und Milchprodukten, *Z. Lebensm. Unters. Forsch.*, 163, 175, 1977.
261. **Kiermeier, F., Weiss, G., Behringer, G., Miller, M., and Ranfft, K.**, Vorkommen und Gehalt an Aflatoxin M_1 in Molkerei-Anbeferungsmilch, *Z. Lebensm. Unters. Forsch.*, 163, 171, 1977.
262. **Shotwell, O. L., Hesseltine, C. W., Burmeister, H. R., Kwolek, W. F., Shannon, G. H., and Hall, H. H.**, Survey of cereal grains and soybeans for the presence of aflatoxin. II. Corn and soybeans, *Cereal Chem.*, 46, 454, 1969.
263. **Shotwell, O. L., Hesseltine, C. W., Vandegraft, E. E., and Goulden, M. L.**, Survey of corn from different regions for aflatoxin, ochratoxin and zearaleonone, *Cereal Sci. Today*, 16, 266, 1971.
264. **Shotwell, O. L., Hesseltine, C. W., and Goulden, M. L.**, Incidence of aflatoxin in Southern corn, 1969-1970, *Cereal Sci. Today*, 18, 192, 1973.
265. **Rambo, G. W., Tuite, J., and Caldwell, R. W.**, *Aspergillus flavus* and aflatoxin in preharvest corn from Indiana in 1971-1972, *Cereal Chem.*, 51, 848, 1974.
266. **Lillehoj, E. B., Kwolek, W. F., Shannon, G. M., Shotwell, O. L., and Hesseltine, C. W.**, Aflatoxin occurrence in 1973 corn at harvest. I. A limited survey in the Southeastern U.S., *Cereal Chem.*, 52, 603, 1975.
267. **Anderson, H. W., Nehring, E. W., and Wichser, W. R.**, Aflatoxin contamination of corn in field, *J. Agric. Food Chem.*, 23, 795, 1975.
268. **Hunt, W. H., Semper, R. C., and Liebe, E. B.**, Incidence of aflatoxin in corn and a field method for aflatoxin analysis, *Cereal Chem.*, 53, 227, 1976.
269. **Zuber, M. S., Calvert, O. H., Lillehoj, E. B., and Kwolek, W. F.**, Preharvest development of aflatoxin B_1 in corn in the U.S., *Phytopathology*, 66, 1120, 1976.

330. **Sekita, S., Yoshihira, K., Natori, S., and Kuwano, H.**, Structures of chaetoglobosins C, D, E and F, cytotoxic indol-3-YL-[13] cytochalasins from *Chaetomium globosum, Tetrahedron Lett.*, (17), 1351, 1976.

331. **Sekita, S., Yoshihira, K., Natori, S., and Kuwano, H.**, Chaetoglobosins G and J, cytotoxic indol-3-YL[13]-cytochalasins from *Chaetomium globosum, Tetrahedron Lett.*, (32), 2771, 1977.

332. **Riche, C., Pascard-Billy, C., Devys, M., Gaudemer, A., Barbier, M., and Bousquet, J.-F.**, Structure cristalline et moleculaire de la phomenone, phytotoxine produite par le champignon *Phoma exigua ver. non oxydabilis, Tetrahedron Lett.*, (32), 2765, 1974.

333. **Moule, Y., Moreau, S., and Bousquet, J. F.**, Relationship between the chemical structure and the biological properties of some eremophilane compounds related to PR toxin, *Chem. Biol. Interact.*, 17, 185, 1977.

334. **Lhoest, G., Bettencourt, A., Mercier, P., and Roberfroid, M.**, Identification by mass spectrometry of a metabolite of aflatoxin B_1, *Pharm. Acta Helv.*, 50, 293, 1975.

335. **Steyn, P. S., Vleggaar, R., Pitout, M. J., Steyn, H., and Thiel, P. G.**, 3-Hydroxyaflatoxin B_1: a new metabolite of in vitro aflatoxin B_1 metabolism by vervet monkey (Cercopithecus acthiops) liver, *J. Chem. Soc. Perkin Trans. I*, 22, 2551, 1974.

336. **Hobson, W., Bailey, J., Kowalk, A., and Fuller, X.**, Structure activity relationships in zearalenones, in *Mycotoxins in Human and Animal Health*, Rodricks, J., Hesseltine, C. W., and Mehlman, M. A., Eds., Pathotox Publishers, Park Forest South, Ill., 1977, 379.

337. **Beno, M. A., Cox, R. H., Wells, J. M., Cole, R. J., Kirksey, J. W., and Christoph, G. G.**, Structure of a new cytochalasin, cytochalasin H or kodo-cytochalasin-1, *J. Am. Chem. Soc.*, 99, 4123, 1977.

338. **Wells, J. M., Cutler, H. G., and Cole, R. J.**, Toxicity and plant growth regulator effects of cytochalasin H isolated from *Phomopsis* sp., *Can. J. Microbiol.*, 22, 1137, 1976.

339. **Eppley, R. M., Mazzola, E. P., Highet, R. J., and Bailey, W. J.**, Structure of satratoxin H, a metabolite of *Stachybotrys atra.* Application of proton and carbon 13 nuclear magnetic resonance, *J. Org. Chem.*, 42, 240, 1977.

340. **Landsen, J. A., Cole, R. J., Dorner, J. W., Cox, R. H., Cutler, H. G., and Clark, J. D.**, A new trichothecene mycotoxin isolated from *Fusarium tricinctum, J. Agric. Food Chem.*, 26, 246, 1978.

341. **Spilsburgh, J. F. and Wilkinson, S.**, The isolation of festuclavine and two new clavine alkaloids from *Aspergillus fumigatus* Fres., *J. Chem. Soc.*, p. 2085, 1961.

342. **Cole, R. J., Kirksey, J. W., Corner, J. W., Wilson, D. M., Johnson, J. C., Jr., Johnson, A. N., Bedell, D. M., Springer, J. P., Chexal, K. K., Clardy, J. C., and Cox, R. H.**, Mycotoxins produced by *Aspergillus fumigatus* species isolated from molded silage, *J. Agric. Food Chem.*, 25, 826, 1977.

343. **Howard, C. C. and Johnstone, R. A. W.**, Fungal metabolites. I. Stereochemical features and mass spectrometry of secalonic acids, *J. Chem. Soc. Perkin I*, p. 2033, 1973.

344. **Howard, C. C. and Johnstone, R. A. W.**, Fungal metabolites. III. Isolation of secalonic acids from *Phoma terrestris, J. Chem. Soc. Perkin I*, p. 2440, 1973.

345. **Harada, M., Yano, S., Watanabe, H., Yamazaki, M., and Miyaki, K.**, Phlogistic activity of secalonic acid A, *Chem. Pharm. Bull. (Japan)*, 22, 1600, 1974.

346. **Hooper, J. W., Marlow, W., Whalley, W. B., Borthwick, A. D., and Bowden, R.**, The chemistry of fungi. LXV. The structures of ergochrysin A, isoergochrysin A and ergoxanthin and of secalonic acids A, B, C and D, *J. Chem. Soc.*, (C), 3580, 1971.

347. **Yamazaki, M., Maebayashi, Y., and Miyaki, K.**, The isolation of secalonic acid A from *Aspergillus ochraceus* cultured on rice, *Chem. Pharm. Bull.*, 19, 199, 1971.

348. **Howard, C. C. and Johnstone, R. A. W.**, Fungal metabolites. VI. Crystal and molecular structure of secalonic acid A, *J. Chem. Soc. Perkin I*, p. 1820, 1976.

349. **Howard, C. C. and Johnstone, R. H. W.**, Isolation of a new secalonic acid, *J. Chem. Soc. Commun.*, p. 464, 1973.

350. **Andersen, R., Buchi, G., Kobbe, B., and Demain, A. L.**, Secalonic acids D and F are toxic metabolites of *Aspergillus aculeatus, J. Org. Chem.*, 42, 352, 1977.

351. **Hatsuda, Y., Hamasaki, T., Ishida, M., Matsui, K., and Hara, S.**, Dihydrosterigmatocystin and dihydrodemethylsterigmatocystin, new metabolites from *Aspergillus versicolor, Agric. Biol. Chem. (Japan)*, 36, 521, 1972.

352. **Lansden, J. A., Clarkson, R. J., Neeiy, W. C., Cole, R. J., and Kirksey, J. W.**, Spectroanalytical parameters of fungal metabolites. IV. Moniliformin, *J. Assoc. Off. Anal. Chem.*, 57, 1392, 1974.

353. **Kriek, N. P. J., Marasas, W. F. O., Steyn, P. S., van Rensburg, S. J., and Steyn, M.**, Toxicity of a moniliformin-producing strain of *Fusarium moniliforme* var. *subglutinans* isolated from maize, *Food Cosmet. Toxicol.*, 15, 579, 1977.

354. **Rabie, C. J., Lubben, A., Louw, A. I., Rathbone, E. B., Steyn, P. S., and Vleggaar, R.**, Moniliformin, a mycotoxin from *Fusarium fusariodes, J. Agric. Food Chem.*, 26, 375, 1978.

355. **Steyn, M., Thiel, P. G., and van Schalkwyk, G. C.**, Isolation and purification of moniliformin, *J. Assoc. Off. Anal. Chem.*, 61, 578, 1978.

356. **Cole, R. J., Cutler, H. G., Doupnik, B. L., and Pecknam, J. C.,** Toxin from *Fusarium moniliforme:* effect on plants and animals, *Science,* 179, 1324, 1973.
357. **Springer, J. P., Clardy, J., Cole, R. J., Kirksey, J. W., Hill, R. K., Carlson, R. M., and Isidor, J. L.,** Structure and synthesis of moniliformin, a novel cyclobutane microbial toxin, *J. Am. Chem. Soc.,* 96, 2267, 1974.
358. **Marasas, W. F. O., Kriek, N. P. J., van Rensburg, S. J., Steyn, M., and van Schalkwyk, G. C.,** Occurrence of zearalenone and deoxynivalenol mycotoxins produced by *Fusarium graminearum* Schwabe, in maize in Southern African, *S. Afr. J. Sci.,* 73, 346, 1977.
359. **Scott, P. M., Panalaks, T., Kanhere, S., and Miles, W. F.,** Determination of high pressure liquid chromatography, and gas liquid chromatography/high resolution mass spectrometry, *J. Assoc. Off. Anal. Chem.,* 61, 593, 1978.
360. **Schroeder, H. W. and Hein, H., Jr.,** A note on zearalenone in grain sorghum, *Cereal Chem.,* 52, 751, 1975.
361. **Mirocha, C. J., Harrison, J., Nichols, A. A., and McClintock, M.,** Detection of a fungal estrogen (F-2) in hay associated with infertility in dairy cattle, *Appl. Microbiol.,* 16, 797, 1968.
362. **Osegovie, L.,** Moldy maize poisoning in pigs, *Veterinaria (Sarajevo),* 19, 525, 1970.
363. **Shreeve, B. J., Patterson, D. S. P., and Roberts, B. A.,** Investigation of suspected cases of mycotoxicosis in farm animals in Britain, *Vet. Rec.,* 97, 275, 1975.
364. **Jemmali, M.,** Presence d'un facteur oestrogenique d'origine fongique la zearalenone ou F-2 comme contaminant naturel, dans du mais, *Ann. Microbiol. (Inst. Pasteur),* 124B, 109, 1973.
365. **Korpinen, E. L.,** Natural occurrence of F-2 and F-2 producing *Fusarium* strains associated with field cases of bovine and swine infertility, in Control of Mycotoxins, IUPAC Symp., Kungälv, Sweden, 21.
366. **Mirocha, C. B., Schauerhamer, S., and Pathra, S. V.,** Isolation, detection and quantitation of zearalenone in maize and barley, *J. Assoc. Off. Anal. Chem.,* 57, 1104, 1974.
367. **Thiel, P. G.,** A molecular mechanism for the toxic action of moniliformin, a mycotoxin produced by *Fusarium moniliforme, Biochem. Pharmacol.,* 27, 483, 1978.
368. **Mantle, P. G.,** Interruption of early pregnancy in mice by oral administration of agroclavine and sclerotia of *Claviceps fusiformis, J. Reprod. Fertil.,* 18, 81, 1969.
369. **Mirocha, C. J. and Pathre, S. V.,** Identification of the toxic principle in a sample of poaefusarin, *Appl. Microbiol.,* 26, 719, 1973.
370. **Rukmini, C., Prasad, J. S., and Rao, K.,** Effects of feeding T-2 toxin to rats and monkeys, *Food Cosmet. Toxicol.,* 18, 267, 1980.
371. **Bamburg, J. R., Riggs, N. V., and Strong, F. M.,** The structure of toxin from two strains of *Fusarium tricinctum, Tetrahedron,* 24, 3329, 1968.
372. **Ishii, K., Sakai, K., Ueno, Y., Tsunoda, H., and Enomoto, M.,** Solaniol, a toxic metabolite of *Fusarium solani, Appl. Microbiol.,* 22, 718, 1971.
373. **Tam, C.,** Chemistry and biosynthesis of trichothecenes, in *Mycotoxins in Human and Animal Health,* Rodricks, J., Hesseltine, C. W., and Mehlman, M. A., Eds., Pathotox Publishers, Park Forest South, Ill., 1977, 209—228.
374. **Pathre, S. V., Mirocha, C. J., Christensen, C. M., and Behrens, J.,** Monoacetoxyscirpenol: a new mycotoxin produced by *Fusarium roseum Gibbosum, J. Agric. Food Chem.,* 24, 97, 1976.
375. **Tidd, B. K.,** Phytotoxic compounds produced by *Fusarium equiseti.* III. Nuclear magnetic resonance spectra, *J. Chem. Soc.,* p. 218—220, 1967.
376. **Tatsuno, T., Morita, Y., Tsunoda, H., and Umeda, M.,** Researches toxicologiques des substances metaboliques du *Fusarium nivale.* VII. La troisieme substance metabolique de *F. nivale,* le diacetate de nevalinol, *Chem. Pharm. Bull.,* 18, 1485, 1970.
377. **Steyn, P. S., Vleggaar, R., Rabie, C. J., Kriek, N. P. J., and Harington, J. S.,** Trichothecene mycotoxins from *Fusarium sulphureum, Phytochemistry,* 17, 949, 1978.
378. **Yagen, B., Joffee, A. Z., Horn, P., Mor, N., and Lutsky, I. I.,** Toxins from a strain involved in ATA, in *Mycotoxins in Human and Animal Health,* Rodricks, J., Hesseltine, C. W., and Mehlman, M. A., Eds., Pathotox Publishers, Park Forest South, Ill., 1977, 329.
379. **Doster, R. C., Sinnhuber, R. O., and Pawlowski, N. E.,** Acute intraperitoneal toxicity of ochratoxin A and B derivatives in rainbow trout *(Salmo gardneri), Food Cosmet. Toxicol.,* 12, 499, 1974.
380. **Peers, F. G. and Linsell, C. A.,** Dietary aflatoxins and liver cancer: a population study based in Kenya, *Br. J. Cancer,* 27, 473, 1973.
381. **Russell, T. E., Watson, T. F., and Ryan, G. F.,** Field accumulation of aflatoxin in cottonseed as influenced by irrigation termination dates and pink bollworm infestation, *Appl. Environ. Microbiol.,* 31, 711, 1976.
382. **Dickens, J. W. and Welty, R. E.,** Fluorescence in pistachio nuts contaminated with aflatoxin, *J. Am. Oil Chem. Soc.,* 32, 448, 1975.
383. **Steyn, P. S. and Holzapfel, C. W.,** The isolation of the methyl and ethyl esters of ochratoxins A and B_1 metabolites of *Aspergillus ochraceus* Wilh., *J. S. Afr. Chem. Inst.,* 20, 186, 1967.

MICROBIAL FOOD TOXICANTS: RUBRATOXINS

M. O. Moss

INTRODUCTION

Two toxigenic molds, *Aspergillus flavus* Link and *Penicillium rubrum* Stoll, were isolated from corn implicated in an outbreak of moldy corn toxicosis,[1] a disease first described in pigs and cattle.[2] During the early stages of work on moldy corn toxicosis *P. rubrum* seemed to be the more toxic mold and it was not until a few years later that the importance of *A. flavus* and the aflatoxins became apparent. Although *P. rubrum* was first associated with the etiology of moldy corn toxicosis in farm animals it soon became evident that this disease had a complex etiology which may involve species of *Penicillium, Aspergillus,* and *Fusarium.*

The disease known as hepatitis X of dogs, first described by Siebold and Bailey,[3] was at one time considered to have the same etiology as moldy corn toxicosis of pigs.[4] However, it was subsequently shown[5] that the symptoms of canine hepatitis X could be reproduced with pure crystalline aflatoxin demonstrating that it was in fact due to *A. flavus.* Wilson et al.[5] have pointed out that, although many of the symptoms of moldy corn toxicosis could be ascribed to the action of aflatoxin, some of the gross hemorrhagic symptoms found in poisoned farm animals could not be reproduced in laboratory conditions using pure aflatoxins. These hemorrhagic symptoms would seem to be the particular feature of poisoning by *P. rubrum* and indeed a hemorrhagic disease of poultry could be reproduced using pure cultures of *P. rubrum,* and the related species *P. purpurogenum* Stoll.[6] Although the direct role of the rubratoxins in the disease of animals is uncertain, the presence of these compounds would undoubtedly influence both the production and toxicology of other mycotoxins.

PENICILLIUM RUBRUM STOLL

This species has for many years been considered as belonging to a well-defined series in the Biverticillata-Symmetrica along with *P. purpurogenum, P. aculeatum* Raper and Fennel, and *P. variabile* Sopp. The series understood by Raper and Thom is characterized by the production of colonies with deep yellow-green to gray-green conidial areas, usually velvety or lanose but sometimes with floccose regions, with the reverse of the colonies being deeply colored red to almost maroon. Sometimes the arial hyphae are pigmented yellow or orange. Growth is usually fairly limited on Czapek agar but more spreading on malt agar.

The conidiophores are typically symmetrical and biverticillate with the characteristic lanceolate phialides of the section. Some of the characters useful in separating members of the series are given in Table 1.

Raper and Thom[7] describe members of this series as having a widespread occurrence as part of the soil mycoflora, the early isolates being particularly associated with damp paper, starch paste, and similar materials. *P. rubrum* is included among the isolates from cereal and legume products listed by Scott[8] and has been isolated from peanut pods and infected kernals, as well as maize, bran, and sunflower seeds.[9-11] *P. purpurogenum* has also been isolated from maize.[12] A survey of the fungal flora of foodstuffs collected in Japanese rural villages produced as many as 35 isolates of *P. purpurogenum* and 12 isolates of *P. purpurogenum* var. *rubri-sclerotium* Thom.[13] One of the strains of *P. purpurogenum* was shown to be toxigenic to HeLa cells and it was demonstrated that this was due to its ability to produce the same toxic metabolite as *P. rubrum,* namely rubratoxin B.[14]

Table 1
CHARACTERS USEFUL IN THE SEPARATION OF MEMBERS OF THE *P. PURPUROGENUM* SERIES

		Conidia			Conidiophores	
Species	Pigment in reverse	Shape	Size (μm)	Surface texture	Size (μm)	Wall texture
P. rubrum	Deep red	Subglobose	2.2—2.8 × 2.0—2.5	Smooth	200 × 2.2—3.0	Smooth to granular
P. purpurogenum	Deep red	Elliptical	3.0—3.5 × 2.5—3.0	Rough	100—150 × 2.5—3.5	Smooth
P. aculeatum	Deep red	Globose	3.0—3.5	Echinulate	50—100 × 3.5—4.0	Granular
P. variabile	Orange red to greenish brown	Strongly elliptical	3.0—3.5 × 2.0—2.5	Smooth	200 × 2.5—3.0	Smooth

Recently the taxonomy and identification of the genus *Penicillium* has been reconsidered by Pitt,[15] who has reduced *P. rubrum* to synonomy with *P. purpurogenum*. This species is included in the series Miniolutea with *P. minioluteum* Dierckx (with which an isolate regarded by Raper and Thom as *P. purpurogenum* var. *rubrisclerotium* is included). *P. pinophilum* Hedgcock (considered as including the type strains of *P. purpurogenum* var. *rubrisclerotium*), *P. lignorum* Stolk (a species requiring a low pH and isolated from wood), *P. diversum* Raper and Fennell, *P. marneffei* Segretain, Capponi and Sureau (isolated as a pathogen of the Vietnamese bamboo rat), *P. funiculosum* Thom, *P. mirabile* Belyakova and Mil'ko, *P. verruculosum* Peyronel (which produces the tremorgenic mycotoxin verruculogen), and *P. aculeatum* Raper and Fennell. *P. variabile* is included by Pitt in his series Islandica which contains a number of important mycotoxin producing species but is considered as a heterogeneous group brought together for convenience.

RUBRATOXINS

Isolation, Structure, and Physicochemical Properties

A crude preparation of the toxic agents of *P. rubrum* was first isolated by Wilson and Wilson,[16] who also demonstrated that their material could produce symptoms in experimental animals comparable to those associated with natural outbreaks of moldy corn toxicosis.[17] The crude toxin was eventually resolved into two crystalline compounds referred to as rubratoxins A and B,[18-20] the structures of which have been shown to be represented by the formula indicated in Figure 1.[21-23] The nine-membered carbocyclic ring and two anhydride groups of rubratoxin B are characteristic of a family of mold metabolites referred to as nonadrides which include glauconic and glaucanic acids produced by *P. purpurogenum*,[24] byssochlamic acid produced by *Byssochlamys fulva*,[24] heveadride produced by *Helminthosporium heveae*,[25] and scytalidin, a metabolite of *Scytalidium* sp.[26] The molecule of rubratoxin B contains six asymmetric centers and, from a study of the X-ray crystallography of a suitable derivative, Büchi et al.[27] were able to deduce the stereochemistry shown in Figure 2.

Rubratoxin B is very soluble in acetone and ethyl acetate, moderately soluble in ethyl alcohol, slightly soluble in diethyl ether, and insoluble in nonpolar solvents such

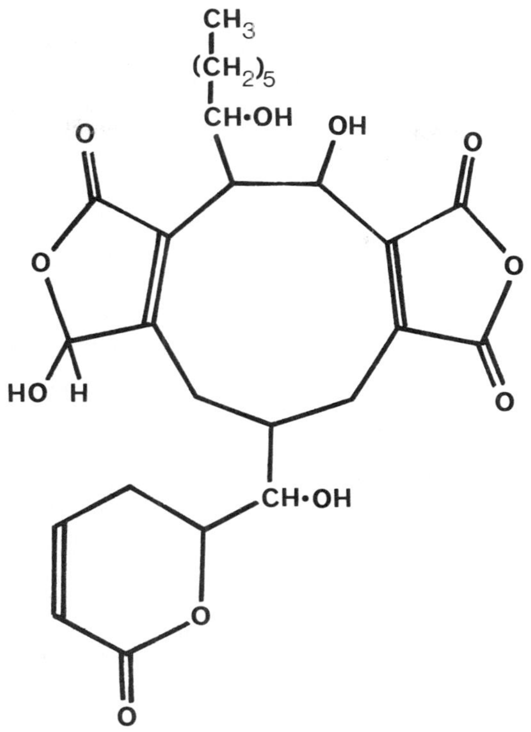

FIGURE 1. Structure of rubratoxin A.

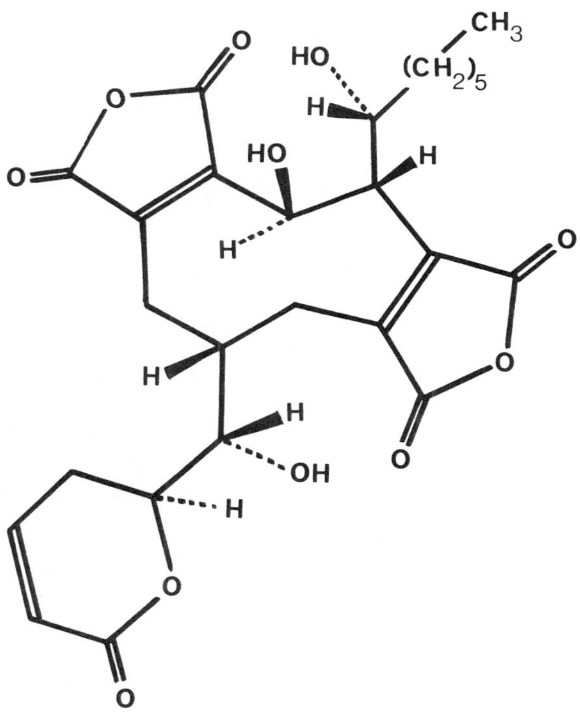

FIGURE 2. The stereochemistry of rubratoxin B.

Table 2
PHYSICOCHEMICAL PROPERTIES OF RUBRATOXINS

	Rubratoxin A	Rubratoxin B
Molecular formula	$C_{26}H_{32}O_{11}$	$C_{26}H_{30}O_{11}$
Mol wt	520	518
Melting point	210—214°C (dec)	168—170°C (dec)
λ max (CH_3CN)	252 nm (4430)	251 nm (9700)
ν max	3400, 1850, 1815, 1770, 1760, 1720, 1695, 960, 925, 895, 815, 750 cm^{-1}	3520, 1958, 1815, 1785, 1755, 1705, 1690, 925, 905, 820, 750, 720 cm^{-1}
$(\alpha)_D^{20}$ (c = 2 acetone)	+ 87°	+ 67°

Modified from Moss, M. O., in *Microbiol Toxins*, Vol. 6, Ciegler, A., Kadis, S., and Ajl, S. J., Eds., Academic Press, New York, 1971.

as petroleum ether. Although it is soluble in distilled water, to the extent of about 4 mg/mℓ, the crystalline compound usually only dissolves after prolonged shaking and it is helpful in preparing aqueous solutions to wet the crystals first with a drop of ethyl alcohol. The physicochemical properties of the rubratoxins are summarized in Table 2.

Biosynthetically the rubratoxins are best considered as derived from the coupling of two C_{13} units, each of which is derived from an acetate-derived decanoic acid derivative and oxaloacetic acid. The background to these ideas is reviewed by Vleggaar and Steyn.[28]

Analysis

Although the rubratoxins are complex compounds they do not appear to lend themselves to a sensitive and specific colorimetric method of analysis. They behave as weak acids and will react with reagents which attack double bonds, but such tests are relatively insensitive and lack specificity. The anhydride groups react with derivatives of hydrazine, such as dinitrophenylhydrazine, to form hydrazide derivatives and it may be possible to adapt such a reaction to form a confirmatory test. Rubratoxin B is relatively stable and can be recovered unchanged after dissolving in weak aqueous alkali. Rubratoxin A, however, is unstable to alkali in which it decomposes on heating to form heptaldehyde and a yellow color. This reaction has been used by Moss and Hill to analyze rubratoxin A in mixtures of the two toxins.[29]

Hayes and Wilson[19] have described a solvent system suitable for the separation of the rubratoxins by thin-layer chromatography (TLC). Using a mixture of methanol-chloroform-acetic acid (20:80:2 by volume) rubratoxin B has an R_f in the region 0.55 to 0.6 on silica gel HF-254. Because of the presence of the chromophore absorbing at 250 nm, the toxins may be seen as dark spots on a fluorescent green background when such plates are examined under short wavelength UV light. Other solvent systems have been suggested such as methanol-chloroform-acetic acid-water (20:80:1:1 by volume)[30] and ethyl acetate-acetic acid (85:15 by volume).[31] It is essential to run markers when using TLC for the analysis of rubratoxins for the R_f values are very sensitive to the activity of the plates and the environmental conditions in which the plates are run. Plates exposed to the atmosphere between spotting and developing usually show an artifact at low R_f for each of the two rubratoxins.[29] Hayes and McCain[30] have since demonstrated that rubratoxin B, but not A, may be converted into a fluorescent derivative when TLC plates are heated at 200°C for 10 min. They report being able to detect as little as 0.5 μg rubratoxin B using this method. With the use of high-pressure liquid chromatography (HPLC) it has been reported that it is possible to detect as little as 5 ng rubratoxin B.[32] A reversed-phase HPLC system using an acetonitrile-water-ethyl acetate eluent and detection by UV absorbance at 254 nm has also been developed for analyzing rubratoxins A and B. They are resolved as sharp peaks in 3 min and the method has been successfully applied to the analysis of urine and plasma samples for rubratoxin B.[33]

A sophisticated approach to the analysis of mycotoxins, but one which cannot be available for general use, is by field desorption mass spectroscopy (MS) which allows the formation and detection of the parent molecular ion of even relatively nonvolatile compounds with hardly any fragmentation of the molecule.[34] The normal mass spectrum of rubratoxin B is complex and the highest peak is usually at m/e 474 (M — CO_2). Using field desorption MS a relatively simple spectrum is obtained in which the parent molecular ion at m/e 518 forms a dominant peak.

The production of radioactively labeled rubratoxin B has been reported by several groups of workers[35,36] and a radioimmune assay for rubratoxin B has been developed by raising an antirubratoxin antibody in rabbits with ovalbumin coupled to rubratoxin.

Biological Activity

General Comments

The studies of Burnside et al.[1] using corn infected with *P. rubrum* showed it to be toxic to pigs, mice, horses, and goats. Pigs fed with this material were lethargic and seemed to be in pain, it being possible to see erythema of the central abdominal skin surface before and after death. Horses became depressed and suffered from a loss of appetite as well as general incoordination. Necropsy of pigs revealed congestion and frequently hemorrhage of most tissues. Field outbreaks of moldy corn toxicosis in both pigs and cattle frequently reveal hemorrhages on postmortem examination, although such field outbreaks may be complicated by the presence of several mycotoxin producing molds and, indeed, *P. rubrum* itself has been reported to produce other mycotoxins such as rugulovasines and chlororugulovasines.[38]

Mice fed with doses of rubratoxin well above the LD_{50} value died within 2 to 4 hr of administration.[18] There was dilation of the s.c. blood vessels and post-mortem examination showed extensive hemorrhaging of the liver, which also had a characteristic mottled appearance. The kidneys were slightly anemic and there was occasional hemorrhaging of the lungs. When mice were killed and examined 7 days after receiving a sublethal dose of toxin, no macroscopic liver damage could be seen, although mice treated with the same doses but killed only 24 hr later had the typically mottled liver, but showed no signs of hemorrhaging or damage in other parts of the body. It seemed that the liver was the primary site of activity and could recover over a period of a week. Using the sleeping time test with sodium pentobarbitone[39] Townsend et al. were able to demonstrate that rubratoxin B had an effect on liver function, even at doses which did not cause macroscopic lesions.[18]

The general picture from these early studies was of a compound which is primarily a hepatotoxin, although, in acute toxicoses, lesions may occur in other parts of the body, especially the kidney and lungs. The generalized effects of acute doses of rubratoxins in most species of higher animals appear to be hemorrhage in a wide range of organs and tissues.

Quantitative and Comparative Toxicology in Higher Animals

Although there is some variation in acute toxicity of rubratoxins among different species of warmblooded animals, this is not as large as the effect of route of administration within any one species, or even the vehicle in which the toxin is administered in some species. Thus the oral LD_{50} for the rat is over 1000 times greater than the value obtained when the toxin is given i.p. In the mouse the LD_{50} is about ten times

Table 3
LD_{50} VALUES OF RUBRATOXIN B

Animal species	Route	Vehicle[a]	LD_{50} (mg/kg)	Ref.
Mouse	i.p.	PG	3.0	41
Mouse	i.p.	DMSO	0.27	42
Mouse	Oral	—	250	43
Cat	i.p.	PG	1.0—1.5	42
Cat	i.p.	DMSO	0.2	42
Dog	i.p.	PG	5	42
Guinea pig	i.p.	DMSO	0.5	42
Rat	i.p.	DMSO	0.35	42
Rat	i.p.	PG	0.36	42
Rat	Oral	DMSO	400—450	42
24-hr rat	Oral	Corn oil	6.38	44
Chicken	i.p.	PG	4.0	42
1-day chick	Oral	0.5 *M* $NaHCO_3$	83.2	47

[a] PG, propylene glycol; DMSO, dimethyl sulfoxide.

greater using propylene glycol as the solvent than it is when dimethylsulfoxide is used as the solvent, and yet in the rat the solvent makes very little difference. The LD_{50} values for rubratoxin B given i.p. in dimethyl sulfoxide are less than 0.5 mg/kg for the rat, mouse, guinea pig, and cat, whereas for the dog and chicken LD_{50} values are greater than 5.0 mg/kg (see Table 3). There are very few studies of the toxicity of rubratoxins to coldblooded vertebrates but tadpoles of the amphibians *Rana temporaria* and *Triturus alpestris* were not affected by any levels of rubratoxin B administered.[40]

From studies of the effect of structural changes in the rubratoxin molecule on its toxicity to the mouse, it has been shown that rubratoxin B itself is the most toxic compound and any changes in this structure result in a loss of toxicity (Table 4).[41] In particular, hydrogenation of the unsaturated δ-lactone resulted in a decrease in toxicity confirming the importance of this part of the molecule in the toxicology of the rubratoxins.

Wogan et al.[42,43] carried out extensive chronic tests to establish whether or not rubratoxin B had any carcinogenicity to rats, which are so sensitive to the carcinogenic activity of aflatoxin. These studies demonstrated that rubratoxin B is not carcinogenic to rats but they did reveal an intriguing type of response to chronic exposure to low levels of toxin. Although 70% of the experimental animals survived for more than 80 weeks and showed no significant lesions after being sacrificed and carefully examined, the 30% which did die did so in a dramatic manner. They appeared to be healthy and showed normal weight gain until just before the onset of symptoms. Then, within a period as short as 3 days, they lost their appetite, rapidly lost body weight, and developed diarrhea and a discharge from the eyes, ears, and nose and then died. The livers of these animals were severely damaged with extensive hemorrhagic necrosis. It seemed as though the liver could withstand, and recover from, a continuous sublethal exposure to rubratoxin, but beyond a certain threshold which depended very much on the individual, the liver was unable to withstand further insult and rapidly degenerated with dramatic consequences for the rest of the body. The presence of other hepatotoxic agents, such as aflatoxin, have synergistic effects on the toxicity of rubratoxin B.[42] This last observation is particularly important for it is probable that field outbreaks of mycotoxicosis will involve more than one species of mold and the presence of more than one toxic metabolite, and yet most of our information is based on studies with pure compounds studied one at a time.

Table 4
LD_{50} VALUES OF DERIVATIVES OF RUBRATOXIN B GIVEN i.p. TO MICE IN PROPYLENE GLYCOL[41]

Compound	Nature of change	LD_{50} (mg/kg)
Rubratoxin B	—	3.0
Rubratoxin A	Reduction of one anhydride to a lactol	6.6
Dihydrorubratoxin B	Hydrogenation of δ-lactone	12
Ketorubratoxin B	Oxidation of one hydroxyl to ketone	ca. 20
Rubratoxin B triacetate	Acetylation of all OH groups	ca. 125
Tetra sodium rubratoxin B	Hydrolysis of anhydrides	12

Another example of synergism with rubratoxin B was demonstrated with ochratoxin during studies on the effects of mycotoxins on infant rats.[44] These studies illustrated another widespread phenomenon, the increased sensitivity of young animals compared with adults (see Table 3). In the case of rubratoxin B, the oral LD_{50} in adult rats is nearly 50 times that in 24-hr-old animals.

In studies on the acute toxicity of penicillic acid and rubratoxin B in dogs, Hayes et al.[45] reported that an additive effect due to a combined administration of both mycotoxins was only seen as an elevation in serum sodium and chloride levels. The same authors studied the effects of aflatoxin B and rubratoxin on dogs and concluded that lesions induced by administering these two mycotoxins were very similar to those reported for dogs suffering from hepatitis X.[46] These studies confirmed that aflatoxin is the primary etiological factor in canine hepatitis X[5] but indicated that rubratoxin B may also be involved.

Studies on the effects of rubratoxin in broiler chickens[47] have indicated a relatively low toxicity in this species (Table 3) and that, on its own, it may not have been responsible for the hemorrhagic anemia associated with *P. rubrum* infested poultry feed.[6] It would also seem that, in broiler chicks, aflatoxin and rubratoxin do not show the synergistic effects found in rats.[48]

Toxicity of Rubratoxins to the Embryo and Tissue Cultures

Single i.p. injections of sublethal doses of rubratoxin B to pregnant mice on any day during the period 6 to 12 days of gestation caused significant increases in embryonic mortality; indeed, doses greater than 0.4 mg/kg on day 8 caused 100% mortality. Many of the surviving fetuses developed abnormalities such as encephalies, malformed pinnae and jaws, umbilical hernias, and open eyes.[49,50] When one considers that the parent animal will metabolize and excrete a fair proportion of the dose given,[51] toxicity to the 8-day fetus must be very high indeed. Evans et al.[52] also demonstrated a dose-related increase in resorptions of the fetus in mice. A reduction in fecundity, but not in mating behavior, has also been reported when male mice were treated with sublethal doses of rubratoxin prior to mating.[53] Phenomena associated with mutagenesis were also described in the offspring of untreated females mated with such rubratoxin-treated males. The teratogenic activity of rubratoxin B has also been demonstrated in chick embryogenesis by injecting 0.005 to 0.007 mg rubratoxin per egg into the air sac of eggs using propylene glycol as the solvent.[54] Gross malformations and cardiac anomalies were observed in treated embryos.

When HeLa cells were used in a study of the biological effects of rubratoxin, the phenomena described included an accumulation of mitotic cells and the appearance of polynuclear cells.[13] Cells treated with 100 μg rubratoxin B per milliliter for 24 hr and then replaced into a control medium recovered and grew at a normal rate. Umeda et al.[13] suggested that rubratoxin B has a directly damaging effect on chromosomes, an observation which fits with the claims that it is mutagenic.[52] However, rubratoxin B does not behave as a mutagen in the Ame's test using *Salmonella*.[55] Rabbit cornea cells become enlarged, irregularly shaped, and packed with tiny spherical refractile inclusions, as well as vacuolate, when treated with 75 μg rubratoxin B per milliliter.[56]

Toxicity to Arthropods

The insecticidal activity of the rubratoxins has been reported by Cole and Rolinson.[57] Using the sheep blowfly (*Lucilia sericata*) at the first instar larval stage, they obtained LC_{50} values of 18 μg/mℓ for rubratoxin A and 200 μg/mℓ for rubratoxin B. It is interesting that rubratoxin A should be the more toxic contrary to observations in higher animals but comparable with observations on protozoa. Rubratoxin B has also been shown to be toxic to the larvae of the fruit fly (*Drosophila melanogaster*)[58] and the crustacean, *Cyclops fuscus*.[59]

Toxicity to Microorganisms

In a general survey of the sensitivity of microorganisms to rubratoxin B, Hayes and Wyatt tested 133 strains of bacteria, protozoa, and algae.[60] Gram-negative bacteria seemed not to be affected at the concentrations used, whereas strains of Gram-positive organisms from the genera *Bacillus, Micrococcus*, and *Staphylococcus* were inhibited by 1.0 mg/mℓ. These authors found *Tetrahymena pyriformis* and *Volvox aureus* to be the most sensitive, having MIC values of 25 and 50 μg/mℓ, respectively. Although one report suggests that rubratoxin B has no effect on the growth of *Chlorella pyrenoidosa*,[61] it has also been reported that a rubratoxin producing strain of *P. rubrum* could produce a zone of inhibition on a plate seeded with *Chlorella*.[62]

The sensitivity of *Tetrahymena pyriformis* has been used for developing a possible bioassay for the rubratoxins,[63,64] as well as for fundamental biochemical studies.[65] Wyatt and Townsend[63] give MIC values of 8.75 μg rubratoxin A per milliliter and 20.5 μg rubratoxin B per milliliter using this protozoan. The value for rubratoxin B is in good agreement with that quoted by Hayes and Wyatt.[60] Although the majority of bacteria are relatively insensitive to rubratoxin B, it is possible that *Bacillus stearothermophilus* may be sensitive enough to form the basis of a fairly nonspecific assay.[66]

Using a disc assay as little as 100 μg rubratoxin B per disc will inhibit the sporulation of several species of mold and cause damage to vegetative hyphae.[67] In liquid media as little as 10 μg rubratoxin B per milliliter will cause swelling, distortion, and increased branching of the mycelium of *A. niger*.[68] As well as inhibiting growth and causing morphological changes, low concentrations of rubratoxin B also affected glucose consumption, respiration, protein secretion, and the cell wall composition of *A. niger* as well as inducing the secretion of secondary metabolites such as pigments.[69] There is a report of the production of a fungicidal agent produced by a strain of *Penicillium rubrum* isolated from a chestnut stump in Yugoslavia.[70] The compound which was not characterized, was particularly active against *Phytophthora parasitica* var. *nicotianae*.

Biochemical Aspects of Toxicity

Using ^{14}C-labeled rubratoxin B to dose both mice and rats, Hayes demonstrated that 40 to 50% of the radioactivity was excreted by both species within 24 hr of administration.[51] The major route of excretion was as respiratory CO_2 (30% in mice and 40% in rats). Radioactivity was found to be most highly concentrated in the liver 1 hr after administration, an observation which is compatible with rubratoxin acting primarily as an hepatotoxin. In further studies on the fate of rubratoxin B in the rat,[71] it was demonstrated that significant quantities of rubratoxin were secreted unchanged in the urine and, although the concentration reached a peak in the liver within 1 hr of i.p.

injection it equilibrated with that in the plasma within a few hours. Using isolated perfused rat liver preparations to study hepatic uptake biliary excretion and metabolism of rubratoxin B, Unger et al.[72] detected both glucuronide and sulfate conjugates of the toxin in the bile. A number of studies have concentrated on the effects of rubratoxin on liver metabolism. Single doses of 1.67 and 2.67 mg rubratoxin B per kilogram body weight, given i.p., resulted in decreases of total liver glycogen and protein and an increase in total liver lipids, although a smaller dose (0.67 mg/kg) caused a stimulation of liver protein synthesis as measured by the incorporation of labeled leucine.[73] One result of the activity of rubratoxin in the liver is the disaggregation of polysomes not apparently mediated by a direct interaction of polysomes with rubratoxin itself. The disruption of polysomes could account for the inhibition of protein biosynthesis observed at higher concentrations.[74]

Although the total serum protein remained unchanged in guinea pigs receiving oral doses of 2 to 10 mg rubratoxin per day over a 3-week period, β-globulin levels decreased, γ-globulin levels increased, prothrombin time increased, and complement activity was suppressed.[75] The suppression of complement activity has been made the basis of a test for the effect of hepatotoxic mycotoxins using radial immunodiffusion.[76]

It has been noticed that rats dying of acute rubratoxicosis appeared to decompose more readily than controls and the livers of those rats which died after prolonged exposure to chronic doses deteriorated incredibly quickly. Such a generalized loss of tissue structure may be associated with the increased activity of lysosomal enzymes and, indeed, Hayes and Hunter[77] have demonstrated an increase in both acid phosphatase and β-glucuronidase in liver tissue after a dose of 0.5 mg rubratoxin B per kilogram given i.p.

It is possible then that rubratoxin causes a breakdown of lysosomes with a consequent release of lysosomal enzymes. There is some evidence that rubratoxin B destroys the complex integrity of another organelle in which membranes play an important part, the mitochondrion. Intact liver mitochondria have little or no ATPase activity which does, however, appear after some structural damage. The low basal level of ATPase in intact mitochondria is stimulated by uncoupling agents such as dinitrophenol, but no such stimulation of the basal level of ATPase of rubratoxin damaged mitochondria was observed.[18] It should be noted, however, that the oligomycin sensitive magnesium dependant ATPase, which is mitochondrial, is itself inhibited by rubratoxin B.[79] Neubert and Merker[80] noticed a swelling of liver mitochondria following a sublethal dose of the crude toxin from *P. rubrum* but they concluded that the mitochondrial lesions, inhibition of respiration, and uncoupling of oxidative phosphorylation were not responsible for the symptoms of rubratoxicosis. Whatever the role of mitochondrial damage is in rubratoxin poisoning there appears to be no doubt that the respiration of isolated liver mitochondria is inhibited[81,82] and this inhibition may involve a particular site near the end of the electron transport system.[82] Using whole homogenates and preparations of mitochondria from several animal species and from males and females Hayes and Hannan[83] demonstrated both species and sex differences in the in vitro uptake of oxygen by citrate and succinate.

Two particular enzyme systems that have been studied intensively are the (Na^+ - K^+) - ATPase and cytochrome P-450 dependent oxygenase of mice. Rubratoxin B would seem to be a potent inhibitor of membrane ATPase for the (Na^+ - K^+) - ATPase of mouse brain microsomes is inactivated (IC_{50} 6.0×10^{-6} *M*) in a Na^+ and K^+ - dependent and reversible manner. Compounds containing thiol groups have a protective effect.[84] Rubratoxin B has also been shown to inhibit the liver P-450 oxygenases such as pentobarbital hydroxylase and ethyl morphine demethylase.[85] Glutathione levels in the liver were also reduced and indeed glutathione offers some protection against rubratoxin poisoning.[86] Several studies have demonstrated that, unlike aflatoxin, rubratoxin

B does not require activation in the animal body but is the molecule directly involved in toxicosis.[87,88]

Although a considerable amount of detailed information is available about the effect of rubratoxin B on a variety of biological processes it is still not possible to describe the sequence of events leading to the death of an animal suffering from rubratoxicosis or to avoid the conclusion that rubratoxicosis in farm animals should not occur with careful animal husbandry. *P. rubrum* and its toxins are the subjects of several recent reviews.[89-91]

REFERENCES

1. **Burnside, J. E., Sippel, W. L., Forgacs, J., Carll, W. T., Atwood, M. B., and Doll, E. R.**, A disease of swine and cattle caused by eating mouldy corn. II. Experimental production with pure cultures of moulds, *Am. J. Vet. Res.*, 18, 817, 1957.
2. **Sippel, W. L., Burnside, J. E., and Atwood, M. B.**, A disease of swine and cattle caused by eating mouldy corn, in *Proc. 90th Annu. Meet., Am. Vet. Med. Assoc., Toronto*, 1953, 174.
3. **Seibold, H. R. and Bailey, W. S.**, An epizootic of hepatitis in the dog, *J. Am. Vet. Med. Assoc.*, 121, 201, 1952.
4. **Bailey, W. S. and Groth, A. H.**, The relationship of hepatitis X of dogs and mouldy corn poisoning of swine, *J. Am. Vet. Med. Assoc.*, 134, 514, 1959.
5. **Wilson, B. J., Teer, P. A., Barney, G. H., and Blood, F. R.**, Relationship of aflatoxin to epizootics of toxic hepatitis among animals in Southern United States, *Am. J. Vet. Res.*, 28, 1217, 1967.
6. **Forgacs, J., Koch, H., Carll, W. T., and White-Stevens, R. M.**, Additional studies on the relationship of mycotoxicoses to the poultry hemorrhagic syndrome, *Am. J. Vet. Res.*, 19, 744, 1958.
7. **Raper, K. B. and Thom, C.**, *A Manual of the Penicillia*, Williams & Wilkins, Baltimore, 1949, 630.
8. **Scott, de B.**, Toxigenic fungi isolated from cereal and legume products, *Mycopathol. Mycol. Appl.*, 25, 213, 1965.
9. **Jackson, C. R.**, Peanut pod mycoflora and kernal infections, *Plant Soil*, 23, 203, 1965.
10. **Joffe, A. Z. and Borut, S. Y.**, Soil and kernal mycoflora of groundnut fields in Israel, *Mycologia*, 58, 629, 1966.
11. **Cantini, G. and Scurti, J. G.**, Mycoflora of maize, bran and sunflower seeds, *Allionia (Turin)*, 11, 29, 1965.
12. **Carlton, W. W., Tuite, J., and Mislivec, P.**, Investigations of the toxic effects in mice of certain strains of *Penicillium, Toxicol. Appl. Pharmacol.*, 13, 372, 1968.
13. **Umeda, M., Saito, A., and Saito, M.**, Cytotoxic effects of toxic culture filtrate of *Penicillium purpurogenum* and its toxic metabolite rubratoxin B, on HeLa cells, comparative study among the effects of rubratoxin B, colcemid and vinblastine, *Jpn. J. Exp. Med.*, 40, 409, 1970.
14. **Natori, S., Sakaki, S., Kurata, H., Udagawa, S., Ichinoi, M., Saito, M., Umeda, M., and Ohtsubo, K.**, Production of rubratoxin B. *Penicillium rubrum* Stoll, *Appl. Microbiol.*, 19, 613, 1970.
15. **Pitt, J. I.**, The Genus *Penicillium* and its Teleomorphic States Eupenicillium and Talaromyces, Academic Press, New York, 1979, 436.
16. **Wilson, B. J. and Wilson, C. H.**, Hepatotoxic substance from *Penicillium rubrum, J. Bacteriol.*, 83, 693, 1962.
17. **Wilson, B. J. and Wilson, C. H.**, Extraction and preliminary characterization of a hepatotoxic substance from cultures of *Penicillium rubrum, J. Bacteriol.*, 84, 283, 1962.
18. **Townsend, R. J., Moss, M. O., and Peck, H. M.**, Isolation and characterization of a hepatotoxic substance from cultures of *Penicillium rubrum, J. Pharm. Pharmacol.*, 18, 471, 1966.
19. **Hayes, A. W. and Wilson, B. J.**, Bioproduction and purification of rubratoxin B, *Appl. Microbiol.*, 16, 1163, 1968.
20. **Moss, M. O., Robinson, F. V., Wood, A. B., and Morrison, A.**, Observations on the structure of the toxins from *Penicillium rubrum, Chem. Ind. (London)*, 755, 1967.
21. **Moss, M. O., Wood, A. B., and Robinson, F. V.**, The structure of rubratoxin A, a toxic metabolite of *Penicillium rubrum, Tetrahedron Lett.*, 5, 367, 1969.
22. **Moss, M. O., Robinson, F. V., Wood, A. B., Paisley, H. M., and Feeney, J.**, Rubratoxin B, a proposed structure for a bis-anhydride from *Penicillium rubrum* Stoll, *Nature (London)*, 220, 767, 1968.
23. **Moss, M. O., Robinson, F. V., and Wood, A. B.**, Rubratoxins, *J. Chem. Soc.*, Sect. C, 619, 1971.

24. **Barton, D. H. R. and Sutherland, J. K.**, The nonadrides. I. Introduction and general survey, *J. Chem. Soc.*, 1769, 1965.
25. **Crane, R. I., Hedden, P., MacMillan, J., and Turner, W. B.**, Fungal products. IV. The structure of heveadride, a new nonadride from *Helminthosporium heveae, J. Chem. Soc. Perkin I*, 194, 1973.
26. **Strunz, G. M., Kakushima, M., and Stillwell, M. A.**, Scytalid in: a new fungitoxic metabolite produced by a *Scytalidium* species, *J. Chem. Soc. Perkin I*, 2280, 1972.
27. **Büchi, G., Snadu, K. M., White, J. D., Gonzontas, J. Z., and Singh, S.**, Structures of rubratoxins A and B, *J. Am. Chem. Soc.*, 92, 6638, 1970.
28. **Vleggaar, R. and Steyn, P. S.**, The Biosynthesis of some miscellaneous mycotoxins, in *The Biosynthesis of Mycotoxins*, Steyn, P. S., Ed., Academic Press, New York, 1980, 395.
29. **Moss, M. O. and Hill, I. W.**, Strain variation in the production of rubratoxins by *Penicillium rubrum* Stoll, *Mycopathol. Mycol. Appl.*, 40, 81, 1970.
30. **Hayes, A. W. and McCain, H. W.**, A procedure for the extraction and estimation of rubratoxin B in corn, *Food Cosmet. Toxicol.*, 13, 221, 1975.
31. **Emeh, C. O. and Marth, E. H.**, Synthesis of macromolecules and rubratoxin by *Penicillium rubrum, Arch. Microbiol.*, 115, 157, 1977.
32. **Engstrom, G. W., Richard, J. L., and Cysewski, S. J.**, High pressure liquid chromatographic method for the detection and resolution of rubratoxins, aflatoxins and other mycotoxins, *J. Agric. Food Chem.*, 25, 833, 1977.
33. **Unger, P. D. and Hayes, A. W.**, High-pressure liquid chromatography of the mycotoxins, rubratoxins A and B, and its application to the analysis of urine and plasma for rubratoxin B, *J. Chromatogr.*, 153, 115, 1978.
34. **Sphon, J. A., Dreifuss, P. A., and Schulten, H. R.**, Field desorption mass spectrometry of mycotoxins and mycotoxin mixtures and its application as a screening technique for foodstuffs, *J. Assoc. Off. Anal. Chem.*, 60, 73, 1977.
35. **Emeh, C. O. and Marth, E. H.**, Incorporation of labelled small molecules into rubratoxin, *Arch. Microbiol.*, 118, 7, 1978.
36. **Davis, R. M. and Richard, J. L.**, Production of (14C) rubratoxin B, *Mycopathologia*, 67, 35, 1979.
37. **Davis, R. M. and Stone, S. S.**, Production of anti-rubratoxin antibody and its use in a radio-immunoassay for rubratoxin B, *Mycopathologia*, 30, 29, 1979.
38. **Dorner, J. W., Cole, R. J., Hill, R., Wicklow, D., and Cox, R. H.**, *Penicillium rubrum* and *Penicillium biforme*, new sources of rugulovasines A and B, *Appl. Environ. Microbiol.*, 40, 685, 1980.
39. **Plaa, G. L., Evans, E. A., and Hine, C. H.**, Relative hepatotoxicity of seven halogenated hydrocarbons, *J. Pharmacol. Exp. Ther.*, 123, 224, 1958.
40. **Reiss, J.**, Toxic effects of mycotoxins on amphibian larvae, *Toxicology*, 8, 121, 1977.
41. **Rose, H. M. and Moss, M. O.**, The effect of modifying the structure of rubratoxin B on the acute toxicity to mice, *Biochem. Pharmacol.*, 19, 612, 1970.
42. **Wogan, G. N., Edwards, G. S., and Newberne, P. M.**, Acute and chronic toxicity of rubratoxin B, *Toxicol. Appl. Pharmacol.*, 19, 712, 1971.
43. **Edwards, G. S. and Wogan, G. N.**, Acute end chronic toxicity of rubratoxin in rats, *Fed. Proc.*, 27, 552, 1968.
44. **Hayes, A. W., Cain, J. A., and Moore, B. G.**, Effect of aflatoxin B_1, ochratoxin A and rubratoxin B on infant rats, *Food Cosmet. Toxicol.*, 15, 23, 1977.
45. **Hayes, A. W., Unger, P. D., and Williams, W. L.**, Acute toxicity of penicillic acid and rubratoxin B in dogs, *Ann. Nutr. Aliment.*, 31, 711, 1977.
46. **Hayes, A. W. and Williams, W. L.**, Acute toxicity of aflatoxin B, and rubratoxin B in dogs, *J. Environ. Pathol. Toxicol.*, 1, 59, 1978.
47. **Wyatt, R. D. and Hamilton, P. B.**, The effect of rubratoxin in broiler chickens, *Poultry Sci.*, 51, 1383, 1972.
48. **Wyatt, R. D., Tung, H. T., and Hamilton, P. B.**, Effect of simultaneous feeding of aflatoxin and rubratoxin to chickens, *Poultry Sci.*, 52, 395, 1973.
49. **Hood, R. D., Innes, J. E., and Hayes, A. W.**, Effects of rubratoxin B on prenatal development in mice, *Bull. Environ. Contam. Toxicol.*, 10, 200, 1973.
50. **Koshakji, R. P., Wilson, B. J., and Harbison, R. D.**, Effect of rubratoxin B on prenatal growth and development in mice, *Res. Commun. Chem. Pathol. Pharmacol.*, 5, 484, 1973.
51. **Hayes, A. W.**, Excretion and tissue distribution of radioactivity from rubratoxin B - ^{14}C in mice and rats, *Toxicol. Appl. Pharmacol.*, 23, 91, 1972.
52. **Evans, M. A., Wilson, B. J., and Harbison, R. D.**, Toxicology and mutagenic effects of rubratoxin B, *Pharmacology*, 17, 248, 1975.
53. **Evans, M. A. and Harbison, R. D.**, Prenatal toxicity of rubratoxin B and its hydrogenated analog, *Toxicol. Appl. Pharmacol.*, 39, 13, 1977.
54. **Gilani, S. H., Bancroft, J., and Reilly, M.**, Rubratoxin B and chick embryogenesis: an experimental study, *Environ. Res.*, 20, 199, 1979.

55. **Kuczuk, M. H., Benson, P. M., Heath, H. E., and Hayes, A. W.**, Evaluation of the mutagenic potential of mycotoxins using *Salmonella typhimurium* and *Saccharomyces cerevisiae, Mutat. Res.*, 53, 11, 1978.
56. **Hayes, A. W.**, Effect of aflatoxin B and rubratoxin B on bacteriophage and rabbit cornea cells, *Bull. Environ. Contam. Toxicol.*, 15, 665, 1976.
57. **Cole, M. and Rolinson, G. N.**, Microbial metabolites with insecticidal properties, *Appl. Microbiol.*, 24, 660, 1972.
58. **Reiss, J.**, Insecticidal and larvicidal activities of the mycotoxins aflatoxin B_1, rubratoxin B, patulin and diacetoxyscirpenol towards *Drosophila melanogaster, Chem. Biol. Interact.*, 10, 339, 1975.
59. **Reiss, J.**, Toxic effects of the mycotoxins aflatoxin B_1, rubratoxin B, patulin and diacetoxyscirpenol on the crustacean *Cyclops fuscus, J. Assoc. Off. Anal. Chem.*, 55, 895, 1972.
60. **Hayes, A. W., and Wyatt, E. P.**, Survey of the sensitivity of microorganisms to rubratoxin B, *Appl. Microbiol.*, 20, 164, 1970.
61. **Sullivan, J. D. and Ikawa, M.**, Variations in inhibition of growth of five *Chlorella* strains by mycotoxins and other toxic substances, *Agric. Food Chem.*, 20, 921, 1972.
62. **Ikawa, M., Ma, D. S., Meeker, G. B., and Davis, R. P.**, Use of *Chlorella* in mycotoxin and phycotoxin research, *Agric. Food Chem.*, 17, 425, 1969.
63. **Wyatt, T. D. and Townsend, R. J.**, The bioassay of rubratoxins A and B using *Tetrahymena pyriformis* strain W, *J. Gen. Microbiol.*, 80, 85, 1974.
64. **Hayes, A. W., Melton, R., and Smith, S. J.**, Effect of aflatoxin B_1, ochratoxin and rubratoxin B on a protozoan, *Tetrahymena pyriformis* HSM, *Bull. Environ. Contam. Toxicol.*, 11, 321, 1974.
65. **Hayes, A. W.**, Antiprotozoal activity of rubratoxin B, *Antimicrob. Agents Chemother.*, 4, 80, 1973.
66. **Reiss, J.**, Mycotoxin bioassay, using *Bacillus stearothermophilus, J. Assoc. Off. Anal. Chem.*, 58, 624, 1975.
67. **Reiss, J.**, Toxicity of rubratoxin B to fungi, *J. Gen. Microbiol.*, 71, 167, 1972.
68. **Moss, M. O. and Badii, F.**, The effect of rubratoxin on growth and morphology of *Aspergillus niger, Bull. Br. Mycol. Soc.*, 12, 121, 1978.
69. **Moss, M. O. and Badii, F.**, Effect of rubratoxin B on growth, metabolism and morphology of *Aspergillus niger, Trans. Br. Mycol. Soc.*, 74, 1, 1980.
70. **Krstic, M. M.**, On the fungistatic and fungicidal properties of an antibiotic produced by *Penicillium rubrum, Plant Dis. Rep.*, 51, 669, 1967.
71. **Unger, P. D. and Hayes, A. W.**, Disposition of rubratoxin B in the rat, *Toxicol. Appl. Pharmacol.*, 47, 585, 1979.
72. **Unger, P. D., Hayes, A. W., and Mehendale, H. M.**, Hepatic uptake, disposition and metabolism of rubratoxin B in isolated perfused rat liver, *Toxicol. Appl. Pharmacol.*, 47, 529, 1979.
73. **Hayes, A. W. and Wilson, B. J.**, Effects of rubratoxin B on liver composition and metabolism in the mouse, *Toxicol. Appl. Pharmacol.*, 17, 481, 1970.
74. **Watson, S. A. and Hayes, A. W.**, Evaluation of possible sites of action of rubratoxin B-induced polyribosomal disaggregation in mouse liver, *J. Toxicol. Environ. Health*, 2, 639, 1977.
75. **Richard, J. L., Thurston, J. R., and Graham, C. K.**, Changes in complement activity, serum proteins, and prothrombin time in guinea pigs fed rubratoxin alone or in combination with aflatoxin, *Am. J. Vet. Res.*, 35, 957, 1974.
76. **Thurston, J. R. and Richard, J. L.**, Confirmation by radial immunodiffusion of depression of the fourth component of complement in guinea pigs fed aflatoxin or rubratoxin, *Am. J. Vet. Res.*, 40, 1206, 1979.
77. **Hayes, A. W. and Hunter, C. E.**, Effects of rubratoxin B on acid phosphatase and β-glucuronidase, *Res. Commun. Chem. Pathol. Pharmacol.*, 6, 759, 1973.
78. **Madhavikutty, K. and Shanmugasundaram, E. R. B.**, Functional state of mouse liver mitochondria in toxicoses due to *Aspergillus flavus* and *Penicillium rubrum, Experientia*, 25, 149, 1969.
79. **Desaiah, D., Hayes, A. W., and Ho, I. K.**, Effect of rubratoxin B on adenosine triphosphatase activities in the mouse, *Toxicol. Appl. Pharmacol.*, 39, 71, 1977.
80. **Neubert, D. and Merker, H. J.**, Some effects of toxins from microorganisms on mammalian cellular metabolism and structure, in *Recent Advances in the Pharmacology of Toxins*, Raudonat, H. W., Ed., Pergamon Press, Oxford, 1965, 17.
81. **Bernard, C. and Dumas, P.**, Effect of rubratoxin B on the respiration of liver mitochondria in the rat, *Mycopathologia*, 55, 53, 1975.
82. **Hayes, A. W.**, Action of rubratoxin B on mouse liver mitochondria, *Toxicology*, 6, 253, 1976.
83. **Hayes, A. W. and Hannan, C. J.**, Effects of rubratoxin and aflatoxin on oxygen consumption of Krebs' cycle intermediates, *Toxicol. Appl. Pharmacol.*, 25, 30, 1973.
84. **Phillips, T. D., Hayes, A. W., Ho, I. K., and Desaiah, D.**, Effects of rubratoxin B on the kinetics of cationic and substrate activation of (Na^+ - K^+) - ATPase and *p*-nitrophenyl phosphatase, *J. Biol. Chem.*, 253, 3487, 1978.

85. **Siraj, M. Y. and Hayes, A. W.**, Inhibition of the hepatic cytochrome P-450 dependent mono-oxygenase system by rubratoxin B in male mice, *Toxicol. Appl. Pharmacol.*, 48, 351, 1979.
86. **Siraj, M. Y. and Hayes, A. W.**, Effects of rubratoxin B on the hepatic mono-oxygenase system in phenobarbital and 3-methyl-cholanthrene induced male mice, *Toxicology*, 17, 17, 1980.
87. **Hayes, A. W. and Ho, I. K.**, Interaction of rubratoxin B and pentobarbital in mice, *J. Environ. Pathol. Toxicol.*, 1, 491, 1978.
88. **Phillips, T. D. and Hayes, A. W.**, Structural modification of polyfunctional rubratoxin B: effects on mammalian adenosine triphosphatase, *J. Environ. Pathol. Toxicol.*, 2, 853, 1979.
89. **Moss, M. O.**, The rubratoxins, in *Microbial Toxins*, Vol. 6, Ciegler, A., Kadis, S., and Ajl, S. J., Eds., Academic Press, New York, 1971, 381.
90. **Newberne, P. M.**, *Penicillium rubrum* - rubratoxins, in *Mycotoxins*, Purchase, I. F. H., Ed., Elsevier, Amsterdam, 1974, 163.
91. **Hayes, A. W.**, Rubratoxins, in *Mycotoxins in Human and Animal Health*, Rodericks, J. V., Hesseltine, C. W., and Mehman, M. A., Eds., Pathotox Publishers, Park Forest South, Ill., 1977, 507.

CITREOVIRIDIN

Yoshio Ueno

INTRODUCTION

Mycotoxin pollution occurs in various areas of Monsoon Asia, where the climatic conditions such as high temperature and humidity are suited for fungal growth. In these areas, rice grains are one of the most important foodstuffs, and the grains infected with toxic fungi induce serious foodborne intoxications of human and farm animals.

Acute cardiac beriberi, which is called "Shoskin-kakke" in Japanese, has been prevalent during the last 3 centuries in rice-eating countries including Japan. This disease is characterized by violent and tragic symptoms such as convulsions, vomition, ascending paralysis, and respiratory arrest. Numerous researchers presented many speculative theories about its cause in the past. Infection, avitaminosis, and intoxication are presented, and the avitaminosis theory is accepted in general without any explanation for sudden disappearance of this violent disease. In 1891, Sakaki[1] suspected that rice grains damaged by infection of fungi are responsible for the development of the acute type of cardiac beriberi, since he succeeded in demonstrating that an ethanol extract of naturally mildewed rice grains contained neurotoxic agent(s) which caused convulsions, paralysis, and respiratory arrest in frogs, rabbits, and mice.

This is the first experimental toxicology on the etiology of cardiac beriberi from the standpoint of the intoxication theory, although both the causal agent(s) and the actual fungal name were not identified at that time.

In 1918, Miyake started the screening of fungi which are responsible for the damage of rice grains. The first report on yellowed rice was published in 1940.[2] One sample was from Taiwan (Formosa) rice produced in 1936, and the remaining nine samples were from domestic rice produced in Northern Japan during 1934 to 1939. The isolated fungal species was originally named *Penicillium toxicarium* Miyake,[3] but later the name was changed to *P. citreo-viride* Biourge by Naito.[4-5]

Since 1942, Uraguchi and co-workers have conducted a series of toxicological observations, in vivo and in vitro, using only the crude extract of *P. citreo-viride*-molded rice, as will be described in detail in a later section. The major symptoms of intoxicated animals were characterized by a ascending paralysis and respiratory arrest, similar to the actual cases of cardiac beriberi. As for the chemical principle of the crude toxin, Hirata[6-8] isolated a yellow pigment, named citreoviridin, and Sakabe et al.[9] determined the chemical structure. The toxicology on the purified toxin was carried out by Uraguchi[10] and the author.

In this chapter, the author aimed to summarize the mycology, chemistry, toxicology, and biochemistry of citreoviridin. The detailed information is also summarized by Uraguchi[11-12] and the author.[13-15]

MYCOLOGY

Citreoviridin Producing Fungi

Citreoviridin was first isolated by Hirata[6-8] from *P. citreo-viride* Biourge and later from *P. ochrosalmoneum* Udagawa (= *Eupenicillium ochrosalmoneum*).[16-17] Nagel and Steyn[18] reported that, besides *P. ochrosalmoneum, P. pulvullorum* produces the mycotoxin on maize.

Since the discovery by Sakaki[1] that an ethanol extract of naturally molded rice grains

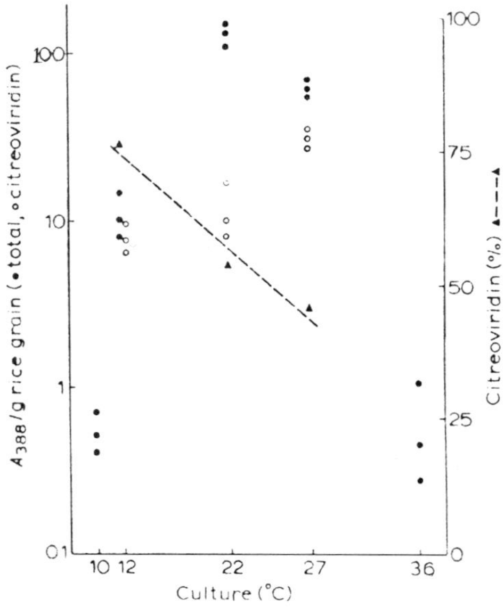

FIGURE 1. Temperature-dependency of citreoviridin production by *P. citreo-viride on rice grains.* (From Ueno, Y., *Jpn. J. Exp. Med.*, 42, 107, 1972. With permission.)

causes neurotoxic effects to animals, Miyake and co-workers, Naito and Tsunoda, carried out an extensive mycological survey on domestic and imported rice grains. Miyake et al.[2] detected *P. citreo-viride* for the first time in rice imported from Formosa, and in domestice rice grains harvested in 1937 to 1939. Thereafter, Tsunoda et al.[19] and Tsuruta and Tsunoda[20] examined the fungal pollution of cereals imported during 1954 to 1957 and found that *P. citreo-viride* was detected in 7.4% of the total samples examined and that a high incidence was observed in samples from Thailand, Burma, Italy, Spain, and others. These findings suggest that *P. citreo-viride* is distributed in rice throughout the world. A recent survey on the fungal pollution revealed that wheat flour, homemade miso (bean-paste), and maize are also contaminated by this fungus, but the incidence was very low in comparison with other toxic fungi.[21-22]

According to the mycological survey of Miyake et al.[2] the rice-parasitic fungi grow on rice during storage after harvest. A scratch on the surface of unpolished rice grains allows the fungi to enter the interior. The infection begins on stored rice grains when its water content reaches 14.6%. When the content is increased as little as 1% or more, other fungi start to grow and overwhelm *P. citreo-viride.* Their observations indicate that an early infection with this species occurs when rice grains are stored at such a narrow moisture content.

Citreoviridin Producibility

Since Miyake's survey indicates a great influence of humidity and temperature for the fungal growth and infection, the author analyzed the ecological behavior of *P. citreo-viride.* The fungus was grown on rice grains under different temperatures, and the lethal toxicity of the molded grains as well as the content of citreoviridin were estimated. As summarized in the publications,[13,23] the lethal toxicity of an ethanol extract of the moldy grains was highly toxic when the fungus was cultured at a rather low temperature. The direct determination of citreoviridin contents in these extracts revealed that the relative content of the neurotoxin was also high when the fungus was fermented at a lower temperature, as shown in Figure 1. Experiments with a liquid

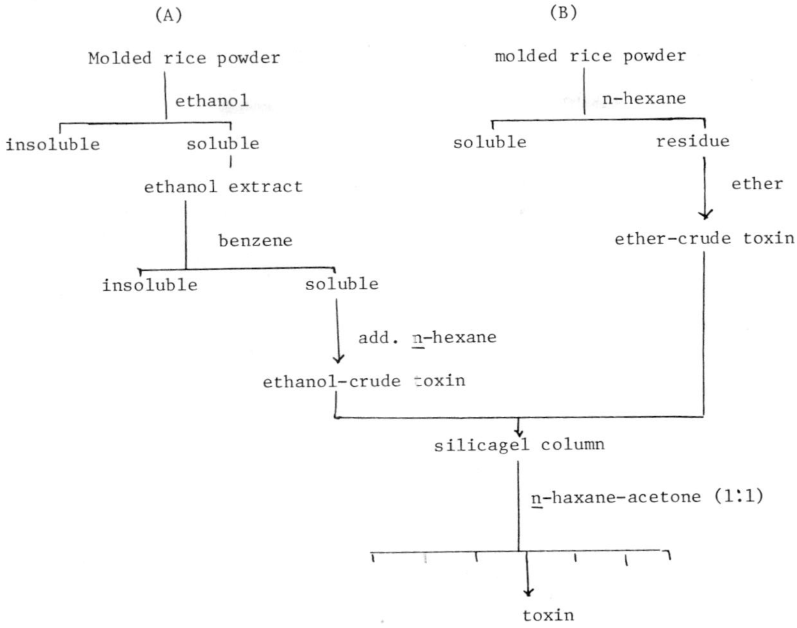

FIGURE 2. Isolation procedure of citreoviridin from *P. citreo-viride*-molded rice grains. (From Ueno, Y. and Ueno, I., *Jpn. J. Exp. Med.*, 42, 91, 1972. With permission.)

medium also revealed that the fungal growth as well as the citreoviridin production were rather high at 20°C.[23]

These ecological approaches support an assumption that the toxic fungus occurs particularly on the so-called "soft grains" produced in the northern part of the Main Island of Japan which faces to the Sea of Japan. In this region the duration of insolation is shorter since the climate is less sunny, and the temperature here is lower than in the regions faced to the Pacific Ocean. From these considerations, Uraguchi[12] has stressed that a more extensive investigation should be carried out on rice grains which are stored in regions where a less sunny and lower temperatural climate are predominant.

CHEMISTRY

Isolation of Citreoviridin

P. citreo-viride produces the mycotoxin either on rice grains or liquid culture media. The author isolated a large amount of citreoviridin from artificially molded rice grains by different procedures, as shown in Figure 2.[13,17]

Polished rice grains were allowed to stand for 30 min in tap water and autoclaved for 20 min at 15 psi. After cooling to room temperature, the grains were inoculated with the fungal spore suspension and incubated in a dark room for 3 weeks at 20 to 24°C. The yellowed rice grains thus obtained were dried at 52 to 53°C for 2 to 3 days, crushed in a mortar, and stored in a dark desiccator until used. In the procedure A, the rice powder was mixed with two volumes of ethanol, and after 4 hr with occasional shaking a deeply yellowed supernatant was decanted. The residue was then extracted with the same volume of ethanol, and the combined ethanol extract was concentrated *in vacuo* to a small volume. After standing at 4°C overnight, the white precipitate was filtered off and the yellow solution was further condensed almost to dryness *in vacuo.*

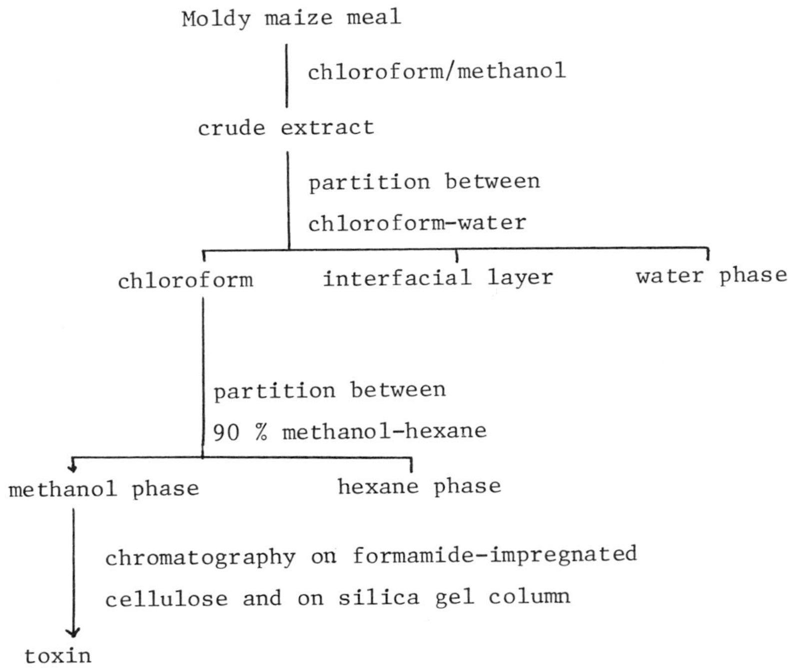

FIGURE 3. Isolation procedure of citreoviridin from *P. pulvillorum*-infected moldy maize meat. (From Nagel, D. W. and Steyn, P. S., *Phytochemistry*, 11, 627, 1972. With permission.)

The oily residue was dissolved in hot benzene and allowed to stand at 4°C for several days to precipitate a benzene-insoluble brown oil. Upon a drop-wise addition of *n*-hexane affords *n*-hexane insoluble precipitate. The thus obtained yellow precipitate was filtered and dried *in vacuo*. The author referred to this extract as "ethanol crude toxin".

In the procedure B, the rice powder was first extracted continuously with *n*-hexane for 2 days in a large Soxhlet extractor, then with ether. The ether-soluble fraction was dissolved in hot methanol and allowed to stand at 4°C overnight, and the clear supernatant was concentrated into dryness. The yellow powder was referred to "ether crude toxin". The crude toxins thus obtained were applied on a Kieselgel column and fractionated with *n*-hexane-acetone (2:1). Citreoviridin was crystallized from methanol.

According to Nagel and Steyn,[18] citreoviridin was isolated from *P. pulvillorum*-infected maize meal by the flow sheet presented in Figure 3. First, the moldy maize was extracted with chloroform-methanol, and the chloroform-soluble fraction was mixed with water and the partition between chloroform and water gave rise the toxic chloroform fraction. After treated with a mixture of methanol-*n*-hexane-water, the methanol phase was fractioned by formamide-impregnated cellulose and silicagel columns. Finally, the toxic was crystallized from methanol.

Synthetic liquid media are preferred for the production of citreoviridin as well as for study of its biogenetic pathway. Of the five synthetic media tested, the author found that Ushinsky medium supports the highest level of production both the mycelium and citreoviridin.[17]

Structure of Citreoviridin

Chemical structure of citreoviridin (Figure 4, I) was determined by Sakabe et al.[9] from the physical data (Table 1) and its degradation products (Figure 5).

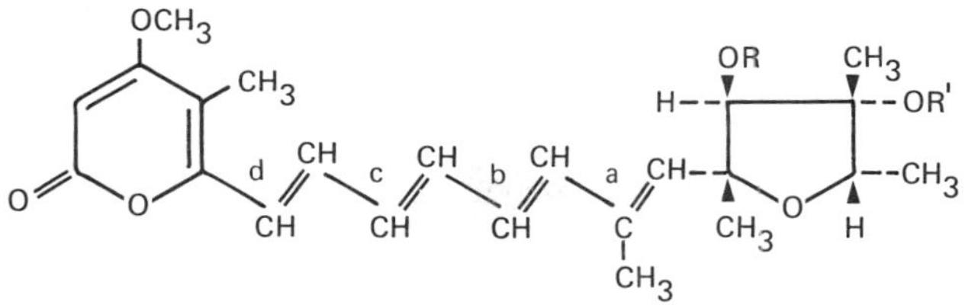

citreoviridin (I) R=R′ = H
citreoviridin monoacetate (II) R=CH_3CO, R′ = H
citreoviridin diacetate (III) R=R′ = CH_3CO

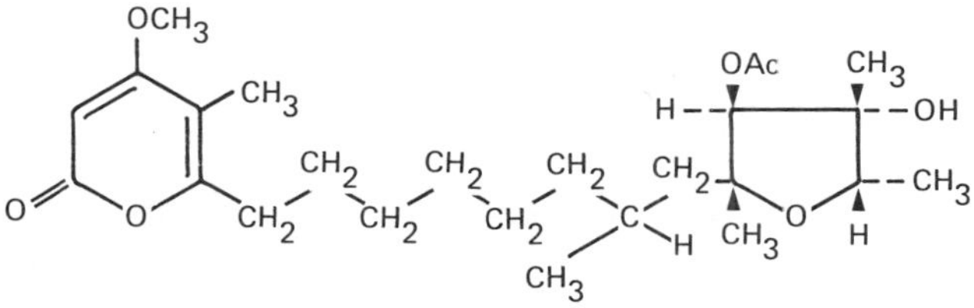

octahydrocitreoviridin monoacetate (IV)

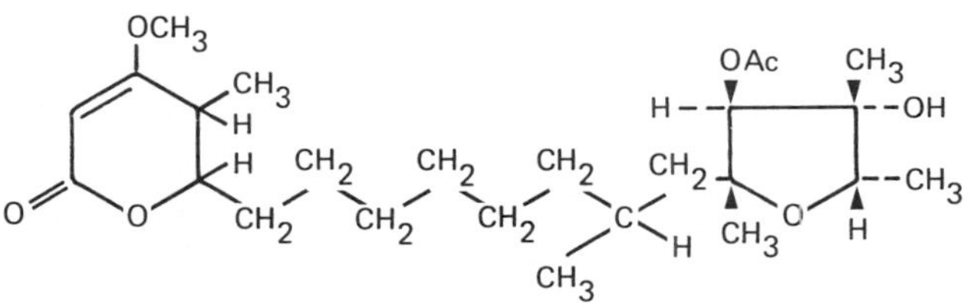

decahydrocitreoviridin monoacetate (V)

FIGURE 4. Citreoviridin and its derivatives.

Table 1
PHYSICOCHEMICAL PROPERTY OF CITREOVIRIDINS

	Citreoviridin				Isocitreoviridin		Citreoviridin monoacetate	Citreoviridin diacetate
Molecular formula	$C_{23}H_{30}O_6$				$C_{23}H_{30}O_6$		$C_{25}H_{32}O_7$	$C_{27}H_{34}O_8$
Mol wt	402.2				402.2		444	486
Melting point (°C)	107—111						99—101	
$[\alpha]^{18.6}_D$ (c = 1 in $CHCl_3$)	−107.8							
UV absorption (nm, δ)	388	(48000)[a]	383	(44000)[b]	382	(43200)	400_{sh}	399_{sh}
	294	(27100)	294	(24800)	297	(28800)	385	383
	286_{sh}	(24600)	285	(22000)	287	(27800)	294	293
	234	(10200)	238	(10500)	232	(18200)	285_{sh}	285_{sh}
	204	(17000)	206	(14600)	206	(24900)	238	230_{sh}
							204	204

[a] λ^{EtOh}_{max} (Ref. 9).
[b] λ^{MeOH}_{max} (Ref. 18).

Citreoviridin, when being crystallized from methanol, forms yellow needles $C_{23}H_{30}O_6 \cdot CH_3OH$, whose methanol is lost during storage. Permanganate oxidation of citreoviridin (I) in pyridine afforded a carboxylic acid (4). The IR (1720, 1630, 1558 cm^{-1}) and the UV spectra (λ^{EtOH}_{max} 294 nm, ε 6030) of the methyl ester suggest that it is

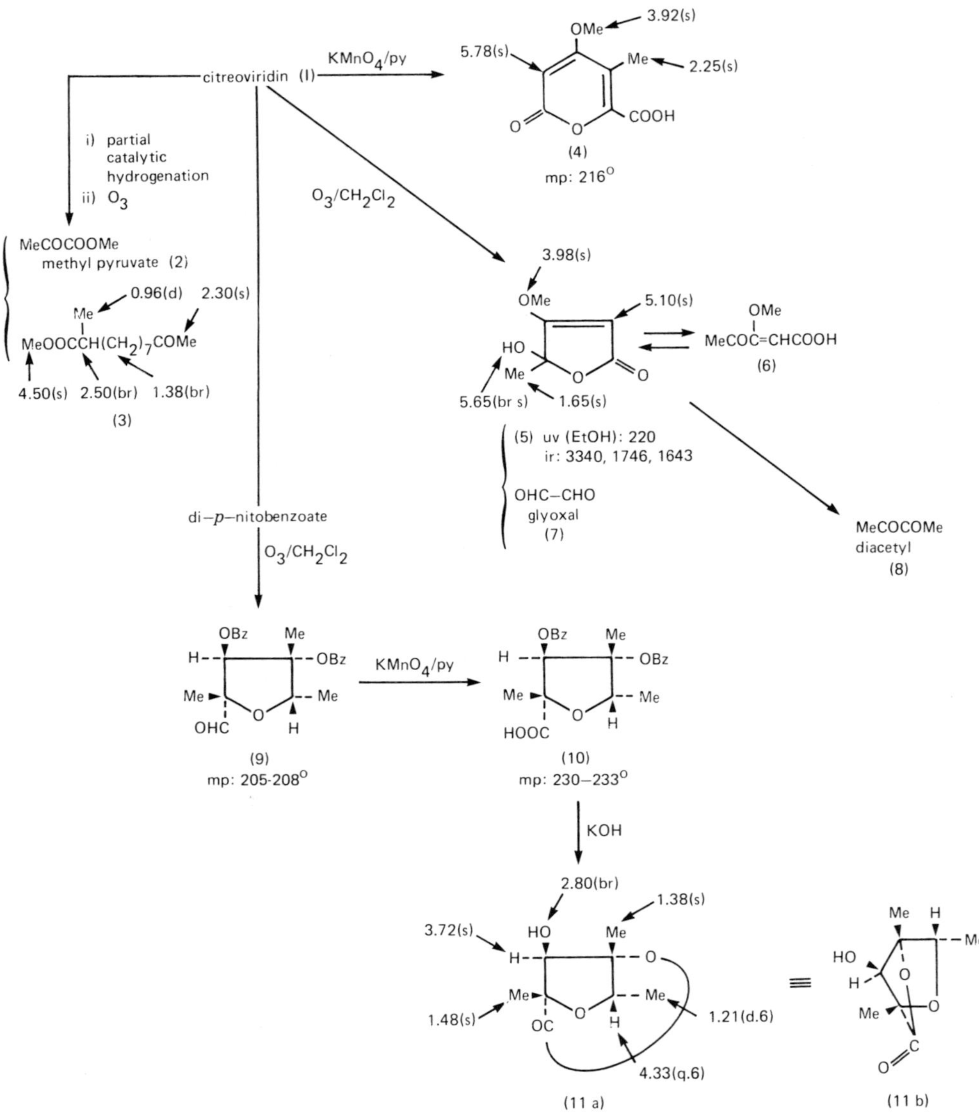

FIGURE 5. Chemical degradation of citreoviridin. (From Sakabe, N., Goto, T., and Hirata, Y., *Tetrahedron Lett.*, 27, 1825, 1964. With permission.)

an α-pyrone and not a γ-pyrone. Ozonolysis of citreoviridin (I) gave glyoxal (7) and diacetyl (8), both were identified as their 2,4-dinitrophenylhydrazones.

Partial catalytic hydrogenation followed by ozonolysis of I afforded methyl pyruvate (2) and an oily methyl ester (3). Ozonolysis of di-*p*-nitrobenzoate of citreoviridin (I) gave a saturated aldehyde (9), and the oxidation of 9 yielded the corresponding acid (10), which gave an oily five-membered lactone on hydrolysis with base followed by acidification (11a and b). From these degradation experiments and spectral analysis, the structure of citreoviridin (I) was proved to be composed from three moieties, α-pyrone, tetraene, and hydrofuran.

Treatment of citreoviridin (I) in diffuse light with a catalytic amount of iodine at room temperature gave a mixture of citreoviridin and a new yellow compound in a ratio of 7:3. The IR and mass spectra of these compounds were very nearly identical and the main difference in their UV spectra, estimated in methanol, was a shift of the

Table 2
R_f VALUES OF CITREOVIRIDIN AND ITS DERIVATIVES

Compounds	Solvents	R_f values	Ref.
Citreoviridin	Acetone-*n*-hexane (1:1)	0.45	17, 23
	Ethylacetate-toluene (1:1)	0.50	17
	Chloroform-methanol (9:1)	0.85	17
	Chloroform-acetone (5:1)	0.33—0.49	24
	Chloroform-methanol-acetone (45:3:2)	0.74	25
Isocitreoviridin	Chloroform-methanol-acetone (45:3:2)	0.79	25
Citreoviridin monoacetate	Chloroform-acetone (5:1)	0.57—0.68	24
Citreoviridin diacetate	Chloroform-acetone (5:1)	0.68—0.89	24

bands at 294 and 285 nm for citreoviridin (I) to 297 and 287 nm for the new compound. This bathochromic shift may be due to a change in the geometrical isomerism of citreoviridin (I) and all other evidence suggests that the two compounds are geometrical isomers. Using molecular models it appears from steric considerations that a change in geometrical isomerism from *trans* to *cis* has occurred at double bond c (Figure 4, I). This compound was named isocitreoviridin.[18] During the isolation of citreoviridin from moldy maize meal, isocitreoviridin was yielded in approximately 0.5% of citreoviridin, but this compound is an artifact since when very pure citreoviridin was submitted to simulated cultural and extraction procedure, it gave a mixture of citreoviridin and isocitreoviridin.[18]

Acetylation of citreoviridin (I) affords a monoacetate (II), mp 99 to 101°C,[9] and diacetate (III).[24] Upon hydrogenation of citreoviridin monoacetate (II) in ethyl acetate over 10% (w/w) Pd per charcoal at 20°C for 20 min at atmospheric pressure, octahydrocitreoviridin monoacetate (IV) and decahydrocitreoviridin monoacetate (V) are obtained. Their UV spectra revealed only 288 and 233 nm, respectively[24] (Table 1 and Figure 4).

As for the solubility of citreoviridin, the mycotoxin is soluble in ethanol, methanol, ether, benzene, chloroform, and acetone, and insoluble in *n*-hexane and water.[17] Upon exposure to light, the yellow color of citreoviridin turns to brown or dark red, and an illumination with UV lamp (366 nm) gives a brilliant yellowish fluorescence.[17]

Chemical Determination of Citreoviridin

Citreoviridin is soluble in alcohol and benzene, and absorbs light in visible region; λ_{max}^{EtOH} 388 nm (48,000). These chemical features were applied by the author to the analytical procedure for citreoviridin in moldy rice grains as well as in biological materials such as tissues and excreta.[17] The author[17] and Engel[25] developed photometric and fluorometric determination of citreoviridin. Table 2 summarized R_f values of the mycotoxin in different developing solvents. The UV maximum is at 388 nm in ethanol solution but is at 360 nm on thin-layer plates. Its emission maximum has been found to be 525 nm both in chloroform and on thin-layer plates. UV-densitometric and fluorodensitometric estimations of citreoviridin are similar in their sensitivity, and 0.1 to 1.0 μg per spot of the mycotoxin is able to estimate quantitatively.

TOXICOLOGY

Acute Toxicity of the Crude Extracts

Uraguchi carried out extensive research on the toxicology of citreoviridin. In an early stage of investigation, he[26] employed a crude extract of rice grains molded artificially with *Penicillium* sp., which was later identified as *P. citreo-viride* Biourge. The *Penicillium* sp. isolated from the Formosan rice by Miyake was inoculated on normal unpolished rice grains and incubated for a month to provide the moldy rice grains, and the extraction with ethanol followed with ether gave a yellowish-brown, malodorous, and viscous oily liquid. The crude toxin thus obtained were given i.p., s.c., or p.o. into seven mammals and five other vertebrates. Early onset of progressive paralysis in the hindlegs and flank, vomiting, and convulsions was followed by gradual respiratory disorder. The nervous symptoms were to be markedly observed in higher mammals such as cats, while less in amphibia. LD_{50} values fluctuated depending on the preparations, and mostly this value in mice ranged in 1.0 to 10 mg/10 g body weight. The i.p., s.c., and p.o. LD_{50} values formed a ratio of 8:10:30, and the times to death were 1.5 to 3 hr (i.p.), 3 to 6 hr (s.c.), and 3 to 8 hr (p.o.). From these observations, Uraguchi[26] came to the conclusion that the crude extract contains toxic principle, acting orally and parenterally, which induces progressive ascending paralysis in vertebrates.

With an aim to disclose the cause of toxicological effect of the crude toxin, Uraguchi[27-30] conducted extensive observations utilizing clinical and physiological techniques, with the following conclusions. Progressive paralysis of the ascending type is ascribed to spinal and medually depression, and the toxin is presumed to attack the motor neurons or internuncial neurons along the spinal cord and motor nerve cells in the medulla oblongata. The inhibition of the respiratory center is the cause of death, and direct inhibition of peripheral nerves and striated muscle ruled out.

Sakai and Uraguchi[31] investigated the long-term effect of the crude toxin to rats. Rats were exposed every day to the crude toxin by the repeated ingestion in doses of 1/100 or 1/6 of oral LD_{50} for 6 to 16 months. In the early stage of intoxication, loss of body weight was marked, and the death rate was 55% in 3 months in the case of the small dose and over 30% in the first 10 days with the large dose. Following oral administration, a histological examination showed only slight atrophy and pleomorphism of the liver cells in an animal dying in the later stage. These chronic experiments revealed that the crude toxin induces a chronic effect to rats but no particular evidences are observed. Long-term feeding experiment on rats with the moldy rice infected with *P. citreo-viride* revealed also no significant pathological changes except anemia.

Toxicity of Citreoviridin

Uraguchi[10] and the author[17] examined the acute toxicity of citreoviridin in mice, rats, cats, and dogs. The LD_{50} of the neurotoxin in male mice was 11 (s.c.), 7.5 (i.p.), and 29 (p.o.) mg/kg, and in female rats, the s.c. LD_{50} was 3.6 mg/kg. After a single dose near the LD_{50}, the poisoned mice died mostly within 0.5 to 6 hr (s.c.), 0.5 to 3 hr (i.p.), and 1 to 4 hr (p.o.). Acute symptoms developed such as lameness of the posterior extremities, impairment of voluntary movement and tremors, and the mice became unable to stand due to advancing paralysis. Later, depression, coma, gasping, and convulsions followed by respiratory arrest occurred. On autopsy, the right ventricle was dilated. In cats and dogs, vomiting occurs shortly after the administration of citreoviridin and followed by progressive ascending paralysis, convulsion, and respiratory arrest.

Subacute experiments showed that, in mice administered toxin orally in doses of 5 to 10 mg/kg/day, about half of the males and females died within 1 week. In mice administered the toxin s.c. at a dose of 3 mg/kg for 7 weeks, the death rate was 14/18 in the male and 2/10 in the female. These experiments show that the cumulative toxicity of citreoviridin is not demonstrated by short-term feeding and that the male is more susceptible than the female to the subacute intoxication by the neurotoxin.

In cats, the CNS is damaged by subacute administration of citreoviridin. A female cat administered i.v. four times in 2.5 hr in a total dose of 15.4 mg/kg developed ascending paralysis in the acute stage and followed by loss of eyesight several days after the administration. Pathological examination after 1 year revealed atrophy of the optic nerve. Another cat, received s.c. in two doses of 30 mg/kg and then in one or two doses of 15 mg/kg/week for 1 month, induced paroxysmal convulsions sporadically, which continued for about 10 min in intervals of 1 to 2 weeks. After this stage, the cat became nervous and abnormal behavior such as anxiety was found. From these cases, the neurotoxic action of citreoviridin impairs the CNS in cats.

Distribution and Fate of Citreoviridin

Employing the spectrophotometry and TLC analytical methods described previously, the author[17] investigated absorption, distribution, and excretion of citreoviridin in rats. The neurotoxin administered s.c. to female rats disappeared rapidly from the injection site, and among the major four organs the liver showed the highest content of mycotoxin; 3, 1.7, and 0.7% of the total dose were recovered as citreoviridin from the liver 8, 20, and 52 hr, respectively, after the administration.

About 3 and 1% of the mycotoxin administered s.c. to rats were detected from the feces at 21 and 45 hr, respectively, and no detectable amount was recovered from the urine. In cats and dogs, a small amount of citreoviridin was detected from the vomit. From these data, it is presumed that the rate of absorption of citreoviridin is rapid and metabolic degradation occurs in vivo.

BIOCHEMISTRY

Citreoviridin has a close structural similarity to the aurovertin B, a metabolite of *Calcarisporium arbuscula*,[32-33] which is a potent inhibitor of the mitochondrial ATP synthesis.[33-34] Experiments with soluble mitochondrial ATPase from ox heart revealed that autovertin B acts on ATP side of the oligomycin-sensitive site and inhibits at a site located on the coupling factor F_1.

Detailed experiments with rat liver and ox heart mitochondria were performed by Beechey and co-workers,[24,36] and the obtained information was as follows: citreoviridin inhibits the mitochondrial energy-linked reactions such as ADP-stimulated respiration, ATP-driven reduction of NAD^+ by succinate, ATP-driven NAD transhydrogenase, and ATPase from submitochondrial particles. The dissociation constant (Kd) calculated by a simple law-of-mass-action treatment for the citreoviridin-ATPase complex was 0.5 to 4.2 μM for ox-heart mitochondrial preparations and 0.15 μM for rat liver mitochondrial preparations. Citreoviridin monoacetate decreased its inhibitory potency (Kd = 2 to 25 μM for ox heart and 0.7 μM for rat liver), and citreoviridin diacetate greatly decreased the inhibitory potency (Kd = 60 to 215 μM for ox heart). Upon hydrogenation of double bond of citreoviridin monoacetate, the resulting octahydro- and decahydro-derivatives were much less effective inhibitors than the parent compound. From these data, it is clear that citreoviridin is a potent inhibitor of ATP synthesis reactions in mitochondria and that the conjugated polyene moiety of citreoviridin derivatives is an important component in the binding of the inhibitors to the ATPase. No significant enhancement of fluorescence was observed when citreoviridin interacted with the mitochondrial ATPase, in contrast to the case of aurovertin. Presumably the conformation in which citreoviridin is bound to the ATPase is such that an increased quantum yield is not obtained.

Recently, Verschoor et al.,[37] Douglas et al.,[38] and Gause et al.[39] investigated the effects of aurovertin and citreoviridin on the subunit of beef-mitochondrial and yeast ATPase. The isolated β-subunit, which possesses no ATPase activity, contained one aurovertin binding site with a dissociation constant of 0.56 *M*, and citreoviridin binds with the β subunit in different binding site. Fluorometrical analysis revealed that the binding of citreoviridin appears to alter protein conformation for ligand binding and influences the positioning of aurovertin within its site. At the present time, no information is available in regards to the relationship between the inhibitory effect of citreoviridin on ATPase and the toxicological feature of the mycotoxin.

As revealed in the preceding section, citreoviridin is a potent neurotoxic agent for vertebrates. The administration of citreoviridin causes ascending paralysis, disturbances in the CNS, convulsion, and respiratory arrest.[10-11] Datta et al.[40-43] also demonstrated the convulsive signs followed by paralysis of the upper limb in rats. Since the glycogen metabolism in brain tissues and convulsion are closely related, Datta et al.[44] investigated the level of glycogen metabolizing enzyme activities in brain tissues of rats after in vivo administration of citeroviridin. In the brain of rats administered with citreoviridin s.c., 48% (0.50 LD_{50}), and 76% (LD_{50}), indicating the dose dependency of its action. Partial recovery was noted during 5- and 10-day withdrawal periods. As for the enzyme activities, there is a significant decrease of glycogen synthetase activity in a dose-dependent manner. Glycogen phosphorylase and glucose-6-phosphatase activities were almost unaltered after citreoviridin treatment indicating the glycogen utilization process remain unimpaired and the reduction of glucogen level is mainly due to decreased glycogen synthesis.

Since some neurotropic agents such as picrotoxin, pentylenetetrazole, and others causes the reduction of glycogen level in brain tissues, the present data does not mean the specific site of citreoviridin. Impairment of the regulation system of glucose level is presumed to be more significant for evaluation of citreoviridin neurotoxicity.

REFERENCES

1. **Sakaki, J.,** Toxicological study on moldy rice, *Z. Tokio Med. Gesselsch.*, 5, 21, 1891.
2. **Miyake, I., Naito, H., and Tsunoda, H.,** Studies on the production of toxin in the Saprophyte-growing rice grains under storage (in Japanese), *Beikokuriyo Kenkyujo Hokoku*, 1, 1, 1940.
3. **Miyake, I.,** *Penicillium toxicarium* growing on the yellow moldy rice (in Japanese), *Nisshin Igaku*, 34, 161, 1947.
4. **Naito, H.,** Revision of *Penicillium toxicarium* Miyake (Rice Yellowsis mold), *Bull. Food Res. Inst.*, 18, 75, 1964.
5. **Naito, H.,** Studies on Molds Injurious to Stored Rice, Doctoral thesis (in Japanese), 1962.
6. **Hirata, Y.,** On the products of mould. I. Poisonous substance from mouldy rice (Part I), Extraction, *J. Jpn. Chem. Soc.*, 68, 63, 1947.
7. **Hirata, Y.,** On the products of mould. II. Poisonous substance from mouldy rice (Part II), Molecular weight, *J. Jpn. Chem. Soc.*, 68, 74, 1947.
8. **Hirata, Y.,** On the products of mould. III, IV (in Japanese), *J. Jpn. Chem. Soc.*, 68, 104, 1964.
9. **Sakabe, N., Goto, T., and Hirata, Y.,** The structure of citreoviridin, a toxic compound produced by *P. citreoviride* molded on rice, *Tetrahedron Lett.*, 27, 1825, 1964.
10. **Uraguchi, K.,** Mycotoxic origin of cardiac beriberi, *J. Stored Prod. Res.*, 5, 227, 1969.
11. **Uraguchi, K.,** Pharmacology of mycotoxins, in *International Encyclopedia of Pharmacology and Toxicology*, Section 71, Raskova, H., Ed., Pergamon Press, New York, 1971, chap. 17.
12. **Uraguchi, K.,** Citreoviridin: *Penicillium toxicarium* Miyake, *Penicillium citreo-viride* Biourge, and *Penicillium ochrosalmoneum* Udagawa, in *Microbial Toxins*, Vol. 6, Ciegler, A., Kadis, S., and Ajl, S. J., Eds., Academic Press, New York, 1971, 299.
13. **Ueno, Y.,** Production of citreoviridin, a neurotoxic mycotoxin of *Penicillium citreo-viride* Biourge, in *Mycotoxins in Human Health*, Purchase, I. F. H., Ed., Macmillan Press, London, 1974, 115.
14. **Ueno, Y.,** Citreoviridin from *Penicillium citreo-viride* Biourge, in *Mycotoxins*, Purchase, I. F. H., Ed., Elsevier, New York, 1974, 283.

15. **Ueno, Y. and Ueno, I.**, Toxicology and biochemistry of mycotoxins, in *Toxicology: Biochemistry and Pathology of Mycotoxins*, Uraguchi, K. and Yamazaki, M., Eds., John Wiley & Sons, New York, 1978, 107.
16. **Udagawa, S.**, Taxonomic studies of fungi on stored rice grains. III. *Penicillium* group (*Penicillia* and related genera) 2, *J. Agric. Sci. Tokyo Nogyo Daigaku*, 5, 10, 1959.
17. **Ueno, Y. and Ueno, I.**, Isolation and acute toxicity of citreoviridin, a neurotoxic mycotoxin of *Penicillium citreo-viride* Biourge, *Jpn. J. Exp. Med.*, 42, 91, 1972.
18. **Nagel, D. W. and Steyn, P. S.**, Production of citreoviridin by *Penicillium pulvullorum*, *Phytochemistry*, 11, 627, 1972.
19. **Tsunda, H., Tsuruta, O., and Takahashi, M.**, Researches for the microorganisms which deteriorate the stored cereals. XVII. Parasites of imported rice (1), *Bull. Food Res. Inst.*, 13, 29, 1958.
20. **Tsuruta, O. and Tsunoda, H.**, Researches for the micro-organisms which deteriorate the stored cereals. XX. An examination on appearance frequency among countries, of *Aspergillacea*, the main organism introductive to mildue putrefaction of imported rice, *Bull. Food Res. Inst.*, 14, 32, 1956.
21. **Ichinoe, M. and Kurata, H.**, Present status of the mycotoxin-producing fungi contaminated with domestic and imported foodstuffs, *J. Food Hyg. Soc. Jpn.*, 17, 337, 1976.
22. **Miyake, K., Yamazaki, M., Horie, Y., and Udagawa, S.**, Studies on the toxigenic fungi found in rice of Chiba Prefecture. IV. On the microflora of rice (3), *Ann. Rep. Inst. Food Microbiol. Chiba Univ.*, 23, 31, 1970.
23. **Ueno, Y.**, Temperature-dependent production of citreoviridin, a neurotoxin of *Penicillium citreoviride* Biourge, *Jpn. J. Exp. Med.*, 42, 107, 1972.
24. **Linnett, P. E., Mitchell, A. D., Osselton, M. D., Mulheirn, L. J., and Beechey, R. B.**, *Biochem. J.*, 170, 503, 1978.
25. **Engel, G.**, Untersuchungen zur Bildung von Mykotoxinen und deren Quantitative Analyse, *J. Chromatogr.*, 130, 293, 1977.
26. **Uraguchi, K.**, Pharmacological studies on the toxicity of the yellowed rice "O-hen-mai". I. Evidence of a toxin in the yellowed rice polluted by *Penicillium* sp., *Nisshin Igaku*, 34, 155, 1947.
27. **Uraguchi, K.**, Pharmacological studies on the toxicity of the yellowed rice "O-hen-mai". II. Physical properties of the yellowed rice toxin, *Nisshin Igaku*, 34, 224, 1947.
28. **Uraguchi, K.**, Pharmacological studies on the toxicity of the yellowed rice "O-hen-mai". IV. Toxicological feature of the yellowed rice toxin in developing acute poisoning, *Nisshin Igaku*, 36, 13, 1949.
29. **Uraguchi, K.**, Toxicological studies on the toxicity of the yellowed rice "O-hen-mai". VIII. Site of action of the yellowed rice toxin and cause of death in the acute poisoning, *Nisshin Igaku*, 42, 690, 1955.
30. **Uraguchi, K., Sakai, F., and Mori, S.**, Pharmacological studies on the toxicity of the yellowed rice "O-hen-mai". VIII. Site of action of the yellowed rice toxin and cause of death in the acute poisoning, *Nisshin Igaku*, 42, 690, 1955.
31. **Sakai, F. and Uraguchi, K.**, Pharmacological studies on the toxicity of yellowed rice "O-hen-mai". VII. Research on the possibility of chronic poisoning through long-term oral administration with the yellowed rice toxin in rats, *Nisshin Igaku*, 42, 690, 1955.
32. **Osselton, M. D., Baum, H., and Beechey, R. B.**, Isolation, purification and characterization of aurovertin B, *Biochem. Soc. Trans.*, 2, 200, 1974.
33. **Mulheirn, L. J., Beechey, R. B., and Leworthy, D. P.**, Aurovertin B, a metabolite of *Calcarisporium arbuscula*, *J. Chem. Soc. Chem. Commun.*, p. 874, 1974.
34. **Lardy, H., Connelly, J. L., and Johnson, D.**, Antibiotics as tools for metabolic studies. II. Inhibition of phosphoryl transfer on mitochondria by oligomycin and aurovertin, *Biochemistry*, 3, 1961, 1964.
35. **Roberton, A. M., Beechey, R. B., Holloway, C. T., and Knight, I. G.**, The effect of aurovertin on a soluble mitochondrial adenosine triphosphatase, *Biochem. J.*, 104, 54c, 1967.
36. **Beechey, R. B., Osselton, D. O., Baum, H., Linnett, P. E., and Michell, A. D.**, Citreoviridin diacetate; a new inhibitor of the mitochondrial ATP-synthesis, in *Membrane Proteins in Transport and Phosphorylation*, Azzone, G. F., Klingenberg, M. E., Quagliariello, E., and Siliprandi, N., Eds., North-Holland, Amsterdam, 1974, 201.
37. **Verschoor, G. J., Sluis, P. R., van der, and Slater, E. C.**, The binding of aurovertin to isolated subunit of F_1 (mitochondrial ATPase). Stoicheiometry of β subunit in F_1, *Biochim. Biophys. Acta*, 462, 438, 1977.
38. **Douglas, M. G., Koh, Y., Ebner, E., Agsteribbe, E., and Schatz, G.**, A nuclear mutation conferring aurovertin resistance to yeast mitochondrial adenosine triphosphatase, *J. Biol. Chem.*, 254, 1335, 1979.
39. **Gause, E. M., Buck, M. A., and Douglas, M. G.**, Binding of citreoviridin to the β subunit of the yeast F-ATPase, *J. Biol. Chem.*, 256, 557, 1981.
40. **Datta, S. C., Sengupta, D., and Ghosh, J. J.**, Toxicity of citreoviridin, a neurotoxic mycotoxin of *Penicillium citreoviride* on glycogen metabolism of rat heart, *Indian J. Pharmacol.*, 11, 84, 1979.

41. **Dattam, S. C., Sengupta, N., and Sengupta, D.,** Citreoviridin induced neurochemical changes in developing brain, in Abstr. 12th Annu. Conf. Nutr. Soc. India, Chandigarh, March 3 to 4, 1979, 16.
42. **Datta, S. C. and Ghosh, J. J.,** Effect of citreoviridin, a neuritoxin of *Penicillium citreoviride,* on some enzymes of rat liver, *Asian J. Pharm. Sci.,* 1, 119, 1979.
43. **Datta, S. C. and Ghosh, J. J.,** The GABA system in brain during acute citreoviridin intoxication, in Abstr. 48th Annu. Conf. Soc. Biol. Chem., Lucknow, October 11 to 13, 1979, 99.
44. **Datta, S. C. and Ghosh, J. J.,** Effect of citreoviridin, a toxin from *Penicillium citreoviride* NRRL 2579, on glycogen metabolism of rat brain, *Toxicon,* 19, 217, 1981.

MICROBIAL FOOD TOXICANTS: OCHRATOXINS

J. Harwig, T. Kuiper-Goodman, and P. M. Scott

INTRODUCTION

Ochratoxin A (OA) derives its name from *Aspergillus ochraceus,* the first mold from which it was isolated. It is the main toxic component in cultures of this mold.[1] Related fungal metabolites are the methyl and ethyl esters of OA, and the dechloro derivative ochratoxin B and its esters.[2,3] These were not discovered as a result of investigations into the etiology of a disease, but as a result of large-scale toxicity screening of molds present in South African foods.[4] Such screening has yielded numerous new compounds referred to as mycotoxins, but the link between these toxins and animal or human disease has remained tenuous in many cases. In contrast, the significance of OA as an agent of disease soon became evident. Following its discovery in 1965, information has become available on the natural occurrence of OA in foods and feeds, its toxicity, the molds producing OA, and its role in a disease of swine and poultry in Scandinavia. Attempts are now being made to link OA to a human kidney disease. The extraordinarily rapid progress achieved in this field is, to a large extent, attributable to the great chemical stability of OA and to its fluorescent properties. These facilitated development of suitable and sensitive analytical techniques. OA and related toxins have been the subject of previous reviews.[5-13]

CHEMISTRY

Physicochemical Data

Physicochemical data on ochratoxins and their structures are presented in Table 1 and Figure 1, respectively. Abbreviations for these compounds are listed in Table 1. Oa and Ob are also often abbreviated as Oα and Oβ. OA consists of a dihydroisocoumarin moiety linked through its 7-carboxyl group to L-β-phenylalanine. Support for this structure is derived from NMR data and mass spectra for the *O*-methylated methyl ester of OA[2] and from optical rotatory dispersion spectra of OA and related compounds.[14,15] It is a colorless crystalline compound with a melting point of about 90°C when crystallized from benzene. When the benzene of crystallization is removed by heating *in vacuo* and the compound is recrystallized from xylene, it melts at 169°C without xylene of crystallization.[2] It is soluble in polar organic solvents such as chloroform and methanol and in dilute aqueous sodium bicarbonate and emits green and blue fluorescence in acid and alkaline media, respectively.[19] The molar absorptivity (ε) of OA may be dependent on the concentration. However, at 333 nm and in the range of 1.9×10^{-6}-4.8×10^{-4}*M*, ε varied only little (Table 1). The UV absorption spectrum varied with pH[6,16,20] and with the solvent polarity;[21] a shift in absorption maximum from 325 to 375 nm was observed in changing the solvent from 96% ethanol to absolute ethanol, for example, reflecting the strong influence of water on the ochratoxin molecule. The same type of shift was found in the fluorescence excitation spectra while the corrected fluorescence emission spectra showed the opposite change with solvent polarity, going from a maximum of 467 nm in 96% ethanol to 428 nm in absolute ethanol.[21] Fluorescence intensity increased with increasing pH for OA, OB, and their ethyl esters.[22] The infrared (IR) spectrum in chloroform[14] included peaks at 3380, 1723, 1678, and 1655 cm^{-1}. OA is optically active.[2]

The mass spectra of OA, OB, and their methyl and ethyl esters have been recorded in the publication "Mycotoxins Mass Spectral Data Bank".[23] However, the spectrum

Table 1
PHYSICOCHEMICAL DATA ON OCHRATOXINS

Name	Abbreviation	Chemical formula	mp(°C)[a]	$\lambda_{(max)}$[b]	$[\alpha]20/_D$[c]	Ref.
Ochratoxin A	OA	$C_{20}H_{18}ClNO_6$	90(benzene), 169(xylene)	213(36,800), 332(6,400) [ethanol]	−118°(1.1)[c]	2, 14
				333(4,680 ± 40) [$1.9 \times 10^{-6} - 4.8 \times 10^{-4} M$ in ethanol]		15
				333(6,100) [pH < 4.0]		6
				380(7,650) [pH > 9.0]		6
Ochratoxin a (hydrolysis product of OA)	Oα	$C_{11}H_9ClO_5$	239, 237—239	212(30,000), 338(5,600) [ethanol]		2, 16
				217(32,090), 336(6,440)		16
Ochratoxin B	OB	$C_{20}H_{19}NO_6$	208—209, 221 (methanol)	218(37,200), 318(6,900) [ethanol]	−35°(0.15)	2, 17
				218(36,200), 318(6,700) [ethanol]		17
				318(6,500) [pH < 5.1]		6
				365(10,400) [pH > 10.0]		6
Ochratoxin b (hydrolysis product of OB)	Oβ	$C_{11}H_{10}O_5$	223	218(32,000), 322(6,300) [ethanol]		14
Ochratoxin C (ethyl ester of OA)	OC	$C_{22}H_{22}ClNO_6$	Amorphous	213(32,700), 331(4,100), 378(2,050)	−100°(1.2)	2, 3
				333(6,500) [pH < 4.0]		6, 17
				380(10,200) [pH > 9.0]		6
Ochratoxin D (4-hydroxy OA)	OD	$C_{20}H_{18}ClNO_7$	216—218 (benzene)	213(32,500), 334(6,400) [ethanol]		18
Ochratoxin A methyl ester	OA-M	$C_{21}H_{20}ClNO_6$		333(6,500) [methanol]	−78°(0.027)	3, 17
Ochratoxin B methyl ester	OB-M	$C_{21}H_{21}NO_6$	134—135 (benzene)	Identical to OB-E	−62°(0.02)	3
				318(6,500) [$CHCl_3$]		17
Ochratoxin B ethyl ester	OB-E	$C_{22}H_{23}NO_6$	102—103 (ether)	218(32,000), 318(5,200), 364(1,250)	−49°(0.04)	3

[a] Solvent used for crystallization indicated in brackets.
[b] UV absorption maxima; corresponding molar absorptivities in parens; solvents, pH, or concentration in brackets.
[c] Specific optical rotation.

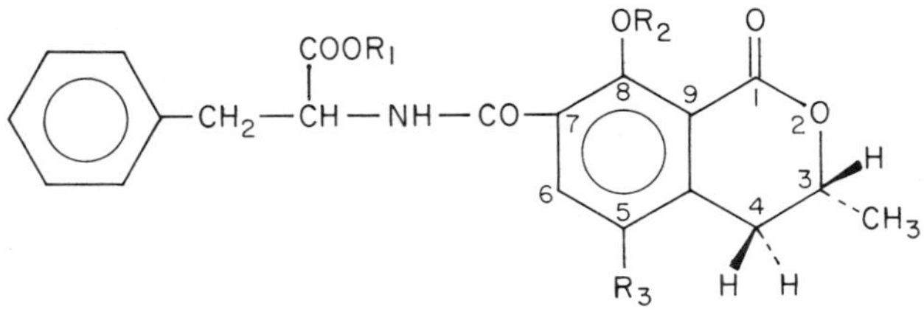

FIGURE 1. Structure of the ochratoxins; OA:$R_1 = R_2 = H$, $R_3 = Cl$; OB:$R_1 = R_2 = R_3 = H$; OC:$R_1 = C_2H_5$, $R_2 = H$, $R_3 = Cl$; OA-M:$R_1 = CH_3$, $R_2 = CH_3$, $R_3 = Cl$; OB-M or −E:$R_1 = CH_3$ or C_2H_5, $R_2 = R_3 = H$.

of OA is incorrect[24] and major peaks should be at m/z 239/241 and 255/257; the parent ion at m/z 403 is of very low intensity.

Although ethanol solutions of OA can be stored in the refrigerator for a year or more,[25] care needs to be taken during analytical procedures. OA solutions should be kept in the dark as photolysis may occur on exposure to fluorescent light.[15]

OA is hydrolyzed by vigorous treatment with acid[2] and more readily by treatment with α-chymotrypsin or carboxypeptidase A, thereby yielding L-β-phenylalanine and the lactone acid 7-carboxy-5-chloro-3,4-dihydro-8-hydroxy-3-methylisocoumarin (Oα).[20,26]

OB is the dechloro derivative of OA and is a colorless crystalline compound. Oβ, the dechloro analog of Oα, is the hydrolysis product of OB and was also detected in culture extracts.[18] Ochratoxin C is the amorphous ethyl ester of OA.[2,3] Ochratoxin D is identical to the nontoxic 4-hydroxyochratoxin A, isolated from *Penicillium viridicatum*.[18,27]

OA is readily esterified with methanol-HCl to yield the methyl ester of OA.[2] A red color reaction with ethanolic ferric chloride shows the presence of a free phenolic hydroxyl group in this ester. Subsequent methylation with diazomethane yields the *O*-methylated ester in which the phenolic group has also been methylated. Physicochemical properties of the *O*-methylated esters of OA and of other ochratoxins have been presented by van der Merwe et al.[2]

The methyl and ethyl esters of OB, which occur in cultures of *A. ochraceus*[3] can be obtained by esterification of OB with BF_3.[17] OA has been chemically synthesized.[14,15,28]

Isolation and Purification

Although OA is now commercially available, there is sometimes a need to isolate and purify the ochratoxins in the laboratory. The first step in this procedure involves extraction of ochratoxins from cultures of a suitable toxin-producing mold. OA can be extracted from acidified culture filtrates with chloroform,[29,30] or by passage through a column of ion-exchange resin followed by elution and subsequent extraction of the eluate.[31] Moldy substrates or mold cultures are extracted with ethyl acetate,[32] methanol,[33] hot chloroform,[27,30] acidified chloroform,[34] chloroform-methanol,[2,29] or chloroform-methanol preceded by hexane extraction.[25]

As a preliminary cleanup step, the acidic components of the extract including OA and OB are transferred into bicarbonate. This is then followed by acidification and extraction with chloroform.[2,17] The next step aims at separation of OA from OB and other acidic compounds and has been achieved by ion-exchange chromatography,[2] column chromatography on silica gel impregnated with 5% oxalic acid,[27] on Sephadex® LH 20,[25] by Sephadex® chromatography followed by silica gel chromatography in benzene-acetic acid (9 + 1),[34] by preparative liquid chromatography (LC),[33] or by preparative thin-layer chromatography (TLC).[35] For final purification, OA is crystallized from benzene[2,17,25,27] and OB from acidic aqueous methanol[2] or methanol.[17] It is important to remove any acetic acid before crystallization of OA from benzene.[34] OA-M, OC, OB-M, and OB-E are obtained by esterification of the corresponding acids with methanolic or ethanolic BF_3.[17]

Analysis

It is appropriate to note here that, like the aflatoxins, the ochratoxins may be produced at high concentration in nonrandomly distributed pockets or single seeds. To maximize the chances of detecting ochratoxins, samples should therefore consist of subsamples taken randomly from various areas of the bulk of a product. The subsamples should be pooled, ground finely, and mixed thoroughly, as suggested for mycotoxins in general.[36] Other general considerations in connection with ochratoxin analysis[6] are (1) the presence of benzene of crystallization in some preparations of OA, necessitating a correction in the quantitation by weight, (2) variation of absorptivity values with pH and solvent, (3) effect of pH and solvent on fluorescence properties of OA on TLC plates, and (4) photodecomposition by UV light.

Semiquantitative methods have been developed for various foods and have been reviewed.[6] Most use TLC for quantitation but several LC methods have been published recently.[37] Solvents and methods of extraction, fluorescence properties, and solvent systems for TLC determination of ochratoxins have been tabulated.[6] Emphasis will therefore be laid here on a few methods that have proven effective in the detection of OA in naturally contaminated products.

A method developed for barley[38] has probably received widest acceptance. It is fast, sensitive, and specific for ochratoxins, and has been tested in collaborative studies. It has been adopted by the Association of Official Analytical Chemists (AOAC) as an Official First Action Method[39] and has received the status of "Recommended method" by the International Union of Pure and Applied Chemists.[8] Samples are acidified with dilute phosphoric acid and extracted with chloroform. The extract is passed through a diatomaceous earth-aqueous bicarbonate column on which the acids are retained. Fats and ochratoxin esters are removed with hexane and chloroform, which are combined for ochratoxin ester analysis. Ochratoxin acids are eluted from the column by elution with formic acid-chloroform. The fraction containing ochratoxin esters is passed through another column consisting of methanol-aqueous bicarbonate-diatomaceous earth and is defatted with hexane-benzene. The esters are then eluted with formic acid-hexane-benzene. Amounts of ochratoxin acids and esters are determined by fluorescence on TLC plates, either visually or by densitometry.[39-42] The OA spots may be eluted with ethanol before fluorescence measurement.[43] For the acids, TLC plates are developed with benzene-methanol-acetic acid; for the esters, hexane-acetone-acetic acid is used.[39] In acidic solvent systems and under long wave UV light, the color of fluorescence is green, green blue, and pale green for OA, OB, and OC, respectively; all change to blue when sprayed with alkaline solutions[19,38] or when exposed to ammonia fumes.[44] This change in fluorescence can be used for confirmation of quantitation. The identity of OA and OB is confirmed by formation of ethyl esters with ethanolic BF_3.[39] These esters have a much higher R_f than the corresponding acids, with benzene-methanol-acetic acid as solvent system, and recently have been formed by derivatization at the origin of the TLC plate.[45] The percentage recovery of OA by the AOAC method, as determined in a collaborative study, was 112%;[46] the detection limit was 12 μg/kg.[38]

The sensitivity of detecting OA on TLC plates may be increased by exposing plates to ammonia.[47] The AOAC method has been used, either in its original or somewhat modified version, for barley,[38] barley and oats,[48] kidneys of swine,[49] coffee beans,[50,51] corn and milo,[52] and wheat and hay.[53]

The second most frequently used method was developed for simultaneous detection of ochratoxins, zearalenone, and aflatoxins in a wide variety of foods.[54] Toxins are extracted with water-chloroform. The chloroform extract is added to a silica gel column which is then washed with hexane and benzene for defatting; after eluting zearalenone and aflatoxins with appropriate eluants, ochratoxins are eluted with benzene-acetic acid. Confirmation of the identity of OA is obtained by TLC with several solvent systems, and by solubility in bicarbonate. Substances interfering with this method are present in barley, Brazil nuts, green coffee, cotton seed, and Capsicum pepper. The method was used for wheat,[55,56] corn,[57-59] and soybeans.[56]

A multitoxin detection method for ochratoxins, aflatoxins, zearalenone, sterigmatocystin, and patulin involves extraction with acetonitrile-aqueous KCl, defatting of the acetonitrile fraction with iso-octane, and transfer of toxins into chloroform. The lowest detection level of OA by TLC ranged from 45 to 100 μg/kg.[60] This method was used for analysis of moldy Canadian wheat, oats, barley, rye, mixed feed, and white beans.[61]

The first method described for OA involved Soxhlet extraction with chloroform-methanol, transfer of acidic ochratoxins into bicarbonate, and reextraction of the acidified bicarbonate solution with chloroform.[19] This method was used for analysis of Danish grain (mostly barley) associated with a kidney disease of swine referred to as mycotoxic porcine nephropathy (MPN).[62]

A simple screening method developed for detection of OA in cereals involved the following steps: (1) acidification of the sample and chloroform extraction, (2) transfer into bicarbonate, (3) extraction with chloroform, and (4) passage through a minicolumn. A fluorescent band in the column, under long wave UV light, indicates the presence of OA; levels of 12 μg/kg were detected.[63]

As OA and OB are hydrolyzed by carboxypeptidase A, the fluorescence of OA and OB in alkaline buffer containing carboxypeptidase A will decrease while that of Oα and Oβ will increase. This is the basis of a spectrophotometric method for detection of OA and OB in barley.[64,65] This loss of OA fluorescence was also used to determine the toxin in pig blood, serum, and plasma, using the pigs as in vivo sample collectors to screen for OA in feed.[66] Other multitoxin methods, including two LC methods, have been described for various foods,[67-83] but these have not yet been tested by laboratories other than the ones in which they were developed.

Most LC analyses of OA have been by reversed phase chromatography with acidic aqueous mobile phases, although Hunt et al.[82] used microparticulate silica and eluted OA with chloroform containing water and acetic acid in a method for foods that had a detection limit of 12.5 μg/kg after initial cleanup by preparative TLC. Josefsson and Möller[83] employed column chromatographic cleanup on silica gel, partition on a column of aqueous sodium bicarbonate-Celite and reversed phase LC to determine OA in grains down to 1 μg/kg (barley, corn, and wheat) with good recoveries. Osborne[84] obtained good sensitivity (0.5 μg/kg for flour) but low recoveries after a membrane dialysis cleanup step and sodium bicarbonate partition. OA could be analyzed in pig kidney following membrane dialysis concurrently with enzymic digestion, then partition into dichloromethane in the presence of formic acid; recoveries were satisfactory, 73 to 88%, and sensitivity of fluorescence detection, used in all LC analytical methods for OA, was increased tenfold by the formation of a post-column derivative with ammonia solution.[85] A rapid method based on partition steps only for cleanup was developed by Schweighardt et al.[86] for use in feed analysis but recoveries (45 to 57%) were again low. The methyl ester of OA has been formed and analyzed by LC for confirmation of OA in cereals and pig kidney.[83,87]

Radioimmunoassay and Immunofluorescence Microscopy

A beginning has been made in the development of methods involving radioimmunoassay (RIA) or immunofluorescence microscopy (IFM) for the detection and localization of minute quantities of OA in small samples of tissue or body fluids. These methods are specific and sensitive and require the use of OA-antisera. As OA is not antigenic, the toxin first needed to be conjugated to proteins or polypeptide carriers by established methods.[88-92] A hapten/carrier ratio of 20 in the conjugate has been obtained by Worsaae.[92] The toxin-protein conjugate was then used for immunization of rabbits.

In the development of an RIA method, OA-bovine serum albumin proved to be a useful antigen provided several booster injections were given.[90] As OA binds strongly with serum albumin, antisera were purified and pure immunoglobulin G (IgG) was obtained. Antibody titers were then determined by dialysis of the purified IgG against OA previously tritiated with tritiated water,[90,93] and by determining the amount of ^3H-labeled OA bound to the IgG. Specificity of the binding reaction of OA to IgG was shown by inhibition of binding by nonlabeled OA. OA, OC, and the tyrosine ethyl ester of OA, a synthetic analog of OA,[94] showed comparable binding activity while OB and Oα failed to compete with OA. Further research is needed before RIA methods can be applied to animal and human tissues and body fluids.

A method describing an IFM procedure for the localization of OA in kidneys involved incubation of cryostat sections with OA-antiserum from rabbits. The toxin-OA-antiserum was then "tagged" by incubation with fluorescein isothiocyanate conjugated to goat antirabbit serum IgG, and localized by fluorescence microscopy. Lack of fluorescence in appropriate control sections indicated that this procedure was specific for OA. OA was localized by this technique in renal tissues of rats and pigs containing 100 to 600 μg/kg.[91] Presumably, the method can be successfully applied to tissues containing levels of OA lower than 100 μg/kg.

OCHRATOXIN-PRODUCING FUNGI

Ochratoxin Production by *Penicillia*

P. viridicatum

Although OA was named after *A. ochraceus,* the scanty mycological data available indicate that *P. viridicatum* is the major mold responsible for OA contamination of foods. OA-producing isolates of *P. viridicatum* were present in naturally contaminated Canadian wheat,[61,95] Danish barley,[7,96] American wheat,[55] and Japanese rice.[97] In laboratory media, some of these isolates also produced citrinin.[61,95] As some naturally contaminated samples contained both toxins, the co-production of OA and citrinin supported the conclusion that *P. viridicatum* was the mold responsible for the contamination. However, in a more recent study on developing Danish barley grain (1977 crop) no *P. viridicatum* strains were isolated; *P. purpurescens* was the dominant OA-producing species with four toxin-positive isolates out of 22 strains.[98] Using an undescribed method, French workers failed to detect *P. viridicatum* in contaminated corn grown in France but did obtain OA-producing isolates of *A. ochraceus.*[99] Successful isolation of *P. viridicatum* from moldy grain containing predominant *Aspergillus* spp. required incubation of the grain under moist conditions at 12°C. *Penicillia* then emerged before *Aspergilli.*[95]

In laboratory experiments, some isolates of *P. viridicatum* from corn grown in Indiana have shown hepatotoxicity and nephrotoxicity which could not be attributed to OA and/or citrinin.[100] Different *P. viridicatum* isolates are known to produce different toxic metabolites, including penicillic acid and viridicatumtoxin, but these were not detected in the Indiana isolates.[101] Their toxicity may be due to a variety of pigments, some of which were identified as xanthomegnin, viomellein, rubrosulfin and viopurpurin.[102] Hepatic lesions induced by feeding mice xanthomegnin or viomellein at dietary concentrations of about 450 μg/kg feed were similar to those induced by crude culture extracts of an Indiana isolate.[103] A high pressure liquid chromatography (HPLC) method was recently developed for the detection of xanthomegnin in corn.[104] It is not known if these pigments play a role in naturally occurring animal disease but reference has been made to the similarity of some renal changes in avian, porcine, and human nephropathy and those occurring in swine fed a culture of an Indiana isolate.[104]

P. viridicatum occurs commonly in foods and feeds.[101,105] Among the molds associated with cheese warehouses, two of six *P. viridicatum* isolates produced OA.[106] *P. viridicatum* is one of the molds causing "blue eye", a condition that may develop over the winter in cribbed corn, as reported in the Iowa area.[107] This condition has also been observed in commercial popcorn from which some *Penicillia* were isolated that simultaneously produced OA and penicillic acid.[108] *P. viridicatum* is referred to as a storage mold because it occurs consistently in stored corn but not in corn at the time it is harvested.[109] The mold is able to germinate and grow under conditions of high water activity (a_w) and low temperature.[110,111] At an a_w of 0.90 and temperature of 12°C, *P. viridicatum* competed successfully with other molds native to wheat and barley and produced high levels of OA and citrinin, although yields were greater at 24°C.[112] An exhaustive study by Northolt et al.[111] confirmed these findings and showed that 24°C and an a_w of 0.97 were the optimum conditions for production of OA on barley meal and that some was also produced at 12°C. The same optimum temperature was observed on agar media and for *P. cyclopium.* Optimum a_w values for both species were 0.95 to 0.99 on the agar media. These observations indicate that OA contamination by *Penicillium* spp. may occur in grain and other agricultural products that have become excessively damp in storage.

Penicillia Other Than P. viridicatum

Other *Penicillium* spp. capable of OA production are *P. commune, P. cyclopium, P. purpurescens,* and *P. variabile;* these were present in European mold fermented sausages.[113] Molds belonging to the *P. viridicatum* series are commonly associated with meat products.[114] *P. verrucosum* var. *verrucosum* strains isolated from German meat products and from various foods and feeds consistently produced OA.[115,116] *P. palitans, P. purpurescens,* and *P. verruculosum* isolates from grains also produced the toxin.[61,98] *P. verrucosum* var. *cyclopium* is probably a potent OA producer in western Canadian grains as this was the main *Penicillium* species isolated from barley and wheat in which OA had formed during storage.[117]

The incidence of OA-producing *Penicillia* relative to the total number of *Penicillia* isolated has been investigated for some foods in addition to the study on Danish barley referred to previously, where 6 out of 87 *Penicillium* isolates formed OA.[98] In a survey of the mycotoxin-producing potential of *Penicillia* from mold-fermented sausages, 17 of 422 isolates produced OA;[118] 12 of the 17 were *P. viridicatum* isolates.[113] Also, 39 *P. verrucosum* var. *verrucosum* strains out of 1465 *Penicillium* isolates from foods and feeds and 46 strains of this species out of 954 isolates from meat products, including Italian, Hungarian, and German salami, produced OA;[115] 4 of 287 *Penicillium* isolates from cheddar cheese,[119] none of 159 isolates from Swiss cheese,[120] and 3 of 87 mold isolates associated with home-stored food produced OA.[121] OA-producing *Penicillia* have therefore a low relative incidence in foods such as cheese and meat. One of the species from cheddar cheese that produced OA was identified as *P. commune.*[122]

Ochratoxin Production by *Aspergilli*

A. ochraceus

A. ochraceus was common in samples of green coffee beans some of which contained low levels of OA.[50] It is therefore possible that *A. ochraceus* was responsible for the presence of OA in this tropical product.

As reviewed recently, molds in the *A. ochraceus* group are found in a wide variety of foods.[7,123] Consistent with their osmophilic properties is their ability to grow in dried and salted fish products in the Orient.[123] OA-producing isolates of this group have been obtained from foods such as cereals, nuts, hops, beans, coffee beans, and pepper.[7,98,124-127] If these products are favorable substrates for OA production, contamination may occur as soon as conditions arise that permit OA-producing strains to become predominant. *A. ochraceus* became predominant in corn stored for several weeks at 25°C and 18.5% moisture[128] and in inoculated wheat incubated at 20 to 25°C and > 16% moisture.[129] However, this species has difficulty invading kernels in which *A. glaucus* is already present. Since *A. glacus* is a common early invader of grain stored in the U.S., this observation may explain the observed low incidence of *A. ochraceus* in stored grain.[130]

A. ochraceus metabolites other than the ochratoxins include penicillic acid, secalonic acid A, mellein, 4-hydroxymellein, xanthomegnin, viomellein, rubrosulfin, viopurpurin, and certain pyrazine derivatives.[7,131-137] Some of these metabolites may add to the toxicity of foods contaminated with *A. ochraceus.*

Aspergilli Other Than A. ochraceus

Some strains of other members of the *A. ochraceus* group are capable of OA production. These are *A. alliaceus, A. melleus, A. ostianus, A. petrakii, A. sclerotiorum,* and *A. sulphureus.*[123,131] With the exception of *A. alliaceus,* these spp. also produce penicillic acid.[131]

PRODUCTION OF OCHRATOXINS IN THE LABORATORY

Synthetic Media

Molds in the *A. ochraceus* group are capable of producing OA in chemically defined media.[29,131,138-141] Glutamic acid or proline[29,138] and ammonium nitrate[139] are suitable N sources. Sucrose and glucose are suitable sources of C.[139,142] OA was produced in these media at levels of 100 to 200 mg/ℓ. Addition of 1% phenylalanine and 2% yeast extract to the medium used by Ferreira[138] raised levels to about 650 mg/ℓ.[32] Addition to synthetic media of appropriate amounts of copper, zinc, or iron[141] and of a combination of seven trace elements[142] stimulated OA production by some members of the *A. ochraceus* group. The increased production observed on addition of yeast extract may therefore be partly due to trace elements present in the extract. Another part of the increase may be due to the presence of stimulating factors such as amino acids. Both phenylalanine and glutamic acid stimulated OA production.[29,32,132,138]

Glutamic acid and proline could not be replaced in the induction of OA synthesis by any one of 25 amino acids and 9 peptides tested.[29] During its early stages of growth, *A. ochraceus* rapidly incorporated labeled glutamate into RNA, protein, and lipids. Only a small part of the label was present in the phenylalanine and Oα moieties of OA; the stimulatory effect of glutamic acid is therefore likely to be indirect. A direct effect may be exerted by a fragment of the glutamic acid molecule.[143]

Semisynthetic Media

Media consisting of various proportions of sucrose and yeast extract have proven effective in screening *Aspergilli* and *Penicillia* for OA production. A medium consisting of 2% yeast extract and 15% sucrose was suitable for members of the *A. ochraceus* group and for *P. viridicatum.*[30,61,95,141,144] *A. ochraceus* yielded a maximum of 490 mg OA per liter after 12 days of incubation at 25°C in a medium consisting of 2% yeast extract and 4% sucrose.[145] The same medium was used in studies determining the optimal incubation time, temperature, and ratio of medium volume to flask volume for small-scale stationary cultures of *A. ochraceus.*[146] It was also suitable for medium-scale production in stationary cultures at 25°C.[31]

Solid Substrate Fermentation

Much higher yields than those obtained with stationary or shaken cultures on liquid media have been achieved by solid substrate fermentation involving cereals.[127,147,148] This process requires determination of optimum conditions of initial moisture content, agitation and aeration rates, fermentation time, inoculum, and volume. Under optimal conditions, yields as high as 4 mg/g were obtained in wheat fermented by *A. ochraceus.*[149]

Foods

In the laboratory, OA production by *A. ochraceus* has been demonstrated in numerous foods, many of which had first been moistened and/or autoclaved. These foods include wheat and rice, bread, cakes and various other cereal products, soybeans, beans, peas, pecans, various oil-containing seeds, apple and grape juices, dry sausage, and country cured ham.[123,139,150-159] These observations indicate that, in the absence of limiting conditions of a_w, temperature, and competing microflora, many foods provide an adequate nutritional substrate for OA production by *A. ochraceus.*

Factors Affecting Production and Stability in Foods and Feeds

Temperature and a_w

In a semisynthetic medium, *A. sulphureus* produced about three times more OA at 28°C than at 20°C.[160] The optimal temperature for OA production by *A. ochraceus* on moistened cereal products was also 28°C; production at 15 and 37°C was much lower.[154] In cracked corn and wheat, production of OA was higher at 28°C than at 20 and 10°C; penicillic acid was produced simultaneously with OA, but its production was highest at 10°C.[131] A similar observation was made in studies with poultry feed: at a_w values greater than 0.85, OA production was greater at 30°C than at 15 and 22°C, but penicillic acid accumulation occurred to a greater extent at 15 and 22°C than at 30°C.[161] The highest yields of OA from *A. ochraceus* grown on whole wheat bread were at 20°C.[156]

Moisture conditions optimal for OA production by *A. ochraceus* in shredded wheat necessitated addition of 40 to 70 mℓ water to 100 g substrate. Under these a_w conditions, 2.5 g OA per kilogram substrate was detected after 19 to 21 days of incubation.[152]

A combination of 30°C and 0.95 a_w was optimal for OA production by *A. ochraceus* in poultry feed; a temperature of 22°C and an a_w of 0.90 were optimal for penicillic acid production by the same isolate. Among the a_w values studied, the lowest a_w values at which OA production occurred were 0.85 at 30°C and 0.95 at 15°C; for penicillic acid, these were 0.80 at 22 or 30°C.[161] Northolt et al.[111] also showed that 31°C was the optimum temperature for OA production by *A. ochraceus* on various agar media, compared to 24°C for two *Penicillium* species. Considerable amounts were also found at 37°C and an a_w of 0.95. At 24 and 31°C the optimal a_w was 0.99. These data indicate that OA contamination of foods by *A. ochraceus* may occur under conditions of temperature and moisture that are common in some tropical and semitropical regions. Penicillic acid production by the same mold may occur under conditions that are subboptimal for OA production.[162]

^{60}Co Irradiation

Low levels of ^{60}Co irradiation of spores of *A. ochraceus* increased production of OA in wheat.[163] It was therefore suggested that mishandling of irradiated foods may lead to increased OA production.

Attempts should be made to determine if the increased production was associated with a genetic change rather than a physiological one and if the change had a beneficial or deleterious effect on the survival of the mold in foods.

Insecticides

The effect of insecticides on OA production by *A. ochraceus* and *P. viridicatum* was studied in wheat treated with one of six fumigants, which was then aerated, and stored at 0°C before it was inoculated and incubated at a temperature favorable for toxin production. The effect of the insecticides varied with the strain of mold and type of insecticide. OA production was either increased or decreased but in no instance was it greatly inhibited. These effects may have no practical significance.[164,165] The organophosphate insecticide dichlorvos, incorporated into autoclaved corn at 100 to 300 mg/kg, inhibited OA production by *A. ochraceus*.[166]

Competing Microflora

When *A. ochraceus* and *P. granulatum* were placed in the same medium, OA production by the former was markedly reduced.[167]

Stability

Information on the stability of OA in foods is useful in predicting the extent to which OA, present originally in raw materials, is destroyed during processing. In moistened oatmeal and rice cereal, about 30% of added OA remained after 0.5 to 3 hr of autoclaving; storage of these products for 12 weeks at 28°C in the dark or light reduced the level by about 60%.[154] Naturally contaminated wheat did not lose OA when heated up to 200°C and the reduction was only 32% at 250 and 300°C for 40 min.[168] About 20% of OA in naturally contaminated pig products was lost during frying at 150 to 160°C and no loss from adipose tissue was observed.[169] These data indicate that OA has moderate stability in foods. However, levels of OA in green coffee beans were reduced by 80 to 90% as a result of simulated roasting procedures,[50,170,171] and there was considerable reduction of OA in white flour heated at 250°C for 40 min.[168]

During canning procedures, approximately 21% of the total OA content in contaminated moldy white beans was lost in the soaking water, 10% in the blanching water, and only 11% during heat processing. Therefore, a significant amount is likely to survive practical canning procedures.[172]

As OA has been detected in malting barley, questions arose as to the occurrence of OA in beer. OA present in barley at levels up to 830 μg/kg could no longer be detected after the barley had germinated.[173] Therefore, a large proportion of OA is likely to be destroyed during the malting stage of the brewing process. If some OA survived the malting process, an additional portion would be removed during subsequent steps of the brewing process. Only 10 to 28% of the total amount of OA added at the time of mashing was present in the final beer[173-175] with 28 to 39% present in the spent grain.[174] However, Tomova[168] found that there was no reduction of OA during pasteurization and boiling, excluding 22% adsorbed by beer ingredients. OA-containing barley invaded by *P. viridicatum* showed decreased viability.[112] Such barley is therefore not likely to be used for malting purposes. These observations indicate that, in practice, OA may not present a major problem in the beer brewing industry. OA was not detected in 130 samples of beer commercially available in the U.S.[176]

OA is moderately stable during the drying of sausage.[159] It is stable in pig kidneys stored for 10 months at −70°C when the contamination is of natural origin but stability is lower when the kidneys are injected with OA.[116]

Ochratoxins have been removed from contaminated grain by treatment with 2% NH_4OH for 1 month at 45°C or 6 weeks at 20°C with little change in nutritive quality.[177]

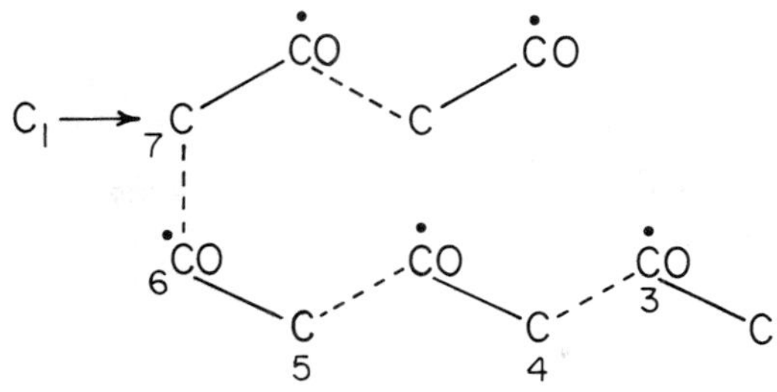

FIGURE 2. Biosynthesis of Oα through head-to-tail condensation of 5 CH_3COOH molecules and incorporation of a C_1 unit from the C_1 pool[178] (the dot indicates the position of the radioactive label).

Biosynthesis

Biogenetic studies on OA have been performed only with *A. ochraceus* cultures. The information available on this topic has been organized below on the basis of the two moieties of the OA molecule.

Isocoumarin Moiety

The first biogenetic scheme proposed for this part of the molecule involved head-to-tail condensation of five acetate units (Figure 2).[178] Evidence consistent with this proposed scheme included the observed exclusive incorporation of radioactivity from (1-^{14}C)-acetate.[179] Also (2-^{14}C)-acetate was incorporated mainly into Oα although some of the label showed up in the phenylalanine moiety,[180,181] possibly because of randomization of label after the long incubation period used.[178] Further evidence for this scheme was derived from the observation that, by degradation of Oα, carbon atom 3 and the methyl carbon at position 3 (Figure 2) could be obtained as acetic acid. One fifth of the total activity incorporated into Oα from (1-^{14}C) was located in carbon 3.[178]

As addition of (2-^{14}C)-malonate to cultures yielded OA with the activity located in Oα, an alternative pathway proposed for the Oα moiety involved condensation of malonate and acetate units.[182]

Theoretically, the carbonyl carbon at position 7 may be derived from a carboxyl group of acetate or from the C-1 pool. A decarboxylation procedure aimed at isolating this carboxyl carbon from Oα, labeled biogenetically from (1-^{14}C)-acetate, yielded inactive carbon dioxide. The first possibility was therefore excluded.[178] Evidence for the second possibility was derived from the following observations: (1) cultures incubated with ($^{14}CH_3$)-methionine yielded labeled OA and the carbon dioxide released as a result of decarboxylation of Oα contained the label,[178] (2) ^{13}C from (^{13}C)-formate was incorporated into the carboxyl carbon at position 7, as determined by NMR spectroscopy.[180,182]

Recently, the five intact acetate units shown in Figure 2 have been completely verified by ^{13}C NMR spectroscopy following incorporation of (1,2-^{13}C)-, (1-^{13}C)- and (2-^{13}C)-acetate.[183,184]

Little information is presently available on the introduction of the Cl atom into Oα. The greatest incorporation of ^{36}Cl into Oa occurred when Na^{36}Cl was added on the second or third day of incubation.[185] The chlorination step is probably brought about by a chloroperoxidase.[186]

Phenylalanine Moiety

As labeled phenylalanine was incorporated exclusively into the amino acid part of the OA molecule,[178-181] phenylalanine may be taken up and incorporated intact into OA. Phenylalanine itself is probably derived through the shikimic acid pathway.[178]

Table 2
NATURAL OCCURRENCE OF OA IN PLANT PRODUCTS

Sample	Incidence	Level (μg/kg)	Location	Ref.
Feeds (suspect, peanut meal excluded)	3/188	Traces	Britain	187
Wheat (moldy, 1976—1977)	10/37	?	Britain	188
Barley (moldy, 1976—1977)	9/31	?	Britain	188
Oats (moldy, 1976—1977)	1/2	?	Britain	188
Corn (food grade)	11/29	<50—500	Britain	189
Cornflour	4/13	<50—200	Britain	189
Soyabeans and soya products	15/53	<50—500	Britain	189
Cocoa beans and cocoa products	15/81	>60—500	Britain	189
Flour	2/9	6,250[a]	Britain	190
Flour, lumps (moldy)	2/7	2,900; 490	Britain	191
Bread (moldy)	1/50	210	Britain	191
Grains (heated)[b]	19/30[c]	20—27,000	Canada	61, 95
Mixed feeds (suspect)	2/7	20—530	Canada	61
Feeds (moldy or suspect)	7/95	30—6,000	Canada	53
White beans	3/?	20—2,100	Canada	61
Moldy peanuts	1 sample	4,900	Canada	61
Feedstuffs (suspect, 1972—1977)	3/2022	?	Canada	192
Feedstuffs (suspect, 1975—1979)	5/475	30—4000	Canada	193
Grains (1976—1979)	6/315	3—8	Canada	194
Barley and oats (associated with MPN)	19/33[d]	28—27,500	Denmark	62
Barley and oats	3/50	9—189	Denmark	62
Horse beans (moldy)	1 sample	?	Denmark	195
Barley	6%	<200	Denmark	196
Barley and oats (associated with MPN)	58%	<200—>1,000	Denmark	196
Cereals and cereal products	19/427	10—50	Denmark	197
Corn (1973 and 1974 crop)	18/924	15—200	France	99
Corn (moldy, 1974 crop)	2/75	?	France	198
Various foods	1/21[e]	50	Japan	199
Moldy rice	?	230, 430[f]	Japan	97
Feed grains (1974 crop)	8/150	50—200	Poland	200
Grains (1975—1977)	67/1121	?	Poland	201
Feeds (1975—1978)	31/622	?	Poland	201
Grains	4/21	5—25	Poland	202
Grain	5—7%	<140	Poland	177
Grain (17—30% H_2O)	?	1,000—3,000	Poland	177
Barley associated with MPN	2/4	20, 4,300	Sweden	116
Barley (food)	3/110	5—11	Sweden	203
Oats (food)	0/79	—	Sweden	203
Barley and oats (feed)	7/84	16—410	Sweden	48
Barley and oats (feed)	8%	<1,000	Sweden	196
Barley (1976 crop)	17/269	2—20	Sweden	204
Grains and feeds (suspect)	6/54	av 750	Sweden	205
Corn (feed, 1975 crop)	50/191	45—5,100	Yugoslavia	206
Grains (food from area with endemic human nephropathy)	6/47	5—90	Yugoslavia	207
Grains (food from area with endemic human nephropathy)	70/768, (153/1428)	<19—140 (2—140)	Yugoslavia	208, 209
Grains (food from nonendemic area)	1/64	14	Yugoslavia	207
Corn (1978)	58%	av 500	Yugoslavia	210
Barley (feed)	1 sample	3,800	Czechoslovakia	211
Corn (moldy)	2/49	18—22	E. Germany	212, 213
Coffee beans (moldy)	2/4	10—90	E. Germany	212, 213

Table 2 (continued)
NATURAL OCCURRENCE OF OA IN PLANT PRODUCTS

Sample	Incidence	Level (μg/kg)	Location	Ref.
Various foods	11/180	30—2,000	India	214
Corn (food)	1/283	110—150	U.S.	59
Export corn	3/293[g]	83—166	U.S.	58
Barley (1971 crop)	23/180	10—37	U.S.	176
Wheat (1970—1973 crops)	11/848	Trace—115	U.S.	55
Corn (poultry feed)	7 cases of illness of poultry	≤16,000	U.S.	215
Green coffee beans (imported)	19/267	≤360	U.S.	50
Coffee beans	2/201	24, 96	U.S.	171

[a] Positive samples were moldy.
[b] Wet grains reaching elevated temperature through microbial activity.
[c] Citrinin in 13 of these 19 samples at 70 to 80,000 μg/kg.
[d] Citrinin in 3 of these 19 samples at 160 to 2000 μg/kg.
[e] The only positive sample was discolored rice.
[f] Citrinin in these 2 samples at 700 and 1,113 μg/kg.
[g] Traces of OB in 2 samples.

The activity of an enzyme referred to as OA synthetase has been demonstrated in cell-free extracts containing adenosine triphosphate (ATP) and $MgCl_2$.[179] The enzyme catalyzes the coupling of phenylalanine with Oα through the amide bond. The overall biosynthesis of the ochratoxin molecule was the subject of a recent review.[186]

NATURAL OCCURRENCE

The first report on the occurrence of OA in an agricultural product appeared in 1969.[57] Since then, OA has been detected in a variety of foods and feeds, mostly from countries with a temperate or continental climate (Table 2). The incidence of OA in grains used as food was generally low and levels were mostly lower than 500 μg/kg. In feed, suspect either because of its moldiness or because of its association with nephropathy of swine or poultry, levels up to 27,500 μg/kg have been reported, with OA occurring in more than 50% of the samples.[196] In some samples of moldy rice[97] and grains,[61,62] OA occurred together with another nephrotoxic mycotoxin citrinin. Perhaps because of its instability in grains,[216] the incidence of citrinin was much lower than that of OA (Table 2) and levels in Danish feeds associated with MPN were also generally lower.[62] Occasionally citrinin levels as high as 80,000 μg/kg have been observed.[61]

OA has also been detected in green coffee beans, cocoa beans, soybeans, visibly moldy peanuts, and dried white beans (Table 2).

OA is sometimes present in tissues of swine and poultry consuming contaminated feed (Table 3). In swine showing macroscopically abnormal kidneys, OA was present at decreasing levels in blood, kidneys, liver, muscle, and fat. Although the kidneys may be rejected because of their abnormal appearance, livers, muscle, and fat may pass the Danish meat inspection.[49] This indicates that human exposure to OA may occur not only through consumption of moldy plant products but also through consumption of animal products. Levels detected in kidneys were 218 μg/kg or lower.[116] Cooking was not an effective method for destroying OA in contaminated meat.[169] Despite the associated severe nephropathy at levels of 100 to 200 μg/kg OA in kidneys, fattening pigs slaughtered at the age of 5 to 6 months otherwise appeared to be

Table 3
NATURAL OCCURRENCE OF OA IN ANIMAL PRODUCTS

Sample	Incidence	Level (μg/kg)	Location	Ref.
Kidneys of swine (suspected to suffer from MPN)	18/19	<67	Denmark	217
Livers of swine (suspected to suffer from MPN)	7/8	?	Denmark	217
Kidneys of swine (all but 2 with MPN)	34/34	32—218	Sweden	116
Livers of above swine	5/5	26—65	Sweden	116
Blood of above swine	3/5	175—335	Sweden	116
Macroscopically abnormal kidneys of swine at slaughterhouses	32/129	2—104	Sweden	218
Macroscopically abnormal kidneys of swine at slaughterhouses	24/90	2—88	Sweden	219
Macroscopically abnormal kidneys of swine at slaughterhouses	21/60[a]	<68	Denmark	49
Breast muscle tissue of poultry with macroscopically abnormal kidneys at slaughterhouses	5/14	4—29	Denmark	220
Muscle tissue of swine	?	?	Yugoslavia	221
Pig blood	47/279 herds	2—187 (μg/ℓ)	Sweden	222

[a] Calculated amounts of OA in livers, muscle, and fat from the swine with OA-containing kidneys were 2 to 47.5, 2 to 29.3, and 2 to 20.3 μg/kg, respectively.[49]

healthy.[116] Analysis of the OA content of pig blood has been used as an indicator of feed contamination[116,222] and could be used to determine the optimum time for slaughter after ochratoxin contamination of feed had occurred. In Denmark, if the concentration of OA in the kidney exceeds 10 μg/kg, the whole carcass is condemned[222] and registry of macroscopically abnormal kidneys at slaughter followed by ochratoxin analysis in order to trace outbreaks of MPN is done in Sweden.[116]

MYCOTOXIC PORCINE, AVIAN, AND HUMAN NEPHROPATHY

Relationship Between Nephropathy and Ochratoxin-Contaminated Swine and Poultry Feed

Recent evidence indicates that OA plays a major causal role in a disease of swine and poultry characterized as nephropathies. In Denmark, MPN was recognized as a disease entity in swine 50 years ago and was reproduced experimentally by feeding swine naturally moldy suspect rye.[223,224] In more recent experimentation, the disease was reproduced by feeding rats and swine barley cultures of a *P. viridicatum* isolate from suspect Danish feed.[96] Examination of its metabolites revealed the presence in the culture medium of two nephrotoxins, oxalic acid and citrinin.[225] Citrinin-induced but not oxalate-induced kidney damage in pigs resembled kidney damage observed in field cases.[226] It was therefore suggested that citrinin, possibly in combination with other mycotoxins, was responsible for MPN. In a Canadian study, OA was recognized as yet another metabolite of *P. viridicatum* and it was therefore speculated that OA contributed to the disease syndrome of MPN.[30] Subsequent detection of OA and citrinin in Danish feeds and of OA in organs and tissues of affected swine provided solid grounds for these suggestions. Since MPN could be reproduced in swine by OA in the feed in the absence of citrinin, and since citrinin showed lower nephrotoxicity, incidence, and levels in Danish feed and is less stable than OA,[62,216] OA was concluded to be the major disease determinant.[227]

Few data are available indicating the economic importance of MPN. It was recognized early that in years of wet climatic conditions the disease could affect up to 7% of the pigs in Danish slaughterhouses.[96] More recent data relating to Danish bacon pigs slaughtered at 5 to 6 months of age indicated a prevalence rate in 1971 of 0.6 to 65.9 cases per 100,000 pigs, which was considered an epidemic year. The prevalence rate was greater in females than in (usually castrated) males.[223] Cases have been recognized also in Norway, Sweden, and Ireland.[116,223,228]

A disease of poultry related to OA-contaminated feed has also been discovered in Denmark.[220] Of 14 samples of breast muscle from chickens, condemned at the slaughterhouse because of a pale and enlarged appearance of the kidneys, 5 contained OA (Table 3). The spontaneous mycotoxic avian nephropathy (MAN) was characterized by tubular degeneration and interstitial fibrous tissue formation. Three outbreaks of a similar disease involving 360,000 turkeys have recently been reported in the U.S. and were traced to OA-contaminated corn.[215] Four outbreaks of a disease characterized by poor growth and poor feed conversion in about two million chickens were also associated with OA-contaminated corn products.[215] The economic impact and public health significance of such outbreaks remain to be investigated.

Experimentally Induced Effects

Swine

Morphological and Functional Renal Changes

Morphological changes attributed to OA in naturally contaminated swine feed were restricted mainly to the kidneys.[116,223,227] Gross examination revealed enlarged and pale kidneys with a mottled surface, increased consistency, and, in advanced cases, numerous small cysts. While the medulla was normal in color, the cortex was pale, with the ventral cortex paler than the dorsal cortex.[227]

At the microscopic level, the initial stage of MPN was characterized by degenerative changes in the proximal tubules and interstitial fibrosis. The proximal tubules showed reduction of the brush border and dilatation and contained desquamated epithelial cells.[116,223,227] Initial degenerative changes were followed by tubular atrophy accompanied by thickened basement membranes, hyalinization of some glomeruli, and cyst formation. Cysts presumably arose from dilated lymphatics and interstitial fibrous tissue formation[223] or from dilated proximal tubules.[116] The difference in color between the ventral and dorsal cortex was attributed to more pronounced interstitial fibrosis in the ventral than in the dorsal region.[223,227] Similar microscopic changes in renal structure were present in experimentally induced MPN using exposure levels for several months of 200 μg OA per kilogram feed, corresponding to those encountered in naturally contaminated feed,[223,227,229-232] and kidney tissues from experimentally induced MPN were therefore used to further characterize the kidney lesions.

Using indirect IFM, OA was seen to be focally localized in about one third of the epithelial cells of the proximal tubules of pigs fed 400 μg OA per kilogram feed for 5 days, in accordance with the histologically observed focal necrosis of proximal tubules.[91]

Ultrastructurally, at early stages of induced MPN, atrophied epithelial cells of the proximal tubules showed reduction of the brush border, large apical vesicles, and a large number of lysosomes. This observation was consistent with the observed increased acid phosphatase activity. The atrophied epithelial cells contained a reduced number of mitochondria, some of which had no cristae. This abnormality was reflected in reduced activity of mitochondrial enzymes.[223,232]

As expected from the structural damage observed, tests performed in the course of an experiment involving OA at 200 to 4000 μg/kg feed revealed impairment of kidney proximal tubular function.[223,227] This was apparent fom the early appearance in the urine of leucine-aminopeptidase, an enzyme normally present particularly in the brush border of the proximal tubules. The urinary increase of this enzyme occurred concomitantly with a decrease in the activity of this enzyme in the renal cortex. Quantitative analysis of this enzyme in the urine and kidney may aid in the diagnosis of MPN related to OA poisoning in pigs.[232] Other signs of tubular function impairment were a decreased ratio between maximal tubular excretion of *p*-aminohippuric acid and inulin clearance, a decreased ability to concentrate urine, and an increased excretion of glucose.[223,227]

Long-term consumption of 25 to 50 μg OA per kilogram body weight (1 mg/kg feed) for up to 2 years induced progressive nephropathy in pigs but terminal renal failure was not reached.[232] The structural and functional changes were similar to those found after feeding pigs for 3 months with naturally contaminated barley containing 4 mg OA per kilogram.[227,232,233] In both experiments, during the first 3 months of exposure, the pigs given OA had a significantly lower growth rate and feed conversion than controls, but in the remaining part of the long-term feeding study these differences disappeared. The long-term exposure seemed to aggravate mainly the decreased renal absorptive capacity for glucose and the ability to produce hyperosmolar urine.[232] Other than this there seemed to be a tissue resistance to further nephrotoxic effects of OA, once renal impairment was established.[232]

At levels of OA high enough to cause swine to be moribund within 6 days, microscopic lesions were not restricted to the kidney, but occurred also in the intestinal tract, liver, and lymphoid tissue.[234,235]

Residues and Excretion of OA

As was observed in field cases of MPN (Table 3), OA residues occurred in decreasing order in the kidneys, liver, muscle, and fat of bacon pigs to which pure OA at a level of 1 mg/kg feed had been administered for 1 month, 3 months, or 2 years[230,231] (Table 4). Maximum OA tissue residues were reached after about 1 month (kidney approximately 25 μg/kg) after which tissue levels of OA remained in the same range (Table 4).

At steady state the feed/kidney tissue ratio for OA was therefore about 40, similar to that observed in a natural outbreak of ochratoxicosis,[116] and slightly higher than observed after short-term administration to pregnant pigs.[236] Serum levels in the latter study were 16 times those in the kidney, similar to the 5- and 13-fold greater levels seen in blood and plasma compared to kidney in a natural outbreak of ochratoxicosis.[116] Plasma levels of OA (μg/ℓ) thus approach the levels of OA in feed (μg/kg) and in one feeding study steady-state plasma levels were up to 1.5 × concentration in feed over a wide range of dosing levels[66] (Table 4). Blood or plasma OA residue levels thus are a good indicator of the level of OA in feed as well as other tissue OA residue levels,[222] and there are fewer sampling problems than when sampling feed.[66]

Determination of disappearance rates from kidney, liver, muscle, and fat indicated half-life values of 3 to 5 days,[230] or slightly longer from blood.[66] Therefore use of OA-free feed for 1 month prior to slaughtering may prevent OA-contaminated meat from reaching the market.[230]

In two pregnant swine both fed OA (0.38 mg/kg body weight) and OB (0.13 mg/kg body weight) from day 21 to 29 of pregnancy, and slaughtered the next day, OA but not OB was present in maternal kidneys, liver, muscle, and placenta but not in the fetuses. OA and OB were present in the blood serum. Since OA and OB had been ingested orally, they had been absorbed from the alimentary tract. A comparison of the ratio of OA and OB present in the feed and that present in the blood serum indicated that OA is absorbed more readily than OB. OA, Oα, and Oβ, but not OB were detected in the urine and feces, indicating partial hydrolysis of OA and complete hy-

Table 4
EXPERIMENTALLY INDUCED RESIDUES OF OCHRATOXIN A IN ANIMAL TISSUES

Species	Level of OA in feed (mg/kg)	Time on test	No. of animals	Blood level of OA (μg/ℓ)	Tissue levels of OA (μg/kg): Kidney	Liver	Muscle: White	Muscle: Red	Fat	Skin	Placenta	Fetus	Milk	Eggs	Ref.
Pig	0.2	120 days	11	—	4[a]	1[a]		ND	ND	—					227
	1	118 days	11	—	14[a]	9[a]		4[a]	7[a]	—					227
	4	136 days	6	—	51[a]	38[a]		5[a]	37[a]	—					227
Pig	1	1 month	5	—	25.7	17.8		11.5	6.0	—					230
	1	3 months	3	—	26.8	11.4		9.5	2.7	—					232
	1	2 yr	6	—	23.5	10.4		5.1	1.2	—					232
	0.009[b]	14 days	2	12	—	—		—	—	—					66
	0.045[b]	14 days	2	55	—	—		—	—	—					66
	0.280[b]	14 days	2	300	—	—		—	—	—					66
	1.455[b]	14 days	2	2200	—	—		—	—	—					66
Pregnant pig	20[c]	8 days[d]	1	—	700	340		130	—	—	60	ND	—		236
	14[c]	8 days[d]	1	6990[e]	430	300		150	—	—	40	ND	—		236
Chicken	1	341 days	7	—	18[a]	5[a]	2[a]								237
	0.3	341 days	7	—	12[a]	2[a]	2[a]								237
Laying hen	0.5	6 weeks	8	—	37	26	ND	8	ND	ND				ND	238
	1	6 weeks	12	—	77	58	5	13	ND	—				ND	238
	4	6 weeks	16	—	107	73	16	21	ND	—				ND	238
	0.5	2 weeks	12	7	124	80	8	7	—	—				3	239
	5	2 weeks	12	7	124	12	3	4	—	—				4	239

Table 4 (continued)
EXPERIMENTALLY INDUCED RESIDUES OF OCHRATOXIN A IN ANIMAL TISSUES

					Tissue levels of OA (μg/kg)										
							Muscle								
Species	Level of OA in feed (mg/kg)	Time on test	No. of animals	Blood level of OA (μg/ℓ)	Kidney	Liver	White	Red	Fat	Skin	Placenta	Fetus	Milk	Eggs	Ref.
Broiler chicken	2	8 weeks	3	—	41	24	—	ND	ND	—					240
Cow	0.3—1.1	11 weeks	1	—	5	—		—	—	—	—	—	ND		241

Note: ND, not detected; —, not determined.

[a] Levels estimated from graph.
[b] Calculated equivalent level of OA fed; daily doses were 0.35 — 53.8 μg/kg b w.
[c] Calculated equivalent level of OA in feed; daily doses were 0.38 and 0.13 mg/kg b w.
[d] Day 21 to 28 of pregnancy.
[e] Serum level.

Table 5
LD$_{50}$ VALUES OF OCHRATOXINS IN AVIAN SPECIES

Species	Toxin	Age	LD$_{50}$	Ref.
Chick embryo	OA[a]	0 days	0.04—0.05 μg/egg	243
	OA[a]	6 days	0.01 μg/egg	243
	OA[a]	18 days	0.05—0.08 μg/egg	243
	OA[b]	2 days	0.7 μg/egg	244
	OA[b]	3 days	3 μg/egg	244
	OA[b]	4 days	7 μg/egg	244
	OA[c]	5 days	17 μg/egg	245
	Oα[c]	5 days	>100 μg/egg	245
	OA[d]	8 days	7 μg/egg	6
	OB[d]	8 days	25 μg/egg	6
	Oα[d]	8 days	25 μg/egg	6
Chicks	OA[e]	1 day	166 μg/chick	246
	OC[e]	1 day	216 μg/chick	246
	OA[f]	1 day	3.3—3.9 mg/kg	247
	OB[f]	1 day	Estimated 54 mg/kg	247
	OA[f]	1 day	2.1 mg/kg	248
	OA[f]	3 weeks	3.6 mg/kg	248
	OA[f]	10 days	10.7 mg/kg	249
	OA[f]	3 days	3.4 mg/kg	250
Turkeys	OA[f]	3 days	5.9 mg/kg	246
Quail	OA[f]	3 days	16.5 mg/kg	246
Duckling	OA[f]	1 day	150 μg/duckling	251
	OA, OC, and OA-M[e]	1 day	135—170 μg/duckling	3

[a] In propylene glycol; air cell and yolk sac injection.
[b] In propylene glycol, air sac injection, embryos examined on day 8.
[c] In aqueous dimethylformamide; mortality determined after 48 hr; yolk sac injection.
[d] Yolk sac injection; mortality observed after 1 week.
[e] OA in water; oral intubation.
[f] OA in bicarbonate; oral intubation.

drolysis of OB in the intestine. Data on the ratios of OA and OB or their hydrolysis products present in the feed, blood serum, feces, and urine showed that most of the Oα, arising from hydrolysis of OA in the intestinal tract, was absorbed from the intestine and subsequently excreted by the kidney. Most of the Oβ may be excreted with the feces.[236] In this study the reproductive tract and fetuses appeared normal.[242]

Avian Species
Chicken

The chick embryo has been used to compare the toxicity of the various ochratoxins (Table 5). After fertilization, embryonic sensitivity to OA increased from day 0 to 6, then decreased.[243] OB, Oα, and Oβ were much less toxic than OA. OA also had a teratogenic effect; injections into eggs yielded malformations related to limb, neck, eye, and brain development, and microscopically cardiac anomalies were observed in some 48-hr embryos.[244,252]

Young chicks have also been used extensively in toxicity tests. Day-old chicks were more sensitive to OA than older chicks. As in the embryo, OB was much less toxic than OA (Table 5).

OA is the most potent mycotoxin studied in the chicken.[248] Acute poisoning was associated with acute nephrosis (MAN), hepatic degeneration, and suppression of hemopoiesis in the bone marrow, in decreasing order of frequency.[124,247,253] Other major effects included damage to the proximal tubules, enteritis accompanied by diarrhea, emaciation, and enlargement of the crop, proventriculus, and gizzard.[247,248,253,254]

Reduced growth rate and enlarged kidneys were evident in chicks ingesting OA at 1 mg/kg feed from hatching to 3 weeks of age.[248] In another study, ingestion by chicks of naturally contaminated feed at 0.3 and 1 mg OA per kilogram feed induced morphological renal changes that were evident only by microscopic examination after 341 days of feeding. These changes were not as pronounced as those observed in MAN occurring spontaneously in Denmark. This difference may be due to the presence of other nephrotoxins in the feed associated with the spontaneous cases or to higher levels of OA than those in the experimental feed.[237] Although macroscopic changes were not observed, levels of OA up to 50 μg/kg were present in kidneys, liver, and muscle, the kidneys containing higher residues than the other tissues (Table 4).[237] Functional changes in the kidneys of these birds were comparable with those observed in MPN.[255]

In laying hens, OA present in feed at levels up to 4 mg/kg caused reduced egg production, delayed sexual maturity, increased mortality, and reduced hatchability.[256] Some of these effects may have been due to other toxic metabolites since wheat cultures of *A. alliaceus* were used to adjust the levels of OA in the feed. Levels of 0.8 to 4.6 mg/kg feed had no effect on protein utilization.[257] Such an effect had been postulated because of an inhibition of carboxypeptidase at very high levels of OA (8000 mg/kg in feed). Decreased egg production, in some cases accompanied by eggshell staining, were also observed in other studies with pure OA at levels up to 4 mg/kg feed for 6 weeks, but without increased mortality and decreased hatchability.[238,253] Residues of up to 107 μg OA per kilogram were detected in the kidneys of laying hens fed pure OA at levels up to 4 mg/kg feed (Table 4), but eggs contained no detectable OA.[238] In these studies the half-life for OA in kidney and liver tissue was estimated to be about 13 hr. Similar levels of OA were seen in another study after 5 mg/kg OA in the feed was consumed for 2 weeks (Table 4), accompanied by a level of only 7 μg/ℓ blood,[239] whereas in studies with pigs the level of OA residues in blood was much greater than that in the other tissues.[236]

A series of experiments was conducted with OA in the feed of young broiler chickens at levels of 0.5 up to 8.0 mg/kg from hatching till 3 weeks of age.[258-266] OA lowered levels of plasma carotenoids.[258] As levels of carotenoids are indicators of carcass pigmentation, the presence of OA in poultry feed may explain the occasional occurrence of underpigmented birds in broiler flocks. This effect also occurs in aflatoxicosis of poultry.[258]

OA increased prothrombin and recalcification times.[259] These effects were less sensitive indicators of OA poisoning than the reduced growth rate.[259] Increased prothrombin times also occurred in laying hens.[238] OA therefore affects the coagulating properties of the blood and may predispose birds to bruising.

At 8 mg/kg in feed, OA induced iron deficiency anemia with a reduction in serum iron accompanied by smaller red blood cells. This effect is opposite to the increase in hemoglobin and packed cell volume seen during ochratoxicosis in beagle dogs and swine.[234,267,268] An inhibition of the GI absorption of iron similar to that seen for carotenoids was postulated.[260]

Leukocytopenia was induced at all dose levels and was primarily due to a decrease in lymphocytes, as well as a decrease in monocytes. A direct effect of OA on the germinal center was postulated. These effects may cause an impairment of cellular immunity.[261] Although the number of circulating heterophil leukocytes was not altered,[261] their phagocytic activity was decreased at 4 and 8 mg/kg in feed.[262] This effect was at the cellular and not at the humoral level, and appeared to be related to an impairment of both directed and undirected locomotion of heterophils inducing a "lazy leukocyte syndrome".[262]

In field outbreaks of avian ochratoxicosis the bones of chickens appeared more pliable and brittle than those of control birds.[263] The experimentally induced ochratoxicosis showed indeed increased fragility of bones at 2 mg/kg or more OA in feed, which was found to be due to a direct effect on intrinsic bone strength independent of bone diameter.[263] The cellular mechanism underlying this effect was not elucidated.

Increased intestinal rupture during processing of chickens was associated with 0.4 mg OA per kilogram in feed.[264] Experimentally the breaking strength of the large intestines was decreased at 2 to 8 mg/kg feed. This was accompanied morphologically by a decrease in stainable collagen and an apparent reduction in the size of collagenous folds.[264]

At 4 and 8 mg/kg in feed there was an increase in liver glycogen which morphologically was seen as a cytoplasmic accumulation of glycogen in the peripheral liver lobules.[265] The decreased glycogen mobilization seen in these chickens was associated with a decrease in cAMP-dependent protein kinase activity and consequent decrease in phosphorylase kinase.[266] This decrease leads to a malfunction of the phosphorylase system of glycogenolysis and therefore causes glycogen accumulation similar to that seen in genetically determined glycogen storage disease type X.[266] Glycogen accumulation as well as depletion has also been observed in other species (see Hepatic Glycogen Accumulation in a following section).

The inhibition of glycogenolysis seen in chickens could account for the increased susceptibility to cold stress seen both in experimentally induced ochratoxicosis and in field cases of the disease.[265,269] Field problems associated with OA ingestion may therefore result from the interacting effects of OA and adverse environmental conditions. This may make diagnosis of OA poisoning and evaluation of its economic impact very difficult.[269]

Serum alkaline phosphatase was increased over control values in chickens fed 5 mg/kg OA in feed for 2 months, but levels of serum sorbitol dehydrogenase and serum glutamate dehydrogenase were not affected,[270] indicating no serious cellular damage of liver or kidney tissue. The same authors reported an increase in serum lipids and proteins at 0.5 mg/kg feed, but a decrease at 5 mg/kg feed. At this higher concentration the content of serum globulins was also reduced.[271]

Turkey and Quail

OA was less toxic to young turkey and quail than to chickens (Table 5). In one study, OA at dietary levels of 4 mg/kg caused marked mortality and reduced weight gain in growing quail[272] whereas in another, involving different feed and mature birds, no mortality was reported.[272,273] At 4 mg/kg feed, OA had remarkably little effect on the production characteristics of quail, but at 16 mg OA per kilogram feed fertility and hatchability of eggs were depressed.[273] It was suggested that the embryotoxic effect of OA was through the presence of OA in the egg, rather than through an alteration of gametes.[274]

Duckling

Initial studies on the toxicity of OA in ducklings yielded an LD_{50} value of 25 μg per duckling. Later studies indicated a value comparable to that for chicks, i.e., 150 μg per duckling. OA, OC, and OA-M were equally toxic (Table 5). A histopathological study revealed primarily liver degeneration.[275] No mention was made of kidney damage.

A Link Between Balkan (Human) Endemic Nephropathy and Ochratoxin A?

Balkan endemic nephropathy (BEN) is a chronic and fatal disease of people living in the rural areas close to the Danube River and its tributaries in Yugoslavia, Romania, and Bulgaria. In villages in these areas, more than 12% of the population may be affected.[276] Clinically, the disease presents itself as a slowly progressing kidney failure that is only rarely accompanied by sodium retention or systemic hypertension. About one third of the patients dying from Balkan nephropathy in Bulgaria and Yugoslavia have been reported also to have papillomas and/or carcinomas of the renal pelvis, ureter, or bladder.[277] In spite of numerous attempts to relate the disease to bacteria, viruses, plant toxins, trace elements, or genetic factors, evidence as to its cause has been inconclusive.[207,276-279]

Contamination of foods with mold toxins as a possible cause was suggested by Dimitrov as early as 1960.[276] This suggestion is supported by the demonstrated correlation between the amount of rain in late summer and fall, and mortality from BEN in the three countries.[276] Possibly, wet harvesting and initial damp storage conditions are conducive to mold growth on plants in the field and in storage, and lead to contamination of foods with nephrotoxins of fungal origin.

P. verrucosum var. *cyclopium* (synonymous with *P. cyclopium*) was found to be a common isolate in food samples collected in an endemic area of Yugoslavia. A corn isolate of this mold, force-fed to rats, induced kidney lesions in the lower part of the proximal convoluted tubules which resembled closely the tubular changes seen in BEN patients.[280] Since the isolate produced OA, the induced kidney lesions were attributed by other workers to OA,[281] and because of the similarity in clinical features and pathology between BEN- and OA-induced nephropathy, these workers proposed that BEN is caused by OA.[221]

In an initial survey involving grains produced in Yugoslavia, OA was found in 12.8% of samples from an endemic area and in 1.6% of those from a nonendemic area.[207] A more extensive survey of 768 samples of cereals and bread locally produced in an area of Yugoslavia where BEN is prevalent showed a mean frequency of OA contamination during the 1972 to 1976 study period of 8.7%, with up to 43% contamination of barley in 1974.[208] The total number of samples from this area analyzed has recently reached 1428 with 10.7% OA contamination (2 to 140 μg/kg).[209] Of the human population in an endemic area, 7% had blood levels of up to 1.8 μg/mℓ OA compared to none in the control group.[13,282] These studies indicate greater exposure to the toxin in the endemic than in the nonendemic area.[207] In the future, sensitive IMF and RIA methods may be applicable to tissues and body fluids from BEN patients, thus providing more information on the etiology of BEN. Possibly, fungal nephrotoxins other than just OA are involved in BEN.

If BEN is indeed caused by fungal toxins, some conditions or practices in the Danube area must be more conducive to mold growth on food products than they are elsewhere. The molds implicated are widespread while the disease has only been recognized in some of the Balkan countries.

EXPERIMENTALLY INDUCED EFFECTS IN BIOLOGICAL SYSTEMS OTHER THAN SWINE AND POULTRY

Effects in Mammals

Rodents

Acute Toxicity

Administration of single oral doses of OA yielded LD_{50} values of 20 to 30.3 and

Table 6
LD$_{50}$ VALUES OF OA IN RODENTS

Species	Route	LD$_{50}$ (mg/kg body weight)	Ref.
Rat[a]	Oral	20—30.3	283—285
	i.p.	12.6	284
	i.v.	12.7	284
Rat (24 hr old)[b]	Oral	3.9	286
Mouse[a]	Oral	58.3	284
	i.p.	40.1	284
	i.v.	33.8	284

[a] In bicarbonate[283,284] or aqueous carboxymethyl cellulose-Na.[285]
[b] In corn oil.

58.3 mg/kg body weight for rats and mice, respectively (Table 6). Mice, therefore, showed lower sensitivity to OA than rats. Neonate rats were more sensitive than adult rats. The main pathological changes in rats associated with fatal doses were severe necrosis of renal tubules and necrosis of periportal cells of the liver.[283] Other studies described hemorrhagic effects in many organs, particularly the kidneys and lungs,[284] and severe enteritis.[285] It was speculated that in the rat OA may be metabolized to Oα in the intestinal tract and that both OA and Oα play a role cooperatively in inducing enteritis.[287]

Simultaneous i.p. injection of 1 mg phenylalanine per mouse completely inhibited the 100% lethal effect of 0.8 mg OA per mouse, but much higher doses of phenylalanine were necessary 30 min after OA injection.[288,289] In guinea pigs, multiple-dose oral administration of OA produced primary toxic effects on the renal tubular epithelium, similar to those observed in other species.[290]

The toxicity of OA in neonate rats was increased by simultaneous dosing with another mycotoxin, rubratoxin A, but this did not occur in a synergistic manner.[286] A combination of citrinin and OA or of penicillic acid and OA, however, evoked synergistic lethal responses in mice injected i.p.[291] These findings underline the possibility of synergistic effects on animals from toxins occurring together in feed.

Short-Term Toxicity

Weanling rats fed a diet containing graded levels of OA for 14 days showed retarded growth and reduced feed consumption. These effects were significant only at levels of 9.6 and 24.0 mg OA per kilogram feed. Degenerative changes involving the entire renal tubular system were observed in rats consuming the diet at 4.8 mg OA per kilogram and at higher levels.[292] A similar experiment in which rats were intubated daily with up to 2 mg OA per kilogram body weight described dose-related increases in volume of urine, in blood total protein and urea levels, and decreases in growth rate, urinary total nitrogen, and blood total lipids and cholesterol. Blood glucose levels remained unaffected.[293] These data indicate a profound disturbance of protein and lipid metabolism. Similarly five to seven i.p. injections once daily of 0.75 or 2.0 mg OA per kilogram body weight in rats led to a decreased body weight, increased urine flow, decreased urine osmolarity, increased urinary protein and urinary glucose, and changes in the transport of organic substances by renal cortical tissue.[294] Rats given daily oral doses of 0.25 to 8.0 mg/kg body weight showed hemorrhaging and ulceration of the digestive system, and hypertrophied, pale kidneys, sometimes with perirenal edema.[295]

In another similar short-term study the hemorrhagic syndrome was seen to be due to an alteration in the hemostatic mechanism. Plasma fibrinogen and thrombocyte and megakaryocyte counts were decreased and an anti-vitamin K-like response was evident as shown by a decrease in factors II, VII, and X.[296,297] This effect was attributed to the structural similarity of OA and anticoagulants of the coumarin series.[296]

Repeated oral dosing with OA in the rat led to degeneration of the convoluted tubules, thickening of basement membranes, and functional impairment of the kidney.[292,298] Mice fed isolates from *A. alliaceus* that produce OA showed similar changes.[299]

Single intratesticular injection of OA in the rat caused dilatation of seminiferous tubules 10 days later. Slightly higher doses resulted in degenerative changes in the spermatogenic epithelium.[300]

Subacute Toxicity

In a 90-day feeding study involving a diet containing 0, 0.2, 1.0, or 5.0 mg OA per kilogram, rats developed pathological changes involving the proximal convoluted tubules of the kidney. These changes were dose dependent and were observed at levels as low as 0.2 mg OA per kilogram feed. No hematological changes were observed in this study, in which the dose given was much lower than other studies reported here. By electron microscopy, basement membrane thickening up to 4 μm was observed in proximal tubular cells. Changes in both rough and smooth endoplasmic reticulum were also observed in these cells. When the animals were given an OA-free diet for 90 days following the experiment, regeneration of kidney lesions was only partial in the rats given 5 mg OA per kilogram feed; changes were reversible at the lower levels.[292]

In mice, weekly i.p. injections of 5 mg OA per kilogram body weight for 6 weeks also caused altered kidney function, accompanied by decreases in plasma albumin and gamma globulin and increases in alpha 1 and 2, and beta globulin.[301] In a similar study hematological changes were extensive with a sixfold increase in clotting time, and large decrease in RBC count and hemoglobin content of blood.[302] These changes are similar to those described for rats.[296]

Long-term oral administration of low levels of OA caused changes in the glycolytic processes of the rat's eye lens but these changes did not affect the lens transparency.[303]

Absorption and Tissue Distribution

Studies involving a single oral administration of OA to rats revealed that OA is absorbed by the stomach[304-307] and small intestine[305] and then appears in the blood and in many organs, particularly the kidneys, liver, and muscle.[304-306,308] As is consistent with its nephrotoxicity, the highest levels of OA accumulated in the kidneys soon after administration.[305,306] OA detected in the blood after oral, i.v., or i.p. administration was predominantly present in the serum albumin fraction, partly in the bound and partly in the free form[308-310] depending on the initial dose.[308] OA did not bind to RBC.[310] OA disappeared very slowly from the blood of rats given an oral or i.v. dose of OA[309,310] (half-life po-68 hr[309]), and more rapidly after an i.p. injection (half-life 12 to 18 hr).[308]

Excretion of OA and its Metabolites

Part of the OA ingested is excreted in the urine and feces as OA and part as its main metabolite, the hydrolysis product Oα;[304,305,307,308,310-312] a third compound excreted is the nontoxic 4-hydroxy derivative of OA.[18,311-313] Other unidentified fluorescent compounds have been observed in the feces and urine of OA-treated rats and these may be metabolites of OA.[304,310] Recovery studies of labeled OA suggest that a large part of the OA ingested is converted to unknown compounds.[6,308]

Attempts have been made to localize the site of hydrolysis of OA giving rise to Oα.

In vitro hydrolytic activity of homogenates of the pancreas, duodenum, and ileum has been demonstrated and Oα was detected in the GI tract and in the excreted serous fluid and mucus of rats dosed orally with OA.[305] Oα in the intestinal tract may therefore be the result of hydrolytic activity of proteolytic enzymes such as intestinal carboxypeptidase.[305] Hydrolytic activity with respect to OA could not be demonstrated in the liver[245] or kidney, or occurred only at very low levels.[305] Other investigators attributed the occurrence of Oα in the feces and intestine to the intestinal microflora.[304] Oα resulting from hydrolysis of OA in the intestinal tract may induce the acute catarrhal enteritis observed in rats given large doses of OA.[285,287,305] After i.v. administration, OA can be detected in rabbit milk.[314]

Hepatic Glycogen Accumulation

Glycogen accumulation is a controversial aspect of OA toxicity. It was observed in hepatocytes of rats given a single oral dose[283] and was hypothesized to result from the interaction of OA with nucleic acids or with enzymes of carbohydrate metabolism. No evidence was found for interaction of OA with nucleic acids but, in vitro, OA inhibited some component of the phosphorylase system of the liver and was therefore postulated to prevent breakdown of glycogen to phosphorylated glucose.[315]

Other investigators have observed depletion of hepatic glycogen instead of accumulation in rats and mice,[285,292,316-318] sometimes accompanied by elevated blood glucose levels.[317,318] The small pool of kidney glycogen was also decreased.[319] Intact and adrenalectomized rats showed depleted levels of hepatic glycogen but elevated levels of cardiac glycogen.[316] Hepatic glycogen depletion was more significant in adrenalectomized rats than in intact rats 4 hr after administration of OA. Hepatic glycogen levels in adrenalectomized rats that had been pretreated with hydrocortisone were more similar to those in intact rats than in adrenalectomized rats that had not received the hormone. In vivo, OA may therefore affect the hormone balance in intact animals.[316] Depletion of hepatic glycogen associated with ochratoxicosis of intact rats was associated with inhibition of glucose transport into liver tissue, inhibition of glycogenesis, and increased glycogenolysis.[317]

In the kidney from rats fed 2 mg OA per kilogram body weight for 2 days gluconeogenesis was decreased. The most important regulatory enzyme in this pathway, renal phosphoenolpyruvate carboxykinase (PEPCK) was selectively inhibited by OA. Liver PEPCK or renal hexokinase were not affected.[319]

Pharmacologic Effects

In a series of experiments mice were injected i.p. with OA (5 to 15 mg/kg body weight, dissolved in propylene glycol) and various pharmacologic parameters were studied.[320-324] Within 30 min OA caused a dose-related hypothermia, which was greater than the hypothermia induced by the vehicle alone.[320] Various mechanisms for this effect were suggested.[320] A strong dose-related CNS depressant effect was noted,[321] as well as muscle relaxant and sedative activity.[321,322] OA prolonged pentobarbitone induced sleeping time[322] more than propylene glycol alone. This effect was postulated to be due to a possible inhibition of mitochondrial respiration.[322] OA had nonnarcotic analgesic activity[323] and was more effective than propylene glycol alone in preventing convulsions induced by pentylene tetrazol.[324] A mono-amine oxidase inhibitor effect of OA was postulated as the underlying mechanism.

The same investigators also studied the effect of OA on steroidogenesis in various organs of rats.[325-328] At 5 mg/kg body weight, i.p. on alternate days for 15 days, OA caused a reduction in the histochemical localization of glucose-6-phosphatase and Δ^5-3β-hydroxysteroid dehydrogenase activity in the testis of both mature and immature rats. This was accompanied by a reduction in the weight of the testis and accessory glands, and by increases in ascorbic acid and cholesterol. These observations suggested that testicular steroidogenesis was decreased.[325] Similarly ovarian steroidogenesis in female rats was impaired as also indicated by an arrest of the estrus cycle,[326] and a delayed onset of puberty.[327] In the adrenal gland, however, OA induced an increase of the same two enzymes and a reduction in cholesterol and ascorbic acid content, suggesting a stimulation of adrenal steroidogenesis.[328]

Immune Response

Guinea pigs given oral doses of 0.45 mg OA per day for 4 weeks did not exhibit altered complement activity or antibody response to *Brucella abortus* antigen.[329] However, in mice given OA (5 mg/kg body weight i.p.) daily for 50 days the circulating antibody titer to *B. abortus* was decreased, but the number of antibody forming cells in the spleen was not affected.[330]

At very low doses (0.005 to 50 μg/kg body weight i.p.) OA given to mice exerted a strong suppression of the immune response to sheep erythrocytes.[331] This effect was prevented if at the same time phenylalanine was injected at twice the OA dose (w/w).[331] The immunosuppressive effects of OA were seen only after i.p. injections, and it was speculated that after oral feeding of OA there may be sufficient phenylalanine present in the feed to protect against the immunosuppressive effects of OA.[289,331]

Teratogenic and Reproductive Effects

OA administered orally or i.p. on various days of gestation had severe embryotoxic effects in rats, mice, and hamsters. These effects included an increased number of dead and resorbed fetuses, decreased fetal body weights, as well as hemorrhages and edema in some fetuses.[332-337] In addition to these effects, various anomalies were observed in the fetuses of OA-treated animals. These included celosomy[333,337] and multiple gross visceral and skeletal malformations in the rat,[335] gross abnormalities not related to skeletal development in the hamster,[334] a variety of skeletal anomalies mostly in the ribs, vertebrae, and skull of mice,[336,338] and necrotic cells in the brain and eyes of fetal mice exposed early or later in gestation;[338,339] towards the end of gestation, when fetuses are generally more resistant to overtly teratogenic and embryolethal effects of OA, the type of damage done by OA in mice could remain undetected until expressed much later as a behavioral deficit.[339] As judged from retarded growth and reduced litter size, the adverse effects of prenatal exposure to OA carried through in the surviving F_1 progeny of OA-treated rats; no effects were observed in the F_2 generation.[340] Craniofacial malformation induced by OA in mice did not increase significantly when T-2 toxin, a mycotoxin of the epoxytrichothecene group, was administered simultaneously.[338]

It is concluded, therefore, that OA affects a variety of fetal organs and tissues. The mechanism by which these effects are brought about is unknown; fetal damage may occur indirectly by some effect of the toxin on the mother such as lowered blood glucose levels[337] and/or directly by an effect on the fetus. The latter possibility is consistent with the observation that OA was present in the maternal blood serum of rats and their embryos. OA can therefore pass through the placenta.[341] Enzyme changes indicating a more aerobic metabolism were seen in rats treated pre- and postnatally with OA.[342] The observed difference in teratogenicity between swine and rodents may be attributable to a species difference or to differences in levels of OA or methods used.

Carcinogenicity and Mutagenicity

No increased tumor incidence was noted in rats or mice given oral or s.c. doses of OA.[343,344] However, these studies may have been performed with too limited a number of animals to draw a valid conclusion.[8] Application of OA to the skin of a skin tumor sensitive strain of mice did not lead to increased papilloma development.[345] No reference has been made to tumor formation in experimental or field cases of OA-poisoned animals and no mutagenic effects were observed in *Bacillus subtilis,* mammalian cell, *Saccharomyces cerevisiae,* and *Salmonella typhimurium* mutagenicity tests,[330,346-348] although after activation a mixture of OA and OB were mutagenic in the Ames *Salmonella* systems.[348] An unconfirmed report indicated that, although OA alone did not produce tumors in 8 months, OA in combination with sterculic acid, a component of cottonseed oil, did produce hepatomas when fed to rainbow trout.[349] A recent study reported neoplastic changes in the kidneys and liver of mice fed an OA-containing diet.[350] The question of carcinogenicity of OA requires further study.

Dogs

As in swine, clinical and pathological features of OA poisoning of young dogs reflected damage to the kidneys and indicated a primary toxic effect on the renal tubular epithelium.[267] In young dogs given daily oral doses of 0.2 to 3.0 mg/kg body weight, pathological changes were induced in the kidneys, lymphoid tissues, and liver. These levels of OA were high enough to cause severe illness within a few days of administration.[268] Ultrastructural changes, observed in dogs given daily oral doses of 0.3 mg/kg body weight, suggested that the endoplasmic reticulum in renal tubular cells is the primary site of nephrotoxic activity.[351]

Combined doses of citrinin and OA also produced clinical signs reflecting renal damage.[352] Cellular degeneration and necrosis occurred in the proximal and distal tubules and in the straight segments and collecting ducts.[353] Ultrastructural changes accompanying administration of these combined toxins have been described.[354]

Ruminants

An investigation into the cause of a disease involving abortion of cows led to the isolation of an OA-producing strain of *A. ochraceus* from the feed.[332] It was therefore suggested that the presence of the toxin in the feed was the cause of this outbreak. Evidence brought forward for this suggestion was the observation that corn cultures of the isolate, incorporated into the diet of pregnant rats, caused fetal death and resorption.[332] However, further research indicated that OA is not likely to be an abortifacient in ruminants.

Cows, given up to 1.66 mg OA per kilogram body weight for 4 to 5 days when 3 to 6 months pregnant, delivered normal calves and remained clinically normal; a pregnant cow given a single oral dose of 13.3 mg/kg body weight showed only short-term illness. The latter level amounted to about 23 times the level accomplished by feeding the most contaminated feed found to date.[355] Intravenous introduction of OA at 1 mg/kg body weight into pregnant ewes caused death in less than 24 hr but no abortion. Death was due to either pulmonary congestion and edema or to massive liver necrosis. Little or no OA entered the sheep's placenta. No OA was detected in the ewe's amniotic fluid and fetal tissues had levels much lower than those in the maternal blood.[356]

Calves appeared to be more sensitive to OA than mature cows. This age difference may be related to hydrolysis of OA to the much less toxic Oα by the functional rumen of mature ruminants.[355] In an experiment with cow's rumen fluid obtained by esophageal sounding, more than half of the OA introduced was hydrolyzed to Oα within 24 hr. The fraction in which hydrolysis occurred contained mainly protozoa. In another experiment with rumen fluid of cows and sheep collected at the slaughterhouse, 80% or more of the OA introduced was recovered intact after 24 hr of incubation; smaller amounts were recovered as Oα and, in some samples of cow's rumen, as the ester of OA, i.e., OC. These differences were attributed to differences in the diet or to lack of feed before slaughtering.[357] Hydrolysis of OA was observed in the contents of all the stomachs of the cows except in those from the abomasum.[358] Because of hydrolysis in the rumen, little OA may enter the blood stream of mature ruminants. However, residues of up to 5 μg OA per kilogram were found in the kidney of a cow fed a diet containing OA for 11 weeks[241] (Table 4); the metabolite Oα was not detected.

OA did not inhibit activity of rumen organisms. At 70 μg/50 mℓ of sheep's rumen fluid, the toxin did not affect the digestion of hay.[359]

The natural occurrence of OA in cow's milk appears unlikely. Oα occurred in the milk and urine only when massive oral doses were given.[355]

In goats, the lethal level produced by feeding daily doses of OA for 5 days was 3 mg/kg body weight.[355]

More than 90% of a single dose of OA fed to goats was excreted within 7 days, with the majority being found in the feces (54%),[360] whereas in rats urinary excretion was found to be the primary route. Just as in rats, unaltered OA was excreted primarily via feces, while OA metabolites were found in the urine. Large amounts of unmetabolized OA in the feces might represent incomplete absorption of OA from the GI tract, together with some OA originating from biliary excretion.[360] A major metabolite in urine, feces, and milk was not identified, but was not Oα or 4-OH-OA.[360] As was found in cows, deposition of OA or its metabolites in milk was insignificant.[360]

Most of the OA accumulated primarily in the microsomes and submicrosomal fractions,[360] which is consistent with the finding that OA causes considerable damage to the endoplasmic reticulum structure in other species.[292,351] A much smaller portion was found in the mitochondrial fraction.[360]

Tissue Culture and Nonmammalian Organisms

Several biological systems have been examined to determine their sensitivity to OA. Such test systems have been used to establish bioassay methods for confirmation of chemical analysis, to screen fungal cultures for toxicity, or to study the mode of action of OA.

Among the systems tested (Table 7), tissue cultures had the greatest sensitivity. In general, epithelial cell-derived cultures were more sensitive than fibroblast-derived cultures.[330] Tracheal cultures showed a toxic response to OA at less than 0.2 μg/mℓ of medium.[361] In monkey epithelial cells, microscopically visible effects included abnormal mitotic figures.[362,363] The cytostatic and cytotoxic effect of OA on hepatoma cells could completely be prevented by the simultaneous addition of phenylalanine to the medium.[365]

Brine shrimp and zebra fish larvae showed mortality at levels of 10 μg OA per milliliter medium or less.[366-369] The ready availability of brine shrimp eggs makes them a convenient bioassay for OA or other mycotoxins. OA incorporated into the diets of silk worm larvae had a lethal effect but the length of exposure needed for mortality may prohibit the use of these larvae as a bioassay.[370] A protozoon, *Tetrahymena pyriformis,* and cotelydon discs of cucumber suffered little or no adverse effects of OA.[371,372]

Most of the bacteria tested showed low sensitivity. Gram-positive bacteria were more sensitive than Gram-negative ones.[375] Incorporation of labeled precursors into protein and RNA by growing (cell wall-free) protoplasts of *Escherichia coli,* a Gram-negative organism, occurred at levels 100- to 1000-fold higher than those necessary for growth inhibition of *Bacillus subtilis,* a Gram-positive organism. The difference in sensitivity of Gram-positive and Gram-negative bacteria to OA is therefore not related to limited permeability of the cell wall of Gram-negative bacteria but to other factors.[378] The degree of toxicity of OA to *B. subtilis* was also dependent on the medium.[377] OA lysed *B. subtilis* cells but had no effect on metabolizing protoplasts in hypertonic medium. OA did inhibit cell wall and protein synthesis, and lysates showed the presence of a

Table 7
TOXICITY OF OA TO TISSUE CULTURES AND VARIOUS NONMAMMALIAN ORGANISMS

System	Toxicity	Ref.
Tracheal organ cultures	Cytotoxicity at 0.175 μg/mℓ	361
Monkey epithelial cells	Abnormal mitosis at 0.1—3.2 μg/mℓ and higher	362, 363
Hela cells	Cytotoxicity at 1—32 μg/mℓ	364
Hepatoma cells	Cytotoxic at 18—36 μg/mℓ	365
Brine shrimp larvae	50% mortality at 10 μg/mℓ (16-hr exposure)	366
	20% mortality at 2 μg/mℓ (24-hr exposure)	367
	50% mortality at 3.9 μg/mℓ (24-hr exposure)	368
Zebra fish larvae	50% mortality at 1.7 μg/mℓ (72-hr exposure)	369
Silk worm larvae	50% mortality after 9 days on a diet containing 2 mg OA/kg	370
Protozoan (*Tetrahymena pyriformis*)	Little or no effect on respiration and growth at 5—400 μg/mℓ	371
Discs of cotelydons of cucumber seedlings	Increased absorption of ^{14}C-leucine and incorporation into protein	372
Proteus and *Bacillus* bacteria	Minimum inhibitory dosage at 250—500 μg/mℓ	373
Bacillus cereus var. *mycoides*	Minimum inhibitory dosage at 1.5 μg/disc	374
Actinomyces spp.	Minimum inhibitory dosage at 100—500 μg/mℓ	373
Streptococcus faecalis	Inhibition of protein and RNA synthesis at 5—10 μg/mℓ, no lethality even at 1 mg/mℓ	375
Bacillus subtilis	Autolysis at greater than 12 μg/mℓ	376
Bacillus subtilis	Inhibition of sporulation at 5.5 μg/mℓ, if present 30 min before end of exponential growth	377

lysis inhibitor. It is therefore possible that, in *B. subtilis,* inhibition of protein synthesis of OA leads to inhibition of mucopeptide synthesis, release of a lysis inhibitor into the medium, and finally to lysis.[376] The relatively high sensitivity of *B. cereus* var. *mycoides* was used to quantitate pure OA and OB dissolved in hexane-chloroform.[374]

The rainbow trout has also yielded useful data on the toxicity of the various ochratoxins (Table 8).[379,380] As in other animals, OA and OC had much greater toxicity than the dechlorinated compound OB and the hydrolysis products Oα and Oβ. Pathological changes induced by injection of OA included degeneration of hepatic parenchymal cells; necrosis in the proximal tubules, hematopoietic tissue and glomeruli of the kidneys; and pycnotic nuclei, cast formation and lipid vacuolation in the renal tubules. Although OB was not lethal at the level used, pathological changes were similar to OA-induced changes.[379] Modification of OA and OB to the corresponding ethyl esters increased the toxicity of these toxins considerably, while modification of OA by substitution of phenylalanine with amino acids of lower molecular weight decreased its toxicity.[380]

Structural Factors Affecting the Toxicity of Ochratoxins

Phenolic Hydroxyl Group

It has been suggested that the presence of the phenolic hydroxyl group in its disso-

Table 8
TOXICITY OF OCHRATOXINS INJECTED INTRAPERITONEALLY INTO RAINBOW TROUT

Toxin	Toxicity	Ref.
OA[a]	LD_{50} of 4.67—5.53 mg/kg	379, 380
OC[b]	LD_{50} at 3.0 mg/kg	380
OB[a]	Nonlethal at 66.7 mg/kg	379, 380
OB-E[b]	LD_{50} at 13.0 mg/kg	380
Oα[a]	Nonlethal at 28.0 mg/kg	379
Oβ[a]	Nonlethal at 26.7 mg/kg	379
Ethyl ester of Oα,[b] alanine and leucine analog of OA[a]	Nonlethal at the molar equivalent of the LD_{50} of OA	380

[a] In bicarbonate.
[b] In corn oil.

ciated form is necessary for toxicity. This suggestion is based on the observation that the toxicity of different ochratoxins correlates with the acid dissociation constants (pK) of their phenolic hydroxyl group. For example, ochratoxins of high toxicity such as OA and OC have a pK near neutrality, while the less toxic ones such as OB and Oα have a more alkaline pK.[6,381] Data on pK values and toxicity of OA analogs, in which the phenylalanine moiety has been substituted, support this suggestion.[94] Also, modification of the hydroxyl group by *O*-methylation reduced the toxicity of OC.[381]

Chlorine Atom

The Cl atom must enhance the toxicity of OA since OB, the dechlorinated compound, is much less toxic than OA (Tables 5 and 8). This effect of the Cl atom may be indirect in that it affects the dissociation of the phenolic hydroxyl group in OA and OC.[381]

Binding with Proteins

In vitro studies on the interaction between OA and bovine serum albumin (BSA), indicate that most ochratoxins bind with BSA but that the binding capacity of the various ochratoxins varies strongly. OB bound less strongly with BSA than did the toxic OA and OC which form a complex of similar nature. The nontoxic Oα and Oβ bound with BSA to a smaller degree than did the other three ochratoxins. Both hydrophobic and ionic bonds were considered important in binding to the albumin molecule, the phenolic hydroxyl group being important in the binding of a second molecule of biologically active ochratoxins.[382,383] The carboxyl group was also suggested to play a role in the interaction of the ochratoxins with proteins.[309] In vivo binding of OA with serum albumin has been observed.[308,309,384]

Molecular Weight

Some evidence involving rainbow trout (Table 8) indicates that changing the molecular weight of OA without altering its ionic nature affects its toxicity to rainbow trout. As determined by i.p. injection, a procedure that circumvents the problem of hydrolysis in the GI tract, the ethyl esters of OA and OB showed greater toxicity than the corresponding parent compounds. The increased toxicity of the esters was attributed to their increased molecular weight. The alanine and leucine analogs of OA had lower toxicity than OA and this lowered toxicity was attributed to the decreased molecular weights of the analogs.[380] It is likely that the ethyl esters have greater, and the amino acid analogs lower, lipid solubility than the parent compounds. The changed toxicity observed in this study may therefore reflect changed permeation rates into cells rather than decreased molecular weights.

Amide Bond

Since the ochratoxins contain an amide bond with an aromatic amino acid as carboxyl terminal, they may be hydrolyzed by proteolytic enzymes such as carboxypeptidase A and α-chymotrypsin.[20] In vitro, the affinity of OA for carboxypeptidase A was much greater than for α-chymotrypsin. OA served as substrate as well as competitive inhibitor of carboxypeptidase A and resembled some dipeptides in its action as inhibitor.[385] Similar hydrolytic reactions may take place in vivo and the different toxicities of the ochratoxins may be related to their different rates of hydrolysis to nontoxic metabolites. In in vitro studies involving tissue extracts of liver and intestines of rats, OB was hydrolyzed much faster than OA. As the hydrolysis products are nontoxic, the difference in hydrolysis rate may partially explain the difference in toxicity of OA and OB.[386] Since Oα has been detected in the urine and feces of OA-treated rats[251,311,312] and swine,[236] in vivo hydrolysis of OA may occur by proteolytic enzymes in the intestine and perhaps the liver. Hydrolysis of OA has been attributed also to microbial activity in the rat's cecum and large intestine[304] and in the cow's rumen.[355,357]

Hydroxylation

A green fluorescent compound present in rat urine was identified as the 4-hydroxy derivative of OA. On injection, this compound was nontoxic to rats.[18] Hydroxylation reactions by liver microsomes may be involved in the excretion of ochratoxins from the body.[313]

Molecular Mechanism of Toxicity

The effect of OA on DNA, RNA, and protein synthesis has been studied mainly in bacteria, yeast, or hepatoma cells, or on cell-free systems derived from these cells, and mammalian cells.

OA had an inhibitory, but nonlethal, effect on *S. faecalis* which was evident at pH 6 to 6.5 but not at higher pH. At 10 $\mu g/m\ell$ medium, OA caused a decreased rate of incorporation of amino acids and of uracil into trichloroacetic acid-precipitable material; as this decrease was not related to inhibition of uptake of precursors, the inhibitory effect was attributed to inhibition of protein and RNA synthesis.[375] The toxin did not affect DNA synthesis. Since protein synthesis was inhibited strongly in the presence of OA and chloramphenicol and RNA synthesis continued for some time, the primary effect of OA was concluded to be inhibition of protein synthesis. Such inhibition could result from interference with transcription or translation. The first possibility was excluded because induction of protein synthesis was not inhibited by OA when present during induction time only.[387] As the toxin did inhibit protein synthesis when added after the time necessary for induction, the toxin affects a step in translation and may inhibit peptide elongation. Consistent with this suggestion was the observation that polysome profiles of protoplasts of *S. faecalis* were stabilized after addition of OA.[387] Similarly with *B. subtilis,* OA prevents sporulation, but again not through an effect on transcription as is common with other sporulation inhibitors.[377]

With cultured hepatoma cells the primary effect of OA was also on protein synthesis, RNA and DNA synthesis being affected much later.[365,388] Citrinin affected first RNA synthesis, followed by protein and DNA synthesis.[389] In combination these two mycotoxins acted cooperatively, the inhibition of protein and DNA synthesis being immediate.[388]

In *B. subtilis*,[378] and in yeast,[390,391] t-RNA synthetase for amino acylation of phenylalanine was inhibited competitively by OA, suggesting that the toxin may act as an analog of phenylalanine in the acylation reaction. In yeast the inhibition by OA of this enzyme[390] and in hepatoma cells the inhibition of protein synthesis was prevented by the presence of phenylalanine when given at the same time as OA and at appropriate concentrations. In vivo phenylalanine, when injected in mice simultaneously with OA, also efficiently counteracted the toxic effects of OA[288] and prevented immunosuppression.[331]

The inhibition of phenyl-alanyl-tRNA synthetase was suggested as the mechanism underlying the inhibition of protein synthesis in vivo, since poly-U-directed peptide synthesis in a cell-free, poly-U-dependent polyphenylalanine synthesizing system derived from *B. stearothermophilus*, was inhibited by OA in the same manner.[392] Similarly in yeast the t-RNA-phenyl-alanyl-charging reaction was inhibited by OA.[390]

Phenylalanyl-t-RNA synthetase also catalyzes the phenylalanine-dependent pyrophosphate-ATP exchange reaction and in yeast this reaction is more inhibited than the amino acylation of t-RNA at the same concentration of phenylalanine and OA.[390] A competitive inhibition by OA of inorganic phosphate exchange[393] leading to an inhibition of ATP-stimulated respiration[394,395] was also seen in isolated rat mitochondria. This was accompanied by a competitive inhibition of the exchange of dicarboxylic acids and adenine nucleotides. Such exchange normally takes place in the mitochondrial transport carriers located in the inner membrane of intact mitochondria. Mitochondrial uptake of OA was saturable and energy dependent, leading to depletion of intramitochondrial ATP. Since OA had only weak affinity for isolated cell membranes, its in vivo effect may chiefly involve the mitochondrial membrane and may, initially, result from interference with the phosphate carrier protein.[395] OB did not compete with OA for mitochondrial uptake, indicating that the chlorine atom is involved in the toxic action of OA on isolated mitochondria.[395]

CONCLUSIONS

In general, it has proved very difficult to find unequivocal proof for the role of specific mycotoxins in diseases of man and animals. OA is an exception. The information on this toxin as a disease determinant fulfills most, if not all, of Koch's postulates, if these are modified to apply to a disease-causing microbial toxin rather than a disease-causing microbe: OA was obtained in pure form from molds associated with feeds that proved toxic to swine or poultry; the presence of the toxin in such feeds was demonstrated, and, when the toxin was administered orally at levels equivalent to those resulting from ingestion of such feeds, significant features of the disease syndrome could be reproduced; finally, measurable residues of the toxin were detected in organs of animals suffering from either experimental or naturally occurring MPN or MAN.

Since data on the natural occurrence of OA indicates its presence in grains of many countries, there remains the challenge of reducing the probability of ingestion of such grains by man and animals. For any given year, analysis of randomly selected grain samples or tissues of swine and poultry is unlikely to reveal a significant incidence of OA for the following reasons: (1) in most major grain producing countries, conventional grain handling practices keep moisture levels of grains well below that needed for growth and toxin production of *Penicillium* spp., (2) the incidence of OA may fluctuate widely from year to year due to great annual differences in climatic conditions, and (3) limited resources usually permit examination of only a few hundred samples, so that the probability of detecting a toxin of low incidence such as OA is low. A biased approach, may, therefore, be more fruitful than a random sampling approach and requires: (1) an awareness of local weather conditions and storage practices that may lead to OA contamination, (2) analysis of samples that are suspect because of moldiness resulting from growth of *Penicillium* spp. or because of occurrences of renal disorders of farm animals, and (3) increased efforts to recognize the gross symptoms of MPN and MAN in kidneys at slaughterhouses in years in which weather conditions and analyses of suspect samples indicate that the chances of OA contamination are higher than usual. Such an approach may lead to the detection and subsequent removal of OA-contaminated plant and meat products before they reach the market.

REFERENCES

1. **Merwe, K. J., van der, Steyn, P. S., Fourie, L., Scott, De B., and Theron, J. J.,** Ochratoxin A, a toxic metabolite produced by *Aspergillus ochraceus* Wilh., *Nature (London),* 205, 1112—1113, 1965.
2. **Merwe, K. J., van der, Steyn, P. S., and Fourie, L.,** Mycotoxins. II. The constitution of ochratoxins A, B, and C, metabolites of *Aspergillus ochraceus* Wilh., *J. Chem. Soc.,* p. 7083—7088, 1965.
3. **Steyn, P. S. and Holzapfel, C. W.,** The isolation of the methyl and ethyl esters of ochratoxins A and B, metabolites of *Aspergillus ochraceus* Wilh., *J. S. Afr. Chem. Inst.,* 20, 186—189, 1967.
4. **Scott, De B.,** Toxigenic fungi isolated from cereal and legume products, *Mycopathol. Mycol. Appl.,* 25, 213—222, 1965.
5. **Applegate, K. L. and Chipley, J. R.,** Ochratoxins, *Adv. Appl. Microbiol.,* 16, 97—109, 1973.
6. **Chu, F. S.,** Studies on ochratoxins, *CRC Crit. Rev. Toxicol.,* 2, 499—524, 1974.
7. **Harwig, J.,** Ochratoxin A and related metabolites, in *Mycotoxins,* Purchase, I. F. H., Ed., Elsevier, New York, 1974, chap. 16.
8. IARC, *Evaluation of Carcinogenic Risk of Chemicals to Man,* Monograph, Vol. 10, International Agency for Research on Cancer, Lyon, 1976, 191—197.
9. **Scott, P. M.,** Mycotoxins in feeds and ingredients and their origin, *J. Food Prot.,* 41, 385—398, 1978.
10. **Steyn, P. S.,** Ochratoxin and other dihydroisocoumarins, in *Microbial Toxins,* Vol. 6, Ciegler, A., Kadis, S., and Ajl, S. J., Eds., Academic Press, New York, 1971, chap. 2.
11. **Krogh, P.,** Ochratoxins, in *Mycotoxins in Human and Animal Health,* Rodricks, J. V., Hesseltine, C. W., and Mehlman, M. A., Eds., Pathotox Publishers, Park Forest South, Ill., 1977, 489—498.
12. **Carlton, W. W. and Krogh, P.,** Ochratoxins, in Conf. Mycotoxins in Animal Feeds and Grains Related to Animal Health, Shimoda, W., Ed., U.S. Department of Commerce, Springfield, Va., 1979, 165—287.
13. **Krogh, P.,** Ochratoxins: occurrence, biological effects and causal role in diseases, in *Natural Toxins, Proc. 6th Int. Symp. Anim. Plant Microbial Toxins, Uppsala,* Eaker, D. and Wadstrøm, T., Eds., Pergamon Press, Oxford, 1980, 673—80.
14. **Steyn, P. S. and Holzapfel, C. W.,** The synthesis of ochratoxin A and B, metabolites of *Aspergillus ochraceus* Wilh., *Tetrahedron,* 23, 4449—4461, 1967.
15. **Neeley, W. C. and West, A. D.,** Spectroanalytical parameters of fungal metabolites. III. Ochratoxin A, *J. Assoc. Off. Anal. Chem.,* 55, 1305—1309, 1972.
16. **Roberts, J. C. and Woollven, P.,** Studies in mycological chemistry. Part XXIV. Synthesis of ochratoxin A, a metabolite of *Aspergillus ochraceus* Wilh., *J. Chem. Soc. Sect. C,* pp. 278—281, 1970.
17. **Nesheim, S.,** Isolation and purification of ochratoxins A and B and preparation of their methyl and ethyl esters, *J. Assoc. Off. Anal. Chem.,* 52, 975—979, 1969.
18. **Hutchison, R. D., Steyn, P. S., and Thompson, D. L.,** The isolation and structure of 4-hydroxyochratoxin A and 7-carboxy-3,4-dihydro-8-hydroxy-3-methylisocoumarin from *Penicillium viridicatum, Tetrahedron Lett.,* No. 43, pp. 4033—4036, 1971.
19. **Steyn, P. S. and Merwe, K. J., van der,** Detection and estimation of ochratoxin A, *Nature (London),* 211, 418, 1966.
20. **Pitout, M. J.,** The hydrolysis of ochratoxin A by some proteolytic enzymes, *Biochem. Pharmacol.,* 18, 485—491, 1969.
21. **Golinski, P. and Chelkowski, J.,** Spectral behaviour of ochratoxin A in different solvents, *J. Assoc. Off. Anal. Chem.,* 61, 586—589, 1978.
22. **Gillespie, A. M., Jr. and Schenk, G. M.,** Fluorescence and phosphorescence of ochratoxins, sterigmatocystin, patulin and zearalenone: quantitation of ochratoxins, *Anal. Lett.,* 10, 161—172, 1977.

23. **Pohland, A. E. and Sphon, J. A.**, Mycotoxins Mass Spectral Data Bank, U.S. Food and Drug Administration, Washington, D.C., 1978.
24. **Gallagher, R. T. and Stahr, H. M.**, Mass spectral confirmation of ochratoxin A, *Appl. Spectrosc.*, 35, 131—132, 1981.
25. **Chu, F. S. and Butz, M. E.**, Spectrophotofluorodensitometric measurement of ochratoxin A in cereal products, *J. Assoc. Off. Anal. Chem.*, 53, 1253—1267, 1970.
26. **Pitout, M. J.**, A rapid spectrophotometric method for the assay of carboxypeptidase A, *Biochem. Pharmacol.*, 18, 1829—1836, 1969.
27. **Scott, P. M., Kennedy, B., and Walbeek, W. van**, Simplified procedure for the purification of ochratoxin A from extracts of *Penicillium viridicatum*, *J. Assoc. Off. Anal. Chem.*, 54, 1445—1447, 1971.
28. **Kraus, G. A.**, A facile synthesis of ochratoxin A, *J. Org. Chem.*, 46, 201, 1981.
29. **Ferreira, N. P.**, The effect of amino acids on the production of ochratoxin A in chemically defined media, *van Leeuwenhoek J. Microbiol. Serol.*, 34, 433—440, 1968.
30. **Walbeek, W. van, Scott, P. M., Harwig, J., and Lawrence, J. W.**, *Penicillium viridicatum* Westling: a new source of ochratoxin A, *Can. J. Microbiol.*, 15, 1281—1285, 1969.
31. **Davis, N. D., Sansing, G. A., Ellenburg, T. V., and Diener, U. L.**, Medium-scale production and purification of ochratoxin A, a metabolite of *Aspergillus ochraceus*, *Appl. Microbiol.*, 23, 433—435, 1972.
32. **Yamazaki, M., Maebayashi, Y., and Miyaki, K.**, Production of ochratoxin A by *Aspergillus ochraceus* isolated in Japan from moldy rice, *Appl. Microbiol.*, 20, 452—454, 1970.
33. **Peterson, R. E. and Ciegler, A.**, Ochratoxin A: isolation and subsequent purification by high pressure liquid chromatography, *Appl. Environ. Microbiol.*, 36, 613—614, 1978.
34. **Bunge, I., Heller, K., and Röschenthaler, R.**, Isolation and purification of ochratoxin A, *Z. Lebensm. Unters. Forsch.*, 168, 457—458, 1979.
35. **Galtier, P.**, Toxines *d'Aspergillus ochraceus* Wilhelm. II. Isolement par chromatographie sur couche mince de l'ochratoxine A obtenue à partir de milieux liquides faiblement concentres en toxine, *Ann. Rech. Vet.*, 5, 155—166, 1974.
36. **Stoloff, L.**, Analytical methods for mycotoxins, *Clin. Toxicol.*, 5, 465—494, 1972.
37. **Scott, P. M.**, Liquid chromatography in the analysis of mycotoxins, in *Trace Analysis*, Vol. 1, Lawrence, J. F., Ed., Academic Press, New York, 1981, 193-266.
38. **Nesheim, S., Hardin, N. F., Francis, O. J., Jr., and Langham, W. S.**, Analysis of ochratoxins A and B and their esters in barley, using partition and thin layer chromatography. I. Development of the method, *J. Assoc. Off. Anal. Chem.*, 56, 817—821, 1973.
39. Association of Official Analytical Chemists, Natural poisons, in *Official Methods of Analysis*, 13th ed., Horwitz, W., Ed., Association Official Analytical Chemists, Washington, D.C., 1980, chap. 26.
40. **Chu, F. S.**, Note on solid state fluorescence emission of ochratoxins A and B on silica gel, *J. Assoc. Off. Anal. Chem.*, 53, 696—697, 1970.
41. **Fritz, W. and Donath, R.**, Zur Bestimmung von Ochratoxin A, *Nahrung*, 18, 827—832, 1974.
42. **Faucon, M., Teherani, A., Tantaoui-Eleraki, A., and Jacquet, J.**, Comparaison de deux méthodes de chromatographie en couche mince sur gel de silice pour le dosage des aflatoxines B_1 et G_2 et de l'ochratoxine A, *Bull. Acad. Vet. Fr.*, 50, 123—139, 1977.
43. **Czierwiecki, L.**, A spectrophotoflurometric method for determination of ochratoxin A in cereals, *Rocz. Panstw. Zakl. Hig.*, 29, 389—394, 1978.
44. **Scott, P. M. and Hand, T. B.**, Method for the detection and estimation of ochratoxin A in some cereal products, *J. Assoc. Off. Anal. Chem.*, 50, 366—370, 1967.
45. **Kleinau, G.**, Zur Derivatisierung von Ochratoxin A auf der Dünnschichtplatte, *Nahrung*, 25, K9—K10, 1981.
46. **Nesheim, S.**, Analysis of ochratoxins A and B and their esters in barley, using partition and thin layer chromatography. II. Collaborative study, *J. Assoc. Off. Anal. Chem.*, 56, 822—826, 1973.
47. **Trenk, H. L. and Chu, F. S.**, Improved detection of ochratoxin A on thin layer plates, *J. Assoc. Off. Anal. Chem.*, 54, 1307—1309, 1971.
48. **Krogh, P., Hald, B., Englund, P., Rutqvist, L., and Swahn, O.**, Contamination of Swedish cereals with ochratoxin A, *Acta Pathol. Microbiol. Scand. B*, 82, 301—302, 1974.
49. **Krogh, P.**, Ochratoxin A residues in tissues of slaughter pigs with nephropathy, *Nord. Vet. Med.*, 29, 402—405, 1977.
50. **Levi, C. P., Trenk, H. L., and Mohr, H. K.**, Study of the occurrence of ochratoxin A in green coffee beans, *J. Assoc. Off. Anal. Chem.*, 57, 866—870, 1974.
51. **Levi, C. P.**, Collaborative study of a method for the determination of ochratoxin A in green coffee, *J. Assoc. Off. Anal. Chem.*, 58, 258—262, 1975.
52. **Shishido, Y.**, Analytical method for ochratoxin A in feeding grains, *Shiryo Kenkyu Hokoku (Tokyo Hishiryo Kensasho)*, 5, 123—126, 1979; *Chem. Abstr.*, 93, 24557j, 1980.
53. **Prior, M. G.**, Mycotoxin determinations on animal feedstuffs and tissues in Western Canada, *Can. J. Comp. Med.*, 40, 75—79, 1976.

54. **Eppley, R. M.**, Screening method for zearalenone, aflatoxin, and ochratoxin, *J. Assoc. Off. Anal. Chem.*, 51, 74—78, 1968.
55. **Shotwell, O. L., Goulden, L., and Hesseltine, C. W.**, Survey of U.S. wheat for ochratoxin and aflatoxin, *J. Assoc. Off. Anal. Chem.*, 59, 122—124, 1976.
56. **Shotwell, O. L., Goulden, M. L., Bennett, G. A., Plattner, D., and Hesseltine, C. W.**, Survey of 1975 wheat and soybeans for aflatoxin, zearalenone, and ochratoxin, *J. Assoc. Off. Anal. Chem.*, 60, 778—783, 1977.
57. **Shotwell, O. L., Hesseltine, C. W., and Goulden, M. L.**, Ochratoxin A: occurrence as natural contaminant of a corn sample, *Appl. Microbiol.*, 17, 765—766, 1969.
58. **Shotwell, O. L., Hesseltine, C. W., Vandegraft, E. E., and Goulden, M. L.**, Survey of corn from different regions for aflatoxin, ochratoxin, and zearalenone, *Cereal Sci. Today*, 16, 266—273, 1971.
59. **Shotwell, O. L., Hesseltine, C. W., Goulden, M. L., and Vandegraft, E. E.**, Survey of corn for aflatoxin, zearalenone, and ochratoxin, *Cereal Chem.*, 47, 700—707, 1970.
60. **Stoloff, L., Nesheim, S., Yin, L., Rodricks, J. V., Stack, M., and Campbell, A. D.**, A multimycotoxin detection method for aflatoxins, ochratoxins, zearalenone, sterigmatocystin, and patulin, *J. Assoc. Off. Anal. Chem.*, 54, 91—97, 1971.
61. **Scott, P. M., van Walbeek, W., Kennedy, B., and Anyeti, D.**, Mycotoxins (ochratoxin A, citrinin, and sterigmatocystin) and toxigenic fungi in grains and other agricultural products, *J. Agric. Food Chem.*, 20, 1103—1109, 1972.
62. **Krogh, P., Hald, B., and Pedersen, E. J.**, Occurrence of ochratoxin A and citrinin in cereals associated with mycotoxic porcine nephropathy, *Acta Pathol. Microbiol. Scand. B*, 81, 689—695, 1973.
63. **Hald, B. and Krogh, P.**, Detection of ochratoxin A in barley, using silica gel minicolumns, *J. Assoc. Off. Anal. Chem.*, 58, 156—158, 1975.
64. **Hult, K. and Gatenbeck, S.**, A spectrophotometric procedure using carboxypeptidase A, for the quantitative measurement of ochratoxin A, *J. Assoc. Off. Anal. Chem.*, 59, 128—129, 1976.
65. **Hult, K., Hökby, E., and Gatenbeck, S.**, Analysis of ochratoxin B alone and in the presence of ochratoxin A, using carboxypeptidase A, *Appl. Environ. Microbiol.*, 33, 1275—1277, 1977.
66. **Hult, K., Hökby, E., Hägglund, U., Gatenbeck, S., Rutqvist, L., and Sellyey, G.**, Ochratoxin A in pig blood: method of analysis and use as a tool for feed studies, *Appl. Environ. Microbiol.*, 38, 772—776, 1979.
67. **Roberts, B. A. and Patterson, D. S. P.**, Detection of twelve mycotoxins in mixed animal feedstuffs, using a novel membrane cleanup procedure, *J. Assoc. Off. Anal. Chem.*, 58, 1178—1181, 1975.
68. **Hagan, S. N. and Tietjen, W. H.**, A convenient thin layer chromatographic cleanup procedure for screening several mycotoxins in oils, *J. Assoc. Off. Anal. Chem.*, 58, 620—621, 1975.
69. **Wilson, D. M., Tabor, W. H., and Trucksess, M. W.**, Screening method for the detection of aflatoxin, ochratoxin, zearalenone, penicillic acid, and citrinin, *J. Assoc. Off. Anal. Chem.*, 59, 125—127, 1976.
70. **Holaday, C. E.**, A rapid screening method for the aflatoxins and ochratoxin A, *J. Am. Oil Chem. Soc.*, 53, 603—605, 1976.
71. **Josefsson, B. G. E. and Möller, T. E.**, Screening method for the detection of aflatoxins, ochratoxin, patulin, sterigmatocystin, and zearalenone in cereals, *J. Assoc. Off. Anal. Chem.*, 60, 1369—1371, 1977.
72. **L'vova, L. S., Kravchenko, L. V., and Shul'gina, A. P.**, Chromatographic method for the simultaneous semiquantitative determination of eight mycotoxins in grain, *Prikl. Biokhim. Mikrobiol.*, 15, 143—149, 1979.
73. **Gimeno, A.**, Thin layer chromatographic determination of aflatoxins, ochratoxins, sterigmatocystin, zearalenone, citrinin, T-2 toxin, diacetoxyscirpenol, penicillic acid, patulin, and penitrem A, *J. Assoc. Off. Anal. Chem.*, 62, 579—585, 1979.
74. **Gimeno, A.**, Improved method for thin layer chromatographic analysis of mycotoxins, *J. Assoc. Off. Anal. Chem.*, 63, 182—186, 1980.
75. **Van Egmond, H. P., Paulsch, W. E., Sizoo, E. A., and Schuller, P. L.**, *Multimycotoxin Methods of Analysis for Feedstuffs*, Rijksinstituut voor de Volksgezondheid, Bilthoven, The Netherlands, 1979.
76. **Gertz, C. and Böschmeyer, L.**, Verfahren zur Bestimmung verschiedener Mykotoxine in Lebensmitteln, *Z. Lebensm. Unters. Forsch.*, 171, 335—340, 1980.
77. **Whidden, M. P., Davis, N. D., and Diener, U. L.**, Detection of rubratoxin B and seven other mycotoxins in corn, *J. Agric. Food Chem.*, 28, 784—786, 1980.
78. **Takeda, Y., Isohata, E., Amano, R., and Uchiyama, M.**, Simultaneous extraction and fractionation and thin layer chromatographic determination of 14 mycotoxins in grains, *J. Assoc. Off. Anal. Chem.*, 62, 573—578, 1979.
79. **Schultz, J. and Motz, R.**, Zum Nachweis von Mykotoxinen in biologischen Versuch, *Monatsh. Veterinärmed.*, 19, 751—754, 1978.

80. **Patterson, D. S. P. and Roberts, B. A.**, Mycotoxins in animal feedstuffs: sensitive thin layer chromatographic detection of aflatoxin, ochratoxin A, sterigmatocystin, zearalenone, and T-2 toxin, *J. Assoc. Off. Anal. Chem.*, 62, 1265—1267, 1979.
81. **Steyn, P. S.**, Multimycotoxin analysis, *Pure Appl. Chem.*, 53, 891—902, 1981.
82. **Hunt, D. C., Bourdon, A. T., Wild, P. J., and Crosby, N. T.**, Use of high performance liquid chromatography combined with fluorescence detection for the identification and estimation of aflatoxins and ochratoxin in food, *J. Sci. Food Agric.*, 29, 234—238, 1978.
83. **Josefsson, E. and Möller, T.**, High pressure liquid chromatographic determination of ochratoxin A and zearalenone in cereals, *J. Assoc. Off. Anal. Chem.*, 62, 1165—1168, 1979.
84. **Osborne, B. G.**, Reversed-phase high performance liquid chromatography determination of ochratoxin A in flour and bakery products, *J. Sci. Food Agric.*, 30, 1065—1070, 1979.
85. **Hunt, D. C., Philp, L. A., and Crosby, N. T.**, Determination of ochratoxin A in pig's kidney using enzymic digestion, dialysis and high performance liquid chromatography with post-column derivatisation, *Analyst*, 104, 1171—1175, 1979.
86. **Schweighardt, H., Schuh, M., Abdelhamid, A. M., Böhm, J., and Leibetseder, J.**, Methode zur Ochratoxin-A-Bestimmung in Lebens - und Futtermitteln mittels Hochdruckflussigkeitschromatographie (HPLC), *Z. Lebensm. Unters Forsch.*, 170, 355—359, 1980.
87. **Hunt, D. C., McConnie, B. R., and Crosby, N. T.**, Confirmation of ochratoxin A by chemical derivatisation and high-performance liquid chromatography, *Analyst*, 105, 89—90, 1980.
88. **Aalund, O., Brunfeldt, K., Hald, B., Krogh, P., and Poulsen, K.**, A radioimmunoassay for ochratoxin A: a preliminary investigation, *Proc. World Vet. Congr.*, 2, 1293—1296, 1975.
89. **Aalund, O., Brunfeldt, K., Hald, B., Krogh, P., and Poulsen, K.**, A radioimmunoassay for ochratoxin A: a preliminary investigation, *Acta Pathol. Microbiol. Scand. C*, 83, 390—392, 1975.
90. **Chu, F. S., Chang, F. C. C., and Hinsdill, R. D.**, Production of antibody against ochratoxin A, *Appl. Environ. Microbiol.*, 31, 831—835, 1976.
91. **Elling, F.**, Demonstration of ochratoxin A in kidneys of pigs and rats by immunofluorescence microscopy, *Acta Pathol. Microbiol. Scand. A*, 85, 151—156, 1977.
92. **Worsaae, H.**, Production of an ochratoxin A antigen with high hapten/carrier molar ratio, *Acta Pathol. Microbiol. Scand. C*, 86, 203—204, 1978.
93. **Chang, F. C. C. and Chu, F. S.**, Preparation of ^{3}H-labelled ochratoxins, *J. Labelled Comp. Radiopharm.*, 12, 231—238, 1976.
94. **Wei, R. D. and Chu, F. S.**, Synthesis of ochratoxins T_A and T_C, analogs of ochratoxins A and C, *Experientia*, 30, 174—175, 1974.
95. **Scott, P. M., van Walbeek, W., Harwig, J., and Fennell, D. I.**, Occurrence of a mycotoxin, ochratoxin A, in wheat and isolation of ochratoxin A and citrinin producing strains of *Penicillium viridicatum, Can. J. Plant. Sci.*, 50, 583—585, 1970.
96. **Krogh, P. and Hasselager, E.**, Studies on fungal nephrotoxicity, *Royal Vet. Agric. Coll. Yearb. (Copenhagen)*, 116, 198—214, 1968.
97. **Sugimoto, T., Minamisawa, M., Takano, K., Sasamura, Y., and Tsuruta, O.**, Detection of ochratoxin A, citrinin and sterigmatocystin from storaged rice by natural occurrence of *Penicillium viridicatum* and *Aspergillus versicolor, J. Food Hyg. Soc. Jpn.*, 18, 176—181, 1977.
98. **Lillehoj, E. B. and Goransson, B.**, Occurrence of ochratoxin and citrinin-producing fungi on developing Danish barley grain, *Acta Pathol. Microbiol. Scand. B*, 88, 133—137, 1980.
99. **Galtier, P., Jemmali, M., Larrieu, G., Yvon, M., and Ponce, G.**, Enquête sur la présence eventuelle d'aflatoxine et d'ochratoxine A dans des mais récoltés en France en 1973 et 1974, *Ann. Nutr. Alim.*, 31, 381—389, 1977.
100. **Carlton, W. W., Tuite, J., and Caldwell, R.**, *Penicillium viridicatum* toxins and mold nephrosis, *J. Am. Vet. Med. Assoc.*, 163, 1295—1297, 1973.
101. **Stack, M. E., Eppley, R. M., and Pohland, A. E.**, Metabolites of *Penicillium viridicatum, in Mycotoxins in Human and Animal Health*, Rodricks, J. V., Hesseltine, C. W., and Mehlman, M. A., Eds., Pathotox Publishers, Park Forest South, Ill., 1977, 543—555.
102. **Stack, M. E., Eppley, R. M., Dreifuss, P. A., and Pohland, A. E.**, Isolation and identification of xanthomegnin, viomellein, rubrosulphin, and viopurpurin as metabolites of *Penicillium viridicatum, Appl. Environ. Microbiol.*, 33, 351—355, 1977.
103. **Carlton, W. W., Stack, M. E., and Eppley, R. M.**, Hepatic alterations produced in mice by xanthomegnin and viomellein, metabolites of *Penicillium viridicatum, Toxicol. Appl. Pharmacol.*, 38, 455—459, 1976.
104. **Stack, M. E., Brown, N. L., and Eppley, R. M.**, High pressure liquid chromatographic determination of xanthomegnin in corn, *J. Assoc. Off. Anal. Chem.*, 61, 590—592, 1978.
105. **Ciegler, A., Fennell, D. I., Sansing, G. A., Detroy, R. W., and Bennett, G. A.**, Mycotoxin-producing strains of *Penicillium viridicatum*: classification into subgroups, *Appl. Microbiol.*, 26, 271—278, 1973.

106. **Boer, E. de and Stolk-Horsthuis, M.**, Sensitivity to natamycin (pimaricin) of fungi isolated in cheese warehouses, *J. Food Prot.*, 40, 533—535, 1977.
107. **Semeniuk, G. and Barre, H. J.**, Molds in stored corn and their control, *Annu. Rep. Iowa Corn Res. Inst.*, 8, 55—57, 1943.
108. **Bullerman, L. B.**, Occurrence of blue-eye condition in commercial popcorn, *Cereal Foods World*, 20, 104—106, 1975.
109. **Mislevic, P. B. and Tuite, J.**, Species of *Penicillium* occurring in freshly-harvested and in stored dent corn kernels, *Mycologia*, 62, 67—74, 1970.
110. **Mislevic, P. B. and Tuite, J.**, Temperature and relative humidity requirements of species of *Penicillium* isolated from yellow dent corn kernels, *Mycologia*, 62, 75—88, 1970.
111. **Northolt, M. D., van Egmond, H. P., and Paulsch, W. E.**, Ochratoxin A production by some fungal species in relation to water activity and temperature, *J. Food Prot.*, 42, 485—490, 1979.
112. **Harwig, J. and Chen, Y.-K.**, Some conditions favoring production of ochratoxin A and citrinin by *Penicillium viridicatum* in wheat and barley, *Can. J. Plant Sci.*, 54, 17—22, 1974.
113. **Ciegler, A., Fennell, D. J., Mintzlaff, H-J., and Leistner, L.**, Ochratoxin synthesis by *Penicillium* species, *Naturwissenschaften*, 59, 365—366, 1972.
114. **Hadlok, R., Samson, R. A., and Schnorr, B.**, Schimmelpilze und Fleisch: Gattung *Penicillium, Fleischwirtschaft*, 55, 979—984, 1975.
115. **Leistner, L. and Eckardt, C.**, Vorkommen toxinogener Penicillien bei Fleischerzeugnissen, *Fleischwirtschaft*, 59, 1892—1896, 1979.
116. **Rutqvist, L., Björklund, N-E., Hult, K., Hökby, E., and Carlsson, B.**, Ochratoxin A as the cause of spontaneous nephropathy in fattening pigs, *Appl. Environ. Microbiol.*, 36, 920—925, 1978.
117. **Abramson, D., Sinha, R. N., and Mills, J. T.**, Mycotoxin and odor formation in moist cereal grain during granary storage, *Cereal Chem.*, 57, 346—351, 1980.
118. **Ciegler, A., Mintzlaff, H.-J., Machnik, W., and Leistner, L.**, Untersuchungen ueber des Toxinbildungsvermögen von Rohwürsten isolierte Schimmelpilze der Gattung *Penicillium, Fleischwirtschaft*, 10, 1311—1314, 1972.
119. **Bullerman, L. B. and Olivigni, E. J.**, Mycotoxin producing-potential of molds isolated from Cheddar cheese, *J. Food Sci.*, 39, 1166—1168, 1974.
120. **Bullerman, L. B.**, Examination of Swiss cheese for incidence of mycotoxin producing molds, *J. Food Sci.*, 41, 26—28, 1976.
121. **Torrey, G. S. and Marth, E. H.**, Isolation and toxicity of molds from foods stored in homes, *J. Food Prot.*, 40, 187—190, 1977.
122. **Pohlmeier, M. M. and Bullerman, L. B.**, Ochratoxin production by a *Penicillium* species isolated from cheese, *J. Food Prot.*, 41, 829, 1978.
123. **Hesseltine, C. W., Vandegraft, E. E., Fennell, D. I., Smith, M. L., and Shotwell, O. L.**, *Aspergilli* as ochratoxin producers, *Mycologia*, 54, 539—550, 1972.
124. **Doupnik, B., Jr. and Peckham, J. C.**, Mycotoxicity of *Aspergillus ochraceus* to chicks, *Appl. Microbiol.*, 19, 594—597, 1970.
125. **Galtier, P., Le Bars, J., Henry, G., and Alvinerie, M.**, Toxines *d'Aspergillus ochraceus* Wilhelm. I. Production d'ochratoxines par des souches isolées de fourrages secs, cultivées sur blé, *Ann. Rech. Vet.*, 4, 487—497, 1973.
126. **Ichinoe, M., Takatori, K., Tanaka, S., Kumata, H., Suzuki, T., and Kurata, H.**,Mycotoxin producibility of the fungi isolated from wheat and barley grains, *J. Food Hyg. Soc. Jpn.*, 16, 381—390, 1975.
127. **Kumata, K., Amano, R., Ichinoe, M., and Uchiyama, S.**, Culture conditions and purification method for large-scale production of ochratoxins by *Aspergillus ochraceus, Shokuhin Eiseigaku Zasshi*, 21, 171—176, 1980; *Chem. Abstr.*, 93, 166089f, 1980.
128. **Lopez, L. C. and Christensen, C. M.**, Effect of moisture content and temperature on invasion of stored corn by *Aspergillus flavus, Phytopathology*, 57, 588—590, 1967.
129. **Christensen, C. M.**, Invasion of stored wheat by *Aspergillus ochraceus, Cereal Chem.*, 39, 100—106, 1962.
130. **Christensen, C. M.**, Mycotoxins, *Crit. Rev. Environ. Control*, 2, 57—80, 1971.
131. **Ciegler, A.**, Bioproduction of ochratoxin A and penicillic acid by members of the *Aspergillus ochraceus* group, *Can. J. Microbiol.*, 18, 631—636, 1972.
132. **Miyaki, K., Maebayashi, Y., and Yamazaki, M.**, On the toxic metabolites of *Aspergillus ochraceus* isolated from the moldy rice, *Annu. Rep. Inst. Food Microbiol. Chiba Univ.*, 23, 41—46, 1970.
133. **Yamazaki, M., Maebayashi, Y., and Miyaki, K.**, The isolation of secalonic acid A from *Aspergillus ochraceus* cultured on rice, *Chem. Pharm. Bull.*, 19, 199—201, 1971.
134. **Durley, R. C., MacMillan, J., Simpson, T. J., Glen, A. T., and Turner, W. B.**, Fungal products. XIII. Xanthomegnin, viomellein, rubrosulphin, and viopurpurin, pigments from *Aspergillus sulphureus* and *Aspergillus melleus, J. Chem. Soc. Perkin I*, pp. 163—169, 1975.

135. **Stack, M. E. and Mislivec, P. B.**, Production of xanthomegnin and viomellein by isolates of *Aspergillus ochraceus, Penicillium cyclopium,* and *Penicillium viridicatum, Appl. Environ. Microbiol.*, 36, 552—554, 1978.
136. **Robbers, J. E., Hong, S., Tuite, J., and Carlton, W. W.**, Production of xanthomegnin and viomellein by species of *Aspergillus* correlated with mycotoxicosis produced in mice, *Appl. Environ. Microbiol.*, 36, 819—823, 1978.
137. **Maebayashi, Y., Sumita, M., Fukushima, K., and Yamazaki, M.**, Isolation and structure of red pigment from *Aspergillus ochraceus* Wilh., *Chem. Pharm. Bull.*, 26, 1320—1322, 1978.
138. **Ferreira, N. P.**, Recent advances in research on ochratoxin (part 2), in *Biochemistry of Some Foodborne Microbial Toxins,* Mateles, R. I. and Wogan, G. N., Eds., Massachusetts Institute of Technology, Cambridge, Mass., 1967, 157—168.
139. **Lai, M., Semeniuk, G., and Hesseltine, C. W.**, Nutrients affecting ochratoxin-A production by Aspergillus spp., *Phytopathology,* 58, 1056, 1968.
140. **Donceva, I.**, A study of the influence of various synthetic food media on the production of ochratoxin A, *Khig. Zdraveopaz.*, 19, 372—277, 1976.
141. **Steele, J. A., Davis, N. D., and Diener, U. L.**, Effect of zinc, copper, and iron on ochratoxin A production, *Appl. Microbiol.*, 25, 847—849, 1973.
142. **Lai, M., Semeniuk, G., and Hesseltine, C. W.**, Conditions for production of ochratoxin A by *Aspergillus* species in a synthetic medium, *Appl. Microbiol.*, 19, 542—544, 1970.
143. **Bacon, C. W., Robbins, J. D., and Burdick, D.**, Metabolism of glutamic acid in *Aspergillus ochraceus* during the biosynthesis of ochratoxin A, *Appl. Microbiol.*, 29, 317—322, 1975.
144. **van Walbeek, W., Scott, P. M., and Thatcher, F. S.**, Mycotoxins from food-borne fungi, *Can. J. Microbiol.*, 14, 131—137, 1968.
145. **Davis, N. D., Searcy, J. W., and Diener, U. L.**, Production of ochratoxin A by *Aspergillus ochraceus* in a semisynthetic medium, *Appl. Microbiol.*, 17, 742—744, 1969.
146. **Sansing, G. A., Davis, N. D., and Diener, U. L.**, Effect of time and temperature on ochratoxin A production by *Aspergillus ochraceus, Can. J. Microbiol.*, 19, 1259—1263, 1973.
147. **Hesseltine, C. W.**, Solid state fermentations, *Biotechnol. Bioeng.*, 14, 517—532, 1972.
148. **Hesseltine, C. W.**, Solid state fermentation — part 1, *Process Biochem.*, 12, 24—27, 1977.
149. **Lindenfelser, L. A. and Ciegler, A.**, Solid-substrate fermentor for ochratoxin A production, *Appl. Microbiol.*, 29, 323—327, 1975.
150. **Escher, F. E., Koehler, P. E., and Ayres, J. C.**, Production of ochratoxins A and B on country cured ham, *Appl. Microbiol.*, 26, 27—30, 1973.
151. **Doupnik, B., Jr. and Bell, D. K.**, Toxicity to chicks of *Aspergillus* and *Penicillium* species isolated from moldy pecans, *Appl. Microbiol.*, 21, 1104—1106, 1971.
152. **Schindler, A. F. and Nesheim, S.**, Effect of moisture and incubation time on ochratoxin A production by an isolate of *Aspergillus ochraceus, J. Assoc. Off. Anal. Chem.*, 53, 89—91, 1970.
153. **Rehm, H.-J. and Schmidt, I.**, Mykotoxine in Lebensmitteln, III. Mitt.: Bildung von Ochratoxinen in verschiedenen Lebensmitteln, *Zentralbl. Bakteriol. Parasitenk. Infektionskr. Hyg.*, 124, 364—368, 1970.
154. **Trenk, H. L., Butz, M. E., and Chu, F. S.**, Production of ochratoxins in different cereal products by *Aspergillus ochraceus, Appl. Microbiol.*, 21, 1032—1035, 1971.
155. **Lindenfelser, L. A., Ciegler, A., and Hesseltine, C. W.**, Wild rice as a substrate for mycotoxin production, *Appl. Environ. Microbiol.*, 35, 105—108, 1978.
156. **Reiss, J.**, The formation of the mycotoxin ochratoxin A by *Aspergillus ochraceus* in bread, *Dtsch. Lebensm. Rundsch.*, 77, 27—32, 1981.
157. **Reiss, J.**, Mycotoxins in foodstuffs. XII. The influence of the water activity (a_w) of cakes on the growth of moulds and the formation of mycotoxins, *Z. Lebensm. Unters. Forsch.*, 167, 419—422, 1978.
158. **Hitokoto, M., Morozumi, S., Wauke, T., Sakai, S., and Kurata, H.**, Fungal contamination and mycotoxin producing potential of dried beans, *Mycopathologia,* 73, 33—38, 1981.
159. **Labie, C. and Tache, S.**, Étude sur les conditions de production d'ochratoxine A dans les saucissons sens, *Bull. Acad. Vet. Fr.*, 52, 553—559, 1979.
160. **Lillehoj, E. B., Aalund, O., and Hald, B.**, Bioproduction of (^{14}C)ochratoxin A in submerged culture, *Appl. Environ. Microbiol.*, 36, 720—723, 1978.
161. **Bacon, C. W., Sweeney, J. G., Robbins, J. D., and Burdick, D.**, Production of penicillic acid and ochratoxin A on poultry feed by *Aspergillus ochraceus:* temperature and moisture requirements, *Appl. Microbiol.*, 26, 155—160, 1973.
162. **Engel, G.**, Untersuchungen zur Bildung von Mykotoxinen und deren quantitative Bestimmung. V. Zur Synthese von Ochratoxin A and Penicillinsaüre durch *Aspergillus ochraceus* in Abhängigkeit von Qualität and Quantität des Kohlenhydratangebotes, *Milchwissenschaft,* 31, 264—267, 1976.
163. **Applegate, K. L. and Chipley, J. R.**, Production of ochratoxin A by *Aspergillus ochraceus* NRLL-3174 before and after exposure to ^{60}Co irradiation, *Appl. Environ. Microbiol.*, 31, 349—353, 1976.

164. **Vandegraft, E. E., Shotwell, O. L., Smith, M. L., and Hesseltine, C. W.**, Mycotoxin production affected by insecticide treatment of wheat, *Cereal Chem.*, 50, 264—270, 1973.
165. **Vandegraft, E. E., Shotwell, O. L., Smith, M. L., and Hesseltine, C. W.**, Mycotoxin formation affected by fumigation of wheat, *Cereal Chem.*, 18, 412—414, 1973.
166. **Wu, M. T. and Ayres, J. C.**, Effects of dichlorvos on ochratoxin production, *J. Agric. Food Chem.*, 22, 536—537, 1974.
167. **Escoula, L. and Larrieu, G.**, Sur la signification d'une double contamination par la patuline et l'ochratoxine, *Ann. Rech. Vet.*, 11, 119—122, 1980.
168. **Tomova, S.**, Stability of ochratoxin in foods, *Zhig. Zdraveopaz.*, 20, 266—270, 1977; *Chem. Abstr.*, 87, 166157c, 1977.
169. **Josefsson, B. G. E. and Möller, T. E.**, Heat stability of ochratoxin A in pig products, *J. Sci. Food Agric.*, 31, 1313—1315, 1980.
170. **Gallaz, L. and Stalder, R.**, Ochratoxin A in Kaffee, *Chem. Mikrobiol. Technol. Lebensm.*, 4, 147—149, 1976.
171. **Levi, C.**, Mycotoxins in coffee, *J. Assoc. Off. Anal. Chem.*, 63, 1282—1285, 1980.
172. **Harwig, J., Chen, Y.-K., and Collins-Thompson, D. L.**, Stability of ochratoxin A in beans during canning, *Can. Inst. Food Sci. Technol. J.*, 7, 288—289, 1974.
173. **Gjertsen, P. and Myken, F.**, Malting and brewing experiments with ochratoxin and citrinin, in Proc. 14th Congr. Eur. Brewery Convention, Salzburg, Austria, 1973, 373—383.
174. **Nip, W. K., Chang, F. C., Chu, F. S., and Prentice, N.**, Fate of ochratoxin A in brewing, *Appl. Microbiol.*, 30, 1048—1049, 1975.
175. **Chu, F. S., Chang, C. C., Ashoor, S. H., and Prentice, N.**, Stability of aflatoxin B_1 and ochratoxin A in brewing, *Appl. Microbiol.*, 29, 313—316, 1975.
176. **Fischbach, H. and Rodricks, J. V.**, Current efforts of the Food and Drug Administration to control mycotoxins in food, *J. Assoc. Off. Anal. Chem.*, 56, 767—770, 1973.
177. **Chelkowski, J.**, Formation and detoxification of mycotoxins (ochratoxins) in cereal grain, *Rocz. Akad. Poln. Poznzniu Rozpr. Nauk*, 100, 1980; *Chem. Abstr.*, 94, 172995k, 1981.
178. **Steyn, P. S., Holzapfel, C. W., and Ferreira, N. P.**, The biosynthesis of the ochratoxins, metabolites of *Aspergillus ochraceus, Phytochemistry*, 9, 1977—1983, 1970.
179. **Ferreira, N. P. and Pitout, M. J.**, The biogenesis of ochratoxin, *J. S. Afr. Chem. Inst.*, 22, S1—S8, 1969.
180. **Maebayashi, Y., Miyaki, K., and Yamazaki, M.**, Application of ^{13}C-NMR to the biosynthetic investigations. I. Biosynthesis of ochratoxin A, *Chem. Pharm. Bull.*, 20, 2172—2175, 1972.
181. **Searcy, J. W., Davis, N. D., and Diener, U. L.**, Biosynthesis of ochratoxin A, *Appl. Microbiol.*, 18, 622—627, 1969.
182. **Yamazaki, M., Maebayashi, Y., and Miyaki, K.**, Biosynthesis of ochratoxin A, *Tetrahedron Lett.*, No. 25, pp. 2301—2303, 1971.
183. **de Jesus, A. E., Steyn, P. S., Vleggar, R., and Wessels, P. L.**, Carbon-13 nuclear magnetic resonance assignments and biosynthesis of the mycotoxin ochratoxin A, *J. Chem. Soc. Perkin I*, pp. 52—54, 1980.
184. **Weisleder, D. and Lillehoj, E.**, Carbon-13 nuclear magnetic resonance assignments and biosynthesis of ochratoxin A, *Tetrahedron Lett.*, No. 21, 993—996, 1980.
185. **Wei, R.-D., Strong, F. M., and Smalley, E. B.**, Incorporation of Chlorine-36 into ochratoxin A, *Appl. Microbiol.*, 22, 276—277, 1971.
186. **Huff, W. E. and Hamilton, P. B.**, Mycotoxins — their biosynthesis in fungi: ochratoxins — metabolites of combined pathways, *J. Food Prot.*, 42, 815—820, 1979.
187. **Shreeve, B. J., Patterson, D. S. P., and Roberts, B. A.**, Investigation of suspected cases of mycotoxicosis in farm animals in Britain, *Vet. Rec.*, 97, 275—278, 1975.
188. **Clarke, J. H. and Niles, E. V.**, Fungi and mycotoxins detected in samples from the Infestation Control Service, 1976—77, and recent laboratory experiments on mycotoxins at PICL, in Proc. 3rd Meet. Mycotoxins Anim. Dis., Pepin, G. A., Patterson, D. S. P., and Shreeve, B. J., Eds., Ministry of Agriculture, Fisheries and Food, England, 1979, 8—12.
189. **Anon.**, Survey of Mycotoxins in the United Kingdom, Her Majesty's Stationery Office, London, 1980.
190. **Richardson, E. A., Flude, P. J., Patterson, D. S. P., MacKenzie, D. W. R., and Wakefield, E. H.**, Ochratoxin A in retail flour, *Lancet*, p. 1366, 1978.
191. **Osborne, B. G.**, The occurrence of ochratoxin A in mouldy bread and flour, *Food Cosmet. Toxicol.*, 18, 615—617, 1980.
192. **Funnell, H. S.**, Mycotoxins in animal feedstuffs in Ontario 1972 to 1977, *Can. J. Comp. Med.*, 43, 243—246, 1979.
193. **Prior, M.**, Mycotoxins in animal feedstuffs and tissues in Western Canada 1975 to 1979, *Can. J. Comp. Med.*, 45, 116—119, 1981.

194. Tarter, E., unpublished results, 1979.
195. **Madsen, A., Mortensen, H. P., Larsen, A. E., Flengmark, P., Hald, B., Krogh, P., and Elling, F.**, Ochratoksinholdige hestebønner til slagterisvin, *Ugeskrift Agron. Hort.*, 6, 100—101, 1973.
196. **Krogh, P.**, Natural occurrence of ochratoxin A, a kidney-toxic mold metabolite in Scandinavian cereals, *Zesz. Probl. Posterpow Nauk. Rolnichzych*, 189, 21—24, 1977.
197. **Pedersen, E. and Hansen, H. N.**, Ochratoxin A in Danish cereals, in Abstr. 4th Int. IUPAC Symp. Mycotoxins and Phycotoxins, Lausanne, August 29 to 31, 1979.
198. **Collet, J.-C., Regnier, J.-M., and Maréchal, J.**, Contamination par mycotoxines des mais conservés en cribs et visiblement altérés, *Ann. Nutr. Alim.*, 31, 447—457, 1977.
199. **Uchiyama, M., Isohata, E., and Takeda, Y.**, A case report on the detection of ochratoxin A from rice, *J. Food Hyg. Soc.*, 17, 103—104, 1976.
200. **Juszkiewicz, T. and Piskorska-Pliszcyznska, J.**, Mycotoxins in grain for animal feeds, *Ann. Nutr. Alim.*, 31, 489—493, 1977.
201. **Juszkiewicz, T. and Piskorska-Pliszczynska, J.**, Occurrence of mycotoxins in animal feeds, in Abstr. 4th IUPAC Symp. Mycotoxins and Phycotoxins, Lausanne, August 29 to 31, 1979.
202. **Chelkowski, J., Godlewska, B., and Radomyska, W.**, Occurrence of mycotoxins in foods and feeds, *Przem. Spozyw.*, 32, 285—286, 1978; *Chem. Abstr.*, 94, 45683g, 1981.
203. **Josefsson, E.**, Ochratoxin i korn- och havreprodukter, *Var Foda*, 28, 206—209, 1976.
204. **Hökby, E., Hult, K., Gatenbeck, S., and Rutqvist, L.**, Ochratoxin A and citrinin in 1976 crop of barley stored on farms in Sweden, *Acta Agric. Scand.*, 29, 174—178, 1979.
205. **Pettersson, H. and Kiessling, K.-H.**, Mycotoxins in Swedish grains and mixed feeds, in Abstr. 4th Int. IUPAC Symp. Mycotoxins and Phycotoxins, Lausanne, August 29 to 31, 1979.
206. **Balzer, I., Bogdanić, C., and Mužić, S.**, Natural contamination of corn (*Zea mais*) with mycotoxins in Yugoslavia, *Ann. Nutr. Alim.*, 31, 425—430, 1977.
207. **Krogh, P., Hald, B., Pleština, R., and Čeović, S.**, Balkan (endemic) nephropathy and foodborne ochratoxin A: preliminary results of a survey of foodstuffs, *Acta Pathol. Microbiol. Scand. B*, 85, 238—240, 1977.
208. **Pavlović, M., Pleština, R., and Krogh, P.**, Ochratoxin A contamination of foodstuffs in an area with Balkan (endemic) nephropathy, *Acta Pathol. Microbiol. Scand. B*, 87, 243—246, 1979.
209. **Pleština, R., Radic, B., Habazin-Novak, V., Čeović, S., and Kralj, Z.**, Ochratoxin A and Balkan endemic nepthropathy (BEN), Survey of foodstuffs for ochratoxin A in an endemic region, in Abstr. 4th Int. IUPAC Symp. Mycotoxins and Phycotoxins, Lausanne, August 29 to 31, 1979.
210. **Balzer, I., Pepeljnjak, S., and Cuturic, S.**, The contamination of corn with mycotoxins in Yugoslavia in the year 1978—79, in Abstr. 4th IUPAC Symp. Mycotoxins and Phycotoxins, Lausanne, August 29 to 31, 1979.
211. **Veselá, D., Veselý, D., Jelínek, S., and Kušak, V.**, A finding of ochratoxin A in fodder barley, *Vet. Med., Praha*, 23, 431—436, 1978.
212. **Fritz, W., Buthig, C., Donath, R., and Engst, R.**, Untersuchungen zur ernahrungsrelevaten Bildung von Ochratoxin A in Getreide und sonstigen Lebensmitteln, *Z. Ges. Hyg.*, 25, 929—932, 1979.
213. **Fritz, W. and Engst, R.**, Survey of selected mycotoxins in food, *J. Environ. Health Sci. Part B*, 16, 193—210, 1981.
214. **Rao, E. R., Basappa, S. C., and Murthy, V. S.**, Studies on the occurrence of ochratoxins in food grains, *J. Food Sci. Technol.*, 16, 113—114, 1979.
215. **Hamilton, P. B., Huff, W. E., Harris, J. R., and Wyatt, R. D.**, Outbreaks of ochratoxicosis in poultry, in Proc. Am. Soc. Microbiol. Annu. Meet., 1977, 248.
216. **Harwig, J., Blanchfield, B. J., and Jarvis, G.**, Effect of water activity on disappearance of patulin and citrinin from grains, *J. Food Sci.*, 42, 1225—1228, 1977.
217. **Hald, B. and Krogh, P.**, Ochratoxin residues in bacon pigs, in IUPAC Symp. Control of Mycotoxins, Kungalv, Sweden, August 21 to 22, 1972, 18.
218. **Rutqvist, L., Björklund, N.-E., Hult, K., and Gatenbeck, S.**, Spontaneous occurrence of ochratoxin residues in kidneys of fattening pigs, *Zentralbl. Vet. Med. Reihe A*, 24, 402—408, 1977.
219. **Josefsson, E.**, Undersökning av ochratoxin A i svinnjurar, *Var Foda*, 31, 415—420, 1979.
220. **Elling, F., Hald, B., Jacobsen, C., and Krogh, P.**, Spontaneous toxic nephropathy in poultry associated with ochratoxin A, *Acta Pathol. Microbiol. Scand. A*, 83, 739—741, 1975.
221. **Krogh, P. and Elling, F.**, Fungal toxins and endemic (Balkan) nephropathy, *Lancet*, July 3, 40, 1976.
222. **Hult, K., Hökby, E., Gatenbeck, S., and Rutqvist, L.**, Ochratoxin A in blood from slaughter pigs in Sweden: use in evaluation of toxin content of consumed feed, *Appl. Environ. Microbiol.*, 39, 828—830, 1980.
223. **Krogh, P. and Elling, F.**, Mycotoxic nephropathy, *Vet. Sci. Commun.*, 1, 51—63, 1977.
224. **Krogh, P.**, Causal associations of mycotoxic nephropathy, in *Medical Mycology, Zentralbl. Bakteriol.* Suppl. 8, Preusser, Eds., Gustav Fischer Verlag, New York, 1980.
225. **Friis, P., Hasselager, E., and Krogh, P.**, Isolation of citrinin and oxalic acid from *Penicillium viridicatum* Westling and their nephrotoxicity in rats and pigs, *Acta Pathol. Microbiol. Scand.*, 77, 559—560, 1969.

226. Krogh, P., Hasselager, E., and Friis, P., Studies on fungal nephrotoxicity. II. Isolation of two nephrotoxic compounds from *Penicillium viridicatum* Westling: citrinin and oxalic acid, *Acta Pathol. Microbiol. Scand. B,* 78, 401—413, 1970.
227. Krogh, P., Axelsen, N. H., Elling, F., Gyrd-Hansen, N., Hald, B., Hyldgaard-Jensen, J., Larsen, A. E., Madsen, A., Mortensen, H. P., Möller, T., Peterson, O. K., Ravnskov, U., Rostgaard, M., and Aalund, O., Experimental porcine nephropathy. Changes of renal function and structure induced by ochratoxin A-contaminated feed, *Acta Pathol. Microbiol. Scand. A,* Suppl. 246, 1—21, 1974.
228. Buckley, H. G., Fungal nephrotoxicity in swine, *Ir. Vet. J.,* 25, 194—196, 1971.
229. Krogh, P., Elling, F., Gyrd-Hansen, N., Hald, B., Larsen, A. E., Lillehøj, E. B., Madsen, A., Mortensen, H. P., and Ravnskov, U., Experimental porcine nephropathy: changes of renal function and structure perorally induced by crystalline ochratoxin A, *Acta Pathol. Microbiol. Scand. A,* 84, 429—434, 1976.
230. Krogh, P., Elling, F., Hald, B., Larsen, A. E., Lillehoj, E. B., Madsen, A., and Mortensen, H. P., Time-dependent disappearance of ochratoxin A residues in tissues of bacon pigs, *Toxicology,* 6, 235—242, 1976.
231. Krogh, P., Elling, F., Friis, C., Hald, B., Larsen, A. E., Lillehoj, E. B., Madsen, A., Mortensen, H. P., Rasmussen, F., and Ravnskov, U., Porcine nephropathy induced by long-term ingestion of ochratoxin A, *Vet. Pathol.,* 16, 466—475, 1979.
232. Elling, F., Ochratoxin A-induced mycotoxic porcine nephropathy: alterations in enzyme activity in tubular cells, *Acta Pathol. Microbiol. Scand. A,* 87, 237—243, 1979.
233. Hyldgaard-Jensen, J., Ochratoksinforgifning hos svin. Aedringer i blodets og urienes enzmindhold hidrørendre fra nyrene, *Aarsberet. Inst. Sterilitetsforsk. K. Vet. Landbohoejskole,* 16, 115—120, 1973.
234. Szczech, G. M., Carlton, W. W., Tuite, J., and Caldwell, R., Ochratoxin A toxicosis in swine, *Vet. Pathol.,* 10, 347—364, 1973.
235. Szczech, G. M. and Hood, R. D., Animal model of human disease: alimentary toxic aleukia, fetal brain necrosis, and renal tubular necrosis, *Am. J. Pathol,* 91, 689—692, 1978.
236. Patterson, D. S. P., Roberts, B. A., and Small, B. J., Metabolism of ochratoxins A and B in the pig during early pregnancy and the accumulation in body tissues of ochratoxin A only, *Food Cosmet. Toxicol.,* 14, 439—442, 1976.
237. Krogh, P., Elling, F., Hald, B., Jylling, B., Petersen, V. E., Skadhauge, E., and Svendsen, C. K., Experimental avian nephropathy, *Acta Pathol. Microbiol. Scand. A,* 84, 215—221, 1976.
238. Prior, M. G. and Sisodia, C. S., Ochratoxicosis in white leghorn hens, *Poultry Sci.,* 57, 619—623, 1978.
239. Frye, C. E. and Chu, F. S., Distribution of ochratoxin A in chicken tissues and eggs, *J. Food Saf.,* 1, 147—159, 1977.
240. Prior, M. G., O'Neil, J. B., and Sisodia, C. S., Effects of ochratoxin A on growth response and residues in broilers, *Poultry Sci.,* 59, 1254—1257, 1980.
241. Shreeve, B. J., Patterson, D. S. P., and Roberts, B. A., The 'carry-over' of aflatoxin, ochratoxin and zearalenone from naturally contaminated feed to tissues, urine and milk of dairy cows, *Food Cosmet. Toxicol.,* 17, 151—152, 1979.
242. Shreeve, B. J., Patterson, D. S. P., Pepin, G. A., Roberts, B. A., and Wrathall, A. E., Effect of feeding ochratoxin to pigs during early pregnancy, *Br. Vet. J.,* 133, 412—417, 1977.
243. Choudhury, H. and Carlson, C. W., The lethal dose of ochratoxin for chick embryos, *Poultry Sci.,* 52, 1202—1203, 1973.
244. Gilani, S. H., Bancroft, J., and Reily, M., Teratogenicity of ochratoxin A in chick embryos, *Toxicol. Appl. Pharmacol.,* 46, 543—546, 1978.
245. Yamazaki, M., Suzuki, S., Sakakibara, Y., and Miyaki, K., The toxicity of 5-chloro-8-hydroxy-3,4-dihydro-3-methyl-isocoumarin-7-carboxylic acid, a hydrolyzate of ochratoxin A, *Jpn. J. Med. Sci. Biol.,* 24, 245—250, 1971.
246. Chu, F. S. and Chang, C. C., Sensitivity of chicks to ochratoxins, *J. Assoc. Off. Anal. Chem.,* 54, 1032—1034, 1971.
247. Peckham, J. C., Doupnik, B., Jr., and Jones, O. H., Jr., Acute toxicity of ochratoxins A and B in chicks, *Appl. Microbiol.,* 21, 492—494, 1971.
248. Huff, W. E., Wyatt, R. D., Tucker, T. L., and Hamilton, P. B., Ochratoxicosis in the broiler chicken, *Poultry Sci.,* 53, 1585—1591, 1974.
249. Galtier, P., Moré, J., and Alvinerie, M., Acute and short-term toxicity of ochratoxin A in 10-day-old chicks, *Food Cosmet. Toxicol.,* 14, 129—131, 1976.
250. Prior, M. G., Sisodia, C. S., and O'Neil, J. B., Acute oral ochratoxicosis in day-old white leghorns, turkeys, and Japanese quail, *Poultry Sci.,* 55, 786—790, 1976.
251. Purchase, I. F. H. and Nel, W., Recent advances in research on ochratoxin, I. Toxicological aspects, in *Biochemistry of Some Foodborne Microbial Toxins,* Mateles, R. I. and Wogan G. N., Eds., Massachusetts Institute Technology Press, Cambridge, 1967, 153—156.

252. Gilani, S. H., Bancroft, J., and O'Rahily, M., The teratogenic effects of ochratoxin A in the chick embryo, *Teratology,* 11, 18A, 1975.
253. Page, R. K., Stewart, G., Wyatt, R., Bush, P., Fletcher, O. J., and Brown, J., Influence of low levels of ochratoxin A on egg production, egg-shell stains, and serum uric-acid levels in leghorn-type hens, *Avian Dis.,* 24, 777—780, 1980.
254. Huff, W. E., Wyatt, R. D., and Hamilton, P. B., Nephrotoxicity of dietary ochratoxin A in broiler chickens, *Appl. Microbiol.,* 30, 48—51, 1975.
255. Svendsen, C. and Skadhauge, E., Renal functions in hens fed graded dietary levels of ochratoxin A, *Acta Pharmacol. Toxicol.,* 38, 186—194, 1976.
256. Choudhury, H., Carlson, C. W., and Semeniuk, G., A study of ochratoxin toxicity in hens, *Poultry Sci.,* 50, 1855—1859, 1971.
257. Kiessling, K.-H. and Pettersson, H., Ochratoxin A and protein utilization *in vivo, Swed. J. Agric. Res.,* 8, 227—229, 1978.
258. Huff, W. E. and Hamilton, P. B., Decreased plasma carotenoids during ochratoxicosis, *Poultry Sci.,* 54, 1308—1310, 1975.
259. Doerr, J. A., Huff, W. E., Tung, H. T., Wyatt, R. D., and Hamilton, P. B., A survey of T-2 toxin, ochratoxin, and aflatoxin for their effects on the coagulation of blood in young broiler chickens, *Poultry Sci.,* 53, 1728—1734, 1974.
260. Huff, W. E., Chang, C. F., Warren, M. F., and Hamilton, P. B., Ochratoxin A-induced iron deficiency anemia, *Appl. Environ. Microbiol.,* 37, 601—604, 1979.
261. Chang, C. F., Huff, W. E., and Hamilton, P. B., A leucocytopenia induced in chickens by dietary ochratoxin A, *Poultry Sci.,* 58, 555—558, 1979.
262. Chang, C.-F. and Hamilton, P. B., Impairment of phagocytosis by heterophils from chickens during ochratoxicosis, *Appl. Environ. Microbiol.,* 39, 572—575, 1980.
263. Huff, W. E., Doerr, J. A., Hamilton, P. B., Hamann, D. D., Peterson, R. E., and Ciegler, A., Evaluation of bone strength during aflatoxicosis and ochratoxicosis, *Appl. Environ. Microbiol.,* 40, 102—107, 1980.
264. Warren, M. F. and Hamilton, P. B., Intestinal fragility during ochratoxicosis and aflatoxicosis in broiler chickens, *Appl. Environ. Microbiol.,* 40, 641—645, 1980.
265. Huff, W. E., Doerr, J. A., and Hamilton, P. B., Decreased glycogen mobilization during ochratoxicosis in broiler chickens, *Appl. Environ. Microbiol.,* 37, 122—126, 1979.
266. Warren, M. F. and Hamilton, P. B., Inhibition of the glycogen phosphorylase system during ochratoxicosis in chickens, *Appl. Environ. Microbiol.,* 40, 522—525, 1980.
267. Szczech, G. M., Carlton, W. W., and Tuite, J., Ochratoxicosis in beagle dogs. I. Clinical and clinicopathological features, *Vet. Pathol.,* 10, 135—154, 1973.
268. Szczech, G. M., Carlton, W. W., and Tuite, J., Ochratoxicosis in beagle dogs. II. Pathology, *Vet. Pathol.,* 10, 219—231, 1973.
269. Huff, W. E. and Hamilton, P. B., The interaction of ochratoxin A with some environmental extremes, *Poultry Sci.,* 54, 1659—1662, 1975.
270. Liker, B., Rupić, V., Bogdanić, Č., Balzer, I., Mužić, S., and Herceg, M., Influence of ochratoxin A on the activity of alkaline phosphatase, sorbitol dehydrogenase and glutamate dehydrogenase in chick blood plasma, *Vet. Arhiv,* 48, 23—32, 1978.
271. Rupić, V., Liker, B., Mužić, S., Bogdanić, Č., and Balzer, I., The effect of ochratoxin A in feed on the blood content of lipids and proteins in chickens, *Arhiv Higijenu Rada I Toksikologiju,* 29, 139—145, 1978.
272. Doster, R. C., Arscott, G. H., and Sinnhuber, R. O., Comparative toxicity of ochratoxin A and crude *Aspergillus ochraceus* culture extract in Japanese quail *(Coturnix coturnix japonica), Poultry Sci.,* 52, 2351—2353, 1973.
273. Prior, M. G., O'Neil, J. B., and Sisodia, C. S., Effects of ochratoxin A on production characteristics and hatchability of Japanese quail, *Can. J. Anim. Sci.,* 58, 29—33, 1978.
274. Prior, M. G., Sisodia, C. S., O'Neil, J. B., and Hrudka, F., Effect of ochratoxin A on fertility and embryo viability of Japanese quail *(Coturnix coturnix japonica), Can. J. Anim. Sci.,* 59, 605—609, 1979.
275. Theron, J. J., Merwe, K. J., van der, Liebenberg, N., Joubert, H. J. B., and Nel, W., Acute liver injury in ducklings and rats as a result of ochratoxin poisoning, *J. Pathol. Bacteriol.,* 91, 521—529, 1966.
276. Austwick, P. K. C., Balkan nephropathy, *Proc. R. Soc. Med.,* 68, 219—221, 1975.
277. Cooper, P., The kidney and ochratoxin A, *Food Cosmet. Toxicol.,* 17, 406—408, 1979.
278. Anon., Balkan nephropathy, *Lancet,* March 26, 683—684, 1977.
279. Berndt, W. O., Hayes, A. W., and Phillips, R. D., Effects of mycotoxins on renal function: mycotoxic nephropathy, *Kidney Int.,* 18, 656—664, 1980.

280. **Barnes, J. M., Carter, R. L., Peristianis, G. C., Austwick, P. K. C., Flynn, F. V., and Aldridge, W. N.**, Balkan (endemic) nephropathy and a toxin-producing strain of *Penicillium verrucosum* var. *cyclopium*: an experimental model in rats, *Lancet,* March 26, 671—675, 1977.
281. **Elling, F. and Krogh, P.**, Fungal toxins and Balkan (endemic) nephropathy, *Lancet,* June 4, 1213, 1977.
282. **Hult, K., Hökby, E., Gatenbeck, S., Plestina, R., and Ceovic, S.**, Ochratoxin A and Balkan endemic nephropathy. IV. Occurrence of ochratoxin A in humans, Abstr. 4th Int. IUPAC Symp. Mycotoxins and Phycotoxins, Lausanne, August 29 to 31, 1979.
283. **Purchase, I. F. H. and Theron, J. J.**, The acute toxicity of ochratoxin A to rats, *Food Cosmet. Toxicol.,* 6, 479—483, 1968.
284. **Galtier, P., More, J., and Bodin, G.**, Toxines d'*Aspergillus ochraceus* Wilhelm, *Ann. Rech. Vet.,* 5, 233—247, 1974.
285. **Kanisawa, M., Suzuki, S., Kozuka, Y., and Yamazaki, M.**, Histopathological studies on the toxicity of ochratoxin A in rats. I. Acute oral toxicity, *Toxicol. Appl. Pharmacol.,* 42, 55—64, 1977.
286. **Hayes, A. W., Cain, J. A., and Moore, B. G.**, Effect of aflatoxin B_1, ochratoxin A and rubratoxin B on infant rats, *Food Cosmet. Toxicol.,* 15, 23—27, 1977.
287. **Kanisawa, M., Suzuki, S., and Moroi, K.**, Ochratoxin α, is it an inducing factor of acute enteritis by ochratoxin A, Abstr. 6th Int. Symp. Anim., Plant, Microbial Toxins, Uppsala, *Toxicon,* 17 (Suppl. 1), 84, 1979.
288. **Creppy, E., Schlegel, M., Röschenthaler, R., and Dirheimer, G.**, Action préventive de la phenylalanine sur l'intoxication aigüe par l'ochratoxine-A, *C. R. Acad. Sci. Paris,* 289, 915—918, 1979.
289. **Creppy, E. E., Schlegel, M., Roschenthaler, R., and Dirheimer, G.**, Phenylalanine prevents acute poisoning by ochratoxin A in mice, *Toxicol. Lett.,* 6, 77—80, 1980.
290. **Thacker, H. L. and Carlton, W. W.**, Ochratoxin A mycotoxicosis in the guinea-pig, *Food Cosmet. Toxicol.,* 15, 563—574, 1977.
291. **Sansing, G. A., Lillehoj, E. B., Detroy, R. W., and Miller, M. A.**, Synergistic toxic effects of citrinin, ochratoxin A and penicillic acid in mice, *Toxicon,* 14, 213—320, 1976.
292. **Munro, I. C., Moodie, C. A., Kuiper-Goodman, T., Scott, P. M., and Grice, H. C.**, Toxicologic changes in rats fed graded dietary levels of ochratoxin A, *Toxicol. Appl. Pharmacol.,* 28, 180—188, 1974.
293. **Hatey, F. and Galtier, P.**, Toxicité à court terme de l'ochratoxine A chez le rat, *Ann. Rech. Vet.,* 8, 7—12, 1977.
294. **Berndt, W. O. and Hayes, A. W.**, *In vivo* and *in vitro* changes in renal function caused by ochratoxin A in the rat, *Toxicology,* 12, 5—17, 1979.
295. **Galtier, P., Bodin, G., and Moré, J.**, Toxines d'*Aspergillus ochraceus* Wilhelm. IV. Toxicité de l'ochratoxine A par administration orale prolongée chez le rat, *Ann. Rech. Vet.,* 6, 207—218, 1975.
296. **Galtier, P., Boneu, B., Charpenteau, J. L., Bodin, G., Alvinerie, M., and Moré, J.**, Physiopathology of haemorrhagic syndrome related to ochratoxin A intoxication in rats, *Food Cosmet. Toxicol.,* 17, 49—53, 1979.
297. Anon., Hemorrhagic effects of ochratoxin A, *Nutr. Rev.,* 38, 348—349, 1980.
298. **Suzuki, S., Kozuka, Y., Satoh, T., and Yamazaki, M.**, Studies on the nephrotoxicity of ochratoxin A in rats, *Toxicol. Appl. Pharmacol.,* 34, 479—490, 1975.
299. **Zimmermann, J. L., Carlton, W. W., Tuite, J., and Fennell, D. I.**, Mycotoxic diseases produced in mice by species of the *Aspergillus ochraceus* group, *Food Cosmet. Toxicol.,* 15, 411—418, 1977.
300. **Moré, J. and Camguilhem, R.**, Effects of low doses of ochratoxin A after intratesticular injection in the rat, *Experientia,* 35, 890—892, 1979.
301. **Gupta, M., Bandyopadhyay, S., Sasmal, D., and Majumder, S. K.**, Effects of ochratoxin A and citrinin on kidney functions, *IRCS Med. Sci.,* 7, 466, 1979.
302. **Gupta, M., Bandyopadhyay, S., Paul, B., and Majumder, S. K.**, Hematological changes produced in mice by ochratoxin A, *Toxicology,* 14, 95—98, 1979.
303. **Rankov, B. G. and Tomova, S.**, Studies on the influence of ochratoxin A on rat lenses, *Albrecht Graefes Arch. Klin. Exp. Ophthal.,* 205, 135—139, 1978.
304. **Galtier, P. and Alvinerie, M.**, Devenir de l'ochratoxine A dans l'organisme animal. II. Distribution tissulaire et élimination chez le rat, *Ann. Rech. Vet.,* 5, 319—328, 1974.
305. **Suzuki, S., Satoh, T., and Yamazaki, M.**, The pharmacokinetics of ochratoxin A in rats, *Jpn. J. Pharmacol.,* 27, 735—744, 1977.
306. **Lillehoj, E. B., Kwolek, W. F., Elling, F., and Krogh, P.**, Tissue distribution of radioactivity from ochratoxin A-^{14}C in rats, *Mycopathologia,* 68, 175—177, 1979.
307. **Galtier, P.**, Contribution of pharmacokinetic studies to mycotoxicology — ochratoxin A, *Vet. Sci. Commun.,* 1, 349—358, 1978.
308. **Chang, F. C. and Chu, F. S.**, The fate of ochratoxin A in rats, *Food Cosmet. Toxicol.,* 15, 199—204, 1977.

309. Galtier, P., Devenir de l'ochratoxine A dans l'organisme animal. I. Transport sanguin de la toxine chez le rat, *Ann. Rech. Vet.*, 5, 311—318, 1974.
310. **Galtier, P., Charpenteau, J.-L., Alvinerie, M., and Labouche, C.**, The pharmacokinetic profile of ochratoxin A in the rat after oral and intravenous administration, *Drug Metab. Dispos.*, 7, 429—434, 1979.
311. **Nel, W. and Purchase, I. F. H.**, The fate of ochratoxin A in rats, *J. S. Afr. Chem. Inst.*, 21, 87—88, 1968.
312. **Walbeek, W., van, Moodie, C. A., Scott, P. M., Harwig, J., and Grice, H. C.**, Toxicity and excretion of ochratoxin A in rats intubated with pure ochratoxin A and fed cultures of *Penicillium viridicatum*, *Toxicol. Appl. Pharmacol.*, 20, 439—441, 1971.
313. **Størmer, F. C. and Pedersen, J. I.**, Formation of 4-hydroxyochratoxin A from ochratoxin A by rat liver microsomes, *Appl. Environ. Microbiol.*, 39, 971—975, 1980.
314. **Galtier, P., Baradat, C., and Alvinerie, M.**, Étude de l'élimination d'ochratoxine A par le lait chez la lapine, *Ann. Nutr. Alim.*, 31, 911—918, 1977.
315. **Pitout, M. J.**, The effect of ochratoxin A on glycogen storage in the rat liver, *Toxicol. Appl. Pharmacol.*, 13, 299—306, 1968.
316. **Suzuki, S. and Satoh, T.**, Effects of ochratoxin A on tissue glycogen levels in rats, *Jpn. J. Pharmacol.*, 23, 415—419, 1973.
317. **Suzuki, S., Satoh, T., and Yamazaki, M.**, Effect of ochratoxin A on carbohydrate metabolism in rat liver, *Toxicol. Appl. Pharmacol.*, 32, 116—122, 1975.
318. **Gupta, M., Bandyopadhyay, S., and Sashmal, D.**, Effects of ochratoxin A and citrinin on liver function and metabolism, *IRCS Med. Sci.*, 7, 320, 1979.
319. **Meisner, H. and Selanik, P.**, Inhibition of renal gluconeogenesis in rats by ochratoxin, *Biochem. J.*, 180, 681—684, 1979.
320. **Gupta, M., Sasmal, D., and Bandyopadhyay, S.**, Hypothermic action of ochratoxin A in mice, *IRCS Med. Sci.*, 7, 457, 1979.
321. **Gupta, M. and Sasmal, D.**, Behavioural pharmacology of ochratoxin A in mice, *IRCS Med. Sci.*, 7, 454, 1979.
322. **Gupta, M. and Sasmal, D.**, Prolongation of pentobarbitone sleeping time by ochratoxin A, *IRCS Med. Sci.*, 7, 367, 1979.
323. **Gupta, M., Sasmal, D., and Bandyopadhyay, S.**, Effects of ochratoxin A in morphine analgesia and acetic acid induced writhing in mice, *IRCS Med. Sci.*, 7, 458, 1979.
324. **Gupta, M., Sasmal, D., and Bandyopadhyay, S.**, Effects of ochratoxin A on pentylene tetrazol induced convulsions in mice, *IRCS Med. Sci.*, 7, 609, 1979.
325. **Paul, B., Deb, C., and Banik, S.**, Testicular steroidogenesis in rats following ochratoxin A treatment, *Indian J. Exp. Biol.*, 17, 121—123, 1979.
326. **Gupta, M., Bandyopadhyay, S., Mazumdar, S. K., and Paul, B.**, Ovarian steroidogenesis in rats following ochratoxin A treatment, *Toxicol. Appl. Pharmacol.*, 53, 515—520, 1980.
327. **Gupta, M., Bandyopadhyay, S., Paul, B., and Mazumdar, S. K.**, Onset of puberty and ovarian steroidogenesis following administration of ochratoxin A, *Endokrinologie*, 75, 292—298, 1980.
328. **Gupta, M., Bandyopadhyay, S., Paul, B., and Mazumdar, S. K.**, Histochemical determination of adrenal steroidogenesis in rat after treatment with ochratoxin A, *Endokrinologie*, 75, 369—372, 1980.
329. **Richard, J. L., Thurston, J. R., Deyoe, B. L., and Booth, G. D.**, Effect of ochratoxin and aflatoxin on serum proteins, complement activity, and antibody production to *Brucella abortus* in guinea pigs, *Appl. Microbiol.*, 29, 27—29, 1975.
330. **Prior, M. G. and Sisodia, C. S.**, Some tissue culture and immunological studies with ochratoxin A, Abstr. 6th Int. Symp. Anim., Plant and Microbiol Toxins, Uppsala, *Toxicon*, 17 (Suppl. 1), 146, 1979.
331. **Haubeck, H. D., Lorkowski, G., Eckehart, K., and Röschenthaler, R.**, Immunosuppression by ochratoxin A and its prevention by phenylalanine, *Appl. Environ. Microbiol.*, 41, 1040—1042, 1981.
332. **Still, P. E., Macklin, A. W., Ribelin, W. E., and Smalley, E. B.**, Relationship of ochratoxin A to foetal death in laboratory and domestic animals, *Nature (London)*, 234, 563—564, 1971.
333. **Moré, J., Galtier, P., Alvinerie, M., and Escribe, M.-J.**, Toxicité de l'ochratoxine A. I. Effet embryotoxique et tératogène chez le rat, *Ann. Rech. Vet.*, 5, 167—178, 1974.
334. **Hood, R. D., Naughton, M. J., and Hayes, A. W.**, Prenatal effects of ochratoxin A in hamsters, *Teratology*, 13, 11—14, 1975.
335. **Brown, M. H., Szczech, G. M., and Purmalis, B. P.**, Teratogenic and toxic effects of ochratoxin A in rats, *Toxicology*, 37, 331—338, 1976.
336. **Hayes, A. W., Hood, R. D., and Lee, H. L.**, Teratogenic effects of ochratoxin A in mice, *Teratology*, 9, 93—98, 1974.
337. **Moré, J., Galtier, P., and Alvinerie, M.**, Embryotoxic and teratogenic effects of ochratoxin A in rats, in Proc. 19th Morphol. Con. Symp. Charles University, Prague, 1978, 321—326.

338. **Hood, R. D., Kuczuk, M. H., and Szczech, G. M.**, Effects in mice of simultaneous prenatal exposure to ochratoxin A and T-2 toxin, *Teratology*, 17, 25—30, 1978.
339. **Szczech, G. M. and Hood, R. D.**, Brain necrosis in mouse fetuses transplacentally exposed to the mycotoxin ochratoxin A, *Toxicol. Appl. Pharmacol.*, 57, 127—137, 1981.
340. **Moré, J., Galtier, P., Brunel-Dubech, N., and Alvinerie, M.**, Toxicité de l'ochratoxine A. II. Effets du traitement sur la descendance (F_1 et F_2) de rattes intoxiquées, *Ann. Rech. Vet.*, 6, 379—389, 1975.
341. **Moré, J., Galtier, P., Alvinerie, M., and Brunel-Dubech, N.**, Toxicité de l'ochratoxine A. III. Effets pendant les stades initiaux de la gestation chez le rat, *Ann. Rech. Vet.*, 9, 169—173, 1978.
342. **Tomova, S., Kereshka, P., and Mikhajlova, A.**, Changes in some liver dehydrogenases after treatment with ochratoxin A, *Khig. Zdraveopaz.*, 21, 595—597, 1978.
343. **Dickens, F. and Waynforth, H. B.**, Survey of Compounds Which have Been Tested for Carcinogenic Activity, Vol. 1968—1969, U.S. Department of Health and Education, Washington, D.C.
344. **Purchase, I. F. H. and Watt, J. J., van der**, The long-term toxicity of ochratoxin A to rats, *Food Cosmet. Toxicol.*, 9, 681—682, 1971.
345. **Lindenfelser, L. A., Lillehoj, E. B., and Milburn, M. S.**, Ochratoxin and penicillic acid in tumorigenic and acute toxicity tests with white mice, in *Developments in Industrial Microbiology*, Vol. 14, American Institute Biological Science, Washington, D.C., 1973, 331—336.
346. **Ueno, Y. and Kubota, K.**, DNA-attacking ability of carcinogenic mycotoxins in recombination-deficient mutant cells of *Bacillus subtilis*, *Cancer Res.*, 36, 445—451, 1976.
347. **Umeda, M., Tsutsui, T., and Saito, M.**, Mutagenicity and inducibility of DNA single-strand breaks and chromosome aberrations by various mycotoxins, *Gann*, 68, 619—625, 1977.
348. **Kuczuk, M. H., Benson, P. M., Heath, H., and Hayes, A. W.**, Evaluation of the mutagenic potential of mycotoxins using *Salmonella typhimurium* and *Saccharomyces cerevisiae*, *Mutat. Res.*, 53, 11—20, 1978.
349. **Doster, R. C., Sinnhuber, R. O., Wales, J. H., and Lee, D. J.**, Acute toxicity and carcinogenicity of ochratoxin in rainbow trout *(Salmo gairdneri)*, *Fed. Am. Exp. Biol.*, 30, 578, 1971.
350. **Kanisawa, M. and Suzuki, S.**, Induction of renal and hepatic tumors in mice by ochratoxin A, a mycotoxin, *Gann*, 69, 599—600, 1978.
351. **Szczech, G. M., Carlton, W. W., and Hinsman, E. J.**, Ochratoxicosis in beagle dogs. III. Terminal renal ultrastructural alterations, *Vet. Pathol.*, 11, 385—406, 1974.
352. **Kitchen, D. N., Carlton, W. W., and Tuite, J.**, Ochratoxin A and citrinin induced nephrosis in beagle dogs. I. Clinical and clinicopathological features, *Vet. Pathol.*, 14, 154—172, 1977.
353. **Kitchen. D. N., Carlton, W. W., and Tuite, J.**, Ochratoxin A and citrinin induced nephrosis in beagle dogs. II. Pathology, *Vet. Pathol.*, 14, 261—272, 1977.
354. **Kitchen, D. N., Carlton, W. W., and Hinsman, E. J.**, Ochratoxin A and citrinin induced nephrosis in beagle dogs. III. Terminal renal ultrastructural alterations, *Vet. Pathol.*, 14, 392—406, 1977.
355. **Ribelin, W. E., Fukushima, K., and Still, P. E.**, The toxicity of ochratoxin to ruminants, *Can. J. Comp. Med.*, 42, 172—176, 1978.
356. **Munro, I. C., Scott, P. M., Moodie, C. A., and Willes, R. F.**, Ochratoxin A — occurrence and toxicity, *J. Am. Vet. Med. Assoc.*, 163, 1269—1273, 1973.
357. **Galtier, P. and Alvinerie, M.**, *In vitro* transformation of ochratoxin A by animal microbial floras, *Ann. Rech. Vet.*, 7, 91—98, 1976.
358. **Hult, K., Teiling, A., and Gatenbeck, S.**, Degradation of ochratoxin A by a ruminant, *Appl. Environ. Microbiol.*, 32, 443—444, 1976.
359. **Pettersson, H. and Kiessling, K.-H.**, Effect of aflatoxin, ochratoxin and sterigmatocystin on microorganisms from sheep rumen, *Swed. J. Agric. Res.*, 6, 161—162, 1976.
360. **Nip, W. K. and Chu, F. S.**, Fate of ochratoxin A in goats, *J. Environ. Sci. Health*, B14, 319—333, 1979.
361. **Cardeilhac, P. T., Nair, K. P. C., and Colwell, W. M.**, Tracheal organ cultures for the bioassay of nanogram quantities of mycotoxins, *J. Assoc. Off. Anal. Chem.*, 55, 1120—1121, 1972.
362. **Engelbrecht, J. C. and Purchase, I. F. H.**, Changes in morphology of cell cultures after treatment with aflatoxin and ochratoxin, *S. Afr. J. Lab. Clin. Med.*, 43, 524—528, 1969.
363. **Steyn, P. S., Vleggaar, R., Du Preez, N. P., Blyth, A. A., and Seegers, J. C.**, The *in vitro* toxicity of analogs of ochratoxin A in monkey kidney epithelial cells, *Toxicol. Appl. Pharmacol.*, 32, 198—203, 1975.
364. **Natori, S., Sakaki, S., Kurata, H., Udagawa, S.-I., Ichinoe, M., Saito, M., and Umeda, M.**, Chemical and cytotoxicity survey on the production of ochratoxins and penicillic acid by *Aspergillus ochraceus* Wilhelm, *Chem. Pharm. Bull.*, 18, 2259—2268, 1970.
365. **Creppy, E.-E., Lugnier, A. A. J., Beck, G., Röschenthaler, R., and Dirheimer, G.**, Action of ochratoxin A on cultured hepatoma cells-reversion of inhibition of phenylalanine, *FEBS Lett.*, 104, 287—290, 1979.
366. **Harwig, J. and Scott, P. M.**, Brine shrimp (*Artemia salina* L.) larvae as a screening system for fungal toxins, *Appl. Microbiol.*, 21, 1011—1016, 1971.

367. **Brown, R. F.**, The effect of some mycotoxins on the brine shrimp, *Artemia salina, J. Assoc. Off. Anal. Chem.*, 46, 119, 1969.
368. **Yamamoto, K.**, Methods for determination of mycotoxins. III. Investigation of bioassay method of mycotoxins with brine shrimp larvae (*Artemia salina*) and its application, *J. Nara Med. Assoc.*, 26, 264—278, 1975.
369. **Abedi, Z. H. and Scott, P. M.**, Detection of toxicity of aflatoxins, sterigmatocystin, and other fungal toxins by lethal action on zebra fish larvae, *J. Assoc. Off. Anal. Chem.*, 52, 963—969, 1969.
370. **Murakoshi, S., Ohtomo, T., and Kurata, H.**, Toxic effects of various mycotoxins to silkworm (*Bombyx mori* L.) larvae in *ad libitum* feeding test, *Shokuhin Eiseigaku Zasshhi*, 14, 65—68, 1973.
371. **Hayes, A. W., Melton, R., and Smith, S. J.**, Effect of aflatoxin B_1, ochratoxin and rubratoxin B on a protozoan, *Tetrahymena pyriformis* HSM, *Bull. Environ. Contamin. Toxicol.*, 11, 321—325, 1974.
372. **White, A. G. and Truelove, B.**, The effects of aflatoxin B_1, citrinin, and ochratoxin A on amino acid uptake and incorporation by cucumber, *Can. J. Bot.*, 50, 2659—2664, 1972.
373. **Arai, T. and Otomo. M.**, Antimicrobial activity of ochratoxin A, *Annu. Rep. Inst. Food Microbiol. Chiba Univ.*, 22, 81—82, 1969.
374. **Broce, D., Grodner, R. M., Killebrew, R. L., and Bonner, F. L.**, Ochratoxins A and B confirmation by microbiological assay using *Bacillus cereus mycoides, J. Assoc. Off. Anal. Chem.*, 53, 616—619, 1970.
375. **Heller, K., Schulz, C., Loser, R., and Röschenthaler, R.**, The inhibition of bacterial growth by ochratoxin A, *Can. J. Microbiol.*, 21, 972—979, 1975.
376. **Singer, U. and Röschenthaler, R.**, Induction of autolysis in *Bacillus subtilis* by ochratoxin A, *Can. J. Microbiol.*, 24, 563—568, 1978.
377. **Wegner, A. and Röschenthaler, R.**, Inhibition of *Bacillus subtilis* sporulation by ochratoxin A, *FEMS Microbiol. Lett.*, 4, 147—150, 1978.
378. **Konrad, I. and Röschenthaler, R.**, Inhibition of phenylalanine tRNA synthetase from *Bacillus subtilis* by ochratoxin A, *FEBS Lett.*, 83, 341—347, 1977.
379. **Doster, R. C., Sinnhuber, R. O., and Wales, J. H.**, Acute intraperitoneal toxicity of ochratoxins A and B in rainbow trout *(Salmo gairdneri), Food Cosmet. Toxicol.*, 10, 85—92, 1972.
380. **Doster, R. C., Sinnhuber, R. O., and Pawlowski, N. E.**, Acute intraperitoneal toxicity of ochratoxin A and B derivatives in rainbow trout *(Salmo gairdneri), Food Cosmet. Toxicol.*, 12, 499—505, 1974.
381. **Chu, F. S., Noh, I., and Chang, C. C.**, Structural requirements for ochratoxin intoxication, *Life Sci.*, 11, 503—508, 1972.
382. **Chu, F. S.**, Interaction of ochratoxin A with bovine serum albumin, *Arch. Biophys.*, 147, 359—366, 1971.
383. **Chu, F. S.**, A comparative study of the interaction of ochratoxins with bovine serum albumin, *Biochem. Pharmacol.*, 23, 1105—1113, 1974.
384. **Galtier, P., Camguilhem, R., and Bodin, G.**, Evidence for *in vitro* and *in vivo* interaction between ochratoxin A and three acidic drugs, *Food Cosmet. Toxicol.*, 18, 493—496, 1980.
385. **Pitout, M. J. and Nel, W.**, The inhibitory effect of ochratoxin A on bovine carboxypeptidase A *in vitro, Biochem. Pharmacol.*, 18, 1837—1843, 1969.
386. **Doster, R. C. and Sinnhuber, R. O.**, Comparative rates of hydrolysis of ochratoxins A and B *in vitro, Food Cosmet. Toxicol.*, 10, 389—394, 1972.
387. **Heller, K. and Röschenthaler, R.**, Inhibition of protein synthesis in *Streptococcus faecalis* by ochratoxin A, *Can. J. Microbiol.*, 24, 466—472, 1978.
388. **Creppy, E.-E., Lorkowski, G., Beck, G., Röschenthaler, R., and Dirheimer, G.**, Combined action of citrinin and ochratoxin A on hepatoma tissue culture cells, *Toxicol. Lett.*, 5, 375—380, 1980.
389. **Lorkowski, G., Creppy, E.-E., Beck, G., Dirheimer, G., and Röschenthaler, R.**, Inhibitory action of citrinin on cultured hepatoma cells, *Food Cosmet. Toxicol.*, 18, 489—491, 1980.
390. **Creppy, E.-E., Lugnier, A. A. J., Fasiolo, F., Heller, K., Röschenthaler, R., and Dirheimer, G.**, *In vitro* inhibition of yeast phenylalanyl-tRNA synthetase by ochratoxin A, *Chem. Biol. Interact.*, 24, 257—261, 1979.
391. **Creppy, E.-E., Lugnier, A. A. J., Heller, K., Röschenthaler, R., Fasiolo, F., and Dirheimer, G.**, Action of ochratoxin A, a mycotoxin from *Aspergillus ochraceus,* on the first step of the acylation reaction catalyzed by eukaryotic phenylalanyl-tRNA synthetase, Abstr. 6th Int. Symp. Anim., Plant and Microbial Toxins, Uppsala, *Toxicon,* 17 (Suppl. 1), 32, 1979.
392. **Bunge, I., Dirheimer, G., and Röschenthaler, R.**, *In vivo* and *in vitro* inhibition of protein synthesis in *Bacillus stearothermophilus* by ochratoxin A, *Biochem. Biophys. Res. Commun.*, 83, 398—405, 1978.
393. **Meisner, H. and Chan, S.**, Ochratoxin A, an inhibitor of mitochondrial transport systems, *Biochemistry,* 13, 2795—2800, 1974.
394. **Moore, J. H. and Truelove, B.**, Ochratoxin A: inhibition of mitochondrial respiration, *Science,* 168, 1102—1103, 1970.
395. **Meisner, H.**, Energy-dependent uptake of ochratoxin A by mitochondria, *Arch. Biochem. Biophys.*, 173, 132—140, 1976.

ACUTE AND CHRONIC BIOLOGICAL EFFECT OF TRICHOTHECENE TOXINS

Mamoru Saito

INTRODUCTION

Since the beginning of the 19th century, there have been serious damages to wheat and other crops caused by Fusaria in Japan. According to Nishikado's extensive research on wheat scab, the disease was prevalent in the regions along the Pacific coast including Kyushu, Kinki, and Tokai Districts, and also in Aomori and Hokkaido.[1] A number of reports on toxicosis in human and animals ingesting grains infected by these fungi appeared during this period. *Fusarium saubinatti* was identified as the causative agent of "Akakabibyo" (red-mold scab disease of wheat) affecting the growth and breeding of military horses.

Meteorological and oceanographic factors, such as fog or salt mist on the Pacific coast, mist in the inland region around large lakes, or sea fog in Hokkaido were suggested by Ishii[2] as responsible for the prevalence of scab disease. Other investigators[3] assumed that *Fusarium* spores were carried from the Asian continent on the seasonal winds to cause the extensive plague of scab on the budding head of wheat when it rained during the heading period.

In October and November 1956, more than 100 persons suffered from food poisoning in a small agricultural training institute in Hokkaido. They showed headache, abdominal pain, nausea, vomiting, diarrhea, chills, and fever after ingestion of noodles made from flour contaminated with *Fusarium* sp. Judging from the symptoms, 12,13-epoxytrichothecens produced by Fusaria may have been the causative agent of this food poisoning.

Following the heavy damage to wheat in the western district of Japan in 1963, the Technical Committee of the Ministry of Agriculture and Forestry decided to start an investigation of the red molds of wheat in several research units. *F. graminearum* was isolated from all the samples and *F. nivale* was isolated from 5 or 6% of samples of infected wheat grains obtained by the Agricultural Experimental Station in Kumamoto prefecture. Toxicity screening test on these fungi by Tsunoda suggested the presence of high toxicity in one strain of *F. nivale.* Toxicity to the culture cells (Hela and Chang cells) was used as the characteristic sign to pursuit the toxin in chemical fractions. Finally nivalenol was isolated by Tatsuno et al.[4] and another new toxic substance, named fusarenon-X, was also isolated from the medium of the stationary culture of *F. nivale* by Ueno[5] et al. Both nivalenol and fusarenon-X are now classified to be 12,13-epoxytrichothecene mycotoxins.

INJURY TO THE ACTIVELY DIVIDING CELLS

Histological examination of the experimental animals administered nivalenol and fusarenon-X revealed marked cytotoxic changes in the tissue with actively dividing cells such as the mucosa of the small intestine, germinal center of the lymph follicles in the spleen, lymph nodes, and other lymph apparatus, thymus, and bone marrow. Crypt cells of the small intestine, especially of the duodenum and jejunum, showed degenerative changes, atypical mitosis, pyknosis, hyperchromatosis of the nuclear membrane, or fragmentation of the nuclei. These cytotoxic injuries of the actively dividing cells are considered as "radiomimetic" biological properties; these changes may be compared with acute death after whole-body irradiation.[6]

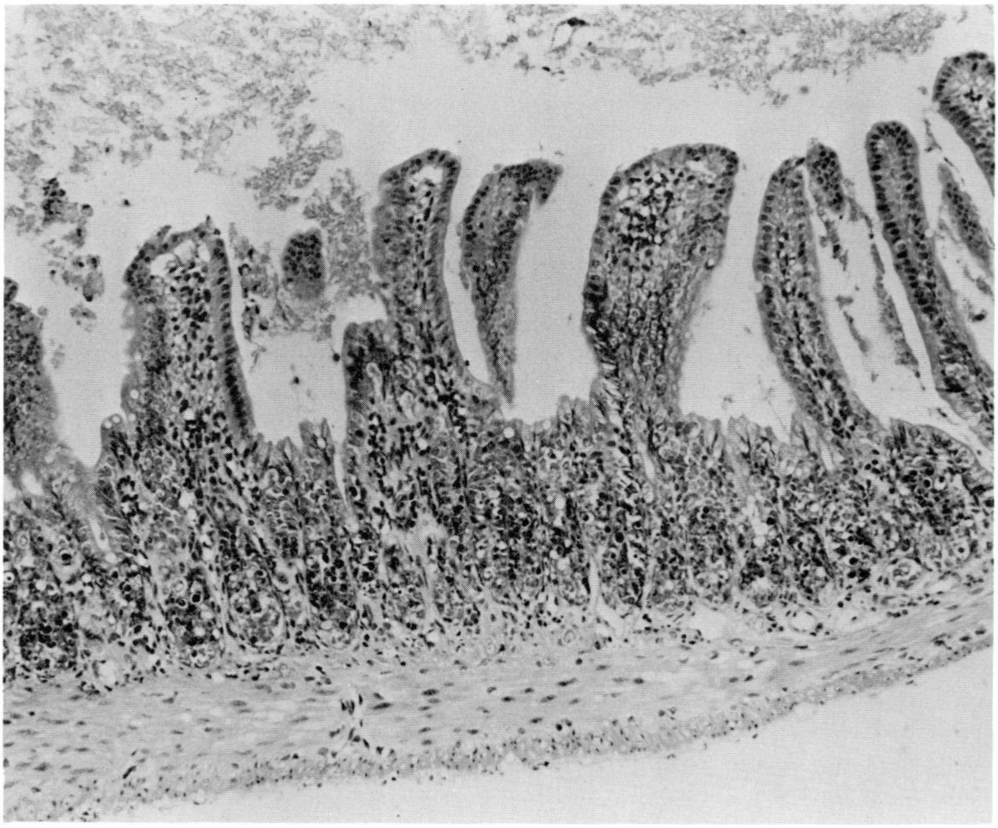

FIGURE 1. Ileum of a mouse showing cytotoxin change of the crypt cells, 6 hr after i.p. administration of fusarenon-X, 6 mg/kg body weight.

As shown in Figure 1 the bottom of the crypts of Lieberkühn of the small intestine was filled with cell debris extruding from the basal cell border into the lumen. The cell damage regularly appeared in the crypts of the small intestine, but the cells of the villi were also involved in the later stage. Some of the mice seemed to have died from intestinal bleeding caused by erosive change and ulceration of the mucosa. The cells of the germinal center of lymph follicles and the cortex of the thymus showed necrosis with karyorrhexis. In the bone marrow, atrophy of the pulp with dilatation of the sinus followed by karyorrhexis of the hematopoietic cells were found in the early stage of toxicosis (Figure 2). The immature cells of the pulp decreased in number and became necrotic later. Compared with changes in the proliferating cells of the tissues described above, striking change was usually not found in the testes. However, cellular components of spermatogenesis were reduced in number; pyknosis and karyolysis of the blastic cells and appearance of multinuclear giant cells were also seen.[6,7]

SKIN IRRITATING EFFECT

Trichothecene mycotoxins including fusarenon-X were highly irritant to the skin tissue of the mouse, rabbit, and guinea pig, causing hemorrhage and necrosis of epidermis and degeneration and necrosis of the hair follicles and dermis. The guinea pig was the highest in skin sensitivity to these toxins among laboratory animals tested.

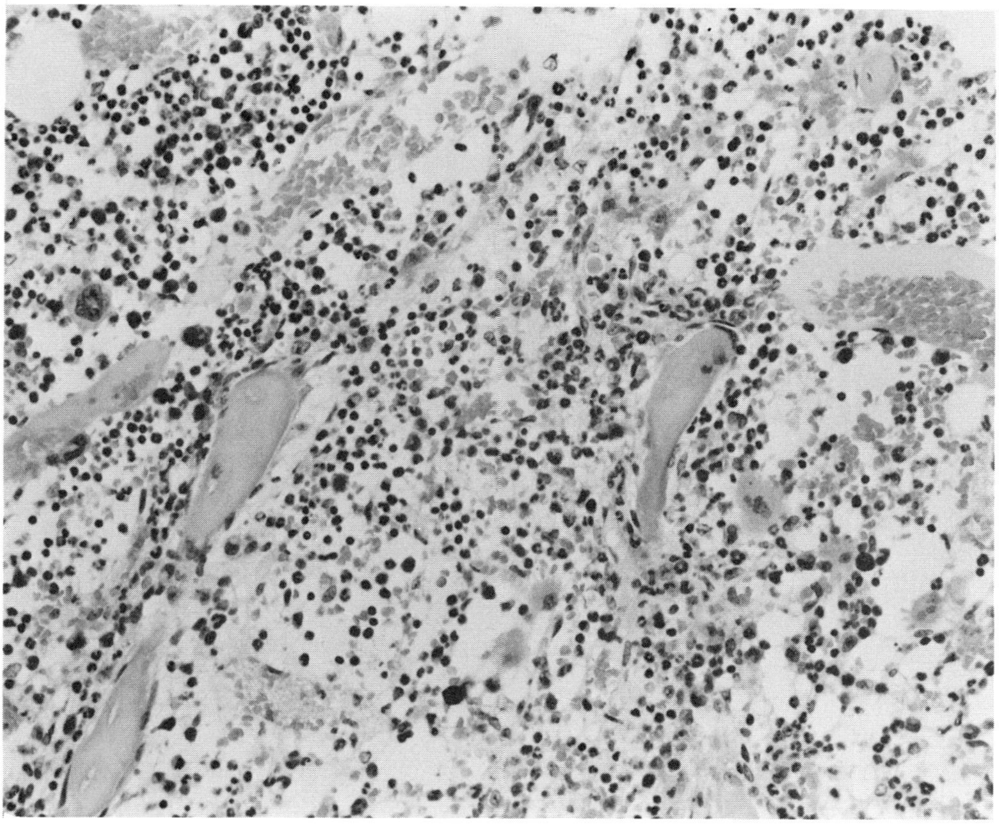

FIGURE 2. Bone marrow of the femur of a mouse 24 hr after i.p. administration of fusarenon-X, 6 mg/kg body weight, shows reduced number of hematopoietic cells in the pulp and dilatation of the sinus.

MUTAGENISTIC ACTIVITY OF THE TRICHOTHECENE MYCOTOXINS

Nagao et al[8] found that, by using Ames' system, fusarenon-X was mutagenic to bacteria, *Salmonella* TA 100, at doses of 0.5 to 1 mg per plate without S9 mixture. This observation indicated that fusarenon-X might be a carcinogen, in conjunction with the pathology of animals of acute trichothecene injuries which resembles that caused by radiation or alkylating agents such as mitomycin C and nitrogen mustard, damaging primarily the GI tract and the hematopoietic system.

CHRONIC EFFECT OF THE TRICHOTHECENE MYCOTOXINS

After feeding rats and mice with rice molded with Fusaria, hyperplasia of the bone marrow, atypical hyperplastic epithelia in the gastric and intestinal mucosa, and proliferation of the intrahepatic bile ducts were observed in several animals. Chronic bronchitis and bronchopneumonia were the frequent cause of death, which may have resulted from a secondary phenomenon of immunological exhaustion to the microorganism of the inflammatory processes in the respiratory organs.

As shown in Table 1, long-term application of moldy rice of *F. nivale* to rats and mice revealed induction of hepatoma, leukemia and adenocarcinoma of the ileum at low incidence, and long-term application of moldy rice of *F. graminearum* revealed a few cases of lymphosarcoma and hemangioma.[9] The toxicity of the molded rice, how-

Table 1
TUMORS DEVELOPED IN ANIMALS BY LONG-TERM APPLICATION OF *F. NIVALE*, *F. GRAMINEARUM*, AND FUSARENON-X

Experimental group	Route of application	Number of animals	Effective no. of animals	Neoplasms
Rats (Donryu, male)				
Control	p.o.	10	9	0
F. nivele (moldy rice)	p.o.	30	22	0
F. graminearum (moldy rice)	p.o.	18	16	1 Lymphosarcoma
Fursarenon-X	Intragastric	20	16	1 Hepatoma
Fusarenon-X	s.c.	18	14	1 Epidermoid tumor of lung
Mice (DDD, male and female)				
Control	p.o.	11		0
	s.c.	25		0
F. nivale (moldy rice)	p.o.	37	32	1 Hepatoma 2 Leukemia 1 Adenocarcinoma of ileum 1 s.c. dermatofibro-myosarcoma
F. graminearum (moldy rice)	p.o.	43	36	1 Hemangiofibroma 1 Hemangioma
Fusarenon-X	s.c.	34	13	1 Leukemia

Table 2
TUMORS DEVELOPED IN RATS FED FUSARENON-X (F-X)

Doses of F-X	Control	50 μg[a] (2 years)	105 μg[a] (1 year)	105 μg[a] (2 years)
No. of rats used	48	49	52	25
Effective no. of rats	33	35	19	14
All tumors (incidence)	8(24.2%)	7(20.0%)	4(21.1%)	2(14.2%)
Pituitary adenoma	1	1	1	1
Thyroid carcinoma/adenoma	4	2	1	
Adrenal tumors				
Cortical adenoma			1	
Pheochromocytoma	1	1	1	
Gastric carcinoma		1		
Urinary bladder carcinoma/ papilloma		1		1
Leydig cell tumor of testis	1			
Leukemia		1		
Plasmocytoma of abdominal cavity	1			

[a] μg/day/rat.

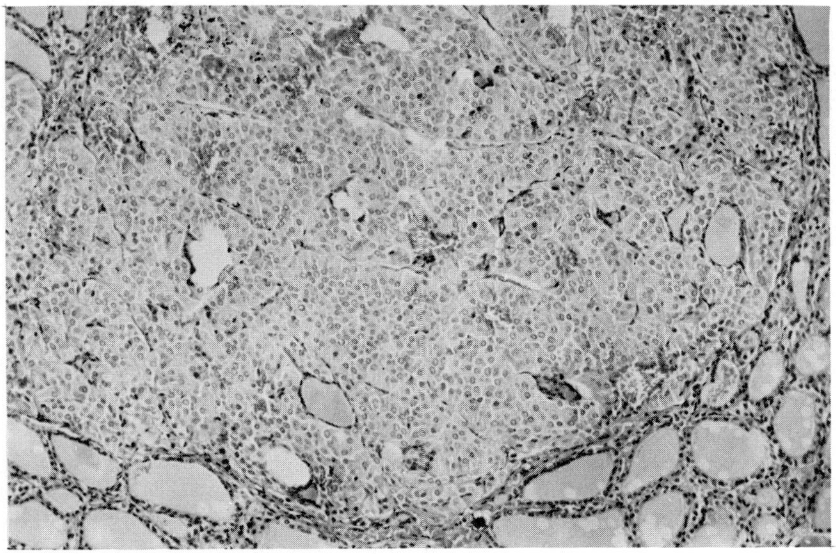

FIGURE 3. Anaplastic tumor of the thyroid of a rat administered 7 ppm of fusarenon-X in diet for 1 year.

ever, varied considerably from lot to lot, making it impossible to determine the dose-response relationship. Results of the study on rats and mice given fusarenon-X weekly by oral administration or s.c. injection (10 to 22 weeks) were similar to those described above. With the latter route, loss of hair was prominent near the injection site. Long-term application of fusarenon-X revealed induction of hepatoma and epidermoid tumor of the lung with destructive growth into the pulmonary tissue at low incidence (Table 1). From the feeding experiment of fusarenon-X to rats (Table 2), it should be pointed out that the high incidence of total tumors including hypophyseal adenomas, thyroid tumors (mostly c-cell carcinoma) (Figure 3), and transitional cell carcinoma of

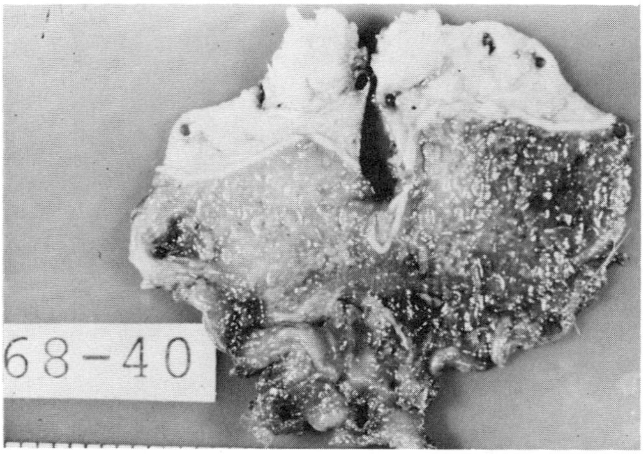

FIGURE 4. Hyperkeratosis of the forestomach of mouse administered diet containing 15 ppm T-2 toxin.

urinary bladder appearing in experimental groups as compared with that of control animals may indicate neoplastic whole-body effect of trichothecene compounds similar to the radiation effect exposed at low level.

A direct irritating effect of trichothecenes was also observed in the forestomach.[10] Mice were given feeds containing 10 and 15 ppm of T-2 toxin for up to 12 months. On opening the stomach at sacrifice, diffuse or localized polypoid-papillary extrusions with hyperkeratosis were found in every mouse (Figure 4). Histological examination of the forestomach revealed hyperkeratosis, acanthosis, and papillomatosis with inflammatory cell infiltration. This change was found within 13 weeks, diffuse with the high dose and localized with the low dose. The lesions were consistently observed during the feeding period of 12 months, but most subsided within 3 months after cessation of feeding. No squamous cell carcinoma of the forestomach developed by the end of 15 months, but one case of adenocarcinoma of the glandular stomach was observed.

CONCLUSION

To avoid intoxication due to ingestion of moldy agricultural commodities, examination of 12,13-epoxytrichothecene is recommended using bioassay of growth inhibitory effect to the culture cell system or the inhibitory effect of protein synthesis of rabbit reticulocytes. We should also examine mutagenicity of the toxin, and if the positive mutagenicity is observed, contaminated food material should be forbidden to be ingested by humankind as well as by livestock.

REFERENCES

1. **Nishikado, Y.**, Studies on the Control of *Fusarium* Scab Disease of Wheat. Data on Agricultural New Technics, No. 97, Japanese Ministry of Agriculture and Forestry, 1958, 59.
2. **Ishii, H.**, *Byogaichu Hasseiyosatsu Hokoku*, 8, 1, 1961.
3. Cited by **Ishii, H.**, *Byogaichu Hasseiyosatsu Hokoku*, 8, 1, 1961.
4. **Tatsuno, T., Fujimoto, Y., and Morita, Y.**, Toxicological research on substances from Fusarium nivale. III. The structure of nivalenol and its monoacetate, *Tetrahedron Lett.*, 33, 2823, 1969.
5. **Ueno, Y., Ueno, I., Tatsuno, T., Okubo, K., and Tsunoda, H.**, Fusarenon-X, a toxic principle of *Fusarium nivale*-culture filtrate, *Experientia*, 25, 1062, 1969.

6. **Saito, M., Enomoto, M., and Tatsuno, T.,** Radiomimetic biological properties of the new scirpene metabolitites of *Fusarium nivale, Gann,* 60, 599, 1969.
7. **Ueno, Y., Ueno, I., Iitoi, Y., Tsunoda, H., Enomoto, M., and Ohtsubo, K.,** Toxicological approaches to the metabolities of *Fusaria.* III. Acute toxicity of fusarenon-X, *Jpn. J. Exp. Med.,* 41, 521, 1971.
8. **Nagao, M., Honda, M., Kiyono, Y., Yahagi, T., Sugimura, T., Hamaki, T., Natori, S., Ueno, Y., and Yamazaki, M.,** Mutagenicities of mycotoxins on *Salmonella, Proc. Jpn. Assoc. Mycotoxicol.,* No. 3-4, 41, 1976.
9. **Saito, M., Enomoto, M., Ohtsubo, K., and Murata, Y.,** Chronic toxic effects of metabolities of *Fusarium* sp. in rats and mice, *Trans. Soc. Pathol. Jpn.,* 60, 79, 1971.
10. **Ohtsubo, K. and Saito, M.,** Chronic effect of trichothecene toxins, in *Mycotoxins in Human and Animal Health,* Rodricks, J. V., Hesseltine, C. W., and Mehlman, M. A., Eds., Pathotox Publishers, Park Forest South, Ill., 1977, 255.

EXAMINING FOOD AND DRINK FOR PARASITIC, SAPROPHYTIC, AND FREE-LIVING PROTOZOA AND HELMINTHS

George J. Jackson

INTRODUCTION

Food infested with parasites is often thought to be a thing of the past or, at most, a hazard associated with low life in places distant from our temperate civilization with its researched, sanitized, prepackaged meals. When the actual situation is examined, even experts are surprised to find that human infections are not limited to underdeveloped tropic or arctic habitations,[19] that parasitic animals cause approximately a third of the GI diseases in our climatic zone,[6] that the traditional foodborne and waterborne parasites have not disappeared,[21] and that several grave ones have been "discovered" only since World War II.[1,4,15,18,24]

Moreover, it seems unlikely that parasites can be eliminated from foods in the near future. Today, the market has become international and so has the distribution of foodborne pathogens.[3,14] Freezing and thawing techniques as well as other food-storage procedures are constantly being improved to preserve taste, but the same innovations may also preserve microbes.[8] Old practices of recycling human and animal waste into the food chain are being revived and could increase the viable content of edibles.[20] Consumer habits may also account for the current importance of parasites as foodborne pathogens, particularly the vogues for raw recipes and undercooking to savor "natural" taste and save heat-labile nutrients. Whether foodborne parasitic diseases really are reappearing among industrialized populations or merely seem more prominent because of the unexpected occurrence is difficult to determine. Certain though is the state of practical food parasitology. It is a neglected discipline compared to food bacteriology. Methods usually have been adapted *ad hoc* from those used for clinical specimens or soil samples. None have yet been standardized rigorously as required by the Association of Official Analytical Chemists.[10]

TYPES OF ANIMALS THAT MAY BE FOODBORNE PARASITES

Many microscopic or small macroscopic animals are consumed live with food and drink. Given the right circumstances, most can survive ingestion at least for short time spans. To consider all as parasites would be impractical. Some of these animals are classified as "free-living", i.e., they normally dwell in soil or water. Others are classified as saprophytes, i.e., they exist on decaying organic matter. In contrast, parasites are defined as being tolerant of existence in or on another living organism; they may even require one host or a series of specific hosts for their survival, development, or reproduction.

Most of the small free-living animals and many of the facultative and obligate parasites of plant or coldblooded animal hosts that are associated with foods cause no pathology when they are ingested live by humans. In due course they are either digested or evacuated with the feces. Exceptions occur. For instance, human infections with high-temperature strains of soil-and-water or sewage amebas (usually by way of water) are among the "new" parasitoses that have been documented since World War II.[24] Gordiid worms ("hair snakes") are free in water as adults and as larvae are parasites of insects; in humans they cause irritation when swallowed accidentally.[5]

It is not exceptional to find that those parasites which require warmblooded animals as "definitive" hosts for their sexual maturation will cause pathology when ingested live with food. Even if, like the fishborne anisakine nematodes of marine mammals, they do not mature or survive long in humans,[17] these organisms nevertheless may be etiologic agents of disease and are commonly listed among our foodborne parasites. Normally not counted as foodborne parasitic diseases are human hookworm and malaria, even though the possibility of infection by ingestion has been demonstrated in experiments.[25] Malaria, of course, is usually acquired from a biting insect, not from one that is accidentally in food; hookworms usually penetrate into the skin from soil, not into the mucosa from ingested soil-grown raw vegetables.

More than 60 genera of invertebrate animals[11] are implicated in human infection through food and drink.[9] Included are unicells (Protozoa) and such Metazoa as tapeworms (cestodes), flukes (trematodes), roundworms (nematodes), and insects (usually larval flies). Despite the variety, relatively few species cause most of the incidents of foodborne and waterborne parasitic disease in industrialized nations of the temperate zone. Some of the more common protozoan diseases that may be acquired through ingestion are amebiasis and giardiasis, which are initiated by cysts of *Entamoeba histolytica* and *Giardia lamblia* transmitted by fecally contaminated water or food, other carriers, or by direct contact; and toxoplasmosis, which is initiated by oocysts from feces (usually cat) or by tissue cysts in meat. Several species of roundworms (nematodes) may be responsible for foodborne diseases in the temperate zone. The larvae of *Trichinella spiralis* are encysted in meat, especially pork and bear; the eggs of *Ascaris* spp. and *Trichuris* spp. survive sewage treatment and contaminate food crops when sewage sludge is used as plant fertilizer; eggs of the pinworm, *Enterobius vermicularis* from the host's perianal region may become airborne and settle onto foods or into drinks. Cestodes (tapeworms) such as *Taenia* spp. and *Echinococcus* spp. are infective as larvae encysted in meat or, in some instances, as eggs from the host's feces.

Infection statistics may be deceptive, however. The evidence is both incomplete and, in some cases, circumstantial. It is incomplete insofar as parasitic diseases are often undiagnosed, misdiagnosed, or unreported to health authorities. It is circumstantial insofar as epidemiological backup for incidence data may be inconclusive or lacking. For instance, although one can assume that cases of trichinosis are caused by the ingestion of undercooked meat or meat products (containing *Trichinella spiralis* larvae), it is not safe to assume that cases of ascariasis are foodborne. The infective eggs of *Ascaris* spp. may be ingested from several different sources, only one of which is fecally contaminated food.

TYPES OF FOODS WITH WHICH PARASITES ARE ASSOCIATED

Water and fresh edibles are the principal items with which drinkborne and foodborne human parasitoses have been associated. The edibles may be of either animal or plant origin. Those plants having their consumable portions in contact with soil or water are the ones most likely to be involved if they are normally eaten unpeeled, uncooked, or undercooked. Raw or undercooked meats are also likely sources of infective stages of parasites.

TYPES OF INFECTIONS CAUSED BY INGESTED PARASITES

Parasites that enter the human body by way of the digestive tract do not necessarily remain confined to the lumen or adjacent tissues.[5] Ingested *T. spiralis* reproduce in the intestine, but the new generation of larvae migrates to the musculature. Ingested eggs of *Ascaris lumbricoides* hatch and release larvae that migrate from the intestine to the liver and lungs. Then they return to the intestine where they produce eggs that are evacuated with the feces. *Entamoeba histolytica* infections may involve just the lumen and tissues of the digestive tract or may spread to other organs and organ systems. There are liver flukes, lung flukes, and intestinal flukes. Adults of *Taenia saginata,* the beef tapeworm of humans, are found in the human intestinal lumen; *T. solium,* the pork tapeworm of humans, may be found in humans either as an adult in the intestinal lumen or as an encysted "bladderworm" (cysticercus) in internal tissues.

Species of parasites vary considerably in where they go and how they get there. Nevertheless, patterns based on host adaptation or, sometimes, on phylogenetic relationships have been discerned. Those parasites that are "adapted" to *Homo sapiens* may migrate to predictable niches of the body by fairly regular routes. Species more adapted to other hosts may "wander." Two closely related liver flukes, *Opisthorchis felineus* and *Clonorchis sinensis,* migrate by way of the bile duct, but *Fasciola hepatica,* not as similar phylogenetically, migrates to the liver through the intestinal wall and abdominal cavity.

It may also be useful to classify infections by the survival and reproductive potentials of the parasites in a particular host. Human infections with anisakine nematodes, although fulminating at first, eventually abort. Human infections with "intestinal" Protozoa are greatly modified by the conditions of a given host's physiological resistance or immunity.[13]

EXAMINATION OF FOOD AND RECOVERY OF PARASITES

One has to know at what and for what one is looking! Examination procedures depend both on the type of food being analyzed and on the sort of parasite suspected of being present. Culture systems for "multiplying" the number of parasites in a sample exist for few species.[23] Seldom can any invertebrate be inoculated directly from food because the growth media are not selective. Associated bacteria tend to overgrow the cultures and the parasites are killed. Only after they have been isolated and cleaned (axenized), will some parasites multiply.

Large parasites, such as tapeworm larvae in meat and fish or larval roundworms in fish, can be seen with the unaided eye or with a low magnification hand lens or dissecting microscope. However, since the parasites may not be on the exposed surface, it might be necessary to dissect a sample or candle it (by holding the sample, pressed between glass or plastic, against a light source).

Even when parasites can be seen by direct examination, more must be done to determine quantity and viability. Washing and scrubbing are sometimes used to free helminth eggs and protozoan cysts from lettuce. Grinding plus digestion in artificial stomach juice are used to isolate *T. spiralis* larvae from pork muscle, but larger parasites, such as *T. solium* larvae, may be destroyed by the grinding. Elution of parasites into warm water, the so-called Baermann Method or Baermannization, was originally[2] designed to free hookworm larvae from soil, but the process also works with an amazing number of other types of samples and parasites. Protozoan cysts and helminth eggs are often concentrated by flotation and/or centrifugation in diverse solutions.[7]

SELECTION OF PARASITES AND ASSESSMENT OF VIABILITY

As mentioned, many of the microscopic and small macroscopic animals associated with foods are free-living soil or water inhabitants or are parasites of plants or cold-blooded animals. They die quickly when they are ingested live by humans. In analyzing food for parasites it would be convenient to eliminate this "living background". No absolute method exists but, presumably, those animals that are infective by way of the GI tract must withstand the digestive juices as well as mammalian temperatures. Procedures[16] often used to select potential parasites from among obscuring invertebrates that do not survive in warmblooded hosts include incubation of infested food or invertebrates from foods in invertebrate Ringer's solution at 35 to 37°C, or digestion in artificial stomach juice at 35 to 37°C.

Motility in water or Ringer's solution is the most commonly used test for parasite viability, but it is useless with the true cyst stages that may occur in food and drink. Staining properties, uptake of a label, excystation by physical or chemical means, and infectivity for a model host are some of the techniques that can be used to ascertain the viability of cysts.

DEAD PARASITES IN FOODS

Dead parasites in foods obviously cannot be infective for the consumer, although, like the cyst stages in parasites' life cycles, it is often difficult to confirm viability or death. In sufficient concentration, dead parasites dilute the food. Before death, they may have caused changes in the food. Although there is as yet no firm evidence that parasites contain harmful substances, even a few parasites may be aesthetically objectionable.[3,12]

MORPHOLOGICAL IDENTIFICATION

After parasites have been recovered from foods, cleaned of adhering particles, prodded for viability, and tested for infectivity, they should be identified morphologically.[9] To aid identification, protozoan cysts and helminth eggs may be fixed and stained with Lugol's iodine.

Nematodes may usually be fixed in steaming 70% ethanol or a solution of glacial acetic acid and 95% ethanol (1:3) and stored in 70% ethanol with a few drops of glycerin. Other fixatives are available. Nematode morphology is studied in temporary mounts after specimens are removed from the storing solution and cleared in glycerin, phenol, or lactophenol. Before the nematodes are stored again, clearing fluid should be washed away with 70% ethanol. Sectioning and staining are sometimes necessary for the detailed identification of nematodes.

Before fixation, both trematode and cestode flatworms should be relaxed in distilled water for 15 min. Trematodes (flukes) can be fixed in boiling 10% formalin, kept 1 day in 10% formalin, then stored in 5% formalin. Cestodes should be immersed briefly and repeatedly in 70°C water and fixed for 1 day in a mixture of ethanol (85 parts), commercial strength formalin (10 parts), and glacial acetic acid (5 parts), then stored in 70% ethanol. All flatworms should be stained and mounted permanently for identification.

Acanthocephala, if alive, should be placed in water to evert the proboscis, then fixed in steaming 70% ethanol with a few drops of glacial acetic acid, and stored in 70% ethanol until they are stained and mounted permanently.

Fleas, lice, mites, copepods, fly larvae, and other parasitic and food-inhabiting arthropods should be fixed in hot water or hot 70% ethanol and stored in 70% ethanol.

In all cases, specimens should be kept in tightly capped vials with identifying data (food source; anatomic location in host, if possible; geographical origin; date of sample collection; date of parasite collection; collector's name; presumed identification; ultimately, the name of the identifying expert) written in indelible pencil on a slip of paper. The paper should be placed in the liquid-filled vial with the parasite. Assistance in identifying properly fixed and stored food parasites may be obtained from FDA (Laboratory of Parasitology, HFF-234, Division of Microbiology, Bureau of Foods, Food and Drug Administration, 200 C Street, S. W., Washington, D.C., 20204) or from laboratories listed by the American Public Health Association.[9]

EXPERIMENTAL IDENTIFICATION

With mice as hosts, inoculation is used to identify *Toxoplasma* cysts. Serological identification for some parasites is carried out at the Centers for Disease Control (CDC), Atlanta, Ga., 30333. Identification of parasites by their chemical profiles is still in the experimental stage.[22]

PREVENTION OF HUMAN INFECTION

Parasites are usually killed when food has been heated thoroughly, as in cooking or as required for canning. Infections occur mostly with underheated or raw foods, and contact contamination of other foods and kitchen utensils is possible with some parasites. Parasitic animals vary considerably from species to species in their susceptibility to freezing, drying, chlorination, salting, and marinating procedures.

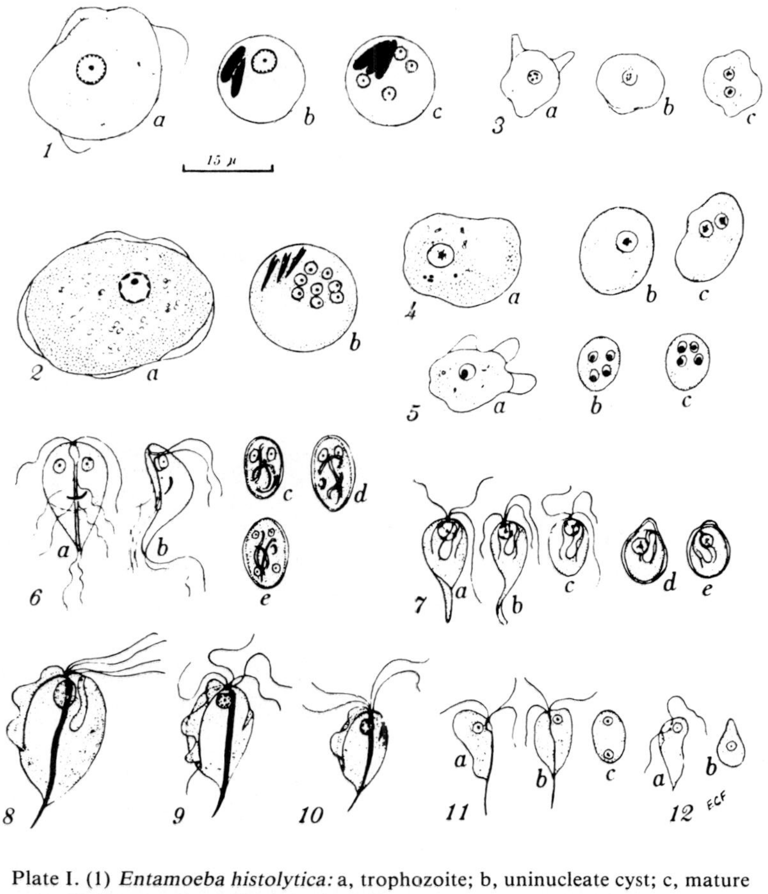

Plate I. (1) *Entamoeba histolytica:* a, trophozoite; b, uninucleate cyst; c, mature cyst; (2) *Entamoeba coli:* a, trophozoite; b, mature cyst; (3) *Dientamoeba fragilis:* a to c, representative trophozoites; (4) *Iodamoeba butschlii:* a, trophozoite; b, c, mature cysts; (5) *Endolimax nana:* a, trophozoite; b, c, mature cysts; (6) *Giardia lamblia:* a, ventral view, and b, lateral view of trophozoite; c, d, immature and e, mature cysts; (7) *Chilomastix mesnili:* a to c, trophozoites; d, e, cysts; (8) *Trichomonas vaginalis:* trophozoite, from the vagina; (9) *Trichomonas hominis:* trophozoite, from the cecum; (10) *Trichomonas tenax:* trophozoite, from the mouth; (11) *Enteromonas hominis:* a, b, trophozoites; c, cyst; (12) *Retortamonas intestinalis:* a, trophozoite; b, cyst. (From Faust, E. C., Beaver, P. C., and Jung, R. C., *Animal Agents and Vectors of Human Disease,* 3rd ed., Lea & Febiger, Philadelphia, 1968. With permission.)

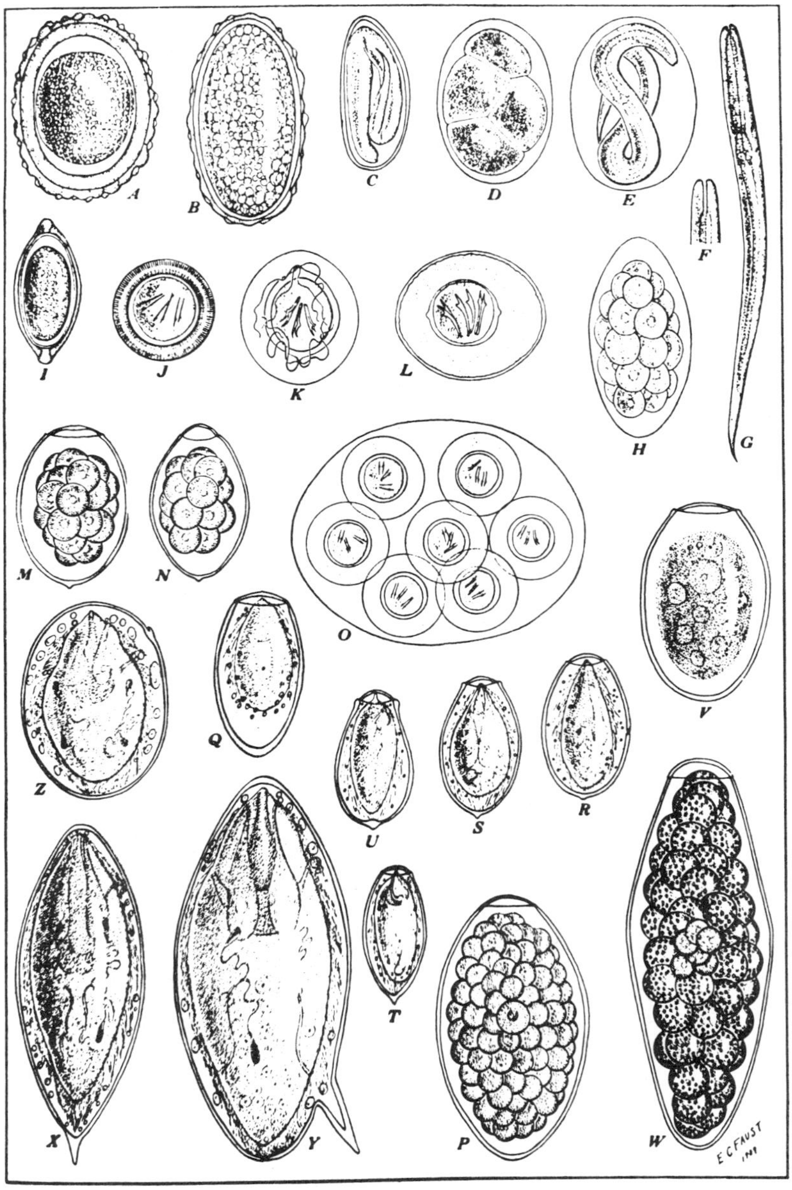

Plate II

Plate II. Eggs and larvae of the more common Helminth parasites of man. (A) *Ascaris lumbricoides* (large intestinal roundworm), unsegmented fertile egg, usually with bile-stained outer shell, passed in feces; (B) *A. lumbricoides*, infertile egg, usually with bile-stained outer shell, passed in feces; (C), *Enterobius vermicularis* (pinworm or seatworm), with completely developed larva, usually deposited by the mother worm on the perianal skin; (D) *Ancylostoma duodenale* ("Old World" hookworm) or *Necator americanus* ("American" hookworm), early cleavage stage, passed in semiformed feces; (E) *A. duodenale* or *N. americanus*, with completely developed first-stage (rhabditoid) larva, passed in constipated stool or developed in feces that have stood 24 to 48 hr in the laboratory; (F) *A. duodenale* or *N. americanus*, anterior extremity of hatched rhabditoid larva, showing long, narrow, buccal cavity (contrast with anterior end of G); (G) *Strongyloides stercoralis* (threadworm), rhabditoid larva passed in feces or obtained by duodenal drainage, showing very short buccal cavity (contrast with F); (H) *Trichostrongylus* sp., characteristic morula-stage egg passed in feces; (I) *Trichuris trichiura* (*Trichocephalus trichiurus* or whipworm), with unsegmented ovum, usually with bile-stained outer shell, passed in feces; (J) *Taenia saginata* (beef tapeworm) or *T. solium* (pork tapeworm), with fully embryonated oncosphere, with dark brown outer shell, passed in feces; (K) *Hymenolepis nana* (dwarf tapeworm), with fully embryonated oncosphere, passed in feces; (L) *Hymenolepis diminuta* (rat tapeworm) with fully embryonated oncosphere, passed in feces; (M) *Diphyllobothrium latum* (fish tapeworm), characteristically unembryonated, as passed in feces; (N) *Spirometra mansoni, S. erinacei, S. houghtoni,* and other species of genus *Spirometra,* characteristically unembryonated, as passed in feces of definitive host; (O) *Dipylidium caninum* (double-pored dog tapeworm), mother egg capsule containing several fully embryonated oncospheres, as passed in feces or expressed from disintegrating gravid proglottid; (P) *Fasciolopsis buski* (giant intestinal fluke) or *Fasciola hepatica* (sheep liver fluke), unembryonated, as passed in feces or obtained by duodenal and/or biliary drainage; (Q) *Dicrocoelium dendriticum*, with developed miracidium, passed in feces or obtained by duodenal or biliary drainage; (R) *Heterophyes heterophyes,* with developed miracidium, passed in feces; (S) *Metagonimus yokogawai*, with developed miracidium, passed in feces; (T) *Opisthorchis felineus,* with developed miracidium, passed in feces or obtained by duodenal or biliary drainage; (U) *Clonorchis sinensis* (Chinese liver fluke), with developed miracidium, passed in feces or obtained by duodenal or biliary drainage; (V) *Paragonimus westermani* (Oriental lung fluke), unembryonated, recovered from sputum or swallowed and passed in feces; (W) *Gastrodiscoides hominis,* unembryonated, passed in feces; (X) *Schistosoma haematobium* (blood fluke), with developed miracidium, passed in urine or at times in feces; (Y) *Schistosoma mansoni* (Manson's blood fluke), with developed miracidium, passed in feces; (Z) *Schistosoma japonicum* (Oriental blood fluke), with developed miracidium, passed in feces. (R, S, T, and U, magnification × 666; all other figures, magnification × 333). (From Faust, E. C., *Human Helminthology,* Lea & Febiger, Philadelphia. With permission.)

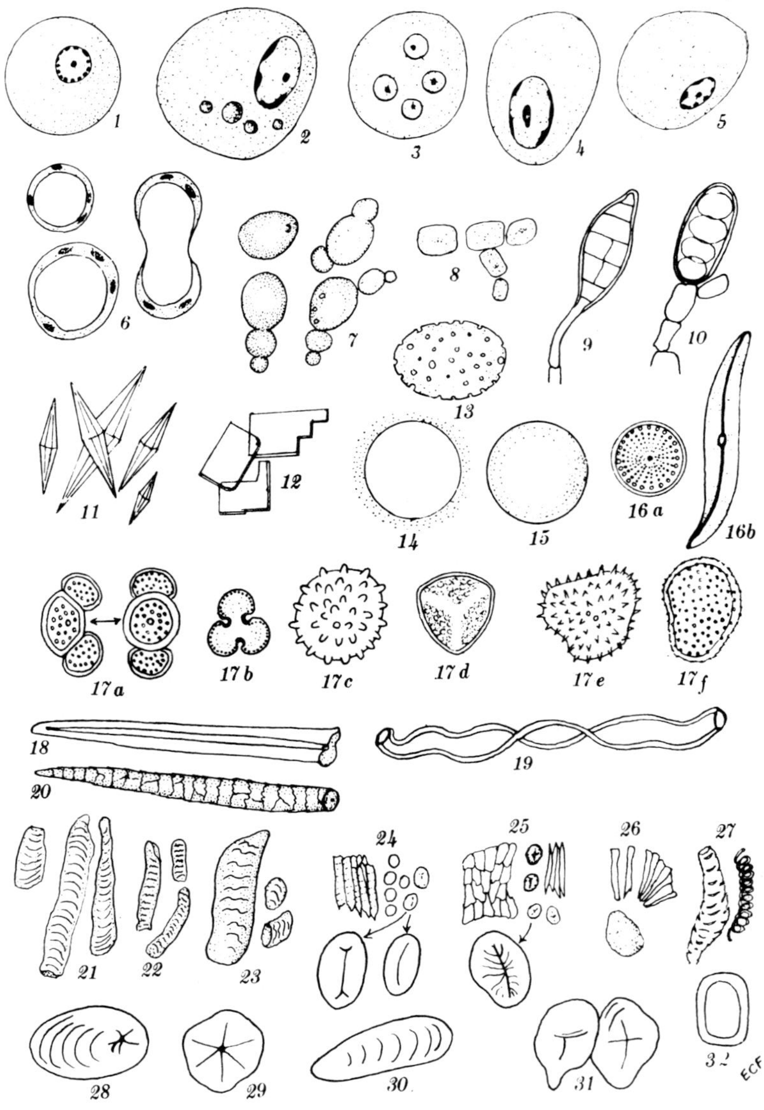

Plate III. Objects in the feces at times causing confusion in parasitologic diagnosis. (1) Precyst of *Entamoeba histolytica,* for comparison with cellular exudate; (2) macrophage; (3) neutrophilic polymorphonuclear leukocyte; (4) squamous epithelial cell from aspirate of the rectum; (5) plasma cell from aspirate of the rectum; (6) *Blastocystis* sp.; (7) yeast cells; (8) units from septate mycelium of *Monilia* spp.; (9, 10) conidia respectively of the fungi *Alternaria* spp. and *Helminthosporium* spp.; (11) Charcot-Leyden crystals; (12) cholesterol crystals; (13) partly digested particle of casein; (14) air bubble; (15) oil droplet; (16a, 16b) diatoms; (17) pollen grains (a, pine; b, African violet; c, hibiscus; d, broom sage; e, ragweed; f, timothy grass); (18) plant hair; (19) fragment of cotton fiber; (20) mammalian hair; (21 to 32) food remnants (21, beef or pork muscle; 22, crab meat; 23, fish; 24, wheat grain; 25, corn kernel; 26, string beans; 27, conducting tubules of fibrovascular bundle; 28, Irish potato starch grain; 29, rice starch; 30, plantain starch; 31, sweet potato starch; 32, woody cell wall). (1 to 12, magnification × 1125; 16a, × 700, 16b, × 200; 17 to 20, × approximately 300; 21 to 27, × approximately 240; 28 to 31, × approximately 750; 32, × 200. (From Faust, E. C., Beaver, P. C., and Jung, R. C., *Animal Agents and Vectors of Human Disease,* 3rd ed., Lea & Febiger, Philadelphia, 1968. With permission.)

REFERENCES

1. **Alicata, J. E. and Jindrak, K.,** *Angiostrongylosis in the Pacific and Southeast Asia,* Charles C Thomas, Springfield, Ill., 1970.
2. **Baermann, G.,** *Eine einfache Methode zur Auffindung von Ankylostomum (Nematoden) Larven in Erdproben,* Meded. geneesk. Lab. Weltvreden, Feestbundel (Batavia), 1917.
3. **Bier, J. W.,** Protozoa and Helminths, Technical Bull. No. 1, Food and Drug Administration, Washington, D.C., 1981, 49.
4. Giardiasis in Residents of Rome, N.Y. and in U.S. Travellers to the Soviet Union, Morbidity and Mortality Weekly Rep. 24, Center For Disease Control, Atlanta, 1975, 366.
5. **Chandler, A. C.,** *Introduction to Parasitology,* John Wiley & Sons, New York, 1952.
6. Epidemological Research Laboratory of the Public Health Laboratory Service and Hospital Laboratories of England, Wales and Northern Ireland, *Communicable Disease Report,* 10, 1, 1975.
7. **Faust, E. C., Sawitz, W., Tobie, J., Odom, V., Peres, C., and Lincicome, D. R.,** Comparative efficiency of various techniques for the diagnosis of protozoa and helminths in feces, *J. Parasitol.,* 25, 241, 1939.
8. **Gordon, R. M., Graedel, S. K., and Stucki, W. P.,** Cryopreservation of viable axenic *Entamoeba histolytica, J. Parasitol.,* 55, 1087, 1969.
9. **Healy, G. R., Jackson, G.J., Lichtenfels, J. R., Hoffman, G. L., and Cheng, T. C.,** Foodborne parasites, in Compendium of Methods for the Microbiological Examination of Foods, M. L. Speck, Ed., American Public Health Association, Washington, D.C., 1976, 471.
10. **Horwitz, W.,** *Official Methods of Analysis,* 5th ed., Association of Official Analytical Chemists, Arlington, Va., 1978.
11. **Hyman, L.,** *The Invertebrates,* Vols. 1 to 6, McGraw-Hill, New York, 1940 to 1967.
12. **Ikari, P. T.,** Parasites and Related Forms. Microscopic-Analytical Methods in Food and Drug Control, Food and Drug Tech. Bull. No. 1, Food and Drug Administration, Washington, D.C., 1960, 177 (out of print, see Ref. 3).
13. **Jackson, G. J., Singer, I., and Herman, R.,** *Immunity to Parasitic Animals,* Vols. 1 and 2, Appleton-Century-Crofts, New York, 1969 and 1970.
14. **Jackson, G. J., Andrews, W. H., Jr., Gorham, J. R., Bier, J. W., Payne, W. L., and Wilson, C. R.,** Survey of microflora and microfauna associated with the Moroccan food snail, *Helix aspersa, Proc. 3rd Int. Congr. Parasitol.,* Vol. 3, Facta Publication, H. Egerman Verlag, Vienna, 1974, 1601.
15. **Jackson, G. J.,** The "new disease" status of human anisakiasis and North American cases, *J. Milk Food Technol.,* 38, 769, 1975.
16. **Jackson, G. J., Bier, J. W., Payne, W. L., Gerding, T. A., and Rude, R. A.,** *Bacteriological Analytical Manual,* 5th ed., Association of Official Analytical Chemists, Arlington, Va., 1978, 1.
17. **Juels, C. W., Butler, W., Bier, J. W., and Jackson, G. J.,** Temporary human infection with a *Phocanema* sp. larva, *Am. J. Trop. Med. Hyg.,* 24, 942, 1975.
18. **Morera, P.,** Life history and redescription of *Angiostrongylus costaricensis* Morera and Cespedes, *Am. J. Trop. Med. Hyg.,* 22, 613, 1971.
19. **Most, H.,** Manhattan: a tropic isle? *Am. J. Trop. Med. Hyg.,* 17, 333, 1968.
20. **Rudolfs, W., Falk, L. L., and Ragotzkie, R. A.,** Contamination of vegetables grown in polluted soil, *Sewage Ind. Wastes,* 23, 253, 1951.
21. Cooperative Parasite Survey, Intestinal Parasite Diagnoses in State (County) Health Department Laboratories: Continuing Series, State Health Laboratories and Centers for Disease Control, Atlanta, 1976, 1977.
22. **Tarr, G. E.,** Comparison of ascarosides from four species of ascarid and one oxyurid, *Comp. Biochem. Physiol.,* 46B, 167, 1973.
23. **Taylor, A. E. R. and Baker, J. R.,** *The Cultivation of Parasites In Vitro,* Blackwell Scientific, Oxford, 1968.
24. **Willaert, E.,** Primary amoebic meningo-encephalitis. A selected bibliography and tabular survey of cases, *Ann. Soc. Belg. Med. Trop.,* 54, 429, 1974.
25. **Yoeli, M. and Most, H.,** Sporozoite-induced infections of *Plasmodium berghei* administered by the oral route, *Science,* 173, 1031, 1971.

FOOD CONTAMINANTS: ENTERIC PATHOGENS IN SHELLFISH

Rudolph Di Girolamo

SHELLFISH — WHAT THEY ARE, AND HOW THEY FEED

When we speak of shellfish, we are referring primarily to a group of bivalve mollusks (oysters, mussels, clams, and cockles) which ingest food by a process termed filter-feeding. This process involves the selective ingestion of small particles of organic matter from large volumes of water drawn into the shellfish in the form of currents. As potential food particles travel across the animal's gills, they become entrapped by mucus strands secreted continuously by the shellfish during pumping. The string-like masses of food material and mucus are directed by ciliary action to the mouth region where the matter is sorted. According to its nature, the entrapped material may be swept, via ciliary action, directly into the mouth or along rejection paths to the outside.[9,45,142] Due to this method of feeding, it is possible for shellfish to accumulate any organic matter suspended in the over-lying water, including such small particles as viruses and bacteria. Hence, shellfish residing in waters receiving sewage pollution can accumulate enteric pathogens and serve as vectors of enteric diseases.

UPTAKE OF ENTERIC BACTERIA BY SHELLFISH

In 1895, Foote was one of the first investigators to show, under laboratory conditions, that oysters would accumulate typhoid bacilli if they were placed in seawater contaminated with these microbes.[41] In field studies, he also isolated typhoid bacilli from oysters growing in sewage polluted waters. Foote concluded that the shellfish absorbed these microorganisms from the surrounding seawater during feeding. These findings were confirmed by Boyce and Herdman in 1896.[10] They reported that shellfish harvested from sewage polluted waters were probable carriers of enteric diseases because the animals could ingest any pathogenic microorganisms present in seawater. Subsequently, Stiles demonstrated the consistent presence of coliform, typhoid, and paratyphoid bacilli in the heavily polluted waters overlying the oyster beds of Jamaica Bay, L.I.[124,125] He believed that the oysters became contaminated because they were feeding upon fecal matter suspended in the water. In later studies, Arcisz and Kelly demonstrated that oysters and clams exposed to polluted waters not only readily ingest bacteria, but also concentrate them to levels higher than that of the surrounding water.[4,69] In field studies with clams and oysters, Vasconcelos and Ericksen showed that both types of shellfish responded to bacteriological changes in water quality and readily ingested coliform and fecal coliform bacteria.[134] In their studies, the clams consistently accumulated greater levels of enteric bacteria than did the oysters.

That these ingested enteric microorganisms would persist in shellfish was first demonstrated in 1905 by Klein.[72] He reported that typhoid and other enteric bacilli would survive for at least 4 days in contaminated shellfish stored at 50 to 55°F This was the temperature range at which shellfish were usually kept for shipment and sale. These findings were later confirmed by Tonney and White.[129]

One of the major problems faced by many of these early investigators was the difficulty of isolating enteropathogens from seawater with the techniques then available. Consequently various researchers, including Round[118] suggested using coliform and fecal coliform counts as indices of pollution. These microorganisms are normal inhabitants of the human intestinal tract. Round noted that counts of these enteric bacteria were consistently high in shellfish and in water subjected to sewage pollution. A standard technique for coliform detection was later adopted by the U.S. Public Health Service,[131] and has become the standard method for determining the sanitary quality not only of shellfish, but also of the water overlying shellfish beds.[2]

SHELLFISH-BORNE ENTERIC BACTERIAL DISEASES

Thus, early research clearly established the fact that shellfish could ingest pathogenic bacteria from polluted water and serve as vectors for enteric bacterial diseases. Conn[22] recorded one of the earliest known outbeaks of typhoid fever attributed to eating oysters (*Crassostrea virginica*) held in an estuary 300 ft below a sewage outfall. The shellfish may not have been contaminated originally, but certainly became so after feeding on raw sewage. Similar outbreaks of gastroenteritis or typhoid fever due to eating raw shellfish were recorded on a regular basis between the years 1902 and 1911.[72,123,125] In every incident, the shellfish responsible for the outbreaks were proven to have been stored in waters receiving raw sewage discharge prior to harvesting and shipment to market. Two of these incidences resulted from eating so-called "Rockway" or "Blue Point" oysters harvested from Jamaica Bay, L.I.[123,124] This bay was known to be heavily polluted by millions of gallons of raw sewage dumped into it daily.[125] Despite this fact, oysters and clams continued to be harvested from this area and sold on the open market until 1925. At that time a major typhoid outbreak, due to eating Jamaica Bay oysters, occurred simultaneously in New York City, Chicago, and Washington, D.C.[88] As a result of this tri-city outbreak, a permanent embargo was finally placed on the sale of shellfish from Jamaica Bay. More stringent and effective methods and measures of controlling the shellfish industry were also adopted by the U.S. Public Health Service.[131] These methods included the monitoring of all shellfish growing areas, control of shellfish harvesting and packing, and the licensing of all reputable interstate shippers.

In recent times, shellfish associated outbreaks of typhoid fever or salmonellosis (*Salmonella* associated gastroenteritis) have been traced either to the contamination of the animals by asymptomatic carriers, or to the illegal harvesting and sale of oysters, clams, or mussels from polluted waters.[54,101,107] Typical of this situation was the outbreak recorded by Old and Gill.[107] These workers reported 94 cases of typhoid fever and 8 of bacterial gastroenteritis among individuals who had eaten raw oysters in St. Charles Parish, La. The oysters were purchased from an asymptomatic carrier whose feces were positive for *Salmonella typhi*. The oysters were not polluted originally, but became so when the carrier defecated in the water adjacent to the shellfish beds some hours prior to actually harvesting the oysters. Similar outbreaks of typhoid fever or salmonellosis due to eating contaminated clams or mussels have been recorded in Europe.[13,120] In each incident the shellfish responsible were first illegally or improperly harvested from heavily polluted waters then either eaten, or sold on the open market. The symptoms of typhoid fever and *Salmonella* induced gastroenteritis (salmonellosis) are shown in Table 1.

SHELLFISH-ASSOCIATED GASTROENTERITIS CAUSED BY BACTERIA OTHER THAN "THE ENTERICS"

A number of outbreaks of gastroenteritis involving bivalve molluscs and crabs have been caused by another agent, *Vibrio parahaemolyticus*[3,6,8] *V. parahaemolyticus* is a true marine bacterium which is apparently pathogenic for humans. Its presence has been demonstrated not only in estuary waters, but also in shellfish.[80] Incidences involving this bacteria have been more frequent in Japan than America. However, this microorganism has been recovered from both the Atlantic and Pacific coasts of the U.S.[68]

Table 1
SOME BASIC SYMPTOMS OF TYPHOID FEVER AND SALMONELLOSIS

Typhoid fever	Salmonellosis
Agent — *Salmonella typhi*	Agent — *Salmonella* spp.
Incubation time — 2 weeks	Incubation time — 24—28 hr
Hyperpyrexia — present, high: 104—105°F	Hyperpyrexia — if present, low: 99—100°F
Chills — present	Chills — sometimes present
Diarrhea — present	Diarrhea — always present, sometimes severe
Descrotic (irregular) pulse	
Bradycardia — present	Bradycardia — usually absent
Erythema — present	Erythema — usually absent. Severe abdominal pain usually present, especially at onset
Mortality rate — fairly high, 75%. Usually death due to complications, i.e., pneumonia	Mortality rate — very low; complications — usually none

SHELLFISH-BORNE ENTERIC VIRUS DISEASES

Early research clearly demonstrated the role of shellfish harvested from sewage polluted waters as vectors of enteric bacterial diseases. However, enteric bacteria are not the only microbial entities released in sewage. Early virus research indicated that certain viral disease, such as poliomyelitis and infectious hepatitis, could be transmitted either in contaminated foodstuffs or via the water route.[18-20,35,46,49,53,65,66,105,115,137] Investigators had also shown the presence of enteric viruses in sewage effluents being pumped into estuaries.[17,67,70] However, the possible association of shellfish with the dissemination of either infectious hepatitis (hepatitis Virus A) or other enteric viral diseases was not considered until 1955. At that time, the first definitive outbreak of infectious hepatitis (IH) was reported by Roos[117] in Sweden. There were a total of 119 victims, but no mortalities. The shellfish responsible for this incident were clams (*Mercenaria mercenaria*) harvested 200 m below a sewage outfall in the bay around Stockholm. The clams were incriminated primarily on the basis of epidemiological evidence. These shellfish were the only unusual food eaten by the victims over a 6-week period, and the time lapse between consumption of shellfish and onset of disease symptoms correlated closely with the incubation time for hepatitis. Roos and co-workers were unable to isolate the causative virus, but they noted that all shellfish examined had high titers of fecal coliforms. The following year, an additional 540 cases of hepatitis in which the vectors were contaminated raw clams and oysters were recorded by Lindberg-Braman[78] and Kjellander.[71] As in the Roos report, incrimination of the shellfish was based upon epidemiological evidence. However, Kjellander made a significant observation: these outbreaks of shellfish-associated hepatitis were more prevalent during winter than any other season of the year. He suggested that the shellfish probably absorbed the hepatitis agent from the raw sewage on which they fed and retained it, in a viable condition, for prolonged periods of time.

Further evidence indicating the role of shellfish which had fed on raw sewage in the transmission of infectious hepatitis was presented by Rindge et al.[115] These workers reported one of the first known shellfish associated outbreaks of IH to occur in the U.S. They proved, epidemiologically, that 15 cases of hepatitis were a direct result of eating polluted clams. The clams in question had been harvested from an estuary channel which fed into the bay of Greenwich, Conn. By using fecal coliform counts, Rindge and co-workers proved the source of pollution to be partially treated sewage. This sewage was released into the estuary from a sewage disposal plant whose disposal facilities proved to be grossly overloaded. During the period of time since the original outbreak recorded by Roos, there have been approximately 11 such shellfish-associated epidemics that have been recorded worldwide. Of those reported in the U.S., at least four incidences were directly related to oysters, clams, or mussels which had fed upon raw sewage.[26,36,37,47,52,93,110] In several outbreaks, the shellfish had either been steamed, or cooked in some fashion prior to being consumed, thus indicating that the hepatitis virus is capable of withstanding common means of food processing.[72,103,119] In two other outbreaks which occurred in 1973 involving 278 persons, the oysters which were the vectors had been harvested from supposedly "safe" waters. These outbreaks and others have been extensively reviewed elsewhere.[11,84,91,110,126,137]

Nature of the hepatitis agent — The virus etiology of infectious hepatitis was first proposed in 1939.[40] However, IH was not made a reportable disease in the U.S. until 1950.[182] No clear-cut technique for detecting shellfish-associated hepatitis was proposed by the U.S. Public Health until 1962.[133] This was due partially to the fact that the virus responsible for IH had not, as yet, been identified. During the period from 1954 to 1968, a number of viral agents were incriminated as the probable agent(s) of hepatitis.[55,62,77,87,95,102,114,116] The actual agent of IH has been identified as Hepatitis A virus (HAV), a 27 mm RNA virus, and techniques for its isolation and titration are being perfected.[50,61,76,104,136,138]

SYMPTOMS OF IH (HEPATITIS A)

Transmission of IH is, typically, via the fecal-oral route. The incubation period for this disease varies from 15 to 50 days with an average of about 30 days. Prodromal symptoms of the disease are recorded as being somewhat similar to poliomyelitis with fever, loss of appetite, nausea, vomiting, and occasional aches in the joints and back of the head.[24,34] As the disease progresses, there is hepatic involvment with the liver first becoming tender then stone-like on palpation. The urine becomes dark, and the feces (stools) of patients clay colored. Due to hepatic insufficiency and the escape of bile salts into the blood, victims show jaundicing first of the "whites" of the eyes, then of the skin. This presence of free bile salts also may lead to a feeling of severe itching (prurescence). Resolution of the disease is slow, usually 6 weeks to 6 months. Fortunately, the mortality rate is very low, less than 2%. Once jaundice is present, IH (or Hepatitis A) caused by eating raw contaminated shellfish has been recorded difficult-to-differentiate from serum hepatitis or Hepatitis B virus (HBV). Some of the more significant differences are shown in Table 2. Of significance is the fact that with shellfish-associated outbreaks of hepatitis, victims may suffer from nonbacterial gastroenteritis prior to actual onset of hepatitis symptoms.[52]

There are at least 100 types of viruses which can be recovered from human sewage. These include agents responsible for respiratory diseases, meningitis, gastroenteritis, and diarrhea. Some of these viruses have been recovered from shellfish; in fact, there is mounting evidence that infectious hepatitis is not the only viral disease transmitted by contaminated shellfish.[12,25,42] Viruses capable of producing diarrhea and gastroenteritis may have been responsible for at least one outbreak of food poisoning involving 35 persons who had eaten cockles.[1] Virus-like particles were recovered from fecal samples of three victims, but not the shellfish. All victims reported rapid onset of symptoms which consisted primarily of vomiting and diarrhea, but no apparent enteropathogenic bacteria were recovered from fecal samples. Consequently, future research may prove that shellfish also serve as vectors for the dissemination of other entero viruses (e.g., rotavirus).

Table 2
SOME EPIDEMIOLOGIC DIFFERENCES BETWEEN INFECTIOUS (HAV) AND "SERUM" HEPATITIS (HBV)

Infectious hepatitis	Serum hepatitis
Average incubation period: 30—36 days	Average incubation period: 60—65 days
Onset usually abrupt	Onset tends to be gradual
Fever 38°C or higher usually present	Fever usually absent
Au antigen/antibody absent	Au antigen/antibody present 30—80 days after exposure

PRESENCE OF ENTERIC VIRUSES IN SEWAGE AND SEAWATER

Faced with the uncertainty as to the nature of the IH virus, early researchers were obliged to use other viruses as models to study the joint problems of viral pollution of seawater and contamination of shellfish. The enteric group of viruses especially poliovirus I (poliovirus lsc-2ab) have been used extensively as models because they are readily cultivated and share some of the characteristics of the hepatitis virus.

Another reason for the choice of enteric viruses as research tools has been their demonstrated presence in sewage effluents and polluted marine bays and estuaries where they can survive for considerable lengths of time.[42,67,70,86]

Kelly and Sanderson[70] were among the first researchers to actually isolate viable polio, coxsackie, and echo viruses from sewage effluents being pumped into Jamaica Bay, L.I. Isolation of these viruses from sewage were subsequently recorded by Clarke and Chang as well as Clarke and Kabler[16,17] who observed that adeno and reo virus could also be recovered, periodically, from sewage effluents. These findings were confirmed by Kallins who observed that primary sewage treatment had no effect on the viability of enteric viruses and must be followed by further treatment.[67]

The actual survival of enteric viruses in marine estuaries has been demonstrated by a number of researchers including Cioglia and Loddo.[15] These workers reported that seawater apparently has an inactivating effect on some viruses. However, polio, echo, coxsackie, and reo viruses were resistant to this inactivating effect and were able to survive in estuaries for at least 2 weeks. They also found virus levels as high as 10^3 to 10^4 plaque-forming units (PFU) per milliliter to be common in some polluted waters during the warmer months of the year. Similar results were recorded by other investigarors,[83,98] who noted that enteric viruses may survive for up to 140 days in natural waters and from 9 to 40 days in the laboratory depending upon experimental conditions. The literature pertaining to the presence and persistance of enteric viruses in sewage and seawater has been extensively reviewed elsewhere.[3,8]

ACCUMULATION AND PERSISTENCE OF ENTERIC VIRUSES IN SHELLFISH

The ability of shellfish to accumulate viruses was first shown experimentally by Crovari[23] using poliovirus 3 as a model. He noted that within a 72-hr period, edible mussels would accumulate up to 90% of the virus present in contaminated seawater. However, when the mussels were placed for the same period of time in flowing seawater at a temperature of 25°C, there followed a log reduction in virus titer. In subsequent research, Hedstrom and Lycke[56] showed that poliovirus 3 was also rapidly accumulated by Eastern oysters placed in a stationary seawater system. These workers attempted, unsuccessfully, to eliminate viruses from the oysters by placing them in chlorinated seawater. Subsequently, Atwood et al.[5] proved that various species of clams would accumulate another enteric virus, coxsackie B-5, but not to the same extent as poliovirus.

Thus, laboratory research clearly established that shellfish readily accumulated enteric viruses. The first actual field recovery of viruses from shellfish growing in a polluted estuary was recorded by Metcalf and Stiles.[96-98] These researchers recovered two enteric viruses, coxsackie B-4 and echo 9, in addition to reo virus and coliphage from Eastern oysters growing in an estuary of Great Bay, N.H. In laboratory studies, Metcalf and Stiles noted that oysters routinely exposed to log $6TCD_{50}$ (tissue culture infective dose) units of coxsackie B-3 and polio I, contained after exposure log $5TCD_{50}$ units (about 90%) of virus per gram of tissue. Further, the virus remained stable in oysters at 5°C for 28 days. In these studies, the greatest recovery of virus was made from the digestive gland, but no viral replication was observed either in this gland or other oyster tissues. Consequently, they interpreted viral contamination of shellfish as a two-phase phenomenon: uptake and persistence.[96]

Metcalf's findings were subsequently confirmed by Liu et al., who studied the fate of poliovirus I in the Northern Quahaug (*Mercenaria mercenaria*).[81,83] In laboratory experiments, they found that the highest level of contamination in shellfish occurred within a few hours after the exposure of the animals to virus. The majority of the accumulated viruses were recovered from the digestive diverticulae, stomach, and hemolymph of the clams. There did not appear to be any adsorption, penetration, or replications of virus in the cells of the test animals. They concluded that the process of viral accumulation by shellfish was both rapid and dynamic in nature. Research by other workers have confirmed that viruses neither replicate nor attach to specific cells of shellfish. In one such study using *Staphylococcus aureus* phage 80 as a model system and the Eastern oyster as a test animal, investigators did report however, that the viral particles appeared more stable in oyster plasma than in seawater.[39]

Studies with West Coast oysters (*Crassostria gigas* and *Ostrea lurida*) revealed that these shellfish readily accumulate virus in the same manner and in approximately the same quantity as East Coast shellfish.[32,33,60] The digestive tract of these animals appeared to be the major site of viral accumulation. In these studies, it was noted that viral uptake occurred more rapidly in the native oyster (*O. lurida*) than in the larger Pacific oyster (*C. gigas*). In subsequent research, these same investigators proved that the actual uptake of viruses by shellfish was due to ionic bonding of viral particles probably to the sulfate radicals on the mucopolysaccharide moiety of shellfish mucus.[33] These findings were subsequently partially confirmed by the research of Bedford et al.,[7] who studied the uptake of radioactively labeled reo virus type III and Semliki Forest virus by the New Zealand Rock Oyster (*Crassostrea glomerate*). In their studies, they found that accumulation of virus was dependent upon viral concentration and that oysters possessed a large, but finite number of sites for viral adsorption. They also noted that the rate at which oysters accumulated, the nonenveloped reo virus was appreciably greater than the rate of uptake for the lipoprotein enveloped Semliki Forest virus. This observation, plus the finding that oyster shells also have the capability of binding virus, led these workers to conclude that surface phenomena, including adsorption to the mucus sheath, are major factors governing viral uptake by filter-feeding shellfish.

Not only are viruses rapidly accumulated by shellfish, but research results indicate that they may persist in these animals for considerable time and under a variety of conditions.[28,47,103] In one study, persistence of poliovirus in chilled, frozen, and processed oysters was studied.[28] The processing procedure used included steaming, baking, frying, and stewing. Results obtained indicated that viruses could survive for a period varying from 30 to 90 days in chilled or frozen shellfish. Virus survival rate in processed shellfish was found to vary from 7 to 10%, depending upon cooking technique and time. This survival rate was directly related to the kinetics of heating. The bulk of viruses accumulated by shellfish are normally recovered from the digestive area.[32,81,96] However, in these investigations, using ordinary cooking procedures, it was observed that this portion of the oyster was never heated for a sufficient length of time to inactivate all the virus present. These results showing virus persistence are consistant with studies indicating the ability of viruses to survive in other food products.[30,31,57,89]

PROBLEMS TO BE SOLVED

Lack of Correlation Between Bacterial Standards and Presence of Viruses

The present federally enforced standards for shellfish growing areas and market shellfish are based upon the use of coliforms and fecal coliforms as indicator organisms of sanitary quality.[2,130] These standards stipulate that shellfish growing waters should have a total coliform count of no more than 70 coliforms per 100 mℓ of seawater. Standards for shellfish meats are 230 coliforms per 100 g and no more than 23 fecal coliforms per gram of meats. Enforcement of these standards plus control and licensing of shellfish growers and packers has virtually eliminated shellfish as vectors in the dissemination of diseases caused by entero pathogenic bacteria.

However, there are mounting data suggesting that these bacterial standards may be valueless as indices of viral contamination to either seawater or shellfish. In recent studies, viruses have been recovered from seawater and shellfish harvested from supposedly "safe" areas.[43] Other research has indicated the presence of viruses in market oysters.[11,12,25] In fact, it has been proven that shellfish grown and harvested under the guidelines of the National Shellfish Sanitation Program were vectors for an outbreak of Hepatitis A involving 273 victims in Houston, Tex. and Calhoun, Ga.[90,110] Faced with the probable inadequacy of bacterial standards, it may be necessary to find another type of indicator organism, develop another form of bacterial standard applicable to viruses, or routinely monitor seawater and shellfish for viral contamination.

Use of coliphage as indicators of viral contamination has been suggested by some investigators.[59,135] However, research results have proven that there is often a disparity or lack of correlation between coliphage and enteric virus levels in seawater and shellfish.[99] Therefore, the development of a viral standard based upon coliphage or bacteriophage titers seems unlikely.

Proposals have been advanced for new bacterial techniques which might offer a more realistic indication of possible viral pollution. One such suggested technique has been to use fecal coliform counts of the marine sediments of shellfish beds; another has been to use fecal streptococci as indicator organisms.[21] The value of fecal coliform counts of sediments as indices of possible viral contamination is based upon three significant findings. First, viruses have been recovered from sediments in concentration up to 1000 times greater than in the overlying waters. In addition, investigators have recovered enteric viruses from marine sediments shellfish beds, and both enteric viruses and fecal coliforms survive for longer periods of time in sediments than in surrounding waters.[48,51] However, a definitive relationship must be established between the presence of fecal coliforms and the presence of enteric, or other viruses before any type of standard can be established or even considered.

The fecal streptococci are known to survive longer in natural waters than do the coliforms or fecal coliforms. They are also considered to be reliable indicators of past pollution to an area. They are known to resist freezing better than fecal coliforms, and their use as indicators of possible viral contamination of frozen shellfish has also been suggested.[27,79,112] However, a definite and consistant relationship between the presence of fecal streptococci and viral pollution of shellfish would have to be established before adopting any form of fecal streptococci standard for seawater and shellfish.

Need for a Standard Method to Assay for Viruses

Obviously, the best method to determine if shellfish or shellfish growing waters are being contaminated with viruses would be to assay for these entities directly. Unfortunately, the routine surveying of shellfish for possible viral contaminants is hampered by one salient factor: no standard techniques have, as yet, been adopted for detecting viruses in shellfish, seawater, or marine sediments. That is, none have been adopted at the time of this writing. Within the past several years, however, a number of excellent techniques have been developed.[58,74,75,121,122] Quite possibly one, or several of these may become recommended for routine assay work. Each of these proposed procedures has certain advantages including varying degrees of reliability in recovering viruses. A disadvantage common to all is the fact that they depend, ultimately, on determining the cytopathogenic effects (CPEs) of viruses on tissue culture monolayers. Consequently, test results are known only after (on the average) 3 to 7 days. Very recently, rapid assay techniques such as enzyme-linked immunassay (ELISA)[141] and radioimmunoassay (RIA)[61] have been developed which do not require tissue culture techniques. At least two of these techniques require no extensive equipment costs and appear reliable in detecting both rotavirus (a causative agent of nonbacterial gastroenteritis) and Hepatitis A. Further perfection of these techniques could ultimately produce the needed tool: a rapid, reliable, and inexpensive assay procedure.

ELIMINATION OF VIRUSES FROM SHELLFISH

Obviously, the best way to avoid contamination of shellfish by enteric pathogens is to prevent these from entering in the first place. Another viable option or adjunct procedure is to treat the shellfish in such a fashion so as to eliminate potential pathogens. Methods to do this include use of ionizing radiation or depuration.

Ionizing Radiation

The use of ionizing radiation has been suggested by some scientists as a viable means of eliminating food spoilage and pathogenic microorganisms (including viruses) from shellfish.[85,92,106] The method is possibly valid for bacteria, but viruses are known to possess a certain degree of resistance to gamma radiation.[94,127,128] Further, it has been shown that the amount of radiation (400 krad) required to inactivate 90% of the virus in experimentally contaminated shellfish renders the animals inedible.[29]

Depuration

The method which holds the greatest promise of removing viruses from shellfish is depuration. The term depuration refers to the ability of bivalve mollusks to cleanse themselves of extraneous matter when they are placed in clean, fresh water.[108,109] Depuration, on a commercial scale, has been employed both in America and Europe to "purify" soft shelled clams and "edible" oysters (*Ostrea edulis*).[44,64,100,139,140] Experimentally, this same process has been shown effective in removing viruses from oysters, mussels, and clams usually within 72 hr. Research has indicated the possibility of some viruses remaining in individual shellfish unless the entire procedure is carefully monitored.[14] However, a number of procedures have been devised which, under experimental conditions, improve the efficiency of depuration.[63]

There are two ways by which shellfish can be depurated: through relay operations, or by use of commercial depuration plants. Relaying, the simplest procedure, involves transporting shellfish from the collection area to clean waters in which they are allowed to reside until complete self cleaning has taken place. The one major objection to this has come from environmentalists who feel continued use of relaying operations would lead to eventual pollution of formerly pristine waters.

Commercial depuration plants have been in operation in certain areas for some years and they do work.[100] However, new plants might have to be designed specifically for elimination of viruses. Objections to the use of such plants could be based upon economics, i.e., building and operational costs might increase market price of shellfish. Other objections could come from certain shellfish farmers who might feel that routine use of depuration would produce adverse publicity as to the safety of the product. There is no doubt, however, that depuration in conjunction with the national shellfish program should assure continued production and sale of a highly nutritious and delectable product.

REFERENCES

1. *Abstr. Hyg.*, 52, 612, 1977.
2. American Public Health Association, Recommended Procedures for the Examination of Sea Water and Shellfish, 4th ed., American Public Health Association, New Yori, 1970.
3. American Public Health Association, Viruses in Water, Berg, G., Bodily, H. L., Lennette, E. H., Melnick, J. L., and Metcalf, T. G., Eds., American Public Health Association, Washington, D.C., 1976.
4. **Arcisz, W. and Kelly, C. B.**, Self Purification of the Self Clam, *Mya Arenaria, Public Health Rep.*, 70, 605—614.
5. **Atwood, R. P., Cherry, J. D., and Klein, J. O.**, Clams and Viruses: Studies with Coxsackie B-5 Virus, Communicable Disease Center Hepatitis Survey, 1964.
6. **Barnard, R.**, *Vibrio parahaemolyticus* newly recognized cause of gastroenteritis in U.S., *J. Environ. Health*, 35, 45—47, 1964.
7. **Bedford, A. J., Williams G., and Bellamy, A. R.**, Virus Accumulation by the Rock Oyster, *Crassostrea glomerata, Appl. Environ. Microbiol.*, 35, 1012—1018, 1978.
8. **Berg, G.**, Virus transmission by the water vehicle. I. Viruses, *Health Lab. Sci.*, 3, 85—91, 1966.
9. **Borradaile, L. A., Eastham, J. T., and Potts, F. A.**, *The Invertebrata*, 4th ed., Cambridge University Press, London, 1962, 625—626.
10. **Boyce, R. W. and Herdmann, W. A.**, On oysters and typhoid, *Rep. Br. Assoc. Adv. Sci.*, 65, 723—726, 1896.
11. **Brisou, J.**, Viral and parasitic pollution of littoral waters and its consequences on public health, *Bull. WHO*, 38, 79—118, 1968.
12. **Brisou, J. and Denis, F.**, Viral Contamination of Seafood Incidences of Human Disease, *Econ. Mediterr. Anim.*, 13, 345—350, 1972.
13. **Buzzanca, E.**, Su d'un episodio di tossinfezione alimentare da *Salmonella enteriditis* nel commune di Corona (Messina), *Igiena Sanitaria Pubblica*, 20, 285—292, 1964.
14. **Canzonier, W. J.**, Accumulation and elimination of coliphage S-13 by the hard clam, *Mercenaria merceneria, Appl. Microbiol.*, 21, 1024—1031, 1971.
15. **Cioglia, E. and Loddo, B.**, Processes of Self Purification in a Marine Environment. III Resistance of Some Enteroviruses, *Nuovi Annali Igiene Microbiol.*, 13, 11—20, 1962.
16. **Clarke, N. A. and Chang, L. A.**, Enteric viruses in water, *J. Am. Water Works Assoc.*, 51, 1299—1303, 1959.
17. **Clarke, N. A. and Kabler, P. W.**, Human enteric viruses in sewage, *Health Lab. Sci.*, 1, 44—47, 1964.
18. **Cliver, D. O.**, Food-associated viruses, *Health Lab. Sci.*, 4, 213—221, 1967.
19. **Cliver, D. O.**, Viral infections, in *Food Borne Infections and Intoxicants*, Riemann, H., Ed., Academic Press, New York, 1969, 73—78.
20. **Cliver, D. O.**, Transmission of viruses through foods, *Crit. Rev. Environ. Control*, 1, 551—579, 1971.
21. **Cohen, J. and Shuval, H. I.**, Coliforms, fecal coliforms, and fecal streptococci as indicators of water pollution, *Water, Air, Soil Bull.*, 2, 89—95, 1973.

22. Conn, H. W., The Outbreak of Typhoid Fever at Wesleyan University, Connecticut State Board of Health Reports, 1894.
23. Crovari, P., Some observations on the depuration of mussels infected with poliomyelitis virus, *Igiene Moderna*, 5, 22—26, 1958.
24. **Davis, D. B., Dulbecco, R., Eisen, H. N., Ginsberg, H. S., and Wood, W. B., Jr.,** *Microbiology*, Harper & Row, New York 1968, 118 and 1276.
25. **Denis, F.,** Coxsackie group A in oysters and mussels, *Lancet*, 6, 1262, 1973.
26. **Dienstag, J. L., Giest, I. D., Lucas, C. R., Wong, D., and Purcell, R. H.,** Mussel associated viral heptatis, type A: serological confirmation, *Lancet*, 1, 561—564, 1976.
27. **Di Girolamo, R., Liston, J., and Matches, J.,** The effects of freezing on the survival of *Salmonella* and *E. coli* in Pacific oysters, *J. Food Sci.*, 35, 13—16, 1970.
28. **Di Girolamo, R., Liston, J., and Matches, J. R.,** Survival of virus in chilled, frozen, and processed oysters, *Appl. Microbiol.*, 20, 58—63, 1970.
29. **Di Girolamo, R., Liston, J., Jr., and Matches, J.,** Effects of irradiation on the survival of virus in West Oysters, *Appl. Microbiol.*, 24, 1005—1006, 1972.
30. **Di Girolamo, R., Wiczynski, L., Daley, M., Miranda, F., and Viehweger, C.,** Uptake of bacteriophage and their subsequent survival in edible West Coast crabs after processing, *Appl. Microbiol.*, 23, 1073—1076, 1972.
31. **Di Girolamo, R. and Daley, M.,** Recovery of bacteriophage from contaminated chilled and frozen samples of edible West Coast crabs, *Appl. Microbiol.*, 25, 1020—1022, 1973.
32. **Di Girolamo, R., Liston, J., and Matches, J.,** Uptake and elimination of poliovirus by West Coast oysters, *Appl. Microbiol.*, 29, 260—264, 1975.
33. **Di Girolamo, R., Liston, J., and Matches, J. R.,** Ionic bonding, the mechanism of viral uptake by shellfish mucus, *Appl. Environ. Microbiol.*, 33, 19—25, 1977.
34. **Di Marco, G.,** *Patologia Generale Umana*, Edizione Mondatori, Milano, 1964, 453—465.
35. **Dingman, J. C.,** Report of a possibly milk-borne epidemic of infantile paralysis, *N. Y. State J. Med.*, 16, 589—590, 1916.
36. **Dismukes, W. E., Bisno, A. L., Katz, S., and Johnson, R. F.,** An outbreak of gastroenteritis and infectious hepatitis attributed to raw clams, *Am. J. Epidemiol.*, 89, 551—561, 1969.
37. **Dougherty, W. J. and Altman, R.,** Viral hepatitis in New Jersey, *Am. J. Med.*, 32, 704—716, 1962.
38. **Earampamoorthy, S. and Koff, R. S.,** Health Hazards of bivalve mollusc ingestion, *Ann. Intern. Med.*, 83, 107—110, 1975.
39. **Feng, J. S.,** The fate of a virus, *Staphylococcus aureus* phage 80, ingested into the oysters, *Crassostrea virginica*, *J. Invert. Pathol.*, 8, 496—498, 1966.
40. **Findley, G. M., McCallum, F. O., and Murgatroyd, F.,** Observations bearing on the etiology of infectious hepatitis (so-called epidemic catarrahal jaundice), *Trans. R. Soc. Trop. Med. Hyg.*, 32, 575—582, 1939.
41. **Foote, C. J.,** A bacteriological study of oysters, with a special reference to them as a course of typhoid infection, *Med. News*, 66, 320—328, 1895.
42. **Fox, J. P.,** Human-associated viruses in water, in Viruses in Water, Berg, G., Bodily, H. L., Lennette, E. H., Melnick, J. L., and Metcalf, T. G., Eds., American Public Health Association, Washington, D.C., 1976, 39—49.
43. **Fugate, K. J., Cliver, D. O., and Hatch, M. T.,** Enteroviruses and potential bacterial indicators in Gulf Coasts oysters, *J. Milk Food Technol.*, 38, 100—104, 1975.
44. **Furfari, S. A.,** Depuration Plant Design, U.S. Department of Health, Education and Welfare Environmental Health Series Publication, Washington, D.C., 1966, 111.
45. **Galtsoff, P.,** The American Oyster, *Crassostrea virginica*, Bull. 64, U.S. Fish and Wildlife Service, Washington, D.C., 1964, 27—235.
46. **Garibaldi, R. A., Murphy, G. D., III, and Wood, B. T.,** Infectious hepatitis outbreak associated with cafe water, *Health Rep.*, 87, 164—171, 1972.
47. **Gaub, J. and Ranek, L.,** Epidemic hepatitis following consumption of imported oysters, *Ugeski Laeger*, 135, 345—348, 1973.
48. **Gerba, C. P. and McLeod, J.,** Effect of sediments on the survival of *Escherichia coli* in marine waters, *Appl. Environ. Microbiol.*, 32, 114—120, 1976.
49. **Goldstein, D. M., McD. Hammon, W., and Viets, H. R.,** An outbreak of polio-encephalitis among navy cadets, possibly food-borne, *JAMA*, 131, 569, 1946.
50. **Grabow, W. O. K.,** Progress in studies on the type A (infectious) hepatitis virus in water, *Water Sanit.*, 2, 20—24, 1976.
51. **Greenberg, A. E.,** Survival of enteric organisms in seawater, *Public Health Rep.*, 71, 77—86, 1956.
52. **Guidon, L. and Pierach, C. A.,** Infectious hepatitis after ingestion of raw clams, *Minn. Med.*, 56, 15—19, 1973.
53. **Hargreaves, E. R.,** Epidemiology of poliomyelitis, *Lancet*, 1, 969—972, 1949.
54. **Hart, J. C.,** Typhoid fever from clams, *Community Health Bull.*, 1944.

55. **Hartwell, W. V., Avernheimer, A. H., and Pearce, G. W.,** Examination by chromatography and immunodiffusion of an adenovirus 3 isolated from humans with infectious hepatitis, *Appl. Microbiol.,* 16, 1859—1862, 1968.
56. **Hedstrom, C. E. and Lycke, E.,** An experimental study on oysters as virus carriers, *Am. J. Hyg.,* 79, 134—138, 1964.
57. **Heidelbaugh, N. D. and Giron, D. J.,** Effect of processing on recovery of poliovirus from inoculated foods, *J. Food Sci.,* 34, 234—237, 1969.
58. **Herrmann, J. E. and Cliver, D. O.,** Methods for detecting food-borne enteroviruses, *Appl. Microbiol.,* 16, 1564—1569, 1968.
59. **Hilton, M. C. and Stotzky, G.,** Use of coliphages as indicators of water pollution, *Can. J. Microbiol.,* 19, 747—751, 1973.
60. **Hoff, J. C. and Becker, R. C.,** The accumulation and elimination of crude and clarified poliovirus suspensions by shellfish, *Am. J. Epidemiol.,* 90, 53, 1969.
61. **Hollinger, B. F., Brodley, D. W., Maynard, J. E., Dreesman, G. R., and Melnick, J. L.,** Detection of hepatitis A viral antigen by radio-immunoassay, *J. Immunol.,* 115, 1464—1466, 1975.
62. **Hsuing, G. D., Isacson, P., and McCollum, R. W.,** Studies on a myxovirus isolated from human blood. I. Isolation and properties, *J. Immunol.,* 88, 284—287, 1976.
63. **Huntley, B. E. and Hammerstrum, B.,** An experimental depuration plant: operation and evaluation, *Chesapeake Sci.,* 12, 231—239, 1971.
64. **Hensen, E. T.,** Depuration, the national posture, in Proc. 5th Nat. Shellfish Sanit. Workshop, Washington, D.C., 1964, 71—74.
65. **Joseph, P. R., Millar, J. D., and Henderson, D. A.,** An outbreak of hepatitis traced to food contamination, *N. Engl. J. Med.,* 273, 188—194, 1965.
66. **Jubb, G.,** A third outbreak of epidemic poliomyelitis at West Kirby, *Lancet,* 1, 67, 1915.
67. **Kallins, S. A.,** The presence of human enteric viruses in sewage and their removal by convention sewage treatment methods, *Adv. Appl. Microbiol.,* 8, 145—151, 1966.
68. **Kaneko, T. and Colwell, R. R.,** Incidence of *Vibrio parahaemolyticus* in Chesapeake Bay, *Appl. Microbiol.,* 30, 251—255, 1975.
69. **Kelly, C. B. and Arcisz, W.,** Survival of enteric organisms in shellfish, *Public Health Rep.,* 69, 1205—1210, 1954.
70. **Kelly, S. M. and Sanderson, W. W.,** Viruses in sewage, *N.Y. Dep. Health News,* 36, 18—22, 1959.
71. **Kjellander, J.,** Hygienic and microbiological viewpoints on oysters as vectors of infection, *Svenska Lakartidn,* 53, 1009—1012, 1956.
72. **Klein, E.,** Report of experiments and observations on the vitality of the bacillus of typhoid and other sewage microbes in oysters and shellfish, in *Science Manscript,* W. Cline and Son, London, 1905, 58—72.
73. **Koff, R. S., Grady, G. F., Chalmers, T. C., Mosley, J. A., Swarz, B. L., and the Boston Inter-Hospital Liver Group,** Viral hepatitis in a group of Boston hospitals. II. Importance of exposure to shellfish in a non-epidemic period, *N. Engl. J. Med.,* 276, 703—710, 1967.
74. **Konowalchuk, J. and Speirs, J. T.,** Enterovirus recovery from laboratory contaminated samples of shellfish, *Can. J. Microbiol.,* 18, 1023, 1029, 1972.
75. **Kostenbader K. D., Jr. and Cliver, D. O.,** Filtration methods for recovering enteroviruses from foods, *Appl. Microbiol.,* 26, 149—154, 1973.
76. **Krejs, G. J., Gassner, M., and Blum, A. L.,** Epidemiology of infectious hepatitis, *Clin. Gastroenterol.,* 3, 277—303, 1974.
77. **Kubelka, V.,** Our knowledge about motol virus isolated from patients suffering with infectious hepatitis, in *Advances in Hepatology,* Vandenbrouch, J., De Groate, J., and Standaert, L. O., Eds., Williams & Wilkins, Baltimore, 1965, 230—235.
78. **Lindberg-Braman, A. M.,** Clinical observations on the so-called oyster hepatitis, *Am. J. Public Health,* 53, 1003—1006, 1956.
79. **Liston, J. and Raj. H.,** Food poisoning problems of frozen seafoods, *J. Environ. Health,* 25, 235—250, 1962.
80. **Liston, J.,** *Vibrio parahaemolyticus,* in *Microbial Safety of Fishery Products,* Chichestes, V. and Graham, D., Eds., AVI Publishing, New York, 1975, 203—213.
81. **Liu, O. C., Seraichekas, H. R., and Murphy, B. L.,** Fate of poliovirus in northern quahaugs, *Proc. Soc. Exp. Biol. Med.,* 121, 601—604, 1966.
82. **Liu, O. C., Seraichekas, H. R., and Murphy, B. L.,** Viral depuration of the northern quahaug, *Appl. Microbiol.,* 15, 307—311, 1967.
83. **Liu, O. C., Seraichekas, H. R., Broshior, D. A., Heffernan, W. P., and Cabelli, V. J.,** The occurence of human enteric viruses in shellfish and estuaries, *Bacteriol. Proc.,* 42, 151, 1968.
84. **Liu, O. C.,** Viral pollution and depuration of shellfish, in Proc. Nat. Specialty Conf. Disinfection, July 1970, New York, 1971, 397—428.

85. **Luizzo, J. A., Farag, M. K., and Novak, A. F.**, Effect of low-level radiation on the proteolytic activity of bacteria in oysters, *J. Food Sci.*, 32, 104—112, 1967.
86. **Lo, S., Gilbert, J., and Hetrick, F.** Stability of human enteroviruses in estuarine and marine waters, *Appl. Environ. Microbiol.*, 32, 245—249, 1976.
87. **Ludberg, R.**, On the nature of the viral agent of hepatitis, *Am. J. Public Health*, 53, 1406—1410, 1956.
88. **Lumsden, T. T., Hazeltine, H. E., Leak, T. P., and Veldee, M. V.**, A typhoid fever epidemic caused by oyster borne infection, *Public Health Rep.*, Vol. 50, 1925.
89. **Lynt, R. K.**, Survival and recovery of enteroviruses from foods, *Appl. Microbiol.*, 14, 218—222, 1966.
90. **Mackowiak, P. A., Caraway, C. T., and Portney, B. L.**, Oyster-associated hepatitis lessons from the Louisiana experience, *Am. J. Epidemiol.*, 103, 181—191, 1976.
91. **Mackowiak, P. A., Caraway, C. T., and Portnoy, B. L.**, Oyster-associated hepatitis lessons from the Louisiana experience, Report, B, *Am. J. Epidemiol.*, 103, 186—191, 1976.
92. **MacLean, D. F. and Welander, C.**, The preservation of fish with ionizing radiation: bacterial studies, *Food Technol.*, 14, 251—254, 1960.
93. **Mason, J. O. and McLean, W. R.**, Infectious hepatitis traced to the consumption of raw oysters, *Am. J. Hyg.*, 75, 90—94, 1962.
94. **McCrea, J. F.**, Ionizing radiation and its effect on animal viruses, *Ann. N.Y. Acad. Sci.*, 83, 692—705, 1960.
95. **McLean, W. D.**, Etiological studies of infectious and serum hepatitis, *Prospectus Virol.*, 3, 184—188, 1959.
96. **Metcalf, T. G. and Stiles, W. C.**, The accumulation of enteric virus by the oyster, *Crassostrea virginica*, *J. Infect. Dis.*, 115, 68—72, 1965.
97. **Metcalf, T. G., Wallis, C., and Melnick, J. L.**, Virus enumeration and public health assessments in polluted surface water contributing to transmission of virus in nature, in *Virus Survival in Water and Wastewater Systems*, Malim, J. F., Jr. and Sagik, B. P., New York, 1974, 57—70.
98. **Metcalf, T. G., Slantz, L. W., and Bartley, C. H.**, Enteric pathogens in estuary waters and shellfish, in *Microbial Safety of Fisheries Products*, AVI Publishing, New York, 1975, 215—234.
99. **Metcalf, T. G.**, Evaluation of shellfish sanitary quality by indicators of sewage pollution, in *Discharge of Sewage from Sea Outfalls*, Gameson, A. L. H., Ed., Pergamon Press, Oxford, England, 1975, 75—84.
100. Ministry of Agriculture, Fisheries and Food, The Purification of Oysters in Installations Using Ultraviolet light, No. 27, Ministry of Agriculture, Fisheries and Food, Fisheries Laboratory, Burnham-on-Crouch, England, 1961.
101. **Mood, E. W.**, First Typhoid Case in Seven Years, Monthly Report, New Haven Department of Health, 1948, 15.
102. **Morris, J. A.**, Viral etiology of hepatitis, in *Primary Hepatoma*, University of Utah Press, Salt Lake City, 1954, 99.
103. **Mosley, J. W.**, Clam Associated Epidemic of Infectious Hepatitis, Hepatitis Surveillance Report, Communicable Disease Center, 18, 14—18, 1964.
104. **Mosley, J. W.**, The epidemiology of viral hepatitis: an overview, *Am. J. Med. Sci.*, 270, 253—270, 1975.
105. **Neefe, J. R. and Stokes, J., Jr.**, Epidemic of infectious hepatitis apparently due to a water-borne agent: epidemiological observations and transmission experiments in human volunteers, *JAMA*, 128, 1963—1970, 1945.
106. **Novak, A. F., Liuzzo, J. A., Grodnerand, R. M., and Lorell, R. T.**, Radiation pasteurization of Gulf Coast oysters, *Food Technol.*, 20, 103—104, 1966.
107. **Old, H. N. and Gill, S. L.**, A typhoid fever epidemic caused by carrier bootlegging oysters, *Am. J. Public Health*, 30, 633—640, 1941.
108. **Petrilli, F. L.**, Recherche sull' autodepurazione dei mitili, *Igiene Moderna*, 31, 309—312, 1938.
109. **Phillips, E. B.** Some experiments upon the removal of oysters from polluted to unpolluted waters, *J. Am. Public Health Assoc.*, 1, 305—310, 1911.
110. **Portnoy, B. L., Mackowiak, P. A., Caraway, C. T., Walker, J. A., McKinley, T. W., and Klein, C. A., Jr.**, Oyster associated hepatitis, failure of shellfish certification programs to prevent outbreaks, *JAMA*, 233, 1065—1068, 1975.
111. **Presnell, M. and Cummins, J.**, The Influence of Seawater Flow Rate on Bacterial Elimination by the Eastern Oyster, Tech. Memo, Gulf Coast Shellfish Sanitation Research Center, 1967, 67—73.
112. **Raj, H., Wiebe, W., and Liston, J.**, Detection and enumeration of fecal indicator organisms in frozen seafoods. II. Enterococci, *Appl. Microbiol.*, 9, 195—200, 1961.
113. **Read, M. R., Bancroft, H., Doull, J. A., and Parker, R. F.**, Infectious hepatitis — presumably foodborne, *Am. J. Public Health*, 36, 367—370, 1946.

114. **Rightsel, W. A.**, Tissue culture cultivation of cytopathogenic agents from patients with clinical hepatitis, *Science*, 124, 226—228, 1957.
115. **Rindge, M. E., Clem, J. David, Linkner, R. E., and Sherman, L. K.**, A Case Study of Infectious Hepatitis by Raw Clams, U.S. Public Health Special Rep., Greenwich, Conn., 1961.
116. **Roane, P. R., Woodford, P., Ward, F. G., and Cross, S.**, Biological and biophysical properties of a virus apparently isolated from a case of human hepatitis, *Bacteriol. Proc.*, 6, 132, 1966.
117. **Roos, B.**, Hepatitis epidemic conveyed by oysters, *Svenska Lakartidn*, 53, 989—1003, 1956.
118. **Round, L. A.**, Contributions to the Bacteriology of the Oyster; The Results of Experiments and Observations Made while Conducting an Investigation Authorized by the Commissioners of Shellfisheries of the State of Rhode Island, Providence State Special Report, 1914.
119. **Rully, S. J., Johnson, R. G., Mosley, J. W., Atwater, J. B., Rossetti, M. A., and Hort, J. C.**, An epidemic of clam-associated hepatitis, *JAMA*, 208, 649, 1969.
120. **Saccani, C. F., Polodeno, D., and Mogliano, E. M.**, L'attivita del centro degli enterobatteri patogeni per L'Italia settentrionale, *Igiene Moderna*, 58, 1066—1070, 1965.
121. **Sobsey, M. D., Wallis, C., and Melnick, J. L.**, Development of a simple method for concentrating enteroviruses from oysters, *Appl. Microbiol.*, 29, 21—26, 1975.
122. **Sobsey, M. D.**, Methods for detecting enteric viruses in water and wastewater, in Viruses in Water, Berg, G., Bodily, H. L., Lennette, E. H., Melnick, J. L., and Metcalf, T. G., Eds., American Public Health Association, Washington, D.C., 1976, 82—127.
123. **Soper, G. A.**, Report on a sporadic outbreak of typhoid fever at Laurence, New York due to oysters, *Med. News*, 86, 241—242, 1905.
124. **Stiles, G. W., Jr.**, Shellfish Contamination from Sewage Polluted Waters and Other Sources, Bull. 136, U.S. Government Printing Office, Washington, D.C., 1911, 1—50.
125. **Stiles, G. W., Jr.**, Sewage Polluted Oysters as a Cause of Typhoid And Other Gastrointestinal Disturbances, Bureau of Chemistry Bull. 156, U.S. Department of Agriculture, Washington, D.C., 1912, 1—44.
126. **Stille, W., Keinkel, B., and Nerger, K.**, Shellfish transmitted hepatitis, *Deutsch Med. Wochenschr.*, 97, 145—147, 1972.
127. **Sullivan, R. and Read, R. B., Jr.**, Method for recovery of viruses from milk and milk products, *J. Dairy Sci.*, 51, 1748—1751, 1968.
128. **Sullivan, R., Fassolitis, A. C., Larkin, E. P., Read, R. B., Jr., and Peeler, J. T.**, Inactivation of thirty viruses by gamma radiation, *Appl. Microbiol.*, 22, 61—65, 1071.
129. **Tonney, F. O. and White, J. L.**, Viability of *Bacillus typhosus* in oysters during storage, *JAMA*, 84, 1403, 1926.
130. Sanitation of Shellfish Growing Waters, Special Rep., U.S. Department of Health, Education and Welfare, Washington, D.C., 1965.
131. The Spread of Typhoid Fever and Other Gastrointestinal Diseases by Sewage Polluted Shellfish, Special Rep., U.S. Public Health Service, Washington, D.C., 1926.
132. U.S. Public Health Service, Viral Hepatitis, a New Disease, No. 5, Communicable Disease Center, Atlanta, Ga., 1951.
133. U.S. Public Health Service, Shellfish Associated Hepatitis, Techniques for Detecting, No. 11, Communicable Disease Center, Atlanta, Ga., 1962.
134. **Vasconcelos, G. J. and Ericksen, T. H.**, Bacterial Accumulation-Elimination Response of Pacific Oysters and Manila Clams Maintained under Commercial Wet Storage, Report of the N.W. Shellfish Sanitation Resource Planning Conference, 1967, 65—78.
135. **Vaughn, J. M., and Metcalf, T. G.**, Coliphages as indicators of enteric viruses in shellfish and shellfish raising estuarine waters, *Water Resour.*, 9, 613—616, 1975.
136. **Waterson, A. P.**, Infectious particles in hepatitis, *Annu. Rev. Med.*, 27, 23—25, 1976.
137. WHO Expert Committee on Fish and Shellfish Hygiene, Fish and Shellfish Hygiene, Committee Report, WHO Technical Rep. Ser., World Health Organization, Geneva, 1974, 3—62.
138. WHO Scientific Group on Virus Diseases, The new program of the World Health Organization in medical virology, *Intervirology*, 6, 133—149, 1976.
139. **Wood, P. R.**, The status of shellfish depuration in the United Kingdom, presented at Nat. Conf. Depuration, Kingston, R.I., 1966.
140. **Wood, B. C.** The Principles and Methods Employed for the Sanitary Control of Molluscan Shellfish, FAO Tech. Conf. Marine Pollut., Effects on Living Resour. Fishing, 1970, 560—565.
141. **Yolken, R. H., Kim, H. W., Clem, T., Wyatt, R. G., Chanock, R. M., Kalica, A. R., and Kapikian, A. Z.**, Enzyme-linked immunosorbent assay (ELISA) for detection of human reovirus-like agent of infantile gastroenteritis, *Lancet*, 2, 263—266, 1977.
142. **Yonge, C. M.**, *Oysters*, Collins Clear Type Press, London, 1967.

ALGAL TOXINS

Nancy H. Shoptaugh and Edward J. Schantz

INTRODUCTION

Toxic algae are found in marine, brackish, and freshwater habitats throughout the world. Until recently, man has not considered seriously the consumption of algae as a source of food and has not been confronted directly with the poisonous products produced by them. The study of toxins produced by algae poses important problems as many species of algae have been cited as potential food supplies for both domestic animals and humans.[18] Field observations from as early as 1878 have recorded the poisoning of wild and domestic animals by the direct consumption of poisonous algae (e.g., *Nodularia spunigena, Gleotorichia echinulata,* and *Coelosphaerium Kuetzingianum*) in the water from lakes and ponds where dense, unialgal growths (blooms) have taken place.[9] The occurrence of poisons is sporadic, even within a particular species and the algal-man interaction can involve gastrointestinal, dermatological, respiratory, and allergic responses.[9,15]

Many species of freshwater and marine organisms are known to be poisonous. However, only a few of these present serious food poisoning problems for man and his domestic animals. With no reports of toxic algae in the phyla of Chlorophyta, Euglenophyta, Rhodophyta, or Phaecophyta,[9] this review has centered on three major taxonomic divisions: (1) the marine algae including the dinoflagellates (Pyrrophyta) that produce poisons causing shellfish and some fish to become poisonous, (2) the blue-green algae (Cyanophyta, Cyanobacteria) that produce poisons in the freshwater lakes throughout certain areas of the world and in some oceans, and (3) the yellow-brown algae (Chrysophyta, phytoflagellates) that occur in brackish water ponds and estuarine waters and produce toxins that kill fish.

PYRROPHYTA

The dinoflagellates produce a variety of toxins. The most lethal is the paralytic type (see "Paralytic Shellfish Poisoning") that has caused considerable sickness and death of humans and also has caused much damage in fish kills along the coast of Florida. The known poisonous dinoflagellates that produce paralytic poisons are marine species of the genus *Gonyaulax,* namely *G. catenella, G. tamarensis* var. *excavata, G. acatenella,* and *G. monilata.*

CYANOPHYTA

Table 1 lists some of the poisonous blue-green and yellow-brown algae along with their usual location and characteristics. Toxins from strains of three species of freshwater cyanobacteria have received the most extensive study to date. The blue-green algae that have caused public health problems and heavy economic loss by killing domestic animals are *Microcystis aeruginosa, Anabaena flos-aquae,* and *Aphanizomenon flos-aquae.* Although the number of animals affected in *An. flos-aquae* poisonings is more dramatic, the geographic distribution of *M. aeruginosa* is wider.[6] These organisms produce blooms in shallow freshwater lakes throughout the world and have killed thousands of cattle, horses, sheep, hogs, fowl, and dogs that drank water containing large numbers of these organisms. Concentrations usually become high on the leeward side of freshwater lakes after heavy algal growths. Outbreaks have been reported in

Table 1
POISONS PRODUCED BY VARIOUS ALGAE

Species	Common location	Biological characteristics	Chemical characteristics
		Cyanophyta; Blue-green Algae	
Microcystis aeruginosa	Shallow freshwater lakes	Toxic to mammals FDF (10)	Cyclic polypeptide of 10 amino acid units; mol wt about 1200 daltons; Water soluble (2)
Anabaena flos-aquae	Shallow freshwater lakes	Toxic to mammals and waterfowl VFDF (10)	Water soluble; $C_{10}H_{15}NO$; mol wt 165 daltons; see structure Figure 1 (7)
Aphanizomenon flos-aquae	Freshwater lakes	Toxic to mammals; physiological action very fast and similar to that of saxitoxin	Water soluble; similar to saxitoxin (12)
Schizothrix calcicola	Pacific Ocean	Produces sign in animals like symptoms of ciguatera poisoning in people	Structure determined called debromoaplysiatoxin; see structure Figure 1 (13)
Caulepra	Western Pacific Ocean	Dizziness in man	Not characterized
Lyngbya majuscula	Pacific Ocean	Contact dermatitis	Structure determined called lyngbyatoxin A; see structure Figure 1 (3)
		Chrysophyta; Yellow-brown Algae; Phytoflagellates	
Prymnesium parvum	Brackish water ponds and estuarine water	Toxic to fish and all gill-breathing animals; potent hemolytic agent	Water soluble, nondialyzable, thermolabile, appears to be a proteo-phospho-lipid (21)

North America, particularly from North and South Dakota, Minnesota, Iowa, Wisconsin, Illinois, Montana, Wyoming, and Colorado in the U.S. and from the provinces of Alberta, Saskatchewan, Manitoba, and Ontario in Canada. Similar blooms of poisonous algae have been reported in Russia, Brazil, Australia, and South Africa.[16]

Of the several poisons that have been identified in blue-green algal cultures, only a few have been isolated in pure form and partially characterized. The cyanophyta toxins are characterized by their effects on mice. One poison, called the fast death factor (FDF), produced by *M. aeruginosa* was extracted with water from decomposed algal cells and subjected to paper electrophoresis, first under acidic conditions and then under alkaline conditions. This process yielded a highly purified toxic component that killed mice in 30 to 60 min at a MLD of 9 μg/20-g mouse.[10] Because the FDF formed only on the decomposition of the algal cells, it is believed to be an endotoxin derived from cellular material. The microcystin toxins have been characterized as peptides of molecular weights ranging from 500 to 1700.[11] One is a cyclic polypeptide, with a molecular weight of about 1200, that has been further characterized to be made up of 10 amino acid residues as follows: 1 residue each of L-aspartic acid, D-Serine, L-valine, and L-ornithine and 2 residues each of L-glutamic acid, L-alanine, and L-leucine.[2] The occurrence of the unnatural amino acid D-serine in the peptide is interesting. The specific mode of action for the FDF in domestic and wild animals is unknown. The FDF is toxic to a great variety of animals except waterfowl, indicating that the waterfowl may have a particular mechanism to protect it against the toxin. In most animals the FDF produces pallor followed by violent convulsions, prostration, and death. Gross signs of poisoning in the liver are manifested by swelling, mottling, and dark coloration as well as the pooling of red blood cells in the liver.[6] In some areas, such as Bangladesh, *M. aeruginosa* in the drinking water has been linked to localized cases of diarrhea. Although no known vector connecting *M. aeruginosa* to the human food chain has been found, the implication of acute hepatitis, jaundice, hepatic degeneration, and cirrhosis in rats[9] alerts investigators and public health officials to the potential hazards of chronic human exposure to blue-green algae. The isolation of a slow death factor (SDF) from this same algal species has been linked to the bacteria associated with the algae.[2] A MLD of SDF causes death in mice within 48 hr.

The poisons produced by *Anabaena flos-aquae*, anatoxins, act more rapidly than the FDF and was termed the very fast death factor (VFDF) by the Canadian workers.[10] The anatoxins are alkaloids, polypeptides, and pteridines[4] Anatoxin-a, an alkaloid of molecular weight 165 daltons, is an an agonist at the nicotinic acid receptor and causes death by respiratory arrest.[5] The VFDF from *An. flos-aquae* will kill waterfowl as well as all mammals tested or observed from accidental poisonings and will kill mice in 15 to 20 min at a MLD with death preceded by paralysis, tremors, and mild convulsions. VFDF has no antigenic nor antibiotic activity.[17]

The poison produced by another blue-green algae, *Aphanizomenon flos-aquae*, is similar in rapidity of action to the VFDF of *An. flos-aquae* and occurs in many lakes and water reservoirs. One toxin from this organism appears similar in its action and chemical characteristics to saxitoxin, produced by one of the poisonous dinoflagellates, *Gonyaulax catenella*[12] (see "Paralytic Shellfish Poisoning"). This similarity is not an anomaly as it has been shown that some poisons from one species or genus within a phylum may be similar or even identical with those from another phylum.[14] As with the other blue-green algal toxins discussed above, *Ap. flos-aquae* toxins show heterogeneity and are being further characterized. These aphantoxins appear to be potent membrane poisons with effects similar to the biotoxins derived from marine dinoflagellates.[1]

Certain species of *Caulapra* and *Schiathrix calcicola* produce poisons that get into the large edible fish through the food chain from smaller plankton-feeding fish. These poisons are among those believed to cause certain disease in people such as ciguatera poisoning.[15] Although acute oral toxicity to humans has not been documented, gastroenteritis and contact irritations from recreational and municipal water supplies have been attributed to lipopolysaccharide endotoxins produced by *S. calcicola* and *An. flos-aquae*.[6] Thus, of the freshwater blue-green algae, the alkaloid neurotoxins may not pose as much of a major human problem as the polypeptide and lipopolysaccharide endotoxins.

The marine blue-green algal toxin from *Lyngbya majuscula* has been associated with contact dermatitis. The indole alkaloid, lyngbyatoxin A, and debromoaplysiatoxin have been isolated from this marine species and both toxins act as tumor promotors and increase epidermal ornithine decarboxylase activity, an activity associated with fast-growing neoplasms.[8] Debromoaplysiatoxin and oscillatoxin A (31-nordebromoaplysiatoxin) have been isolated from the marine blue-green algae *S. calcicola* and *Oscillatoria nigroviridis*.[13] The epiphytic association of the toxic Oscillatoriaceae with edible algae poses a potential danger to man although there is no direct evidence for a causal relationship between human cancer and the toxins of Oscillatoriaceae.[13]

CHRYSOPHYTA

Another poison of considerable commercial interest is one produced by the yellow-brown algal *Prymnesium parvum* that grows in the brackish water ponds and estuary waters in certain areas of the world. The toxin, prymnesin, is composed of at least six separable toxic components.[22] The chemical properties of pyrmnesin are similar to

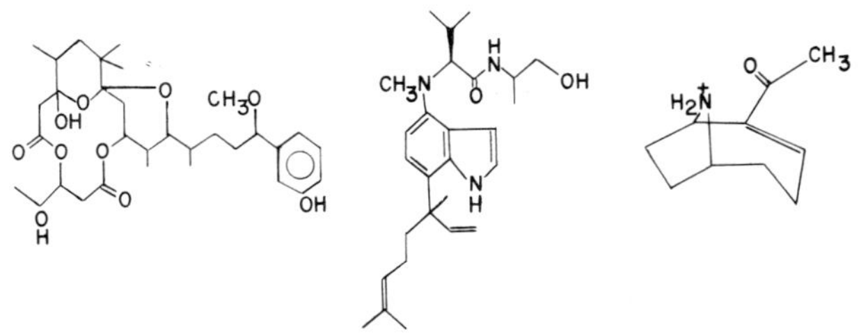

FIGURE 1. Structure of some algal toxins; debromoaplysiatoxin from *S. calcicola* lyngbyatoxin A from *L. majuscula* anatoxin A from *An. flos-aquae.*

acidic polar lipids and have been characterized as containing 15 amino acids, unidentified fatty acids, 0.47% phosphate, and 10 to 12% hexose sugars.[23] Speculation that the proteo-phospho-lipid toxin might be a membrane precursor suggests that the rare toxicity in nature is due to an over-synthesis of a normal membrane intermediate.[21] The broad biological spectrum of toxic activity includes cytotoxic, hemolytic, neurotoxic, and ichthyotoxic activities. The activity expressed depends on the type of lipid-protein complex that the toxin is exhibiting at a given time.[20] The amphipathic properties of the toxin molecule are the basis for the molecular action of the toxin on biological membranes. The toxin alters gill permeability of teleost species, aquatic invertebrates, and the gill-breathing stage of amphibians and is lethal because it inhibits the transfer of oxygen across the gill membrane. The toxin is nondialyzable, thermolabile, and believed to be a saponin-like compound that is a potent hemolytic agent.[19] Prymnesin has been an endemic problem in the commercial carp tanks in Israel and other countries where fish are raised in ponds. Other toxic Chrysophyceae include various species of *Ochromonas* which have icthyotoxic and hemolytic components and *Phaeocystis pouchetii,* whose toxic component, acrylic acid, is responsible for the absence of GI microflora in the antarctic penguin.[20]

REFERENCES

1. Alam, M., Ikawa, M., Sasner, J. J., Jr., and Sawyer, P. J., *Toxicon,* 11, 65, 1973.
2. Bishop, C. T., Amet, E. F. L. J., and Gorham, P. R., *Can. J. Biochem. Physiol.,* 37, 453, 1959.
3. Cardellina, J. H., Marner, F. J., and Moore, R. E., *Science,* 204, 193, 1979.
4. Carmichael, W. W. and Gorham, P. R., *Mitt. Int. Ver. Limnol.,* 21, 285, 1978.
5. Carmichael, W. W., Biggs, D. F., and Peterson, M. A., *Toxicon,* 17, 229, 1979.
6. Carmichael, W.W., in *The Water Environment: Algal Toxins and Health,* Carmichael, W. W., Ed., Plenum Press, New York, 1981, 1—13.
7. Devlin, J. R., Edwards, O. E., Gorham, P. R., Hunter, N. R., Pike, R. K., and Stavrick, B., *Can. J. Chem.,* 55, 1367, 1977.
8. Fujiki, H., Mori, M., Nakayasu, M., Terada, M., Sugimura, T., and Moore, R. E., Submitted for publication, 1980.
9. Gentile, J. H., *Microbial Toxins,* Vol. 7, Kadis, S. Ciegler, A., and Ajl, S., Eds., Academic Press, New York, 1971, 27—66.
10. Gorham, P. R., *Am. J. Public Health,* 52, 2100, 1962.
11. Gorham, P. R. and Carmichael, W. W., *Prog. Water Technol.,* 12, 189, 1980.
12. Jackim, E. and Gentile, J., *Science,* 162, 915, 1968.
13. Moore, R. E., in *The Water Environment: Algal Toxins and Health,* Carmichael, W. W., Ed., Plenum Press, New York, 1981, 15—23.
14. Russell, F. E., *Fed. Proc.,* 26, 1206, 1967.

15. Schantz, E. J., *Biochemistry of Some Foodborne Microbial Toxins,* Mateles, R. I., and Wogan, G. N., Eds., MIT Press, Cambridge, 1967, 51—65.
16. Schantz, E. J., in *Properties and Products of Algae,* Zajic, J. E., Ed., Plenum Press, New York, 1970, 83—96.
17. Schantz, E. J., in *Toxic Constituents of Animal Foodstuffs,* Liener, O. E., Ed., Academic Press, New York, 1974, 111—130.
18. Schwimmer, M. and Schwimmer, D., *The Role of Algae and Plankton in Medicine,* Grune & Stratton, New York, 1955.
19. Shilo, M. and Rosenberger, M., *Ann. N.Y. Acad. Sci.,* 90, 866, 1960.
20. Shilo, M., *Microbial Toxins,* Vol. 7, Kadis, S., Ciegler, A., and Ajl, S., Eds., Academic Press, New York, 1971, 67—103.
21. Shilo, M., in *The Water Enmironment: Algal Toxins and Health,* Carmichael, W. W., Ed., Plenum Press, New York, 1981, 37—47.
22. Ulitzur, S., *Verh. Int. Ver. Limnol.,* 17, 771, 1969.
23. Ulitzur, S. and Shilo, M., *Biochim. Biophys. Acta,* 201, 350, 1970.
24. Wolk, C. P., *Bacteriol. Rev.,* 37, 32, 1973.

Foodborne Diseases

FOODBORNE DISEASES: VIRAL INFECTIONS

Dean O. Cliver

INTRODUCTION

The transmission of viruses through foods has been known at least since 1914.[1] Viruses capable of being foodborne are typically those transmitted by an "anal-oral" cycle, which may include direct anal-oral contract,[2] other "contact" transmission, or indirect transmission such as that in which food and water serve as vehicles. These viruses enter the body by ingestion, are generally produced in the intestine (though they may not cause symptoms there), and are shed with feces. Fecal contamination of foods is known to occur both directly and by way of fecally contaminated water.

Compilation from figures available for the U.S. for the years 1974 to 1978 indicates that outbreaks of proven foodborne viral disease comprised 2.5% of the foodborne outbreaks of known etiology and included 3.2% of the cases of illness in such outbreaks. The majority of outbreaks during the period were of undetermined etiology, and some of these were probably viral gastroenteritis. Many viruses that might be transmissible through foods are likely to cause disease involving organs remote from the digestive tract and thus not be recognized as foodborne by clinicians, and much foodborne disease that takes place is not reported through official channels. In all, it seems likely that a great deal more foodborne viral disease is occurring in the U.S. than would be indicated by the average reported totals of approximately 5 outbreaks and 190 cases per year, but it is important to note that the preeminent mode of transmission for these viruses is almost certainly by contact, rather than via food. Comparable figures for other countries do not seem to be available.

FOODBORNE VIRAL DISEASE

A look at the record of foodborne viral disease in humans should serve to justify some of the generalizations made above.

Poliomyelitis

The first recorded foodborne viral disease seems to have been poliomyelitis.[1] At least ten outbreaks have been recorded;[3] the vehicle in eight of these was milk, either consumed without pasteurization or contaminated after pasteurization. The three serotypes of poliovirus are members of the enterovirus group, which comprises small (about 28 nm), round viruses that contain single-stranded RNA, lack a lipid envelope, and withstand acidity in the range of pH 3 to 5 for reasonable periods of time. Classical paralytic poliomyelitis is the result of viral infection of the CNS, but the viruses are transmitted by the anal-oral route and infect the small intestines or (with large doses of virus) the oropharynx after ingestion; the infection may extend to the CNS in only a small proportion of cases, the rest of which are subclinical or inapparent. In recent years, clinical poliomyelitis is virtually unknown in the U.S. (though it is quite common among children in developing countries) because of the success of the vaccines that are now in use, and foodborne poliomyelitis has not been recorded in the U.S. since 1949. The oral, attenuated vaccine is used almost exclusively in the U.S., whereas some other developed countries have used only the formalin-inactivated vaccine. Either can eliminate the disease and, with sufficiently diligent use, the "wild-type" virus; but random, year-round use of the oral vaccine, as is practiced in the U.S., guarantees the presence of the virus in the sewage of many communities to the extent that the vaccine polioviruses have been proposed as indicators of virus contamination in water[4] and, perhaps, food. The oral vaccine has been relatively unsuccessful in developing countries to date, so the polioviruses (and other enteroviruses) infect people early in life and are probably common in the environment. Even in developed countries, the polioviruses and other enteroviruses are frequently detected in the water and sediments of estuaries and in the shellfish that grow in the estuaries.

The only enterovirus, among established members of the group other than the polioviruses, that has been implicated in an outbreak of human disease is echovirus 4, which seems to have caused 80 cases of aseptic meningitis among persons who ate cole slaw at a picnic.[5] The mode of contamination of the cole slaw was not determined.

Hepatitis A

The viral disease most frequently reported to have been transmitted through foods is Hepatitis A (HA). It was formerly called "infectious hepatitis" and was not recognized as a distinct clinical entity until the time of World War II. The disease is unusual in that its incubation period ranges from 15 to 50 days, with a median of 28 days. Anal-oral transmission is the rule, with direct transmission from person to person most common; however, more than 80 food-associated outbreaks have now been recorded. Viral replication in the liver frequently leads to jaundice, though milder forms and inapparent infections are common. Many of those affected are debilitated for long periods of time, but chronic carriers and shedders of the HA virus (HAV) are not known. The virus is worldwide in distribution and is quite specific for the human species; antibody against HAV has been found in nonhuman primates that are not known to have been exposed to people, but there is no evidence that any nonhuman species plays a significant role in harboring HAV or transmitting it to humans.

Much recent progress has been made in characterizing HAV, which shows many of the properties of the enterovirus group. It has only recently been replicated demonstrably in cell cultures,[6] and it does not cause visible deterioration of the host cells when replicating in vitro. Such propagation may eventually lead to the production of a vaccine, which is a very attractive prospect because only one serotype of the virus is known to exist, and worldwide eradication may be possible. For the time being, the main use of virus derived from cell cultures will probably be as an antigen in the serological methods (radioimmunoassay, RIA, and enzyme immunoassay, EIA) that are now being used in the diagnosis of the disease. Not all field strains of HAV will replicate in cell cultures, so the cell culture procedure itself will probably find limited application in diagnosis unless further improvements are achieved.

The first specifically described foodborne outbreak of HA took place in Scotland in 1943[7]; raw milk was the vehicle. A major stride was the description of the transmission of the disease through shellfish: a large outbreak involving oysters in Sweden was observed late in 1955.[8,9] The majority of outbreaks now on record have been reported in the U.S., but there is every reason to believe that other countries are at least as commonly the sites of such incidents. The record makes it clear that any food mishandled by an infected person can serve as the vehicle for HAV if it is not cooked subsequent to contamination. Salads, fruits, luncheon meats, sandwiches, and many other foods have served in this role. However, the premier vehicle for HAV through the years has been shellfish (bivalve mollusks). Among the factors contributing to this are (1) the increasing pollution with human excrement of many waters in which shellfish grow, (2) the ability of shellfish to concentrate viruses from their environmental waters, although they do not actually become infected, (3) the fact that the shellfish digestive tract, where the majority of the virus probably remains, is eaten along with the rest of the soft tissues of the body, (4) the common practice of eating shellfish raw or with minimal cooking, and (5) the apparently exceptional protection afforded by shellfish to viruses that are present in them during heating.[10] Perhaps surprisingly, shellfish were implicated as the vehicle in only 1 of the 20 HA outbreaks reported in the U.S. during the years 1974 to 1978.

The food handler is the principal known source of HAV in foods, although polluted water has been known to contaminate foods other than shellfish. HAV is shed in the feces of an infected person as early as 14 days before onset of symptoms, peaks at 8 to 10 days before onset, and declines rapidly after symptoms begin. Thus, those who handle food (even nonprofessionally) during their incubation period or an inapparent infection are a threat to those who consume the food without further cooking, at least if the food handler has not scrupulously maintained proper food handling practices and personal hygiene throughout the period.

On the basis of observations in experimentally infected chimpanzees and marmosets, it has been suggested that HAV replicates solely in the liver, and not in the intestine.[11] This would be quite significant to the environmental transmission of HAV if true, for it would explain the virtual absence of the virus from urine and respiratory secretions and would suggest that HAV present in feces as a result of shedding via the common bile duct would not be associated with sloughed cells of the intestinal mucosa, as seems likely to be true of the other enteroviruses. On the other hand, it is extremely hard to imagine how HAV could infect perorally, with the efficiency that it apparently does, if no cells capable of replicating the virus are directly exposed to virus passing down the digestive tract with the ingesta.

Gastroenteritis

In recent years, a great variety of viruses have been proven to be capable of causing acute infectious nonbacterial gastroenteritis (AING) in humans. The AING viruses seem to vary in size, presence of a lipid envelope, and nucleic acid type; a property that many of them share is apparent inability to replicate in laboratory cell cultures. The majority of foodborne disease reported in the U.S. and probably in other countries is of undetermined etiology, and most of this is gastroenteritis because diseases involving parts of the body remote from the digestive tract are unlikely to be suspected of being foodborne. The only viruses definitely implicated as causes of gastroenteritis transmitted through foods are small (about 28 nm), round viruses resembling the "Norwalk agent".[12] These are being incriminated on the basis of electron microscopic examination of the stools of affected persons, though the detection of similar particles in the extract of an oyster from one outbreak has been reported.[12] The vehicles in most of the documented outbreaks in the U.K. have been shellfish, particularly cockles.[13] Where other foods have been involved, a substantially lower proportion of those affected have been shown to be shedding the small, round viruses in their stools than in outbreaks where the vehicle was a shellfish.[12] An outbreak of oyster-associated AING apparently caused by similar viruses has affected over 1000 people in Australia.[14] Diagnosis has been found by the Australian investigators to be reliable when antibody to the virus is sought, rather than the virus itself.[15] The small, round viruses of AING, including the "Norwalk agents", have been tentatively assigned to various groups, most recently the calicivirus group.[16] In that the methods available for the detection of these viruses are based rather on their size and antigenicity than on their infectivity, these procedures are much more likely to succeed in demonstrating virus in stool samples than in foods, where considerable dilution of the fecal contaminant would have taken place before testing. Therefore, the detection of echovirus 8 (an enterovirus) and a reovirus in a batch of oysters from an outbreak of small, round virus-induced AING in Australia is significant, not because the viruses detected could have caused the symptoms, but because these viruses are indicators of the viral contamination of the oysters and might afford a basis for future screening of shellfish to determine whether they are safe.[17] No source of food contamination by the AING viruses other than wastewater has yet been implicated, but it is reasonable to suppose that contamination through mishandling by an infected person is as likely to occur as with HAV. Therefore, it should be noted that the detection of other viruses as an indirect indication of contamination by the AING viruses is only likely to be valid where the source of the contaminant is wastewater that will contain the pooled excreta of many persons — a single food handler is unlikely to harbor more than one virus at a time, so the presence, after mishandling, of a virus detectable in cell culture is unlikely to signal the presence of an AING virus.

Other Agents

Other Human Enteric Viruses

No human viruses not of intestinal origin are known to have been transmitted through foods. In addition to the agents of specific disease that originate in the human intestines, other human enteric viruses are sometimes detected in foods, as was reported for the reovirus found in contaminated Australian oysters. The reoviruses are medium-sized (about 80 nm) viruses that have a double protein coat but no lipid envelope, and contain double-stranded RNA that breaks into ten segments under some analytical conditions. At least three serotypes of reoviruses are known to infect humans; some of these seem to be potentially transmissible among species, but none has been definitely associated with frank disease in any of the species that it infects. Adenoviruses and coronaviruses that replicate poorly or not at all in cell cultures have been identified in diarrheal stool specimens and are, therefore, likely to contaminate foods upon occasion, either directly or by way of water polluted with feces; these agents have not been seen in foods nor shown epidemiologically to be transmitted by foods to the present time.

Zoonoses

Several viral and rickettsial agents of food-source animals are also capable of causing disease in humans. Any of these zoonotic agents may cause illness in those who are directly in contact with diseased animals (e.g., farmers and veterinarians); fortunately, few of these are known to have been transmitted through foods to consumers. Two nonviral agents are discussed in this section because they are obligate intracellular parasites and are, therefore, studied by the same procedures as are used with viruses.

The tick-borne encephalitis (TBE) viruses of central Europe infect dairy animals as a result of tick bites and can then be shed via the mammary gland.[18] Those who drink the milk or milk products without complete pasteurization are liable to be infected perorally. Milk from goats and, more recently, sheep, has been implicated epidemiologically, but cows that were infected experimentally with the virus have also been shown to shed it with their milk.

Q-fever, a disease caused by the rickettsia, *Coxiella burnetti,* affects cattle and other food-source species, some of which may become infected as a result of tick bites. The agent is essentially completely destroyed by proper pasteurization,[19] but it has been known to cause disease in those who drink raw milk from infected cows.[20]

Chlamydia psittaci, the agent of ornithosis, causes respiratory disease occasionally in workers in poultry slaughter plants, especially those engaged in killing and evisceration.[21] Turkeys and, to a lesser extent, chickens, are the species most often involved in the U.S., whereas ducks and geese are more frequently implicated in Europe. Persons who dress infected poultry in their homes are apparently also at risk, but the agent has not been known to infect perorally.

Viruses of Animal Diseases

Several viruses that cause diseases in animals, but essentially do not affect man, have been found in foods or have been shown epidemiologically to be transmitted among animals by way of the human food supply system. The viruses of foot-and-mouth disease are very important in this connection; they are capable of withstanding extremely rigorous treatments in some animal and milk products and of being transmitted internationally in these products to countries such as the U.S. from which the disease has been eradicated.[22] The foot-and-mouth disease viruses are manifestly no threat to human health, but they have a very significant effect upon human well-being as a result of the direct costs of the losses of animal productivity that they cause and of the indirect costs of the restraints to international trade in animal products that are imposed in an attempt to prevent the spread of the disease. Other viruses that cause economically significant animal disease might also be mentioned here, but none of them is nearly as important by the above criteria as foot-and-mouth disease.

A second important class of animal disease agents that sometimes occurs in foods is the group of viruses associated with cancer in animals (oncogenic viruses). These are of particular concern because of the public's special sensitivity to the presence in the food supply of anything regarded as a cause of cancer. Fortunately the oncogenic viruses of animals appear to be quite host specific and do not seem to be capable of affecting humans either through contact or through the consumption of food products containing the viruses. For example, quite intensive surveys of persons exposed strongly to the bovine leukosis agent, which may be present in the meat and the milk of affected animals, has shown that people probably cannot be infected with this virus.[23] The viruses of avian leukosis are present in tumor tissue, but also in apparently normal tissue of infected birds, and can be transmitted vertically to the next avian generation through eggs;[24] despite the presence of these viruses in many foods consumed by humans, there is no indication that people have ever been infected. One agent that does not originate in a food-source species is the polyoma virus. It infects mice, sometimes with production of a variety of tumors, and is shed for long periods in the urine of infected animals.[25] Stored grain to which mice had access has been found contaminated with the polyoma virus; but again, the agent is not known to infect humans, and no transmission via food has been observed.

PREVENTING VIRUS TRANSMISSION THROUGH FOODS

The viral particles that sometimes contaminate food are small (about 28 to 100 nm in diameter) and roughly sperical. They consist of RNA or DNA (and sometimes a few enzymes) coated with protein and, rarely, an outer lipid envelope. The particles are produced exclusively in susceptible living host cells, most often, as has already been noted, those of the human intestines. These particles are capable of withstanding the rigors of the environment outside the host, including that in some foods, for periods of time; however, they are totally inert as they occur in food and cannot multiply or carry out any life processes. *Coxiella burnetii* and *Chlamydia psittaci,* which have also been mentioned, are bacteria that originate in food-source animals and resemble viruses in being obligate intracellular parasites. Concern over the presence of these agents in foods stems from their potential to infect perorally and, perhaps, to produce disease in consumers.

Contamination of Food

The presence of any virus in food is, obviously, undesirable; but the degree of threat to the consumer that viral contamination of food represents must be evaluated on the basis of the kind and quantity of virus that is present. Viruses that infect humans following ingestion must usually establish an infection first in the small intestines. Illness need not result from the viral infection, but if it does, it will either stem from the local infection in the intestinal tract or from a secondary infection of another part of the body. The viruses of AING produce illness rapidly (often within 24 to 48 hr) and are often associated with high attack rates among those who consume a contaminated food. Viruses such as HAV that produce disease only when they have succeeded in reaching and infecting an organ remote from the digestive tract are associated with longer incubation periods before onset of symptoms and with higher rates of inapparent infection. In any case, the probability of producing an infection in any given consumer will be low unless a good deal of virus is present in the food; the further probability that overt disease will result from the infection is a function of host factors including, but not limited to, specific preexisting immunity, as well as the inherent virulence of the virus.

It has already been stated that the most frequent source of foodborne viruses is the human intestines and that the most significant mode of contamination is through mishandling by an infected person. Such generalizations are based largely upon the extensive record of HA transmission through foods, over half of which was reported from the U.S. It is possible that some other foodborne viral diseases may not be transmitted in identical fashion and that some other countries, particularly developing countries, may not show identical patterns of foodborne disease transmission. Despite these reservations, an extensive record does exist, and it should be of some use for the lessons it teaches. The HAV may be shed by an infected individual for as long as 14 days before illness begins, and peak levels of fecal virus are now known to occur well before the onset of symptoms. It is obviously undesirable for ill persons to handle foods, and in the case of foodborne HA, this has certainly been shown to have happened quite often; but the damage may well have been done far enough in advance of illness that little is gained by excluding the person from handling food once illness begins. This observation, and the fairly frequent incidence of inapparent infection with HAV, underscores the importance of good personal hygiene and proper food sanitation among those who handle food, whether professionally or not. If food is cooked or otherwise treated after contamination, there is a fair prospect that any contaminating virus will be inactivated (deprived of its potential to produce infection) as will be discussed below; therefore, the nearer to the time of being eaten the food is contaminated, the more critical the contamination is. It is not surprising, then, that the majority of food handlers who have been implicated in HA outbreaks have been engaged in food service or final preparation. The mode of contamination in these instances would appear to have been from feces, via fingers, to the food. Sometimes the number of cases that occur in an outbreak due to food mishandling denotes quite a degree of fecal contamination; for example, a food handler who apparently carried his concept of organic food to an extreme was implicated in having caused a total of 140 cases among customers of two restaurants featuring such foods at which he worked.[26] Clearly, contamination of food by way of fecally soiled fingers should be avoided; however, studies in the author's laboratory[27] have not revealed a preparation, including soap and water, that would guarantee the removal of virus-containing fecal soil from fingers to a degree that would prevent contamination as a result of a touch. Nevertheless, hand washing is better than none. Disposable plastic gloves, properly used, can prevent passage of fecal virus to or from the fingers. Whatever precautions are instituted, they must be used without fail if they are to protect the consumer.

The second most common mode of contamination of food, based upon the record of HA outbreaks, is by way of water contaminated with human feces. The food most often contaminated in this way has been shellfish, for the reasons enumerated above. However, shellfish have also been known to become contaminated through mishandling, and other foods have been known to become contaminated as a result of exposure to sewage. For example, a backup of sewage into a delicatessen contaminated cold cuts and touched off an outbreak that eventually included 67 cases of HA.[28] Several additional incidents are discussed elsewhere.[29] Though contamination of vegetables, via their roots, with virus from polluted irrigation water has been proposed, there is no solid experimental epidemiological evidence that this is of practical significance.

In addition to water as a proximate source of fecal viruses contaminating foods, one might well imagine that fomites and mechanical vectors would also play a role. Compartmented cafeteria trays, with which the food served on them came in direct contact, were contaminated in drying by an infected person and served as a fomes in a foodborne outbreak of HA.[30] However, no outbreak of HA that was clearly attributable to contamination by vectors such as flies that had access to human feces has yet been recorded. Food exposed as bait at the site of an epidemic of poliomyelitis attracted flies and became contaminated with poliovirus,[31] but it could not be demonstrated directly that any of those who had contracted the disease had gotten it as a result of eating fly-contaminated food. Thus, it seems reasonable to suppose that mechanical vectors such as flies may sometimes play a role in contaminating foods with enteric viruses, but no instance in which this clearly happened has yet been recorded. Insects that harbor enteric viruses may well be more common in developing countries than in the U.S., but so, probably, are waterborne and direct human contamination of foods.

Decontamination of Foods and Inactivation of Foodborne Viruses

Once a food has been contaminated with an enteric virus, consumer infections are quite likely if the food is eaten immediately. However, several things may occur or may be done intentionally to prevent virus transmission through food even after contamination has taken place.

Shellfish that harbor viruses as a result of having siphoned water that is contaminated with human feces do not become infected and do not replicate the virus.[32] As long as the animals are alive, they may rid themselves of the contaminant if they are given virus-free water to siphon.[33] However, individual shellfish eliminate virus at different rates, so a few may continue to harbor viruses after the majority have cleansed themselves completely.[34] One shellfish purification or decontamination process entails holding the animals on commercial premises for periods of time, while exposing them to water that has been decontaminated with UV light or with chlorine — this is called depuration. In a recent study with human volunteers, depurated oysters produced two outbreaks of Norwalk virus gastroenteritis.[35] Another shellfish purification process is called "relaying"; it involves moving shellfish that have been harvested in polluted waters to waters that are naturally free of pollution and leaving them there, sometimes for weeks, to cleanse themselves. This seems a less critical approach, in that time is not so costly but what it can be used liberally; however, no studies on the antiviral effectiveness of relaying seem to have been done. No comparable process is available for the decontamination of any other virus-contaminated food.

The viral load on surface-contaminated fruits and vegetables may be reduced by ordinary washing with water, assuming that the water used is free of viruses. Unfortunately, viral contamination of water may be common at locations where viruses on the surfaces of fruits and vegetables are a concern, and the water itself may be responsible for contaminating the surfaces of foods that were otherwise virus free. Even the best of water cannot be expected to remove all virus from a contaminated food surface; chemical disinfection processes for this purpose are currently under study, under the sponsorship of the U.S. Army Natick Laboratories. UV light is of some value in the antiviral decontamination of vegetable surfaces,[36] but has not been tested extensively in this application.

A contaminating virus is more likely to be present within the substance of other contaminated foods, either as a result of admixture or of penetration through a permeable surface from without. Virus inside a food will be inactivated only by some factor that is capable of penetrating the food; the most common of these is heat. Indeed, viruses will probably be inactivated at some rate at any temperature above absolute zero, but the inactivation rate will be virtually imperceptible for most enteric viruses in most foods at the temperature of a household freezer (about −18°C). The rate of inactivation of foodborne virus will generally increase with increasing temperature, but the chemical composition of the immediate environment of the virus has a strong influence upon the inactivation rate;[37] no general model for the interaction of food constituents, as they affect thermal inactivation of virus, has yet been devised. A further complication to the prediction of the effect of temperature upon inactivation rate is that the coat protein of the virus is the most labile moiety at some temperatures, whereas the nucleic acid is more vulnerable at others.[38] Nonetheless, it appears that most viruses will be completely inactivated within 30 min at a temperature of 55°C in milk;[39] important exceptions to this are the viruses of foot-and-mouth disease, which under some conditions of study show some residual infectivity after full pasteurization.[40] Thermal stability of viruses in foods seems to be favored by high levels of proteins and fats,[41,42] and added salts may afford a degree of protection to the virus,[43] but shellfish in particular seem to protect virus from very high temperatures despite the absence of anything in their general composition that might be expected to exert this effect. The report that a substantial portion of poliovirus accumulated experimentally by oysters was still infectious after frying[10] was lent some health significance by the epidemiologic observations that HAV had on occasion been transmitted by minimally cooked shellfish, such as stewed oysters, cooked just until their edges curled[44] and steamed clams,[45] and that apparently viral (though no virus was isolated from those affected) AING had been associated with cockles that had been "boiled 4 min."[46] When HAV was injected into oysters, substantial but not complete inactivation took place within 19 min at 60°C, as determined by feeding the heated suspension to marmosets.[47] Clearly, any heat that is applied to a food will inactivate at least some virus, assuming that the contamination occurred before the food was heated; of course, heat is of little use in the case of foods that are customarily eaten raw.

Like heat, ionizing radiation (e.g., cobalt-60 γ rays) is capable of penetrating contaminated foods. The dose needed to inactivate approximately 90% of an enterovirus contaminant is in the range of 4 to 5 kGy (0.4 to 0.5 Mrad).[48,49] The food in which a virus is suspended exerts some influence upon the antiviral effect of a given radiation dose, probably at least partly through the scavenging of free radicals.[50,51] Many food irradiation processes have been developed; foods are wholesome after processing with doses at least as high as 10 kGy (1 Mrad).[52] Perhaps most pertinent in the present context is the finding that irradiation at doses likely to be used in food processing probably will not cause mutations in viruses that may be present.[53]

Enteroviruses are generally subject to inactivation upon desiccation. At least 99% inactivation of experimentally inoculated enteroviruses was observed during freeze-drying of foods on a laboratory scale.[50,54] Various constituents of low-moisture foods, but not the level of moisture itself, influenced the stability of virus during subsequent storage.[54] Several viruses are extensively inactivated during air-drying of foods;[55,56] however, the foot-and-mouth disease viruses are not totally inactivated in foods by such processes as vacuum evaporation and hot-air drying of milk and milk products[40,57] or spray-drying of milk and of blood products to be used in foods.[58]

In contrast to the chemical constituents of foods that were described above, which stabilize viruses against thermal inactivation, some other constituents may tend to cause direct inactivation of contaminating virus. The action of the acid produced during rigor mortis against the viruses of foot-and-mouth disease in muscle is well known, and acid conditions in foods generally are probably antiviral when compared to foods whose pH is near neutral.[54] The enteric viruses tend to withstand acid better than alkali, but they are most stable at neutral pH. Food additives such as sodium bisulfite and ascorbic acid may also exert a direct antiviral effect under appropriate conditions.[59,60]

Another potential means of inactivation of foodborne viruses is by microorganisms or their enzymes in food. Some microbial enzymes and organisms have been shown to be capable of inactivating enteroviruses directly;[61] virus-containing foods are likely to contain microbes indigenous to the food, organisms that entered the food with the viral contaminant, or both. However, neither bacterial cultures used in food processing or normal spoilage organisms have yet been shown to exert a significant antiviral effect in experimentally contaminated foods.[62-65]

Foodborne viruses will be inactivated, gradually or rapidly, in any food that is stored above freezing temperatures; but the persistence of virus relative to the shelf life of the food becomes significant only if contamination takes place well before the time that the food is served and consumed. Foods are "stable" if they have been thermally processed to "commercial sterility" and then stored in a hermetically sealed container (in which case viral contamination is unlikely to have occurred unless spoilage organisms and perhaps bacterial pathogens are also introduced) or if their water activity is so low as to preclude the growth of spoilage organisms (in which case virus *may* be inactivated before the food is consumed). Foods that are "perishable" may either be marketed and consumed quickly, to prevent spoilage, or they may be stored at refrigerator or freezer temperatures, to extend their shelf lives; in either case, the time during which a contaminating virus persists is likely to exceed the shelf life of the food. Thus, only thermal processing or thorough cooking of food affords a reliable barrier against virus transmission; perishable foods are "safe" only if contamination is prevented.

Preventive Measures in Context

It has been shown that the transmission of viruses through foods may be prevented either by avoiding contamination of the food in the first place or by removing or inactivating the contaminating virus before the food is eaten. The experiences upon which the discussions are based are those from industrialized nations (principally the U.S.) that already have well-established systems of food and water sanitation and community hygiene. In such places, the anal-oral transmission of viruses is limited both by the sanitation measures themselves and by the relatively low incidence of virus carriage that results from good sanitation. That is, when there is a lapse in sanitation in the U.S., virus is less likely to be transmitted than in many other locations because the fecal pollutants are less likely to have viruses in them. Although it is almost certainly true that the majority of transmission of enteric viruses in industrialized nations is by direct contact, transmission via food or water should be avoided because it may lead to the introduction of unique serotypes of viruses into populations that have no preexisting immunity to them, with noteworthy results. On the other hand, it should perhaps be noted that most outbreaks of viral disease that have been associated with transmission through food or water have apparently been self-limiting before a great number of secondary cases occurred, so community-wide epidemics stemming from introduction of a new viral serotype by food or water have not been seen. Even allowing for sporadic cases of foodborne viral disease that are not part of recognized outbreaks, this mode of viral transmission is quite rare in industrialized nations.

The situation with regard to virus transmission through foods in developing countries is a matter for speculation; no authoritative data seem to be available on the subject. What is known is that enteric virus infections are common from early childhood in some countries and that these infections are almost certainly among the causes of the high infant mortality experienced in these places. Those who survive infection with a given virus are likely to be immune to that particular virus for the rest of their lives, but they may also show some permanent sequellae. Recent surveys have shown that "infantile paralysis", an early term for poliomyelitis, is very descriptive of the syndrome as it still occurs in many developing areas; children who have been infected with the poliomyelitis viruses at an early age may be found with residual paralysis many years after, and possibly for life. Thus, while viral infections are very prevalent in developing countries, they are not merely to be tolerated or taken for granted. Assuming that viruses do contaminate foods in these settings on occasion and that infections sometimes result, one cannot assert confidently that the elimination of all virus transmission via foods would have a perceptible effect upon the overall rate of viral infection or disease. Where rates of infection are high, feces disposal is minimally supervised, and the means for good overall community hygiene are often lacking, so many alternative modes of virus transmission exist that the total prevention of transmission via the food route alone probably would have little immediate impact. The first goals in reducing virus infections in a developing country should probably be making adequate provision for feces disposal and upgrading general community hygiene; whereas in industrialized nations, direct emphasis on food hygiene may produce significant results if the right problems are addressed.

DETECTION OF FOODBORNE VIRUSES

There are many occasions on which one might wish to determine whether a food has virus in it, either before any of the food has been consumed or in the aftermath of a possible foodborne outbreak of viral disease. Laboratory virology is expensive, so testing of food should be undertaken only when human illness has occurred or when there is a reasonably high probability to suspect that contamination has occurred. Otherwise, one may expend a great deal of effort testing foods without detecting any virus that represents a threat to human health.[66]

Detection Based on Infectivity

Viruses were first recognized as infectious agents, and for many years diagnostic virology was based primarily upon the detection of viral infectivity in clinical specimens. The detection of foodborne viruses has been based upon procedures adapted from diagnostic virology. In this context, one can say that virus has been detected when a perceptible infection has been produced in a living host system. Detection procedures based on infectivity are still used when possible because when they can be used at all they are sensitive to smaller quantities of virus than any competing approach and because it is, after all, the infectivity of the virus that may be present in food that causes concern.

The requirement for a living host system in which to detect viruses presents some problems. At times in the history of virology, it has been necessary for humans to serve in this role, but this is generally undesirable because of ethical concerns and because humans present a great variety of uncontrolled variables that make them poor research subjects. Laboratory animals, such as guinea pigs and suckling mice, have been used to some extent, as have embryonated eggs, but virology has come into its own as a science and as a branch of preventive medicine largely since the development of cell culture procedures. The advent of means for producing large numbers of cultures of essentially uniform cells for use in detecting, replicating, and quantitating viruses has enabled the production of enormous amounts of information which, when published, have been applied successfully by laboratories all around the world. That is, the reproducibility of results obtained with cell cultures is unequaled by those based on infection of whole host organisms.

Cells for culture are derived either directly from the tissues of an animal's body ("primary" cultures) or from other cell cultures after repeated passage ("established" cultures). In either case, the species of origin of the cells is usually one of the primates (including man) if viruses of human origin are to be detected. Cells are most often grown and maintained as monolayers on an inner surface of a glass or plastic vessel and bathed in a medium that maintains the pH and ionic balance and supplies the nutrient needs of the cells. Details will not be given here; many good texts on the subject are in print, but the methods can best be learned first in a laboratory where they are already in use. One might test a food sample for viral contamination simply by feeding it to a human or animal subject, but procedures for testing in cell cultures are somewhat more involved. First, the food sample must be liquefied, if it is not already a liquid. Second, food solids, and anything else in the suspension that may be deleterious to the cell cultures, must be removed from the suspension, either by physical or chemical means. Finally, excess fluid is often removed from the sample suspension so that testing can be done in the smallest possible number of cultures and still include enough of the original food sample to afford a reasonable prospect of obtaining a positive test result. This final concentration step is justified only if the savings, in terms of the cost of the cell cultures that are not to be used, exceed the cost of the concentration procedure. More extensive presentations on detection methods in food virology can be found elsewhere.[29,67,68] For the present discussion, it will suffice to say that every step in the procedures described is critical from the standpoint of keeping the virus, in infectious form, in the fraction that is saved, rather than in the fraction that is discarded. Even if one succeeds in producing a small-volume extract that contains all of the virus that was in the original food sample, the virus will only be detected if the extract is inoculated into the proper type of cell culture (no virus of concern in foods destined for human consumption will express itself in all of the types of cell cultures that might be used for testing, and some of the most important viruses do not express themselves in any known type of cell culture), which is then properly maintained and observed.

Infectivity procedures that use cell cultures to detect foodborne viruses present three major problems. First, the cost of the cultures and of preparing the samples for inoculation into them is large, though perhaps not as large as if other living host systems were used instead. Second, the type of cell culture that is inoculated may not be susceptible to the virus that is present. Finally, the replication rates of animal viruses are so slow that days to weeks will often have elapsed before the results of the test are known. These points are underscored principally to emphasize the statement, made earlier, that testing foods for viral contamination should only be undertaken for very good reasons.

Detection Based on Serology

It has already been stated that some of the most important of foodborne viruses (namely, HAV and the small, round viruses of AING) either do not replicate in any known type of cell culture or replicate without producing any discernible effect. Therefore, following the lead of diagnostic virology, one might best undertake to detect these agents on the basis of their antigens. Serological detection is a reasonable second choice after infectivity from the standpoint of sensitivity, because the other alternatives are physical or chemical detection methods, and viruses are not so unique in biologic contexts that the particles could be detected at low levels in foods on any other basis than the antigens of their coat proteins.

Diagnostic procedures for these viruses are now focused on the presence of the viruses in stools (where they occur at much higher levels than are likely to be found in foods) and sometimes in tissues, or on the demonstration of antiviral antibody in the blood serum or the stools of affected persons. The serologic procedures used have included EIA and RIA, which have already been mentioned,[11,15] and immune electron microscopy,[15] which improves upon the the specificity of electron microscopy in detecting the viruses in stool extracts by demonstrating a serologic interaction between what appear to be viral particles and known antibody to a given virus. If such procedures are to be used to detect foodborne viruses, they will succeed only where the levels of contaminating virus are quite high and excellent extraction methods are used in preparing the sample for the serologic test. Thus, shellfish, with their known ability to concentrate virus from their environmental waters, are probably most likely to yield positive tests. Other foods are likely to be positive only if they have been directly contaminated to an extreme degree.

Even though the serologic tests will only rarely detect viral contaminants in foods, a great deal is being learned about HAV and the small, round AING viruses, and their transmission through foods, as a result of the increased use of the serologic methods to confirm diagnoses, for example, during apparent outbreaks of foodborne disease. The procedures designed to detect antibody against the virus are applicable in such epidemiologic investigations, whereas they are of no use in monitoring contamination of food.

The sensitivity of the serological procedures could be improved greatly if there were cell cultures available that would reliably replicate HAV or the small, round AING viruses as they occurred in environmental samples, even though the cell showed no effect of the viral infection. Such cell cultures would serve in the role of an "enrichment", simply increasing the amount of viral antigen that was available to be demonstrated by the serologic reaction. If this enrichment step could be carried out fairly rapidly (e.g., within 1 to 2 days), it would not greatly postpone the results of the test; rapidity is one of the greatest advantages of serologic testing, as contrasted to infectivity testing, which may often take weeks. Unfortunately, present methods for the replication of HAV take weeks and are not successful with some specimens that are known to contain virus.[6]

A further property of serologic tests that might be mentioned here as both an advantage and a disadvantage is specificity. Specificity is an advantage in that a positive test result (assuming that proper controls were included) is definite proof that the virus in question was present in the sample that was tested, though of course not necessarily in infectious condition. On the other hand, the specificity of the test procedure requires that a separate test be performed for each of the viral types that one hopes to be able to detect. Furthermore, a test that is based on the use of known antiserum will never permit the detection of an unknown virus; there are ways to surmount this, but they are indirect and potentially less reliable than the procedures that were described first.

C. Indicators of Viral Contamination

The foregoing discussion indicates that viruses are unusually complicated to detect in foods, even when compared to other potentially foodborne pathogens. Therefore, it is not unreasonable to look for some other, perhaps bacterial, indicator whose presence would signal a high probability of viral contamination. Many bacterial indicator systems have been examined from this standpoint, especially those that are supposed to indicate fecal contamination because most foodborne viruses that threaten human health are of intestinal origin. It seems possible that the presence of fecal indicator organisms (e.g., fecal coliforms or fecal streptococci) in food might suggest the occurrence of recent, direct fecal contamination that might have included viruses; however, contamination that was indirect, especially by way of saline water, might not be accompanied by significant levels of fecal indicator bacteria. Furthermore, the persistence of viruses in contaminated foods seems to differ markedly from that of indicator bacteria.

For these and other reasons, no useful bacterial or other microbial indicator system that would obviate direct testing for foodborne viruses has yet been identified.[69] Virus detectable by infectivity testing may under certain circumstances serve to suggest the presence in food of virus that could otherwise only be detected by the less sensitive serologic procedures,[17] but this kind of association is too fortuitous to afford a reliable basis for monitoring food for viral contamination, even if infectivity testing were much easier and less costly than it is.

SUMMARY

Several viral diseases are sometimes transmitted through foods. At present, the diseases most frequently reported to be foodborne are HA and the AING that is caused by small, round viruses. The foods most often implicated as vehicles have been shellfish, which are able to concentrate viruses from their environmental waters, but any other food may transmit viruses on occasion, as the outbreak record shows. There are no "cures" for viral diseases — those affected either survive or (rarely) not — so anything that can be done to prevent viral infections is certainly worth considering.

The viruses that are transmitted through foods are largely produced in the human intestines and are transmitted more often by direct contact than via the food vehicle. Contamination of foods either takes place directly from the hands of an infected person or indirectly, by way of feces-polluted water or perhaps insects. Viral contamination of foods, at least with agents that are a threat to the health of human consumers, appears to be a relatively rare event in industrialized nations such as the U.S., although it is clear that not all of the foodborne viral disease that occurs is part of recognized outbreaks.

Even if food does become contaminated with virus, the virus may be inactivated before it can cause an infection if the food is not eaten immediately. Heat is probably the most practical general means of inactivating viruses in foods, though of course not all foods are cooked. On the other hand, cold storage of foods can be expected to preserve any virus that may be present. Many components of the environment of a virus within a food appear to modify the effects of temperature upon virus persistence.

Some viruses occurring in some foods can be detected without the necessity of first making someone ill, but routine monitoring for viral contamination of foods is probably not feasible. HAV and the small, round viruses of AING are unlikely to be detected in foods by the serologic detection methods that are presently available, even in most cases where levels of contamination are quite sufficient to cause consumer illnesses. Microbial indicators of fecal contamination can be monitored, but the absence of such indicators certainly does not guarantee the absence of virus from a food.

Some human error is inevitable, so transmission of viral disease through foods will probably never be entirely prevented. In industrialized nations such as the U.S., the foci of prevention should be on the maintenance of established good food handling practices and on good personal hygiene of food handlers, especially those employed in the food service industry. In developing countries, first attention should probably be paid to safer feces disposal and to general community hygiene, in that food is probably only one of several means by which viruses are transmitted in communities where the rate of endemicity is high.

All foodborne disease is theoretically preventable, but not all viral disease transmitted through foods can be prevented. The incidence of virus transmission through foods can certainly be minimized by good sanitation, which does not necessarily entail precautions that are specific for viruses alone.

64. **Kantor, M. A. and Potter, N. N.,** Persistence of echovirus and poliovirus in fermented sausages. Effects of sodium nitrite and processing variables, *J. Food Sci.,* 40, 968, 1975.
65. **Kalitina, T. A.,** Persistence of poliomyelitis virus and some other enteric viruses in cottage cheese, *Vopr. Pitan.,* 30(2), 78, 1971.
66. **Kostenbader, K. D., Jr., and Cliver, D. O.,** Quest for viruses associated with our food supply, *J. Food Sci.,* 42, 1253, 1977.
67. **Cliver, D. O.,** Food-borne viruses, in Compendium of Methods for the Microbiological Examination of Foods, Speck, M. L., Ed., American Public Health Association, Washington, D.C., 1976, 462.
68. **Sobsey, M. D.,** Procedures for the virologic examination of shellfish, sea water and sediment, in Recommended Procedures for the Examination of Sea Water and Shellfish, 5th ed., Hunt, D., Ed., American Public Health Association, Washington, D.C., in press.
69. **Berg, G., Ed.,** *Indicators of Viruses in Water and Food,* Ann Arbor Science Publishers, Ann Arbor, Mich., 1978.

CLOSTRIDIUM PERFRINGENS FOOD POISONING

Betty C. Hobbs

THE ORGANISM

Clostridium perfringens is an important member of the group of organisms called by the generic name *Clostridium*. They are sporing organisms which will grow only in the absence of oxygen; some species are more exacting in their requirements for reduced oxygen than others. Under the microscope the morphological appearance varies from the typical large rod-shaped bacilli with rounded or square ends *(C. perfringens)* to irregular forms including boat- and lemon-shaped organisms, swollen rods, and filaments. Mostly they are arranged singly, but some occur in pairs or in chains and bundles.

Except for *C. perfringens* and a few saprophytic species, they are motile by flagellae growing all around the cell. Individual members of the group vary in their ease of sporulation in laboratory media; with *C. sporogenes* and *C. bifermentans*, sporulation is prolific, whereas *C. perfringans* is slow to spore both in laboratory media and in cooked food. Special media are described for encouraging sporulation, although the properties of the spores so formed may not be quite the same as those produced under natural conditions. The spores of clostridia are widely distributed in soil, dust, feces, flies, and raw foodstuffs such as meat, poultry, fish, and vegetables; they are assumed to be fecal in origin. Small numbers of *C. perfringens* and other anaerobic organisms may be found in cooked foods also.

C. perfringens is a normal inhabitant in the large intestine of man and animals, where it spores readily and so the spores reach soil, dust, and foods from excreta; the dehydrated spores survive a long time.

C. perfringens is the organism commonly associated with gas gangrene. It does not multiply in healthy tissues, but grows rapidly in damaged or devitalized and therefore anaerobic tissue. Sometimes the contamination of damaged tissue with aerobic organisms increases oxygen uptake and anaerobiosis. Tissue damaged in wartime by gunshot, bayonet, and shrapnel wounds and in road accidents may be contaminated with soil and dust containing spores. After a surgical operation the site may be contaminated with clostridia from the patient's own bowel or skin.[1-2] After abortions intestinal clostridia reach necrotic or dead tissue in the uterus and initiate dangerous infections which may be followed by septicemia. Other clostridia may be implicated in gas gangrene also.

There are five types of *C. perfringens*, A to E, distinguished according to the various combinations of toxins they produce; α-toxin, the enzyme lecithinase (phospholipase), is the most important. It kills cells by damaging the cell membranes including red cells thus causing hemolysis. There are more than ten toxins (beta, gamma, delta, epsilon, theta, kappa, and others). Some have lethal or necrotizing actions, others are less pathogenic. During infection, absorption of toxins into the circulation causes toxemia and shock. The typical appearance of gas gangrene is due to the formation of much gas and edema at the site of infection and also the breakdown of hemoglobin.

An important characteristic of the organism is the resistance to heat of the spores. The degree of resistance varies, but some spores are well able to withstand most cooking procedures. Meat ensures an anaerobic medium and during the cooling stage the organisms germinate and multiply to produce enormous numbers of vegetative cells within a few hours of cooking. Survival of spores is ensured by initially heavy contamination and by inadequate heat penetration throughout even normal cooking procedures. Germination of spores is activated by the heat shock of cooking and encouraged by the anaerobic environment of cooked meats and other foods. Growth is enhanced by abundance of soluble nutrients in a progressively anaerobic environment.

All these conditions for growth of the organism are fulfilled in the kitchen and particularly in large-scale catering. It is difficult to disillusion cooks that the heat treatment of cooking does not necessarily destroy all living matter; only when heat under pressure is used can there be assurance that spores are destroyed.

FOOD POISONING

Typical symptoms of food poisoning, nausea, abdominal pain, and acute diarrhea will occur only after the consumption of food heavily contaminated with the vegetative cells of *C. perfringens.* The organism is frequently present in meat, fish, and vegetables as spores and vegetative cells and there is no assurance that it will be killed by cooking. Some method other than cooking must be used to ensure the safety of food which is not immediately eaten freshly cooked and hot; small numbers of surviving spores or even vegetative cells will be harmless. Full outgrowth of spores and rapid multiplication of vegetative cells must be prevented by rapidly reducing the temperature to a level below which the organism cannot grow. Also the food must be kept cold until required to be eaten cold or, if necessary reheated when, for safety's sake, it should be boiled through to the center in order to destroy any cells which have grown.

Special cooling rooms incorporated into the design of kitchens would help to lower the temperature of large masses of meat, poultry, and other foods to a level unsuitable for the growth of *C. perfringens* and other food poisoning organisms. Such small rooms should be maintained at a temperature of approximately 50°F (10°C) and fitted with means for the extraction of steam and to keep a circulation of cold air.

The majority of kitchens have no means to ensure a rapid drop in temperature through the zone known to encourage the growth of bacteria. The multiplication of bacterial agents of food poisoning in certain foods left at temperatures conducive for growth is almost certainly the greatest hazard unknowingly perpetuated in kitchens. The examination of samples of food taken from bulks suspected to have caused food poisoning shows that almost invariably they contain large numbers of the agent responsible for the illness. This evidence confirms not only the identity of the infective agent, but also that there have been faulty storage conditions responsible for growth of bacteria. Long hours of unrefrigerated storage between cooking and eating is not usually regarded as unhygienic practice; "hygiene" is more often considered to be actual manipulation of food with contaminated hands or dirty utensils. Perhaps neither of these faults, undesirable though they are, will cause harm unless followed by a laxity of storage which allows contaminating organisms to reach numbers constituting an effective dose for infection or intoxication. Sutton et al.[3] suggested that large numbers of *C. perfringens* were required to initiate symptoms of food poisoning, not only to allow growth and toxin production in the intestine, but also to permit survival of organisms through the acid regions of the stomach. Unless this growth factor is understood and measures taken to facilitate cooling by the architectural design of kitchens, food poisoning will continue. *C. perfringens* is not the only common agent of bacterial food poisoning, when present in food in large numbers. There are other organisms also such as staphylococci, many salmonella serotypes, *Escherichia coli,* and the vibrios which are usually effective disease agents in large doses only.

Nevertheless, the numbers of organisms responsible for causing illness will vary according to the susceptibility of the individual. Children, aged, and sick persons are at greatest risk and the dose level of organisms for disease may be far lower than that required to initiate symptoms in healthy adults. The virulence or infectivity of bacteria

Table 1
COUNTS OF *C. PERFRINGENS* IN FOODS FROM 24 INCIDENTS OF FOOD POISONING IN GREAT BRITAIN

Count of *C. perfringens*/g	Number of incidents	% of incidents
Less than 10^4	0	—
10^4—9.9 10^4	3	12
10^5—9.9 10^5	3	12
10^6—9.9 10^6	7	29
10^7—9.9 10^7	10	42
Greater than 10^8	1	4

Note: Range 1.3 10^4—1.2 10^8/g: median 5 10^6/g.

Table 2
COUNTS OF *C. PERFRINGENS*, AEROBIC COLONY COUNTS, AND FOODS FROM 6 INCIDENTS OF FOOD POISONING

Incident	Food	Anaerobic count of *C. perfringens*/g	Aerobic count/g
1	Shepherd's pie	4—10^6	Less than 500
2	Braised beef	1.8—10^7	3.5—10^4
3	Cottage pie	2—10^7	8—10^7
4	Boiled salt beef	2.3—10^7	9.8—10^3
5	Stewed steak	4—10^7	1.3—10^6
6	Roast pork sandwiches	1.2—10^8	6.2—10^8

may be enhanced by rapid passage from person to person in overcrowded adult and pediatric wards and more particularly in maternity units. One acute case of bacillary dysentery or salmonellosis excreting vast numbers of organisms in fluid and watery stools may disseminate the organisms and cause cross-infection to an extent dependent on the care of both nurses and doctors.

Tables 1 and 2 give the counts of *C. perfringens* found in foods giving rise to food poisoning. Table 2 shows the foods usually implicated in the food poisoning caused by this organism.[4]

Examples of Outbreaks of *C. perfringens* Food Poisoning

The events in the kitchen leading to *C. perfringens* food poisoning are almost the same in every instance, principally, prolonged storage between cooking and eating, whether or not the food is eaten cold or warmed up. In extreme instances, like the "Harvest Supper" outbreak, there is gross neglect of storage principles. A large turkey was cooked and stored overnight in the slowly cooling oven, it was thus under cover and out of the way. Next day it was left for a period of some hours on top of the oven. Optimum growth conditions were reached again when warm gravy was poured over the sliced turkey meat, and the plates of sliced meat kept warm in the oven pending supper. A fatal case among the 100 or more ill persons the next day highlighted the consequence of ignorance concerning the habits and characteristics of *C. perfringens.* While ineffective thawing of frozen birds leading to inadequate and slow cooking will encourage survival and even growth of bacteria during cooking, the more active multiplication is likely to occur during storage periods after cooking. Many hospital outbreaks, particularly in geriatric wards, have been caused by the growth of *C. perfringens* in minced meat cooked far ahead of requirements and warmed up. Leftover portions from one meal may be added to the new batch prepared for the next day, so that there is a gradual buildup of numbers of contaminants. Many of the spores already present in the meat, or picked up from containers and equipment, survive cooking and grow out after the heat shock of cooking has initiated germination. Refrigerated storage space is often inadequate and pans of cooked meat may be left at kitchen temperature for a few hours and even overnight. The practice is reprehensible, but could be saved from calamity if special cooling rooms were available. In spite of persistant warnings, hospitals and schools are known to have had repeated outbreaks of *C. perfringens* food poisoning, committing the same mistakes each time.

Brisket of beef is frequently a vehicle of *C. perfringens* food poisoning. The chunks of meat are boiled and may be left in the liquor overnight in the boiler; even if taken out and covered they may remain in the kitchen overnight.

Pies made with precooked meat, which are stored overnight and warmed up for the purpose of browning the pastry are not infrequently a cause of trouble. Spores germinate and the vegetative cells multiply during the cooling period; the subsequent heat is evidently too low to ensure the death of even the vegetative cells.

Powdered spices are a prolific source of spores; meat curries kept warm for hours during festive occasions or for picnics and conferences are often responsible for acute diarrhea. The attack rate was high among a coach load of Indians and a few Europeans who ate a picnic lunch of curried chicken prepared during the night and merely warmed through before distribution. The following year the same picnic excursion repeated the experience.

''Meals-on-Wheels'' have not escaped and many recipients have suffered food poisoning because of the delay at warm temperatures between cooking and service.

Meat and poultry are not the only food vehicles; there are reported outbreaks of *C. perfringens* food poisoning from fish, for example, frozen whole salmon, on at least two occasions. The fish were thawed, cooked, and allowed to remain in their liquor overnight without refrigeration.

Even vegetables, misused after cooking, may constitute an ideal medium for the growth of *C. perfringens.* Black beans in gravy sauce and dispensed from large bowls along with other foods were thought to be the food vehicle responsible for acute diarrheal disease affecting hundreds of persons attending an evening function during a conference. The meal was served late in the evening and the various items could have been standing at warm ambient temperature for several hours. It may have been the practice to add each days leftover portion to the next batch prepared. The excursion planned for the next day was marred by the frequent stoppages required by people in great discomfort and by the complete lack of appetite for the mountain lunch.

Other heat-resistant sporing organisms, such as *Bacillus cereus,* may be involved as agents of food poisoning also. As it is well nigh impossible to eliminate these organisms, although strict cleanliness will reduce the numbers in the environment, it is necessary to teach the simple but essential preventive measure of keeping food hot or cold but never warm. Other examples of outbreaks are given by Hobbs and Gilbert.[5]

Statistics

In England and Wales from 1973 to 1975,[6] *C. perfringens* type A, was the second most common cause of food poisoning, between 15 and 32% of the annual totals from all bacterial causes. Outbreaks tended to be larger than those due to other organisms with an average of 50 persons per outbreak; in some instances more than 300 persons were affected. In 1976, strains of *C. perfringens* from 96 outbreaks were investigated in one laboratory.[7] Reports from the U.S. indicate that *C. perfringens* is the commonest agent of food poisoning, the recorded outbreaks surpassing those for salmonella food poisoning.

Statistics for enteritis necroticans due to *C. perfringens* type C strains are not available. There are reports of outbreaks from Germany and Papua, New Guinea, but rarely from other countries; it is possible that the illness is not recognized.

Both in the U.K. and in the U.S., outbreaks of *C. perfringens* food poisoning occur throughout the year without particular prevalence in any season. This may be explained by the fact that the organism multiplies in food in the kitchen, where the ambient temperature is much the same throughout the year. Also the essential germination of spores and outgrowth occurs during the cooling of the cooked food.

The symptoms include abdominal pain, nausea, and diarrhea without fever; they occur 8 to 24 hr after eating food grossly contaminated with the organism. The duration of illness is short, 1 to 2 days. There are few fatalities only, among elderly and debilitated persons. It has been suggested that death in malnourished children with diarrhea may be due to *C. perfringens* septicemia following the increased numbers in the intestine.[24] Parry[8] reported post-mortem findings on geriatric patients who had succumbed to *C. perfringens* food poisoning; there were changes in the large and small bowel including congestion and infection. In the years 1980 and 1981 in England and Wales, *C. perfringens* was responsible for about one quarter of the general outbreaks of "food poisoning".

The Mechanism of Infection and Intoxication

It was some years after the causal agent was incriminated that the mechanism of the illness was revealed. The toxin or enterotoxin responsible for the illness is produced in quantities of clinical significance in vivo only; it is released in the large intestine during active sporulation of the organism. The association between sporulation and enterotoxin production was confirmed by ligated gut experiments in rabbits and lambs, coupled with monkey and human volunteer feeding tests.[9-12] It is thought that the enterotoxin induces increased capillary permeability, vasodilation, and intestinal mobility, which leads to fluid accumulation and diarrhea.[13-15] The secretion of water, sodium, and chloride by the rat ileum was increased by enterotoxin, and the absorption of glucose was inhibited.[16] Although *C. perfringens* spores readily in the intestine, the potential for sporulation in cooked foods is poor and the toxin unlikely to be detected. In media specially designed to encourage sporulation toxin can be detected in vegetative cells 3 hr after inoculation. When spores are formed the toxin increases and can be detected outside the cell in the culture filtrate after 10 or 11 hr. The production peak coincides with the release of free mature spores from the sporangia.[9,17-18] The demonstration of fluid accumulation in the ligated ileal loops of rabbits was used to indicate the action of enterotoxin.[9]

The crude enterotoxin may be purified. It is thought to be a protein of molecular weight 36,000 ± 4000, with an iso-electric point of 4.3.[13] It contains 19 amino acids, among which asparatic acid, serine, leucine, and glumatic acid are predominant. It is heat labile with D_{60} of 4 min, which is the decimal reduction time for the destruction of 90% of toxin at the constant temperature of 60°C. The minimum infective dose for mice is 2000 per milligram N, and lethal doses of toxin caused death within 20 min.[19] Hauschild[20] reported factors of similarity between the toxin of *C. perfringens* and that of *V. cholerae*. Their mode of action may be similar; it is described as enhanced adenocyclase and inhibited phosphodiesterase activities with a chloride and sodium imbalance.

Torres-Anjel[21] demonstrated high antienterotoxic titers in a group of people and suggested that *C. perfringens* enterotoxin was continually produced in the intestine. Antibody against the enterotoxin was found in 82% of 116 serum samples from Brazilians and 65% of 81 serum samples from Americans.[22] All of 141 blood samples from Brazilian slaughterhouse workers showed titers against the enterotoxin, and those who worked on the killing floor had the highest titers. Nevertheless, even *C. perfringens* induced food poisoning, after the administration of large numbers of the organism, did not appear to confer immunity.

Source and Distribution of *C. perfringens*

C. perfringens is commonly found in human and animal excreta and in foods both for human and animal consumption. Because it is prevalent in the environment and also in foods, many doubted its role as a food poisoning organism, although it was well known as the principle agent of gas gangrene. However, it was shown that it was only when the numbers in foods and feces rose to abnormal limits that illness occurred. The essence of prevention is in efforts to keep the numbers low. Mere demonstration of the organism is possible in almost all human stool samples; the counts are usually in the range of 10^3 to 10^4 per gram of feces. In the diarrheal phase of illness the counts rise to 10^6 and higher of *C. perfringens* per gram of feces. In foods the normal count of *C. perfringens* is very low and the organisms may be found through enrichment cultures only. Foods responsible for outbreaks may have many millions of *C. perfringens* per gram of cooked food.

Although the organism is so frequently present in the human intestinal tract, it is considered that human excretors play a very minor role in the spread of *C. perfringens* in the kitchen. Workers should not be penalized when outbreaks occur unless they are active cases of diarrhea. A diarrheal case in the kitchen is a hazard not only from *C. perfringens* but also from the spread of other intestinal pathogens. The likelihood of spread is enhanced when stools are fluid in consistency, and also because of the vastly increased numbers of pathogenic organisms excreted during the acute stage of food poisoning.

Attention should be directed to methods of preparation, cooking, cooling, and storage of foods. Many workers have reported high rates of isolation of *C. perfringens* from raw foods. Roberts[23] found *C. perfringens* in 63% of 183 poultry carcasses. Taniguti isolated *C. perfringens* from 65% of fish samples purchased from retail shops in England. In Japan, he isolated the organism from 67 to 93% of samples from the body surface of seafoods examined from June to September.[24] The isolation rates for heat-resistant strains from samples of pork, beef, veal, mutton, and lamb varied from 1.5 to 42.7%.[25-26] McKillop[27] found *C. perfringens* in 72% of raw meat, fish, and poultry, 20% of cooked foods were contaminated also. Reviews of work on the isolation of *C. perfringens* from foods are given by Bryan,[28,29] Smith and Holdeman,[30] Willis,[31] and Walker.[32].

There are many references to the isolation of *C. perfringens* from animals. Smith and Crabb[33] found *C. perfringens* in the alimentary tract of most farm and domestic animals including dogs, cats, pigs, and cows. Tsai et al.[34] found *C. perfringens* in 88% of chicken feces and in 80% of samples of feces from beef cattle. Narayan[35-36] and Narayan and Takacs[37] isolated *C. perfringens* including type C strains, from muscle tissue and various organs of healthy pigs and cows. The infection rate was enhanced in animals not starved before slaughter and by hasty slaughter of tense and exhausted animals. There is no lack of evidence for the prevalence of *C. perfringens* in other living creatures, including flies[25] and the environment generally. It has been found in

90% of samples of kitchen dust.[27] Thus it must be concurred that the organism will inhabit kitchens to a greater or lesser extent depending on the general state of cleanliness (hygiene). The greater the prevalence and the higher the numbers of *C. perfringens* the greater the risk of contamination and the more rapid the growth of the organism in cooked food. Nevertheless, the growth of the organism in cooked food usually starts from its position in the raw food itself.

Characteristics in Relation to Food Poisoning

A consideration of the characteristics of the organism with regard to rate of growth and optimum and minimum temperatures for growth is helpful in understanding the hazard it presents from food poisoning. The optimum temperatures for growth in cooked meat are 43 to 47°C[38] and the limits for active growth are about 50°C and 15 to 20°C;[39] White and Hobbs[40] could not demonstrate growth at temperatures lower than 6.5°C up to 7 days. Thus cooked food should be cooled rapidly through the temperature zone of optimum growth. Barnes et al.[39] could not recover the organism from cooked meat after freezing at −5 and −20°C for 6 months; destruction of spores was more rapid at −5°C. The vegetative cells are sensitive to cold diluent.[41]

The generation time, that is the time for the division of each vegetative cell, is unusually short, 10 to 12 min, at optimum growth temperatures and in favorable food media.[41a] Thus, the organism can multiply rapidly and reach high numbers in a comparatively short space of time, even within a few hours of cooking depending on the original numbers present. Measurements of D values for spores of *C. perfringens* type A, give a wide range of times and temperatures of heat resistance. Survival of spores may be a few minutes to hours at 100°C, but some spores survive a few seconds only at 90°C. Spores in cooked meat appear to be protected and even more resistant to heat than those suspended in dilutent.

The heat shock of cooking is essential for the germination of many spores. All these characteristics predispose the organism to thrive in cooked foods, mostly meat and poultry. When large numbers are swallowed, they inhabit the gut, sporulate, and release the toxin responsible for diarrhea.

Investigation

The typical history of an outbreak including the incubation period, clinical symptoms, and methods used in food preparation and storage usually directs attention to the possibility of *C. perfringens* food poisoning. Confirmation is dependent on the isolation of the organism in large numbers from the suspected food or foods and also from samples of feces from patients. The identification of the organsm as *C. perfringens* is not difficult, but it is necessary to use finer typing methods in order to ensure that the organisms from food and feces are the same and to find the origin of the particular type or types.

The serological typing of *C. perfringens*, type A, has been used for many years.[42,25] An International Typing Scheme now combines the serotypes described in the U.K., U.S., and Japan.[43-44a] The number of serotypes is at present 75, and in 1976 a causative serotype was established in 73 of the 96 outbreaks investigated.[7] It has been shown that certain serotypes occur more frequently than others as agents of food poisoning. Table 3 reproduced from Stringer et al.[7] gives the distribution of serotypes of *C. perfringens* from outbreaks of food poisoning, and feces and samples of food not associated with food poisoning. *C. perfringens* type 3,4 was responsible for 79 (35%) of 224 of those outbreaks in which the causative serotypes was established. Type 3,4 was associated with outbreaks involving various meats and poultry; outbreaks caused by type 1 strains were mostly associated with beef as the food vehicle. Type 41 was mostly isolated from pork and chicken outbreaks.[7]

Table 3
SEROTYPES OF *CLOSTRIDIUM PERFRINGENS* AND FOOD MOST FREQUENTLY INVOLVED IN OUTBREAKS OF FOOD POISONING (1971 — JUNE 1977)

	Number of outbreaks associated with						
Serotype	Beef and mince	Pork	Lamb	Chicken	Turkey	Other food or not specified	Total
1	11	0	1	1	1	4	18
3,4[a]	36	4	4	9	6	20	79
11 11,13 7,11,13[a]	2	0	1	4	0	4	11
29	10	4	2	0	2	2	19
41	2	6	0	6	0	4	18
Other	29	5	3	11	7	24	79
Not determined[b]	24	8	7	20	7	84	150
Total	114	27	18	51	22	142	374

[a] Isolates were agglutinated by antisera to one or more types.
[b] Refers to outbreaks where cultures were not typable with antisera available at that time and to outbreaks where insufficient cultures were examined.

From Stringer, M. F., Shah, N., and Gilbert, R. J., *Proc. IAMS Meet., Szcecin, Poland, 1977.*

The distribution of *C. perfringens* serotypes in freshly slaughtered animals, and the relationship between serotype and enterotoxin production are being investigated. The type-specific antigens reside in the capsulan polysaccharides.[45-47]

HISTORICAL REVIEW

The historical background to the present day concepts of *C. perfringens* food poisoning is usually found in the beginning; here it appears at the end. It is thought that those who study the subject may better understand the work of the early workers and in the context of the later work decide where next to place the emphasis for investigation.

C. perfringens was associated with outbreaks of mild but chronic diarrhea at the end of the 19th century. Klein[48] and Andrewes[49] reported that abdominal pain was common but vomiting rare. Dunham[50] reported five cases of infection with *"Bacillus aerogenes capsulatus (welch)"*. He gave a time and temperature of 1 min at 98°C for the heat resistance of the spores. Wild[51] said that the spores were more heat resistant in stools than in cultures. Von Hibler[52] and Rodella[53] stated that the spores of *C. perfringens* would survive 1 hr at 100°C. This was vital information in relation to the survival of spores through cooking processes, but its significance was not apprehended. Simonds[54] also made pertinent statements about the sporulation of *C. perfringens.* He claimed that in feces it was related to intestinal disturbances, later work on the association between sporulation and enterotoxin production in the large intestine proved him to be right. Simonds made other claims about sporulation which in later years could not be confirmed. Investigations on toxin (other than the enterotoxin) production and identification in relation to classification were carried out and published by Oakley[55,56] and Oakley and Warrack.[57,58]

Knox and MacDonald[59] described outbreaks in which children were ill after school meals; they found that gravy made the previous day was heavily contaminated with anaerobic sporing bacilli including *C. perfringens.* McClung[60] described four outbreaks of food poisoning resulting from the consumption of chickens steamed the previous day and left to cool slowly. The predominant symptoms were abdominal cramps, nausea, and diarrhea; *C. perfringens* was isolated from the cooked chicken. 1946 records give the earliest results of human volunteer experiments with cultures and filtrates of *C. perfringens*; animals were administered filtrates or living broth cultures orally, also.[61] Cultures from McClung's strains induced symptoms similar to those commonly found in food poisoning in both man and animals. Symptoms from sterile filtrates were not typical of food poisoning. Subsequent experiments showed that most volunteers fed with living cultures actively growing in meat broth media developed typical symptoms of abdominal pain and diarrhea with nausea but rarely vomiting.[25,62,63]

In 1948 there was an outbreak in Hamburg involving at least 400 cases of an intestinal disease described as enteritis necroticans.[64-66] The clinical and bacteriological findings were carefully described in a series of papers.[56,67,68] The symptoms were predominantly abdominal pain and diarrhea; the strain of *C. perfringens* isolated from the food and patients survived 100°C for 1 to 4 hr, and produced much β-toxin. It was described as a new type at first, but later regarded as an atypical type C. These investigations awoke suspicion that outbreaks of a similar nature may be caused by *C. perfringens* or other sporing organisms. Osterling[62] claimed that *C. perfringens* was the predominant organism in food products suspected to be the vehicle of infection in 15 of 33 outbreaks of food poisoning. Hobbs et al.[25] described epidemiological studies and laboratory investigations of outbreaks of food poisoning due to atypical strains of *C. perfringens* type A, which were similar in heat resistance, but dissimilar in the toxicology and colonial appearance, to those described by the German workers. It was observed that the food vehicle was almost invariably a cold or warmed-up meat dish cooked at a temperature not greater than 100°C for 2 to 3 hr the previous day or even a few hours before required, and then allowed to cool slowly in the kitchen or larder. The meat as eaten appeared to be normal in appearance, taste, and smell, although occasionally there were reports of gas bubbles from stews and pies. McKillop[27] reported hospital outbreaks of food poisoning after the consumption of precooked chickens in which the predominant organisms were *C. perfringens* of less heat resistance than hitherto described. Boiled chickens in stock were decanted while hot into deep metal receptacles. Whether the spores and bacilli were in the surface dust or spores had survived cooking, ideal conditions for multiplication were provided in the warm mass of chicken meat and stock. As more and more outbreaks were investigated it was found that strains of *C. perfringens*, type A, of varying heat resistance were involved in outbreaks although the majority were caused by strains capable of surviving the more rigorous cooking procedures.[69-71] Later, Hauschild, Duncan, and Strong and their colleagues[69-71] began to publish the results of their work on the mechanism of infection and intoxication. The purification and characteristics of the enterotoxin were described by Hauschild and Hilsheimer.[19] The work that followed has already been described.

This organism, its action in food and in the human intestinal tract has been the subject of research in many countries and for many years; there are few other organisms which have received such detailed attention. In summary, typical features of *C. perfringens* food poisoning are given by Collee[72] as follows:

1. Pre-cooked or bulk-cooked meat is usually involved.
2. There is an incubation period of about 12 hr.
3. Symptoms include abdominal pain and diarrhea. Fever and vomiting are not typically present.

4. Symptoms usually subside after 24 to 48 hr.
5. The causative organism can be isolated from the food concerned and from the feces of victims.

The food is subjected to a prolonged cooling period within a suitable temperature range which allows rapid and continued bacterial multiplication. Victims therefore ingest the equivalent of a cooked meat broth culture containing millions of viable *C. perfringens.* The organisms sporulate in the gut and release an enterotoxin at this stage. The symptoms of abdominal pain and diarrhea occur 8 to 24 hr or perhaps longer after ingestion of the meal.

Meals served in bulk in schools, hospitals, and canteens tend to be mainly involved in *C. perfringens* outbreaks. The occurrence of this food poisoning is a serious reflection on the cooling and cold storage facilities of the catering establishment concerned.

REFERENCES

1. **Parker, M. T.**, Clostridial sepsis, *Br. Med. J.*, 2, 698, 1967.
2. **Ayliffe, G. A. J. and Lowbury, E. J. T.**, Sources of gas gangrene in hospital, *Br. Med. J.*, 2, 333—337, 1969.
3. **Sutton, R. G. A., Ghosh, A. C., and Hobbs, B. C.**, Isolation and Enumeration of *Clostridium welchii* from Food and Faeces, in *Isolation of Anaerobes.* Soc. Appl. Bact. Tech. Ser. No. 5, Shapton, D. E. and Board, R. G., Eds., Academic Press, London, 1971, 39—47.
4. **Hobbs, B. C., and Gilbert, R. J.**, Microbiological counts in relation to food poisoning, *Proc. IV Int. Congr. Food Sci. Technol.*, 3, 159—169, 1974.
5. **Hobbs, B. C. and Gilbert, R. J.**, *Food Poisoning and Food Hygiene,* 4th ed., Edward Arnold, London, 1978.
6. **Vernon, E.**, Food poisoning and Salmonella infections in England and Wales, 1973—1975. An analysis of reports to the Public Health Laboratory Service, *Public Health (London),* 91, 225—235, 1977.
7. **Stringer, M. F., Shah, N., and Gilbert, R. J.**, Serological typing of *Clostridium perfringens* and its epidemiological significance in the investigation of food poisoning outbreaks, in *Proc. IAMS Meet., Szcecin, Poland, 1977.*
8. **Parry, W. H.**, Outbreak of *Clostridium welchii* food-poisoning, *Br. Med. J.*, 2, 1616—1619, 1963.
9. **Duncan, C. L., and Strong, D. H.**, Experimental production of diarrhea in rabbits with *Clostridium perfringens, Can. J. Microbiol.*, 15, 765—770, 1969.
10. **Duncan, C. L. and Strong, D. H.**, *Clostridium perfringens* Type A food poisoning. I. Response of the rabbit ileum as an indication of enteropathogenicity of strains of *Clostridium perfringens* in monkeys, *Infect. Immunol.*, 3, 167—170, 1971.
11. **Hauschild, A. H. W., Niilo, L., and Dorward, W. J.**, Enteropathogenic factors of food-poisoning *Clostridium perfringens* Type A, *Can. J. Microbiol.*, 16, 331—338, 1970.
12. **Strong, D. H., Duncan, C. L., and Perna, G.**, *Clostridium perfringens* Type A food poisoning, *Infect. Immunol.*, 3, 171—178, 1971.
13. **Stark, R. L. and Duncan, C. L.**, Transient increase in capillary permeability induced by *Clostridium perfringens* Type A enterotoxin, *Infect. Immunol.*, 5, 147—150, 1972.
14. **Niilo, L.**, Mechanism of action of the enteropathogenic factor of *Clostridium perfringens* Type A, *Infect Immunol.*, 3, 100—106, 1971.
15. **Niilo, L., Hauschild, A. H. W., and Dorward, W. J.**, Immunization of sheep against experimental *Clostridium perfringens* Type A enteritis, *Can. J. Microbiol.*, 17, 391—395, 1971.
16. **McDonel, J. L.**, In vivo effects of *Clostridium perfringens* enteropathogenic factors on the rat ileum, *Infect. Immunol.*, 10, 1156—1162, 1974.
17. **Duncan, C. L.**, Time of enterotoxin formation and release during sporulation of *Clostridium perfringens,* Type A, *J. Bacteriol.*, 113, 932—936, 1973.
18. **Uemura, T., Sakaguchi, G., and Riemann, H. P.**, In vitro, production of *Clostridium perfringens* enterotoxin and its detection by reversed passive hemagglutination, *Appl. Microbiol.*, 26, 381—385, 1973.
19. **Hauschild, A. H. W. and Hilsheimer, R.**, Purification and characteristics of the enterotoxin of *Clostridium perfringens* Type A, *Can. J. Microbiol.*, 17, 1425—1433, 1971.

20. Hauschild, A. H. W., Food poisoning by *Clostridium perfringens, Can. Inst. Food Sci. Technol. J.*, 6, 106—110, 1973.
21. Torres-Anjel, M. J., Ph.D. thesis, University of California, Davis. 1974.
22. Uemura, T., Genigeorgis, C., Riemann, H. P., and Franti, C. E., Antibody against *Clostridium perfringens* Type A enterotoxin in human sera, *Infect. Immunol.*, 9, 470—471, 1974.
23. Roberts, D., Observations on procedures for thawing and spit-roasting frozen dressed chickens, and post-cooking care and storage, with particular reference to food-poisoning bacteria, *J. Hyg. Camb.*, 70, 565—588, 1972.
24. Taniguti, T. and Zenitani, B., Incidence of *Clostridium perfringens* in fishes (II) on the detection rate of *Clostridum perfringens* Type A and heat-resistance of isolated strains, *J. Food Hyg. Soc. Jpn.*, 10, 266—271, 1969.
25. Hobbs, B. C., Smith, M. E., Oakley, C. L., Warrack, G. H., and Cruickshank, J. C., *Clostridium welchii* food poisoning, *J. Hyg. Camb.*, 51, 75—101, 1953.
26. Sylvester, P. K., and Green, J., The effect of different types of cooking on artificially infected meat, *Med. Offr.*, 105, 231—235, 1961.
27. McKillop, E. J., Bacterial contamination of hospital food with special reference to *Clostridium welchii* food poisoning, *J. Hyg. Camb.*, 57, 31—46, 1959.
28. Bryan, F. L., What the sanitarian should know about *Clostridium perfringens* foodborne illness, *J. Milk Food Technol.*, 32, 381—389, 1969.
29. Bryan, F. L., *Clostridium perfringens* in relation to meat products, in Proc. 25th Annu. Reciprocal Meat Conf., 1972, 323—341.
30. Smith, L. Ds., and Holdeman, L. V., in *The Pathogenic Anaerbic Bacteria*, 1st ed., Charles C Thomas, Springfield, Ill., 1968.
31. Willis, A. T., in *Clostridia of Wound Infection*, Butterworths, London, 1969.
32. Walker, H. W., Food borne illness from *Clostridium perfringens, CRC Crit. Rev. Food Sci. Nutr.*, 7, 71—104, 1975.
33. Smith, H. W. and Crabb, W. E., The faecal bacterial flora of animals and man: its development in the young, *J. Pathol. Bacteriol.*, 82, 53—63, 1961.
34. Tsai, C. C., Torres-Angel, M. J., and Riemann, H. P., Characteristics of enterotoxigenic *Clostridium perfringens* Type A isolated from cattle and chicken, *J. Formosan Med. Assoc.*, 73, 501—510, 1974.
35. Narayan, K. G., Studies on Clostridia-incidence in the beef cattle, *Acta Vet. Acad. Sci. Hung.*, 16, 65—72, 1966.
36. Narayan, K. G., Incidence of Clostridia in pigs, *Acta Vet. Acad. Sci. Hung.*, 17, 179—182, 1967.
37. Narayan, K. G. and Takacs, J., Incidence of Clostridia in emergency-slaughtered cattle, *Acta Vet. Acad. Sci. Hung.*, 16, 345—349, 1966.
38. Colee, J. G., Knowlden, J. A., and Hobbs, B. C., Studies on the growth, sporulation and carriage of *Clostridium welchii* with special reference to food poisoning strains, *J. Appl. Bacteriol.*, 24, 326—339, 1961.
39. Barnes, E. M., Despaul, J. E., and Ingram, M., The behavior of a food poisoning strain of *Clostridium welchii* in beef, *J. Appl. Bacteriol.*, 26, 415—427, 1963.
40. White, A. and Hobbs, B. C., Refrigeration as a preventative measure in food poisoning, *R. Soc. Health J.*, 83, 111—114, 1963.
41. Sanousi, S. M. El., Ph.D. thesis, University of Bristol, 1975.
41a. Mead, G. C., Growth and sporulation of *Clostridium welchii* in breast and leg muscle of poultry, *J. Appl. Bacteriol.*, 32, 86—95, 1969.
42. Henderson, D. W., The somatic antigens of the *Cl. welchii* group of organisms, *J. Hyg. Camb.*, 40, 501—512, 1940.
43. Stringer, M. F., Turnbull, P. C. B., Hughes, J. A., and Hobbs, B. C., An international serotyping system for *Clostridium perfringens (welchii)* Type A in the near future. Joint OIE-IABS Symp. on clostridial products in veterinary medicine, Paris, 1975, *Dev. Biol. Stand.*, 32, 85—89, 1976.
44. Hughes, J. A., Turnbull, P. C. B., and Stringer, M. F., A serotyping system for *Clostridium welchii (C. perfringens)* Type A, and studies on the type-specific antigens, *J. Med. Microbiol.*, 9, 475—485, 1976.
44a. Stringer, M. F., Turnbull, P. C. B., and Gilbert, R. J., Application of serological typing to the investigation of outbreaks of Clostridium perfringens food poisoning, *J. Hyg. Comb.*, 84, 443—456, 1970 to 1978.
45. Baine, J. and Cherniak, R., Capsular polysaccharides of *Clostridium perfringens* Hobbs 5, *Biochemistry (N. Y.)*, 10, 2949—2952, 1971.
46. Lee, L. and Cherniak, R., Capsular polysaccharides of *Clostridium perfringens* Hobbs 10, *Infect. Immunol.*, 9, 318—322, 1974.
47. Button, J. A., M. Phil. thesis, University of London, 1975.

48. **Klein, E.**, On a pathogenic anaerobic intestinal bacillus, *Bacillus enteritidis sporogenes, Zentbl. Bakt. Parasitkde. I. Abt. Orig.*, 18, 737—743, 1895.
49. **Andrewes, F. W.**, On an outbreak of diarrhoea in the wards of St. Bartholomew's Hospital, probably caused by infection of rice pudding with *Bacillus enteritidis sporogenes, Lancet*, 1, 8—12, 1899.
50. **Dunham, E. K.**, Report of five cases of infection by the bacillus aerogenes capsulatus (welch), *Bull. Johns Hopkins Hosp.*, 8, 68—74, 1897.
51. **Wild, O.** Beitrag Zur Kenntnis des *Bacillus enteriditis sporogenes, Zentbl. Bakt. Parasitkde. I Abt. Orig.*, 23, 913—917, 1898.
52. **von Hibler, E.**, Ueber die Differential diagnose der pathogenen Anaerobiae, *Zentbl. Bakt. Parasitkde. I Abt. Ref.*, 37, 545, 1906.
53. **Rodella, A.** Studien uber Darmfaulnis 6 Faulnisvermogen des acholischen stuhles, *Wien Klin. Wocschr.*, 23, 1383, 1910.
54. **Simonds, J. P.**, Studies in *Bacillus welchii* with Special Reference to Classification and its Relation to Diarrhoea, No. 5, Monographs Rockefeller Institute, 1915.
55. **Oakley, C. L.**, The toxins of *Clostridium welchii, Bull. Hyg. (London)*, 18, 781—806, 1943.
56. **Oakley, C. L.**, The toxins of *Clostridium welchii*, Type F, *Br. Med. J.*, 1, 269—270, 1949.
57. **Oakley, C. L. and Warrack, G. H.**, The ACRA test as a means of estimating hyaluronidase, desoxyribonuclease and their antibodies, *J. Pathol. Bacteriol.*, 63, 45—55, 1951.
58. **Oakley, C. L. and Warrack, G. H.**, Routine typing of *Clostridium welchii, J. Hyg. Camb.*, 51, 102—107, 1953.
59. **Knox, R. and MacDonald, E. J.**, Outbreaks of food poisoning in certain Leicester institutions, *Med. Offr.*, 69, 21—22, 1943.
60. **McClung, L. S.**, Human food poisoning due to growth of *Clostridium perfringens (C. welchii)*, in freshly cooked chicken: preliminary note, *J. Bacteriol.*, 50, 229—231, 1945.
61. **Cravitz, L. and Gillmore, J. D.**, The Role of *Clostridium perfringens* in Human Food Poisoning, Project 1-756, Rep. No. 2, Naval Medical Research Institute, Bethesda, 1946.
62. **Osterling, S.**, Matforgiftninger orsakade av *Clostridium perfringens (welchii), Nord. Hyg. Tidskr.*, 33, 173—179, 1952.
63. **Dische, F. E. and Elek, S. D.**, Experimental food poisoning by *Clostridium welchii, Lancet*, pp. 71—74, 1957.
64. **Ernst, O.**, Epidemiologischer Betrag zur Enteritis Necroticans, *Dte Gesundhwes.*, 3, 262, 1948.
65. **Schütz, F.**, Darmbrand: Neue Erkenntnisse zu Singer Pathogenese und Therapie, *Dtsch Med. Wschr.*, 73, 176, 1948.
66. **Marcuse, K. and König, I.**, Bakteriologische Befunde bei Jejunitis Necroticans, *Zentbl. Bakt. Parasitkde. I. Abt. Orig.*, 156, 107—118, 1950.
67. **Zeissler, J. and Rassfeld-Sternberg, L.**, Enteritis necroticans due to *Clostridium welchii* Type F, *Br. Med. J.*, 1, 267—269, 1949.
68. **Hain, E.**, Origin of *Clostridium welchii* Type F infection, *Br. Med. J.*, 1, 271, 1949.
69. **Hall, H. E., Angelotti, R., Lewis, K. H., and Foter, M. J.**, Characteristics of *Clostridium perfringens* strains associated with food and food-borne disease, *J. Bacteriol.*, 85, 1094—1103, 1963.
70. **Taylor, C. E. D. and Coetzee, E. F. C.**, Range of heat resistance of *Clostridium welchii* associated with suspected food poisoning, *Mon. Bull. Minist. Health*, 25, 142—144, 1966.
71. **Sutton, R. G. A. and Hobbs, B. C.**, Food poisoning caused by heat-sensitive *Clostridium welchii*: a report of five recent outbreaks, *J. Hyg. Camb.*, 66, 135—146, 1968.
72. **Collee, J. G.**, *Applied Medical Microbiology*, (Basic Microbiology 3,), Blackwell Scientific, Oxford, 1976, 77.
73. **Hobbs, B. C.**, *Clostridium perfringens* gastroenteritis, in *Foodborne Infections and Intoxications*, 2nd ed., Rieman, H. and Bryan, F. L., Eds., Academic Press, London, 1979, 131—171.
74. **Torres-Anjel, M. J.**, personal comment.

FOODBORNE DISEASES DUE TO *VIBRIO CHOLERAE*

D. Barua

INTRODUCTION

The species *Vibrio cholerae* at present conta' s 60 known serovars, or serogroups, among which *V. cholerae* 0-1 is the etiologic agent of cholera.[1] The so-called NAG (nonagglutinable) vibrios or NCV (noncholera vibrios) have also been included in the species *V. cholerae* by decision of the International Subcommittee on the Taxonomy of Vibrios in 1971 and are now distinguished only by serovar number.[2] While *V. cholerae* other than serovar 0-1 can sometimes cause cholera-like diarrhea, they do not possess the same epidemic propensity as the true *V. cholerae*; the two clinical entities will therefore be described separately.

CHOLERA

The word cholera continues even today to evoke panic in the minds of both health administrators and the public because it is possibly the most rapid killer of all the communicable diseases and one that can spread extremely fast. The areas where cholera thrives are also those where modern treatment facilities are not readily available; there, an individual in perfect health can be struck by cholera and die within as short a time as 4 hr. The unabated fury of the seventh pandemic of cholera that began in 1961 demonstrates the vulnerability of the modern world in the face of the vigorous ability to spread of the etiologic agent of the disease.

Fear is justified only when cholera is considered in isolation, but when it is viewed as just one of the many acute diarrheal diseases thriving in the same environment, having similar epidemiological and clinical features, causing similar biochemical derangements and requiring similar therapeutic interventions and control measures, it is clear that cholera constitutes only a very small fraction of the total mass of acute diarrheal diseases.[3]

Etiology and Epidemiology

Causative Agent

The etiologic agent of the current seventh pandemic is a biotype of *V. cholerae* 0-1 that used to be called the El Tor vibrio (after the El Tor quarantine camp in the Sinai Peninsula where it was first discovered) and is at present described as *V. cholerae* biotype eltor. The "classical" *V. cholerae* has been almost completely replaced by the *eltor* biotype and is now only rarely isolated from occasional cases in the Indian subcontinent.

Both the classical and *eltor* biotypes are similar in their many characteristics, including the antigenic composition of their two serotypes (Inaba and Ogawa) while their differentiating features are considered to be infrasubspecific and also not very stable as several *eltor* strains with atypical characteristics have now been documented. However, some of the features of the *eltor* biotype are of considerable epidemiological significance: they appear to be more sturdy than the classical type and can survive longer in the environment so that they can be detected in night-soil and latrine samples,[4-5] they are relatively more resistant to antibiotics,[6] and they are excreted over a longer period by untreated cases.[7] However, both the classical and *eltor* biotypes of the cholera vibrio are rather delicate compared to the members of the family Enterobacteriaceae and other serovars of *V. cholerae* as regards their resistance to drying and chemicals. Their only natural host is man.

In the laboratory, *eltor* biotype is differentiated by its ability to cause hemagglutination of chicken, sheep, human, and rabbit cells and by its resistance to polymgxin B and cholera-phage group IV.[8,36]

Asymptomatic Infection

The *eltor* biotype has been found to cause many more mild and asymptomatic infections than typical cases; the ratio of severe cases to asymptomatic or mild infections has been found to vary from 1:10 to 1:100 depending on the density of the population, the hygienic habits of the people, the degree of environmental contamination, and to a very important extent on the thoroughness of surveillance; with classical *V. cholerae* the ratio has been found to be about 1:4.[8] Thus, the visible tip of the iceberg in *eltor* cholera is very small indeed. The few examples of cases that have become long-term carriers after recovery have all been due to the *eltor* biotype.

International Spread

The circumstances that helped the *eltor* biotpye of *V. cholerae* to break out from its endemic focus in Sulawesi (Celebes) in Indonesia in 1960 to 1961 have remained unknown. The ability of this sturdier vibrio to cause many inapparent infections and mild cases whose mobility is not restricted, and the availability of improved, faster transport systems that have freed small towns and villages from isolation in the post-war years, might have contributed to the spread of the infection.[8] At first, it traveled in a fairly predictable fashion to contiguous countries, following population movements by sea and road, and sometimes by air. In 1970, however, cholera made an unanticipated appearance in Guinea in West Africa, having presumably made an enormous jump over hitherto unaffected territory. By mid 1981, 91 countries, including a few in Western Europe and in North America, have so far become involved in the current pandemic, eight of them in 1978, and it is showing no sign of losing its vigor. Predictions of an extension to South and Central America have, however, remained unfulfilled.

Endemicity and Epidemicity

Cholera has now become endemic in many countries of Asia and Africa. The high rate of inapparent infection makes it possible for the disease to persist in a community where conditions for transmission are favorable without the occurrence of typical cases over long periods. It becomes difficult to differentiate the new introductions from recrudescences of infection when cases appear in such communities at long intervals.

It is well known from the long history of cholera in the Indo-Pakistan subcontinent that epidemics of the disease appear after periodic intervals, usually of 2 to 4 years. No clear explanation for this is available, but the appearance of a large number of nonimmune individuals, either by migration or by birth, is usually considered to be the factor responsible for such periodicity. During explosive epidemics of cholera, large numbers of cases appear over a short period of time when there is a single or multiple common source or vehicle of infection. Sometimes cholera outbreaks follow a rather protracted pattern, persisting for several weeks, when only a few cases occur per day or week and the means of transmission remains difficult to define.

Seasonal Incidence

Cholera also exhibits a characteristic seasonal pattern which can vary from one country or region of a country to another. In Calcutta, for example, the cholera incidence rises to a peak during the hot dry months and ends with the monsoon, whereas in Dacca, Bangladesh, only a short distance away, the cholera season starts after the monsoon rains and ceases during the hot dry months. In some areas of the Philippines incidence is highest during the rainy season. In Africa, the disease reaches its peak during the dry period and declines during the rainy months. The seasonal patterns are, however, changing in some areas over the years. The factors that trigger off the seasonal rise and fall remain largely unknown.

Transmission

Cholera is not a highly contagious disease. Studies in volunteers have shown that about 10^{11} organisms are required to produce choleraic symptoms unless the gastric acid is neutralized by bicarbonate when the infective dose can be reduced to 10^5 or 10^6 organisms.[9] The magnitude of the infective dose reduces the chances of direct transmission although Félix has brought evidence of multiplication of *V. cholerae* in sweat and of direct spread from person to person in desert-like, dry areas of Africa.[10]

The higher incidence of asymptomatic infection in studies among close family contacts as compared to that in the community may be due to the sharing of food and water rather than direct contact, although it is difficult to separate the two possibilities.

The role of water in the spread of cholera was conclusively shown by Snow in 1855 long before the discovery of the causative agent or, for that matter, before the germ theory of the disease was known.[11] Experience during subsequent years and in many countries has only confirmed and expanded those observations. *V. cholerae* does not multiply in water though the possibility of some multiplication in water grossly contaminated with sewage cannot be excluded. Its survival in water varies from 1 or 2 days to 17 days at room temperature or up to 42 days at 5 to 10°C, and is also dependent upon pH and salinity as well as the presence of organic matter and competitive bacterial flora. Thus, although water is the common means of spread of the infection, it has to be repeatedly contaminated by human excreta containing the organism for maintenance of infection in the community.

Food also plays an important role in transmission although it is difficult on occasion to separate food from drink. It has only rarely been possible to isolate the causative agent in thousands of cooked or uncooked food samples gathered from the kitchens and market places in endemic situations. The period of survival of the cholera vibrio in various foodstuffs depends on many factors and is generally short at room temperature but longer under refrigeration; vibrios have been found to multiply, to a limited extent, only in milk and milk products.

However, there is convincing epidemiological evidence of many foodborne outbreaks of cholera following festivals and feasts, and particularly after funerals of person dying of cholera. Such funeral gatherings, where the custom of washing and touching the dead body and of eating and drinking facilitates the transference of the organism directly by soiled hands to food and drinks, played a significant role in the spread of cholera in Africa. Teng has provided interesting reports of cholera outbreaks in Hong Kong which were connected with eating in restaurants or with food bought from them.[12]

Contaminated foods like small raw shrimps — called alamang or hippon — in the Philippines, vegetables irrigated with contaminated sewage or freshened with contaminated water in Israel and several other countries, cockles, mussels, and other bivalves in Italy and Portugal, and shellfish and other seafood in Malaysia, Guam, the Gilbert Islands, and the U.S. have been incriminated as the cause of large and small outbreaks of cholera.[15-19] The relation of these foods with water is, however, obvious. These instances have stimulated studies on the survival and growth of the cholera vibrio in shellfish at different temperatures under natural conditions.

The role of food in transmission was also very well demonstrated when 40 bacteriologically confirmed cases occurred among 374 passengers and 19 crew members of a flight arriving in Sydney from London in November 1972. Among 37 passengers who continued their journey to New Zealand, there were 3 cases, 1 of which was fatal. Epidemiological investigation suggested that a dish of hors d'oeuvres prepared in the air company kitchen in Bahrain and served only to economy class passengers was the source of infection. All the cases were among the economy class passengers and the water taken aboard at Bahrian was negative for *V. cholerae.*[20]

Since 1963, a few sporadic cases of cholera have occurred almost every year in Singapore although more than 90% of the residential premises of the cases and carriers have piped water and over 50% of them have modern sanitation facilities. Bacteriological examination of imported and local foods including shellfish failed to isolate the vibrio and extensive epidemiological investigations were unable to demonstrate a link between the cases and a common food source. However, the health authorities believed they had sufficient evidence to point to contaminated food sold by hawkers as the source of infection.[21]

Host Susceptibility

In newly affected areas cholera is predominantly a disease of adults, particularly males where they are more mobile and subject to greater exposure than the females; in endemic areas it is more commonly seen in children, which is possibly an expression of acquired immunity in older age groups due to repeated infections. Studies in Bangladesh have demonstrated that with increasing age there is a rise in serum vibriocidal antibody titer which is associated with a fall in case rate.[22] In a rural endemic area in Bangladesh, the case rate in children aged 1 to 5 years is about ten times that seen in adults.[23] Young adults, however, constitute a big segment of cholera patients even in endemic areas. Cholera is rare in children below 1 year of age though mild and inapparent infection is quite common, especially in artificially fed infants. The sex distribution appears to depend on the degree of exposure to the source of infection. In pregnant women, particularly in the third trimester, cholera is usually a serious disease with a high case fatality and miscarriage rate unless treated early and vigorously.

One of the most important factors in host resistance that has been well elucidated is the role of gastric acidity. Individuals with lower or absent gastric acidity have been found to have an increased risk of infection, a shorter incubation period and a greater chance of developing a more serious disease.[24-26]

Nutritional Status and Host Susceptibility

The close association between malnutrition and susceptibility to acute diarrheal disease has been repeatedly reported but no study has yet succeeded in demonstrating such an association with cholera. Investigators in Bangladesh in the mid 1960s failed to incriminate any nutritional factor when they looked for thiamine, B-12, sodium folate, and ascorbic acid deficiency.[27,29] Nor could any correlation be found between cholera and the severity of histological changes in the intestinal epithelium associated with the malabsorption syndrome.[28,30] Experimental studies on a limited number of monkeys were also noncontributory.[31] However, the duration of clinical disease has been documented to be longer in malnourished adults and children.[32,33] In pediatric cholera, this difference was apparent only in individuals treated with antibiotics.

Pathogenesis and Pathophysiology

All known symptoms and signs in cholera are due to the biochemical and metabolic derangements brought on by the rapid and copious loss of fluid and electrolytes from the gut. The causative agent, if it succeeds in passing through the gastric acid barrier and reaching the small intestine, multiplies (the actual site of multiplication in the small intestine is not known), adheres to the mucous membrane, and produces a protein enterotoxin. The cholera vibrio generally does not invade the intestinal tissue, nor has the enterotoxin been found to have any direct effect on any organ or tissue other than the small intestine in naturally occurring disease. There is no histological evidence of damage to the intestinal mucosa.

The toxin combines with a receptor (GM_1 ganglioside) and the complex is then probably transported to an active site in the cell where it activates the adenylate cyclase with consequent increase in intracellular cyclic AMP. All the available experimental evidence today supports the concept that c = AMP is directly involved in the mediation of enterotoxin-induced gut fluid secretion in cholera.[34]

The fluid lost is isotonic in nature with a very low protein content (less than 200 mg/100 mℓ) and in adults has the following electrolyte concentrations in *M* Eq. per liter: sodium 126 ± 9, potassium 19 ± 9, bicarbonate 47 ± 10, and chloride 95 ± 9; in children the composition is sodium 56 ± 11, potassium 25 ± 4, bicarbonate 14 ± 4, and chloride 55 ± 11.

The loss of this fluid, sometimes at the rate of 1ℓ/hr in adults, rapidly leads to hypovolemic shock and metabolic acidosis with the typical clinical manifestations. Laboratory investigations reveal evidence of hemoconcentration, a drop in plasma bicarbonate levels and in arterial pH, and a marked elevation of plasma protein and of plasma specific gravity. In children, however, the plasma specific gravity is lower in severe cholera than in adults with disease of similar severity. Although the activity of intestinal disaccharidases including lactase may be impaired, glucose absorption and consequently glucose-linked sodium absorption, which form the basis of oral rehydration, are usually preserved.

Clinical Features and Spectrum

The incubation period of cholera varies from 12 hr to less than 5 days; in its typical form the disease is characterized by profuse painless and effortless watery diarrhea that quickly assumes a "rice water" appearance and is frequently associated with vomiting. The patient soon becomes thirsty and oliguric and complains of cramps in the muscles of the extremities and sometimes of the abdomen. There is no fever except in the case of young children. The subject rapidly becomes weak, exhibits all the signs and symptoms of dehydration and acidosis, and eventually collapses.

Like many other infectious diseases, cholera exhibits a very broad clinical spectrum ranging from the above-mentioned manifestation of cholera gravis to a very mild disease which is clinically indistinguishable from mild diarrheas of other origins. The illness may therefore last from 12 hr to 7 days.

The *eltor* biotpye is known to cause many more cases of mild disease than its classical counterparts although severe illness caused by the *eltor* biotype is clinically indistinguishable from that caused by the classical *V. cholerae.*

Delayed or inadequate treatment may result in acute renal failure and problems associated with hypokalemia. Management of acute renal failure with the pathological features of acute tubular necrosis in cholera patients is more complicated because of the associated severe degree of metabolic acidosis due to gastrointestinal losses of bicarbonate. Associated hypokalemia, particularly in pediatric patients, also creates a serious problem requiring careful clinical judgment.

Hypoglycemia is occasionally seen in children, causing convulsions and prolonged coma.[35] The inclusion of glucose in i.v. or oral fluid should prevent such a complication.

Diagnosis

Clinical diagnosis of typical cases particularly during an epidemic is very easy, although cholera-like illness may be caused by organisms other than *V. cholerae*. Laboratory techniques for the diagnosis of cholera are simple; a provisional diagnosis can be made within a few minutes to 2 hr using dark field or fluorescent microscopy, and can be confirmed by cultural procedures within 18 hr.[36] The institution of therapy, however, should not wait for laboratory diagnosis as the same therapeutic approaches are required and are effective in almost all acute diarrheas.

Clinical Management

Cholera is probably the most rewarding disease to treat. No patient should die of cholera if the appropriate treatment can reach him while he is still alive. In well-organized treatment centers the case fatality rate has been reduced to less than 1% in all ages. The basis of successful treatment is prompt and adequate replacement of water and electrolytes lost in the stool and vomitus, correction of acidosis and potassium depletion, maintenance of hydration until diarrhea ceases by replacing fluid and electrolyte losses as they occur by i.v. and oral routes, and reduction of the volume and duration of diarrhea with suitable antibiotics. Administration of antibiotics is not essential for therapy but is extremely advantageous as it shortens the disease and the period of vibrio secretion, and thus reduces the need for i.v. fluid which is expensive and often difficult to obtain. Tetracycline is generally used and is effective. Hirschhorn et al. have described in detail the treatment in adults and Mahalanabis et al. in children.[37,38]

Commercially available Ringer's lactate (Hartman's solution) has been used widely with satisfaction for i.v. rehydration. An oral solution consisting of NaCl (3.5 g), KCl (1.5 g), Na_2HCO_3 (2.5 g), and glucose (20 g) per liter of drinking water has been found adequate for maintenance of hydration as well as for rehydration of mild to moderate cases. Untrained family members and assistants with minimal medical supervision used this oral solution to treat about 4000 adults and children suffering from cholera in a Bengali refugee camp in 1971, when the mortality rate was only slightly greater than that observed in the well-established treatment centers.[39] This same fluid has also been shown to be effective for the treatment of other acute diarrheas in adults and children.

WHO-supported studies on the feasibility, acceptability, and effectiveness of oral rehydration therapy supported by education on proper dietetic management for acute diarrheal diseases in children at the community level, delivered through subprofessional personnel and mothers with some training, have demonstrated not only the effectiveness of the procedure but also a better weight gain — a nutritional benefit — among those so treated as compared with the controls over a 6-month period of observation.[40] The mothers volunteered the information that they would continue to use the fluid because the children receiving it both felt and ate better and bothered them less. Encouraged by these observations in several countries, WHO is developing a program for diarrheal diseases control with the immediate task of promoting oral rehydration.

Immunity and Vaccines

In volunteer studies it has been shown that convalescents from induced cholera acquire serotype-specific immunity lasting for about 14 months;[41] recent studies indicate that the immunity may be shorter and not so type-specific.[42] Several field trials with parenterally administered whole-cell killed vaccines have shown them to offer approximately 60 to 70% protection for about 2 to 3 months, declining rapidly to about 30% in 6 months. All the trials were carried out in endemic areas and mostly with one dose. Two doses were found to give better results in children, who probably had less previous exposure to cholera antigens. In two recent trials, a single dose of a similar vaccine with aluminum hydroxide or phosphate as adjuvant was found to protect children better and for a relatively longer period, but the number of cases of cholera was rather small for a firm conclusion.[43a,43b]

Current information on the role of cholera enterotoxin in pathogenesis has stimulated the development of cholera toxoid which was found to provide highly significant but short-lived protection in experimental animals. A field trial in Bangladesh with commercially prepared, gluteraldehyde-treated toxoid has failed to demonstrate effectiveness.[44a] A more effective vaccine is likely to be a combination of cholera toxoid with vibrio somatic antigen and an appropriate adjuvant; one such vaccine containing B subunit (binding component of cholera toxin) have already been developed[44b] and is being field tested.

An attenuated live oral cholera vaccine that is able to colonize the small bowel and stimulate the host defense mechanism would be ideal. Finkelstein et al. developed such a mutant which did not produce toxin,[45] but the strain was found to be rather unstable. The same group of workers has now developed a laboratory mutant producing only the B subunit of the toxin. This mutant has offered high protection to volunteers against challenge. Further work on safety and mechanism of delivery are currently in progress.

Prevention and Control

Despite the development of very effective treatment, the case fatality rate due to cholera has remained high in many endemic areas and in newly affected areas, particularly in the early stages of epidemics. This is usually a reflection of the logistic problems created by the occurrence of large numbers of cases in an area with already stretched public health facilities. A high death rate invariably causes panic, flight of the population, and spread of the disease. Public health administrators are therefore calling for intensified efforts to develop improved methods of prevention and control.

The most effective method remains the improvement of water supply and sanitation. The disappearance of cholera and most of the acute diarrheas from the industrialized countries, and the absence of cholera in those parts of a town or village ravaged by cholera epidemics where people have facilities permitting better hygienic habits, demonstrates the effectiveness of sanitation and better personal hygiene. However, the present rate of socioeconomic and other improvements in the developing world — where the problem lies — clearly indicates that even the slight improvement in water supply and sewage disposal that might suffice to reduce the transmission of *V. cholerae* may not be an attainable goal in the near future.

In spite of the fact that the cholera vaccine that is generally available is of unreliable potency, and even if potent, gives incomplete and short-lived protection *(vide supra)*, vaccination is widely used for cholera control. Vaccine does not reduce the rate of inapparent infection and thus cannot prevent transmission and introduction of cholera into a new area or country. The considered opinion is that provision of adequate treatment and enforcement of hygienic practices to the extent that available resources permit is much more beneficial for the control of an outbreak than vaccination.

Chemoprophylaxis or prophylactic treatment of close contacts, and sometimes of the whole community, is also resorted to frequently. The effectiveness of mass treatment — often begun too late and done without adequate supervision or surveillance for potential side effects — has not been properly evaluated. Generally, 2 to 3 g of tetracycline in divided doses over 2 to 3 days, or 300 mg of doxycycline in a single dose, is advised for adults and smaller doses for children. Sulfadoxine in a single dose has been extensively used in many countries in Africa.

This approach may be appropriate when cholera strikes a defined population group such as a refugee camp, military barracks, a boarding house, or a mental asylum where there is little prospect of providing treatment or improving sanitation and water supply. In such a situation vaccination with a potent vaccine may also be advisable.

Even in the most desperate situations a great deal can be achieved in cholera control by *ad hoc* improvement of water supply and excreta disposal by simple methods supported by extensive health education to improve personal hygiene.

DIARRHEA CAUSED BY *V. CHOLERAE* OF SEROVARS OTHER THAN 0-1

Yajnik and Prasad seem to have been the first to produce strong epidemiological evidence of the diarrheagenic potential of these vibrios previously called NAG (nonagglutinable) or NCV (noncholera) vibrios.[46] This was subsequently confirmed by other observations in India and also in Bangladesh.[47-50] In the meantime, Dutta and colleagues produced experimental evidence of the same in the infant rabbit.[51] A large waterborne outbreak of gastroenteritis due to these vibrios, resembling a cholera outbreak, has been recorded in Sudan.[52] The report of an explosive outbreak of gastroenteritis due to these vibrios transmitted through potato salad and affecting 56 out of 180 workers in an automobile training center in Czechoslovakia is particularly interesting;[53] 60 additional persons had mild symptoms but did not report sick. Several other foodborne outbreaks due to these organisms have been documented.

More recently two groups of workers have independently shown that these vibrios isolated from cases of severe diarrhea can produce an enterotoxin very similar to that of the true cholera vibrio and which can also be neutralized by the antiserum prepared against purified cholera toxin.[54,55]

The development in the early 1960s of selective media for the diagnosis of cholera has helped tremendously in the isolation of all types of vibrios from cases, carriers, animals, and the environment, particularly water. It may be noted that vibrios other than serovar 0-1 have been found in the environment, particularly in surface water, wherever they have been looked for, irrespective of the presence or absence of cholera in the community. In a WHO collaborative study, nearly 1000 vibrios of serovars other than 0-1 isolated from cases of diarrhea in the absence of other enteropathogens in different countries have been serotyped at the National Institute of Health, Tokyo, by Sakazaki, but no particular serovar(s) could be identified as being particularly diarrheagenic.

REFERENCES

1. **Shimada, T. and Sakazaki, R.,** Additional serovars and inter-o antigenic relationship of *V. cholerae, Jpn. J. Med. Sci. Biol.,* 30, 275—277, 1977.
2. International Committee on Systematic Bacteriology, Subcommittee on Taxonomy of Vibrios, Minutes of the meeting, July 22, 1971, Washington, D.C., *Int. J. Syst. Bacteriol.,* 22, 189—190, 1972.
3. **Barua, D.,** WHO activities in the control of acute diarrhoeal diseases including cholera, in *Proc. 43rd Nobel Symp. on Cholera and Related Diarrheas — Molecular Aspects of a Global Health Problem,* S. Karger AG, Basel, 1979, in press.
4. **Van de Linde, P. A. M. and Forbes, G. I.,** Observations on the spread of cholera in Hong Kong, 1961—63, *Bull. WHO,* 32, 515—530, 1965.
5. **Sinha, R., Deb, B. C., De, S. P.,** Abou-Gareeb, A. H., and Shrivastava, D. L., Cholera carrier studies in Calcutta in 1966—67, *Bull. WHO,* 37, 89—100, 1967.
6. **Kuwahara, S., Goto, S., Kimura, M., and Abe, H.,** Drug sensitivity of El Tor vibrio strains isolated in the Philippines in 1964 and 1965, *Bull. WHO,* 37, 763—771, 1967.
7. **Woodward, W. E. and Mosley, W. H.,** The spectrum of cholera in rural Bangladesh. II. Comparison of El Tor Ogawa and classical Inaba infection, *Am. J. Epidemiol.,* 96, 342—351, 1972.
8. **Cvjetanović, B. and Barua, D.,** The seventh pandemic of cholera, *Nature (London),* 239, 137—138, 1972.

9. Cash, R. A., Music, S. I., Libonati, J. P., Snyder, M. J., Wenzel, R. P., and Hornick, R. B., Response of man to infection with *V. cholerae*. I. Clinical, serologic and bacteriologic responses to a known inoculum, *J. Infect. Dis.*, 129, 42—52, 1974.
10. Félix, H., Le développement de l' épidémie de choléra en Afrique de l'Ouest, *Bull. Soc. Pathol. Exot.*, 64, 561—580, 1971.
11. Snow, J., On the mode of communication of cholera, in *Snow on Cholera,* Richardson, B. W. and Frost, W. H., Eds., Commonwealth Fund, New York, 1936, 1—139.
12. Teng, P. H., The role of foods in the transmission of cholera, in Proc. Cholera Research Symp., Honolulu, Hawaii, U.S. Public Health Service Publ. No. 1328, U.S. Government Printing Office, Washington, D.C., 1965, 328—332.
13. Joseph, P. R., Tamayo, J. F., Mosley, W. H., Alvero, M. G., Dizon, J. J., and Henderson, D. A., Studies of cholera El Tor in the Philippines. II. A retrospective investigation of an explosive outbreak in Bacolod City and Talisay, November 1961, *Bull. WHO,* 33, 637—643, 1965.
14. Cohen, J., Schwartz, T., Klasmer, R., Pridau, D., Ghalayini, H., and Davies, A. M., Epidemiological aspects of cholera El Tor outbreaks in a non-endemic area, *Lancet,* 2, 86—89, 1971.
15. Baine, W. B., Zampieri, A., Mazzoti, M., Angioni, G., Greco, D., Giola, M. D., Izz, E., Gangarosa, E. J., and Pocchiari, F., Epidemiology of cholera in Italy in 1973, *Lancet,* 2, 1370—1381, 1974.
16. Blake, P. A., Rosenberg, M. L., Costa, J. B., Ferreira, P. S., Guimaraes, C. L., and Gangarosa, E. J., Cholera in Portugal, 1974. I. Mode of transmission, *Am. J. Epidemiol.*, 105, 337—343, 1977.
17. Dutt, A. K., Alwi, S., and Velauthan, T., A shellfish-borne cholera outbreak in Malaysia, *Trans. R. Soc. Trop. Med. Hyg.*, 65, 815—818, 1971.
18. Merson, M. H., Martin, W. T., Craig, J. P., Morris, G. K., Blake, P. A., Craun, G. F., Feeley, J. C., Camacho, J. C., and Gangarosa, E. J., Cholera on Guam, 1974, *Am. J. Epidemiol.*, 105, 349—361, 1977.
19. Follow-up on *V. cholerae* infection — Louisiana, Morbidity and Mortality Weekly Rep. 27, Center for Disease Control, Atlanta, 1978, 367.
20. Sutton, R. G. A., An outbreak of cholera in Australia due to food served in flight on an international aircraft, *J. Hyg.*, 72, 441—451, 1974.
21. Goh, K. T., El Tor cholera in Singapore, *Epidemiol. News Bull.*, 3, 27—28, 1977.
22. Gangarosa, E. J. and Mosley, W. H., Epidemiology and surveillance of cholera, in *Cholera,* Barua, D. and Burrows, W., Eds., W. B. Saunders, Philadelphia, 1974, 381—403.
23. Mosley, W. H., The role of immunity in cholera. A review of epidemiological and serological studies, *Tex. Rep. Biol. Med.*, 27, 227—241, 1969.
24. Gitelson, S., Gastrectomy, achlorhydria and cholera, *Isr. J. Med. Sci.*, 7, 663—667, 1971.
25. Schiraldi, O., Benvestito, V., Di Bari, C., Moschetta, R., and Pastore, G., Gastric abnormalities in cholera: epidemiological and clinical considerations, *Bull. WHO,* 51, 349—352, 1974.
26. Nalin, D. R., Levine, R. J., Levine, M. M., Hoover, D., Bergqvist, E., McLaughlin, J., Libonati, J., Alam, J., and Hornick, R. B., Cholera, non-vibrio cholera, and stomach acid, *Lancet,* 2, 856—859, 1978.
27. Rosenberg, I. H., Greenough, W. B., III, Lindenbaum, J., and Gordon, R. S., Jr., Nutritional studies in cholera. Influence of nutritional status on susceptibility to infection, *Am. J. Clin. Nutr.*, 19, 384—389, 1966.
28. Sprinz, H., Sribhibhadh, R., Gangarosa, E. J., Benyajati, C., Kundel, D., and Halstead, S., Biopsy of small bowel of Thai people, with special reference to recovery from Asiatic cholera and to an intestinal malabsorption syndrome, *Am. J. Clin. Pathol.*, 38, 43—51, 1962.
29. Lindenbaum, J., Alam, A. K. M. J., and Kent, T. H., Subclinical small intestinal disease in East Pakistan, *Br. Med. J.*, 2, 1616—1619, 1966.
30. Gangarosa, E. J., Beisel, W. R., Benyajati, C., Sprinz, H., and Piyaratn, P., The nature of the gastrointestinal lesion in Asiatic cholera and its relation to pathogenesis: a biopsy study, *Am. J. Trop. Med.*, 9, 125—135, 1960.
31. Benenson, E. S., Cholera, in *Infectious Agents and Host Reactions,* Mudd, S., Ed., W. B. Saunders, Philadelphia, 1970, 285—302.
32. Palmer, D. L., Koster, F. T., Alam, A. K. M. J., and Islam, M. R., Nutritional status: a determinant of severity of diarrhoea in patients with cholera, *J. Infect. Dis.*, 134, 8—14, 1976.
33. Lindenbaum, J., Greenough, W. B., III, and Islam, M. R., Antibiotic therapy of cholera in children, *Bull. WHO,* 37, 529—538, 1967.
34. Carpenter, C. C. J., Jr., Greenough, W. B., III, and Gordon, R. S., Jr., Pathogenesis and pathophysiology of cholera, in *Cholera,* Barua, D. and Burrows, W., Eds., W. B. Saunders, Philadelphia, 1974, 129—141.
35. Hirschhorn, N., Lindenbaum, J., Greenough, W. B., III, and Alam, S. M., Hypoglycaemia in children with acute diarrhoea, *Lancet,* 2, 128—133, 1966.

36. **Barua, D.** Laboratory diagnosis of cholera, in *Cholera*, Barua, D. and Burrows, W., Eds., W. B., Saunders, Philadelphia, 1974, 85—126.
37. **Hirschhorn, N., Pierce, N. F., Kobari, K., and Carpenter, C. C. J., Jr.**, The treatment of cholera, in *Cholera*, Barua, D. and Burrows, W., Eds., W. B. Saunders, Philadelphia, 1974, 235—252.
38. **Mahalanabis, D., Watten, R. H., and Wallace, C. K.**, Clinical aspects and management of pediatric cholera, in *Cholera*, Barua, D. and Burrows, W., Eds., W. B. Saunders, Philadelphia, 1974, 221—233.
39. **Mahalanabis, D., Choudhuri, A., Bagchi, N. G., Bhattacharya, A. K., and Simpson, T. W.**, Oral fluid therapy of cholera among Bangladesh refugees, *Johns Hopkins Med. J.*, 132, 197—205, 1973.
40. International Study Group, A positive effect on the nutrition of Philippine children of an oral glucose-electrolyte solution given at home for the treatment of diarrhoea. Report of a field trial, *Bull. WHO*, 55, 87—94, 1977.
41. **Music, S. I., Libonati, J. P., Wenzel, R. P., Snyder, M. J., Hornick, R. B., and Woodward, T. E.**, Induced human cholera, *Antimicrob. Agents Chemother. — 1970*, pp. 462—466, 1971.
42. **Levine, M. M.**, Immunity in cholera evaluated in human volunteers, in *Proc. 43rd Nobel Symp. on Cholera and Related Diarrheas — Molecular Aspects of a Global Health Problem*, S. Karger AG, Basel, 1979, in press.
43a. **Sulianti Saroso, J., Bahrawi, W., Witjaksono, H., Budiarso, R. L. P., Brotowasisto, Benčić, Z., Dewitt, W. E., and Gomez, C. Z.**, A controlled field trial of plain and aluminium hydroxide-adsorbed cholera vaccines in Surabaja, Indonesia during 1973—1975, *Bull. WHO*, 56, 619—627 and 995, 1978.
43b. **Pal, S. C., Deb B. C., Sen Gupta, P. G., De, S. P., Sircar, B. K., Sen, D., and Sikdar, S. N.**, A controlled field trial of an aluminium phosphate-absorbed cholera vaccine in Calcutta, *Bull. WHO*, 58, 741—745, 1980.
44a. **Curlin, G., Levine, R., Aziz, K. M. A., Rahman, A. S. M. M., and Verwey, W. F.**, Field trial of cholera toxoid, in Proc. 11th Joint Conf. on Cholera of the U.S.-Japan Cooperative Medical Science Program, National Institutes of Health, Bethesda, 1975, 314—329.
44b. **Holmgren, J., Svennerholm Ann-Mari, and Lönnroth, I.**, Development of improved cholera vaccine based on subunit toxiod, *Nature (London)*, 269, 602—604, 1977.
45. **Finkelstein, R. A., Vasil, M. L., and Holmes, R. K.**, Studies on toxinogenesis in *V. cholerae*. I. Isolation of mutants with altered toxinogenicity, *J. Infect. Dis.*, 129, 117—123, 1974.
46. **Yajnik, B. S. and Prasad, B. G.**, A note on vibrios isolated in Kumbh Fair, Allahabad, 1954, *Indian Med. Gaz.*, 89, 341—349, 1954.
47. **McIntyre, O. R., Feeley, J. C., Greenough, W. B., III, Benenson, A. S., Hassan, S. I., and Saad, A.**, Diarrhoea caused by non-cholera vibrios, *Am. J. Trop. Med. Hyg.*, 14, 412—418, 1965.
48. **Lindenbaum, J., Greenough, W. B., III, Benenson, A. S., Oseasohn, R., Rizvi, J., and Saad, A.**, Non-vibrio cholera, *Lancet*, 1, 1081—1083, 1965.
49. **Carpenter, C. C. J., Barua, D., Wallace, C. K., Sack, R. B., Mitra, P. P., Werner, A. S., Duffy, T. P., Oleinick, A., Khanra, S. T., and Lewis, G. W.**, Clinical and physiological observations during an epidemic outbreak of non-vibrio cholera-like disease in Calcutta, *Bull. WHO*, 33, 665—671, 1965.
50. **Chatterjee, B. D., Gorbach, S. L., and Neogy, K. N.**, *V. parahaemolyticus* and diarrhoea associated with non-cholera virbrios, *Bull. WHO*, 42, 460—463, 1970.
51. **Dutta, N. K., Panse, M. V., and Jhala, H. I.**, Cholerangenic property of certain strains of El Tor, non-agglutinable, and water virbrios confirmed experimentally, *Br. Med. J.*, 1, 1200—1203, 1963.
52. **Kamal, A. M. and Zinnaka, Y.**, Outbreak of gastroenteritis by nonagglutinable (NAG) vibrios in the Republic of Sudan, *J. Egypt. Public Health Assoc.*, 40, 125—174, 1971.
53. **Aldova, E., Láznĭcková, K., Štěpánková, E., and Lietova, J.**, Isolation of nonagglutinable vibrios from an enteritis outbreak in Czechoslovakia, *J. Infect. Dis.*, 118, 25—31, 1968.
54. **Zinnaka, Y. and Carpenter, C. C. J., Jr.**, An enterotoxin produced by noncholera vibrios, *Johns Hopkins Med. J.*, 131, 403—411, 1972.
55. **Ohashi, M., Shimada, T., and Fukumi, H.**, *In vitro* production of enterotoxin and hemorrhagic principle by *Vibrio cholerae*, NAG, *Jpn. J. Med. Sci. Biol.*, 25, 179—194, 1972.

FOODBORNE DISEASES: BRUCELLOSIS

Robert I. Wise

INTRODUCTION

Brucellosis (undulant fever) spreads to people from animals. Spread from person to person rarely, if ever, occurs. In contrast, the infection in animals is contagious, for infected animals readily infect other animals, and their illness may become chronic with shedding of the microorganisms into the environment, often for many years. The process of domestication of animals, herding them together and moving them from one area to another, has produced conditions which are unnatural and increase the ease by which the bacteria survive and are transmitted.

RESERVOIRS OF BRUCELLOSIS

Brucellae are small Gram-negative, aerobic, nonmotile, nonspore-forming coccobacilli. Three species, which have characteristic affinity for specific animal hosts, cause most human brucellosis. *Brucella abortus* causes the disease in cattle, *B. suis* in swine, and *B. melitensis* in sheep and goats; however, cross-infections of animals can occur with all species. Brucellae can infect deer, elk, bison, moose, and other wild animals and infection has been reported in chickens, turkeys, and birds; however, poultry and birds are not significant sources of the infection in people.[1] A newly described species, *B. canis* causes infections of dogs; however, human infection with this organism has rarely been reported.[2]

ACQUISITION OF BRUCELLOSIS

Animals acquire brucellosis by eating, licking, or drinking infected or contaminated materials. The pregnant uterus may become infected with resulting abortion or birth of infected young animals. In suckling, the viable young animals ingest large numbers of *Brucella,* which are shed in the milk of infected animals and become infected.[3]

Brucellosis in people is contracted from infected animals or their products; it is foodborne by raw meat and dairy products such as butter, cream, ice cream, cheese, yogurt, etc. For some people the disease is an occupational hazard, particularly for farmers, livestock workers, dairy people, veterinarians, meat inspectors, packing plant workers, rendering plant employees, butchers, hunters, cooks, and others, who work with or consume contaminated food.[4-10]

Human beings may acquire the infection through various portals of entry. The Brucellae may enter through abrasions or cuts of the skin when there is contamination, by inhalation of aerosols of the bacteria through the respiratory tract, by contamination of the eyes, and by ingestion of food which contains the microorganisms. Unwashed hands, contaminated from infected raw meat or milk or their products, may convey the Brucellae to food which is then eaten, or to abrasions, cuts of the skin, or to the eyes.

The problem in abattoirs is a unique one, for the employees are exposed to a disease for which there is no adequate means of detection, and methods of prevention are not satisfactory. Cows and pigs, which are found to be reactors, are sent to the packing plants for slaughter. Studies have revealed the presence of Brucellae in the meat of reactor animals.[11] There is potential exposure of all people who handle the meat or meat products.

The dairy industry is involved for milk from infected cows and goats contains the Brucellae. Fortunately these microorganisms do not survive the process of pasteurization; however, many people drink unpasteurized milk and eat butter, ice cream, cheese, and yogurt which has been prepared from raw milk. Travelers to countries which do not have adequate controls may contract brucellosis by eating unpasteurized milk products which contain viable Brucellae.

PATHOPHYSIOLOGY

The Brucellae invade mucous membranes, enter the lymphatic system, localize in regional lymph nodes, subsequently invade the blood stream, and disseminate throughout the body. Secondary localization occurs particularly in organs and tissues, which possess abundant reticuloendothelial cells, spleen, liver, bone marrow, lymph nodes, and kidneys. The microorganisms localize in the cytoplasm of the phagocytic mononuclear cells of tissues with resultant formation of granulomatous lesions. Necrosis of tissue may be slight or absent, but abscess formation and caseation may occur. The host response is one of phagocytosis by mononculear and polymorphonuclear cells and the production of both humoral and cellular immunity. A state of hypersensitivity may be a prominent feature in human beings.[12,13]

MANIFESTATIONS

After an incubation period which usually varies from 1 to 3 weeks, the infected person develops malaise, fatigue, chills, fever, sweats, and weakness. Many patients complain of muscle aches, headache, and anorexia. Weight loss may range from 2 to 20 kg. Cough is present in approximately 20% of cases. Other manifestations include nervousness, backache, insomnia, mental depression, nausea, vomiting, pain in the back of the neck, abdominal pain, constipation, diarrhea, visual disturbances, neuritic pain, testicular swelling and pain, dysuria, tinnitus, and dizziness. Physical abnormalities may include fever, splenomegaly, and lymphadenopathy as the most frequent signs; however, some patients demonstrate abdominal tenderness, hepatomegaly, cardiac abnormalities, neurological changes, tenderness over the spine, skin lesions, funduscopic changes, orchitis, jaundice, and tender joints. The illness may vary in severity from a mild episode, which persists for a few days to a fatal result. The mortality rate in the U.S. is approximately 2%.[14]

Most patients, who are diagnosed as having brucellosis and are treated, have an uneventful recovery; however, complications may be severe.[15] There may be a diffuse or localized encephalopathy with debilitating neuropsychiatric disorder,[16] chronic arthritis, particularly spondylitis osetomyelitis and bursitis,[17] endocarditis,[18] hepatitis,[19] splenitis with hypersplenism and associated thrombocytopenic purpura and hemorrhage,[20] infections of the genitourinary tract, orchitis, epididymitis, and pyelonephritis.[21] *Brucella* pneumonia has been described with pleural effusions and empyema.[22,23] Visual disturbances are common manifestations and are due to involvement of the optic nerve and other structures of the eye.[24]

DIAGNOSIS

The diagnosis of brucellosis in human beings depends upon recognition of the manifestations, which are not specific for brucellosis, but may be characteristic of other febrile diseases, and the occupational history of the patient. The patient may be associated with an animal industry (beef, dairy, swine, goat, sheep) or a consumer of raw meat, milk, butter, ice cream, cheese, etc. The illness may vary in severity from a low-

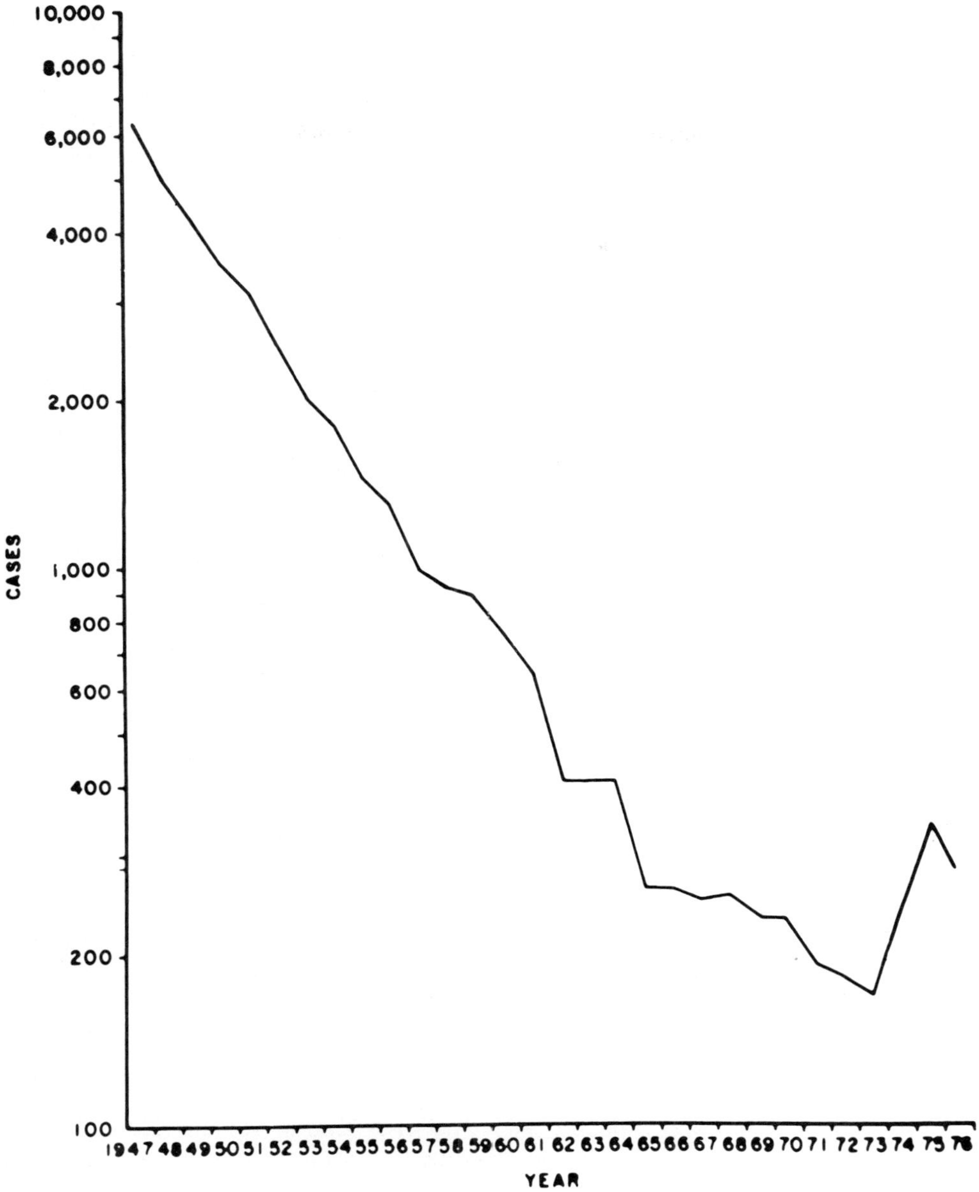

FIGURE 1. Human brucellosis in U.S., 1947 to 1976.

grade fever, which is ill defined, to a febrile state with chills, sweating and prostration. The slide agglutination test provides serological information rapidly. The tube agglutination test supplemented by the complement fixation test on the convalescent serum is the most dependable serologic diagnostic procedure, but requires longer time and greater expense than the rapid slide agglutination. Isolation of *Brucella* from blood or tissues establishes the diagnosis as brucellosis. Blood cultures are reported to be positive in approximately 50% of acute cases when performed by competent methods.[25]

TREATMENT

Therapy with tetracycline for 3 weeks or with tetracycline for 3 weeks plus strepto-

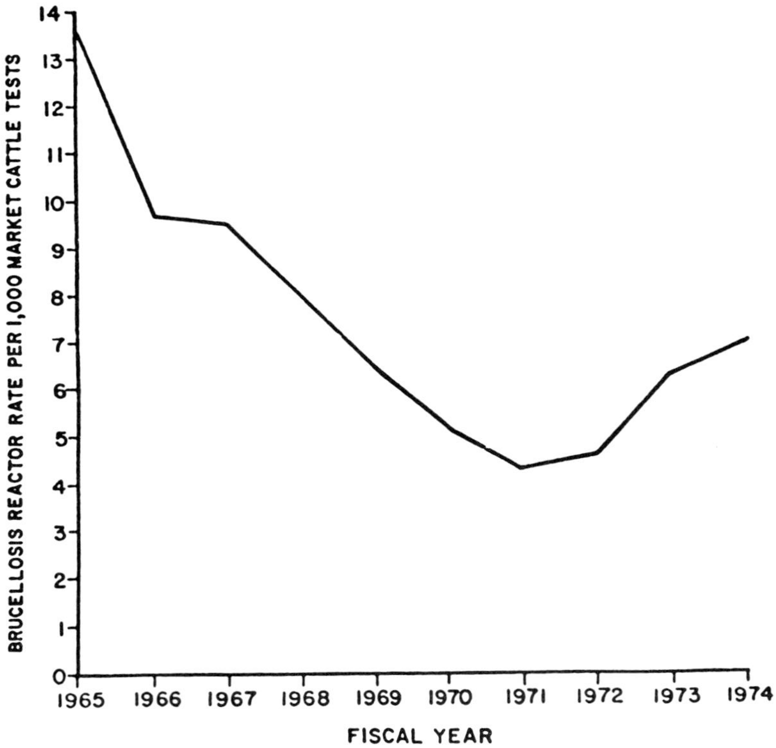

FIGURE 2. Bovine brucellosis reactor rates found in market cattle testing in U.S. for fiscal years 1965 to 1974. (From Cooperative State-Federal Brucellosis Eradication Program Statistical Tables, U.S. Department of Agriculture, Washington, D.C., 1965 to 1974.)

mycin for the first 2 weeks results in cure in approximately 80% of the cases.[26] Chloramphenicol has been used successfully, but has had limited trials in therapy of human brucellosis because of disadvantages in its use.[27] Trimethaprim and sulfamethoxazole in combination has produced remissions in acute brucellosis.[28] A problem in the treatment of brucellosis is that relapse may occur.

IMMUNITY AND RESISTANCE

There may be some degree of immunity following an attack of human brucellosis, but second attacks occur. Subsequent attacks may be mild; however, hypersensitivity may exist and result in severe illness.[5,13] At present, there are no safe, effective methods of enhancing resistance to human brucellosis by the use of vaccines.

In contrast, a broad spectrum of vaccines, viable and killed, virulent and attenuated, have been used in attempts to enhance resistance in animals, particularly cattle, swine, and goats. The use of Strain 19, a viable vaccine of *B. abortus*, and *B. abortus* killed 45/20 vaccine are helpful in enhancement of resistance to bovine brucellosis and are used in many national programs to control or eradicate the disease.[29,30] Since only a partial degree of immunity is developed by vaccination, other measures must be used concurrently with vaccination if eradication of the disease from a herd is to be accomplished.[31] There are no practical methods of immunization available in the control of swine brucellosis. Vaccines of *B. meletensis* have been found to provide some protection in goats.[32]

PREVENTION

Prevention of brucellosis in people is dependent upon eradication of the disease in animals. Many countries have developed programs to prevent the spread of brucellosis among animals, from animals to people, and to eradicate the disease from the area. Attempts to prevent and eradicate brucellosis consists of methods which are designed to eliminate the reservoirs of the infection from local areas, states, provinces, and countries, and to prevent reentry of the infection into the area. These methods consist of the identification of every animal, immunization of animals, prevention of cross-infection from infected animals to healthy animals and from infected herds to uninfected herds, the detection of infection in animals and herds, slaughter of infected animals, pasteurization of milk and milk products, sanitary practices in the processing and handling of meat and meat products, the cooking of meat and meat products, the surveillance and reporting of infections of people and animals to appropriate public and animal health authorities, and the education of every person who has responsibility in the prevention of this disease.[31] Such programs have been effective in reducing the incidence of human disease. Figure 1 demonstrates the reduction of reported cases of human brucellosis in the U.S. from 6321 in 1947 to 175 in 1973.[2] The increase in incidence of human brucellosis reported in 1974 to 1976 followed an increase in the brucellosis reactor rates of market cattle in the U.S. which began in 1972 as seen in Figure 2.[33]

INCIDENCE

Seven countries have reported the eradication of bovine brucellosis. They are in order of the last occurrence of reported bovine brucellosis: Norway, 1952; Sweden, 1957; Finland, 1960; Denmark, 1962; Switzerland, 1963; Czechoslovakia, 1964; and Japan, 1974. Brucellosis is present in all other countries of the world. The Netherlands, West Germany, and Austria report ''exceptional occurrence''; Belgium, Italy, Russia, Yugoslavia, and New Zealand report ''low sporadic incidence''. The disease still exists in Great Britain, Spain, Canada, and the U.S., though it is ''much reduced''. Ireland, France, Portugal, Mexico, and Australia report ''moderate incidence'' of infection. Most countries of Central and South America report moderate to high levels of bovine brucellosis. The countries of Africa report ''low sporadic'' to ''high incidence'', and ''high incidence'' is reported in India, Mongolia, and Hong Kong.[34]

REFERENCES

1. **Spink, W. W.**, The reservoir of brucellosis, in *The Nature of Brucellosis,* Spink, W. W., Ed., The University of Minnesota Press, Minneapolis, 1956, 65—79.
2. *Brucellosis Surveillance, Annu. Summary, 1976, Center for Disease Control, Atlanta, October 6, 1977, 6.*
3. **Cunningham, B.**, A difficult disease called brucellosis, in *Bovine Brucellosis An International Symposium,* Crawford, R. P. and Hidalgo, R. J., Texas A&M University Press, College Station, 1977, 11—20.
4. **Borts, I. H.**, Some observations regarding the epidemiology, spread and diagnosis of brucellosis, *J. Kan. Med. Soc.,* 46, 399—405, 1945.
5. **Buchman, T. M., Faber, L. C., and Feldman, R. A.**, Brucellosis in the United States, 1960—1972, an abatoir-associated disease, *Medicine,* 53, 403—439, 1974.
6. **Hendricks, S. L., Borts, I. H., Heeren, R. H., Hausler, W. J., and Held, J. R.**, Brucellosis outbreak in an Iowa packing house, *Am. J. Public Health,* 52, 1166—1178, 1962.

7. **Hutchings, L. M., McCullough, N. B., Dunham, C. R., Eisele, C. W., and Bunnel, D. E.,** The viability of *Br. meletensis* in naturally infected hams, *Public Health Rep.*, 66, 1402—1408, 1951.
8. **McCullough, N. B., Eisele, C. W., and Byrne, A. F.,** Incidence and distribution of *Br. abortus* in slaughtered Bangs reactor cattle, *Public Health Rep.*, 66, 341—345, 1951.
9. **McCullough, N. B., Eisele, C. W., and Pavelcheck, E.,** Survey of brucellosis in slaughtered hogs, *Public Health Rep.*, 66, 205—208, 1951.
10. **McNutt, S. H.,** Incidence and importance of *Brucella* infection of swine in packing houses, *J. Am. Vet. Med. Assoc.*, 86, 183—191, 1935.
11. **Sadler, W. W.,** Present evidence on the role of meat in the epidemiology of human brucellosis, *Am. J. Public Health,* 50, 504—514, 1960.
12. **Spink, W. W.,** The pathogenesis of brucellosis, in *The Nature of Brucellosis,* Spink, W. W., Ed., The University of Minnesota Press, Minneapolis, 1956, 112—144.
13. **Spink, W. W.,** The significance of bacterial hypersensitivity in human brucellosis: studies on infection due to Strain 19 *Brucella abortus, Ann. Int. Med.*, 47, 861—874, 1957.
14. **Spink, W. W.,** The natural course of brucellosis, in *The Nature of Brucellosis,* Spink, W. W., Ed., The University of Minnesota Press, Minneapolis, 1956, 145—170.
15. **Spink, W. W.,** The complications of brucellosis, in *The Nature of Brucellosis,* Spink, W. W., Ed., The University of Minnesota Press, Minneapolis, 1956, 171—190.
16. **DaJong, R. N.,** Central nervous system involvement in undulant fever, with report of a case and a survey of the literature, *J. Nerv. Ment. Dis.*, 83, 430—442, 1936.
17. **Lowbeer, L.,** Brucellotic osteomyelitis of the spinal column in man, *Am. J. Pathol.*, 24, 1949.
18. **Call, J. D., Baggenstoss, A. H., and Merritt, W. A.,** Endocarditis due to *Brucella:* report of two cases, *Am. J. Clin. Pathol.*, 14, 508—518, 1944.
19. **Chaikin, N. W. and Schwemmer, D.,** Hepatitis in the course of *Brucella* infection, *Rev. Gastroenterol.*, 10, 130—132, 1943.
20. **Weed, L. A., Dohlin, D. C., Pugh, D. G., and Ivins, J. I.,** *Brucella* in tissues removed at surgery, *Am. J. Clin. Pathol.*, 22, 10—21, 1952.
21. **Forbes, K. A., Lowry, E. C., Gibson, T. E., and Soanes, W. A.,** Brucellosis of the genitourinary tract: review of the literature and report of a case in a child, *Urol. Surg.*, 4, 1954.
22. **Harvey, W. A.,** Pulmonary brucellosis, *Ann. Int. Med.*, 28, 768—781, 1948.
23. **McDonald, R. H.,** Acute empyema with *Brucella abortus* as the primary causative agent: case report, *J. Thorac. Surg.*, 9, 92—93, 1939.
24. **Puig Solanes, M., Heatley, M. J., Arenas, F., and Guerrero Ibanra, G.,** Ocular complications in brucellosis, *Am. J. Ophthalmol.*, 36, 675, 1953.
25. **Spink, W. W.,** The diagnosis of brucellosis, in *The Nature of Brucellosis,* Spink, W. W., Ed., The University of Minnesota Press, Minneapolis, 1956, 191—215.
26. **Spink, W. W.,** The therapy of brucellosis, in *The Nature of Brucellosis,* Spink, W. W., Ed., The University of Minnesota Press, Minneapolis, 1956, 216—258.
27. **Knight, V., Ruiz-Sanchez, F., and McDermott, W.,** Chloramphenicol in the treatment of the acute manifestations of brucellosis, *Am. J. Med. Sci.*, 219, 627—638, 1950.
28. **Hassan, A., Erian, M. M., Farid, Z., Hathaut, S. C., and Sorensen, K.,** Trimethaprim-sulphamethazole in acute brucellosis, *Br. Med. J.*, 3, 159—160, 1971.
29. **Godfrey, A.,** Experiences with *Brucella* vaccines, in *Bovine Brucellosis an Int. Symp.*, Crawford, R. P. and Hidalgo, R. J., Eds., Texas A&M University Press, College Station, 1977, 209—217.
30. **Nelson, C.,** Immunity to *Brucella abortus,* in *Bovine Brucellosis an Int. Symp.*, Crawford, R. P. and Hidalgo, R. J., Ed., Texas A&M University Press, College Station, 1977, 177—188.
31. **Anderson, R. K., Berman, D. T., Berry, W. T., Hopkin, J. A., and Wise, R. I.,** Rep. Natl. Brucellosis Tech. Comm., prepared for the Animal and Plant Inspection Service, U.S. Department of Agriculture and U.S. Animal Health Association, August 28, 1978.
32. **Spink W. W.,** The prevention of brucellosis, in *The Nature of Brucellosis,* Spink, W. W., The University of Minnesota Press, Minneapolis, 1956, 259—280.
33. *Cooperative State-Federal Brucellosis Eradication Program Statistical Tables, U.S. Department of Agriculture, Washington, D.C., 1965 to 1974.*
34. **Konigshofer, H. O.,** Animal Health Yearbook 1977, Food and Agriculture Organization, World Health Organization, International Office of Epizootics, Geneva.

FOODBORNE DISEASES: AFLATOXICOSIS

D. S. P. Patterson

INTRODUCTION

In 1960 an outbreak of acute liver disease ("Turkey X disease") occurred among poultry in the U.K. and some 100,000 birds died. The cause was traced to particular shipments of imported groundnut meal used in the manufacture of poultry feed and eventually the specific liver poison was shown to be a fungal metabolite, which became known as aflatoxin. Historic detail will be found in several reviews on this subject.[1-4]

Four closely related toxins occur naturally (Figure 1) but aflatoxin B_1 is the most common of the group: it is an acute liver toxin[1-4] and a carcinogen.[5] It is produced by certain strains of only *Aspergillus flavus* and *A. parasiticus*. These fungi are ubiquitous and the potential for contamination of human food and animal feed is therefore considerable. However, certain commodities are especially prone to fungal attack and, in general, this is more likely to occur in tropical and subtropical countries than in temperate ones.

Moisture content of the substrate and temperature are the main factors regulating the growth of *A. flavus* and aflatoxin formation. These and other factors are considered fully by Hesseltine.[6] Although molding is predominantly a storage problem, it can also occur on some crops in the field before harvest. And, in view of its relatively common occurrence (usually in the microgram per kilogram range) in certain cereals, oil seeds, and tree nuts, aflatoxin is now often considered to be an unavoidable contaminant of these commodities and their products. However, concentrations can be kept low by good farming practice and, in the food and feedstuffs industries, the technology is available to mechanically sort contaminated nuts and kernels, monitor raw material by simplified methods of chemical analysis, and even to decontaminate low grade commodities before use in the manufacture of compounded animal feeds.[7,229]

OCCURRENCE OF AFLATOXIN

A wide variety of crops are susceptible to contamination but corn (maize), oil seed (groundnut or peanuts, cottonseed), and tree nuts (Brazils, pistachios) are most frequently attacked by *A. flavus* and eventually contaminated with aflatoxin (see Table 1).

In addition to direct contamination caused by fungal attack from *A. flavus*, aflatoxin may also be "carried over" from feeds to the tissues of food-producing animals and so to human food. Groundnut meal, cottonseed, or corn (maize) are usually the primary sources of the toxin and, when these commodities are included in dairy rations, a small proportion of the ingested toxin appears in cows' milk as the metabolite M_1. The concentration of this metabolite in the milk is directly proportional to the total daily amount of aflatoxin B_1 ingested (see reviews by Kiermeier,[11] Patterson,[4] and Purchase[12]).

Aflatoxin B_1 has been detected at a concentration of 0.012 μg/kg in naturally contaminated pig liver,[12a] but there appear to be no reports of aflatoxin occurring naturally in other animal products. However, Rodricks and Stoloff[13] have reviewed the literature of experimental observations showing that aflatoxin could be "carried-over" into meat, liver, and eggs (see Table 2).

Thus, milk is the major secondary source of aflatoxin for man and there have been many attempts to measure the level of aflatoxin M_1 in milk since the discovery of this

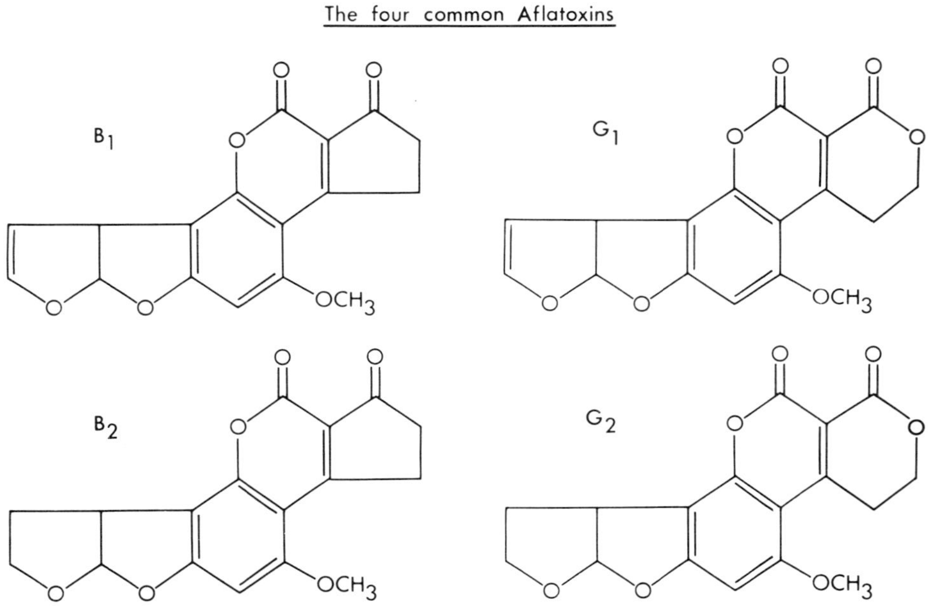

FIGURE 1. Four commonly occurring aflatoxins. The B-series of toxins fluoresce blue-green and the G-series greenish-yellow when separated by silica gel thin-layer chromatography and viewed in UV light.

Table 1
CROPS, FOODS, AND FEEDINGSTUFFS KNOWN TO HAVE BEEN NATURALLY CONTAMINATED WITH AFLATOXIN[8-10,25]

Barley
Brazil nuts*, pistachio nuts, pecans, almonds*, walnuts*, filberts, hazel nuts*, cashew nuts
Copra
Corn (maize)* and corn products
Cotton seed*
Dairy products: milk*, cheeses
Dried figs
Fruit, including apples, oranges, lemons, peaches
Meat products: fermented sausages
Mixed anial feeds (usually those containing oil seed meal or corn)
Peanuts*, groundnut meal*
Peppers, spices, e.g., nutmeg
Rice
Rye
Sorghum
Soya bean
Tomatoes
Various unrefined seed oils
Wheat

Note: The most frequently implicated commodities are marked * and levels of contamination sometimes reach the milligram per kilogram range. Working limits of about 5 μg aflatoxin B_1 per kilogram are now commonly applied in the food industry however.

Table 2
THE "CARRY-OVER" OF AFLATOXIN B_1 FROM ANIMAL FEEDSTUFFS TO EDIBLE ANIMAL TISSUES, EGGS, AND MILK

Food	Aflatoxin residue present	Conc. as approximate percentage of the level in animal feedstuffs[a]
Cows' milk	M_1	0.33
Pig liver	B_1	0.13
Chicken liver	B_1	0.08
Hens' eggs	B_1	0.05
Beef liver/muscle	B_1	0.01

[a] Adapted from a review by Rodricks and Stoloff[13] of experimental data. Milk is the best known *naturally* contaminated food of animal origin: see Table 3.

excretion pathway by Allcroft and Carnaghan[14] in 1963. However, since then analytical methods have become so sensitive that detection limits of about 0.03 μg/ℓ are commonly achieved, and several surveys worldwide have now shown that, while levels of up to 13 μg aflatoxin M_1 per liter are possible, concentrations of 0.1 μg/ℓ or less are more usual, particularly where the use of contaminated dairy rations is properly controlled (see Table 3).

METABOLISM AND MODE OF ACTION

Most of our knowledge of this subject derives from studies in vitro using liver tissues from duck, mouse, rat, and the rainbow trout (see reviews[4,28,29]).

Aflatoxin B_1 is metabolized in the liver to a number of closely related compounds and with one exception (aflatoxin 2,3 oxide, see below), each biotransformation involves the conversion to a hydroxylated metabolite (Figure 2). Only one such derivative, aflatoxin M_1, has appreciable acute oral toxicity[30] however. This and two other metabolites, aflatoxin P_1 and Q_1, also undergo conjugation with glucuronic acid and are excreted in the bile and urine,[31-34] which, although not truly a detoxification reaction, permits the more rapid clearance of these metabolites from the body. The 2,3-oxide would undergo true detoxification when converted by epoxide hydrase to the corresponding dihydrodiol (see below) or conjugated with glutathione[34a,34b] by glutathione *S*-transferase.

Most of these metabolic reactions are catalyzed by hepatic microsomal enzymes (mixed function oxidases) but a few are associated with the NADP-linked dehydrogenases located in the cytoplasm[35-38] (see Figure 2).

Four metabolites of aflatoxin B_1 are thought to be specially important:

1. Aflatoxin M_1 — Although it is somewhat less toxic than the parent toxin it is excreted in cows' milk and therefore constitutes a potential environmental hazard to man (see "Occurrence of Aflatoxin").
2. Aflatoxin B_1 2,3-oxide — When metabolically activated, aflatoxin B_1 is believed to be converted by microsomal mixed function oxidases of the liver to its 2,3-oxide or epoxide[39,40,40a] (Figure 2). This is a short-lived highly reactive molecule which binds covalently to the guanine bases of DNA[41-43] with possible mutagenic and carcinogenic consequences. The resulting aflatoxin-DNA complex and any unchanged epoxide are further metabolized by microsomal enzymes[43a-43c] to the corresponding dihydrodiol (2,3-dihydroxy 2,3-dihydro aflatoxin B_1).

Table 3
SELECTED SURVEYS OF COWS' MILK FOR AFLATOXIN M_1 CONTAMINATION[a]

Country	Total no. samples analyzed	No. containing aflatoxin M_1	Range of conc. μg/ℓ or kg	Ref.
		Liquid		
Belgium	68	42	0.02—0.2	15
E. Germany	36	4[b]	1.7—6.5	16
W. Germany	61	28	0.01—0.25	17
W. Germany	419	79	Trace-0.54	18
W. Germany	260	118[b]	0.05—0.33	19
India	21	3	Up to 13.3	20
Netherlands	95	74	0.09—0.5	21
U.K.	278	85[b,c]	0.03—0.52	22
U.S.	302	177	0.1-more than 0.5	23
		Dried		
E. Germany	18	1[d]	6.4	16
W. Germany	166	8	0.67—2.0	24
W. Germany	55	36	Trace—4.0	25
W. Germany	120	7[b]	0.05—0.13	26
South Africa	56	0	—	27

[a] Surveys carried out during or since 1972 but not including data that imply excessively contaminated dairy rations: detection limits approximately 30 ng/l.
[b] Seasonal effect observed, i.e., concentration depends upon the level of concentrate feeding.
[c] 92.5% contained no more than 0.1 μg aflatoxin M_1 per liter.
[d] Contained aflatoxin B_1.

3. Aflatoxin B_1 dihydrodiol — This may be regarded as the second activated form of aflatoxin B_1. Like the closely related hemiacetal[36,41,44-48] (Figure 2) for which it was at first probably confused,[43a] it forms Schiff bases nonspecifically by covalent interaction with proteins and when this involves key enzyme systems it probably leads to disruption of metabolism and eventually liver cell death.[36,43a] This pathway is well developed in the duckling and some other species that are particularly susceptible to acute aflatoxin poisoning.[36,43a,44,48]
4. Aflatoxicol — Shortly after the discovery that aflatoxin was reversibly reduced in the liver cell cytoplasm, it was suggested[36,48] that this constituted a "metabolic reservoir" of aflatoxin and, where it was a quantitatively important pathway, that it might increase an animal's susceptibility to the toxic effect of aflatoxin. This proposal has since been taken further by Hsieh et al.[49] and Salhab and Edwards,[50] who have produced data which they believe suggest that human liver can be grouped with those of other species that are known to be relatively resistant to the carcinogenic effects of aflatoxin B_1 (see Table 4) and in which the reversible aflatoxicol pathway is relatively poorly developed.

[In the above paragraphs, the metabolism of aflatoxin B_1 only has been considered. This is largely because most data have been obtained for this particular toxin but it will be obvious from an inspection of Figure 2 and comparison of the relevant struc-

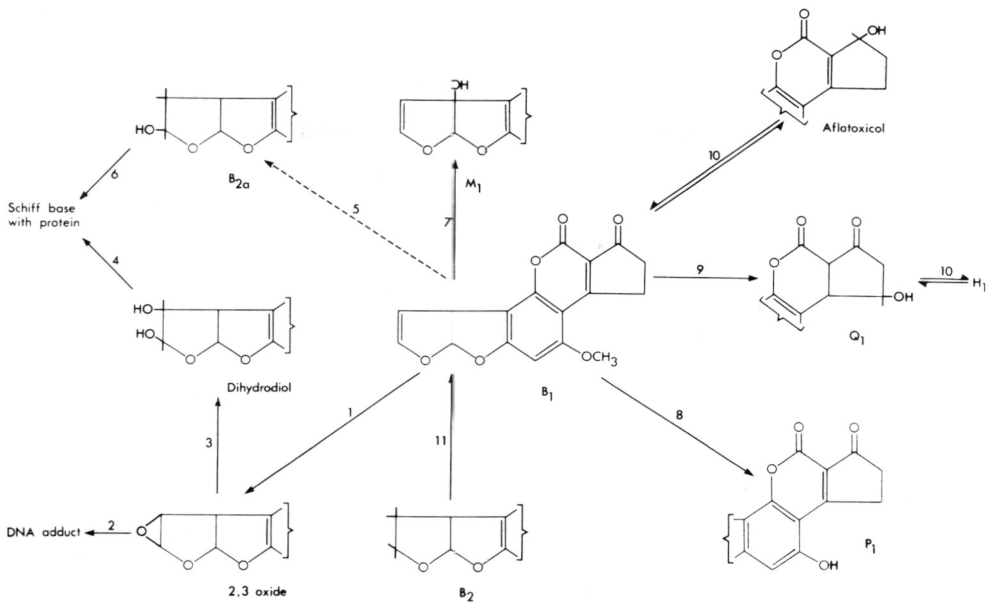

FIGURE 2. Generalized scheme for the metabolism of aflatoxin B_1 in the liver. Aflatoxin B_1 may be metabolically activated by the mixed function oxidases (MFOs)to form the 2,3-oxide or epoxide (1), which either binds covalently to DNA forming an N^7-guanyl adduct (2), or is hydrolyzed by epoxide hydrase to the dihydrodiol (3). The aflatoxin-DNA adduct also spontaneously yields the dihydrodiol. The latter metabolite, in its dialdehydric phenolate form at physiological pH, forms Schiff bases with proteins (6). The other metabolites are hydroxylated derivatives. Aflatoxin M_1 (7), aflatoxin P_1 (8), and aflatoxin Q_1 (9) are produced by MFOs but aflatoxical (10) is formed reversibly by NADP-dependent cytoplasmic dehydrogenases. Aflatoxical H_1 is similarly derived from aflatoxin Q_1: corresponding derivatives of other metabolites of aflatoxin B_1 (11). Reactions 1 to 8 are also open to aflatoxin G_1 and reactions 7, 8, and 11 to aflatoxin G_2.

Not shown in this scheme are the conjugation reactions forming glucuronides of the hydroxylated metabolites and the glutathione conjugate of aflatoxin B_1 2,3-oxide.

tural formulae (Figure 1) that certain pathways are also available to aflatoxins B_2, G_1, and G_2.]

The mode of action of aflatoxin expressed in terms of interactions with liver cell organelles and biochemical systems is summarized in Table 5, but at the molecular level, the acute and chronic action of aflatoxin can probably best be explained in terms of interactions with protein and nucleic acid mentioned above. In either case aflatoxin would appear to be metabolically activated first.

As described above, metabolic pathways in the liver include detoxification as well as activation and these processes tend to differ among the species, from breed to breed, and between the sexes. Metabolism is also influenced by age, hormonal status, and the composition of the diet. So, these factors also largely determine the outcome of ingesting aflatoxin.

ACUTE AFLATOXICOSIS

Aflatoxin B_1 is an acute liver poison and, measured in terms of oral LD_{50} values, there are appreciable species differences in susceptibility (Table 6): the other commonly occurring aflatoxins, B_2, G_1, G_2, and M_1 are intrinsically less toxic[77,77a] (Table 7). Dis-

Table 4
COMPARATIVE AFLATOXIN METABOLISM: THE POSSIBLE CORRELATION BETWEEN THE REVERSIBLE AFLATOXICOL PATHWAY[35,36] IN LIVERS OF DIFFERENT SPECIES AND THEIR SUSCEPTIBILITIES TO AFLATOXIN-INDUCED CARCINOGENESIS

		Reduction of aflatoxin to aflatoxicol in vitro	
Species	Susceptibility to mutagenic/ carcinogenic action of aflatoxin	Aflatoxicol as a[a] percentage of all metabolites formed	Reversibility[b] ratio of aflatoxicol formed to aflatoxin reformed
Rabbit	?	—	6.7
Trout	Susceptible	—	3.3
Duck	Susceptible	17.1	—
Rat	Susceptible	8.0	0.1
Mouse	Relatively resistant	6.0	0.2
Monkey	Relatively resistant	4.1	0.2
Man	?	2.1	0.1

[a] Calculated from the data of Hsieh et al.[49]
[b] Approximate ratios calculated from graphical data of Salhab and Edwards.[50]

tinctive pathological features of aflatoxin-induced hepatotoxicity in most animal species include liver cell necrosis, vacuolation, and hemorrhage with fatty infiltration, fibrosis, and bile duct proliferation: see reviews by Butler,[3] Edds,[78] and Newberne and Butler.[79] In the following accounts of acute aflatoxicosis, clinical and pathological effects relate to experimental or accidental poisoning by aflatoxin B_1.

Laboratory Animals

Much published toxicological data relate to the rat and the duckling, the former, probably because of its well-established position in the toxicology laboratory, and the latter, because the day-old bird is an extremely susceptible subject and was used in bioassay procedures[80,81] before reliable chemical methods of analysis had been developed for aflatoxin.

Aflatoxin induces acute liver injury, widespread hemorrhages, and death in the rat although there are quantitative strain and sex differences in response;[3] the same histological lesion is to be seen in livers of rats of either sex and different strains however. Periportal zone necrosis develops slowly over a period of a few days and is accompanied by marked bile duct proliferation. At nonlethal doses, recovery from this lesion is slow.

Day-old ducklings fail to grow, develop s.c. hemorrhages, and die following the oral administration of 10 to 20 μg aflatoxin B_1. The liver undergoes fatty changes and periportal zone necrosis. In survivors, extensive biliary proliferation occurs and reaches a peak at 3 days before regressing. This characteristic histological response can be used to detect as little as about 1 μg aflatoxin B_1[3,81] in a standardized bioassay procedure.

There are few published accounts of naturally occurring acute aflatoxicosis among laboratory animals but, prior to the discovery of aflatoxin, outbreaks of noninfectious hepatitis and edema in guinea pigs were reported[82,83] and, when a sample of hepatotoxic feed was examined some years later, Allcroft et al.[84] identified aflatoxin as the

Table 5
INTERACTION BETWEEN AFLATOXIN OR ITS METABOLITES AND THE ORGANELLES AND BIOCHEMICAL SYSTEMS OF THE LIVER CELL

Locus	Interaction of biochemical response	Ref.
Nucleus	Mutagenic, carcinogenic: binds covalently to DNA[a,b,d]	5, 40a, 43
	Inhibits DNA-dependent RNA polymerase[a,c]	51
	Stimulates DNA repair synthesis[a]	52
Mitochondria	Increases membrane permeability[b]	53
	Inhibits protein synthesis[a-c]	54
	Interrupts electron transport[b]	55, 56
Lysosomes	Increases membrane permeability[a]	57
	Free acid hydrolases released[a]	58
Endoplasmic reticulum	"Degranulation" occurs[a]	59
	Mimicks steroid sex hormones at polysome building sites[b]	60
	Binds covalently to r RNA[a,b,d]	43
	Inhibits mixed function oxidases[b,c]	61
	Inhibits protein synthesis including blood clotting factors[a,b]	62, 63
	Inhibits glucose 6-phosphatase[a,c]	64
	Inhibits fatty acid and phospholipid synthesis[a,c]	65, 66
	Feedback control of cholesterol synthesis lost[a,f]	67
	Inhibits sialyl transferases[b,c,f]	68
	Inhibits lipid peroxidation[a-c]	69, 70
Cytoplasm	Initially stimulates glycogenolysis/pentose shunt activity[a]	64
	Substrate/inhibitor for 17-hydroxy steroid dehydrogenase (NADP linked)[b]	71
General	Increase capilliary permeability[a]	72
	Aflatoxin B_1 dihydrodiol forms Schiff bases with free amino-groups (amino-acids, peptides, and proteins) and inhibits metabolism	43a

[a] In vivo effect.
[b] Action of aflatoxin B_1 in vitro.
[c] Possibly due to a metabolite.
[d] Probably due to aflatoxin B_1-2,3-oxide.
[e] Cf. other carcinogens.
[f] Precancerous change.

probable cause of one such outbreak. Also in the 1950s, Seibold and Bailey[85,86] described an epizootic of "hepatitis X" among dogs in the southeastern states of the U.S. A commercial dog food contained a toxic factor[87] and later work suggests that this may have been aflatoxin.[88,89]

Farm Animals

In an agricultural context, outbreaks of acute aflatoxicosis can result in economic losses for the farmer but, as aflatoxin residues can accumulate in milk and, at least theoretically, in meat and eggs of food-producing animals, the other important aspect of aflatoxin contamination of animal feedstuffs is that it can also be regarded as a potential health hazard to the community at large.

Most aspects of the acute disease of cattle, sheep, swine, and poultry have already

Table 6
LD_{50} VALUES FOR ACUTE AFLATOXIN B_1 POISONING IN DIFFERENT ANIMAL SPECIES[3,73-76]

Species	mg/kg body weight
Rabbit	0.3
Duckling	0.34
Dog	0.5—1.0
Cat	0.55
Pig	0.62
Guinea pig	1.4
Baboon	2.0
Sheep	2.0
Chicken	3.4—6.8[a]
Rat	7.2—17.9[b]
Macaque monkey	7.8
Mouse	9.0
Hamster	10.2

[a] Strain differences.
[b] Sex differences.

Table 7
RELATIVE POTENCIES OF DIFFERENT AFLATOXINS

Aflatoxin	Toxicity relative to B_1[a]	Ref.	Mutagenicity relative to B_1[b]	Ref.
B_1	100	—	100	—
M^1	72	30	3.2	77a
G_1	62	77	3.3	77a
B_2	41	77	0.2	77a
G_2	26	77	0.1	77a

[a] Calculated from acute LD_{50} values in the day-old duckling.
[b] Calculated from the rates of formation of a bacterial mutagen.

been reviewed[78,90-96] and much of this accumulated information derives from experimental oral dosing or even administration by parenteral routes. It seemed more appropriate here to consider effects of feeding aflatoxin-contaminated rations however, and Tables 8 and 9 summarize reports of naturally occurring and experimental aflatoxicosis in cattle and swine induced in this way. From the same tables it will be seen that clinical effects may probably be observed when dietary levels of aflatoxin B_1 exceed about 100 μg/kg in cattle but only about 50 μg/kg in swine. Adult cattle appear to be fairly resistant and no effects are apparent with dietary levels of up to 300 μg/kg,[93] but calves are more susceptible and death occurs at 1800 μg/kg.[97]

Very little appears to have been reported on the effects of dietary aflatoxin on sheep. One long-term feeding experiment[106] has shown that, ignoring for the present any chronic effects of aflatoxin, levels of about 1750 μg/kg are readily tolerated by sheep, there being no appreciable effects on feed intake or signs of ill health. Salivation, fever, inappetance, rumenal atonia, and icterus have been observed as early clinical signs when aflatoxin B_1 was administered parenterally.[73]

Much more is known of the toxic effects of feeding aflatoxin-contaminated rations

Table 8
ACUTE AFLATOXICOSIS IN CATTLE INGESTING CONTAMINATED FEED

Aflatoxin B_1 conc. (μg/kg)[a]	Natural or experimental	Description of effect	Ref.
		Clinical Effects	
80—200	NE	Low milk yield, aflatoxin M_1 in milk, lowered resistance to infection	10, 14
110	N	Reduced daily weight gain, scouring (calves)	10
300[b]	E	Raised serum alkaline phosphatase activity	93
700	E	Low feed consumption and efficiency	93
1800	E	Anorexia, loss of condition, death (calves)	97
		Pathology	
700	E	Grossly abnormal grey enlarged rubbery liver	93
1000[c]	N	Proliferation of bile duct epithelium; chronic endophlebitis of centrolobular and hepatic veins; variation in size and shape of parenchymel cells, many with dense basophilic nuclei; diffuse fibrosis; variable steatosis	98
1800	E	Ascites, visceral edema	97

[a] Lowest reported level producing the stated effect.
[b] Levels of up to about 300 μg/kg appear to be well tolerated.
[c] Value estimated by the present author.

to poultry. This is partly due to the fact that aflatoxin was discovered during the investigation of extensive outbreaks of acute liver disease in poultry[4,96] and partly as a result of the indefatigable research effort of one American team of poultry scientists.

One of the first acute effects of aflatoxin B_1 ingestion is to retard growth and this is seen more readily in ducks and chickens than in turkeys. There is a drop in egg production by laying birds followed by general malaise and inappetance. In turkeys diarrhea has been observed. Subcutaneous hemorrhage or an increased tendency to bruise is also sometimes a feature. Nervous signs include ataxia, opisthotonus, and convulsions. At sufficiently high dietary levels, death follows within a few days of the onset of clinical signs although asymptomatic deaths have been recorded in turkeys.[96] At post-mortem examination, the liver and kidneys are generally seen to be grossly affected: hepatic necrosis, hemorrhage, and fibrosis are major features although engorgement and congestion of the kidneys of turkey poults has also been stressed.[107]

Acute fatal disease is usually experienced with dietary levels of aflatoxin B_1 of about 1000 μg/kg in poults but five to ten times that figure in broilers[108]. Subclinical or clinical effects can be expected at levels above about 200 μg/kg in poults but at higher levels in broilers.[108] Table 10 summarizes some of these biochemical or pathophysiological responses and clinical effects and indicates the lowest reported levels of dietary aflatoxin that caused them.

Table 9
ACUTE AFLATOXICOSIS IN PIGS INGESTING CONTAMINATED FEED

Aflatoxin B_1 conc ($\mu g/kg$)[a]	Natural or experimental	Description of effect	Ref.
		Clinical Effects	
"Mouldy peanut meal"	N	Acute illness, icterus, and death; increased serum levels of liver enzymes, bilirubin, and globulins, decreased A:G ratio, albumin, PCV and hemoglobin	99
51	B	Plasma levels of NPN, urea N, adenine nucleotides decreased (in addition to above biochemical responses); (at *higher* doses, levels of enzymes, vitamin A, glycogen, and total N are decreased in the liver)	100
200	E	Impaired growth and feed conversion	101
1000	E	Clinically normal but unexpected deaths occur	102
1500	E	Chromosomal changes in bone marrow cells and peripheral leukocytes	103
1750[b]	N	Apathy, jaundice, death	104
		Pathology	
1750	NE	Generalized icterus, petichial and ecchymotic hemorrhages, subendocardial and subserosal hemorrhages, gastric ulceration. Liver shows karyomegaly, ductule proliferation, fibrosis, and steatosis	99, 105

[a] Lowest reported level producing the stated effect.
[b] Value estimated by the present author.

Man

Possible or proved associations between the ingestion of contaminated food and the occurrence of acute aflatoxicosis in man are of two kinds: those involving individuals who had accidentally consumed a single identifiable contaminated food and outbreaks of acute liver disease where field investigations indicate the probable ingestion of an aflatoxin-contaminated diet. Additionally, in some other studies, milk, urine, and liver or other tissues taken post-mortem have been analyzed and shown to contain aflatoxin. This confirms that, in many parts of the world, man is exposed to diets containing the toxin, and suggests that acute or chronic forms of aflatoxin poisonings are possible consequences.

Tables 11 and 12 list some of the published investigations of acute aflatoxicosis in man and of the occurrence of aflatoxin B_1 and M_1 in human tissues and body fluids. The most convincing account of acute aflatoxicosis in man is probably that reported

Table 10
SOME SUBCLINICAL AND CLINICAL EFFECTS IN POULTRY CAUSED BY THE INGESTION OF EXPERIMENTALLY CONTAMINATED FEED

Dietary conc of aflatoxin B_1 (μg/kg)	Biochemical/pathophysiological response	Clinical effect	Ref.
220[a]	Aflatoxin residues in eggs		13
250	Impaired immunogenesis	Weight gain decreased (poults)	108—111
625	Resistance to salt, humidity and thermal stresses reduced; free acid hydrolases released from hepatic lysosomes; altered lipid transport; hemolytic anemia induced; prothrombin synthesis impaired; capillary fragility increased	Resistance to infection diminished; increased tendency to bruise and bleed	66, 112—116
700		Feed intake reduced	117
1000		Acute illness, death (poults)	108
1250	Liver weight increased		118
1300	Liver fat content increased; liver vitamin A decreased; RNA/DNA decreased		119
1500		Weight gain decreased (broilers)	120
2500	Egg yolk and serum carotenoids increased; hepatic fatty acid synthesis impaired; cell mediated immunity (graft against host) and delayed hypersensitivity impaired; serum bloods of immunoglobulins lowered; further effects on blood clotting factors	Av. egg weight decreased and egg production lowered	113, 121—123
5000	Spleen weight increased; bursa of Fabricius smaller; calcium absorption inhibited; bone strength diminished	Hatchability of fertile eggs low; tendency to bone fracture; acute illness, death (broilers)	108, 118, 120 124—126
10000	Compensatory increased in intenstinal absorption of methionine and glucose		127

[a] Average reported level; all other levels are the lowest reported experimental levels producing the stated effect(s).

from India in 1975[130-132] (see Table 11). In this epidemic, 397 village dwellers presumably consumed corn (maize) contaminated with up to 15,000 μg aflatoxin B_1 per kilogram, became ill with acute hepatitis, and 106 people died. It was estimated that up to 6 mg aflatoxin B_1 per person was consumed. Village dogs, feeding on scraps of kitchen waste, also died during this outbreak of disease.*

Reye's syndrome[144-146] (encephalopathy with fatty degeneration of the viscera) may

* Since then, an outbreak of acute aflatoxicosis among domestic dogs has been reported from Kenya.[255]

Table 11
SOME PUBLISHED REPORTS OF ACUTE AFLATOXICOSIS IN MAN[a]

Country	No. cases of liver disease reported	Age group	Suspected food: Description	Suspected food: Approximate aflatoxin B_1 content (μg/kg)	Nature of the disease	Ref.
Germany	1	Adult	Brazil nut	Not determined: association presumed	Acute fatal illness: subject already suffering from hemosiderosis	128
India	20	1.5—5 years	Peanut meal	300	Hepatomegaly, cirrhosis: 3 fatal cases liver failure	129
India	397	All ages	Corn (maize)	250—15,600	Acute hepatitis: 91 deaths	130—132
Senegal	2	4—6 years	Peanut meal	500—1,000	Hepatitis investigated after known exposure in infancy	133
Taiwan	29	All ages	Rice	200	Acute liver disease, 3 children died	134
Uganda	1	15 years	Cassava	1,700	Acute fatal hepatitis	135

[a] Excluding Reye's Syndrome cases: for discussion, see text.

be a particular manifestation of aflatoxin poisoning in children and young adolescents. Clinical illness is characterized by vomiting and convulsions with hypoglycemia and hyperammoniemia; it runs a short course of 1 or 2 days ending in coma and death. The involvement of aflatoxin was first suggested in the investigation of a single case of the disease in Thailand[147] when a 3-year-old boy ate moldy but apparently still palatable leftover cooked rice which, on analysis, was found to contain 10,000 μg total aflatoxin per kilogram. Since then several authors have reported the presence of aflatoxin B_1 in food[146-148] and autopsy specimens of liver.[142,149,150] Additionally aflatoxin B_1 has been detected in the blood[151] and aflatoxins B_1 and M_1 in the urine[147] of Reye's syndrome patients. Consequently this disease appears to be associated with the ingestion of aflatoxin-contaminated food, but it has been suggested that there may also be other viral or possibly nutritional factors in its etiology.[148,152,153] It is consistent with an intoxication hypothesis that a syndrome, in many ways similar to Reye's syndrome, has been produced in macaque monkeys by the acute oral administration of aflatoxin B_1.[76] However, some aspects of the syndrome are difficult to explain solely on this basis[146] and unlike the close association between the generally high dietary levels of aflatoxin in certain regions of Africa and Southeast Asia and the incidence of liver cancer (see "Man" in following section) Reye's syndrome appears to have a much wider geographical distribution, the disease having first been described in New Zealand and later investigated not only in Thailand but also in Czechoslovakia and the U.S.

CHRONIC AFLATOXICOSIS

Aflatoxin is a potent hepatocarcinogen[3,5] and, as discussed in the section "Metabolism and Mode of Action", it is generally believed that the toxin is first metabolized in the liver by mixed function oxidases before interacting with DNA and so initiating

Table 12
HUMAN EXPOSURE TO AFLATOXIN: OCCURRENCE IN TISSUES AND URINE

Country	Age group	Samples examined	No. samples containing aflatoxin/ total no. examined	Other details	Ref.
Czechoslovakia	68 years	Lung	1/1[a]	Carcinoma of the lungs: worker in factory processing groundnut meal	136
France	Adults	Liver	3/50[a]	2 Alcoholic cirrhosis and 1 stomach cancer cases	137
France	Adults	Liver	3/50[a]	3 Cardiopathy cases	138
India	Adults	Milk	3/43[b]	Breast milk from mothers of cirrhotic children	
India	Children	Urine	18/43[b]	Hepatic cirrhosis	139
India	Children	Urine	8/17[b]	Healthy subjects	
Phillipines	Children	Urine	6/36[c]	Healthy subjects: presence of aflatoxin M_1 associated with the ingestion of peanut butter	140
Phillipines	Children and adults	Urine	4/5[b]	Bulked urine from groups of individuals with differing dietary habits	141
Thailand	Children	Liver	11/15[a]	Children dying from causes other than Reye's syndrome	142
USA	56 years	Liver	1/1[a]	Carcinoma of rectum and liver	143

[a] Aflatoxin B_1.
[b] Aflatoxin-like material.
[c] Aflatoxin M_1 detected.

tumor development. The mechanism of this initial interaction and subsequent early phases of aflatoxin-induced carcinogenesis have been investigated and reviewed by several authors (e.g., see References 154 to 157).

Laboratory Animals

That aflatoxin may have been a carcinogen probably suggested itself to early investigators when it was observed that, following the induction of acute lesions, hyperplastic changes developed in the livers of ducklings and other animal species[3,78,79] Carnaghan[158] was first used experimentally to induce hepatic tumors in ducklings by feeding a diet containing as little as 35 μg aflatoxin B_1 per kilogram. Since then it has been shown that dietary aflatoxin B_1 causes liver cell cancer to develop in three species of fish, the rat, and the pig (see Table 13). Additionally, the administration of aflatoxin B_1, orally or by injection, achieved the same result in three species of primate, the mouse, and the ferret. The rainbow trout embryo was found to be extremely susceptible, in that the mere exposure to aflatoxin B_1 in the water (0.5 mg/kg) for only 60 min was sufficient to induce cancers some 10 months later.[172]

As shown in Table 13 there are species differences but breed, age, sex, hormonal status, and diet (to be discussed below) also appear to play a part in determining an animal's susceptibility to the carcinogenic action of aflatoxin B_1. Thus, different breeds of rat responded differently to similar dietary levels of aflatoxin B_1,[162-165] neonatal mice[167] developed liver tumors but older mice did not,[157] and the female rat is strikingly less susceptible to the effects of aflatoxin than the male.[174]

Table 13
HEPATOCARCINOGENICITY OF AFLATOXIN B_1 IN EXPERIMENTAL ANIMALS

Aflatoxin B_1 fed in the diet			Aflatoxin administered in other ways	
Conc range (μg/kg range)	Species	Ref.	Species	Ref.
35 or less	Duckling	158	Ferret	164
35 or less	Rainbow trout	159	Marmoset	166
35 or less	Salmon	160	Mouse[a]	167
1000—6000	Guppy	161	Rhesus monkey	168—171
1000—6000	Rat	162—165	Rainbow trout embryo	172
1000—6000	Pig	102	Tree shrew	173

[a] Dietary aflatoxin at levels of up to 150 mg/kg were ineffective[157] but, administered i.p., the toxin induced tumors in neonatal mice at a dose of 6 mg/kg body weight.[167]

Dose-response relationships have been demonstrated in the rat[163] and the rainbow trout,[175] the former with dietary levels of up to 100 μg aflatoxin B_1 per kilogram, the latter up to about 30 μg/kg. Concentrations of aflatoxin B_1 in feed necessary to induce a 10% incidence of liver tumors (compared with none in appropriate control groups) were 1 μg/kg in the rat and about 0.1 μg/kg in rainbow trout. Other aflatoxins are less potent than aflatoxin B_1,[77] presumably because of differences in the balance between (1) conversion in vivo to the appropriate carcinogenic metabolite, and (2) detoxification or elimination from the body (see "Metabolism and Mode of Action").

The liver is generally the site of action of aflatoxin but tumors have been induced in other organs also. Renal epithelial neoplasms have been reported in male Wistar rats fed diets containing 250 to 1000 μg aflatoxin B_1 per kilogram,[165] colon cancers developed particularly when vitamin A deficient rats were treated with aflatoxin B_1,[79,176,177] and carcinomas of the glandular stomach were observed in rats fed rather high dietary concentrations (3000 to 4000 μg aflatoxin B_1 per kilogram).[178] Other tumor sites include the lacrimal glands,[179,180] the tongue,[181] and the esophagus.[180]

Farm Animals

There are at least two reports of tumor induction in farm animals. In a prolonged feeding trial,[106] a group of five ewes tolerated a ration of groundnut meal, fish meal, and maize (corn) gluten containing 1000 to 1750 μg aflatoxin B_1 per kilogram and, although fertility was not high, they produced normal lambs. Among these ewes and their progeny, one wether developed liver carcinoma and died after 3.5 years on this diet, and nasal chondromas were found in two ewes killed after 4 to 5 years. The latter observation is probably unique and may have arisen from continuous exposure to contaminated groundnut dust. In another more recent report,[102] four brood sows were fed aflatoxin for 28 to 30 months (1000 μg aflatoxin B_1 plus G_1 per kilogram in the diet), farrowed successfully litters of normal piglets, remained clinically healthy, but at autopsy it was found that nodular liver cell adenomas and hepatic cell hyperplasia had developed in all four of them.

Man

The epidemiological investigations of Peers and Linsell,[182,183] Shank et al.,[184,185] and

Table 14
PROBABLE AND POSSIBLE ASSOCIATIONS BETWEEN AFLATOXIN EXPOSURE AND CANCER IN MAN

Presumed route	Estimated exposure level	Associated disease	Incidence	Ref.
		Naturally Contaminated Food		
Diet	3.5 to 222.4 ng aflatoxin B_1/kg body weight ingested daily (from food-on-the-plate data)	Primary liver cell cancer	1.2—13.0/10^5 total population anually (incidence and exposure correlated statistically)	182—189
		Occupational Exposure		
Inhaled peanut meal dust (in mills)	160—395 μg Aflatoxin B_1/m^3 air/man/week	Cancer of liver and other organs	11/67 men exposed: 2 cases of liver cancer	190
Inhaled peanut meal dust (in mills)	No quantitative data	Lung adenoma (lung tissue contained aflatoxin B_1)	1 Case report	136
Inhaled/ingested laboratory dust: pure aflatoxin B_1 on TLC plates	No quantitative data	Colon adenocarcinoma	2 Case reports	191

Van Rensberg et al.[186] are too well known for detailed discussion to be required here. During their studies of an apparent association between the dietary intake of aflatoxin and the high incidence of primary liver cancer among the populations of Africa and Southeast Asia, it emerged that a statistically valid relationship exists between exposure and incidence[187,188] and although this points to aflatoxin as a factor in the etiology of liver cancer in these regions, it has been stressed that a causal relationship cannot be inferred. Indeed, it is known that in many of the same regions there is also a high incidence of hepatitis B antigenemia[189] and the disease may therefore also have a viral component. Table 14 summarizes the epidemiological data and also refers to three reports of possible associations between occupational exposure to aflatoxin and cancer in man.

MODIFYING EFFECTS OF NUTRITIONAL AND OTHER EFFECTS

The toxic effects of aflatoxin as a dietary contaminant depend, among other factors, upon the nutritional status of the animal or person consuming it. This subject has already been considered in some detail by earlier reviewers,[154,157,192-196] but listed in Table 15 will be found many of the known effects of dietary manipulation on acute or chronic experimental aflatoxicosis.

As already discussed in an earlier section aflatoxin is not only detoxified by hepatic enzyme systems but also metabolically "activated" before it causes liver damage. Consequently, the balance between the flux of aflatoxin molecules along these opposing pathways can be expected to influence if not actually determine the nature and intensity of the toxic response. It is unlikely that such a variety of nutritional factors as mentioned in Table 15 would alter this balance but some probably do. In other cases it is not difficult to suggest possible specific protective mechanisms. Thus, to give a few

Table 15
SOME MODIFYING EFFECTS OF NUTRITIONAL FACTORS ON THE COURSE OF EXPERIMENTAL AFLATOXICOSIS

	Interactions with	
Composition of diet	Acute effects of aflatoxin B_1	Development of hepatomas
Gross composition		
Protein	Deficiency enhances (rat,[197] monkey[198]); increased levels (up to 30% protected chickens[201]	Deficiency protects rats[199,200]
Fat		
Total lipid content	Fat up to 18% total diet protects turkeys[202] and chickens[201]	High levels protect rats[203]
Fatty acid components	Unsaturated but not saturated fats correct growth retardation in chickens[201]	Cyclopropenoid fatty acids act as co-carcinogens in trout[159,204]
Trace elements		
Copper	Possible protection in hamsters[205]	—
Selenium	Supplement (1 mg/kg) protects, higher levels enhance (rats[206])	Supplements up to 5 mg/kg ineffective but kidney damage enhanced (rats[207])
Vitamins		
A	Deficiency enhances in male but not female rats;[208] fourfold dietary excess ineffective in chickens;[209] carotene protects rats[210]	Deficiency does not influence liver tumor development but enhances colon carcinogenesis[176,177] 13-*cis* retinoic acid but not vitamin A protects against colon tumors[177]
B_1 (thiamin)	Deficiency protects chickens[209]	—
B_2 (riboflavin)	Deficiency enhances (chickens[209])	Pretreatment enhances (rats[210])
B_6 (pyridoxine)	—	Deficiency possibly enhances (man[194])
B_{12}	Supplements (13.2 μg/kg) ineffective in chickens[211]	Supplements (50 μg/kg) enhance (rats[212])
Lipotropes (choline methionine-folate)	Marginal deficiency (with high fat) protects against effects of single, enhances effects of multiple doses of aflatoxin in rats[213,214]	Marginal deficiency enhances, severe deficiency decreases incidence (rats[194]); choline protects[215]
D_3	Deficiency enhances (ducklings,[216] chickens[209])	—
E	Supplement (up to 18 IU/kg) ineffective (chickens[209,211])	—
K	Fourfold dietary excess ineffective (chickens[209])	—

examples, microsomal enzymes, usually concerned with the detoxification of xenobiotics, are known to respond to changing levels of dietary protein[218] and unsaturated fatty acids may play important structural and functional roles in hepatic microsomal drug metabolism,[219] but selenium, an integral part of the enzyme glutathione peroxidase, is a biological antioxidant[220] that would be essential for the disposal of any lipid peroxides that may be generated and thiamine deficiency might protect by stimulating the oxidation of fats stored in the liver in acute aflatoxicosis.

Detailed explanations for many of these empirically determined nutritional interactions must await the outcome of future research, but meanwhile we are led to the rather obvious general conclusion that the well-nourished subject will probably resist the toxic effects of aflatoxin better than his undernourished fellow.

Other modifying factors include hormonal status, environmental chemicals including drugs and other mycotoxins, concurrent disease whether infectious or not, and the physical environment, notably the effect of UV radiation. Again some examples may be cited by way of illustration: male rats are more susceptible than females to the acute[208,217] and chronic[208] aflatoxin poisoning; DDT, a typical environmental chemical and phenobarbitone, a frequently used drug, are both inducers of microsomal mixed function oxidases,[221] and experimentally, pretreatment has been shown to lower the incidence of and delay liver tumor development in the rat, presumably by selectively promoting detoxification of aflatoxin; the widespread occurrence of viral infection in certain human populations with a high incidence of liver cancer has already been mentioned and experimental aflatoxin injury plus viral hepatitis has been shown to induce a more severe reaction than does aflatoxin alone;[166] and, when exposed to UV radiation, rats become more susceptible to acute aflatoxin poisoning,[210,222] an observation that suggests that the aflatoxin molecule may be activated in the skin in addition to the liver.

In general, naturally molded crops, feeds, and food cannot be expected to bear a pure culture of a single fungal species and although the analyst may be examining his samples for only one mycotoxin, simultaneous contamination with several mycotoxins must be regarded as the norm. Groundnut is frequently attacked by *A. flavus* or *A. parasiticus* in the field or post-harvest and consignments often contain one or more of the aflatoxins, but the presence of other mycotoxins have seldom been reported. Corn (maize) is commonly contaminated with aflatoxin and/or the *Fusarium* toxin zearalenone and other mixed contaminations occur in cereal crops such as wheat or barley. Thus, any modification of aflatoxin toxicity due to interactions with other mycotoxins are most likely when intoxication follows the ingestion of cereals and corn but least likely in the case of groundnut or its products.

A simple illustration of a possible interaction would be the co-contamination of barley with aflatoxin B_1 and ochratoxin A.[10] The latter, a metabolite of the *A. ochraceus* group, is a nephrotoxin[223] and its presence in a food or feed might cause damage to the kidneys sufficient to impair the excretion of aflatoxin metabolites and so to enhance the toxicity of the latter mycotoxin. The question of mycotoxin-mycotoxin interactions in vivo deserves serious consideration[224] particularly at the low levels of contamination that usually occur.

SECONDARY AFLATOXICOSIS

The "carry-over" of aflatoxin contamination from feedstuffs to the tissues of food-producing animals has been considered briefly in an earlier section on occurrence. Although aflatoxin B_1 or its metabolite aflatoxin M_1 can at least theoretically be transferred to eggs, liver, and meat (Table 2) it is only the secondary contamination of milk that appears to occur naturally to any significant extent (Table 3). This has important public health implications because, in the U.K. and probably in other countries, milk and milk products constitute the largest constant component of the daily national diet.

Groundnut meal and corn (maize) are excellent sources of protein and are frequently used in the formulation of dairy feedstuffs even though contamination of the former commodity with aflatoxin B_1 is virtually unavoidable. Levels of contamination are usually controlled by law or some form of trade agreement however. In the U.K. and other EEC countries the maximum allowable concentration of aflatoxin B_1 is 20 μg/kg in dairy concentrates[225] and, assuming a "carry-over" of 0.33% (Table 2) it can then be anticipated that the resulting maximum concentration of aflatoxin M_1 in milk from a single herd would be about 0.1 μg/ℓ. Bulked milk from several farms would probably contain much less.

It is difficult to make a satisfactory assessment of the risk to people of consuming milk with this level of contamination as no cases of human disease actually or possibly caused by milkborne aflatoxin M_1* have ever been reported but a few comparisons may be useful here. In the first place, ignoring the toxicological differences between aflatoxins M_1 and B_1, a concentration of 0.1 μg aflatoxin M_1 per liter is merely 1/50 the statutory limit of many countries for the contamination of human food with aflatoxin B_1. Secondly, while the daily intake by man from consuming 0.5 ℓ of milk of this quality would be only 50 ng aflatoxin M_1, the same quantity of aflatoxin B_1 could be obtained from as little as 25 g peanuts containing a not improbable 2 μg/kg. Furthermore, the daily ingestion of a 0.5 ℓ of this milk by a 70-kg man would result in a dosage of about 0.7 ng aflatoxin M_1 per kilogram body weight per day. And, this is equivalent to one fifth the daily intake of aflatoxin B_1 in one section of the Kenyan population surveyed by Peers and Linsell[182] and in whom the incidence of liver cancer (0.7 in 100,000 population) was no higher than that in many Western countries.

Thus, secondary aflatoxicosis is more a concept than an identifiable disease problem but nonetheless there is a potential, but perhaps slight, hazard to public health associated with the "carry-over" of aflatoxin residues into milk. Interestingly, a recent report shows that milk substitutes may be contaminated with aflatoxin B_1.[226]

CONCLUSION AND RECENT DEVELOPMENTS

Aflatoxin-producing strains of *A. flavus* and *A. parasiticus* are virtually ubiquitous and contamination of cereals, fruit, nuts, and their products (see Table 1) appear to be unavoidable. Aflatoxin B_1 is the most commonly occurring of the aflatoxins and is a potent carcinogen (see section "Chronic Aflatoxicosis"). In particular, groundnut crops are almost invariably contaminated but this commodity constitutes such a valuable source of protein that it continues to be in demand both as a human and animal food despite some 20 years' research reports on the toxicology of aflatoxin.

Unless control measures are taken to reduce sufficiently the aflatoxin content of animal feedstuffs the health of livestock may be impaired (see "Farm Animals" in sections on acute aflatoxicosis and chronic aflatoxicosis) but perhaps more importantly, even at levels that do not affect the health or performance of farm animals, aflatoxin, usually as its metabolite M_1, may be "carried over" into milk[227] and animal tissues destined for human food (see Tables 2 and 3 and "Secondary Aflatoxicosis.") Although with continuing improvements in analytical methods, it is now known that traces of aflatoxin M_1 are frequently present in milk, the public health significance of this degree of contamination has not been properly evaluated.

The entry of aflatoxin B_1 into the human food chain and pathways taken by the parent toxin and metabolites like M_1 are summarized schematically in Figure 3 taking the groundnut as the typical source of aflatoxin contamination. Control of the aflatoxin problem can be exerted at four points in this chain: (1) fungal attack of groundnuts may be minimized by selecting plant varieties that are naturally resistant to insect damage, by controlling insects, by using appropriate fungicides, and by avoiding physical damage during harvest,[228,229] (2) storage conditions can be chosen that do not encourage molding,[6,229] (3) the use of groundnut meal for animal feedstuffs can be restricted by legislation (e.g., Reference 225) or trade agreement; excessively contaminated groundnut meal can be decontaminated by chemical means;[229,230] contamination of milk is eliminated altogether when cows are at pasture, and (4) statutory or "working" limits can be imposed on the level of aflatoxin contamination in peanuts for human consumption[25,231-234] and to some extent roasting can reduce the level of contamination.

* Although Dvorackova et al.[148] have reported the apparent association between feeding milk powder containing B_1 and Reyes' syndrome in children, the absence of aflatoxin M_1 indicates that this contamination was not "carried-over" from cows' feedstuffs.

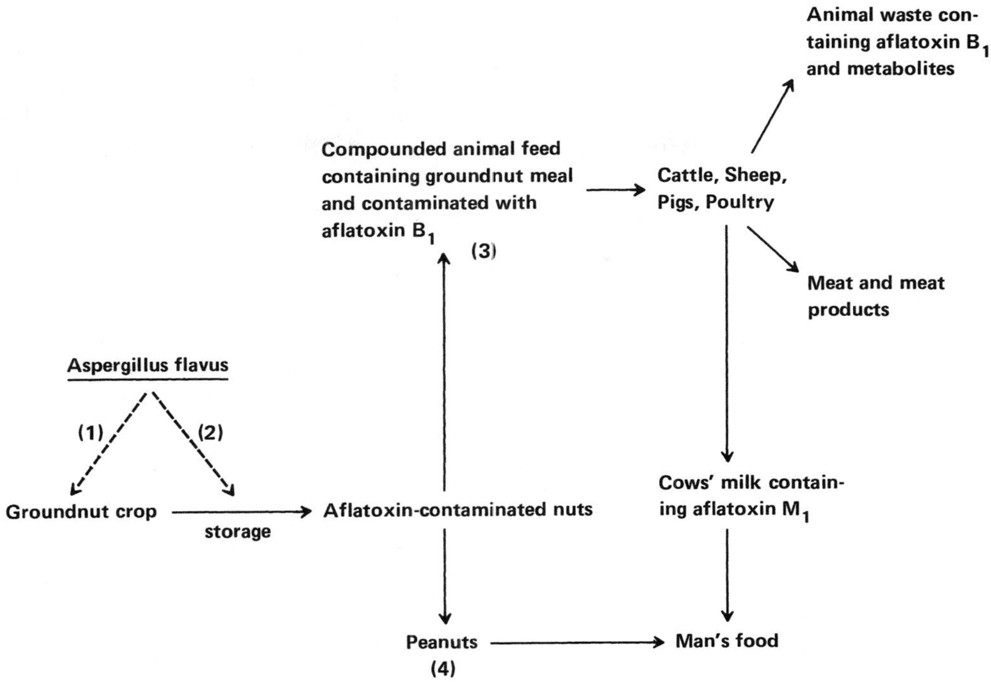

FIGURE 3. Aflatoxin in the food chain, showing how man's food can be directly or indirectly contaminated with traces of aflatoxin B_1 or its metabolite M_1. Control is possible at the four points indicated (see text).

There are two gaps in our detailed knowledge of the distribution of aflatoxins in the environment as outlined in Figure 3 however. One concerns the fate of aflatoxin and its metabolites once excreted by livestock. This amounts to a substantial proportion of any aflatoxin B_1 ingested by cattle and sheep,[235-237] and possibly much is detoxified by ammonia formed from the voided urine. Secondly, little is known of the possible "carry-over" of aflatoxin metabolites other than M_1 into animal products used as human food but in a recent study of the rat it has been shown that water-soluble metabolites and low-molecular-weight conjugates are virtually nontoxic when fed to a second animal.[238]

There have been many other developments since this chapter of the handbook was first written. Our understanding of aflatoxin metabolism has altered somewhat (see for example Reference 43a) and this has necessitated a slight change in that section of the original text. Otherwise, little updating has been attempted, partly because the continuing, almost exponential, growth of the literature made a comprehensive revision such a daunting task, but mainly because the essential public and animal health problem stated above remains qualitatively the same. Of course, additional data on the occurrence of aflatoxin in food, its toxicity to laboratory and farm animals, its metabolism and mode of action, control measures, and the detoxification of aflatoxin-contaminated feedstuffs have been reported in the proceedings of several recent conferences[239-243] and reviewed elsewhere.[244-249] The U.K. Ministry of Agriculture, Fisheries and Food sponsored a survey of human food and has published what is probably a unique account of the experience of one country with dietary mycotoxins.[250] And, soon afterwards, stricter control[251] was imposed on the importation of raw materials used for the manufacture of dairy and other feeds so as to minimize the "carry-over" of aflatoxin M_1 into cow's milk. A task group was also asked by the World Health Organization to evaluate the hazard of foodborne mycotoxins and their report[252] was chiefly concerned with aflatoxin. At about the same time the U.S. Food and Drug Administration made a critical assessment of the carcinogenic risk from aflatoxin in corn (maize) and peanuts.[253] The FDA subsequently carried out a survey of milk for aflatoxin M_1 contamination, found levels of up to about 0.5 $\mu g/\ell$ and now, in common with the WHO, are committed to the principle of reducing milk contamination as far as practicable.[254]

REFERENCES

1. **Goldblatt, L. A., Ed.,** *Aflatoxin: Scientific Background, Control, and Implications,* Academic Press, New York, 1969.
2. **Detroy, R. W., Lillehoj, E. B., and Ciegler, A.,** Aflatoxin, in *Microbial Toxins,* Vol. 6, Ciegler, A., Kadis, S., and Ajl, S. J., Eds., Academic Press, New York, 1971, 3.
3. **Butler, W. H.,** Aflatoxin, in *Mycotoxins,* Purchase, I. F. H., Ed., Elsevier, Amsterdam, 1974, 1.
4. **Patterson, D. S. P.,** Aflatoxin and related compounds, in *Mycotoxic Fungi, Mycotoxins, Mycotoxicoses: an Encyclopedic Handbook,* Vol. 1, Wyllie, T. D. and Morehouse, L. G., Eds., Marcel Dekker, New York, 1977, 131, 144, 156, 159.
5. **IARC,** Aflatoxins, in *IARC Monographs on the Evaluation of the Carcinogenic Risk of Chemicals to Man,* Vol. 10, IARC Working Group, International Agency for Reserach on Cancer, Lyon, 1976, 51.
6. **Hesseltine, C. W.,** Conditions leading to mycotoxin contamination of foods and feeds, in *Mycotoxins and Other Fungal Related Food Problems,* Adv. Chem. Ser. No. 149, Rodricks, J. V., Ed., American Chemical Society, Washington, D.C., 1976, 1.
7. **Goldblatt, L. A., and Dollear, F. G.,** Detoxification of contaminated crops, in *Mycotoxins in Human and Animal Health,* Rodricks, J. V., Hesseltine, C. W., and Mehlman, M. A., Eds., Pathotox Publishers, Park Forest South, Ill., 1977, 139.
8. **Stoloff, L.,** Occurrence of mycotoxins in foods and feeds, in *Mycotoxins and Other Fungal Related Food Problems,* Adv. Chem. Ser. No. 149, Rodricks, J. V., Ed., American Chemical Society, Washington, D.C., 1976, 23.
9. **Jones, B. D.,** Aflatoxin and related compounds, occurrence in foods and feed, in *Mycotoxic Fungi, Mycotoxins, Mycotoxicoses: an Encyclopedic Handbook,* Vol. 1, Wyllie, T. D. and Morehouse, L. G., Eds., Marcel Dekker, New York, 1977, 190.
10. **Patterson, D. S. P., Roberts, B. A., Shreeve, B. J., Wrathall, A. E., and Gitter, M.,** Aflatoxin, ochratoxin and zearalenone in animal feedstuffs: some clinical and experimental observations, *Ann. Nutr. Aliment.,* 31, 643, 1977.
11. **Kiermeier, F.,** The significance of aflatoxins in the dairy industry, *Annu. Bull. Int. Dairy Fed.,* No. 98, 1, 1977.
12. **Purchase, I. F. H.,** Aflatoxin residues in food of animal origin, *Food Cosmet. Toxicol.,* 10, 531, 1972.

12a. **Hayes, A. W., King, R. E., Unger, P. D., Phillips, T. D., Hatkin, J., and Bowen, J. H.,** Aflatoxicosis in swine, *J. Am. Vet. Med. Assoc.,* 172, 1295, 1978.

13. **Rodricks, J. V. and Stoloff, L.,** Aflatoxin residues from contaminated feed in edible tissues of food producing animals, in *Mycotoxins in Human and Animal Health,* Rodricks, J. V., Hesseltine, C. W., and Mehlman, M. A., Eds., Pathotox Publishers, Park Forest South, Ill., 1977, 67.
14. **Allcroft, R. and Carnaghan, R. B. A.,** Groundnut toxicity: an examination for toxin in human food products from animals fed toxic groundnut meal, *Vet. Rec.,* 75, 259, 1963.
15. **Van Peé, W., Van Brabant, J., and Joostens, J.,** The detection and concentration of aflatoxin M_1 in milk and milk products, *Rev. Agric. (Bruxelles),* 30, 403, 1977.
16. **Fritz, W., Donath, R., and Engst, R.,** Determination and occurrence of aflatoxin M_1 and B_1 in milk and dairy produce, *Nahrung,* 21, 79, 1977.
17. **Kiermeier, F.,** Aflatoxin M_1 secretion in cows' milk depending on the quantity of aflatoxin B_1 ingested, *Milchwissenschaft,* 28, 683, 1973.
18. **Kiermeier, F., Weiss, G., Behringer, G., Miller, M., and Ranfft, K.,** On the presence and content of aflatoxin M_1 in milk shipped to a dairy plant, *Z. Lebensm. Unters. Forsch.,* 163, 171, 1977.
19. **Polzhofer, K.,** Determination of aflatoxins in milk and milk products, *Z. Lebensm. Unters. Forsch.,* 163, 175, 1977.
20. **Paul, R., Kalra, M. S., and Singh, A.,** Incidence of aflatoxins in milk and milk products, *Indian J. Dairy Sci.,* 29, 318, 1976.

21. Schuller, P. L., Verhülsdonk, C. A. H., and Paulsch, W. E., Aflatoxin M_1 in liquid and powdered milk, *Zesz. Probl. Postepow. Nauk. Roln.*, 139, 255, 1977.
22. Patterson, D. S. P., unpublished data, 1977.
23. Anon., FDA is preparing to set action level for aflatoxin in fluid milk, *Food Chem. News*, 19 (37), 38, 1977.
24. Neumann-Kleinpaul, A. and Terplan, G., The presence of aflatoxin M_1 in dried milk products, *Arch. Lebensm.*, 23, 128, 1972.
25. Hanssen, E. and Jung, M., Control of aflatoxins in the food industry, *Pure Appl. Chem.* 35, 239, 1973.
26. Jung, M. and Hanssen, E., The occurrence of aflatoxin M_1 in dried milk products, *Food Cosmet. Toxicol.*, 12, 131, 1974.
27. Luck, H., Steyn, M., and Wehner, F. C., A survey of milk powder for aflatoxin content, *S. Afr. J. Dairy Technol.*, 8, 85, 1977.
28. Shank, R. C. Metabolic activation of mycotoxins by animals and humans: an overview, *J. Toxicol. Environ. Health*, 2, 1229, 1977.
29. Patterson, D. S. P., Metabolism of aflatoxin and other mycotoxins in relation to their toxicity and the accumulation of residues in animal tissues, *Pure Appl. Chem.*, 49, 1723, 1977.
30. Holzapfel, C. W., Steyn, P. S., and Purchase, I. F. H., Isolation and structure of aflatoxins M_1 and M_2, *Tetrahedron Lett.*, 25, 2799, 1966.
31. Bassir, O. and Osiyemi, F., Biliary excretion of aflatoxin in the rat after a single dose, *Nature (London)*, 215, 882, 1967.
32. Dalezios, J. I., Wogan, G. N., and Weinreb, S. M., Aflatoxin P_1: a new aflatoxin metabolite in monkeys, *Science*, 171, 584, 1971.
33. Dalezios, J. I. and Wogan, G. N., Metabolism of aflatoxin B_1 in rhesus monkeys, *Cancer Res.*, 32, 2297, 1972.
34. Masri, M. S., Haddon, W. F., Lundin, R. E., and Hsieh, D. P. H., Aflatoxin Q_1: A newly identified major metabolite of aflatoxin B_1 in monkey liver. *J. Agric. Food Chem.*, 22, 512, 1974.
34a. Degen, G. H. and Neumann, H. G., The major metabolite of aflatoxin B_1 in the rat is a glutathione conjugate, *Chem. Biol. Interact.*, 22, 239, 1978.
34b. Lotlikar, P. D., Insetta, S. M., and Lyons, P. R., Inhibition of microsome-mediated binding of aflatoxin B_1 to DNA by glutathione S-transferase, *Cancer Lett.*, 9, 143, 1980.
35. Patterson, D. S. P. and Roberts, B. A., The *in vitro* reduction of aflatoxins B_1 and B_2 by soluble avian liver enzymes, *Food Cosmet. Toxicol.*, 9, 829, 1971.
36. Patterson, D. S. P. and Roberts, B. A., Aflatoxin metabolism in duck liver homogenates: the relative importance of reversible cyclopentenone reduction and hemiacetal formation, *Food Cosmet. Toxicol.*, 10, 501, 1972.
37. Salhab, A. S. and Hsieh, D. P. H., Aflatoxicol H_1: a major metabolite of aflatoxin B_1 produced by human and monkey livers *in vitro*, *Res Commun. Chem. Pathol. Pharmacol.*, 10, 419, 1975.
38. Salhab, A. S., Abramson, F. P., Geelhoed, G. W., and Edwards, G. S., Aflatoxicol M_1, a new metabolite of aflatoxicol, *Xenobiotica*, 7, 401, 1977.
39. Garner, R. C., Miller, E. C., Miller, J. A., Garner, J. V., and Hanson, R. S., Formation of a factor lethal for *S. typhimurium* TA1530 and TA1531 on incubation of aflatoxin B_1 with rat liver microsomes, *Biochem. Biophys. Res. Commun.*, 45, 774, 1971.
40. Garner, R. C., Miller, E. C., and Miller, J. A., Liver microsomal metabolism of aflatoxin B_1 to a reactive derivative toxic to *Salmonella* typhimurium TA1530, *Cancer Res.*, 32, 2058, 1972.
40a. Garner, R. C., Carcinogenesis by fungal products, *Br. Med. Bull.*, 36, 47, 1980.
41. Swenson, D. H., Miller, J. A., and Miller, E. C., 2,3-Dihydro-2,3-dihydroxy-aflatoxin B_1: an acid hydrolysis product formed by hamster and rat liver microsomes *in vitro*, *Biochem. Biophys. Res. Commun.*, 53, 1260, 1973.
42. Swenson, D. H., Miller, J. A., and Miller, E. C., The reactivity and carcinogenicity of aflatoxin B_1-2,3-dichloride, a model for the putative 2,3-oxide metabolite of aflatoxin B_1, *Cancer Res.*, 35, 3811, 1975.
43. Lin, J-K., Miller, J. A., and Miller, E. C., 2,3-Dihydro-2-(guan-7-yl)-3-hydroxy-aflatoxin B_1, a major acid hydrolysis product of aflatoxin B_1-DNA or -ribosomal RNA adducts formed in hepatic microsome-mediated reactions and in rat liver *in vivo*, *Cancer Res.*, 37, 4430, 1977.
43a. Neal, G. E., Judah, D. J., Stirpe, F., and Patterson, D. S. P., The formation of 2,3-dihydroxy-2,3-dihydro-aflatoxin B_1 by the metabolism of aflatoxin B_1 by liver microsomes isolated from certain avian and mammalian species and the possible role of this metabolite in the acute toxicity of aflatoxin B_1, *Toxicol. Appl. Pharmacol.*, 58, 431, 1981.
43b. Neal, G. E. and Colley, P. J., The formation of 2,3-dihydro-2,3-dihydroxy aflatoxin B_1 by the metabolism of aflatoxin B_1 *in vitro* by rat liver microsomes, *FEBS Lett.*, 101, 382, 1979.
43c. Wang, T-C.V. and Cerutti, P., Spontaneous reactions of aflatoxin B_1 modified DNA *in vitro*, *Biochemistry*, 19, 1692, 1980.

44. **Patterson, D. S. P. and Roberts, B. A.** The formation of aflatoxins B_{2a} and G_{2a} and their degradation products during the *in vitro* detoxification of aflatoxin by livers of certain avian and mammalian species, *Food Cosmet. Toxicol.*, 8, 527, 1970.
45. **Gurtoo, H. L. and Campbell, T. C.**, Metabolism of aflatoxin B_1 and its metabolism dependent and independent binding to rat hepatic microsomes, *Mol. Pharmacol.*, 10, 776, 1974.
46. **Ashoor, S. H. and Chu, F. S.**, Interaction of aflatoxin B_{2a} with amino acids and proteins, *Biochem. Pharmacol.*, 24, 1799, 1975.
47. **Chipley, J. R., Mabee, M. S., Applegate, K. L., and Dreyfuss, M. S.**, Further characterization of tissue distribution and metabolism of ^{14}C-aflatoxin B_1 in chickens, *Appl. Microbiol.*, 28, 1027, 1974.
48. **Patterson, D. S. P.**, Metabolism as a factor in determining the toxic action of the aflatoxin in different animal species, *Food Cosmet. Toxicol.*, 11, 287, 1973.
49. **Hsieh, D. P. H., Wong, Z. A., Wong, J. J., Michas, C., and Ruebner, B. H.**, Comparative metabolism of aflatoxin, in *Mycotoxins in Human and Animal Health,* Rodricks, J. V., Hesseltine, C. W., and Mehlman, M. A. Eds., Pathotox Publishers, Park Forest South, Ill., 1977, 37.
50. **Salhab, A. S. and Edwards, G. S.**, Comparative *in vitro* metabolism of aflatoxicol by liver preparations from animals and humans, *Cancer Res.*, 37, 1016, 1977.
51. **Pong, R. S. and Wogan, G. N.**, Time course and dose-response characteristics of aflatoxin B_1. Effects on rat liver RNA polymerase and ultrastructure, *Cancer Res.*, 30, 294, 1970.
52. **Stich, H. F. and Laishes, B. A.**, The response of *Xeroderma pigmentosa* cells and controls to the activated mycotoxins, aflatoxins and sterigmatocystin, *Int. J. Cancer,* 16, 266, 1975.
53. **Bababunmi, E. A. and Bassir, O.**, Effects of aflatoxin B_1 on the swelling and ATPase activities of mitochondria isolated from different tissues of the rat, *FEBS Lett.*, 26, 102, 1972.
54. **Belt, J. A. and Campbell, T. C.**, Effect of aflatoxin on mitochondrial protein synthesis, *Fed. Proc. Fed. Am. Soc. Exp. Biol.*, 34, (Abstr. 41), 226, 1975.
55. **Doherty, W. P. and Campbell, T. C.**, Inhibition of rat liver mitochondria electron-transport flow by aflatoxin B_1, *Res. Commun. Chem. Pathol. Pharmacol.*, 3, 601, 1972.
56. **Doherty, W. P. and Campbell, T. C.**, Aflatoxin inhibition of rat liver mitochondria, *Chem. Biol. Interact.*, 7, 63, 1973.
57. **Adekunle, A. A. and Elegbe, R. A.**, Lysosomal activity in aflatoxin B_1-treated avian embryos, *Lancet,* 1, 991, 1974.
58. **Pokrovsky, A. A., Kravchenko, L. V., and Tutelyan, V. A.** Effect of aflatoxin on rat liver lysosomes, *Toxicon,* 10, 25, 1972.
59. **Butler, W. H.**, Further ultrastructural observations on injury of rat hepatic parenchymal cells induced by aflatoxin B_1, *Chem. Biol. Interact,* 4, 49, 1971 to 1972.
60. **Williams, D. J. and Rabin, B. R.**, Disruption by carcinogens of the hormone dependent association of membranes with polysomes, *Nature (London),* 232, 102, 1971.
61. **Fairclough, D. L., Fox, J. P., and Campbell, T. C.**, Effect of aflatoxin preincubation and addition to rat liver microsomes on mixed function oxidase activity, *Fed. Proc. Fed. Am. Soc. Exp. Biol.*, 34, (Abstr. 3197), 784, 1975.
62. **John, D. W. and Miller, L.**, Effect of aflatoxin B_1 on net synthesis of albumin, fibrinogen, and $alpha_1$-acid glycoprotein by the isolated perfused rat liver, *Biochem. Pharmacol.*, 18, 1135, 1969.
63. **Bababunmi, E. A. and Bassir, O.**, Effect of aflatoxin on blood clotting in the rat, *Br. J. Pharmacol.*, 37, 497, 1969.
64. **Shankaran, R., Raj, H. G., and Venkitasubramanian, T. A.**, Effect of aflatoxin on carbohydrate metabolism in chick liver, *Enzymologia,* 39, 371, 1970.
65. **Donaldson, W. E., Tung, H-T., and Hamilton, P. B.**, Depression of fatty acid synthesis in chick liver by aflatoxin, *Comp. Biochem. Physiol.*, 41B, 843, 1972.
66. **Tung, H-T., Donaldson, W. E., and Hamilton, P. B.** Altered lipid transport during aflatoxicosis, *Toxicol. Appl. Pharmacol.*, 22, 97, 1972.
67. **Horton, B. J., Horton, J. D., and Sabine, J. R.**, Metabolic control in precancerous liver. II. Loss of feedback control of cholesterol synthesis measured repeatedly *in vivo* during treatment with the carcinogens N-2 fluorenylacetamide and aflatoxin, *Eur. J. Cancer,* 8, 437, 1972.
68. **Bernacki, R. J. and Gurtoo, H. L.**, Differential inhibition of rat liver sialyl transferases EC.2.4.99.1 by various aflatoxins and their metabolites, *Res. Commun. Chem. Pathol. Pharmacol.*, 10, 681, 1975.
69. **Raj, H. G., Santhanam, K., Gupta, R. P., and Venkitasubramanian, T. A.**, Oxidative metabolism of aflatoxin B_1 by rat liver microsomes *in vitro* and its effect on lipid peroxidation, *Res. Commun. Chem. Pathol. Pharmacol.*, 8, 703, 1974.
70. **Wells, P., Aftergood, L., Parkin, L., and Alfin-Slater, R. B.**, Effect of dietary fat upon aflatoxicosis in rats fed torula yeast containing diet, *J. Am. Oil Chem. Soc.*, 52, 139, 1975.
71. **Patterson, D. S. P. and Roberts, B. A.** Steroid sex hormones as inhibitors of aflatoxin metabolism in liver homogenates, *Experientia,* 28, 929, 1972.

72. **Tung, H-T., Smith, J. W., and Hamilton, P. B.,** Aflatoxicosis and bruising in the chicken, *Poultry Sci.*, 50, 795, 1971.
73. **Armbrecht, B. H., Shalkop, W. T., Rollins, L. D., Pohland, A. E., and Stoloff, L.,** Acute toxicity of aflatoxin B_1 in wethers, *Nature (London)*, 225, 1062, 1970.
74. **Carnaghan, R. B. A., Hebert, C. N., Patterson, D. S. P., and Sweasey, D.,** Comparative biological and biochemical studies in hybrid chicks. II. Susceptibility to aflatoxin and effects on serum protein constituents, *Br. Poultry Sci.*, 8, 279, 1967.
75. **Peers, F.G. and Linsell, C. A.,** Acute toxicity of aflatoxin B_1 for baboons, *Food Cosmet. Toxicol.*, 14, 227, 1976.
76. **Shank, R. C., Johnsen, D. O., Tanticharoenyoz, P., Wooding, W. L., and Bourgeois, C. H.,** Acute toxicity of aflatoxin B_1 in the macaque monkey, *Toxicol. Appl. Pharmacol.*, 20, 227, 1971.
77. **Wogan, G. N., Edwards, G. S., and Newberne, P. M.,** Structure-activity relationships in toxicity and carcinogenicity of aflatoxins and analogs, *Cancer Res.*, 31, 1936, 1971.
77a. **Hayes, A. W.,** Mycotoxins: a review of biological effects and their role in human disease, *Clin. Toxicol.*, 17, 45, 1980.
78. **Edds, G. T.,** Acute aflatoxicosis: a review, *J. Am. Vet. Med. Assoc.*, 162, 304, 1973.
79. **Newberne, P. M. and Butler, W. H.,** Acute and chronic effects of aflatoxin on the liver of domestic and laboratory animals: a review, *Cancer Res.*, 29, 236, 1969.
80. **Sargeant, K., O'Kelly, J., Carnaghan, R. B. A., and Allcroft, R.,** The assay of a toxic principle in certain groundnut meals, *Vet. Rec.*, 73, 1219, 1961.
81. **Legator, M. S.,** Biological assays for aflatoxins, in *Aflatoxin: Scientific Background, Control, and Implications,* Goldblatt, L. A., Ed., Academic Press, New York, 1969, 107.
82. **Paget, G. E.** Exudative hepatitis in guinea pigs, *J. Pathol. Bacteriol.*, 67, 393, 1954.
83. **Stalker, A. L., and McLean, D. L.,** The incidence of oedema in young guinea pigs, *J. Anim. Tech. Assoc.*, 8, 18, 1957.
84. **Paterson, J. S., Crook, J. C., Shand, A., Lewis, G., and Allcroft, R.,** Groundnut toxicity as the cause of exudative hepatitis (oedema disease) of guinea pigs, *Vet. Rec.*, 74, 639, 1962.
85. **Seibold, H. R. and Bailey, W. S.,** An epizootic of hepatitis in the dog, *J. Am. Vet. Med. Assoc.*, 121, 201, 1952.
86. **Seibold, H. R.,** Hepatitis X in dogs, *Vet. Med.*, 48, 242, 1953.
87. **Newberne, J. W., Bailey, W. S., and Seibold, H. R.,** Notes on a recent outbreak and experimental reproduction of hepatitis X in dogs, *J. Am. Vet. Med. Assoc.*, 127, 59, 1955.
88. **Newberne, P. M., Russo, R., and Wogan, G. N.,** Acute toxicity of aflatoxin B_1 in the dog, *Pathol. Vet.*, 3, 331, 1966.
89. **Chaffee, V. W., Edds, G. T., Himes, J. A., and Neal, F. C.,** Aflatoxicosis in dogs, *Am. J. Vet. Res.*, 30, 1737, 1969.
90. **Allcroft, R.,** Aflatoxicosis in farm animals, in *Aflatoxin: Scientific Background, Control, and Implications,* Goldblatt, L. A., Ed., Academic Press, New York, 1969, 223.
91. **Keyl, A. C. and Booth, A. N.,** Aflatoxin effects in livestock, *J. Am. Oil Chem. Soc.*, 48, 599, 1971.
92. **Newberne, P. M.,** The new world of mycotoxins in animal and human health, *Clin. Toxicol.*, 7, 161, 1974.
93. **Keyl, A. C.,** Aflatoxicosis in cattle, in *Mycotoxic Fungi, Mycotoxins, Mycotoxicoses: an Encyclopedic Handbook,* Vol. 2, Wyllie, T. D. and Morehouse, L. G. Eds., Marcel Dekker, New York, 1977, 9.
94. **Armbrecht, B. H.,** Aflatoxicosis in sheep, in *Mycotoxic Fungi, Mycotoxins, Mycotoxicoses: an Encyclopedic Handbook,* Vol. 2, Wyllie, T. D. and Morehouse, L. G., Eds., Marcel Dekker, New York, 1977, 189.
95. **Armbrecht, B. H.,** Aflatoxicosis in swine, in *Mycotoxic Fungi, Mycotoxins, Mycotoxicoses: an Encyclopedic Handbook,* Vol. 2, Wyllie, T. D. and Morehouse, L. G., Eds., Marcel Dekker, New York, 1977, 227.
96. **Austwick, P. K. C.,** Aflatoxicosis in poultry, in *Mycotoxic Fungi, Mycotoxins, Mycotoxicoses: an Encyclopedic Handbook,* Vol. 2, Wyllie, T. D. and Morehouse, L. G., Eds., Marcel Dekker, New York, 1977, 279.
97. **Allcroft, R. and Lewis, G.,** Groundnut toxicity in cattle: experimental poisoning of calves and a report on clinical effects in older cattle, *Vet. Rec.*, 75, 487, 1963.
98. **Loosmore, R. M. and Markson, L. M.,** Poisoning of cattle by Brazilian groundnut meal, *Vet. Rec.*, 73, 813, 1961.
99. **Adams, L. G.,** Spontaneous epizootic of toxic hepatitis in swine attributed to aflatoxicosis, *Rev. Inst. Colomb. Agropecu.*, 9, 31, 1974.
100. **Gumbmann, M. R. and Williams, S. N.,** Biochemical effects of aflatoxin in pigs, *Toxicol. Appl. Pharmacol.*, 15, 393, 1969.

101. **Armbrecht, B. H., Wiseman, H. G., Shalkop, W. T., and Geleta, J. N.,** Swine aflatoxicosis. I. An assessment of growth efficiency and other responses in growing pigs fed aflatoxin, *Environ. Physiol.*, 1, 198, 1971.
102. **Shalkop, W. T. and Armbrecht, B. H.,** Carcinogenic response of brood sows fed aflatoxin for 28 to 30 months, *Am. J. Vet. Res.*, 35, 623, 1974.
103. **Petrickova, V., Lojda, L., Rubes, J., and Stavikova, M.,** Effect of aflatoxin B_1 on the chromosomal pattern and reproduction of rats and pigs, *Proc. 8th Int. Congr. Anim. Reprod. Art. Insem., Krakow*, Vol. 1, Abstr. 199, 1976.
104. **Loosmore, R. M. and Harding, J. D. J.,** A toxic factor in Brazilian groundnut causing liver damage in pigs, *Vet. Rec.*, 73, 1362, 1961.
105. **Harding, J. D. J., Done, J. T., Lewis, G., and Allcroft, R.,** Experimental groundnut poisoning in pigs, *Res. Vet. Sci.*, 4, 217, 1963.
106. **Lewis, G., Markson, L. M., and Allcroft, R.,** The effect of feeding toxic groundnut meal to sheep over a period of five years, *Vet. Rec.*, 80, 312, 1967.
107. **Stevens, A. J., Saunders, C. N., Spence, J. B., and Newnham, A. G.,** Investigations into "disease" of turkey poults, *Vet. Rec.*, 72, 627, 1960.
108. **Pier, A. C.,** Biological effects and diagnostic problems of mycotoxicoses in poultry, Proc. 25th Western Poult. Dis. Conf., 1976, 76.
109. **Thaxton, P. and Hamilton, P. B.,** Immunosupression in broilers by aflatoxin, *Poultry Sci.*, 50, 1636, 1971.
110. **Thaxton, P., Tung, H-T., and Hamilton, P. B.,** Immunosupression in chicks by aflatoxin, *Poultry Sci.*, 53, 721, 1974.
111. **Edds, G. T. and Simpson, C. F.,** Cecal coccidiosis in poultry as affected by prior exposure to aflatoxin B_1, *Am. J. Vet. Res.*, 37, 65, 1976.
112. **Hamilton, P. B. and Harris, J. R.,** Interaction of aflatoxicosis with *Candida albicans* infection and other stresses in chickens, *Poultry Sci.*, 50, 906, 1971.
113. **Doerr, J. A., Wyatt, R. D., and Hamilton, P. B.,** Impairment of coagulation function during aflatoxicosis in young chicks, *Toxicol. Appl. Pharmacol.*, 35, 437, 1976.
114. **Tung, H-T., Smith, J. W., and Hamilton, P. B.,** Aflatoxicosis and bruising in the chicken, *Poultry Sci.*, 50, 795, 1971.
115. **Wyatt, R. D., Thaxton, P., and Hamilton, P. B.,** Interaction of aflatoxicosis with heat stress, *Poultry Sci.*, 54, 1065, 1975.
116. **Tung, H-T., Cook, F. W., Wyatt, R. D., and Hamilton, P. B.,** The anaemia caused by aflatoxin, *Poultry Sci.*, 54, 1962, 1975.
117. **Rajion, M. A. and Farrell, D. J.,** Energy and nitrogen metabolism of diseased chickens: aflatoxicosis, *Br. Poultry Sci.* 17, 79, 1976.
118. **Boonchuvit, B. and Hamilton, P. B.,** Interaction of aflatoxin and paratyphoid infections in broiler chickens, *Poultry Sci.*, 54, 1567, 1975.
119. **Carnaghan, R. B. A., Lewis, G., Patterson, D. S. P., and Allcroft, R.,** Biochemical and pathological aspects of groundnut poisoning in chickens, *Pathol. Vet.*, 3, 601, 1966.
120. **Smith, J. W., and Hamilton, P. B.,** Aflatoxicosis in the broiler chicken, *Poultry Sci.*, 49, 207, 1970.
121. **Hamilton, P. B. and Garlich, J. D.,** Aflatoxin as possible cause of fatty liver syndrome in laying hens, *Poultry Sci.*, 50, 800, 1971.
122. **Huff, W. E., Wyatt, R. D., and Hamilton, P. B.,** Effects of dietary aflatoxin on certain egg yolk parameters, *Poultry Sci.*, 54, 2014, 1975.
123. **Giambrone, J. J., Ewart, D. L., Wyatt, R. D., and Eidson, C. S.,** Effect of aflatoxin on the humoral and cell-mediated immune systems of the chicken, *Am. J. Vet. Res.*, 39, 305, 1978.
124. **Huff, W. E., Chang, C-F., Garlich, J. D., and Hamilton, P. B.,** Aflatoxicosis in chickens fed diets varying in calcium and phosphorus, *Poultry Sci.*, 56 (Abstr.), 1724, 1977.
125. **Huff, W. E., Doerr, J. A., and Hamilton, P. B.,** Decreased bone strength during ochratoxicosis and aflatoxicosis, *Poultry Sci.*, 56 (Abstr.), 1724, 1977.
126. **Howarth, B., and Wyatt, R. D.,** Effect of dietary aflatoxin on fertility, hatchability, and progeny performance of broiler breeder hens, *Appl. Environ. Microbiol.*, 31, 680, 1976.
127. **Ruff, M. D. and Wyatt, R. D.,** Intestinal absorption of L-methionine and glucose in chickens with aflatoxicosis, *Toxicol. Appl. Pharmacol.*, 37, 257, 1976.
128. **Bosenberg, H.,** Diagnostic possibilities for the determination of aflatoxin intoxications, *Zentralbl. Bakteriol. Parasitkd. Infektionskr. Hyg. Abt. 1: Orig. A.*, 220, 252, 1972.
129. **Amla, I., Kamala, C. S., Gopalakrishna, G. S., Jayaraj, A. P., Sreesnivasamurthy, V., and Parpia, H. A. B.,** Cirrhosis in children from peanut meal contaminated by aflatoxin, *Am. J. Clin. Nutr.*, 24, 609, 1971.
130. **Krishnamachari, K. A. V. R., Bhat, R. V., Nagarajan, V., and Tilak, T. B. G.,** Hepatitis due to aflatoxicosis: an outbreak in Western India, *Lancet*, 1, 1061, 1975.

131. **Bhat, R. V. and Krishnamachari, K. A. V. R.**, Follow-up study aflatoxic hepatitis in parts of Western India, *Indian J. Med. Res.*, 66, 55, 1977.
132. **Tandon, B. N., Krishnamurthy, L., Koshy, A., Tandon, H. D., Ramalingaswami, V., Bhandari, J. R., Mathur, M. M., and Mathur, P. D.**, Study of an epidemic of jaundice, presumably due to toxic hepatitis, in Northwest India, *Gastroenterology*, 72, 488, 1977.
133. **Payet, M., Cros, J., Guenman, C., Sankale, M., and Monlanier, M.**, Two observations of children who had consumed for a prolonged period flours contaminated with *Aspergillus flavus*, *Presse Med.*, 74, 649, 1966.
134. **Ling, K-H., Wang, J. J., Wu, R., Tung, T-C., Lin, C. K., Lin, S-S., and Lin, T-M.**, as cited by **Shank, R. C.**, in *Mycotoxins and other Fungal Related Food Problems*, J. V. Rodricks, Ed., Adv. Chem. Ser. No. 149, American Chemical Society, Washington, D.C., 1976, 51.
135. **Serck-Hanssen, A.**, Aflatoxin-induced fatal hepatitis? A case report from Uganda, *Arch. Environ. Health*, 20, 729, 1970.
136. **Dvorackova, I.**, Aflatoxin inhalation and alveolar cell carcinoma, *Br. Med. J.*, 1, 691, 1976.
137. **Richir, C., Morard, J-L., Larcebau, S., and Pujol, J.**, Demonstration of aflatoxin in the human liver, *Arch. Fr. Mal. Appar. Dig.*, 63, 391, 1974.
138. **Richir, C., Paccalin, J., Larcebeau, S., Faugeres, J., Morard, J-L., and Lamant, M.**, The presence of aflatoxin B_1 in the human liver, *Cah. Nutr. Diet.*, (3/4), 223, 1976.
139. **Robinson, P.**, Infantile cirrhosis of the liver in India with special reference to probable aflatoxin etiology, *Clin. Pediatr. (Bologna)*, 6, 57, 1967.
140. **Campbell, T. C., Caedo, J. P., Bulatao-Jayme, J., Salamat, L., and Engel, R. W.**, Aflatoxin M_1 in human urine, *Nature (London)*, 227, 403, 1970.
141. **Campbell, T. C., Sinnhuber, R. O., Lee, D. J., Wales, J. H., and Salamat, L.**, Hepatocarcinogenic material in urine specimens from humans consuming aflatoxin, *J. Natl. Cancer. Inst.*, 52, 1647, 1974.
142. **Shank, R. C., Bourgeois, C. H., Keschamras, N., and Chandavimol, P.**, Aflatoxins in autopsy specimens from Thai children with an acute disease of unknown aetiology, *Food Cosmet. Toxicol.*, 9, 501, 1971.
143. **Phillips, D. L., Yourtee, D. M., and Searles, S.**, Presence of aflatoxin B_1 in human liver in the United States, *Toxicol. Appl. Pharmacol.*, 36, 403, 1976.
144. **Reye, R. D. K., Morgan, G., and Baral, J.**, Encephalopathy and fatty degeneration of the viscera: a disease entity in childhood, *Lancet*, 2, 749, 1963.
145. **Pollack, J. D., Ed.**, *Reye's Syndrome*, Grune & Stratton, New York, 1975.
146. **Shank, R. C.**, Mycotoxicoses of man: dietary and epidemiological considerations, in *Mycotoxic Fungi, Mycotoxins, Mycotoxicoses: An Encyclopedic Handbook*, Vol. 3, Wyllie, T. D. and Morehouse, L. G., Eds., Marcel Dekker, New York, 1978, 1.
147. **Bourgeois, C. H., Shank, R. C., Grossman, R. A., Johnsen, D. O., Wooding, W. L., and Chandavimol, P.**, Acute aflatoxin B_1 toxicity in the macaque and its similarities to Reye's syndrome, *Lab. Invest.*, 24, 206, 1971.
148. **Dvorackova, I., Kusak, V., Vesely, D., Vesela, J., and Nesnidal, P.**, Aflatoxin and encepalopathy with fatty degeneration of viscera (Reye), *Ann. Nutr. Aliment.* 31, 977, 1977.
149. **Becroft, D. M. O. and Webster, D. R.**, Aflatoxins and Reye's disease, *Br. Med. J.*, 4, 117, 1972.
150. **Chaves-Carballo, E., Ellefson, R. D., and Gomez, M. R.**, An aflatoxin in the liver of a patient with Reye-Johnson syndrome, *Mayo Clin. Proc.*, 51, 48, 1976.
151. **Hogan, G. R., Ryan, N. J., and Hayes, A. W.**, Aflatoxin B_1 and Reye's syndrome, *Lancet*, 1, 561, 1978.
152. **Becroft, D. M. O.**, Encephalopathy and fatty degeneration of the viscera, *Am. J. Dis. Child.*, 115, 750, 1968.
153. **Glick, T. H., Likosky, W. H., Levitt, L. P., Mellin, H., and Reynolds, D. W.**, Reye's syndrome: an epidemiological approach, *Pediatrics*, 46, 371, 1970.
154. **Wogan, G. N.**, Mycotoxins and other naturally occurring carcinogens, in *Environmental Cancer. Advances in Modern Toxicology*, Vol. 3, Kraybill, H. F. and Mehlmann, M. A., Eds., John Wiley & Sons, New York, 1977, 263.
155. **Butler, W. H. & Neal, G. E.**, Mode of action and human health aspects of aflatoxin carcinogenesis, *Pure Appl. Chem.*, 49, 1747, 1977.
156. **Moulé, Y.**, Mode of action of mycotoxins, *Pure Appl. Chem.*, 49, 1733, 1977.
157. **Wogan, G. N.**, Aflatoxin carcinogenesis, in *Methods in Cancer Research*, Busch, M., Ed., Academic Press, New York, 1973, 309.
158. **Carnaghan, R. B. A.**, Hepatic tumours in ducks fed on low level of toxic groundnut meal, *Nature (London)*, 208, 308, 1965.
159. **Sinnhuber, R. O., Wales, J. H., Ayres, J. L., Engebrecht, R. H., and Amend, D. L.**, Dietary factors and hepatoma in rainbow trout (*Salmo gairdneri*). I. Aflatoxins in vegetable protein feedstuffs, *J. Natl. Cancer Inst.*, 41, 711, 1968.

160. **Wales, J. H. and Sinnhuber, R. O.**, Hepatomas induced by aflatoxin in the sockeye salmon *(Oncorhynchus nerka)*, *J. Natl. Cancer Inst.*, 48, 1529, 1972.
161. **Sato, S., Matsushima, T., Tankaka, N., Sugimura, T., and Takashima, F.**, Hepatic tumours in the guppy (*Lebistes reticulatus*) induced by aflatoxin B_1, dimethylnitrosamine, and 2-acetylaminofluorene, *J. Natl. Cancer Inst.*, 50, 767, 1973.
162. **Svoboda, D., Grady, H. T., and Higginson, J.**, Aflatoxin B_1 injury in rat and monkey liver, *Am. J. Pathol.*, 49, 1023, 1966.
163. **Wogan, G. N. and Newberne, P. M.**, Dose-response characteristics of aflatoxin B_1 carcinogenesis in the rat, *Cancer Res.*, 27, 2370, 1967.
164. **Butler, W. H.** Aflatoxicosis in laboratory animals, in *Aflatoxin: Scientific Background, Control. and Implications*, Goldblatt, L. A., Ed., Academic Press, New York, 1969, 223.
165. **Epstein, S. M., Bartus, B., and Farber, E.**, Renal epithelial neoplasms induced in male Wistar rats by oral aflatoxin B_1, *Cancer Res.*, 29, 1045, 1969.
166. **Lin, J. J., Liu, C., and Svoboda, D. J.**, Long-term effects of aflatoxin B_1 and viral hepatitis on marmoset liver, *Lab. Invest.*, 30, 267, 1974.
167. **Vesselinovitch, S. D., Mihailovich, N., Wogan, G. N., Lombard, L. S., and Rao, K. V. N.**, Aflatoxin B_1, a hepatocarcinogen in the infant mouse, *Cancer Res.*, 32, 2289, 1972.
168. **Gopalan, C., Tulpule, P. G., and Krishnamurthi, D.**, Induction of hepatic carcinoma with aflatoxin in the rhesus monkey, *Food Cosmet. Toxicol.*, 10, 519, 1972.
169. **Tilak, T. B. G.**, Induction of cholangiocarcinoma following treatment of a rhesus monkey with aflatoxin, *Food Cosmet. Toxicol.*, 13, 247, 1975.
170. **Adamson, R. H., Correa, P., and Dalgard, D. W.**, Occurrence of a primary liver carcinoma in a rhesus monkey fed aflatoxin B_1, *J. Natl. Cancer Inst.*, 50, 549, 1973.
171. **Adamson, R. H., Correa, P., Sieber, S. M., McIntire, K. R., and Dalgard, D. W.**, Carcinogenicity of aflatoxin B_1 in rhesus monkeys: two additional cases of primary liver cancer, *J. Natl. Cancer Inst.*, 57, 67, 1976.
172. **Sinnhuber, R. O. and Wales, J. H.**, Aflatoxin B_1 hepatocarcinogenicity in rainbow trout embryos, *Fed. Proc. Fed. Am. Soc. Exp. Biol.*, 33, 247, 1974.
173. **Reddy, J. K., Svoboda, D. J., and Rao, M. S.**, Induction of liver tumors by aflatoxin B_1 in the tree shrew *(Tupaia glis)*, a non human primate, *Cancer Res.*, 36, 151, 1976.
174. **Newberne, P. M. and Wogan, G. N.**, Sequential morphologic changes in aflatoxin B_1 carcinogenesis in the rat, *Cancer Res.*, 28, 770, 1968.
175. **Halver, J. E.**, Aflatoxicosis and trout hepatoma, in *Aflatoxin: Scientific Background, Control and Implications*, Goldblatt, L. A., Ed., Academic Press, New York, 1969, 265.
176. **Newberne, P. M. and Rogers, A. E.**, Rat colon carcinomas associated with aflatoxin and marginal vitamin A, *J. Natl. Cancer Inst.*, 50, 439, 1973.
177. **Newberne, P. M. and Suphakarn, V.**, Preventive role of vitamin A in colon carcinogenesis in rats, *Cancer (Brussels)*, 40, 2553, 1977.
178. **Butler, W. H. and Barnes, J. M.**, Carcinoma of the glandular stomach in rats given diets containing aflatoxin, *Nature (London)*, 209, 90, 1966.
179. **Goodall, C. M. and Butler, W. H.**, Aflatoxin carcinogenesis: inhibition of liver cancer induction in hypophysectomized rats, *Int. J. Cancer*, 4, 422, 1969.
180. **Butler, W. H., Greenblatt, M., and Lijinsky, W.**, Carcinogenesis in rats by aflatoxins B_1, G_1 and B_2, *Cancer Res.*, 29, 2206, 1969.
181. **Ward, J. M., Sontag, J. M., Weisburger, E. K., and Brown, C. A.**, Effect of lifetime exposure to aflatoxin B_1 in rats, *J. Natl. Cancer Inst.*, 55, 107, 1975.
182. **Peers, F. G. and Linsell, C. A.**, Dietary aflatoxins and liver cancer — a population based study in Kenya, *Br. J. Cancer*, 27, 473, 1973.
183. **Peers, F. G., Gilman, G. A., and Linsell, C. A.**, Dietary aflatoxins and human liver cancer. A study in Swaziland, *Int. J. Cancer*, 17, 167, 1976.
184. **Shank, R. C., Wogan, G. N., Gibson, J. B., and Nondasuta, A.**, Dietary aflatoxins and human liver cancer. II. Aflatoxins in market foods and foodstuffs of Thailand and Hong Kong, *Food Cosmet. Toxicol.*, 10, 61, 1972.
185. **Shank, R. C., Gordon, J. E., Wogan, G. N., Nondasuta, A., and Subhamani, B.**, Dietary aflatoxins and human liver cancer. III. Field survey of rural Thai families for ingested aflatoxins, *Food Cosmet. Toxicol.*, 10, 71, 1972.
186. **Van Rensburg, S. J., Van der Watt, J. J., Purchase, I. F. H.**, Pereira Coutinho, L., and Markham, R., Primary liver cancer rate and aflatoxin intake in a high cancer area, *S. Afr. Med. J.*, 48, 2508a, 1974.
187. **Van Rensburg, S. J.**, Role of epidemiology in the elucidation of mycotoxin health risks, in *Mycotoxins in Human and Animal Health*, Rodricks, J. V., Hesseltine, C. W., and Mehlman, M. A., Eds., Pathotox Publishers, Park Forest South, Ill., 1977, 699.

188. **Peers, F. G. and Linsell, C. A.**, Dietary aflatoxins and human primary liver cancer, *Ann. Nutr. Aliment.*, 31, 1005, 1977.
189. **Linsell, C. A. and Peers, F. G.**, Field studies on liver cell cancer, in *Origins of Human Cancer*, Hiatt, H. H., Watson, J. D., and Winsten, J. A., Eds., Cold Spring Harbor Laboratory, Cold Spring Harbor, New York, 1977, 549.
190. **Van Nieuwenhuize, J. P., Herber, R. F. M., De Bruin, A., Meyer, P. B., and Duba, W. C.**, Cancers in men following a long-term exposure to aflatoxins in indoor air, *Tijdschr. Soc. Geneeskd.*, 51, 754, 1973.
191. **Deger, G. E.**, Aflatoxin — human colon carcinogenesis? *Ann. Intern. Med.*, 85, 204, 1976.
192. **Newberne, P. M.**, Mycotoxins: toxicity, carcinogenicity and the influence of various nutritional conditions, *Environ. Health Perspect.* 9, 1, 1974.
193. **Newberne, P. M.**, Environmental modifiers of susceptibility to carcinogenesis, *Cancer Detect. Prev.*, 1, 129, 1976.
194. **Newberne, P. M. and Gross, R. L.**, The role of nutrition in aflatoxin injury, in *Mycotoxins in Human and Animal Health*, Rodericks, J. V., Hesseltine, C. W., and Mehlman, M. A., Eds., Pathotox Publishers, Park Forest South, Ill., 1977, 51.
195. **Newberne, P. M. and Rogers, A. E.**, Nutritional modulation of carcinogenesis, in *Fundamentals in Cancer Prevention*, Magee, P. N., Ed., University Park Press, Baltimore, 1976, 15.
196. **Hamilton, P. B.**, Interrelationships of mycotoxins with nutrition, *Fed. Proc. Fed. Am. Soc. Exp. Biol.*, 36, 1899, 1977.
197. **Madhavan, T. V. and Gopalan, C.**, Effect of dietary protein on aflatoxin liver injury in weanling rats, *Arch. Pathol.*, 80, 123, 1965.
198. **Madhavan, T. V., Suryanarayano Rao, K., and Tulpule, P. G.**, Effect of dietary protein level on susceptibility of monkeys to aflatoxin liver injury, *Indian. J. Med. Res.*, 53, 984, 1965.
199. **Madhavan, T. V. and Gopalan, C.**, The effect of dietary protein on carcinogenesis of aflatoxin, *Arch. Pathol.*, 85, 133, 1968.
200. **Newberne, P. M. and Wogan, G. N.**, Potentiating effects of low-protein diets on effect of aflatoxin in rats, *Toxicol. Appl. Pharmacol.*, 12, 309, 1968.
201. **Smith, J. W., Hill, C.H., and Hamilton, P. B.**, The effect of dietary modification on aflatoxicosis in the broiler chicken, *Poultry Sci.*, 50, 768, 1971.
202. **Hamilton, P. B., Tung, H-T., Harris, J. R., Gainer, J. H., and Donaldson, W. E.**, The effect of dietary fat on aflatoxicosis in turkeys, *Poultry Sci.*, 51, 165, 1972.
203. **Rogers, A. E.**, personal communication, 1978.
204. **Sinnhuber, R. O., Lee, D. J., Wales, J. H., Landers, M. K., and Keyl, A. C.**, Hepatic carcinogenesis of aflatoxin, M_1 in rainbow trout *(Salmo gairdneri)* and its enhancement by cyclopropene fatty acids, *J. Natl. Cancer Inst.*, 53, 1285, 1974.
205. **Thomen, E. and Llewellyn, G. C.**, The effects of copper acetate and mixed aflatoxins on male Syrian hamsters, *Assoc. Southeast. Biol. Bull.*, 24 (Abstr.), 90, 1977.
206. **Newberne, P. M. and Conner, M. W.**, Effect of selenium on acute response to aflatoxin B_1, in *Trace Substances in Environmental Health VIII*, Hemphill, D. D., Ed., University of Missouri, Columbia, 1974, 323.
207. **Grant, K. E., Conner, M. W., and Newberne, P. M.**, Effect of dietary sodium selenite upon lesions induced by repeated small doses of aflatoxin B_1, *Toxicol. Appl. Pharmacol.*, 41, 166, 1977.
208. **Reddy, G. S., Tilak, T. B. G., and Krishnamurthi, D.**, Susceptibility of vitamin A-deficient rats to aflatoxin, *Food Cosmet. Toxicol.*, 11, 467, 1973.
209. **Hamilton, P. B., Tung, H-T., Wyatt, R. D., and Donaldson, W. E.**, Interaction of dietary aflatoxin with some vitamin deficiencies, *Poultry Sci.*, 53, 871, 1974.
210. **Newberne, P. M., Chan, W. C. M., and Rogers, A. E.**, Influence of light, riboflavin and carotene on the response of rats to the acute toxicity of aflatoxin and monocrotaline, *Toxicol. Appl. Pharmacol.*, 28, 300, 1974.
211. **Hamilton, P. B. and Garlich, J. D.**, Failure of vitamin supplementation to alter the fatty liver syndrome caused by aflatoxin, *Poultry Sci.*, 51, 688, 1972.
212. **Temcharoen, P., Anukarahanonta, T., and Bhamarapravati, N.**, Influence of dietary protein and vitamin B_{12} on the toxicity and carcinogenicity of aflatoxins in rat liver, *Cancer Res.*, 38, 2185, 1978.
213. **Rogers, A. E. and Newberne, P. M.**, Diet and aflatoxin B_1 toxicity in rats, *Toxicol. Appl. Pharmacol.*, 20, 113, 1971.
214. **Rogers, A. E.**, Variable effects of a lipotrope-deficient, high fat diet on chemical carcinogenesis in rats, *Cancer Res.*, 35, 2469, 1975.
215. **Newberne, P. M. and Rogers, A. E.**, Aflatoxin carcinogenesis in rats: dietary effects, in *Mycotoxins and Human Health*, Purchase, I. F. H., Ed., Macmillan, London, 1971, 195.
216. **Ferrando, R., Murthy, T. R. K., and Henry, N.**, Aflatoxicosis and vitamin D_3 deficiency in the duckling, *Rev. Med. Vet. (Toulouse)*, 126, 1259, 1975.

217. **Butler, W. H.,** Acute toxicity of aflatoxin B_1 in rats, *Br. J. Cancer,* 18, 756, 1964.
218. **McLean, A. E. M. and McLean, E. K.,** Diet and toxicity, *Br. Med. Bull.,* 25, 278, 1969.
219. **DiAugustine, R. P. and Fouts, J. R.,** The effects of unsaturated fatty acid on hepatic microsomal drug metabolism and cytochrome P-450, *Biochem. J.,* 115, 547, 1969.
220. **Hoekstra, W. G.,** Biochemical function of selenium and its relation to vitamin E, *Fed. Proc. Fed. Am. Soc. Exp. Biol.,* 34, 2083, 1975.
221. **Conney, A. H.,** Pharmacological implications of microsomal enzyme induction, *Pharmacol. Rev.,* 19, 317, 1967.
222. **Joseph-Bravo, P. I., Findley, M., and Newberne, P. M.,** Some interactions of light, riboflavine, and aflatoxin B_1 *in vivo* and *in vitro, J. Toxicol. Environ. Health,* 1, 353, 1976.
223. **Krogh, P.,** Ochratoxins, in *Mycotoxins in Human and Animal Health,* Rodricks, J. V., Hesseltine, C. W., and Mehlman, M. A., Eds., Pathotox Publishers, Park Forest South, Ill., 1977, 489.
224. **Tulpule, P., Manabe, M., Jemmali, M., Hamilton, P., and Patterson, D.,** Panel on aflatoxins, in *Mycotoxins in Human and Animal Health,* Rodricks, J. V., Hesseltine, C. W., and Mehlman, M. A., Eds., Pathotox Publishers, Park Forest South, Ill., 1977, 181.
225. **Anon.,** *Fertiliser and Feedingstuffs (Amendment) Regulations,* Statutory Instrument No. 840, Her Majesty's Stationery Office, London, 1976.
226. **Hayes, A. W., Unger, P. D., Stoloff, L., Trucksess, M. W., Hogan, G. R., Ryan, N. J., and Wray, B. B.,** Occurrence of aflatoxin in hypoallergenic milk substitutes, *J. Food Prot.,* 41, 974, 1978.
227. **Patterson, D. S. P., Glancy, E. M., and Roberts, B. A.** The 'carry-over' of aflatoxin M_1 into milk of cows fed rations containing a low concentration of aflatoxin B_1, *Food Cosmet. Toxicol.,* 18, 35, 1980.
228. **Dickens, J. W.,** Aflatoxin occurrence and control during growth, harvest and storage of peanuts, in *Mycotoxins in Human and Animal Health,* Rodricks, J. V., Hesseltine, C. W., and Mehlman, M. A., Eds., Pathotox Publishers, Park Forest South, Ill., 1977, 99.
229. **Goldblatt, L. A. and Dollear, F. G.,** Modifying mycotxin contamination in feeds — use of mold inhibitors, ammoniation, roasting, in *Interactions of Mycotoxins in Animal Production,* Proc. Symp., National Academy of Sciences, Washington, D.C., 1979, 167.
230. **Beckwith, A. C., Vesonder, R. F., and Ciegler, A.,** Chemical methods investigated for detoxifying aflatoxin in foods and feeds, in *Mycotoxins and Other Fungal Related Food Problems,* Rodricks, J. V., Ed., Adv. Chem. Ser. No. 149, American Chemical Society, Washington, D.C., 1976, 58.
231. **Kensler, C. J. and Natoli, D. J.,** Processing to ensure wholesome products, in *Aflatoxin: Scientific Background, Control and Implications,* Goldblatt, L. A., Ed., Academic Press, New York, 1969, 334.
232. **Tiemstra, P. J.,** Aflatoxin control during food processing of peanuts, in *Mycotoxins in Human and Animal Health,* Rodricks, J. V., Hesseltine, C. W., and Mehlman, M. A., Eds., Pathotox Publishers, Park Forest South, Ill., 1977, 121.
233. **Krogh, P.,** Mycotoxin tolerances in foodstuffs, *Ann. Nutr. Aliment.,* 31, 411, 1977.
234. **Rodricks, J. V.,** Regulatory aspects of the mycotoxin problem in the United States, in *Mycotoxic Fungi, Mycotoxins, Mycotoxicoses: An Encyclopedic Handbook,* Vol. 3, Wyllie, T. D. and Morehouse, L. G., Eds., Marcel Dekker, New York, 1978, 161.
235. **Allcroft, R., Roberts, B. A., and Lloyd, M. K.,** Excretion of aflatoxin in a lactating cow, *Food Cosmet. Toxicol.,* 6, 619, 1968.
236. **Nabney, J., Burbage, M. B., Allcroft, R. and Lewis, G.,** Metabolism of aflatoxin in sheep: excretion pattern in the lactating ewe, *Food Cosmet. Toxicol.,* 5, 11, 1967.
237. **Stoloff, L., Dantzman, J., and Armbrecht, B. H.,** Aflatoxin excretion in wethers, *Food Cosmet. Toxicol.,* 9, 839, 1971.
238. **Jaggi, W., Lutz, W. K., Luthy, J., Zweifel, U., and Schlatter, C.,** *In vitro* covalent binding to rat liver DNA of macromolecule bound aflatoxin B_1. A relay toxicity study, *Experientia (Basel),* 35, 950, 1979.
239. **Poiger, H., Ed.,** Health Hazards from Aflatoxin: a Workshop, University of Zurich, Switzerland, 1978.
240. **Pepin, G. A., Patterson, D. S. P., and Shreeve, B. J., Eds.,.,** Proc. 3rd Meet. Mycotoxins in Anim. Dis., Ministry of Agriculture, Fisheries and Food, Pinner, Middlesex, U.K., 1979.
241. **Anon.,** *Interactions of Mycotoxins in Animal Production,* Proc. Symp., National Academy of Sciences, Washington, D.C., 1979.
242. **Preusser, H-J., Ed.,** Medical Mycology, Proc. Mycol. Symp. 12th Int. Congr. Microbiol., *Zbl. Bakt.,* Suppl. 8, 1980.
243. **Eaker, D. and Wadström, T., Eds.,** *Natural Toxins,* Pergamon Press, New York, 1980.
244. **Uraguchi, K. and Yamazaki, M., Eds.,** *Toxicology, Biochemistry and Pathology of Mycotoxins,* John Wiley & Sons, New York, 1978.

245. **Heathcote, J. G. and Hibbert, J. R.,** *Aflatoxins: Chemical and Biological Aspects,* Elsevier, New York, 1978.
246. **Busby, W. F. and Wogan, G. N.,** Food-borne mycotoxins and alimentary mycotoxicosis, in *Food Borne Infections and Intoxications,* 2nd ed., Riemann, H. and Bryan, F. L., Eds., Academic Press, New York, 1979, 579.
247. **Bullerman, L. B.,** Significance of mycotoxins to food safety and human health, *J. Food Prot.,* 42, 65, 1980.
248. **Stoloff, L.,** Aflatoxin M in perspective, *J. Food Prot.,* 43, 226, 1980.
249. **Patterson, D. S. P.,** Mycotoxins, in *Environmental Chemistry,* Vol. 2, Bowen, H. J. M., Ed., Royal Society of Chemistry, London, in press, 1981.
250. **Anon.,** Survey of Mycotoxins in the United Kingdom, Food Surveillance Paper No. 4, Her Majesty's Stationery Office, London, 1980.
251. **Anon.,** Fertilisers and Feeding Stuffs (Amendment) Regulations, Statutory Instrument No. 10, Her Majesty's Stationery Office, London, 1981.
252. WHO, Mycotoxins, Environmental Health Criteria No. 11, World Health Organization, Geneva, 1979 (also published in French, 1980).
253. FDA, Assessment of Estimated Risk Resulting From Aflatoxins in Consumer Peanut Products and other Food Commodities, Bureau of Foods, Food and Drug Administration, Washington, D.C., 1978.
254. **Robens, J. F.,** Health risk and regulatory aspects of aflatoxin, *Vet. Hum. Toxicol.,* 22, 264, 1980.
255. **Price, J. E. and Heinonen, R.,** An epizootic of canine aflatoxicosis, *Kenya Vet.,* 2, 45, 1978.

FOODBORNE DISEASES: ALIMENTARY TOXIC ALEUKIA

Abraham Z. Joffe

INTRODUCTION

The studies reported in this chapter were carried out by the author in the Russian Confederate Republic of U.S.S.R., in the Orenburg district in order to elucidate the factors conducive to the occurrence and development of the very severe, so-called Alimentary Toxic Aleukia (ATA) disease. These findings have been supplemented with recent work in Israel in the Laboratory of Mycology and Mycotoxicology of the Hebrew University of Jerusalem by studying the toxicity of authentic ATA-producing *Fusarium* species to animals and plants and also the chemical nature of their toxic compounds. The author hopes that this published material will be of great interest and use to biologists, chemists, toxicologists, physicians, veterinarians, and pharmacologists.

The studies described in this chapter constitute a comprehensive investigation of the causes and problems of ATA, a widespread disease among the population of the Soviet Union, especially the Orenburg district, during World War II and the postwar years up to 1947.

In the years 1942 to 1947, ATA occurred widely as a very serious and in most cases lethal disease accompanied by various symptoms, which have been fully described in Russian literature[47-49,75,94,95,169,170,171,232] and by the author.[104-109,113-116,118,120,124]

Occurrence of this disease could be related to the near-famine condition prevailing in some parts of the Soviet Union at that time. Under these conditions the population was driven to collecting even grains that had been left in the field and passed the winter under snow cover. Consumption of such cereal grains by the population in the Orenberg district and in some other areas of the U.S.S.R. led to serious and frequently lethal outbreaks of a disease designated at "alimentary toxic aleukia" since it was characterized by a progressive leukopenia and often led to a subsequent stage known as "septic angina".

Many different names have been used for the description of the clinical aspects of ATA: septic angina,[41,47-49,79,81,82,92,95,132,133,144,150,154,155,159,161,162,169,201,202,227,233,236,286,292,339-341] alimentary hemorrhagic aleukia,[198,222,307] alimentary aleukia,[143] agranulocytosis,[94,134,167] acute myelotoxicosis,[170] alimentary panhematopathy,[344] aplastic mesenchymatopathy,[134,143] and endemic panmyelo-toxicosis.[177,178]

The isolation and identification of the toxic agents produced by the authentic strains of *F. poae* and *F. sporotrichioides* isolated from overwintered grains in Soviet Union is established with trichothecenes, chiefly T-2 toxin as the cause of ATA.[129]

In this chapter the author describes the history of the development of ATA in the U.S.S.R., the epidemiological background, etiology and biology of the disease, detailed studies on toxicity of overwintered cereal grains, bioassay methods and procedures with personal analyses of toxic fungi and the chemical nature of their toxicity, taxonomic problems of *Fusarium* species of the Sporotrichiella section, general clinical syndromes, and pathology in man and animals and also effect of *Fusarium* toxins on plants.[114-120,122,124,127,130]

DISTRIBUTION OF ATA IN THE U.S.S.R.

ATA or septic angina, has been recorded in Russia from time to time, probably since the 19th century. It was noted in eastern Siberia and in the Amur region as a food intoxication in 1913.[22,118,339-341] After 3 decades the areas of outbreaks widened and in spring of 1932, ATA appeared suddenly in endemic form in several districts of western Siberia[48,72,78,80,132,246] and in Kazakhstan.[22,247] In May and June of 1934, the disease was recorded again in western Siberia.[307] After a few years ATA reoccurred in Ryazan, Molotov, Sverdlovsk, Omsk, Novosibirsk,[198] Altai territory,[94,254,255,339,340] and some counties of Kazakhstan and Kirghiz S.S.R.[154,155] The disease became widespread in 1942 and, according to the data presented in the report of Beletskij,[22] appeared again at the beginning of World War II in different republics and districts, including Molotov, Kirov,[204] Saratov,[287] Gorkov, Yaroslavsk, Kuybyshev, Chelabinsk, Orenburn of the Ural,[2,47,48,49,116,118,161,171,172,182-238] and also in the Udmurt A.S.S.R.,[288] as well as in the Tatar A.S.S.R. [63,64,162] and Bashkir A.S.S.R.[92,286,292,308,309,344,345] In 1943 the disease appeared in the Leningrad,[201] Uljanovsk,[309] and Stalingrad districts and also in the Moldavian S.S.R. and Mari A.S.S.R.[22] In 1944 the disease spread considerably; it appeared in the northwest and southeast sections of the Ural,[22,47,48,79,182-187,253] and approached regions of the Volga River.[22] It occurred in the Ukrainian S.S.R., in central regions of the European Soviet, and also in central Asia and in far eastern Siberia.[250]

All in all, in 1944 there were outbreaks of ATA in 34 districts and counties. In that year the food situation deteriorated further and much of the population was forced to collect grains that had been left in the fields throughout the winter months. In 1945 the disease occurred in the Voronezh district,[202] in Komi A.S.S.R., and in 12 other districts and regions.[250] In 1946, ATA was observed for the first time in the Dostromsk district, in Kabardino-Balkarsk A.S.S.R., and in other areas, for a total of 19 districts and counties.[251] In 1947, ATA appeared in 23 regions, such as Tomsk, Omsk, Arkhangelsk, and Novosibirsk, among others. The frequency of occurrence of the disease in the U.S.S.R. was found to be highest between longitudes of 40 to 140°E and latitudes of 50 to 60°N.[178] In this zone the ATA disease recurred several times. For example, in the Altai territory, the outbreaks were described every year for 14 years;[340-341] in the Molotov district, for 8 years; in Bashkir A.S.S.R., 7 years;[308,309,345] and in 6 out of 31 regions affected by ATA, it appeared only once, e.g., in Komi A.S.S.R.[250] Later in 1952, 1953, and 1955 severe cases of the disease appeared again in various regions of the U.S.S.R.[21,244,246,247]

EPIDEMIOLOGY

The epidemiological investigations proved that ATA occurred in families which gathered various grains from the fields in the spring after the snow thawed. In certain sporadic cases, the disease occurred in different seasons because people purchased the overwintered toxic grains, meal, or products prepared from this grain in the markets or in the villages and towns. In 1945 to 1947 and 1948 to 1949 the disease occurred comparatively seldom and was mainly caused by instances in which overwintered toxic cereal crops were sold.[114,118,124] The mortality in 1942 to 1944 was high, and whole families, or even entire villages, were affected, mostly in agricultural areas.[115,116]

Initially research into the nature and cause of the condition was carried out simultaneously by different investigators without any central planning or coordination. The fact that the typical clinical syndrome of ATA could not be reproduced in laboratory animals hindered research. In 1932 the nature of the disease was still unknown, and outbreaks were erroneously labeled as diptheria or cholera.

In view of the sudden outbreaks and high mortality ATA was considered in the beginning to be an epidemic disease of infectious origin; however, epidemiological and bacteriological studies did not confirm this hypothesis. Moreover, the fact that none of the medical staff who took care of the patients was ever affected by ATA led to rejection of the contagious hypothesis.

For quite a long period ATA was considered to be due to a dietary deficiency of vitamins B_1, C, and riboflavin.[236,337] This hypothesis, like others relating ATA to bacterial infection, was rejected. Although a deficiency of riboflavin was detected in experimental animals suffering from severe aleukia, agranulocytosis, and anemia, the diet of patients with ATA was not deficient in this vitamin.[91,236,237,245]

Belief in these theories delayed recognition of the true nature of ATA. Eventually, however, it was realized that the disease was caused by the consumption of fungal contaminated overwintered grains, or other agricultural products, which had formed the staple diet of the peasant population in the agricultural areas in Russia and which had been infected by toxic fungi.[112,115-118,120,124]

The most important question at this stage was to determine which specific fungi were responsible for the intoxication. Eventual identification of the cause of the disease led to the introduction of proper prophylactic measures.

The studies described here deal especially with the Orenburg district where the author worked for 8 years for the Institute of Epidemiology and Microbiology of the U.S.S.R. Ministry of Health and headed the Mycological Division of the Institute. This institute established a special ATA laboratory (septic angina) for the investigation of all aspects of the disease. The following were investigated: the role of overwintered cereal crops, the mycoflora of these grains, the toxic properties of cryophilic fungi developing at especially low temperatures, climatic and ecological conditions for toxin production in grains, characteristics of these toxins in animals, the chemistry of toxins, and mainly, the clinical symptoms and pathological findings in man. The studies were subsequently continued by the author in the Department of Botany, in the Laboratory of Mycology and Mycotoxicology, the Hebrew University of Jerusalem, Israel.

The first outbreaks in the Orenburg district were in 1924 and in 1934.[78,79] In 1942 and 1943 the disease reappeared on a large scale. The Orenburg district has a conterminal boundary with the Kuybyshev, Saratov, and Chelabinsk districts, Tatar and Bashkir A.S.S.R., and also with Kazakhstan S.R.R. (see Figure 1). In 1942 ATA appeared in 15 counties in the northwest and in 4 counties in the central portions of the district. In the spring of 1943, the disease affected 30 out of 50 counties in the north, west, and south of the district. The outbreaks peaked in 1944, when 47 out of 50 counties were affected (only three easterly counties were not affected). In that year the population in the Orenburg and other districts of Soviet Russia suffered enormous casualties. It can be seen from Figure 1 that more than 10% of the population was affected, and many fatalities occurred in 9 of the 50 counties of the district.[108,115,116,120,124] In 1945, 1946, and 1947 the disease declined and affected 14, 8, and 12 counties, respectively. In 1948 and 1949 the disease was almost absent.

The outbreak of ATA was not uniform in the entire district. It was limited to some counties, or even to a single small village within that county.[120,124]

The spread of ATA was especially marked in districts where it occurred early in the spring season. It is therefore a seasonal disease, the main outbreak of which is evident during April to May until mid June. In July to August, the rate of spread declines, and in autumn and winter the disease does not appear at all. In 1942, the disease appeared in Orenburg later than usual, i.e., in July to August, but in most cases it appears where spring starts early and the snow melts early. Only in a few cases was ATA recorded in other seasons of the year, in urban patients who received overwintered grains later than they were available to the peasants in the agricultural areas.

ETIOLOGY

The seasonal occurrence of ATA, its endemicity, and the composition of the affected population suggested the importance of climatic and ecological factors in outbreaks

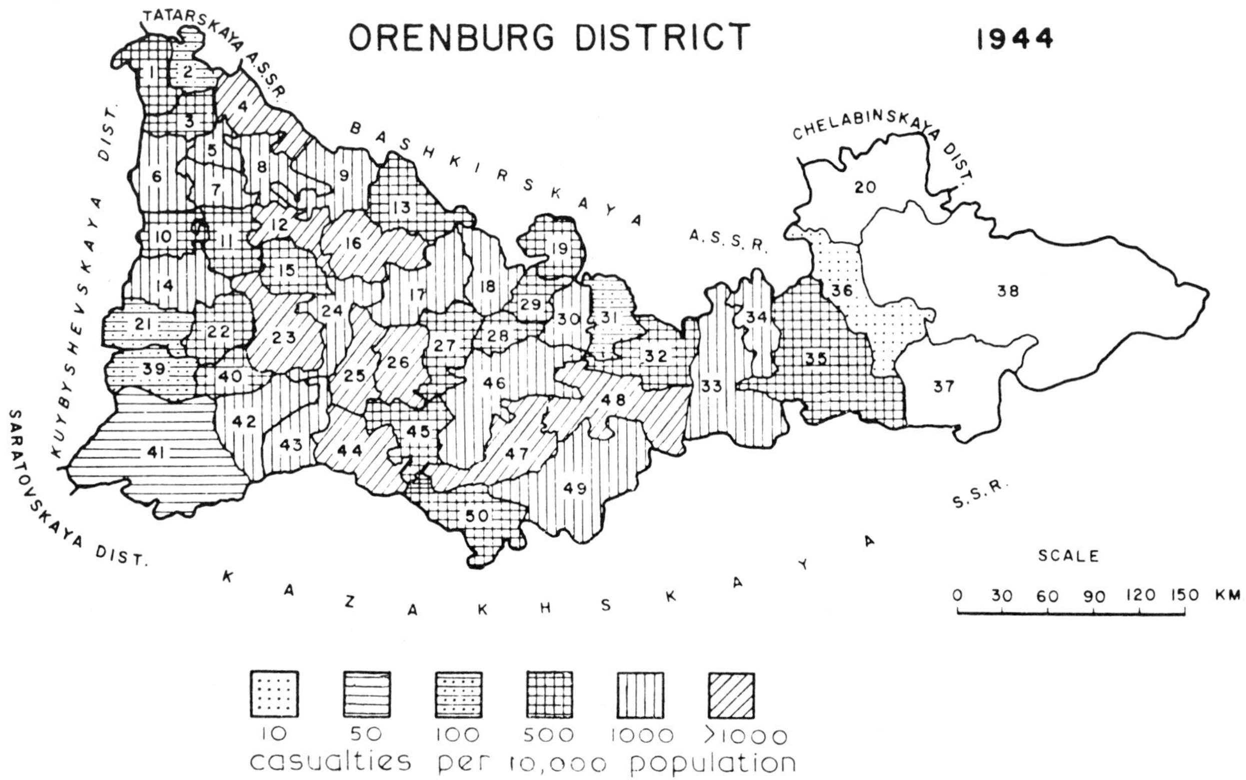

FIGURE 1. The outbreaks of ATA disease in various counties of the Orenburg district in 1944. (Figure courtesy of Marcel Dekker, Inc.)

of the disease. Surveys in the affected areas showed that the major nutrients of the patients consisted of cereal grains in the form of wheat, prosomillet, barley, oats, buckwheat, and leguminous plants. It was thus assumed that there was a correlation between the food of these rural peasants and development of the disease.

The peak outbreaks occurred in the spring when overwintered grains which had been under snow cover for the whole autumn and winter were consumed by the rural peasants. This led investigators to look for changes in the grains as a result of their being under snow cover throughout the winter. Considerable changes were found in the mycoflora of the overwintered cereals, and this suggested that ATA was caused by certain fungi which developed well under the cover of snow.

In discussing the etiology of ATA, we shall first survey the mycoflora of overwintered grains. After identifying the fungi which are responsible for ATA, we will describe the biological and chemical properties of these fungi and show that the toxin which they produce is the direct cause of the disease. The conclusion that toxin produced by *Fusarium* species which developed on overwintered grains in the field is the cause of ATA will be supported by comparing the clinical syndrome in man with the picture produced in various laboratory animals and by identification of the toxic principles responsible for the ATA disease.

The occurrence of the disease in man was found to be influenced by the following etiological factors:

Quantity of overwintered grain ingested — ATA disease usually appeared after the consumption of at least 2 kg of food prepared from toxic overwintered grains.

Duration of feeding on the toxic grain — Lesions in the hematopoietic system were produced as a result of accumulation of toxic material in the body. Symptoms usually appear 2 to 3 weeks after consumption of toxic grains (prosomillet, wheat, barley, rye, oats, buckwheat, and other products). For example, when a patient consumed toxic grain in sufficient quantity at the end of April, he became sick with the disease of ATA in the middle of May. Death occurred 6 to 8 weeks after the initial eating of the toxic grain.[49,170,171,186,202,236,341]

Concentration of toxin in the food — Great variations were found in the toxicity of overwintered grains, even within samples of one field. Families who fed on nontoxic overwintered grains were not affected. Entire families were found who had ingested highly toxic grains and who were all affected by ATA.

Kind of cereal ingested — Prosomillet and wheat were the most toxic.[107-109,115,192,193]

Sensitivity of the individual to toxin — Populations fed on balanced diets were less sensitive to the toxin than undernourished persons. Those whose diet consisted almost entirely of overwintered cereals were more severely affected.[118,124]

Season of harvesting — Grains harvested during the thaws were toxic. Grains which had been harvested in autumn and winter before the snow melted were either nontoxic or slightly toxic, but grains harvested after a less severe winter with abundant snow, followed by frequent alternate freezing and thawing in spring, were conducive to growth of the toxic fungi.[115]

Age — Breast-fed babies less than 1 year old were not affected since the toxic substance was not secreted in the milk of the sick mother.[49,170,279,281] The disease occurred in infants 1 year of age and older if they had eaten products of toxic overwintered cereals.[162,166-168,341] Disease occurred most frequently and caused the highest mortality between the ages of 8 and 50 years.[49,238]

Sex — The literature did not mention any sex differences in the mortality of ATA. We also suggested that there was no difference in the incidence of the disease in men

FIGURE 2. Sample of overwintered prosomillet affected by *F. sporotrichioides* consumed by two members of a family who died. (Figure courtesy of Marcel Dekker, Inc.)

and women, although Talayev et al.[307] noticed a greater incidence of the disease in middle-aged women.

Mycological and Mycotoxicological Studies

Initially it was thought that prosomillet was the most dangerous source of ATA, since many people fell ill after eating this grain. This cereal crop was an extensive source of the disease; it was widely grown in the Orenburg district and in other parts of the Soviet Union, and ripened late, producing an abundant harvest so that large amounts were left to overwinter in the field. Later it was shown that wheat and barley were the main causes of the disease, as well as oats, rye, and buckwheat.

All these cereals were left unharvested during the winter under snow, and the climatic, meteorologic, and ecologic conditions were favorable for enhancing toxic properties. The role of toxic fungi in causing ATA was studied extensively in our laboratory. According to Joffe[115,120,124] the great number of deaths in the Orenburg district and in other places in the U.S.S.R. were caused by *Fusarium* fungi. Researchers who worked with the ATA problem suggested that this disease was associated with a toxic origin, especially with overwintered grains contaminated by toxic fungi. This hypothesis was indeed correct and important in the etiology of ATA and was supported by many investigators.[25,62,71,108,112-116,118-120,124,131,134,140,152,192,193,197,223,224,238-241,245-247,249,253-255,317] Samples were taken from the most dangerous foci for mycological and toxicological analyses. The studies on the overwintered and normal grains, mycoflora,

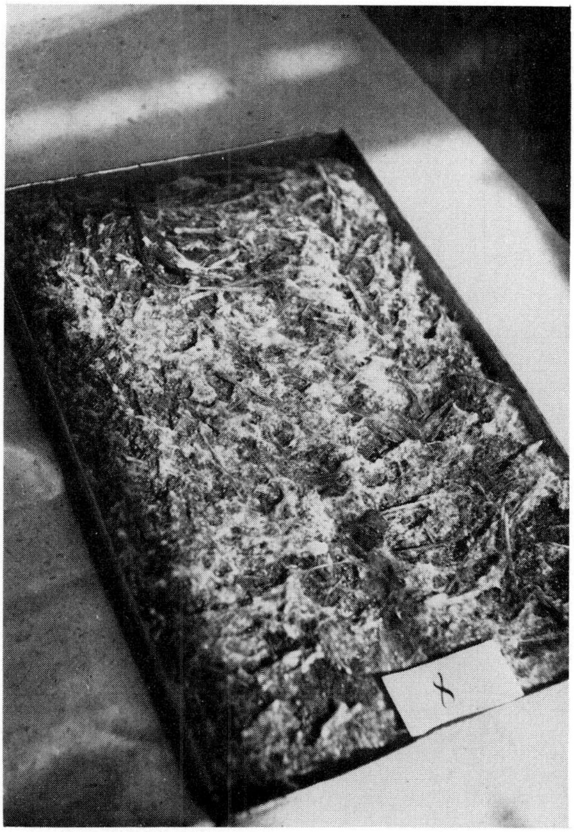

FIGURE 3. Wheat ear overwintered in the field and affected by *F. poae*, which caused the death of three persons after consumption. (Figure courtesy of Marcel Dekker, Inc.)

and the toxicity of the various isolated species enabled us to understand more clearly the etiology of ATA (Joffe[115,116,118,120,124]). The fungi most frequently associated with the toxic grain causing ATA belonged to the Sporotrichiella section of the genus *Fusarium* and included the following species: *F. poae, F. sporotrichioides, F. sporotrichioides* var. *tricinctum* and *F. sporotrichioides* var. *chlamydosporum*.[108,112,113,121,123,128]

The identification of toxic fungi which caused ATA disease was established by the following steps: The regional offices of the U.S.S.R. Ministry of Health gave notification of death caused by ATA. Samples of the grain and products consumed by the deceased were taken from their homes. These samples were carefully investigated with the help of mycological analysis (Figures 2 and 3); the fungal extracts obtained were then examined by the skin test method on rabbits and at the same time were fed to animals. The toxins produced by *F. poae* and *F. sporotrichioides,* which cause ATA, were investigated in detail on various animals by Joffe.[108,112,113,116,118,120,124] When these animals exhibited symptoms resembling those observed in man, the toxic principle was considered to be inherent in the samples.

The toxins of *F. poae* and *F. sporotrichioides* were also isolated from different overwintered grains and soil collected from various lethal fields of the Orenburg district.

In addition to all these investigations, we carried out experiments from 1943 to 1949 on 39 variously treated trial plots in the Orenburg district. These experiments contrib-

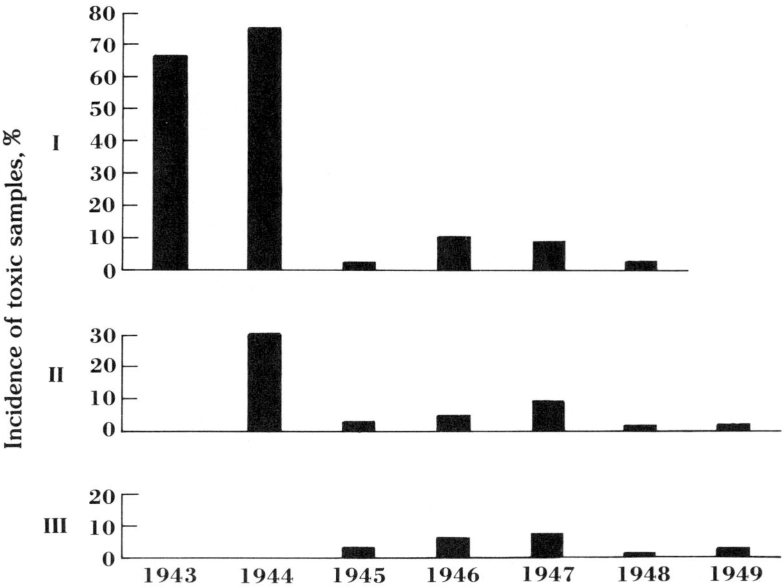

FIGURE 4. Toxic samples collected from overwintered cereal crops from (I) counties of the Orenburg district, (II) 39 experimental plots in Orenburg district, and (III) the experimental plot of the Orenburg district, shown as percentages of total samples examined from each situation. (Adapted from Joffe, A. Z., *Plant Soil,* 18, 31, 1963. With permission.)

uted much to the determination of the conditions for toxin production in overwintered grain.[115]

Material and Methods

For this study, 39 experimental plots were set up under normal field conditions in different counties of the Orenburg district. The plots varied in size from 100 m^2 to 0.5 ha. The experimental crops included the most commonly grown cereals, namely millet, wheat, and barley.

At harvest time the crop from one half of each experimental plot was cut and arranged in stacks, while the other half was left uncut. Samples of the stacked cereals were taken from the upper layer of the stack as well as from the bottom layer, at soil level. Plant and soil samples were usually collected from the experimental plots, twice a month, from August or September of each year to the following May. The samples of cereals were threshed and then dried at 45°C; the soil samples were similarly dried. Samples of dried and threshed grain weighing 30 to 50 g, samples of vegetative parts (stems, leaves, ears, panicles, husks, etc.) weighing 15 to 30 g, and soil samples of not less than 200 g were soaked in ether or alcohol. After 3 to 5 days of soaking with repeated shaking, the ether or alcohol was driven off in a distilling apparatus. The residue obtained after evaporation was assayed for toxicity by application to the shaved skin of a rabbit. The method of testing has been fully described elsewhere.[112-114,126,192,193] A total of 4702 tests with cereal and soil samples were performed on 528 rabbits.

At the beginning of each month, a summary was made of the preceding month's weather condition, for each of the experimental plots. These records included:

1. Readings taken at 1:00, 7:00, 13:00, and 19:00 hr of air temperature, relative

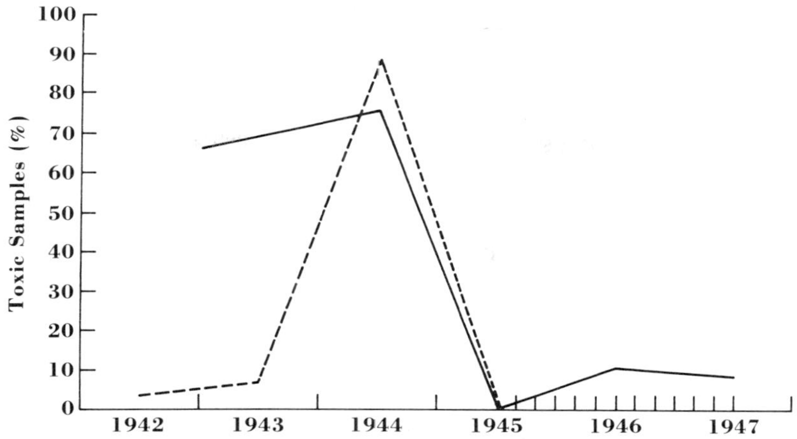

FIGURE 5. Incidence of ATA compared with incidence of toxic samples collected from overwintered cereal crops in 1942 to 1947. (From Joffe, A. Z., *Plant Soil,* 18, 31, 1963. With permission.)

humidity, and soil temperature at depths of 5, 10, 15, and 20 cm; soil temperatures were recorded until the establishment of a permanent snow cover
2. Mean daily maximum and minimum temperatures and minimum temperatures of the soil surface and of the surface of the snow
3. Moisture content of the upper soil horizon, determined every 10 days up to the formation of the snow cover and again at the time of intensive thaw; when snow was present, these measurements were taken only once a month
4. Amount of precipitation
5. Depth of freezing of the soil
6. Average depth of the snow cover, measured every 10 days, and general characteristics of the snow

The material of the 6-year-long investigation of the experimental fields consisted of 1336 samples of grain and vegetative parts, and 327 soil samples, as well as 300 samples of cereals and 122 soil samples collected at normal harvest times for control on the sites which subsequently served as experimental plots. Thus, 2085 cereal and soil samples were assayed for toxicity during the experimental period.

Concurrent investigations during the years 1943 to 1949 were conducted on 1931 samples of overwintered cereals and 230 soil samples collected from all counties of the Orenburg district. Control consisted of 296 samples of cereal and 160 soil samples collected during the normal summer harvesting period. Thus a total of 2617 cereal and soil samples were studied for toxicity. The material for investigation consisted of millet, wheat, barley, oats, and buckwheat as well as occasional samples of sunflower seed, acorns, legumes, etc., gathered in the fields.

Yearly Variability of Toxicity, 1943 to 1949*

Comparison of the toxicity of cereals from field samples and from the experimental plots in different years (Figures 4 and 5) showed that samples collected in 1945, 1948, and 1949 were mostly nontoxic. The years 1946 and 1947 hold an intermediate position, inasmuch as the toxicity of samples investigated in those years, even though not

[a] In cases where data on years of observation are mentioned, as for instance 1944 to 1949, they are to be read from autumn 1944 to spring 1949.

Table 1
TOXICITY OF CEREAL AND SOIL SAMPLES FROM EXPERIMENTAL PLOTS FOR THE PERIODS 1943—1944 AND 1944—1949

Years	Samples	No. examined	Toxic		Slightly toxic		Nontoxic	
			No.	%	No.	%	No.	%
1943—1944	Cereal samples	287	89	31.0	101	35.2	97	33.8
	Soil samples	36	14	38.9	4	11.1	18	50.0
1944—1949	Cereal samples	1049	35	3.3	55	5.2	959	91.4
	Soil samples	291	8	2.8	24	8.2	259	89.0

Adapted from Joffe, A. Z., *Plant Soil,* 18, 31, 1963.

high, was quite significant. The years 1943 and 1944 were characterized by exceptionally potent toxicity of the grains. Samples during the period 1943 to 1944 almost invariably produced a positive skin reaction. Particularly toxic were samples collected in the spring of 1944 in different counties: 75.6% of the tested samples exhibited a strong toxic activity while in the remaining 24.4% toxicity was less pronounced. The number of toxic samples from the experimental plots in spring 1944 exceeded the number of toxic samples collected during the whole period from the autumn of 1944 to the spring of 1949. This is clearly indicated in Table 1.

It seemed of interest to compare the toxicity of cereals gathered in different years on the same experimental plot. A plot on the grounds of the District Experiment Station (Orenburg county), which was under observation for 5 consecutive years (since 1945), may serve as an example (Figure 4).

It is evident that the toxicity of cereals on the same plot varied from year to year. Thus, both the collated material for the period 1943 to 1949 and the results obtained for the same plot in different years indicate that the degree of toxicity of overwintered cereals varies according to the year. On the basis of the presented data, it can be assumed that only in certain years are environmental conditions favorable for the formation and accumulation of toxic substances in cereals overwintering under the snow.

Relation Between Toxicity of Cereals and Incidence of ATA

The correlation between the incidence of ATA and the toxicity of overwintered cereals, in a number of years, is presented in Figure 5. In 1943 the percentage of toxic samples was high but was not paralleled by high incidence of disease: this is due to the fact that in 1943 food in Russia was not yet so scarce as to force large numbers of people to collect overwintered grains. However, in 1944 the high prevalence of toxicity in the samples coincided with extreme scarcity of food, and large parts of the population could only subsist by searching for overwintered grain; this resulted in a very high incidence of disease in that year. In the following years, the toxicity of samples decreased and the food situation improved, so that incidence of the disease was greatly reduced.

Role of the Soil in the Process of Toxin Formation

The role of the soil was studied during the years 1944 to 1945 in specially designed experimental blocks consisting of nine plots in all. For three of the plots (No. 1 to 3), soil was obtained from a field in which the cereals harvested in spring 1944 had been responsible for many, frequently fatal cases. The transported soil was spread out in a 15-cm thick layer over one half of each plot, leaving the other half as a control. Prosomillet panicles and wheat ears gathered in August 1944 were spread in a thin layer over the surface of the plots. On each of the 6 remaining plots (No. 4 to 9), 20 poles were put up prior to the beginning of the experiment and a small sheaf of wheat was tied to each of the poles on plots 4, 6, and 8 in such a way that it did not come in contact with the soil. Spikes from another sheaf were arranged around the pole, directly on the ground. An analogous arrangement was used on plots 5, 7, and 9 using sheaves of millet. Before the initiation of the experiment, spikes which remained in the fields on which the experimental plots were later established were collected once monthly from August to October and assayed for toxicity. During the experiment, samples from the experimental sheaves were taken for determination of toxic properties twice monthly from December to May from plots 4 and 5, and from November to May from plots 6 to 9.

Of the 36 grain samples collected from those halves of experimental plots 1 to 3, which had been covered with soil from the field suspected of harboring the toxic principle, strong toxicity was displayed by 4 samples and slight toxicity by 3 samples; however, not a single instance of toxicity was detected in the 36 control samples taken from the untreated halves of the same plots. Samples from plots 4 to 9, which had not been in contact with the soil during the winter-spring period, were also assayed, but none of the 45 samples examined showed signs of toxicity. The 47 sheaves which had been lying on the ground in these same plots yielded 4 toxic and 3 slightly toxic samples. Of 24 soil samples taken from these plots, 2 proved to be toxic.

The fungi most commonly encountered in the soil samples were representatives of the genus *Penicillium,* with lesser numbers of *Mucor, Cladosporium, Fusarium, Alternaria,* and *Rhizopus.* These fungi varied greatly in their degree of toxicity, those of the genus *Fusarium,* and especially *F. poae* and *F. sporotrichioides,* being the most highly toxic.

A relationship has been established between the fungal flora of overwintered cereals and the soil mycoflora in the years of 1944 and 1945.[77,108,112,113,116,118,120,121,192,193] The resemblance between the mycoflora of overwintered cereals and of the soils points to the part played by the soil in fungal infection of cereals. It should also be mentioned that the mycoflora of the experimental fields, as determined in spring of the years 1947 to 1949, did not differ materially in generic composition from the fungal population of samples received from the counties.[108]

Toxicity in Soil, Grains, and Vegetative Cereal Parts

A comparison of the toxicity determined in the samples of soil, grain, and vegetative parts examined during the period 1944 to 1949 from experimental field plots is given in Table 2. The tabulated data show that toxicity occurred most frequently in soil (11.2%) and somewhat less prevalent in grain (10.7%), while the vegetative parts of cereals were the least toxic (6.2%).

The detection of 11.2% toxic soil samples points to the presence of a toxic principle in the soil. This soil factor apparently spreads to the cereal plant, affecting first the vegetative parts and then the grain. The conditions prevailing in the latter are evidently particularly favorable for toxin accumulation. Hence, toxicity was more frequent in grain than in the vegetative parts. In the light of this reasoning, the findings described below and their practical value can be understood.

Of a total of 92 samples of vegetative parts gathered between September and December of 1943, 17 samples (18.4%) were toxic. Of a total of 136 assayed samples collected

FIGURE 6. (a) Overwintered prosomillet grains taken from an experimental field which caused severe cases of ATA. Growth of *F. sporotrichioides* was heavy. (b) Grains of nontoxic normal prosomillet gathered in autumn 1943 from experimental field. (Figure courtesy of Marcel Dekker, Inc.)

The findings from samples collected from a 0.5-ha experimental field plot during the different seasons of the 1943 to 1944 millet are given in Table 6. All the spring samples of millet were toxic, as compared to a much smaller proportion of toxicity among samples gathered in the autumn and winter.

Persistence of Toxicity in Stored Grains

With a view toward determining whether the toxicity of overwintered cereals is modified by prolonged storage, strongly toxic samples of millet, from the years 1943 and 1944, were reinvestigated in 1949 and 1950. The results demonstrated that the toxic

Table 6
TOXICITY OF OVERWINTERED MILLET GRAIN FROM ONE EXPERIMENTAL PLOT SAMPLED IN AUTUMN, WINTER, AND SPRING OF 1943—1944

	Autumn		Winter		Spring	
Toxicity	No. of samples	%	No. of samples	%	No. of samples	%
Toxic	16	20.5	7	11.2	38	100.0
Slightly toxic	18	23.1	19	30.1	—	—
Nontoxic	44	56.4	37	58.7	—	—
Total no. of samples examined	78		63		38	

Adapted from Joffe, A. Z., *Plant Soil*, 18, 31, 1963.

ingredient contained in the grain was not affected by 6 or 7 years of storage and that toxicity remains unchanged.[114,118,120,124]

However, concurrent mycological investigations of these samples failed to isolate *F. poae* and *F. sporotrichioides*, which had obviously perished. These findings may account for the difficulties encountered in attempts to isolate toxic fungi from cereal grains after a prolonged period of storage.

Environmental Conditions and the Incidence of ATA

To verify the extent to which meteorological factors may have influenced toxin development in overwintered cereals, an analysis was made of the meteorological data pertaining to the years 1939 to 1949. This analysis showed that conditions in 1943 to 1944, the year in which ATA reached maximum proportions, deviated from the normal in at least four important respects:

1. The rainfall total in September 1943 was only 15.2 mm, as compared with a 10-year average of 57.0 mm.
2. The mean temperature of the months January to February 1944 was only −8.3°C as compared with the much lower 10-year average of −14.7°C for these months.
3. The average depth of the snow cover in January to March 1944 amounted to 79 cm (in March 108 cm), as compared with a 10-year average of only 41 cm for the January to March period (48 cm for March alone).
4. The depth to which the soil was frozen in February to March 1944 reached only 19 cm, as compared with 77 cm for a 6-year average.

It has been shown in laboratory studies[114] that toxin accumulation in cultures of fungi isolated from overwintered grain is intensified by fluctuations of temperature such as those obtained by alternate freezing and thawing and that cultures of toxin-producing species of *F. sporotrichioides* grew well at temperatures as low as −2 to −7°C. Toxin formation is most active at temperatures just below zero, i.e., a few degrees above the growth minimum.

There were years in which overwintered grains were ingested but no outbreak of ATA was recorded because the climatic and ecological conditions were not favorable for toxin production. The conditions prevailing in spring 1944 appear to have favored toxin formation by the relatively high temperatures and especially by the dense snow cover which prevented the soil from freezing to its usual depth. Alternate thawing and freezing occurred in the spring of that year more frequently than in other years, and this favored development of toxin-producing fungi on overwintering grain.

Though not substantiated by laboratory work, it is believed that the light rainfall in autumn 1943 may have been a contributory factor. The rain-free and relatively cold autumn nights, together with ample dew, may well have favored development of the toxin-producing fungi on vegetative cereal parts, where they proceeded to attack the grain under the favorable conditions of the following spring.

A comprehensive analysis of the connection between various environmental conditions prevailing in autumn and winter and the incidence of ATA in the 50 counties of the Orenburg district of the U.S.S.R. was carried out over the 7 years 1941/1942 to 1947/1948 (see Table 7).

These dates show that the year of heaviest disease incidence, 1943 to 1944, was characterized by higher temperatures in January and February, by the depth of snow cover in March, and by the depth at which the soil was frozen, which was less in autumn and especially low in March.

During the 30 years preceding the outbreaks of ATA in 1943/1944, similar combinations of low September-October rainfall (up to 25 mm) and relatively mild January-February temperatures (−9.5 to 12.5°C) were recorded in the Orenburg district only in 1924/1925 and 1934/1935. It is of interest to note that, according to Geminov, [78,79] cases of poisoning occurred in considerable numbers in the rural population of this district in those 2 years.

The weather prevailing in 1943/1944 thus greatly favored accumulation of toxin in the cereals overwintering under the snow, and thus caused the disastrous outbreaks of ATA in 1944.

Another factor affecting the degree of toxicity of overwintered cereals was brought out by a survey of the incidence of ATA, carried out annually from 1943 to 1949, in each of the 50 counties of the Orenburg district. In 1944, the year with the most numerous cases in which 47 counties were affected, not a single case was recorded in the 3 most easterly counties (Figure 1), despite the fact that the population of this area consumed overwintered cereals no less than in other counties. The distinguishing characteristic of the three unaffected countries is their high altitude, which generally exceeds 350 to 400 m above sea level. In these counties, the minimal winter temperatures are considerably lower than in the other counties. All the other counties, in which the number of cases was moderate to very numerous, are at an altitude of 70 to 200 m above sea level. Yefremov [339,340] and Mitiukevich[196] also stated that overwintered grains were more toxic in low areas.

To verify this effect of altitude, two experimental plots were set up in the autumn of 1948 in the eastern counties. All the cereal and soil samples taken from this plot were found to be nontoxic. Samples collected on experimental plots situated in the Orenburg county at a height of 108 m served as control. Among the control samples, 5 proved to be toxic and 12 slightly toxic. These data supported the conclusion that the toxicity of overwintered cereals is affected by the altitude of the locality concerned.

Comparison Between the Toxicity of Fungal Cultures and Cereal Samples

It was interesting to compare toxicity of cultures with the toxicity of the cereal samples from which they had been isolated. Highly and mildly toxic fungi were detected in 1945 to 1949, both on toxic and on nontoxic overwintered cereals, whereas in 1944, when ATA was widely occurring in the Orenburg district, mildly and highly toxic cultures were isolated almost exclusively from toxic samples of overwintered cereals.

On comparing the type of reaction obtained from highly toxic fungi with that obtained from the respective cereal samples from which they had been isolated (samples

Table 7

RELATION BETWEEN AUTUMN AND WINTER WEATHER AND THE INCIDENCE OF ATA IN THE ORENBURG DISTRICT OF U.S.S.R., 1941—1942 TO 1947—1948

Weather		1941—1942	1942—1943	1943—1944	1944—1945	1945—1946	1946—1947	1947—1948
Temp (°C)	January	−17.7	−18.8	−8.3	−16.2	−12.4	−16.6	−7.0
	February	−18.7	−15.7	−8.0	−20.8	−5.4	−15.1	−12.6
Snow cover (in cm)	December	14	40	29	5	11	4	8
	January	22	53	50	5	29	6	15
	February	23	50	79	6	50	25	21
	March	25	43	108	17	70	18	28
Depth to which soil was frozen (cm)	November	52	21	12	25	40	33	NR
	December	NR[a]	36	14	30	50	75	41
	January	NR	NR	27	NR	50	106	68
	February	100	NR	29	80	NR	123	118
	March	NR	NR	10	80	91	NR	124

Incidence of ATA in 50 counties

Number of counties in which the population was affected by ATA at the following rates per 10000 head

No disease	31	20	3	36	42	38	0
0—50 cases	12	17	3	14	8	11	0
50—500 cases	7	13	19	0	0	0	0
500—100 cases	0	0	16	0	0	0	0
More than 1000 cases	0	0	9	0	0	0	0

[a] NR, not recorded.

Adapted from Joffe, A. Z., in *Handbook of Mycotoxins and Mycotoxicoses,* Vol. 2, Marcel Dekker, New York, 1978. With permission.

gathered in spring 1944), it was noteworthy that toxic *Cladosporium* cultures, producing a reaction of the leukocytic edematous type, and *Fusarium* cultures, giving an edematous-hemorrhagic or necrotic reaction, had been isolated only from highly toxic cereal samples. Thus it is most probable that the decisive part in the development of toxicity in overwintered cereals was played by those *Fusarium* and *Cladosporium* species which gave these types of reaction.

Strukov and Mironov[303] studied the histology of the skin to which several toxic fungi from our material had been applied. These authors indicated that the tissue changes which developed following an application of ether extracts of these fungi to the skin of rabbits were similar to reactions caused by extracts from toxic millet.

As we have stated elsewhere[112,116] toxic species of *Fusarium* were always highly toxic when inoculated on sterilized grain.

Myasnikov,[199] who tested *F. poae* in our laboratory, found that its toxic ingredient was similar to that contained in overwintered cereals. Similar results were obtained from our material by Olifson,[205-209,211] who studied the chemical composition of millet after it had been experimentally infected with pure cultures of *F. poae* and *F. sporotrichioides.*

MYCOFLORA OF OVERWINTERED CEREALS AND THEIR TOXICITY

We studied a large number of cereal samples which were collected from trial plots after being covered with snow all winter,[115] samples of soils on which the investigated plants had been grown, as well as overwintered cereals from various areas of the Orenburg district. Summer harvested samples collected in the same fields and grain samples from large government storehouses served as controls.

Methods of Mycological Study

The superficial flora of the grains was studied by culturing on various media. For study of the internal flora of the grains, the latter were surface sterilized and then rinsed and cultured. A variety of synthetic culture media were used as well as substrates prepared from sterile normal prosomillet, prosomillet husks, barley, wheat, and rice.

Fungi which were found to produce toxins on sterilized millet and wheat were invariably found to produce their toxins also on agar or on a liquid medium. Thus, throughout the year of investigations the toxic cultures were grown both on agar and on natural substrates.

The number of genera found on samples of overwintered cereals was proportionately much larger than that found on summer-harvested samples. Thus, on overwintered cereals, the genera represented by numerous isolates were *Penicillium, Fusarium, Cladosporium, Alternaria,* and *Mucor.* On normal samples only the three genera — *Pencillium, Mucor,* and *Alternaria* — were present as a rule, while *Fusarium* and *Cladosporium* were found in a few cases only or were absent altogether.

We isolated 3549 cultures belonging to 42 genera and to 192 species (Joffe[112,113]) from more than 1000 selected samples of overwintered toxic and nontoxic cereals from various regions of Orenburg district and from the experimental field plots.

The investigation of the mycoflora of overwintered grains, their toxic properties, and toxicity of various fungi enabled us to understand more clearly the etiology of ATA disease.

Isolation of *Fusarium* Strains from Different Sources

From each overwintered sample 100 to 200 grains were dipped in 0.1% solutions of mercuric chloride ($HgCl_2$) or 2 to 5% hypochlorite for 1 to 3 min, washed four to five times in sterile distilled water and plated on potato-dextrose-agar (PDA), mainly to 5 grains for each petri dish. The antibiotic, added just before the plates were poured, effectively inhibited bacterial growth. The plates were incubated at 24 or 18°C for 11 or 15 days, respectively, and the pure cultures maintained as monoconidial isolates were stored on PDA slants in test tubes and on sterilized soil at 3°C. The *Fusarium* strains were identified according to the taxonomic system published by Joffe[121,123,124] and Joffe and Palti.[128] Microcultures also were frequently used for identification of the fungi of *Fusarium* and other species.

The pure isolates were grown on various liquid and solid media at different temperatures (1, 6, 12, 18, 24, 35, and 40°C) for a period of 6 to 76 days, and sometimes from +4 to −5, −7, and −10°C on natural media for a period of 30 to 90 days or longer.

When using natural media, prosomillet, wheat, or barley grains were prepared in 150-, 250-, 500-mℓ, or 1ℓ flat flasks containing 10, 30, 50, or 100 g grain, and 15, 45, 75, or 140 mℓ tap water, respectively. The agar cultures were sometimes placed in a 3-ℓ flask on whose walls a layer of agar had been deposited by rolling. A variety of liquid media were used (Joffe[119,122]). The cultures were incubated in 1000- or 2000-mℓ Fernbach flasks containing 400 to 800 mℓ of liquid medium, respectively.

The liquid media were as follows:

1.	Substrate I:	2g KNO_3 1 g KH_2PO_4, 0.5 g KC1, 0.5 g $MgSO_4 \cdot 7H_2O$, traces of $FeSO_4$, 10 g sucrose, 1000 mℓ distilled water
2.	Substrate II:	first 5 compounds as in substrate I, 5 g soluble starch, 2.5 g dextrose, 2.5 g sucrose, 1000 mℓ distilled water
3.	Substrate III:	25 g malt extract, 1000 mℓ distilled water
4.	Substrate IV:	30 g sucrose, 10 g glucose, 1 g peptone, 1000 mℓ distilled water (carbohydrate-peptone medium)
5.	Substrate V:	2 g $NaNO_3$, 1 g K_2HPO_4, 0.5 g KC1, 0.5 g $MgSO_4 \cdot 7H_2O$, traces of $FeSO_4$, 30 g sucrose, 1000 m ℓ distilled water
6.	Substrate VI:	200 g potato, 20 g dextrose, 1000 mℓ tap-water
7.	Substrate VII:	2 g $NaNO_3$, 1 g K_2HPO_4, 0.5 g KCl, 0.5 g $MgSO_4 \cdot 7H_2O$, 0.001 g $FeSO_4$, 25 g soluble starch, 1000 cc distilled water
8.	Substrate VIII:	0.5 g K_2HPO_4, 0.25 g $MgSO_4 \cdot H_2O$, 0.5 g KCl, 5 g glucose, 3 g yeast extract, 1000 mℓ distilled water

The media were autoclaved at 0.5 to 1 atm for 20 to 30 min or were sometimes steam sterilized two or three times for 1 hr. The pH of the media varied depending on the time of cultivation and the kind of medium.

Preparation of Crude Extract

After incubation at 5 or 12°C for 46 or 21 days, respectively, different isolates of *F. poae* and *F. sporotrichioides* incubated on natural media, wheat, or millet were taken out, dried overnight at 45 to 50°C, and extracted with ethyl alcohol 96%. They were again extracted 1 to 2 min in a Waring® blender. The wheat or millet slurry was filtered under vacuum through filter paper and the residue was washed twice with ethyl alcohol. Evaporation of the solvents yielded a biologically active oil. To purify the oil from water-soluble compounds, it was extracted five times with absolute ethyl alcohol and the combined alcoholic extracts were concentrated to about 1 mℓ. Presence of T-2 toxin in the sample of crude extract was shown by TLC and GLC.

After cultivation the liquid media infected by strains of *Fusarium* were filtered through filter paper or passed through a Seitz filter and then extracted with 96% ethanol or ether and concentrated in a vacuum rotary evaporator.

The Role of *Fusarium* Species in Toxin Producing and Toxicity of Other Fungi

For the evaluation of the role of *Fusarium* in toxin production, it was necessary to study the toxicity of isolated cultures. The mycotoxic properties of the *Fusarium* crude extracts were assessed by skin test on rabbits. Details of this test and estimation of intensity of the skin reaction are described in "Bioassay Method and Procedures".

Among the isolates studied, 80 of *F. poae* (out of 82), 77 of *F. sporotrichioides* (out of 83), 8 of *F. sporotrichioides* var. *tricinctum* (out of 32), and 7 of *F. sporotrichioides* var. *chlamydosporum* (out of 15) were toxic. The results of toxicity and characteristics of the *Fusarium* strains according to substrate, place and time of isolation are given in Tables 8 to 11.

During 5 years (1944 to 1948) we also isolated 49 strains belonging to various species of *Fusarium* from overwintered cereals and soil: 20 of them were highly toxic, the remaining 29 slightly toxic. The data are given in Table 12.

All species of *Fusarium* isolated from overwintered toxic and nontoxic grain and summer harvested cereals and soil collected in experimental fields or received from various parts of Orenburg district are listed in Table 13.

Among 546 *Fusarium* cultures isolated from the above-mentioned substrates, 40.5% showed toxicity of varying degrees; 20.7% were highly toxic, and 19.8% slightly toxic (see Table 14). Highly and mildly toxic cultures were much less common in *Cladosporium,* and still less in *Mucor, Alternaria,* and *Penicillium.* The results of different kinds of fungi other than *Fusarium* which can produce toxic compounds, isolated from overwintered toxic and nontoxic grains of various cereals, are also very essential and important (Table 15).

In view of the supposition that the fungi borne on the grain may give rise to disease, each of the fungus isolates was tested for its toxicity. In Table 15 the fungus species which were isolated and their respective degrees of toxicity are listed.

As evident from the data (see Tables 8 to 13 and 15) cited, 199 toxic and strongly toxic cultures and 309 mildly toxic cultures were isolated from overwintered cereals and their soils. No toxic cultures were found on grain or on vegetative parts of summer-harvested plants.[124]

Further investigations aimed at assessing the genera most likely to produce toxins. It was assumed that the toxicogenic properties of different genera of fungi might be estimated from the frequency of their occurrence on overwintered cereals and from the appearance among them of highly toxic strains. From toxic and highly toxic cultures 13 genera were isolated and from mildly toxic cultures of 17 genera. The most frequent occurrence of toxic fungi belonged to the genera *Alternaria, Mucor, Penicillium,* and especially *Fusarium* and *Cladosporium,* each of these represented by many species. *F. poae* and *F. sporotrichioides,* both of very common occurrence on overwintered cereals in all those parts of the Orenburg district from which material had been collected, were represented in the majority of cultures. Also common were *C. epiphyllum* and *C. fagi.* Cultures of *Alternaria tenuis, Mucor hiemalis, M. racemosus, Penicillium brevi-compactum, P. steckii,* and others showed considerable toxicity (Figures 7 and 8).

At the same time many mycologists working in a parallel direction obtained various amounts of toxic *Fusarium* species and other fungi from overwintered cereal grains in relation to ATA. Murashinskij[197] isolated some strains of *Alternaria* and *Fusarium* from Siberian toxic wheat grain, and Sirotinina[287] isolated some *Alternaria* from toxic prosomillet of the Saratov district. Kvashnina[152] isolated 1227 strains belonging to 83 species among which she found *Aspergillus caliptratus, Phoma* spp., *Hymemopsis* spp., 14 cultures of *F. poae,* and 32 of *F. sporotrichioides,* very toxic from 107 samples of wheat, rye, oat, barley, pea, and sunflower gathered in the Altai territory, Bashkir and Tatar A.S.S.R., Byelorussian S.S.R., and the Ivanov, Saratov, Kuybyshov, Tambov, Chkalov, and Yaroslav districts.

Table 8
CHARACTERISTICS OF *F. POAE*(PECK) WR. STRAINS

No. of culture	Toxicity of culture: L	E	N	Year of isolation	Culture medium	Toxicity of original host	Original host	County of collection
345		+	+	1944	Millet, barley, AP, M, C	Highly toxic	Millet	N.-Sergievskij
400	+			1944	Millet, barley, PD, M, C	Toxic	Millet	N.-Sergievskij
401	+	+		1944	Millet, wheat, PD, C	Weakly toxic	Wheat	Orenburgskij
585	+			1945	Millet, barley, AP, M	Weakly toxic	Barley	Orenburgskij
619	+			1945	Millet, barley, AP	Weakly toxic	VP of P barley	Orenburgskij
655		+		1945	Millet, wheat, PD, C	Nontoxic	VP of P wheat	Orenburgskij
658	+			1945	Millet, barley, PD, C	Nontoxic	Wheat	Zianchurinskij
688		+		1945	Millet, wheat, PD	Nontoxic	Wheat	Orenburgskij
692	+			1945	Millet husks, AP, PD	Nontoxic	Millet	Orenburgskij
763	+			1946	Millet, wheat, PD, C	Nontoxic	Wheat	Orenburgskij
809		+		1947	Millet, barley, PD, M	Nontoxic	Barley	Orenburgskij
848	+			1947	Millet, barley, AP, C	Nontoxic	Millet	Orenburgskij
860	+			1947	Millet, wheat, AP, M	Nontoxic	Millet	Orenburgskij
916	+			1947	Millet, wheat, AP, M	Toxic	Oats	Zianchurinskij
954	+			1947	Millet, barley, C	Toxic	Rye	Zianchurinskij
973	+			1947	Millet, PD, C	Toxic	Millet	N.-Sergievskij
980		+		1947	Millet, PD, M	Nontoxic	Wheat	Orenburgskij
991	+			1947	Millet, PD, M	Toxic	Rye	Zianchurinskij
1059		+		1947	Millet, barley, AP	Nontoxic	VP of P barley	Orenburgskij
1164		+		1947	Millet, barley, C	Nontoxic	Barley	Orenburgskij
1185		+		1947	Millet, barley, C	Nontoxic	Barley	Orenburgskij
1191	+			1947	Millet, PD, M	Nontoxic	Wheat	Orenburgskij
1221		+		1947	Millet, barley, CP, C	Nontoxic	Soil	Orenburgskij
1226	+			1947	Millet, barley, C	Nontoxic	Barley	Orenburgskij
1240	+			1947	Millet, barley, C	Weakly toxic	Barley	Orenburgskij
1333	+			1947	Millet, C	Weakly toxic	Millet	Orenburgskij
1377		+		1948	Millet, barley, CP, C	Nontoxic	Soil	Orenburgskij
1547	+			1948	Millet, AP, PD	Nontoxic	Millet	Ak.-Bulakskij
1557		+		1948	Millet, barley	Nontoxic	Barley	Orenburgskij
1686		+		1948	Millet, CP, C	Nontoxic	Soil	Sakmarskij

Table 8 (continued)
CHARACTERISTICS OF *F. POAE*(PECK) WR. STRAINS

No. of culture	Toxicity of culture L	E	N	Year of isolation	Culture medium	Toxicity of original host	Original host	County of collection
2125		+		1948	Millet, barley, C	Nontoxic	Barley	Orenburgskij
2126		+		1948	Millet, barley, C	Nontoxic	Barley	Orenburgskij
2364	+			1948	Millet, wheat, PD, M	Weakly toxic	Wheat	Asekeevskij
2375		+ 1		1948	Millet, barley, PD, M	Nontoxic	Wheat	Asekeevskij
60/9		+ + +	+ +	1944	Millet husks, barley, AP, C	Highly toxic	Millet	Sol-Iletskij
344		+ + +	+	1944	Millet, barley, AP, C	Highly toxic	Millet	N.-Sergievskij
346		+ +	+ +	1944	Millet, barley, AP, C	Highly toxic	Millet	N.-Sergievskij
348		+ +	+ +	1944	Millet, barley, AP, C	Highly toxic	Millet	N.-Sergievskij
350		+ +		1944	Millet, barley, AP, C	Highly toxic	Millet	N.-Sergievskij
395		+ + +	+	1944	Millet husks, PD, M	Toxic	Millet	Orenburgskij
396		+ + +	+ + +	1944	Millet husks, PD, C	Toxic	Millet	Orenburgskij
426	+ + +			1944	Millet, barley, AP	Toxic	Barley	Orenburgskij
467	+ +			1944	Millet, barley, PD	Toxic	Barley	Orenburgskij
792	+	+ + +	+ +	1946	Millet, barley, AP	Weakly toxic	Barley	Orenburgskij
856		+ + +	+ +	1947	Millet, barley, PD, C	Nontoxic	Millet	Orenburgskij
914		+ + +		1947	Millet, wheat, AD, C	Highly toxic	Wheat	Sol-Iletskij
923		+ + +		1947	Millet, barley, C	Toxic	Rye	Zianchurinskij
928	+ +	+ + +		1947	Millet husks, barley, AP, C	Weakly toxic	Millet	Sol-Iletskij
958	+ +	+ +		1947	Millet, wheat, PD, C	Toxic	Wheat	Sakmarskij
975	+	+ + +	+ +	1947	Millet husks, PD, M	Weakly toxic	Millet	Orenburgskij
985	+ +	+		1947	Millet husks, PD, M	Nontoxic	VP of P millet	Orenburgskij
986		+ +	+	1947	Millet, wheat, PD, C	Toxic	Wheat	Orenburgskij
988	+ +	+ +		1947	Millet, barley, C	Toxic	Rye	Zianchurinskij
994		+ + +	+ + +	1947	Millet, wheat, PD, C	Nontoxic	Wheat	Orenburgskij
1190		+ +		1947	Millet, barley, AP	Nontoxic	Barley	Orenburgskij
1370		+ + +	+	1948	Millet, wheat, PD	Nontoxic	Wheat	Orenburgskij
1372		+ +	+	1948	Millet, wheat, PD	Nontoxic	Wheat	Orenburgskij
1374		+ + +	+ +	1948	Millet, CP, C	Nontoxic	Soil	Orenburgskij
1539		+ + +	+ +	1948	Millet, wheat, PD, C	Nontoxic	Wheat	Ak.-Bulakskij
1633	+ +	+		1948	Millet, CP, C	Nontoxic	Soil	Orenburgskij
1825		+ +	+ +	1948	Millet, barley, C	Nontoxic	Barley	Orenburgskij

1858		+ +	+ +	1948	Millet, barley, C	Nontoxic	Barley	Orenburgskij
1903		+ + +	+	1948	Millet, barley, C	Nontoxic	Barley	Orenburgskij
1912		+ + +	+ +	1948	Millet, barley, C	Nontoxic	Barley	Orenburgskij
1916		+ + +	+ +	1948	Millet, barley, C	Nontoxic	Barley	Dombarovskij
1941	+ +			1948	Millet, wheat, PD, C	Weakly toxic	Wheat	Orenburgskij
1960		+ +		1948	Millet, barley, C	Nontoxic	Barley	Orenburgskij
2001		+ + +	+	1948	Millet, barley, C	Nontoxic	Barley	Orenburgskij
2106		+ + +	+	1948	Millet, barley, C	Nontoxic	Barley	Orenburgskij
2121		+ +	+	1948	Millet, barley, C	Nontoxic	Barley	Orenburgskij
2128		+ + +	+ + +	1948	Millet, wheat, PD, C	Weakly toxic	Wheat	Orenburgskij
2158		+ + +	+ + +	1948	Millet, barley, PD, C	Nontoxic	Barley	Orenburgskij
2189		+ +		1948	Millet, wheat, PD, C	Weakly toxic	Wheat	Orenburgskij
2272		+ + +	+ +	1948	Millet, wheat, PD, C	Nontoxic	Wheat	Orenburgskij
2274		+ +	+	1948	Millet, barley, PD, C	Nontoxic	Barley	Orenburgskij
2275		+ + +	+ +	1948	Millet, barley, PD, C	Nontoxic	Barley	Orenburgskij
2340		+ +	+	1948	Millet, barley, PD, C	Nontoxic	Barley	Orenburgskij
2380	+ +	+		1948	Millet, wheat, PD, C	Nontoxic	Wheat	Orenburgskij
2520	+ +			1949	Millet, wheat, AP, M, C	Nontoxic	Sunflower	Kurmanaevskij
2627		+ + +	+ +	1949	Millet, barley, PD, C	Weakly toxic	Barley	Dombarowskij

Note: L, leukocytorrhea; E, edematous reaction; N, necrotic reaction; AP, acid potato agar; C, Czapek agar; M, malt agar; CP, carbohydrate-peptone agar; PD, potato-dextrose agar; VP of P, vegetative parts of plant; +, mildly toxic; + +, toxic; + + +, strongly toxic.

Adapted from Joffe, A. Z., *Bull. Res. Counc. Isr.*, 8D, 81, 1960.

Table 9
CHARACTERISTICS OF *F. SPOROTRICHIOIDES* SHERB. STRAINS

No. of culture	Toxicity of culture			Year of isolation	Culture medium	Toxicity of original host	Original host	County of collection
	L	E	N					
332	+	+		1944	Millet, barley, AP, M, C	Highly toxic	Millet	N.-Sergievskij
358		+	+	1944	Millet, barley, M, C	Toxic	Millet	Orenburgskij
375	+			1944	Millet, wheat, PD, M, C	Toxic	Millet	Sorochinskij
625	+			1945	Millet, wheat, PD, M	Nontoxic	Soil	Orenburgskij
650	+			1945	Millet, wheat, CP	Toxic	Wheat	Orenburgskij
684		+		1945	Millet, PD, C	Nontoxic	Millet	Orenburgskij
690		+		1945	Millet, wheat, CP,C	Toxic	Barley	Orenburgskij
752		+		1946	Millet, wheat, PD	Toxic	Wheat	Orenburgskij
806	+			1946	Millet, wheat, CP	Nontoxic	Barley	Orenburgskij
816	+			1946	Millet, barley, AP, M	Nontoxic	Barley	Orenburgskij
901	+	+		1946	Millet, barley, C	Nontoxic	Rye	Buguruslanskij
903	+			1946	Millet, barley, AP, C	Nontoxic	Rye	Buguruslanskij
943	+	+		1946	Millet, wheat, AP, M	Toxic	Wheat	Sol-Iletskij
956	+	+		1947	Millet, wheat, PD	Nontoxic	Wheat	Orenburgskij
964	+			1946	Millet, barley, AP, C	Weakly toxic	Millet	Ak.-Bulakskij
976	+	+		1947	Millet, barley, AP, M, C	Weakly toxic	Millet	Orenburgskij
977	+	+		1947	Millet, barley, AP, M, C	Weakly toxic	Millet	Orenburgskij
998		+		1947	Millet, barley, AP, C	Toxic	Wheat	Zianchuninskij
1049		+		1947	Millet, wheat, C	Nontoxic	Wheat	Orenburgskij
1067	+	+		1947	Millet, CP, C	Nontoxic	Soil	Orenburgskij
1103	+			1947	Millet, barley, AP, C	Nontoxic	Soil	Orenburgskij
1184		+		1947	Millet, wheat, PD	Nontoxic	Wheat	Orenburgskij
1206		+	+	1947	Millet, barley, C	Nontoxic	Millet	Orenburgskij
1207	+			1947	Millet, wheat, PD, M	Nontoxic	Wheat	Orenburgskij
1264		+		1947	Millet, barley, PD, C	Nontoxic	VP of P wheat	Orenburgskij
1307	+	+		1947	Millet, wheat, PD	Nontoxic	Wheat	Orenburgskij
1682	+	+		1948	Millet, wheat, PD	Toxic	Wheat	Orenburgskij
1684		+		1948	Millet, CP, C	Weakly toxic	Soil	Dombarovskij
1698		+		1948	Millet, wheat, PD, C	Weakly toxic	Soil	Dombarovskij
1873	+	+		1948	Millet, wheat, PD, M	Nontoxic	Wheat	Dombarovskij
2223	+			1948	Millet, wheat, PD, M	Nontoxic	Wheat	N.-Orskij

2384	+	+		1948	Millet, barley, AP, C	Nontoxic	Barley	Orenburgskij
60/10		+ +	+ +	1944	Millet, barley, AP, M, C	Highly toxic	Millet	Sol-Iletskij
341	+ +			1944	Millet, barley, AP, M, C	Highly toxic	Millet	N.-Sergievskij
342	+ +	+		1944	Millet, barley, AP, M, C	Highly toxic	Millet	N.-Sergievskij
343	+ +	+ +		1944	Millet, barley, AP, M, C	Highly toxic	Millet	N.-Sergievskij
347		+ +	+ +	1944	Millet, barley, AP, M, C	Highly toxic	Millet	N.-Sergievskij
349	+ +			1944	Millet, barley, AP, M, C	Highly toxic	Millet	N.-Sergievskij
351		+ +	+	1944	Millet, barley, AP, M, C	Highly toxic	Millet	N.-Sergievskij
738		+ + +	+ +	1946	Millet, barley, C	Highly toxic	Millet	Alexandrovskij
740		+ +		1946	Millet, barley, C	Highly toxic	Millet	Alexandrovskij
855		+ + +	+ + +	1947	Millet, barley, AP, M, C	Nontoxic	VP of P millet	Orenburgskij
920		+ +		1947	Millet, barley	Nontoxic	Millet	Sakmarskij
921		+ + +	+ +	1947	Millet, barley	Highly toxic	Rye	Zianchurinskij
955	+	+ +		1947	Millet, barley, AP, C	Toxic	Barley	Orenburgskij
1000		+ + +	+ +	1947	Millet, barley, AP, C	Weakly toxic	Millet	S.-Karmalinskij
1012		+ + +	+ +	1947	Millet, barley, AP, M, C	Nontoxic	Millet	Orenburgskij
1070		+ + +	+ +	1947	Millet, wheat, PD, C	Nontoxic	Wheat	Orenburgskij
1072		+ + +	+ + +	1947	Millet, wheat, PD, C	Nontoxic	Wheat	Orenburgskij
1172	+ +	+		1947	Millet, wheat, PD, C	Nontoxic	Wheat	Orenburgskij
1139		+ +		1947	Millet, wheat, PD, C	Nontoxic	Wheat	Orenburgskij
1182		+ +	+ +	1947	Millet, wheat, PD, C	Weakly toxic	Wheat	Orenburgskij
1193		+ +		1947	Millet, wheat, PD, C	Weakly toxic	Wheat	Orenburgskij
1208		+ +	+	1947	Millet, wheat, PD, D	Weakly toxic	Wheat	Orenburgskij
1225		+ +	+	1947	Millet, barley, AP, M, C	Nontoxic	Millet	Orenburgskij
1229		+ + +	+ +	1947	Millet, CP, C	Nontoxic	Soil	Orenburgskij
1369		+ +	+	1948	Millet, barley, AP, M, C	Nontoxic	Millet	Orenburgskij
1464		+ +	+ + +	1948	Millet, barley, AP, M, C	Nontoxic	Millet	Orenburgskij
1530		+ + +	+	1948	Millet, wheat, PD, M, C	Nontoxic	Wheat	Orenburgskij
1555		+ + +		1948	Millet, CP, C	Nontoxic	Soil	Orenburgskij
1656		+ + +	+	1948	Millet, barley	Nontoxic	Millet	Orenburgskij
1670		+ + +		1948	Millet, wheat, AP	Nontoxic	Wheat	Dombarovskij
1823		+ + +	+ +	1948	Millet, barley, C	Nontoxic	Barley	Orenburgskij
1830		+ +	+	1948	Millet, barley, C	Nontoxic	Barley	Orenburgskij
1869		+ + +	+ +	1948	Millet, barley, C	Nontoxic	Barley	Orenburgskij
1883		+ +		1948	Millet, wheat, PD, C	Weakly toxic	Wheat	Orenburgskij
1919		+ +	+	1948	Millet, barley, C	Nontoxic	Barley	Dombarovskij
1933		+ +	+	1948	Millet, barley, C	Nontoxic	Barley	Orenburgskij
2127		+ + +	+ +	1948	Millet, barley, AP, M	Nontoxic	Barley	Dombarovskij

Table 9 (CONTINUED)
CHARACTERISTICS OF *F. SPOROTRICHIOIDES* SHERB. STRAINS

No. of culture	Toxicity of culture			Year of isolation	Culture medium	Toxicity of original host	Original host	County of collection
	L	E	N					
2193		+ +		1948	Millet, barley, C	Nontoxic	Barley	Orenburgskij
2270		+ +	+	1948	Millet, barley, C	Nontoxic	Barley	Orenburgskij
2273		+ +	+	1948	Millet, barley, AP, C	Weakly toxic	Barley	Orenburgskij
2276		+ + +	+ +	1948	Millet, barley, C	Nontoxic	Barley	Orenburgskij
2335		+ +	+	1948	Millet, barley, C	Nontoxic	Barley	Orenburgskij
2342		+ +	+ +	1948	Millet, barley, PD	Nontoxic	Wheat	Orenburgskij
2465		+ + +	+	1949	Millet, barley, AP, C	Nontoxic	Barley	Orenburgskij

Note: L, leukocytorrhea; E, edematous reaction; N, necrotic reaction; AP, acid potato agar; C, Czarek agar; M, malt agar; CP, carbohydrate-peptone agar; PD, potato-dextrose agar; VP of P, vegetative parts of plant; +, mildly toxic; + +, toxic; + + +, strongly toxic.

Adapted from Joffe, A. Z., *Bull. Res. Counc. Isr.*, 8D, 81, 1960. With permission.

Table 10
CHARACTERISTICS OF *F. SPOROTRICHIOIDES* VAR. *TRICINCTUM*(CDA.) RAILLO STRAINS

No. of culture	Toxicity of culture			Year of isolation	Culture medium	Toxicity of original host	Original host	County of collection
	L	E	N					
1227		+ +	+	1947	Millet, barley, AP, M	Toxic	Wheat	Orenburgskij
1268		+		1947	Millet, wheat, PD, C	Nontoxic	Wheat	Orenburgskij
1292	+			1947	Millet, PD, M	Nontoxic	Millet	Orenburgskij
1344	+			1948	Millet, wheat, AP, M	Nontoxic	Wheat	Orenburgskij
1868		+		1948	Millet, barley, AP, C	Nontoxic	Wheat	Dombarovskij
1895	+	+		1948	Millet, barley, PD, C	Nontoxic	Barley	Asekeevskij
2332		+		1948	Millet, PD, C	Nontoxic	Wheat	Asekeevskij
2457		+	+	1949	Millet, PD, C	Nontoxic	Millet	Orenburgskij

Note: L, leukocytorrhea; E, edmatous reaction; N, necrotic reaction; AP, acid potato agar; C, Czapek agar; M, malt agar; CP, carbohydrate-peptone agar; PD, potato-dextrose agar; +, mildly toxic; + +, toxic; + + +, strongly toxic.

Table 11
CHARACTERISTICS OF *F. SPOROTRICHIOIDES* VAR. *CHLAMYDOSPORUM*(WR. AND RG.). JOFFE COMB. NOV. STRAINS

No. of culture	Toxicity of culture L	Toxicity of culture E	Toxicity of culture N	Year of isolation	Culture medium	Toxicity of original host	Original host	County of collection
367	+ +	+		1944	Millet, barley, PD, M, C	Weakly toxic	Millet	N.-Sergievskij
1174		+ + +	+ +	1947	Millet, barley, PD, C	Nontoxic	Millet	Orenburgskij
1255	+			1947	Millet, barley, PD, M	Nontoxic	Millet	Asekeevskij
2067		+		1948	Millet, wheat, PD, C	Nontoxic	Soil	Dombarovskij
2371	+	+		1948	Millet, wheat, PD, C	Weakly toxic	Wheat	Asekeevskij
2388	+	+		1948	Millet, wheat, PD, M	Weakly toxic	Barley	Asekeevskij
2458	+	+		1949	Millet, wheat, PD, C	Nontoxic	Barley	Asekeevskij

Note: L, leukocytorrhea; E, edmatous reaction; N, necrotic reaction; AP, acid potato agar; C, Czapek agar; M, malt agar; CP, carbohydrate-peptone agar; PD, potato-dextrose agar; +, mildly toxic; + +, toxic; + + +, strongly toxic.

Table 12
CHARACTERISTICS OF STRAINS OF *FUSARIUM* SPECIES*

No. of culture	Name of fungus	Toxicity of culture L	Toxicity of culture E	Toxicity of culture N	Year of isolation	Culture medium	Toxicity of original host	Original host	County of collection
369	*F. nivale* (Fr.) Ces.	+			1944	Millet, AP, C	Weakly toxic	Millet	N. Sergievski
944	*F. nivale* (Fr.) Ces.	+	+		1947	Millet, AP, PD	Weakly toxic	Millet	S. Iletskij
996	*F. nivale* (Fr.) Ces.		+		1947	Millet, AP, PD	Weakly toxic	Millet	S. Iletskij
809	*F. arthrosporioides* Sherb.		+		1946	Millet	Nontoxic	Barley	Orenburgskij
1163	*F. avenaceum* (Fr.) Sacc.		+	+	1947	Millet, barley, AP, C	Nontoxic	Wheat	Orenburgskij
137	*F. avenaceum* (Fr.) Sacc.	+			1944	Millet, barley, AP	Weakly toxic	Wheat	Orenburgskij
1448	*F. avenaceum* (Fr.) Sacc.		+	+	1948	Millet, barley, AP	Nontoxic	VP of P millet	Orenburgskij
1018	*F. avenaceum* (Fr.) Sacc.		+ + +	+ +	1947	Millet, PD, C	Nontoxic	Barley	Orenburgskij
1403	*F. avenaceum* (Fr.) Sacc.		+ +		1948	Millet, PD, C	Weakly toxic	Barley	S. Iletskij
1061	*F. avenaceum* (Fr.) Sacc.	+	+ +		1947	Millet, barley, AP	Nontoxic	Barley	Orenburgskij
1378	*F. semitectum* Berk. and Rav.		+		1948	Millet, wheat, M	Nontoxic	Millet	Orenburgskij
1265	*F. semitectum* Berk. and Rav.		+ +	+	1947	Millet, wheat, M	Nontoxic	Barley	Orenburgskij
2466	*F. semitectum* Berk. and Rav.	+	+	+ +	1949	Millet, wheat, M	Nontoxic	Millet	Orenburgskij
1559	*F. semitectum* Berk. and Rav. var. *majus*	+	+		1948	Millet, barley, PD	Nontoxic	Wheat	N. Orskij
1040	*F. equiseti* (Cda.) Sacc.	+	+		1947	Millet, barley, AP	Nontoxic	Millet	S. Iletskij
1106	*F. equiseti* (Cda.) Sacc.	+			1947	Millet, barley, PD, M	Nontoxic	Soil	Orenburgskij
1860	*F. equiseti* (Cda.) Sacc.	+	+		1948	Millet, wheat, PD, M	Nontoxic	Soil	Orenburgskij
1890	*F. equiseti* (Cda.) Sacc.		+		1948	Millet, wheat, PD, M	Nontoxic	Barley	Orenburgskij
687	*F. equiseti* (Cda.) Sacc.		+ +		1945	Millet, barley, AP	Nontoxic	Wheat	Orenburgskij
1900	*F. equiseti* (Cda.) Sacc.		+ +	+	1948	Millet, barley, AP	Nontoxic	Wheat	Orenburgskij
2385	*F. equiseti* (Cda.) Sacc.	+	+ +	+ +	1948	Millet, barley, AP	Nontoxic	Barley	Orenburgskij
2387	*F. equiseti* (Cda.) Sacc.		+ +	+	1948	Millet, barley, AP	Nontoxic	Barley	Orenburgskij
2315	*F. equiseti* (Cda.) Sacc. var. *acuminatum* (El. and Ev.) Bil.		+ +	+	1948	Millet, barley, PD, M	Nontoxic	Barley	Orenburgskij

Table 12 (CONTINUED)
CHARACTERISTICS OF STRAINS OF *FUSARIUM* SPECIES*

No. of culture	Name of fungus	Toxicity of culture L	E	N	Year of isolation	Culture medium	Toxicity of original host	Original host	County of collection
1140	*F. equiseti* (Cda.) Sacc. var. acuminatum (El. and Ev.) Bil.	+ +	+		1947	Millet, barley, PD, M	Nontoxic	Millet	Orenburgskij
359	*F. equiseti* (Cda.) Sacc. var. caudatum (Wr.) Joffe	+	+	+ +	1944	Millet, barley, PD, M	Nontoxic	Millet	Orenburgskij
765	*F. graminearum* Schwabe			+	1946	Millet, barley, AP	Toxic	Wheat	Orenburgskij
843	*F. culmorum* (W.G.Sm.) Sacc.	+	+		1947	Millet, wheat, PD	Highly toxic	Wheat	S. Iletskij
2328	*F. sambucinum* Fuckel	+			1948	Millet, AP	Nontoxic	Wheat	Orenburgskij
1213	*F. culmorum* (W.G.Sm.) Sacc.	+ +	+ +		1947	Millet, wheat, PD	Weakly toxic	Wheat	Orenburgskij
1540	*F. culmorum* (W.G.Sm.) Sacc.	+	+ +		1948	Millet, wheat, PD	Nontoxic	Millet	Ak.-Bulakskij
1904	*F. sambucinum* Fuckel		+ +	+ + +	1948	Millet, AP, C	Nontoxic	Barley	Orenburgskij
1767	*F. lateritium* Nees		+		1948	Millet, barley, CP, PD	Nontoxic	Soil	Orenburgskij
2330	*F. lateritium* Nees		+		1948	Millet, barley, CP, PD	Nontoxic	Barley	Orenburgskij
2638	*F. lateritium* Nees	+	+		1949	Millet, barley, CP, PD	Nontoxic	Wheat	Dombarovskij
1421	*F. lateritium* Nees		+ +		1948	Millet husks, AP	Nontoxic	VP of P millet	Orenburgskij
2381	*F. lateritium* Nees		+	+ +	1948	Millet husks, AP	Nontoxic	Wheat	Orenburgskij
989	*F. moniliforme* Sheld.	+	+		1947	Millet, CP, PD, C	Weakly toxic	Wheat	Sakmarskij
995	*F. moniliforme* Sheld.	+	+		1947	Millet, CP, PD, C	Weakly toxic	Soil	Orenburgskij
1863	*F. moniliforme* Sheld.	+	+		1948	Millet, CP, PD, C	Nontoxic	Barley	Orenburgskij
2269	*F. moniliforme* Sheld.		+		1948	Millet, AP, C	Nontoxic	Barley	Orenburgskij
1857	*F. moniliforme* Sheld.		+ +	+	1948	Millet, AP, C	Nontoxic	Barley	Orenburgskij
847	*F. oxysporum* Schlecht.	+			1947	Millet, PD, C	Nontoxic	Soil	Orenburgskij
1465	*F. oxysporum* Schlecht.		+		1948	Millet, PD, C	Nontoxic	Soil	Orenburgskij
1637	*F. oxysporum* Schlecht.		+	+	1948	Millet, PD, C	Highly toxic	Wheat	Orenburgskij
2341	*F. oxysporum* Schlecht.		+ +	+	1948	Millet, PD, C	Nontoxic	Wheat	Orenburgskij

2317	*F. oxysporum* Schlecht. var. *redolens* (Wr.) Gordon	+ +		1948	Millet, PD, C	Nontoxic	Wheat	Orenburgskij
965	*F. solani* (Mart.) App. and Wr.	+		1947	Millet, C	Toxic	Wheat	Sakmarskij
990	*F. solani* (Mart.) App. and Wr.	+	+	1947	Millet, C	Highly toxic	Wheat	Sakmarskij
1664	*F. solani* (Mart.) App. and Wr.		+	1948	Millet, C	Highly toxic	Millet	Orenburgskij

Note: L, leukocytorrhea; E, edmatous reaction; N, necrotic reaction; AP, acid potato agar; C, Czapek agar; M, malt agar; CP, carbohydrate-peptone agar; PD, potato-dextrose agar; VP of P, vegetative parts of plant; +, mildly toxic; + +, toxic.

Adapted from Joffe, A. Z., *Bull. Res. Counc. Isr.*, 8D, 81, 1960.

Table 13
TOXICITY OF *FUSARIUM* FUNGI ISOLATED FROM OVERWINTERED CEREALS, THEIR SOIL, AND SUMMER-HARVESTED CEREALS

Fungi	Overwintered cereals			Soil			Summer-harvested cereals (Nontoxic)
	Toxic	Mildly toxic	Nontoxic	Toxic	Mildly toxic	Nontoxic	
Arachnites section							
Fusarium nivale (Fr.) Ces.		3	20				
Sporotrichiella section							
F. poae (Pk.) Wr.	44	31	2	2	3		
F. sporotrichioides Sherb.	42	28	4	2	5		2
F. sporotrichioides Sherb. var *tricinctum* (Cda.) Raillo	1	7	19			5	
F. sporotrichioides var. *chlamydosporium* (Wr. and Rg.) Joffe	2	5	8				
Roseum section							
F. avenaceum (Fr.) Sacc.	3	3	26			10	3
F. arthrosporioides Sherb.		1	7			2	5
Arthrosporiella section							
F. semitectum Berk. and Rav.	2	1	26				
F. semitectum Berk. and Rav. var. *majus*		1					
Lateritium section							
F. lateritium Nees	2	2	24		1	3	
Liseola section							
F. moniliforme Sheld.	1	3	22		1	10	
Gibbosum section							
F. equiseti (Cda.) Sacc.	4	1	19		2	6	18
F. equiseti (Cda.) Sacc. var. *acuminatum* (El. and Ev.) Bil.	2	1	5				
F. equiseti (Cda.) Sacc. var. *caudatum* (Wr.) Joffe	1		17			3	9

Discolor section						
F. graminearum Schw.		1	2			
F. sambucinum Fuckel.	1	1	14			
F. culmorum (W.G.Sm.) Sacc.	2	1	13			
Elegans section						
F. oxysporum Schl.	1	2	16	1	13	2
F. oxysporum Schl. var. *redolens* (Wr.) Gordon	1					
Martiella section						
F. solani (Mart.) Sacc.		3	16		5	
F. jaavicum Koord.			8			5

Adapted from Joffe, A. Z., in *Handbook of Mycotoxins and Mycotoxicoses,* Vol. 3, Marcel Dekker, New York, 1978.

Table 14
GENERA OF FUNGI ASSOCIATED WITH TOXIN PRODUCTION IN OVERWINTERED GRAIN

Genus	Total no. of isolates	Highly toxic (%)	Mildly toxic (%)
Fusarium	546	20.7	19.8
Cladosporium	480	5.4	8.5
Alternaria	506	2.8	5.3
Penicillium	830	1.6	3.8
Mucor	335	3.0	7.2

Adapted from Joffe, A. Z., in *Microbial Toxins*, Vol. 7, Academic Press, New York, 1971.

Pidoplichka and Bilai[223,224] and Bilai[23,25] examined about 1400 strains belonging to 23 genera and 160 species isolated from 765 grain samples (of prosomillet, wheat, oat, buckwheat) obtained in Bashkir A.S.S.R. and Ukrainian S.S.R. The most toxic isolates were *Mucor hiemalis* and *M. albo ater* together with *Piptocephalis freseniana, Mortierella polycephala, M. candelabrum,* var. *minor, F. lateritium, Gliocladium, Ammoniophilum, Trichoderma lignorum,* 34 *F. poae* (only 14 of them toxic) and 36 *F. sporotrichioides* of which only 12 were toxic.

From cultures of *F. poae* and *F. sporotrichioides,* which had been isolated from overwintered cereals such as wheat, prosomillet, and oats, Bilai[23,25] found 11.5, 7.5 and 4.7% to be toxic, respectively. In our isolates, the incidence of toxicity was usually higher.

Tables 8 to 13 list the different species, especially the *Fusarium* fungi, which had been isolated from overwintered and summer-harvested cereals and soil. The results showed that the group most frequently associated with the overwintered cereal grains and which caused ATA was *Fusarium* of the Sporotrichiella section, principally *F. poae* and *F. sporotrichioides.* According to Joffe,[112,116,118,120,124] Joffe and Palti,[127,128] Sarkisov,[252,253,255] Bilai,[23-32] Pidoplichka and Bilai,[223,224] Rubinstein[238-241,245,246] and Rubinstein and Lyass,[249] one of the characteristic biological properties of *F. poae* and *F. sporotrichioides* compounds was their inflammatory and irritative action on rabbit skin.

A relationship was found between the nature of the toxic *Fusarium* cultures and the toxicity of the samples from which they had been isolated. Some of the *Fusarium* cultures caused reactions on the skin of rabbits which were analogous to those produced by the action of toxic cereals which had passed the winter under snow.[112,116] A detailed description of morphological and cultural properties of the toxic fungi of the Sporotrichiella section has been given by Joffe.[104-106,114,121,123]

TAXONOMIC PROBLEMS OF THE *FUSARIUM* OF THE SPOROTRICHIELLA SECTION

The taxonomy of species of the Sporotrichiella section may once have appeared a subject of purely academic interest, but this no longer holds true. Ever since species of this section have been proven to induce serious disorders in humans, who died after consuming overwintered grain,[112-116,118,120,124] and various diseases in animals,[39,40,96,100,124,130,166,173,175,313,319,337,338] the correct identity of these species has become of acute importance. Establishment of this identity will enable us to relate, at least from a taxonomic angle, a large body of toxicological work carried out in recent

Table 15
TOXIC FUNGI DETERMINED BY RABBITS SKIN TEST ISOLATED FROM OVERWINTERED CEREALS, SOILS, AND SUMMER-HARVESTED CEREALS

Fungi	Overwintered cereals			Soils			Summer-harvested cereals (Nontoxic)
	Toxic	Mildly toxic	Nontoxic	Toxic	Mildly toxic	Nontoxic	
Phycomycetes							
Chaetocladium grefeldii v. Tiegh, et. L. M.		2	3		1	2	
Mucor corticola Hag.	1	1	19		1	6	8
M. hiemalis Wehm.	3	4	34		2	13	20
M. humicola Raillo	1	1	9		1	8	
M. racemosus Fres.	2	3	33		1	20	46
M. albo-ater Naum.			15		1	8	
M. corticola Hag.	1	1	19		1	6	8
M. dispersus Hag.		1	8			4	
M. fumosus Naum.	1		9			6	
M. globosus Naum.	1		21		1	7	
M. griseo-ochraceus Naum.		1	5			5	
M. heterosporus Fisch.		1	11		1	5	
M. murorum Naum.					1	2	
M. oblongisporus Naum.		1	7			4	
M. sciurinus Naum.			7	1		6	
M. silvaticus Hag.			13		1	3	16
Piptocephalis freseniana D. B. et W. with *Mucor albo-ater* Naum.	1	1	4			14	
Rhizposu nigricans Ehr.	1	2	2		1	16	54
Thamnidium elegans Link	1	3	33		8	11	6
Ascomycetes							
A. fumigatus Fres.		1	4				
A. niger v. Tiegh		1	6			6	4
Penicillium brevicompactum Dier.	2	1	23				
P. chrysogenum Thom	1	1	40			8	7
P. cyclopium Westl.	1	1	24			3	

Table 15 (continued)
TOXIC FUNGI DETERMINED BY RABBITS SKIN TEST ISOLATED FROM OVERWINTERED CEREALS, SOILS, AND SUMMER-HARVESTED CEREALS

Fungi	Overwintered cereals			Soils			Summer-harvested cereals (Nontoxic)
	Toxic	Mildly toxic	Nontoxic	Toxic	Mildly toxic	Nontoxic	
P. nigricans Bain.	2	2	34	1	1	11	
P. notatum Eestl.	1	2	23			7	19
P. steckii Zal.	2	1	11			7	
P. umbonatum Sopp.	1	1	25			4	
P. viridicatum Westl.	1	1	27		1	4	
P. cyaneo-fulvum Biour.		1	6				
P. albidum Sopp.			5	1	10		
P. aurantio-virens Biour.		2	30		5		
P. bioregeianum Zal.		1	30				
P. bevi-compactum Dier.	2	1	23				5
P. citreo-roseum Dier		2	20				
P. crustosum Thom.		1	43		17		12
P. griseo-roseum Dier.		1	33		9		
P. howardii Thom.		1	9		1	11	
P. jenseni Zal.			7	1		12	
P. martensii Biour.		2	15			3	
P. miczynskii Zal.		2	17				
P. palitans Westl.		1	11				
P. purpurogenum Fler.-stoll		1	17			3	19
P. restrictum Gilm. et a.			11		1	8	17
P. westlingi zal.		1	18			1	
Fungi imperfecti							
Alternaria humicola Oud.	2	4	40			9	3
A. tenuis Nees.	10	21	350	2	2	58	81
Botrytis cinerea Pers.		1	1				
Cladosporium epiphyllum (Pers.) Mart.	8	11	73	2	2	25	2

C. exoasci Link	2		15			3	1
C. fagi Oud.	4	1	33			9	
C. fuligineum Bon.	1	1	17			6	
C. gracile Cda.	3		19			4	
C. herbarum (pers.) Link	2	12	35		1	24	11
C. molle Cke.	2		29			2	
C. penicillioides Preuss	1	4	36		1	5	2
C. pisi Cug. et March.		4	14			8	
C. epiphyllum (Pers.) Mart	8	11	73	2	2	25	2
C. graminum Cda.		2	11			6	2
C. grumosum (Pers.) Link		2	22			3	
C. spherospermum Penz.	1		9			5	
Coremium glaucum Link	1						
Gliocladium penicillioides Cda.		1	1				
Gonatobotrys flava Bon.		3	5				
Macrosporium commune Rabenh		1	3				
Trichoderma lignorum Tode (Harz)	1		18			10	2
Trichothecium roseum Link	2	3	26			5	
Verticillium lateritium Rabenh.	1	1	8				4
Actinomycetes							
A. griseus Kras.				1		10	
A. globisporus Kras.					1	12	

Adapted from Joffe, A. Z., *Bull. Res. Counc. Isr.*, 9D, 101, 1960.

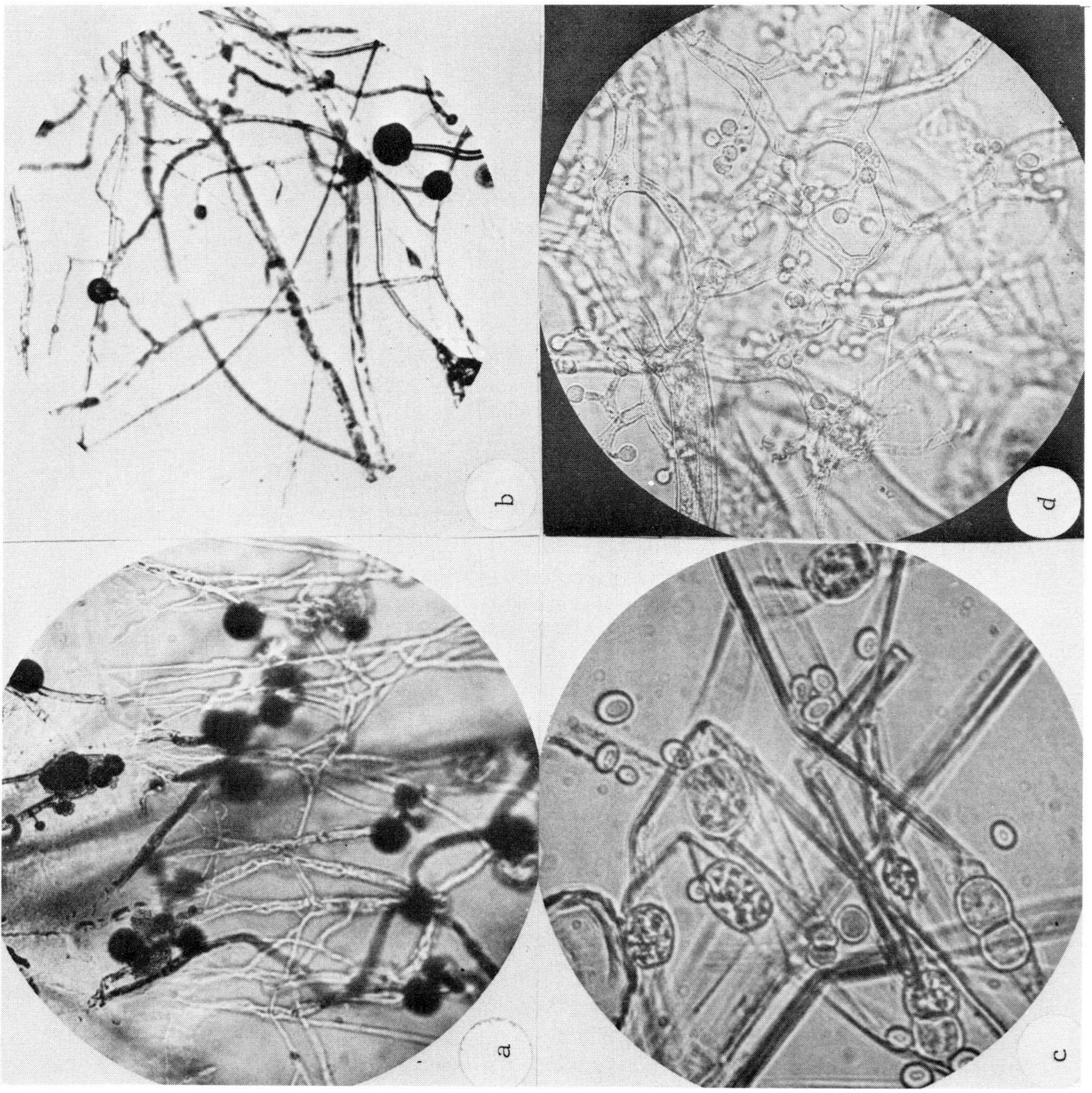
b
d
a
c

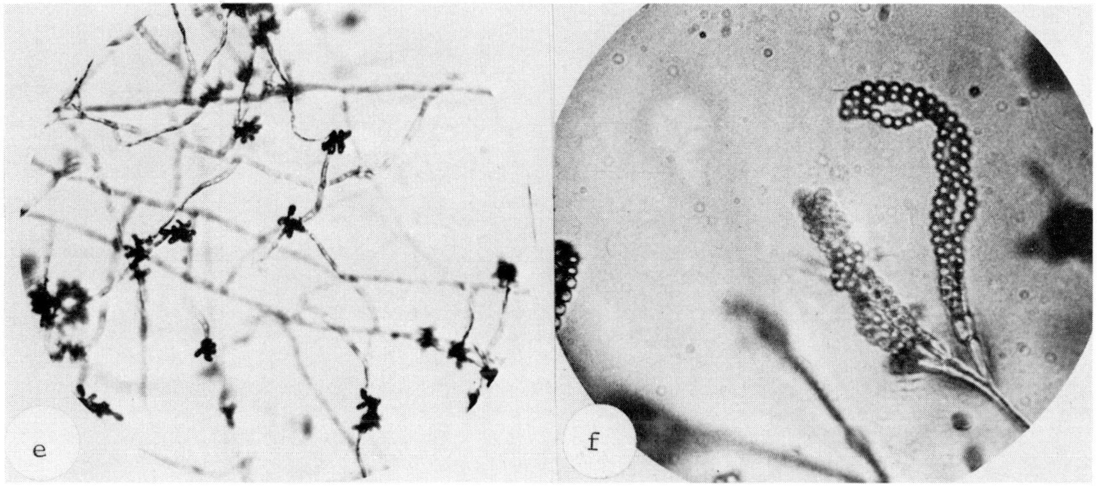

FIGURE 7. (a) *Mucor corticola* Hag., isolated from overwintered wheat, showing sporangia developed on hyphae, magnification × 205; (b) *M. globosum* Fisher, isolated from overwintered barley, showing sporangia developed on hyphae, magnification × 205; (c) *M. racemosus* Fres., isolated from overwintered millet, showing chlamydospores developed in hyphae, magnification × 400; (d) *Thamnidium elegans* Link., isolated from overwintered wheat, showing sporangiophore with numerous lateral sporangiola, magnification × 400; (e) *Trichoderma roseum* Link., isolated from overwintered wheat, showing conidia on aerial hyphae, magnification × 107; (f) *Penicillium steckii* Zal., isolated from overwintered millet, showing characteristic conidiophore and conidia, magnification × 533. (Adapted from Joffe, A. Z., *Mycopathol. Mycol. Appl.*, 16, 201, 1962. With permission.)

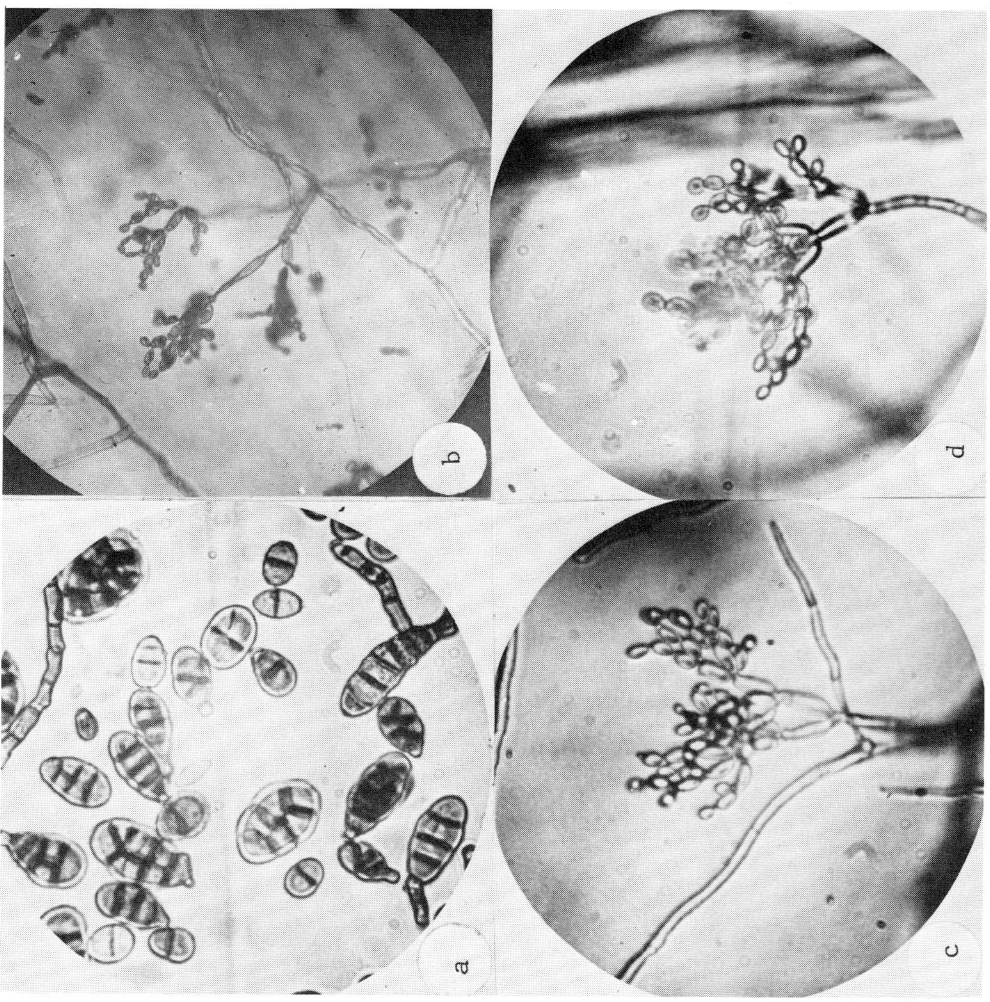
b
d
a
c

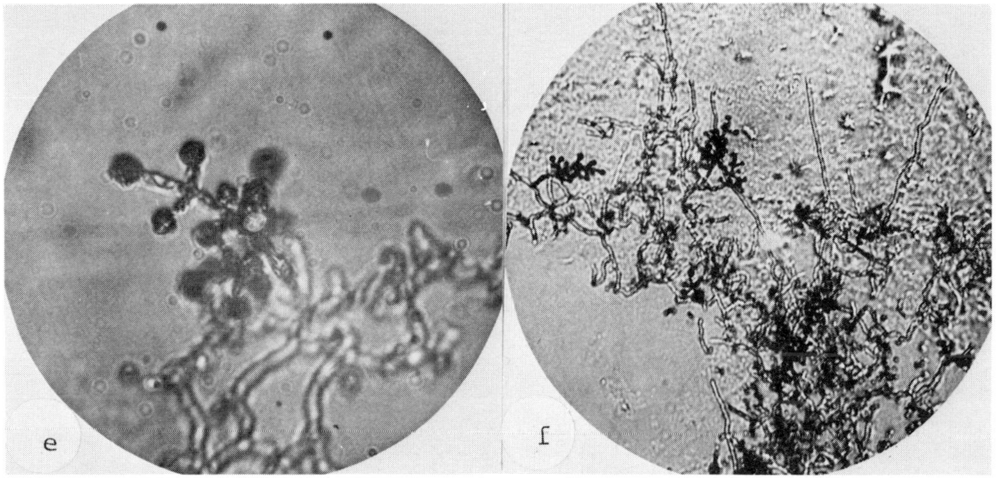

FIGURE 8. (a) *Alternaria tenuis* Nees., isolated from overwintered wheat, showing conidia, magnification × 533; (b) *Cladosporium fagi* Oud., isolated from overwintered millet, showing conidiophores with type of sporulation, magnification × 515; (c) *C. epiphyllum* (Pers.) Mart., isolated from overwintered millet, showing conidiophore with type of sporulation, magnification × 533; (d) *C. penicillioides* Preuss., isolated from overwintered wheat, showing long conidiophore with type of sporulation, magnification × 533; (e) *Trichoderma lignorum* (Tode) Harz, isolated from soil, showing sterigmata of the conidiophore with conidia, magnification × 450; (f) *T. viride* Pers., isolated from overwintered wheat, showing growth habit with type of sporulation, magnification × 107. (Adapted from Joffe, A. Z., *Mycopathol. Mycol. Appl.*, 16, 201, 1962. With permission.)

Table 16
REIDENTIFICATION OF SPOROTHRICHIELLA STRAINS OF *FUSARIUM* USED IN RECENT TOXICOLOGICAL STUDIES IN THE U.S.

No. and designation of strain	Supplied by	Country of origin	Reidentified as
NRRL 3249 *F. tricinctum*	C. W. Hesseltine	U.S.	*F. sporotrichioides*
NRRL 5508 *F. tricinctum*	C. W. Hesseltine	U.S.	*F. sporotrichioides*
NRRL 5509 *F. tricinctum*	C. W. Hesseltine	U.S.	*F. sporotrichioides* var. *tricinctum*
NRRL 3299 *F. tricinctum*	C. W. Hesseltine	U.S.	*F. poae*
NRRL 3287 *F. poae*	C. W. Hesseltine	U.S.	*F. poae*
2061-C *F. tricinctum*	C. J. Mirocha	U.S.	*F. sporotrichioides*
YN-13 *F. tricinctum*	C. J. Mirocha	U.S.	*F. sporotrichioides*
T-2 *F. tricinctum*	W. F. O. Marasas	France	*F. poae*

Adapted from Joffe, A. Z. and Palti, J., *Appl. Microbiol.*, 29, 575, 1975. With permission.

years, mostly in the U.S. and Japan, to the earlier work carried out principally in the U.S.S.R. and Israel.

The taxonomy of the species of the Sporotrichiella section is a matter of dispute. In the U.S., beginning in the 1940s Snyder and Hansen[293] and later collaborators[294,316] began to publish fundamentally different and extremely simplified concepts of *Fusarium* taxonomy. For an example, for reasons never adequately explained, they decided to group all the species of Sporotrichiella section together under one species, *F. tricinctum*. This section has attracted wide attention from research scientists all over the world because of their mycotoxic potential and capacity to cause severe disease in man and animals.

The most prominent of this section are *F. poae* and *F. sporotrichioides,* which developed in the Soviet Union on overwintered cereals and caused the often fatal ATA disease in man, and *F. tricinctum,* which grows in the U.S. on corn, wheat, fescue hay, and other substrates, causing disease of animals.

Most strains of *F. tricinctum* which had been used in U.S. for comprehensive mycotoxicological studies on animals were kindly supplied by colleagues and institutions to my laboratory and were re-identified and determined according to our taxonomic system as *F. poae* (NRRL 3299, NRRL 3287, T-2) and *F. sporotrichioides* (NRRL 3249, NRRL 5908, 2061-c, and YN-13). Only one isolate, NRRL 3509, belonged to *F. sporotrichioides* var. *tricinctum* (Table 16).[121,123,128]

The author,[112,114,116,119,121,123] and Gerlach[83] and Seemüller[278] have proved that these species differ in their morphological and cultural properties from what has generally been called *F. tricinctum* according to Snyder and Hansen.[293]

Table 17 gives the classification of Sporotrichiella section published by various researchers. The morphological, cultural, and toxicological characteristics of the species and varieties in this section as established by the author are given in Table 18. Our concepts approximate those of Gordon[88-91] and Seemüller[278] with some slight variations, and we distinguish between two species of *F. poae* (Peck.)Wr. and *F. sporotrichioides* Sherb. and two varieties, *F. sporotrichioides* Sherb. var. *tricinctum* (Corda)Raillo and *F. sporotrichioides* Sherb. var. *chlamydosporum* (Wr. and Rg.) Joffe.[121,123] The most prominent of these are *F. poae* and *F. sporotrichioides,* which in the Soviet Union grew on overwintered cereals and caused the fatal disease of alimentary toxic aleukia in man (Joffe[108,112-116,118,120,130]), and *F. tricinctum* (according to Snyder and Hansen[293]), which appears in the U.S. on moldy corn, wheat, fescue hay,

Table 17
CLASSIFICATION OF SPOROTHRICHIELLA FUSARIA

Wollenweber and Reinking (1935)	Jamalainen (1943)	Snyder and Hansen (1945)	Raillo (1950)	Bilai (1955, 1970, 1977)	Gordon (1952)	Seemüller (1968)	Joffe (1974, 1977)	Booth (1971, 1975)
F. poae	*F. citriforme*	—	*F. poae*	*F. sarcochroum, F. sporotrichiella* var. *poae*	*F. poae* *F. poae*	*F. poae*	*F. poae*	
F. sporotrichioides	*F. sporotrichioides*	—	*F. sporotrichioides*	*F. sporotrichioides*	*F. sporotrichiella*	*F. sporotrichioides*	*F. sporotrichioides*	*F. sporotrichioides*
F. sporotrichioides var. *minus*	—	—	*F. sporotrichioides* subsp. *minus*	*F. sporotrichiella* var. *sporotrichioides*	—	*F. sporotrichioides* var. *minus*	—	—
F. chlamydosporum	—	—	—	—	—	*F. chlamydosporum*	*F. sporotrichioides* var. *chlamydosporum*	*F. fusarioides*
F. tricinctum	—	*F. tricinctum*	*F. sporotrichioides* var. *tricinctum*	*F. sporotrichiella* var. *tricinctum*	—	*F. tricinctum*	*F. sporotrichioides* var. *tricinctum*	*F. tricinctum*
	—	—	—	*F. sporotrichiella* var. *anthophilum*	—	—	—	—

Note: Booth includes *F. sporotrichioides* and *F. fusarioides* in section Arthrosporiella.

Table 18
CULTURAL, MORPHOLOGICAL, AND TOXICOLOGICAL CHARACTERISTICS OF THE SPECIES AND VARIETIES IN THE SECTION SPOROTRICHIELLA

	F. poae	*F. sporotrichioides*	*F. sporotrichioides* var. *tricinctum*	*F. sporotrichioides* var. *chlamydosporum*
Aerial mycelium				
Color	White, yellow, red or red-brown	White, whitish, rose, or red	White, carminered to purple	White, light yellow to carmine-brown, or intensive rose
Consistency	Felted, somewhat powdery	Downy	Weely	Downy, flocculose
Growth rate	7—8 cm	4—4.5 cm	2.7—3 cm	3—3.5 cm
Microconidia				
Shape	Oval, globose, spherical with basal papilla, rarely pear-shaped	Globose, pear-shaped, ellipsoid, elongate	Lemon-and also pear-shaped, oval	Spindle-shaped or elongate, rarely oval-ellipsoidal
Occurrence in aerial mycelium	Single or false heads	Singly, rarely in chain or false heads	In false heads	Singly, rarely in groups
Measurements (μm)				
0-septate	6.8—9.5 × 4.6—8.2	6—11 × 4.0—7.0	7—9.5 × 2.8—7.2	7.5—10.5 × 2.6—3.2
1-septate	10.5—15 × 4—7.4	9—20 × 3.8—7.5	10—19 × 3—5.5	11—14 × 3.0—3.8
Abundance in relation to macroconidia	More abundant than macroconidia	Often as numerous as macroconidia	More abundant than macronidia	Usually more numerous than macroconidia
Macroconidia				
Shape	Falcate, sometimes lightly curved without foot cell	Falcate to curved with or without foot cell	Falcate, elliptical or more strongly curved, with well-marked foot cell	Curved with narrowly painted apex, well-marked foot cell
Measurements (μm)				
3-septate	19—38 × 3.5—6.0	28—35 × 3.2—4.6	23—45 × 3.4—4	28—35 × 3.2—3.8
5-septate	19—38 × 3.5—6.0	37—44 × 3.5—4.8	34—51 × 3.6—4.4	38—44 × 3.5—4.5

Frequency of				
3-septate	Preponderant	Numerous	Preponderant	Preponderant
5-septate		Numerous	Rare	Rare
Location	Aerial mycelium only	Aerial mycelium sporodochia, rarely in pionnotes	Aerial mycelium, sporodochia	Aerial mycelium
Chlamydospores	Intercalary, singly or in pairs or chains; rare, sometimes absent; hyaline to light brown	Intercalary, singly or in pairs, knots, or chains; sometimes terminal; smooth-walled, hyaline to light brown	Intercalary, singly or in chains; rarely terminal; smooth-walled, brown	Terminal, singly or in pairs or chains; smooth-walled or rough, brown
Plectenchymatous sclerotia	Absent	Occasionally present	Present	Absent, or occasionally present in young cultures
Stroma on PDA	Red or ochre-yellow sometimes violet, rarely colorless	Blood-red, yellow, purple, brown, or light carmine	Carmine-purple, ochre-brown, rarely colorless	Red, brown-red carmine, purple, or light violet
Toxicity to rabbit skin				
Frequency of toxic isolates	High	High	Low	Low to very low

Adapted from Joffe, A. Z., in *Handbook of Mycotoxins and Mycotoxicoses,* Vol. 2, Marcel Dekker, New York, 1978.

and other substrates (Bamburg,[11,13,14] Bamburg et al.,[12,15,16,18] Bamburg and Strong,[17,19] Gilgan et al.,[85] Kosuri et al.,[140,141] Smalley et al.,[290] Szathmary et al.,[306] Yates et al.,[336,337]) and produces toxic symptoms in many animals (Burmeister,[38] Burmeister and Hesseltine,[39] Burmeister et al.,[40] De Nicola et al.,[56] Grove et al.,[96] Joffe,[112,113,125] Joffe and Yagen,[130] Lutsky et al.,[166] Mirocha and Pathre,[180] Mirocha and Christensen,[181] Schoental,[261,263,265-273] Schoental and Joffe,[260] Schoental et al.,[262,264,274-276] Scott and Somer,[277] Smalley[289]) and in Japan (Saito et al.,[259] Ueno et al.,[319-321]).

Characterization of the Fungi of the Sporotrichiella Section

The basis for classification of this section is as follows: variability in shape, microconidia one or two celled, abundant, lemon or pear-shaped, globose, elipsoid, elongate to fusoid, dispersed in aerial mycelium. They formed on simple or branching conidiophores, and sometimes in false heads. Macroconidia are sparse, small, oblong, narrowly fusoid to falcate, spindle or sickle shaped, pedicellate, formed in aerial mycelium, less often in sporodochia or pionnotes. Chlamydospores are intercalary, terminal, in chains or knots, formed chiefly in the mycelia and sometimes in macroconidia. The stroma on PDA is red, yellow-brown, or carmine, sometimes uncolored. The aerial mycelia are white, red, or carmine.

The description of morphological and cultural properties of the species and varieties is given below.

F. poae (Peck) Wr. (Figures 9 to 11, 958)

Synonyms are *F. citriforme* Jamalainen,[103] *F. tricinctum* (Cda.) Sacc. emend. Snyder and Hansen,[293] *F. tricinctum* (Cda.) Sacc. emend. Snyder and Hansen *F. poae* (Pk.),[293] *F. poae* (Pk.) Wr. *F. pallens* Wr.,[326] and *F. sporotrichiella* Bilai var. *poae* (Pk.) Bilai.[26] Cultures are white yellow, rose, carmine, or red-brown. Aerial mycelium hairy, cobwebby to felted, or somewhat powdery.

Microconidia are abundant in relation to macroconidia, scattered over the mycelium, formed singly or in false heads, broadly oval, round or spherical with basal papilla, lemon-shaped, ellipsoid, rarely pear-shaped. Conidiophores are well developed and microconidia formed in lateral, small, broad, phialides.

Macroconidia are sparse, curved, falcate, small, only in aerial mycelium, formed on narrow, elongated phialides.

Sporodochia absent, pionnotes very rare or absent.

Measurements (micrometers) shape, and frequency of conidia:

0—sept., 92% oval, spherical	6.4 to 9 × 4.2 to 8.0
0—sept., 1% elongated, spindly	8 to 14 × 2.5 to 4.2
1—sept., 6% oval, pear-shaped	10.0 to 15 × 4.2 to 7.2
3-sept., 2% curved falcate	17 to 36 × 4.0 to 6.6

Growth rate is 8.0 cm. Chlamydospores are rare, intercalary, single or in pairs, knots, or chains, or absent. Sclerotia are absent. Stroma are carmine, ocher-yellow, and sometimes violet.

F. sporotrichioides Sherbakoff (Figures 9 to 11, 921)

Synonyms are *F. tricinctum* (Cda.) Sacc. emend. Snyder and Hansen,[293] *F. tricinctum* (Cda.) Sacc. emend. Snyder and Hansen *F. poae* (Pk.),[293] *F. sporotrichiella* Bilai,[26] *F. sporotrichiella* Bilai var. *sporotrichioides* (Sherb.) Bilai,[26] and *F. sporotrichioides* Sherb. var. *minus* Wr.[278] Cultures are white, white-rose, red, purple, sometimes light brown. Aerial mycelium downy, floccose.

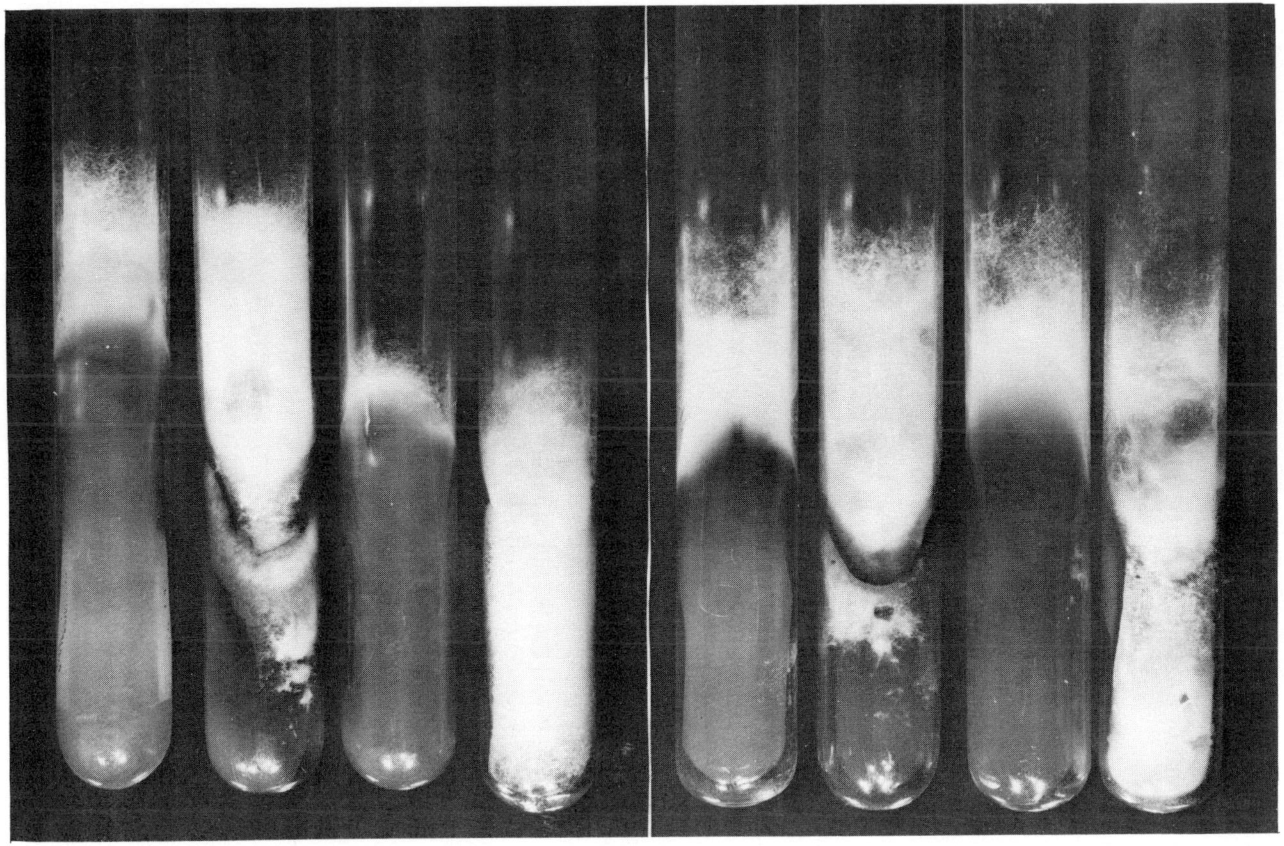

FIGURE 9. Cultural appearance of *F. sporotrichioides* #921, *F. poae* #958, *F. sporotrichioides* var. *tricinctum* #1227, *F. sporotrichioides* var. *chlamydosporum* #4337, PDA slants (from left to right). (Adapted from Joffe, A. Z., in *Handbook of Mycotoxins and Mycotoxicoses*, Vol. 3, Marcel Dekker, New York, 1978. With permission.)

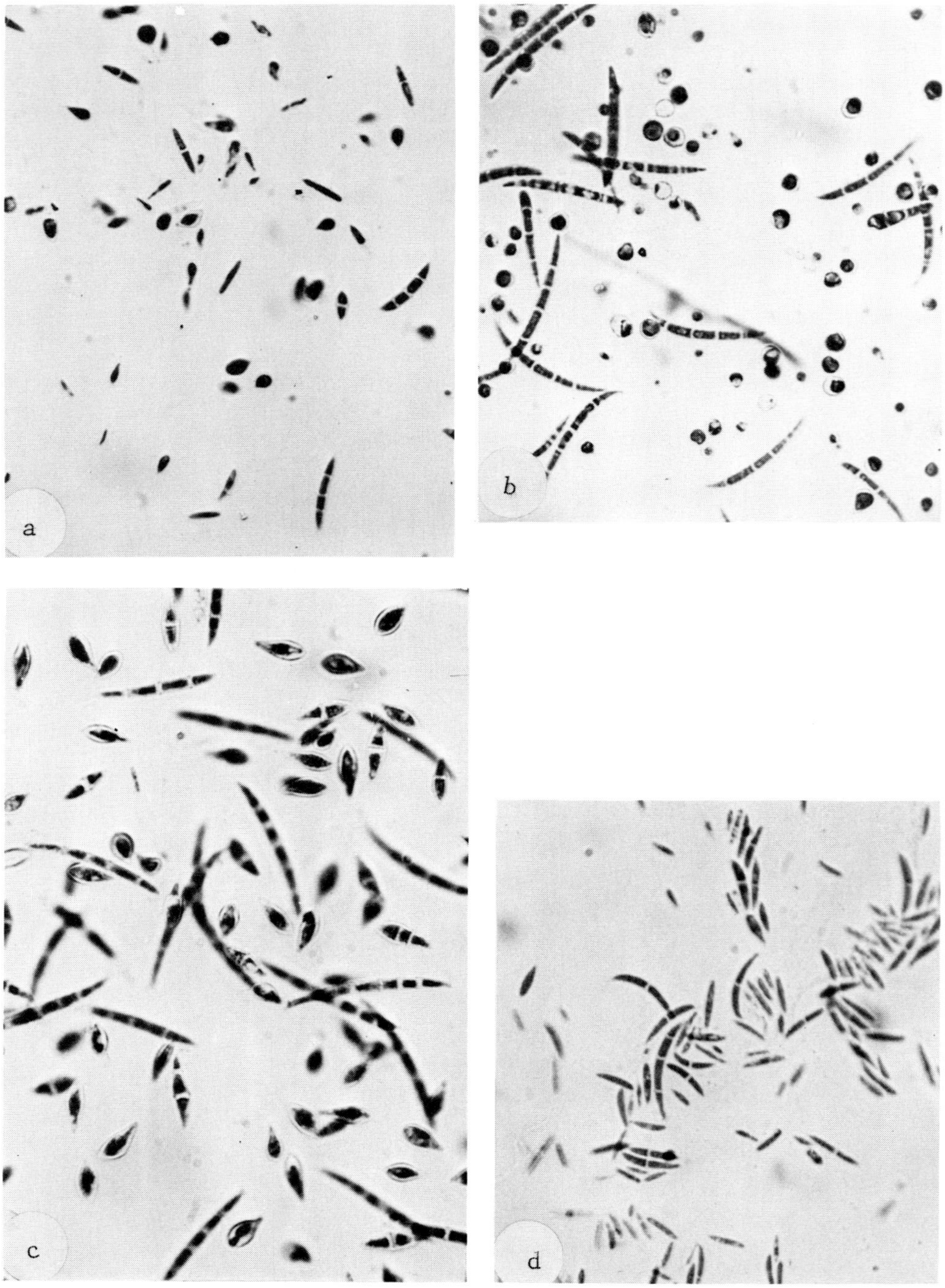

FIGURE 10. (a) *F. poae* #958; (b) *F. sporotrichioides* #921; (c) *F. sporotrichioides* var. tricinctum #1227; (d) *F. sporotrichioides* var. *chlamydosporum* #4337. (a) macroconidia; (b) microconidia; (d) chlamydosporum.

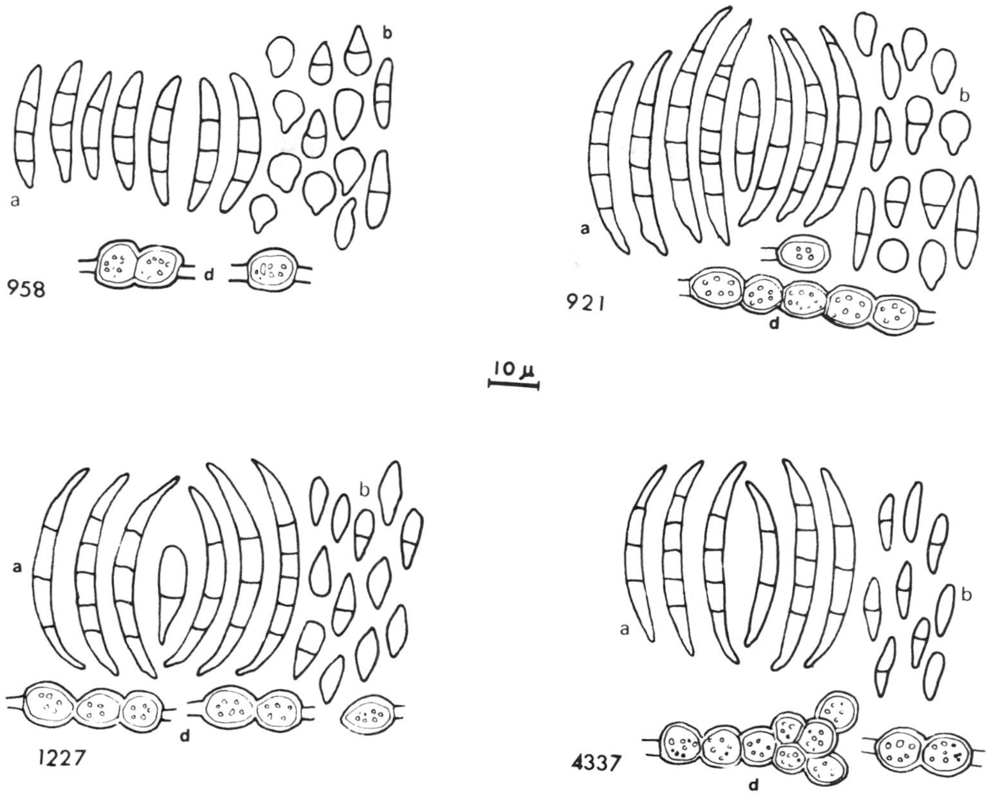

FIGURE 11. (a) *F. poae* #958; (b) *F. sporotrichioides* #921; (c) *F. sporotrichioides* var. *tricinctum;* (d) *F. sporotrichioides* var. *chlamydosporum* #4337.

Microconidia are often as numerous as macrononidia, formed singly or in short chains, globose, pyriform, ellipsoid, lemon-shaped, elongated, slightly falcate, dispersed in aerial mycelium; formed on apical cylindrical phialides of branched conidiophores.

Macroconidia are falcate to curved, oblong, and narrowly fusoid, scattered in aerial mycelium or in sporodochia and formed on elongated phialides.

Measurements (micrometers) shape, and frequency of conidia:

0—sept., 19% pear-shaped	5.4 to 9.5 × 5.2 to 7.0
0—sept., 5% ellipsoidal	7.5 to 12 × 4.4 to 7.5
0—sept., 9% elongated or sickle-shaped	9 to 12 × 2.6 to 4.0
0—sept., 2% spindle-shaped	8 to 15 × 2.7 to 3.8
1—sept., 17% pear-shaped	9 to 16.5 × 4.2 to 7.5
1—sept., 7% ellipsoidal	10 to 19 × 3.6 to 7.8
1—sept., 8% elongated or sickle-shaped	11 to 18 × 3.2 to 4.0
1—sept., 6% spindle-shaped	13 to 24 × 2.4 to 4.0
3—sept., 13% sickle- or spindle-shaped	23 to 34 × 3.8 to 5.0
4—sept., 7% sickle- or spindle-shaped	35 to 42 × 3.4 to 4.6
5—sept., 7% sickle- or spindle-shaped	38 to 47 × 4.2 to 5.4

Growth rate is 4.0 cm. Chlamydospores are intercalary, rarely terminal, single or in pairs, chains or knots, globose, hyaline, or light brown, smooth-walled. Sclerotia are red-brown, occasionally present. Stroma are red, yellow, purple-red, or dark carmine, sometimes brown.

F. sporotrochioides var. tricinctum (Corda) Raillo (Figures 9 to 11, 1227)

Synonyms are *F. tricinctum* (Cda.) Sacc. emend. Snyder and Hansen,[293] *F. sporotrichiella* Bilai var. *tricinctum* (Cda.) Bilai,[26] and *F. tricinctum* (Cda.) Sacc.[26,278,324]

Cultures are white, rose, red, purple, or carmine. Aerial mycelium are abundant, cottony, floccose.

Microconidia are more abundant than macroconidia, formed on conidiophores bearing cylindrical phialides; dispersed, usually singly, in aerial mycelium, or in false heads, pyriform to clavate, lemon-shaped, oval, cylindrical, ellipsoidal or spindly elongated, slightly falcate.

Macroconidia are usually formed in sporodochia on small, slightly curved phialides; falcate or elliptical, strongly curved with well-marked foot cell.

Measurements (micrometers) shape, and frequency of conidia:

0—sept., 39% pear-shaped	8 to 10 × 4.8 to 7.4
0—sept., 16% elongated	7.5 to 15 × 3.0 to 3.7
1—sept., 1% pear-shaped	9 to 17 × 4.5 to 5.2
1—sept., 8% elongated	13 to 21 × 3.0 to 4.0
3—sept., 27% elongated, curved falcate	22 to 39 × 3.4 to 4.2
4—sept., 3% elongated, curved falcate	31 to 45 × 3.6 to 4.4
5—sept., 6% elongated, curved falcate	37 to 52 × 3.5 to 4.6

Growth rate 4.5 cm. Chlamydospores are rare, globose, smooth-walled, intercalary, single or in chains. Sclerotia are white, purple to brown, sometimes absent. Stroma are carmine, purple, ochre-brown, or violet, rarely colorless.

F. sporotrichioides var. chlamydosporum (Wr. and Rg.) Joffe (Figures 9 to 11, 4337)

Synonyms are *F. tricinctum* (Cda.) Sacc. emend. Snyder and Hansen,[293] *F. chlamydosporum* Wr. and Rg.,[278,324] and *F. fusarioides* (Frag. and Cif.) Booth.[35] Cultures are rose, carine, rarely white, yellow, or brown. Aerial mycelium are floccose, cottony.

Microconidia are usually less numerous than macroconidia, formed only singly in aerial mycelium on irregular conidiophores; they are small, narrow to clavate, lemon-shaped, usually one-celled, with rounded apex, spindle-ellipsoid or oblong, rarely puriform or oval-ellipsoid.

Macroconidia develop on phialides of distinct conidiophores in the aerial mycelium. Macroconidia are falcate, curved with narrowly rounded to pointed apical cell and marked foot cell. Sporodochia are absent or very rare.

Measurements (micrometers) shape, and frequency of conidia:

0—sept., 31% spindle-elongated	7.5 to 11 × 2.5 to 3.3
1—sept., 8% spindle-elongated	11.5 to 15 × 2.8 to 3.8
3—sept., 54% curved	28 to 35 × 3.2 to 4.0
4—sept., 5% curved	35 to 40 × 3.4 to 4.2
5—sept., 2% curved	38 to 44 × 3.5 to 4.4

Growth rate is 3.5 cm. Chlamydospores are terminal or intercalary, abundant, single or in pairs, knots or long chains, smooth or slightly rough-walled, light brown. Sclerotia are purple-brown to brown, or absent.

Criteria for distinguishing between species of the Sporotrichiella section have recently been discussed by Joffe [121,123] and are summarized in Table 18.[124]

BIOLOGY OF TOXIC *FUSARIUM* FUNGI

The distribution of *Fusarium* fungi of the Sporotrichella section is fairly widespread in plants, soils, and other substrates. The remnants of vegetative parts of cereals in the field, as well as grains which are left after harvesting, constitute a good medium for the development of fungi, as do cereals which have been mowed and heaped in the field and then wet by rain or harvested late in autumn after the rains have already started.

Since toxin may already be found in vegetative parts in the autumn,[104,115] the danger exists that the grains of cereals harvested in the spring will also be toxic. The toxin is not equally distributed in the grain. In prosomillet grains, more toxin was found in the glumes than in the grain itself. It was also observed that there are light and heavy grains. By separating them with 10 to 25% sodium chloride solution, it was found that the light grains, which floated in the solution, were toxic, whereas the heavy grains were not toxic or less toxic.[183-185] This observation led to the assumption that toxin does not invade the grain from without, but is produced within the grain. If the light and toxic grains are pressed slightly, they turn into powder in contrast with the heavy nontoxic grains which cannot be easily ground. The powder which is derived from the grains contaminated with fungi contains the highest concentration of toxin.

The toxic fungus is believed to develop first in the embryo of the grain, and later the mycelium spreads through the whole grain. This is why the percentage of germination of overwintered grains infected with toxic fungi is much less than that of normal grains.[119]

Climatic and ecological conditions are very important for the development of the fungus. If the winter is mild and temperatures are not too low, development of the *Fusarium* fungi is possible. The condition of the soil also influences development of the fungi; the thicker the layer of snow and the less frozen the soil, the better the fungi develop. When the layer of snow is thinner and the soil freezes, fungi will not develop and toxin will not be produced.

Differential Response of Toxic and Nontoxic Strains of the Same Fungal Species to Temperature

In investigating the cryophilic properties of the isolated toxic fungi, we used mainly toxic cultures of *F. poae* and *F. sporotrichioides*. For comparison, investigations were conducted at the same time on nontoxic cultures isolated from normal, high-quality, non-overwintered cereals. These cultures included *F. sporotrichioides, Cladosporium epiphyllum,* and others.[114] Growth under these specified conditions was determined daily. Observations indicated that for the most part the toxic fungi grew well at the temperatures prevailing in the refrigerator and in the cellar at temperatures of 0 to 1°C. It should be stressed that satisfactory growth also took place at 23 to 25°C. However, cultures isolated from normal cereal samples produced luxuriant growth at 23 to 25°C, while at 0 to 2°C their growth, if any occurred, was as a rule exceedingly scanty.

Additional experiments were also carried out in which toxic cultures of *F. poae, F. sporotrichioides, Cl. epiphyllum, Cl. fagi, Alternaria tenuis, Penicillium brevi-compactum, Mucor hiemalis,* and *M. racemosus* were grown on the same nutrient media at temperatures of −2 to −7°C. Nontoxic cultures of the same species with the excep-

Table 19
ACCUMULATION OF TOXIN DUE TO SHARP TEMPERATURE FLUCTUATIONS

	Toxin accumulation[a]								
	Fusarium poae			*Cladosporium epiphyllum*			*Fusarium + Cladosporium*		
Conditions	L	E	N	L	E	N	L	E	N
Room temperature (18°C)	—	+	—	+	+	—	—	—	—
In snow	+	+	+ +	+	+ +	—	—	+ + +	+
Room temperature-in snow	—	—	—	—	—	—	+ + +	+	—
In snow — room temperature	—	—	—	—	—	—	—	+	+ +
Room temperature — freezing — room temperature	—	+ + +	+ + +	+	+ + +	—	—	—	—
In snow — freezing (−2) (10°C)	—	+ + +	+ + +	+ +	+	—	—	—	—
Room temperature — freezing — in snow	—	—	—	—	—	—	—	+ + +	—
In snow — freezing — in snow	—	—	—	—	—	—	—	+ + +	—
In snow — freezing — room temperature	—	—	—	—	—	—	—	+ + +	+ +
Alternating room temperature — freezing	—	—	—	—	—	—	—	+ + +	+
Alternating in snow — freezing	—	—	—	—	—	—	—	+ +	—

[a] Reactions: L, leukocytic; E, edematous; N, necrotic. Degree of toxicity: +, mildly toxic; + +, toxic; + + +, very toxic.

Adapted from Joffe, A. Z., *Mycopathol. Mycol. Appl.*, 16, 201, 1962.

tion of *F. poae,* derived from normal non-overwintered cereals, were grown for comparison. Toxic cultures of these fungi developed in 19 to 47 days at −2 to −7°C. It should be stressed in this connection that the development of these fungi took place on media which were in a frozen state. Nontoxic cultures did not show any growth during the entire experimental period, which extended over 72 days.

These results indicated that toxic cultures of *F. poae, F. sporotrichioides, Cl. epiphyllum, Cl. fagi, M. hiemalis, A. tenuis,* etc. are cryophilic, while nontoxic cultures of these species are not. There is ample evidence confirming the tolerance to low temperatures of fungi belonging to the genera we have studied.

Effect of Temperature and Substrate on Growth Associations of Some Toxic Fungi

In view of the fact that in most instances *Fusarium* and *Cladosporium* were found in association on overwintered cereals, it was considered of interest to investigate the effect of the temperature factor on the growth associations of these fungi. Experiments were set up with *F. poae, F. sporotrichioides, Cl. epiphyllum,* and *Cl. fagi.*

The study of the effect of temperature regimes on the growth of *F. poae* and *Cl. epiphyllum* in mixed culture showed that active growth of *F. poae* takes place at temperatures from +25°C down to −7°C. At temperatures lower than −7°C the growth of *Fusarium* was arrested. *Cl. epiphyllum* developed actively at low temperatures from −2 to −10°C.

The experiments conducted with *F. sporotrichioides* showed that the growth of these cultures took place at temperatures down to −2°C, while cultures of *Cl. epiphyllum* grew also between −2 and −10°C. *F. sporotrichioides* and *Cl. fagi* yielded analogous results. It appears from this experiment that *F. poae* in mixed culture grows at temperatures from 25°C down to −7°C, but not from −7 to −10°C. *Cladosporium* grows at temperatures from 25 to −10°C but its growth was weaker than that of *Fusarium* in all tests.

The results of our experiments are in agreement with observations conducted under natural conditions. Various forms of growth associations of fungi were observed at different times of the year, both in samples supplied from counties of the Orenburg district and from experimental plots. During the autumn-winter period, copious growth of *Cladosporium* was in evidence on the ears of cereals, while the amount of *Fusarium* was insignificant. However, *Fusarium* predominated on infected ears during the spring.

Effect of Temperature Toxin Formation

In order to determine the effect of temperature factors on the formation of toxin, a number of experiments were set up with *F. poae* and *Cl. epiphyllum.* Pure and mixed cultures of these fungi were sown on potato-agar, potato-acid, and carbohydrate-peptone media, and also on sterile moistened millet. Growth took place at temperatures listed in Table 19. Within a month of initiation of the experiment, ether extracts were prepared from dried experimental cultures and were transferred onto the skin of a rabbit.

The results in Table 19 show that cultures of *Fusarium* and *Cladosporium* grown at low temperatures were more toxic than cultures maintained at room temperature. Sharp fluctuations of temperature greatly increased the toxicity of extracts; application of such extracts to the skin of rabbits resulted in acute edema and necrosis. Pure cultures of *F. poae* and mixed cultures of *F. poae* and *Cl. epiphyllum* gave rise to a more intense skin reaction than *Cl. epiphyllum* alone. *Cl. epiphyllum* is characterized by its edemo-leukocytic reaction, *F. poae* by its edemo-necrotic reaction. Mixed cultures were characterized as a rule by edema or edemo-necrotic reaction. Three replications of the tests produced analogous results.

Table 20
EFFECT OF TEMPERATURE AND OF THE DEVELOPMENT STAGE OF *F. POAE* ON TOXIC ACCUMULATION

Rearing temperature (°C)	Development stage of the fungus	Effect of s.c. application[a] to mice at the following doses (mℓ)			Toxin accumulation[b] (rabbit skin reaction)								
					Evaporated liquid			Ether extract from liquid			Fungal mass		
		0.2	0.5	1.0	L	E	N	L	E	N	L	E	N
40 days at 23—25	Before sporification	A	A	A	—	—	—	—	+	—	—	—	—
	Sporification	A	A	A	—	—	—	—	+ +	+	+	—	—
	Senescence	A	A	A	—	—	—	—	+	+	—	—	—
40 days at 1 to −2	Before sporification	A	A	A	—	—	+	—	+	—	—	—	—
	Sporification	D	D	D	—	+	—	—	+ + +	+ +	—	+ +	—
	Senescence	A	A	A	—	—	—	—	+	—	—	—	—
10 days at 0— + 5 and 30 days at −7 to −10; continued at 0— + 5 for 4 days	Before sporification	A	A	A	—	—	—	—	+	+	—	—	—
Sporification	D	D	D	—	+ +	—	—	+ + +	+ + +	—	+ +	—	
	Senescence	A	A	A	—	—	+	—	+ +	+	+	+	—
40 days at −7 to −10	Before sporification	D	D	D	—	—	—	—	+ +	+ +	—	—	—
	Sporification	D	D	D	—	+ +	—	—	+ + +	+ + +	+	+ +	—
	Senescence	A	A	A	—	—	—	—	+ +	+	—	—	—

[a] Condition of mice: A = alive; D = dead.
[b] Rabbit skin reactions: L, leukocytic; E, edematous; N, necrotic. Degree of toxicity: +, slightly toxic; + +, toxic; + + +, very toxic.

Adapted from Joffe, A. Z., *Mycopathol. Mycol. Appl.*, 16, 201, 1962.

Table 21
EFFECT OF TEMPERATURE AND THE DEVELOPMENT STAGE OF *F. SPOROTRICHIOIDES* ON TOXIC ACCUMULATION

Rearing temperature (°C)	Development stage of the fungus	Effect of s.c. application[a] to mice at the following doses (mℓ)			Toxin accumulation[b] (rabbit skin reaction)								
					Evaporated liquid			Ether extract from liquid			Fungal mass		
		0.2	0.5	1.0	L	E	N	L	E	N	L	E	N
40 days at 23—25	Before sporification	A	A	A	—	—	—	—	—	—	—	—	—
	Sporification	A	A	D	—	—	—	+	+	+	+	+	—
	Senescence	A	A	A	—	—	—	+	—	—	—	—	—
40 days at 1 to −2	Before sporification	A	A	A	—	—	—	—	—	—	—	—	—
	Sporification	D	D	D	+	—	—	+ +	+ +	+	+	+	—
	Senescence	A	A	A	—	—	—	—	—	—	—	—	—
10 days at 0 to + 5 30 days at −7 to −10 continued at 0 to + 5 for 4 days	Before sporification	A	A	D	—	—	—	+	+	—	—	—	—
	Sporification	D	D	D	+	+ +	+	+ +	+ + +	+	+	+ +	+
	Senescence	A	A	A	—	—	—	—	+	—	—	—	—
40 days at −7 to −10	Before sporification	A	A	A	—	—	—	—	—	—	—	—	—
	Sporification	D	D	D	+ +			+ +	+ +	+ +	+ +	+	—
	Senescence	A	A	A	—	—	—	—	+	—	—	—	—

[a] Condition of mice: A = alive, D = dead.

[b] Rabbit skin reactions: L, leukocytic; E, edematous; N, necrotic. Degree of toxicity: +, slightly toxic; + +, toxic; + + +, very toxic.

Adapted from Joffe, A. Z., *Mycopathol. Mycol. Appl.*, 16, 201, 1962.

The effects of alternate freezing and thawing on toxin formation by *F. poae* were tested under laboratory conditions.[112,113,118] Cultures grown on potato-dextrose agar, potato-dextrose acid agar, and carbohydrate-peptone media were alternately kept at room temperature (18°) and at various freezing temperatures down to −10°, and this was repeated 4 to 5 times. Results clearly showed that sharp fluctuations of temperature greatly increased the toxicity of extracts from these cultures. Application of such extracts to the skin of rabbits resulted in acute edema, hemorrhage, and necrosis.

Temperature in Relation to the Stage of Fungal Development

In order to elucidate the dependence of toxin formation on temperature and the developmental stage of the fungus, numerous experiments were conducted in 1946 and again in 1952. Pure cultures of *F. poae* and *F. sporotrichioides* sown on liquid medium (synthetic with starch and carbohydrate-peptone) were grown at different temperatures. At each of the three stages of development, i.e., prior to sporification, at the time of abundant sporification, and at senescence, toxicity assays were carried out with native filtrate, evaporated filtrate, extracts of the fungal mass, and ether extracts of the liquid media.

The toxicity of evaporated and native liquid, of the ether extract, and of the extracts of the fungal film were investigated by means of skin tests on 22 rabbits. The liquid substrates were passed through a Seitz filter, and the sterile filtrates obtained in this way were tested for toxicity in white mice by s.c. injections of 0.2, 0.5, and 1.0 mℓ. The results obtained with *F. poae* #60/9 are given in Table 20 and with *F. sporotrichioides* #60/10 in Table 21. It is evident from the tables that injection of filtrates of cultures obtained at different stages of development at the temperature of 23 to 25°C did not kill white mice.

In the case of extracts of cultures maintained at low temperatures, as well as those kept at 0 to 5°C with intervening freezing, death of the mice occurred within 12 to 48 hr, depending on the dosage and the fungal species. Death was due to systemic toxemia. Post-mortem examinations in every case disclosed necroses in the digestive system and other organs. The highest toxicity was displayed by filtrates of *Fusarium* obtained during the stage of abundant sporification from cultures grown at the temperature of −2 to −7°C and prior to sporification from the −7 to −10°C series, while extracts obtained at an advanced stage of senescence were considerably less toxic.

Heating for 30 min at 100°C of filtrates passed through an asbestos filter did not result in any reduction of toxicity. In certain instances, the presence of toxin in the filtrate could be detected at a very early stage of development. This appears to be due to the fact that spores were occasionally present during the first stage, even though in much smaller numbers than during the second stage.

Application of liquid ether extracts of *F. poae* and *F. sporotrichioides* to the skin of rabbits invariably produced a distinct skin reaction, even in the case of cultures grown at incubator temperatures of 23 to 25°C. This circumstance seems to be connected with the high concentration of the toxin concerned. It should be mentioned, however, that in every instance cultures grown at low temperature produced more pronounced reactions than incubator-reared cultures. Highest toxicity was associated with material obtained at the stage of abundant spore formation.

The results of investigations on the toxicity of native liquid and of ether extracts thus proved that active accumulation of toxin takes place under low temperature conditions, especially at the stage of abundant sporification. Passing the native liquid through a Seitz filter did not affect its toxicity in any way.

The tabulated data bring out yet another interesting characteristic, namely, that the application of ether extracts of the liquid substrate to rabbit skin produced a stronger response than did extracts of the fungal film of *F. poae* and *F. sporotrichioides* cultures. This indicates that the toxins of *F. poae* and *F. sporotrichioides* are excreted into the surrounding medium and thus act as exotoxins.

Variability of Toxic Strains and Species of *Fusarium*

In the course of studies on the toxicity of isolated cultures, we encountered instances of morphological and cultural variability of strains of *Fusarium* and also weakening of their toxic properties. The variability within the genus *Fusarium* has been mentioned by many authors. According to Appel and Wollenweber,[10] Brown,[36] and Brown and Horne,[37] cultures of *Fusarium* readily mutate. Similar observations have been reported by Leonian,[156] who obtained around 50 varieties from a single species. He noted a considerable variability in cultures of the genus *Fusarium* as regards ability to produce spores and he stated that "the only thing constant in *Fusarium* is its variability." The great variability of morphological characters within the genus *Fusarium* has also been indicated by Raillo[234] and Bilai.[26,30,32]

The character of growth of the fungal mycelium type of sporification, amount of pigmentation, morphology of conidia and sclerotia, and other attributes vary according to environmental conditions and according to the derivation and age of the culture.

We have established that the morphological and cultural properties of fungi isolated from cereals overwintered under snow cover display great variability under the influence of unfavorable ecological conditions. This variability is also associated with changes in physiological behavior.[110,111] Changes in morphological and cultural properties of toxic cultures of *Fusarium* were often seen in connection with frequent transplantations into liquid and solid media. These changes were associated with loss of toxicity of the cultures, which was proved by applications to skin of rabbits and by animal feeding experiments. Pure cultures of *Fusarium* sown on liquid and carbohydrate-peptone agar medium and on sterile normal millet, under room temperature conditions underwent marked changes of morphological characteristics and a weakening of toxic properties. The degree of toxicity was determined by biological rabbit skin tests and s.c. injections of liquid substrates of cultures in mice. Applications to the skin of rabbits resulted in weak reactions.

Only in the toxic and strongly toxic strains isolated from overwintered cereals (e.g., *F. poae* and *F. sporotrichioides*) is an unimpaired capacity for high-level production of T-2 toxin retained. This phenomenon is certainly connected with the specific environmental conditions obtaining in Orenburg district in the years when these strains were originally isolated (and ATA outbreaks were widespread).

Effects of Substrate on Growth of Toxic Fungi

As shown by our observations, the extent of toxin formation by cultures of *F. poae, F. sporotrichioides, Cl. epiphyllum,* and *Alternaria tenuis* is determined not only by temperature, but also by nutrition. Both the character of growth and the toxin-forming propensities of certain mold fungi may be modified in relation to the composition of the media.

We used various natural media (millet, wheat, barley, oats, rice, potato, etc.) as well as solid and liquid synthetic media for culturing toxic fungi (Joffe[108,112,113]) and investigated numerous sources of nitrogenous and carbohydrate nutrients and their effect on the formation of toxin by fungal cultures. The following sources of nitrogen and carbon were tested: peptone, casein, glycocoll, cystine, albumin, asparagin, alanine, arginine, histidine, tryptophan, tyrosine, glutamic acid, urea, ammonium sulfate, sodium nitrate, sodium nitrite, ammonium nitrate, arabinose, dextrose, galactose, glucose, saccharose, maltose, lactose, D-mannose, D-fructose, mannitol, starch, cellulose, sodium citrate, sodium acetate, and sodium oxalate. These tests involved 274 cultures of toxic fungi.

Table 22
EFFECTS OF SUBSRTATE AND SOURCE OF ISOLATE ON PRODUCTION OF TOXINS (MEAN OF RESULTS OBTAINED AT 8 TEMPERATURES)

		Mean grade of toxicity per isolate					
		F. poae 9 isolates		*F. sporotrichioides* 9 isolates		*F. sporotrichioides* var. *tricinctum* 3 isolates	
	Liquid substrates	liquid	thallus	liquid	thallus	liquid	thallus
No. IV.	Carbohydrate-peptone	1.2[a]	0.5[a]	1.4[a]	0.8[a]	1.5[b]	0.7[b]
No. V.	Czapek's	1.4	0.7	1.5	0.9	1.5	0.7
No. VIII.	Starch	1.6	0.8	1.8	1.0	1.7	1.0
	Mean	1.4	0.7	1.6	0.9	1.6	0.8
		18 isolates		18 isolates		3 isolates	
	Grain substrates	light	dark	light	dark	light	dark
	Prosomillet	1.3	2.2	1.5	2.0	1.6	2.0
	Wheat	1.3	1.8	1.3	1.9	1.4	1.9
	Barley	1.7	2.3	1.7	2.1	1.1	1.8
	Mean	1.4	2.1	1.5	2.0	1.4	1.9
	Sources of isolates						
	Prosomillet	1.3	2.0	1.4	1.9	1.5	2.1
	Wheat	1.7	2.2	1.2	1.7		
	Barley	1.2	1.8	2.2	2.8	1.1	1.4
	Rye	1.4	2.1	1.7	1.9		
	Soil	2.1	2.6	1.1	1.5		

[a] Each figure in this column represents the mean of 144 cultures.
[b] Each figure in this column represents the mean of 48 cultures.

Adapted from Joffe, A. Z., *Mycopathol. Mycol. Appl.*, 54, 35, 1974.

The best nutrient sources among organic substances for *F. poae* and *F. sporotrichioides* proved to be carbohydrates (starch, glucose), and the best suppliers of nitrogen were peptone and asparagine. Best results among inorganic substances were obtained with ammonium sulfate and sodium nitrate. Very meager growth of *Fusarium* was obtained with the use of organic acids. It is noteworthy that satisfactory development and production of toxic substances were obtained on filter paper with cultures of *F. poae* and *F. sporotrichioides.*

Certain problems relating to the nutritional physiology of *F. sporotrichioides* were investigated by Sarkisov[253,255] and Kvashnina.[152] Among the substances tested, the best sources of nitrogen and carbon were found to be peptone, casein, asparagine, glucose, starch, and mannitol. Bilai[23,25] states that aspartic acid, glutamic acid, and its amides, alanine, glycocoll, and ammonium carbonate, as well as gaseous ammonia, all provide suitable sources of nitrogen for *Fusarium* species of the section Sporotrichiella.

Substrates, Sources of Isolates, and pH

Effects of substrate on overall toxicity production were first studied by growing the *F. poae* and *F. sporotrochioides* at the six temperatures from 2 to 35°C on three grain substrates (wheat, barley, prosomillet) and on the three liquid substrates designated as Substrates IV, V, and VII (see "Isolation of *Fusarium* Strains from Different Sources").

Table 23
EFFECTS OF SUBSTRATES AND TEMPERATURES ON TOXICITY PRODUCED BY ISOLATES OF *F. POAE* AND *F. SPOROTRICHIOIDES* AND THE CHANGES INDUCED BY THESE ISOLATES IN THE pH VALUES OF THE SUBSTRATE (MEANS FOR ALL ISOLATES TESTED)

Substrate		*F. poae*						*F. sporotrichioides*					
		7 toxic isolates				2 nontoxic isolates		7 toxic isolates				2 nontoxic isolates	
		8°C		25°C		8°C	25°C	8°C		25°C		8°C	25°C
No.[a]	pH	T[b]	pH	T	pH	pH	pH	T	pH	T	pH	pH	pH
I	3.8	0.6	4.0	0.1	4.4	3.7	3.5	1.1	3.9	0.4	4.2	3.6	3.8
	5.6	3.7	6.3	1.9	6.7	5.6	5.5	2.1	6.5	1.2	7.0	6.0	6.3
	7.2	2.4	7.1	1.2	7.3	7.0	7.2	1.6	7.3	0.5	7.6	7.1	7.1
III	3.8	1.7	4.5	0.4	5.0	4.0	4.3	0.5	4.2	0.2	4.5	4.1	4.2
	5.6	2.2	6.6	0.5	7.0	5.6	6.1	1.4	6.3	0.5	7.3	6.0	6.2
	7.2	1.7	7.4	0.3	7.7	7.1	7.2	1.6	7.5	0.8	8.0	7.3	7.4
IV	3.8	1.1	3.8	0.3	4.3	3.3	3.4	0.6	3.7	0.3	4.1	3.6	3.2
	5.6	2.6	6.1	0.6	6.7	5.5	6.1	1.7	6.1	0.6	6.9	5.7	5.8
	7.2	1.3	7.2	0.3	7.3	6.6	7.2	1.6	7.1	1.1	7.4	6.8	7.1
V	3.8	0.8	3.5	0.5	4.1	3.5	3.4	0.9	3.6	0.5	3.9	3.7	4.0
	5.6	2.2	5.8	1.6	6.3	5.7	5.7	1.9	5.8	1.2	6.5	6.0	6.0
	7.2	1.6	7.0	0.8	7.4	6.8	6.9	1.3	7.2	0.8	7.5	7.4	7.5
VI	3.8	1.7	4.3	0.9	5.0	4.0	4.1	0.6	4.4	0.4	5.0	4.1	4.2
	5.6	2.7	6.5	2.0	7.2	6.1	6.5	1.9	6.5	1.4	7.2	6.1	6.3
	7.2	1.6	7.4	1.0	8.1	7.3	7.5	2.9	7.6	0.9	8.0	7.2	7.3
VII	3.8	2.1	4.4	1.7	4.9	4.1	4.5	1.1	4.6	0.7	5.2	4.1	4.6
	5.6	3.3	7.1	2.4	7.3	6.1	6.4	3.1	6.6	1.8	7.5	6.2	6.7
	7.2	1.7	7.9	1.5	8.3	7.3	7.4	2.2	7.9	1.6	8.7	7.5	7.8
VIII	3.8	1.0	4.0	0.4	4.4	3.7	3.8	0.9	3.9	0.2	4.4	4.3	4.5
	5.6	1.9	6.0	1.2	6.5	5.7	6.1	1.4	6.2	0.7	6.7	6.5	6.5
	7.2	1.5	7.3	1.4	7.5	6.7	6.6	1.1	7.2	0.9	7.9	7.2	7.5

[a] Composition of substrates is detailed in the section on methods.
[b] Toxicity rating (0—4).

Adapted from Joffe, A. Z., *Mycopathol. Mycol. Appl.*, 54, 35, 1974.

In the case of the liquid cultures, toxicity was determined separately from the mycelial mass (thallus) grown on the liquid and from the substrate after removal of that mass. Results are summarized in Table 22.

These figures show that toxicity derived from the liquid substrate was always higher than that from the thallus grown on that substrate. Overall toxin production was constantly strongest on the starch substrate, and in the case of *F. poae* it was somewhat stronger on Czapek's than on the carbohydrate peptone substrate.

Among the grain substrates, barley yielded growth with higher overall toxicity of *F. poae* and *F. sporotrichioides* both in light and in darkness. The favorable effect of darkness on toxin production was evident on all three of these substrates.

As regards the source of isolates, those of *F. poae* were most highly toxic when isolated from soil, those of *F. sporotrichioides* when isolated from barley. Here again, darkness favored toxin production regardless of the source of the isolate.

Further studies with liquid substrates at three pH levels were carried out with *F. poae* and *F. sporotrichioides* at 8 and 25°C. All the eight substrates listed in an earlier section on methods were used, but since Substrates I and II gave closely similar results,

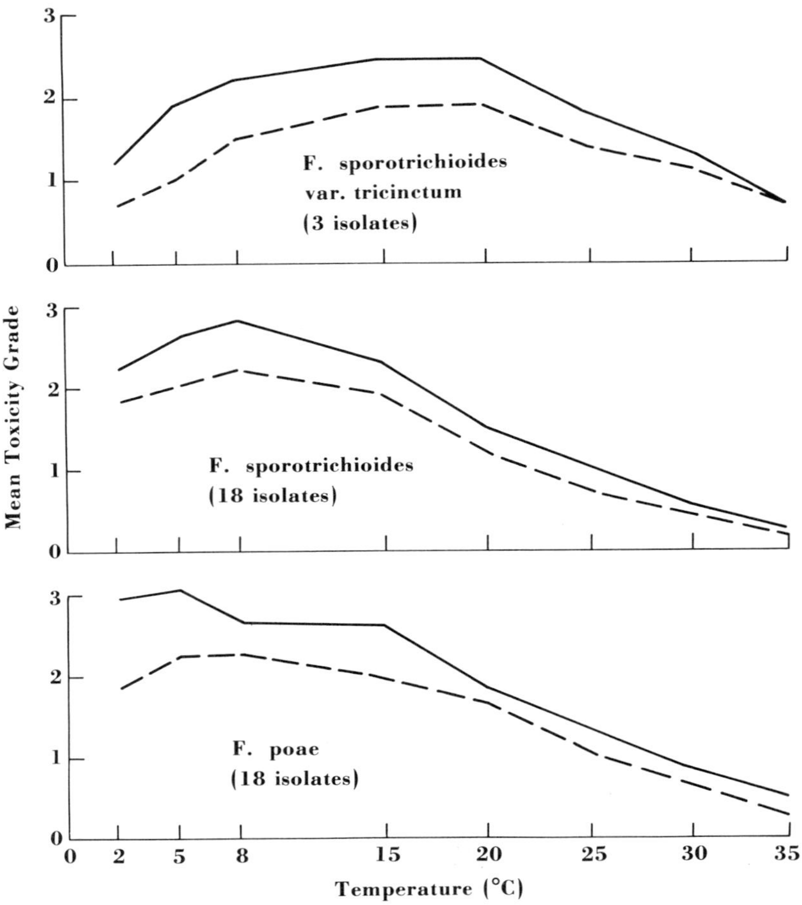

FIGURE 12. Effect of temperature and light on toxicity of the Sporotrichiella section isolates; dark, solid line; light, broken line. (Adapted from Joffe, A. Z., *Mycopathol. Mycol. Appl.*, 54, 35, 1974. With permission.)

those obtained with Substrate II are omitted. The purpose of these studies was to ascertain effects of the various substrates on the toxicity rating of toxic isolates, and the extent to which toxic and nontoxic isolates changed the pH level of substrates on which they were grown.

Results are summarized in Table 23. They show that *F. poae* and *F. sporotrichioides* produced higher toxicity ratings at 8°C than at 25°C. This rating on all substrates was highest at pH 5.6, and mostly lower at pH 3.8 than at 7.2. Top toxicity ratings at pH 5.6 were produced by both fungi on Substrates I and VII, but *F. sporotrichioides* produced an exceptionally high rating at pH 7.2 on Substrate VI.

With regard to changes of pH levels induced by isolates on their substrates, the changes on substrates originally at pH 3.8 were rather small, mostly upwards up to pH 4.5, and only in a few cases (all at 25°C) up to pH 5.0 to 5.2. The changes induced on substrates originally at pH 5.6 were larger, especially at 25°C, and frequently reached pH 7.0, or on Substrate III, VI, and VII, even pH 7.2 to 7.3. On substrates originally at pH 7.2, the changes induced were small, and at 25°C they reached pH 8.0 to 8.3 only on Substrates III, VI, and VII, and as high as pH 8.7 for *F. sporotrichioides* on Substrate VII. It may be assumed that toxin formation is also conditioned by the acidity of the medium. The most suitable pH values were found to be 4.6 to 5.4 for *Fusarium*.

Temperature and Light

The overall toxicity produced by extracts from wheat cultures grown in light and darkness at eight temperatures, from 2 to 35°C, is shown in Figure 12.

F. poae and *F. sporotrichioides* produced highest toxicity both in light and in darkness, at 5 and 8°C, and *F. sporotrichioides* var. *tricinctum* at 15 and 20°C. Though toxicity of the latter fungus also lessened at high temperatures, it yet developed, at 30 and 35°C, twice as much toxicity as the other two species. Darkness clearly favored development of toxicity of all three fungi at all temperatures.

Closely similar results regarding the temperature and light effects on production of toxicity by the species were obtained in additional test series, in which the fungi were grown on liquid substrate.[122]

As far as we are aware, no studies have been published on the relationship between environmental and substrate factors and the extent of toxin production by fungi of the Sporotrichiella section.

There is hardly any literature on the relationships between toxic properties of fungal species and the temperatures under which they were grown. Only Sarkisov and Kvashnina[257] found that cultures of *F. sporotrichioides* were more toxic when grown at 1.5 to 4°C than when grown at 22 to 25°C.

With regard to effects on growth, Lacicowa[153] found that development of *F. poae* on potato dextrose agar and maltose agar media was best at 24 to 25°C, which is in agreement with our results. This author states that the fungus developed equally well in daylight and darkness.

Seemüller,[278] working with two to six of the following fungi, determined temperature relationships as follows: *F. poae* had its minimum growth at 2.5°C, optimum at 22.5 to 27.5°C and maximum at 32.5°C, the latter being appreciably lower than in our work (35°C). *F. sporotrichioides* and what Seemüller called *F. tricinctum* had lower minima (0°C), optima at 27.5 and 22.5°C, respectively, and maxima at 35 and 32.5°C, respectively. Seemüller's *F. chlamydosporum* had a higher minimum (5°C), an optimum at 27.5°C and maximum at 35°C or a little above that value. This agrees with our observation that *F. sporotrichioides* var. *chlamydosporum* made better growth at 35°C than any of the other fungi.

Growth Characteristics Possibly Associated with Fungal Toxicity

Repeated investigations have shown that low temperatures promote rapid accumulation of toxin in cultures of *F. poae* and *F. sporotrichioides,* despite the slow growth of the mycelia. Cultures grown at high temperatures, while producing luxuriant mycelial growth, were only slightly toxic or nontoxic. It is of interest to point out that absence of sporification was occasionally observed in cultures of *F. poae* and *F. sporotrichioides* grown at room temperature, 18 to 20°C. Such cultures when subjected to rabbit skin tests were shown to be nontoxic or slightly toxic. Cultures of the toxic fungi *F. poae* and *F. sporotrichioides* grown at low temperature or under conditions of alternating freezing and thawing were characterized by prolific production of spores and high toxicity.

The presence of abundant nonsporifying aerial mycelium usually coincided with the absence of toxicity. However, scanty aerial mycelium with a large number of spores is normally associated with high toxicity. Toxicity of these cultures is apparently associated with intensive production of spores.

It should be noted, in the Sporotrichiella section, that the presence of pigmentation provides an indication of toxicity in fungi. Not infrequently, *Fusarium* cultures from various other sections with an insignificant amount of pigment proved to be highly toxic, while strongly pigmented cultures sometimes had low toxicity. The germination rate of spores of various toxic fungi was also found to be dependent on temperature.[114]

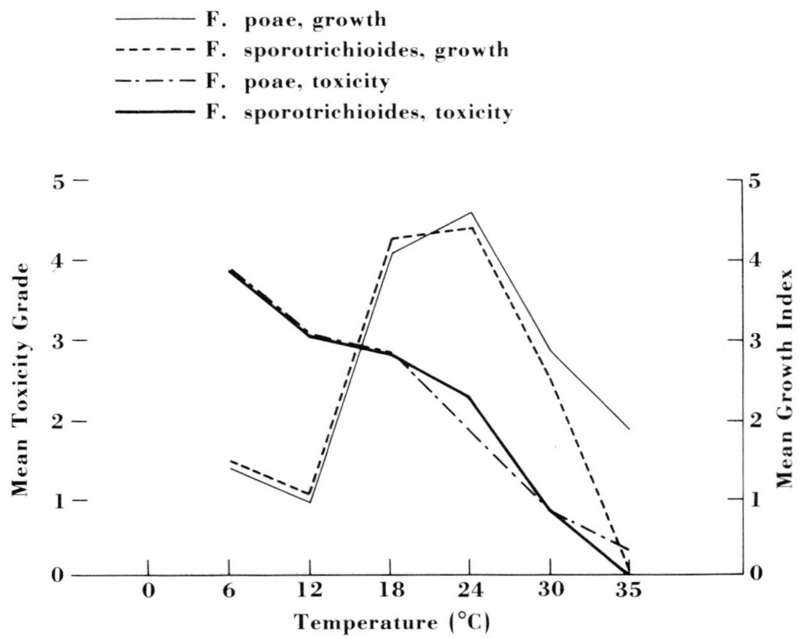

FIGURE 13. Relation between growth in vitro and grade of overall toxicity of *F. poae* (five isolates) and *F. sporotrichioides* (six isolates) at various temperatures. (Adapted from Joffe, A. Z., *Mycopathol. Mycol. Appl.*, 54, 35, 1974. With permission.)

Vigor of Growth and Toxigenicity

Is there any relation between the growth of these fungi in vitro and between the degree of toxicity of their culture extract to rabbit skin? The following conclusions can be drawn from results obtained with five isolates of *F. poae* and six of *F. sporotrichioides* showing strong toxicity.[122] The peak for growth of both species was at 18 to 24°C, where toxicity was moderate; but toxicity was strongest at 6 to 12°C, where growth was limited. At temperatures of 30 to 35°C growth was moderate and toxicity weak. Thus there seems to be no relation between vigor of growth at any of the above temperatures and the toxicity of these fungi. This is clearly illustrated in Figure 13.

Antigenic Properties of Toxic Fungi

Following our proof of the role played by certain fungi in the etiology of ATA, our material was used by Mironov and Alisova[187] in work on immunization of rabbits with toxic cultures of *F. poae* and *Cl. epiphyllum* with the object of determining their antigenic properties. For immunization of rabbits, these authors employed filtrates of cultures, as well as suspensions of mycelium and spores. On the basis of their experiments, Mironov and Alisova concluded that it is possible to obtain specific immunization sera for toxic cultures of *F. poae* and *Cl. epiphyllum*. They further found that in order to obtain these sera, it is necessary to secure immunization of rabbits by means of filtrates of cultures started at the stage of intensive sporification and cultured at a temperature range of −2 to +2°C.

It is necessary to stress the difference in the degree of toxicity between extracts obtained from sterile millet grain infected with toxic fungi and extracts of the mycelial mass grown on starch or carbohydrate-peptone medium. Extracts derived from sterile millet grain infected with toxic fungal cultures displayed more pronounced toxic action than extracts from dry mycelial film.

FIGURE 14. General view of rabbits treated with toxic extract of *F. poae, F. sporotrichioides, F. sporotrichioides* var. *tricinctum,* and *F. sporotrichioides* var. *chlamydosporum* 24 hr after application. White squares on left are controls. (Photo courtesy of Marcel Dekker, Inc.)

BIOASSAY METHODS AND PROCEDURES

The toxic grain responsible for outbreaks of ATA was investigated chiefly by skin tests on rabbits. This skin test is, at present, a generally accepted laboratory test for toxicity of overwintered cereal grains and also for toxins produced by various fungi, especially by *Fusarium* of the Sporotrichiella section, developed on overwintered grains in the field which cause ATA.

Toxicity Test on Rabbits

The mycotoxins of *Fusarium poae, F. sporotrichioides, F. sporotrichioides* var. *tricinctum,* and other isolates were assessed by skin tests performed on male and female rabbits weighing 1.5 to 2 kg. Only rabbits with nonpigmented skin were used because the skin was thinner and they were more sensitive to toxins.[112,113,118,124]

Each rabbit was kept in a separate cage and was maintained on laboratory feed. On the back and sides of each rabbit, squares of skin measuring 3 × 3 to 4 or 5 cm were carefully cleared of hair so that four to five rows with five to seven squares were obtained, and the skin looked like a chessboard (Figure 14). The skin tests were made 24 hr after depilation. The extracts from grains or fungi were applied by a platinum loop, by micropipet or by Hamilton syringe to the skin of the rabbit twice within a 24 hr interval. Two squares of each rabbit served as a control and were treated with ethanol extract of autoclaved normal wheat grain not infected by any fungus. The

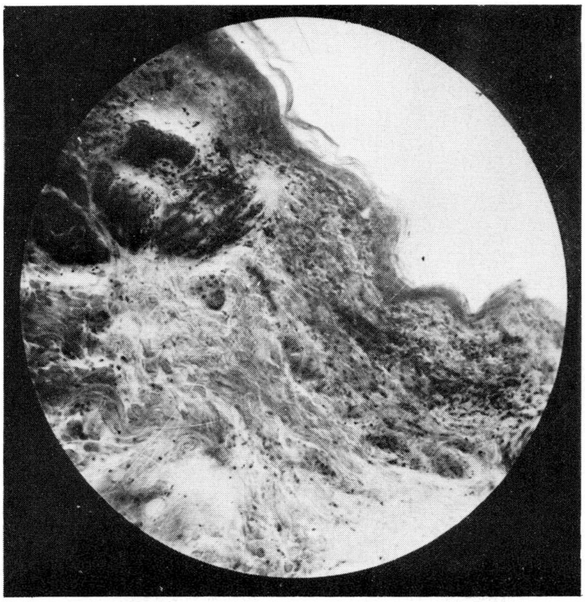

FIGURE 15. Necrosis of both epidermis and subepidermal tissue following application of toxic *F. poae* extract. (Adapted from Joffe, A. Z., *J. Stored Prod. Res.*, 5, 211, 1969. With permission.)

reaction was recorded for 48 hr, but the rabbits were kept under observation for at least 6 to 8 days after the first application.

Prior to being subjected to the experimental application, each rabbit was given two control treatments, one with an extract of known toxicity and another with the extract of unaffected grain.[112,113,120] During the experiment stiff collars of cardboard prevented the rabbits from licking the test squares.

There were various types of skin reactions, the evaluation of which was based on their principal components; leukocytic film, edema, hemorrhage, and necrosis (Figures 15 and 16).

The leukocytic reaction was characterized by the formation on the skin surface of a whitish cream, an easily detachable film which consists of infiltration of mass leukocytes accumulating in the horny layer of the epithelium or dermis. The intensity of the leukocytic components was estimated according to the massiveness of the superficial film.

The edema or acute edema reaction sometimes caused severe inflammatory changes in the epidermis; the intensity of the edema was assessed by the thickness of the skin fold. In this case there was no leukocytic film, or sometimes the leukocytic infiltration of the dermis was accompanied by edema and necrosis of the skin.

The appearance and intensity of hemorrhage reaction was connected also with inflammatory changes in the skin and estimated by the quantity of visible extravasations. The intensity of necrosis, as judged by the massiveness of the scab and the time of shedding, was determined on the 8th day; the other components were recorded on the 3rd day. The intensity of the skin reaction produced by each culture of *F. poae* and *F. sporotrichioides* and others was assessed on a scale of grades by Joffe[112,113,122] and Joffe and Palti.[127]

The presence of edema, hemorrhage, and necrosis was regarded as evidence of marked toxicity of the fungus or of the grain samples. A pronounced leukocytic film,

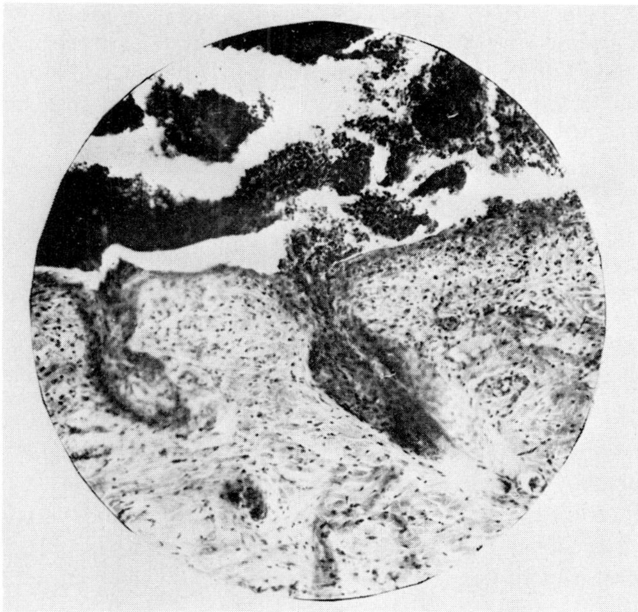

FIGURE 16. Membrane containing leukocytes, following application of toxic *F. sporotrichioides* extract to rabbit skin (superficial leukocyte infiltration). (From Joffe, A. Z., *J. Stored Prod. Res.*, 5, 211, 1969. With permission.)

unaccompanied by any of the other components, was considered as an indication of the toxicity of the experimental material; a reaction consisting of a thin leukocytic film, scattered vesicles, reddening, and desquamation was assessed as weak and doubtful.[112,113,116,118,127]

Skin reactions caused by the action of *Fusarium* extracts were thus distinguished both by their external appearance and by histological changes. The areas of skin which were in contact with toxins of *F. poae, F. sporotrichioides,* and *F. sporotrichioides* var. *tricinctum* were fixed with 4% formaldehyde. Paraffin sections were cut and stained with hematoxylin and eosin.

The toxin of *Fusarium* extracts often had a general toxic effect on the rabbits, manifested by a loss of appetite and weight, sleepiness, and changes in the blood composition and in the organs; in not infrequent cases, the animals died after the administration of the toxin.

Other Tests

Davydova[54] and Mironov and Davydoya[188] found that the skin of cats, sheep, dogs, cows, and horses was also sensitive to *Fusarium* toxins and overwintered cereals. Similar results were obtained by pipetting toxic extracts on the conjuctiva of rabbits. The skin of guinea pigs, white mice, and rats was less sensitive. In contradistinction to these results, Schoental et al.[262,265] established that small doses applied to the skin of mice and rats have a strong local cytotoxic effect.

Other tests were also suggested for the recognition of toxic grain. Thus, Kretovich and Skripkina[149] and Mishustin et al.[194] proposed a quick test which was based on the observation that the ethanol extract of toxic grain of prosomillet, wheat, and other crops suppressed the fermentative power of *Saccharomyces cerevisiae.*[147] Elpidina[63,64] suggested a phytotoxicological test in which a drop of toxic grain or ethanol extract of *F. poae* on leaves caused necrotic dark brown spots on plant tissue of leaves and the wilting of the plant. Joffe[119] also put different plants (tomatoes, beans, peas) into liquid media containing toxic metabolites of *F. poae* and *F. sporotrichioides.* After 4 hr the turgor had already decreased and the plant had begun to droop. After 20 to 24 hr the plants were completely dry. Seedling mortality was caused also by inoculation of fungi of the Sporotrichiella section onto field and vegetable crops.[119] Toxic extracts of these *Fusarium* species caused more severe changes than toxic extracts of prosomillet and wheat overwintered in the field.

Elpidina[63,65,66] detected that two drops of the toxic extract of prosomillet infected by *F. poae* reduced the antitoxic power of diphtheria antiserum, and the toxin named poin obtained from the *Fusarium* inhibited the growth of Ehrlich's adenocarcinoma and Croker's sarcoma.[66-69]

Tkachev[310] suggested that dough made of flour from overwintered grains, when boiled in milk, coagulated the milk; this did not happen with meal of normal nontoxic overwintered grain. Manoilova,[172] Kozin and Yershova,[144] and Olifson[204] proposed color change methods in the extract of toxic overwintered grain and toxic *Fusarium* strains of the Sporotrichiella section.

Rubinstein[246,247] found an effective method for examination of toxic and nontoxic *Fusarium* of the Sporotrichiella section. With the help of a specialized fatty medium toxic *Fusarium* was shown to express the properties of fluorescence while nontoxigenic strains did not show fluorescence after growth.

Bioassay with Toxic *Fusarium*

In feeding experiments on various animals, we used different bioassay methods. The culture filtrates, mycelium (dry fungi), infected grains, ethanol, or ether extracts of toxic *F. poae, F. sporotrichioides,* or *F. sporotrichioides* var. *tricinctum* were administered p.o. and culture filtrate and crude extract were administered s.c. or i.p. in various doses to laboratory animals.[112,113,118,120,125,130,166] We also used gastric fistulas in experiments with dogs[99] and intragastric intubation with mice and rats.[260,275]

The following animals were used in our studies for the biological assay test with toxic strains of *F. poae, F. sporotrichioides,* and *F. sporotrichioides* var. *tricinctum* (including observations on loss of weight, mortality, and histopathological changes in the organs and tissues): frogs, mice, rats, guinea pigs, rabbits, dogs, cats, horses, chickens, ducklings,[7-9,99,112,113,120,130,166,260] and also some protozoa.[58,59,112] All the animals that died were autopsied and tissues were stained with hematoxylin and eosin for histological investigation.

Sarkisov[253,255] carried out experiments by the p.o. method on mice, rats, guinea pigs, rabbits, dogs, and cats as well as on horses, cattle, sheep, and pigs. He fed them toxic overwintered grain infected by *F. sporotrichioides.* Bilai[23-25,27] and Pidoplichka and Bilai[223] fed young rabbits and guinea pigs grain infected by *F. poae* and *F. sporotrichioides* and used aqueous extracts by s.c. injections for the assessment. Rubinstein[240,242,245] and Rubinstein and Lyass[249] carried out studies on feeding mice, rats, cats, and monkeys p.o. with grain infected by *F. sporotrichiodes.* Getsova[84] used i.p. injections on mice with toxins of ATA. The effect of *Fusarium* toxins in animals has been reviewed by Joffe.[112,113,118,120,124]

CHEMISTRY OF TOXINS FROM AUTHENTIC STRAINS INVOLVED IN ATA DISEASE

Much attention has been paid to the study of the chemistry of the *F. poae* and *F. sporotrichioides* isolated from overwintered cereal grains. Numerous investigations have been reported and published in the U.S.S.R. of the structure and chemical properties associated with overwintered grain and toxicity of the *Fusarium* belonging to the Sporotrichiella section.

Composition of the Toxins

Gubarev and Gubareva[97] isolated two fractions from ether extracts of toxic grains. One fraction contained derivatives of fatty acids that caused the local skin inflammatory reaction in rabbits. The second was a fraction of nonsaponifiable material which produced a necrotic skin reaction. Barer[20] related toxins of overwintered grains to steroid type aromatic polycyclic compounds, and Okuniev[203] isolated two separate fractions: the first was a steroid resembling coumarol and vitamin E, which associated with ATA, and the second a toxic product of hydroxy fatty acids which caused only local irritation of the mucous membranes.

Kretovitch[145] and Kretovitch and Bundel[146] found that the toxins are present in overwintered grains as oxidation products of unsaturated fatty acids. Kretovitch and Sosedov[148] found that toxic grains contain more nonprotein nitrogen and amino nitrogen and less starch than nontoxic grains. Toxic grains showed increased activity of dextrinogenic amylase and decreased activity of peroxidase and oxidase.

According to Kolosova[138] the toxic materials were unsaturated acids, whereas Gabel[74] concluded that they were sapogenins. Gabel,[74] Myasnikov,[200] and Svojskaja[305] found that these toxins had acidic properties, but they also believed that there were other toxins in the form of neutral lactones and compound ethers. Zavyalova[342] isolated lipoproteins from wheat contaminated with *Fusarium* species causing ATA disease.

Mironov,[182] Mironov and Myasnikov,[186] and Yefremov[339-341] considered that the toxic extracts obtained from overwintered cereal grains infected with *Fusarium* toxic strains contained only one component that produced both the rabbit skin reaction and the ATA disease in man.

Kozin[144] considered that the lipid fraction isolated from the overwintered grains serves only as a solvent for the toxic substance.

The studies of Pidoplichka and Bilai[223,224] indicated that *F. poae* and *F. sporotrichioides* possess various enzymes which enable the fungi to utilize nutritional sources of nitrogen, carbon, and minerals in the grain. The fermentative system of fungi which operates in stored grains was studied in our laboratory and we concluded that toxic fungi secrete enzymes which act on the components of the grain and thus render the grain toxic under suitable ecological conditions.[130a]

It is obvious from the many theories that there was no unanimous opinion by Russian researchers on the chemical composition of *Fusarium* toxins isolated from overwintered toxic cereals because no one has yet succeeded in isolating a pure toxin and determining its structure.

Many details have been added by the studies by Olifson,[206,207] who found a method for purification and isolation of toxins from overwintered grains and from normal grains previously autoclaved and then infected with toxic fungi *F. poae* and *F. sporotrichioides* of varying degrees of toxicity. Olifson's basic assumption was that the toxin is found in the lipid fraction of the grains. When such a lipid extract is applied to the skin of rabbits, it causes a strong inflammatory reaction. Cats, mice, and guinea pigs which were fed on a diet mixed with this lipid fraction died. Olifson[204,205,208-214] also determined the physicochemical constants for these lipids. The analysis indicated an increase of acid value from a normal of 14.1 to 17.8 to 121.4 for toxic lipids and also an increase in peroxide value from a normal of 1.3 to 1.8 to 6.8 to 8.4 in the nonsaponifiable residue. On the other hand, the refractive index decreased and the iodine value decreased from 132.4 to 66.6 for toxic lipids. If these constants for free and bound lipids are compared, it is evident that free lipids yield a higher acid value (121.4) and a lower peroxide value (6.8 to 8.4) than bound lipids whose acid value and peroxide value are 15.8 and 29.1, respectively. It thus seems that a big difference exists in the chemical composition of free and bound lipids extracted from prosomillet grains contaminated with *F. poae* and *F. sporotrichioides*.

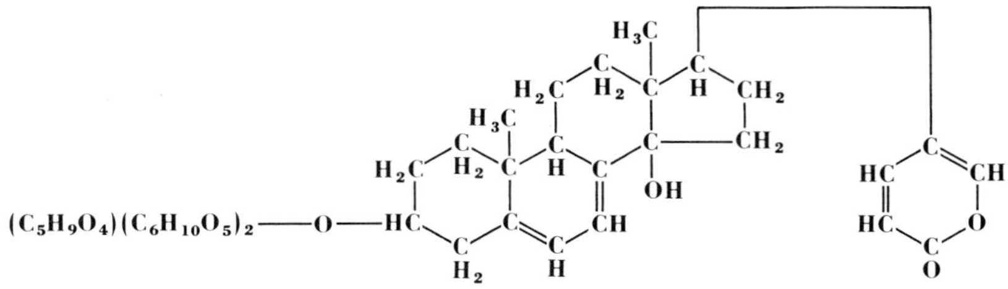

FIGURE 17. Structure of sporofusarin.[211]

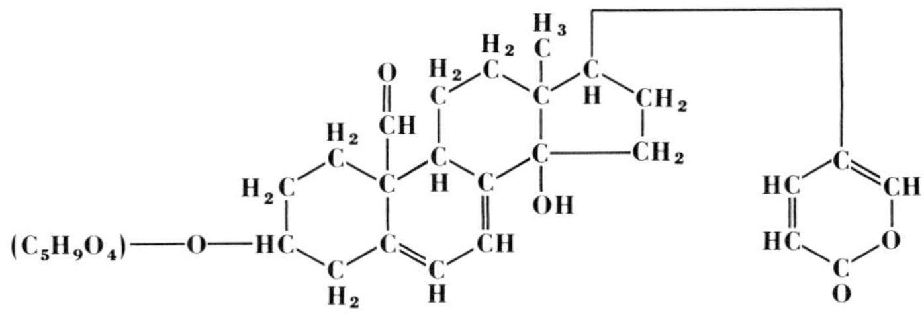

FIGURE 18. Structure of sporofusariogenin.[211]

Under the influence of the *Cladosporium* fungi and *Mucor* the content of lipids was reduced; also *Alternaria* and *Mucor* reduced by 35% the proteins in prosomillet inoculated with these fungi.

A comparison was also made between the chemical composition of prosomillet grains intentionally infected with toxic cultures of *F. poae* and *F. sporotrichioides*, and overwintered grains which became toxic in the field. The grains infected with pure cultures had been grown for some days at −10 to −15°C and then transferred to +3 to +5°C for 1 month. The cultures were then autoclaved and after the infected grains had completely dried in the air, tests for humidity, ash content, proteins, cellulose, acid value, and iodine value were performed. The infected grains showed an increased ash content and acid value and a decreased iodine value. Olifson isolated a neutral fraction from both overwintered prosomillet and normal prosomillet inoculated with *F. poae* and *F. sporotrichioides*, and an acid fraction from millet infected with *Cl. epiphyllum* and *Cl. fagi*. Olifson[211] found that the toxicity of the neutral fraction was more marked for *F. poae* and *F. sporotrichioides* than for some other toxigenic fungi, e.g., species of *Cladosporium*.

Olifson[206,207,211] isolated a saponinic steroid, which he named sporofusarin, from prosomillet which had been infected with *F. sporotrichioides* (Figure 17). Its empirical formula is $C_{65}H_{96}O_{25}$ and melting point 246 to 248°C. Hydrolysis of 4 hr with 5% H_2SO_4 yielded a saponin with an empirical formula of $C_{24}H_{31}O_4$; it was given the name sporofusariogenin (Figure 18).

Olifson also isolated a monoglucoside from *F. poae*, poaefusarin (Figure 19), with an empirical formula of $C_{35}H_{39}O_{12}$. This glycoside contained xylose and a steroidal aglycone. The aglycone was designated poaefusariogenin ($C_{24}H_{28}O_5$) and differed from sporofusariogenin by the presence of an aldehyde group instead of a methyl group (Figure 20). These two derivatives were tested on cats and compared with the effect of a lipid material, called lipotoxol by Olifson, which was isolated from the lipid fraction of toxic overwintered prosomillet.

All these derivatives gave a similar syndrome in cats, characterized mainly by a constant leukopenia. The oral lethal dose was 0.5 mg and all animals died within 15 to 16

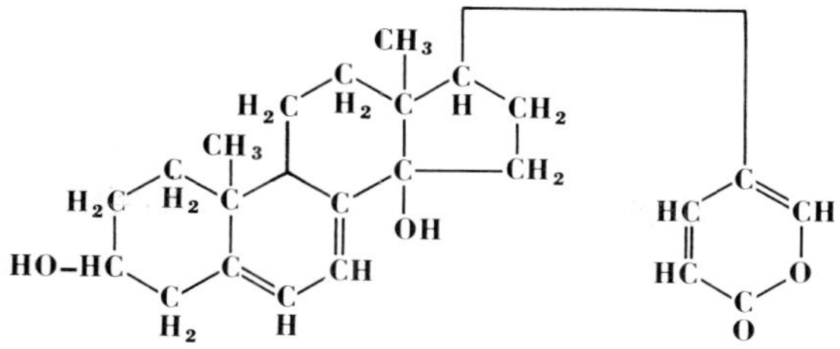

FIGURE 19. Structure of poaefusarin.[211]

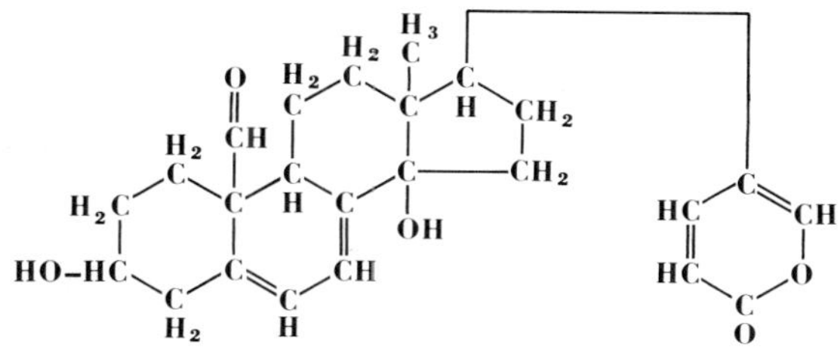

FIGURE 20. Structure of poaefusariogenin.[211]

days. In cats and dogs lipotoxol inhibited the normal action of the heart. The lethal dose for mice was 0.06 to 0.07 mg of 1% lipotoxol extracted from the fatty fraction of prosomillet. On the skin of rabbits lipotoxol caused a typical hemorrhagic-edematous reaction. Lipotoxol resembles in structure and properties both sporofusariogenin and poaefusariogenin which are both steroids, and since the syndrome produced in cats was very much like ATA in man, Olifson concluded that the cause of the disease was the toxin secreted by the two fungi, *F. sporotrichioides* and *F. poae*, in overwintered grains.

It is also suggested by Olifson[211] that intermediate products of these final derivatives such as unsaturated fatty acids, oxyacids, and steroidal lactones are associated with the ATA syndrome. According to Olifson, the two steroidal glycosides, poaefusarin and sporofusarin, and their aglycones are C_{24} steroids that have a doubly unsaturated six-membered lactone ring and carry a 14β-hydroxyl group. He considered that these toxins contributed to the ATA disease. Recently Olifson[209,212] and Misiurenko[195] reported a method for the isolation and production of poaefusarin, further affirming its biological activity.

It should be noted that the structure of these compounds is reminiscent of the cardioactive steroidal lactones, which act specially on heart muscle,[70] and are not listed as skin irritants or compounds that damage the bone marrow. Only Ueno et al.[319] working with our authentic strain NRRL 3510 isolated the following compounds from the crude extract of this culture: neosolaniol, HT-2 toxin, butenolide, and chiefly T-2 toxin. This strain appeared in our protocol under #738, was isolated from overwintered millet grains in the Orenburg district, and according to the request of Ellis, was sent to the Fermentation Laboratory of the U.S. Department of Agriculture, Peoria, Ill., on July 31, 1969 and delivered to Ueno.

Identification of Olifson's toxic "poaefusarin sample" was performed by Mirocha and Pathre[180] and they determined the following compounds: T-2 tetrol, neosolaniol, zearalenone, and mainly T-2 toxin. Later, Bamburg et al.[11-19] and Szathmary et al.[306] studied *F. tricinctum, F. poae,* and *F. sporotrichioides* from eastern Europe and isolated from their extracts zearalenone, T-2 toxin, neosolaniol, HT-2 toxin, and again, mainly T-2 toxin.

The last-named scientists together with Ueno et al.[319] stated that extracts from *F. poae* and *F. sporotrichioides* did not contain any poaefusarin or sporofusarin; they also could not confirm the presence of a steroid-type compound in the toxic poaefusarin sample of Olifson.[211]

All these above-mentioned chemists have found some metabolites of trichothecenes and indicated that chiefly T-2 toxin having a characteristic trichothecene structure and causing severe hematopoietic damage may be one of the principal features of ATA in man.

In experiments on animals the trichothecenes cause necrosis of the actively dividing cells in tissues, of the epithelium in the GI tract, thymus, lymph nodes, and the bone marrow and also produce a skin irritating or necrotizing effect on laboratory animals.[318-320]

In view of the fact that the Russian scientists have persisted in their standpoint concerning toxicity of the mycotoxins "poaefusarin" and "sporofusarin", the author, together with Yagen, again undertook a study with all the original and authentic strains of *F. poae* and *F. sporotrichioides* isolated by the author from overwintered grains collected at the time of the fatal ATA outbreaks in the Orenburg district in the Soviet Union. We have examined the toxic metabolites from 131 isolates (106 *F. sporotrichioides* and 25 *F. poae*) cultivated at low temperatures in our laboratory.[331] These toxic *Fusarium* species were obtained from monoconidial cultures and maintained on sterilized soil and on standard PDA medium at 3°C. The cultures were inoculated and grown on sterilized wheat or millet grains.

The first screening report on the distribution of T-2 toxin-producing *Fusarium* fungi isolated in the U.S.S.R., and associated with ATA, showed that more than 95% of the *F. poae* and *F. sporotrichioides* isolated produced T-2 toxin in various quantities. Among the isolates in our collection there were many which produced gram quantities of T-2 toxin when grown on 1 kg wheat or millet at 12 or 5°C for a period of 21 and 45 days, respectively. Those were isolates involved in the more severe cases of ATA in the Orenburg district.

Crude extracts applied to the skin of rabbits resulted in essentially the same injuries as those produced by pure T-2 toxin, i.e., edema, hemorrhage, and necrosis of the epidermis, dermis, and hair follicles. The results of the rabbit skin screening show that 71 out of 106 isolates of *F. sporotrichioides* and 20 out of 25 cultures of *F. poae* produced very strong or strong dermatitic reactions.

The identification of the isolated compounds and chiefly of T-2 toxin was determined by TLC, GLC, spectroscopic analyses, and also by bioassay test on rabbit skin.

A good correlation was demonstrated between T-2 toxin detection by TLC and inflammatory skin reactions of rabbits, and the comparison of amounts of T-2 toxin determined by GLC also corresponded to the rabbit skin response. The most toxic among all *Fusarium* isolates were *F. sporotrichioides* #921 and *F. poae* #958 (Figure 21).[129]

F. sporotrichioides #921 was isolated in 1947 by the author of this chapter from overwintered rye grains which served as a general food source for two families of four and five persons, respectively. Three members from each family died within 6 to 8 weeks of consuming products prepared from the toxic grains.

This strain was also inoculated and cultured on sterile wheat or millet at 5, 12, and 29°C for 45, 21, and 10 days, respectively. The isolation of the metabolites was accomplished by extraction of the infected grain with ethyl alcohol, and toxicity was established by rabbit skin test. It was found that the extract obtained from 29°C incubation was much less active than the extract obtained from 5 and 12°C incubation. In order to analyze the metabolites produced by *F. sporotrichioides* #921 we cultivated this strain on 1 kg of millet divided among 14 1-ℓ flasks at 12°C for 21 days.[129,332]

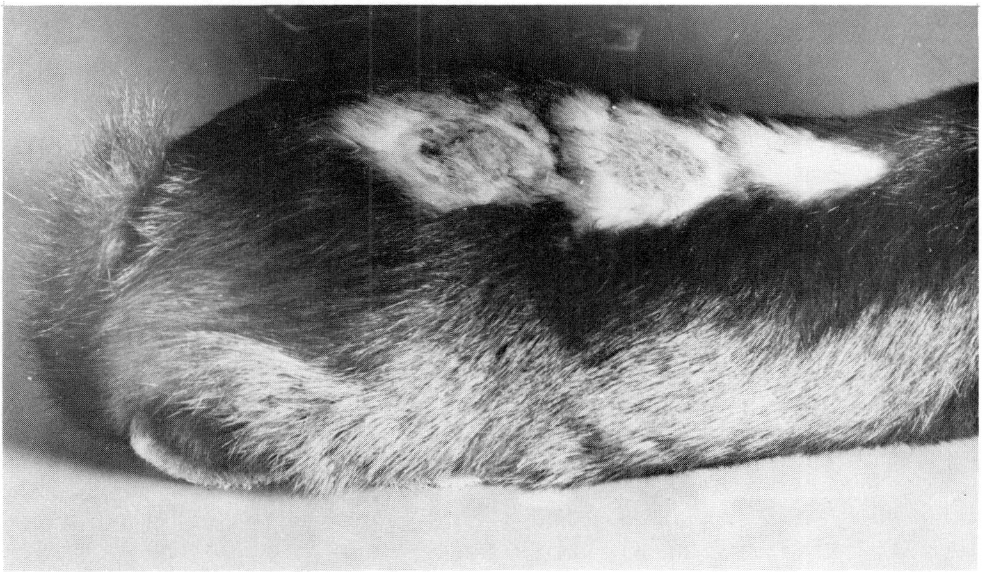

FIGURE 21. Edematous necrotic reaction on rabbit skin 24 hr after application of *F. poae* #958 (left), *F. sporotrichioides* #921 (middle) and control (spot on right). (Photo courtesy of Marcel Dekker, Inc.)

The ethanolic extracts were concentrated by evaporation under reduced pressure to one fifth of the volume, diluted with water, and extracted with cold hexane, ether, ethyl acetate, chloroform, and methylene chloride. The organic extracts were washed with water, dried over $MgSO_4$, and evaporated to dryness. The hexane fraction (12.8 g) was the major one. The ether fraction after two recrystallizations yielded 2.8 g of white crystals, mp 151°, which were identified at T-2 toxin (Figure 22) by elemental analysis, NMR, IR, and mass spectra and optical rotation data.[333]

The mother liquids from the recrystallization of the ether fraction were added to the hexane extracts. The hexane extracts contained fatty acids which interfered with further purification. These acids were converted to their methyl esters with diazomethane by the usual procedure. The resulting oil was subjected to chromatography on alumina with acetone-petroleum ether (6:4).

From crude extract of 22.6 g isolated from 1 kg wheat infected with *F. sporotrichioides* #921 were isolated the following fractions: palmitic, oleic and linoleic fatty acids, β-sitosterol, campesterol, stigmasterol, ergosterol,[70,102] and a novel sterol metabolite of 12β-acetoxy-4,4-di-methyl-24-methylene-5α-cholesta-8,14-diene-3β,11α-diol.[334] The last-named fraction contained an oil which is in the process of being identified. Apart from these sterols, some trichothecenes — neosolaniol, HT-2 toxin, and chiefly T-2 toxin — were also isolated (Figure 23).[333]

Thin-layer analysis showed that the ethyl acetate, chloroform, and methylene chloride extracts contained a similar complicated mixture of compounds; therefore they were combined. The polar fractions from the above chromatography were added to this mixture, which was further purified by column chromatography on alumina using an ethyl acetate-methyl alcohol mixture for elution. The fractions obtained were further purified by preparative TLC. The identification of the isolated compounds was done by GLC combined gas chromatography-mass spectrometry (GC-MS), MS, NMR, and IR spectrometry.

The structure of the known compounds was confirmed by comparing MS, NMR, and IR spectral data to the spectra of the authentic samples or to the data published

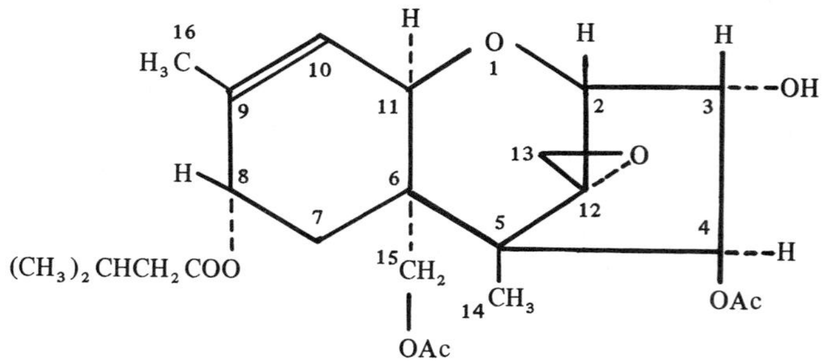

FIGURE 22. Structure of T-2 toxin.[319]

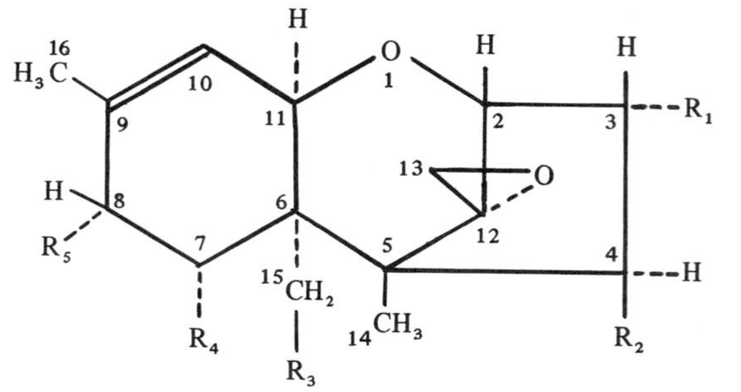

	R_1	R_2	R_3	R_4	R_5
T-2 toxin	OH	OAc	OAc	H	$OCOCH_2CH(CH_3)_2$
HT-2 toxin	OH	OAc	OAc	H	$OCOCH_2CH(CH_3)_2$
Neosolaniol	OH	OAc	OAc	H	

FIGURE 23. Trichothecenes from *F. sporotrichioides* #921.[319]

by Pouchert et al.[228,229] Bamburg et al.,[17] and Ishii et al.[101] We have shown in experiments on animals (mainly on cats and New Hampshire chicks)[130,166] that T-2 toxin (Figure 22) a trichothecene (3α-hydroxy-4β,15-diacetoxy-8α-(3-methyl-butyryloxy)-12,13-epoxytrichothec-9-one), caused local inflammation, hemorrhage, and necrosis in skin and in the GI tract, lymph nodes, and bone marrow, and also severe hematopoietic damage, mainly leukopenia, a drastic decrease of leukocytes.

A total of 4.1 g T-2 toxin was isolated from 1 kg of infected millet. The six isolated steroids, a total of 0.15 g, were neither toxic nor skin irritants, and have not the structure of the poaefusarin or sporofusarin of Olifson.[211] Since they are not skin irritants and are present in such relatively small amounts, we assume that this steroid part of the crude *Fusarium* extract does not contribute in any significant way to intoxication. The fatty acid fractions isolated from the crude extract were neither skin irritants nor toxic.[57] The contribution of HT_2-toxin and neosolaniol to the intoxication must be very small, because they present in the crude extract in relatively small amounts.[333]

We also examined *F. poae* #958 using the same methods and analysis as for *F. sporotrichioides* #921 and a total of 2.8 g of T-2 toxin was isolated from 1 kg of infected millet.[333]

Using TLC, GLC, and MS we analyzed 131 isolates of *F. sporotrichioides* and *F. poae* associated with the clinical ATA toxicoses in the Soviet Union, and found that 95% produce T-2 toxin in varying amounts.[332] On the basis of these chemical analyses we conclude that the intoxication seen in ATA is primarily a result of T-2 toxin poisoning.

We also undertook a comparative study of the amount of T-2 toxin produced by *F. poae*, *F. sporotrichioides*, and *F. sporotrichioides* var. *tricinctum* obtained from overwintered cereal grains involved in ATA disease in man with the yield of this toxin from different sources.[129] These isolates were obtained from fescue hay, corn, and wheat in various areas of the U.S. which were associated with outbreaks of severe toxicity in farm animals. These isolates have been referred to by American and Japanese investigators mostly as *F. tricinctum*.[128] Closer study has shown them to belong to various species of the Sporotrichiella section, especially *F. poae* and *F. sporotrichioides*.[119,121,123,128]

There were, in general, well-defined differences in the yield of T-2 toxin between the strains isolated from overwintered grain sources in the U.S.S.R. and strains isolated in the U.S. The results are summarized in Table 24. Thus the recent studies carried out in the U.S.[180,306] Japan,[319] and also Israel[129,332,333] could not confirm conclusions reached in the Russian studies concerning the role of steroidal lactones in causing ATA disease.

The following interesting works were carried out in the U.S. and Japan with *F. tricinctum:* Goldfredsen et al.,[87] Gilgan et al.,[85] Bamburg et al.,[12,15,16] Bamburg and Strong,[17,19] Marasas et al.,[173-175] Smalley et al.,[289-291] Burmeister,[38] Burmeister and Hesseltine,[39] Burmeister et al.,[40] Chu,[52] Mirocha and Pathre,[180] Ellison and Kotsonis,[60,61] Grove et al.,[96] Yates,[337] Yates et al.,[335,336] Hsu et al.,[100] Scott and Somers,[277] Tookey et al.,[312] Ueno et al.,[318-320] Ishii et al.,[101] Sato et al.,[259] Yamazaki,[334] and Ciegler.[53]

These studies were carried out with U.S. strains and determined as *F. tricinctum* (Corda) Sacc. emend. by Snyder and Hansen. We have reason to assume, according to our mycological investigation,[128] that they should more properly be called *F. poae* (Peck.) Wr. and *F. sporotrichioides* Sherb.

CLINICAL CHARACTERISTICS OF ATA DISEASE

The clinical findings were described by Chilikin,[47-49] Lyass,[167] Manburg,[170] Manburg and Rachalskij,[171] Myasnikov,[198] Nesterov,[202] Romanova,[238] and Yefremov.[338]

The clinical features of ATA are usually divided into four stages. If the disease is diagnosed during the first stage and even at the transition from the second to third stages, early hospitalization may still enable the patient's life to be saved. If, however, the disease is only detected during the third stage, the patient's condition is usually desperate and in most cases death cannot be prevented. Only very few patients in the third stage survive.

First Stage

The characteristic symptoms of this stage appear a short time after ingestion of the toxic grain. They may sometimes appear after a single meal of overwintered toxic grains and disappear completely even if the patient continues to eat the grain. The characteristics of the first stage include primary changes in the oral cavity and GI tract. Shortly after eating food prepared from toxic grain, the patient feels a burning sensation in the mouth, tongue, throat, palate, esophagus, and stomach as the toxin acts on the mucous membranes. The tongue may feel swollen and stiff, and the mucosa of the oral cavity may be hyperemic. Inflammation of the gastric and intestinal mucosa results in vomiting, diarrhea, and abdominal pain. In most cases excessive salivation, headache, dizziness, weakness, fatigue, and tachycardia accompany this stage, and there may be fever and sweating. The leukocyte count may already decrease in this stage to levels of $2000/mm^3$ with relative lymphocytosis, and there may be an increased erythrocyte sedimentation rate.[73,341,342]

Table 24
THE TOXIC STRAINS USED, THEIR ORIGIN, AND T-2 TOXIN YIELD

Strain	Origin		Supplied by	Rabbit skin bioassay	TLC assay[a]	Yield of T-2 toxin (mg/10 g wheat grain)[b]
			Fusarium poae			
60/9	Millet	U.S.S.R.	A. Z. Joffe[c]	+ + + +	Very strong	20.0
396	Millet	U.S.S.R.	A. Z. Joffe	+ + + +	Very strong	21.0
792	Barley	U.S.S.R.	A. Z. Joffe	+ + +	Strong	6.2
958	Wheat	U.S.S.R.	A. Z. Joffe	+ + +	Strong	8.0
NRRL 3287	Unknown	U.S.	C. W. Hessel-tine[d]	+	Slight	1.0
NRRL 3299	Corn	U.S.	C. W. Hessel-tine	+ + +	Strong	5.8
T-2	Corn	France	W. F. O. Marasas[f]	+ +	Medium	3.4
			F. sporotrichioides			
60/10	Millet	U.S.S.R.	A. Z. Joffe	+ + +	Strong	10.3
347	Millet	U.S.S.R.	A. Z. Joffe	+ + + +	Very strong	23.2
351	Millet	U.S.S.R.	A. Z. Joffe	+ + + +	Very strong	15.2
738	Millet	U.S.S.R.	A. Z. Joffe	+ + +	Strong	7.8
921	Rye	U.S.S.R.	A. Z. Joffe	+ + + +	Very strong	24.0
1182	Wheat	U.S.S.R.	A. Z. Joffe	+ + +	Strong	10.2
1823	Barley	U.S.S.R.	A. Z. Joffe	+ + +	Strong	4.6

Table 24 (continued)
THE TOXIC STRAINS USED, THEIR ORIGIN, AND T-2 TOXIN YIELD

Strain	Origin		Supplied by	Rabbit skin bioassay	TLC assay[a]	Yield of T-2 toxin (mg/10 g wheat grain)[b]
NRRL 3249	Tall fescue	U.S.	C. W. Hesseltine	+ +	Medium	4.0
NRRL 5908	Tall fescue	U.S.	H. R. Burmeister[d]	+	Slight	0.6
2061-C	Corn cobs	U.S.	C. J. Mirocha[e]	+	Slight	1.5
YN-13	Corn	U.S.	C. J. Mirocha	+	Slight	2.0
		F. sporotrichioides var. *tricinctum*				
NRRL 3509	Unknown	U.S.	C. W. Hesseltine	+	Slight	0.5

[a] The intensity of gray-brown spot at RF of T-2 toxin.
[b] As determined by GLC.
[c] Department of Botany, The Hewbrew University of Jerusalem, Israel.
[d] Northern Regional Research Lab., USDA, Peoria, Ill., 61604.
[e] Department of Plant Pathology, University of Minnesota, St. Paul, Minn., 55101.
[f] Plant Protection Research Institute, Pretoria, South Africa.

Adapted from Joffe, A. Z., in *Handbook of Mycotoxins and Mycotoxicoses,* Vol. 2, Marcel Dekker, New York, 1978.

The danger exists that this stage may not always be detected because it appears and disappears relatively quickly; the patient may become accustomed to the toxin and a quiescent period may follow while the toxin accumulates and the patient enters the second stage. The first stage may last from 3 to 9 days.

Second Stage

This stage is often called the latent stage[47-49] because the patient feels well and is capable of normal activity. Sometimes it is also called the leukopenic stage[170,171,238] because its main features are disturbances in the hematopoietic system characterized by a progressive leukopenia, a granulopenia, and a relative lymphocytosis. In addition, there is anemia and a decrease in the platelet count. The decrease in the number of leukocytes lowers the resistance of the body to bacterial infection. In addition to changes in the hematopoietic system, there are also disturbances in the central and autonomic nervous systems. Weakness, headache, palpitations, and mild asthmatic

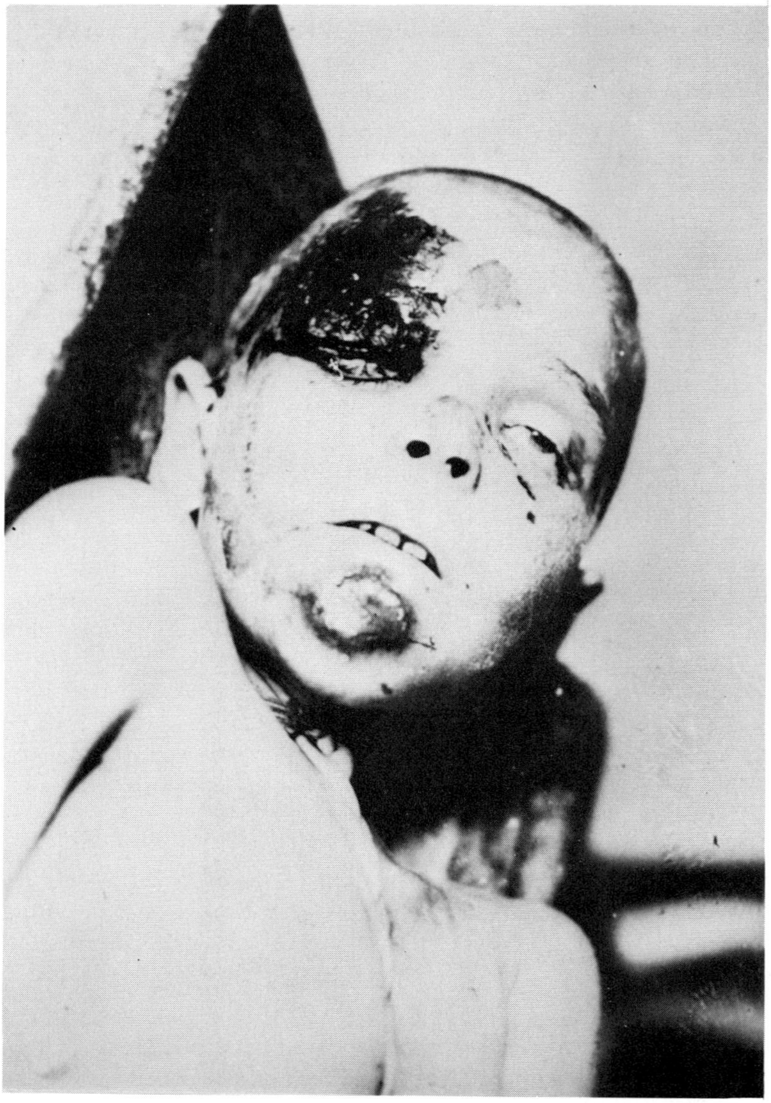

FIGURE 24. Necrotic lesions around the eye and face of a child who died from ATA disease after consuming overwintered prosomillet infected with *F. sporotrichioides.* (Photo courtesy of Marcel Dekker, Inc.)

symptoms may occur. The skin and mucous membranes may be icteric, the pupils dilated, the pulse soft and labile, and the blood pressure decreased. The body temperature does not exceed 38°C and the patient may even be afebrile. There may be diarrhea or constipation.

The normal duration of this stage is usually from 3 to 4 weeks, but it may extend over a period of 2 to 8 weeks. If consumption of toxic grain continues, the symptoms of the third stage rapidly develop.

Third Stage

The transition from the second to the third stage is sudden. At this stage the patient's resistance is already low, and violent symptoms may be present, especially under the influence of stress associated with physical exertion and fatigue. The first visible sign of this stage is the appearance of petechial hemorrhages on the skin of the trunk, in the axillary and inguinal areas, on the lateral surfaces of the arms and thighs, and in serious cases, also on the face and head. The petechial hemorrhages vary from a few millimeters to larger areas a few centimeters in diameter.[48,49]

As a result of increased capillary fragility, any light trauma may cause the hemorrhages to increase in size. Hemorrhages may also be found on the mucous membranes of the mouth and tongue and on the soft palate and tonsils. Nasal, gastric, and intestinal hemorrhages may occur.[49,341]

Necrotic changes soon appear in the throat, with difficulty and pain on swallowing. The necrotic lesions may extend to the vulva, gums, oral mucosa, larynx, and vocal cords and are usually contaminated with a variety of avirulent bacteria. The necrotic areas are an excellent loci for bacterial infection, which can result from lowered resistance of the body due to the damage to the hematopoietic and reticuloendothelial systems. Bacterial infection causes an unpleasant odor from the mouth due to the enzymatic activity of bacteria on proteins. Areas of necrosis may also appear on the lips and on the skin of the nose, jaws, and eyes (see Figure 24).

The regional lymph nodes are frequently enlarged. The submandibular and cervical lymph nodes may become so large and the adjoining connective tissue so edematous that the patient experiences difficulty in opening his mouth. Esophageal lesions may occur, and involvement of the epiglottis may cause laryngeal edema and aphonia (loss of voice). In such cases, death may occur by strangulation. Death of about 30% of the patients was directly related to stenosis of the glottis.[230]

The blood abnormalities observed initially in the first and second stage intensify during the third stage. The leukopenia increases to counts of 100 or even fewer leukocytes per cubic millimeter. The lymphocytes may constitute 90% of the white cells present; the number of thrombocytes decreases below 5000 cells/mm^3 and the erythrocytes below 1 million/mm^3.

The blood sedimentation rate is increased. The prothrombin time ranges between 20 and 56 sec, and the clotting time is usually not much prolonged. There may be a deficiency in fibrinogen in severe cases.[82,151,322] Some investigators found that patients suffer an acute parenchymatous hepatitis accompanied by jaundice. Bronchopneumonia, pulmonary hemorrhages, and lung abscesses are frequent complications.

Fourth Stage

This is the stage of convalescence, and its course and duration depend on the intensity of the toxicosis. Therefore the duration of the recovery period is variable. Only about 3 to 4 weeks of treatment, or sometimes longer, is needed for the disappearance of necrotic lesions and hemorrhagic diathesis and also the bacterial infections. Usually 2 months or more elapse before the blood-forming capacity of the bone marrow returns to normal: as a general rule, first the leukocytes, then the granulocytes, the platelets, and subsequently the erythrocytes.[72,82,92,133,285,309]

Pathologic Findings in Man

The toxic metabolites derived from authentic strains of *F. poae* and *F. sporotrichioides* isolated from overwintered cereals, when ingested by people, caused fatal outbreaks of ATA in the U.S.S.R. Therefore, the determination of the pathological findings of organs and tissues by this disease is very useful and important.

The local action was manifested by clinical burning sensations in the mouth, soft palate, and tongue. These phenomena usually passed when the patient put an end to eating products made of toxic grains. If he continued to consume the fatal grains, then the first signs of toxicosis appeared in the pharynx, esophagus,[81] and stomach,[170] and later various hemorrhagic syndromes appeared, characterized by hemorrhagic rash on

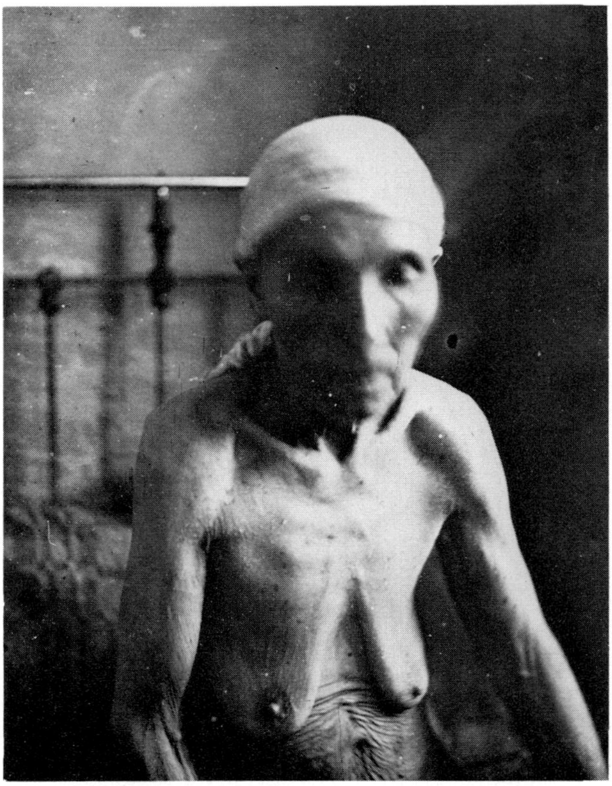

FIGURE 25. Petechial spots, first small and red, later blue or dark, caused by intradermal or submucous hemorrhage on chest and left arm. (Photo courtesy of Marcel Dekker, Inc.)

the skin of the chest (Figure 25), trunk, abdomen, legs, and arms, and even on the face. The hemorrhagic petechiae appeared most abundantly at the start of the development of necrotic angina.[49,150,171] Hemorrhagic and necrotic lesions developed also in the stomach and intestines, including all of the digestive tract, causing changes in intestinal function.[49,55,170,171] Hemorrhage and necrosis of the appendix and cecum and inflammation of the rectum also developed.[49] Hemorrhages were also present in different visceral organs, in the adrenal and the thyroid glands, gonads, uterus, and chiefly in the pleura. Pulmonary changes of bronchopneumonia and severe hemorrhages in the lung tissue with pneumonic abscesses frequently developed.[49] Marked changes were observed in the heart accompanied by vascular insufficiency and arterial hypotonia. Pressure was very low, and sometimes thrombophlebitis and endocarditis were indicated.[202] These changes developed as a result of the serious inflammation of the blood vessels and the condition of the endocrine system. In the third stage of the disease hemorrhages in the liver were revealed with acute parenchymatous hepatitis accompanied by jaundice,[49,202,238,340,341] and various changes in the glucose, protein, mineral, and other metabolisms of the liver were observed.[49,170] Hemorrhages and necrosis in the kidneys appeared as well (Figure 26). In the necrotic angina stage, severe changes were observed in the lymph nodes, which became edematous and were characterized by disappearance of the lymphoid formation. The entire lymphatic and reticuloendothelial systems were affected and showed proliferation of the red blood cells and of the endothelial capillaries and sinuses. Various investigators have observed severe changes in the central and autonomic nervous systems by ATA,[98,142,230] such as im-

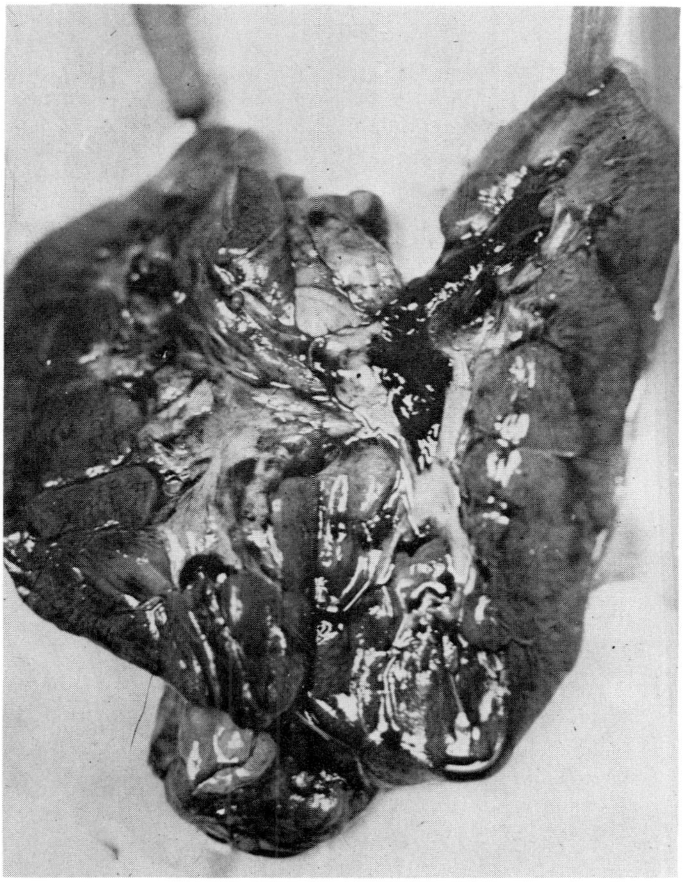

FIGURE 26. Kidney of fatal case of ATA, showing marked hemorrhage in pelvic mucosa. (Photo courtesy of Marcel Dekker, Inc.)

paired nervous reflexes, meningism, general depression and hyperesthesia, encephalitis, cerebral hemorrhages, and destructive lesions in nervous and sympathetic ganglia.[98,137,282,283]

According to Tomina,[311] toxins from overwintered cereals which were absorbed in the stomach and intestines had a cumulative effect in various organs and tissues. The most severe effects were on the hematopoietic system which correlated with the development of the different stages of the disease and resulted in depression of leukopoiesis, erythropoiesis, and thrombopoiesis.[133,143,285,309] Progressive leukopenia appeared (the leukocyte count dropped to 100/mm^3 or even lower) as well as lymphocytosis (to 90%) and a decrease in erythrocytes (which dropped to 1 million/mm^3). The hemoglobin content dropped to 8% and the granulocytes completely disappeared; at the same time, the sedimentation rate increased and showed a deficiency of prothrombin as a result of the irritation of the bone marrow.[47,48,133] Frequently, in severe cases, destructive and hemorrhagic lesions appeared in the blood circulation, which caused thrombosis in the blood vessels of different organs.

The destruction, devastation, and sometimes atrophy of the bone marrow[49,55,339-341] showed serious changes in the organism after consumption over a long period of time of overwintered grain contaminated by the toxic strains of *F. poae* and *F. sporotrichioides*. Hemorrhagic diathesis, necrotic angina, sepsis, and severe hematological changes developed as well.[47,55,95,170,171,341]

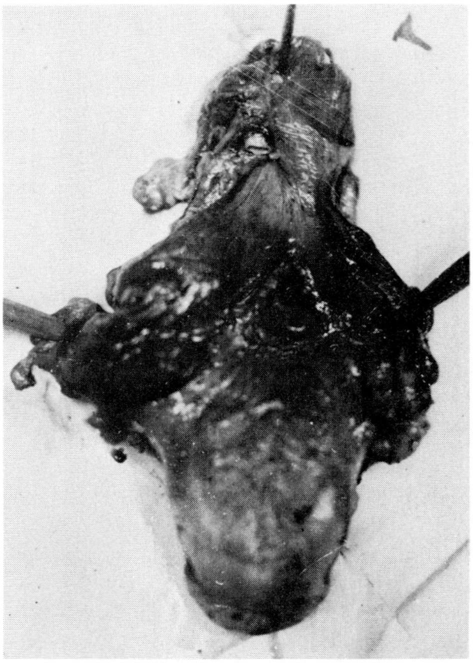

FIGURE 27. Necrosis of pharyngeal mucosa caused by *F. poae* which was isolated from prosomillet wintered under snow cover. The toxic prosomillet grain was eaten by this patient. (Photo courtesy of Marcel Dekker, Inc.)

An important contribution to the pathogenesis of ATA was that of Strukov[302] and Strukov and Tishchenko,[301,304] who showed that disturbances of the hematopoietic system were reversible and did not lead to the destruction of bone marrow. Strukov,[302] Aleshin and Eyngorn,[2] and Aleshin et al.[3] thought that toxins of overwintered cereals did not act primarily on the bone marrow but on an extramedullary apparatus which regulated the hematopoietic, autonomic nervous, and endocrine systems.

The necrotic changes in the final stage of ATA developed initially in the pharyngeal tonsils, and later in so-called necrotic angina, in the throat, and even in the esophagus; severe gangrenous pharyngitis occurred as well (Figure 27). The necrotic angina in the throat brought about an increase of fever. In severe cases signs of glottis edema with asphyxia were observed.[220,242] The necrotic appearance caused weakness, apathy, and damage of the leukocytic, phagocytic, and also reticuloendothelial functions of the organism. Necrotic lesions were present along the entire GI tract and also in other organs.

The strong and profuse menstrual and nose bleeding (Figure 28) were fatal in many cases.

Recovery depended on the amount of consumed toxic grain, on the type of therapy, and chiefly on the presence or absence of additional complications.

Sometimes the disease recurred with hemorrhagic and necrotic syndromes after consumption of infected grain was resumed, or after physical strain.[48,49,202]

PROPHYLAXIS AND TREATMENT

The most important prophylactic measure is to refrain from eating overwintered

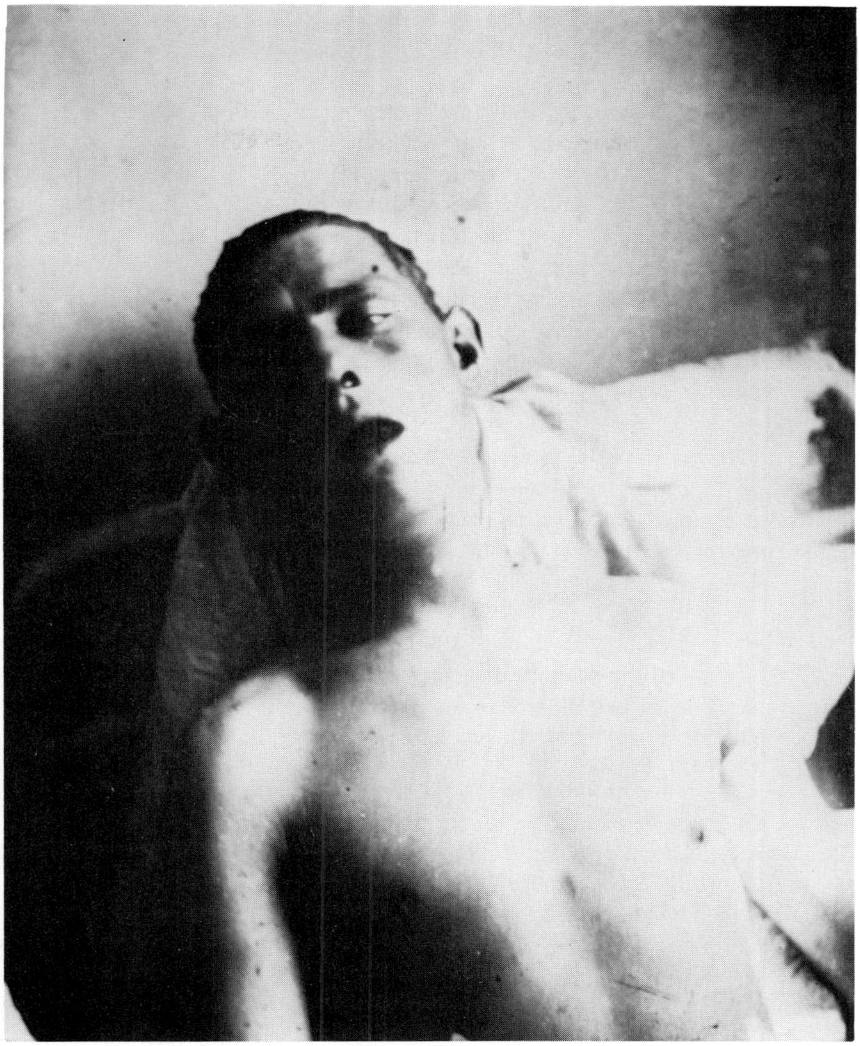

FIGURE 28. General views of a patient with a severe form of ATA: nosebleed (epistaxis), respiratory distress, hemorrhage on left arm. (Photo courtesy of Marcel Dekker, Inc.)

toxic grain. The primary preventive measure, therefore, consists in educating the rural population as to the etiology and clinical symptoms of ATA.[33,309] Such measures have greatly reduced outbreaks of ATA. When the outbreaks of the disease were first reported, medical teams were sent to the affected areas and the population was examined clinically and hematologically.[226,286]

Grain samples should be examined for toxicity by skin test. At the same time, toxic overwintered cereal grain collected by the population should be replaced by wholesome grain. When ATA was detected in the second stage, the treatment recommended at the time included blood transfusion[150,157] and administration of nucleic acid and calcium preparations, antibiotics,[158] Bogomolt's antireticular cytotoxic serum,[136,179] sulfonamides,[135,136] and vitamins C an K.[72,157] When the number of leukocytes declined below 3000/mm^3, hospitalization was recommended.[132,133]

The measures employed in the second stage were also used in the third stage, but treatment is more intensive. Following recovery, a rich diet was given for 1 month and the patient remained under periodic hematological checkup.[157,169]

EFFECTS OF TOXINS ISOLATED FROM *F. POAE* AND *F. SPOROTRICHIOIDES* ON ANIMALS

The toxins of *Fusarium* cultures have both a localized and general toxic effect. The localized effect is first apparent in an inflammatory reaction and is accompanied by subsequent skin necrosis at the site of toxin application. The general effect is apparent in defective hemopoiesis and acute degeneration in the internal organs, as well as in extreme hyperemia especially of the digestive tract.

The symptoms in animals vary according to the potency and quantity of the toxin, the route of administration, and the sensitivity of the animal. General effects of the toxins have been studied in a variety of animals and results of experiments carried out by us and others will be briefly summarized.

Protozoa

The effect of dilute (1:1000) alcohol extracts of millet grain experimentally infected with *F. poae* and *F. sporotrichioides* was studied on certain protozoa and a toxic effect demonstrated in *Paramecium caudatum, Stylonychia mytilus, Opalina ranarum* and *Nyctotheras cordiformis*.[58,59]

Frogs

Frogs were tested in order to establish their susceptibility to *F. poae* #60/9. Following repeated feeding with powdered dry fungi the animals died at 4 to 14 days depending on the dose. The effect of the toxin was cumulative. Post-mortem dissection showed extreme hyperemia, hemorrhages, and edema of the digestive tract and in other organs and tissues.[118]

Fish

Very little work has been done on the effects of fusariotoxins on fish. The possibility that these toxins might contaminate fish food pellets manufactured from maize led Marasas et al.[173] to investigate the effect of a toxic metabolite of *F. tricinctum* related to T-2 toxin. Rainbow trout were fed the contaminated pellets for 12 days. The LD_{50} of the toxin was estimated to be 6.1 mg/kg. Fish offered pellets containing 200 mg/kg toxin or more refused the feed after the first feeding; fish given pellets containing 4 mg/kg ate the food. All fish consuming the toxin exhibited identical pathological symptoms which included shedding of intestinal mucosa and severe edema with accumulation of fluid in body cavities and behind the eyes causing ventral swelling and bulging eyes.

Mice and Rates

The toxic properties of various *F. poae* and *F. sporotrichioides* were tested by the author on 288 experimental and 192 control white mice. These *Fusarium* species were separately incubated on solid and liquid Czapek media and millet grain at low temperatures with alternate freezing and thawing. The mice died 2 to 7 days after having been fed p.o. varying doses of the 12 *Fusarium* strains. The data are presented in Table 25.

Experiments with extracts, liquid medium, and fungal mass of *F. poae* #60/9 and *F. sporotrichioides* #60/10 were carried out on mice and repeated with *F. poae* #958 and *F. sporotrichioides* #921 (see Tables 20 and 21). Subcutaneous injections of fungus filtrates prepared during their abundant sporulation produced a lethal effect after 13 to 24 hr.[112,113,118]

In recent work, Schoental and Joffe[260] and Schoental et al.[262-264,274-276] have studied the long-term effects of extracts and pure T-2 toxin isolated from *F. poae* #958 and *F. sporotrichioides* #921 on mice and rats.

Table 25
LETHAL EFFECT OF CRUDE TOXIN FROM *F. POAE* AND *SPOROTRICHIOIDES*

		Crude toxin							
			Millet grain		Millet extract		Filtrate of Czapek medium		
Species	No. of strain	Czapek agar (20 mg)	50 mg	150 mg	5 mg	8 mg	0.2 mℓ	0.3 mℓ	Dry mycelium (10 mg)
F. poae	60/9	3/2[a]	3/2	3/3	3/2	3/3	3/1	3/2	3/0
	24	3/2	3/3	3/3	3/1	3/3	3/2	3/3	3/0
	396	3/3	3/3	3/3	3/3	3/3	3/2	3/3	3/1
	792	3/3	3/3	3/3	3/3	3/3	3/2	3/3	3/1
	958	3/3	3/3	3/3	3/3	3/3	3/2	3/3	3/1
F. sporotrichioides	60/10	3/2	3/1	3/3	3/1	3/2	3/1	3/2	3/1
	347	3/3	3/3	3/3	3/2	3/3	3/2	3/3	3/1
	351	3/2	3/1	3/3	3/1	3/2	3/1	3/2	3/0
	738	3/3	3/3	3/3	3/2	3/3	3/2	3/3	3/1
	921	3/3	3/3	3/3	3/3	3/3	3/3	3/3	3/1
	1182	3/2	3/1	3/3	3/1	3/2	3/1	3/1	3/0
	1823	3/3	3/3	3/3	3/2	3/3	3/2	3/3	3/1

[a] Number of mice used/died.

Crude concentrated alcoholic culture extracts were applied to rodents by various routes. Extracts from both fungi appeared to produce similar effects in mice and in rats. Single doses, of the order of 0.1 mℓ, were toxic to young animals, which died within a few days. Smaller doses allowed the animals to survive longer so that dosage could be repeated at varying time intervals. High doses, which are lethal in 1 day (or more), cause depletion of the lymphoid tissues, as a consequence of which infections develop in various organs, including heart and kidneys.

When extracts were applied to the skin, the treated site became congested and edematous within 24 to 48 hr depending on the efficacy of the solution. Keratinization followed, the scab falling off within 2 to 3 weeks. When the applications were repeated weekly or less often, some of the hair follicles became atrophic, ulceration was slower to heal, and the treated areas remained depilated for an increasingly longer time, possibly permanently.

Microscopically, the treated skin exhibited disorganized architecture and areas of desquamation and regeneration occurred next to each other; changes were also present in the muscle, which sometimes became edematous, containing foci of infection and infiltration by inflammatory cells (Figures 29 and 30). When treatment was interrupted, the necrotic lesions tended to disappear, but the hyperplastic changes appeared to persist and may have become progressive. Local application of nonlethal doses of the *F. poae* and *F. sporotrichioides* extracts to the skin or by intragastric intubation to the esophagus and stomach caused irritation, edema and ulceration, and desquamation and healing, accompanied by basal cell hyperplasia of the squamous epithelium.

Figure 31 shows a small papilloma in the skin of a mouse, which developed in the course of 10 weeks after a few applications of *F. sporotrichioides*. Figure 32 shows the hyperplasia and an invagination of the esophagus in a rat that received *F. poae* extract by intragastric tube. The regeneration of the basal cells suggests that the lesions may become neoplastic in due course.

Crude extracts of *F. poae* and *F. sporotrichioides* were used to test long-term effects in rodents when administered in various ways. The authors drew attention to the chronic lesions which developed in animals surviving for several months; these suggested that the *Fusarium* metabolites may have been carcinogenic and immunosup-

FIGURE 29. Skin of mouse dying 10 weeks after beginning of approximately weekly applications of *F. poae* extract: areas of desquamation of keratinized epithelium, hyperplasia of basal cells, and foci of infection, some surrounded by inflammatory cells. HE × 70. (Photo courtesy of Marcet Dekker, Inc.)

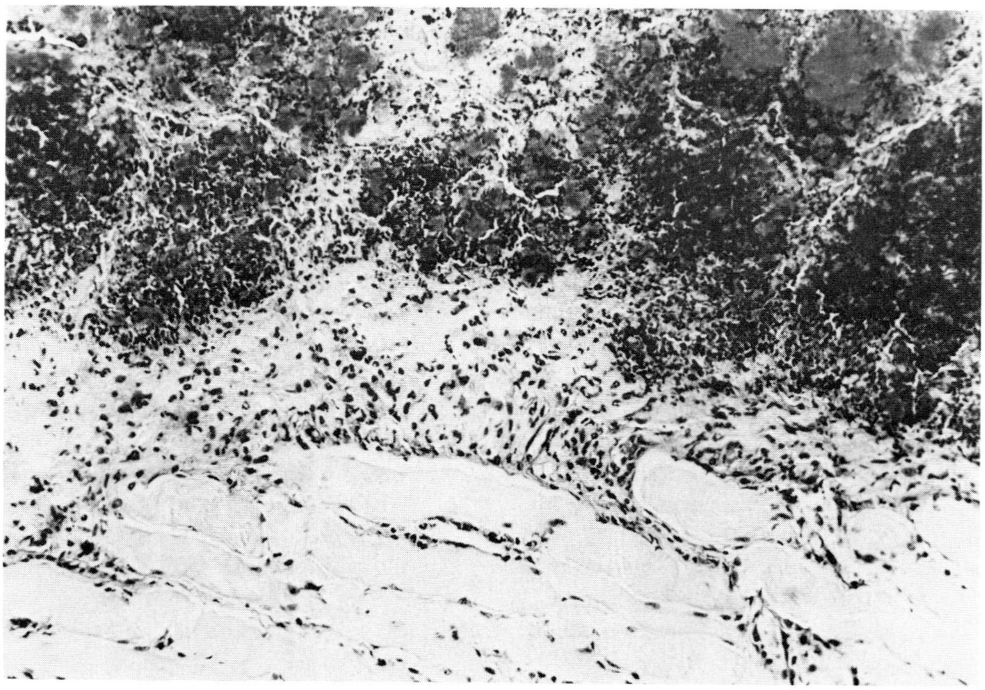

FIGURE 30. Subcutaneous foci of infection, some surrounded by inflammatory cells; mouse died 10 wks after beginning of approximately weekly skin application of *F. poae* extract. (H.E.; magnification × 120.) (From Schoental, R. and Joffe, A. Z., *J. Pathol.*, 112, 37, 1974. With permission.)

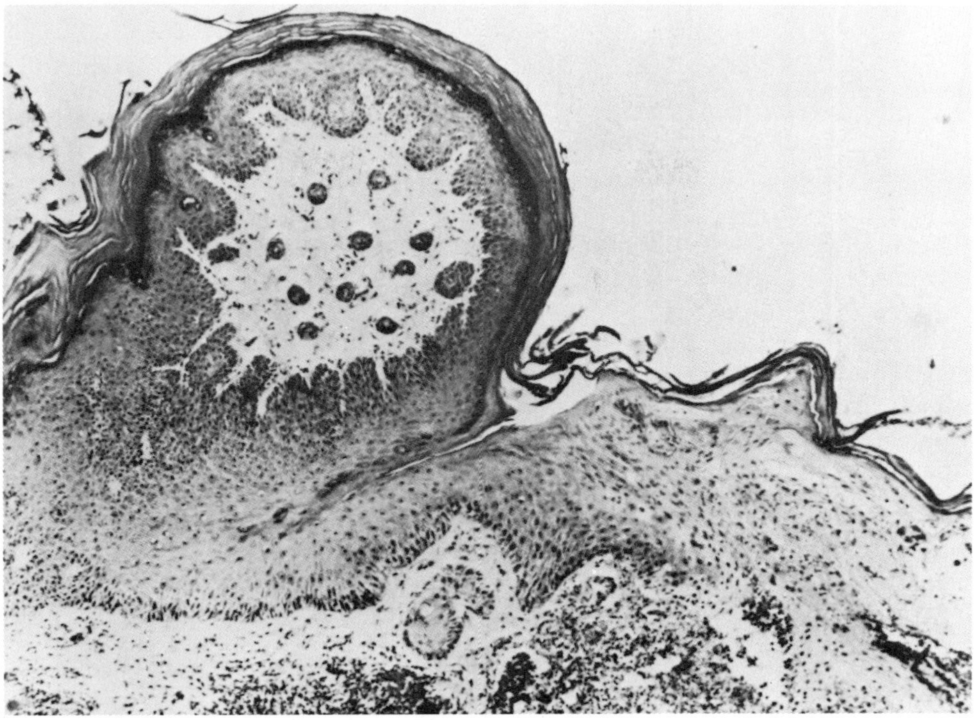

FIGURE 31. Small papilloma on skin of mouse after application of extract of *F. sporotrichioides* #921. (H.E.; magnification × 76.)

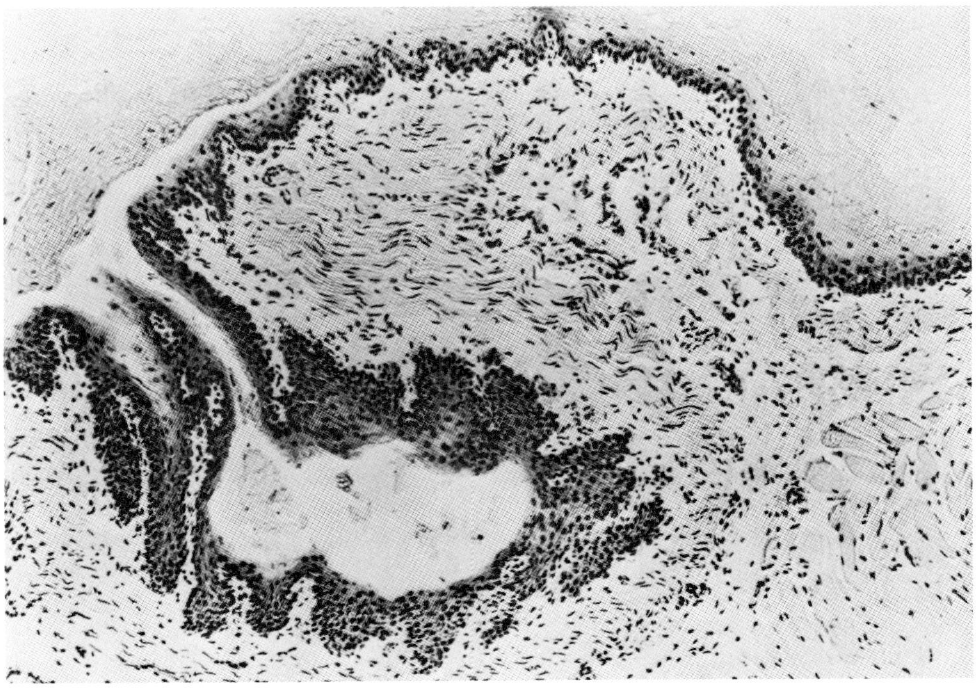

FIGURE 32. Invagination and basal cell hyperplasia of squamous epithelium of the esophagus in a rat to which extract of *F. poae* #958 was applied by stomach tube. (H.E.; magnification × 100.) (Photo courtesy of Marcel Dekker, Inc.)

pressive. The effects of the extracts on thymus, spleen, and lymph glands, and the development of infective foci in various tissues, including the kidneys, strongly suggest immunosuppressive action.[264]

Schoental et al.[274] also carried out long-term experiments on rats, using pure crystalline T-2 toxin isolated from crude extract of *F. sporotrichioides* #921. This mycotoxin is one of the most toxic and irritant among the trichothecene type metabolites of *Fusarium*. T-2 toxin caused severe cardiovascular lesions and various tumors, benign and malignant, in rats. Lesions and tumors were similar to those found among rats that survived when treated with crude ethanol extracts isolated from cultures of *F. poae* and *F. sporotrichioides*.[260,264]

After the first or one of the subsequent treatments with T-2 toxin, alone or in conjunction with nicotinamide, about two thirds of the experimental rats died within a few days. At autopsy, stomach and small intestine were greatly distended with soft, often blood-stained content, hemorrhagic petechia, and erosions in the stomach; lymph glands and spleen appeared enlarged, the thymus appeared small, lungs were congested, heart engorged, and blood vessels congested, particularly conspicuously in the brain.

Microscopically, the stomach mucosa were usually denuded; there was glandular atrophy, cellular infiltration, and striking submucosal edema in the stomach and duodenum. The pancreas showed interlobular edema; there was depletion of the lymphoid elements in the hemorrhagic spleen and lymph glands; thymus showed involution and foci necrosis; lungs were often edematous. Necrotic changes were present in the gonads. Details of the main degenerative or vascular and neoplastic lesions found in rats that survived 12 to 27.5 months after the first, and several weeks or months after the last of several doses of T-2 toxin are given in Table 26 and illustrated in Figures 33 to 45.

Guinea Pigs

In feeding tests, 12 guinea pigs were given p.o. liquid filtrates or dry mycelium of toxic *F. poae* #60/9 and *F. sporotrichioides* #60/10. The animals died within 5 to 13 days of receiving 1.0 to 2.0 $m\ell$ liquid culture or 50 to 70 mg of dry fungus, those receiving the toxic fungi in liquid form being more severely affected. Effects included hemorrhages in organs and tissues, and mainly hyperemia of adrenal gland and necrosis in the GI tract.[112] In guinea pigs eating overwintered millet grains contaminated by the above Fusaria the number of leukocytes declined to 500/mm^3.[4,5]

T-2 toxin isolated from *F. sporotrichioides* #921, administered orally to guinea pigs, caused severe hemorrhages and necrosis in organs, mainly in the duodenum (Figure 46), cecum (Figure 47), and spleen (Figure 48).

The acute and chronic effects on guinea pigs of T-2 toxin isolated from an authentic strain of *F. sporotrichioides* #921 associated with ATA disease [113,129,331,332] were investigated by De Nicola et al.[56] Single oral doses of 2.5 or 5.0 mg/kg T-2 toxin caused gastric and cecal hyperemia and hemorrhage. The animals exhibited water-fluid distension of the cecum, edematous intestinal lymphoidal tissue, and hyperemia of the adrenal gland. Histological examination revealed typical radiomimetic symptoms: necrosis of lymphoidal tissues, bone marrow, and testis, and necrosis and ulceration of the GI tract (particularly the stomach and cecum). The oral LD_{50} for guinea pigs was estimated at 3.06 mg/kg.

Subacute oral doses of 0.5 mg/kg for 21 days, followed by 0.75 mg/kg for 21 days, caused no clinical changes or gross or microscopic lesions. The only symptoms were a moderate leukopenia and an absolute lymphopenia. Slightly higher doses, 0.9 mg/kg for 27 days, also produced no clinical or histological effects, but the erythrocyte count fell and again there was leukopenia and an absolute lymphopenia. Morphological

Table 26
DEGENERATIVE OR VASCULAR AND NEOPLASTIC, IN RATS THAT SURVIVED 12 TO 27.5 MONTHS AFTER THE FIRST OF 3 TO 8 DOSES OF T-2 TOXIN GIVEN I.G. ALONE OR IN CONJUNCTION WITH I.P. INJECTIONS OF NICOTINAMIDE

				Organs affected										
							Stomach			Pancreas				
No.	Sex	Survival (months)	No. of doses	Heart	Arteries	Kidneys	Squamous	Glandular	Duodenum	Exocrine	Islet cell	Brain	Pituitary	Others
1	M	19.5 D	3 + NA	***	***	*	+		+			+		
2	M	23 K	3	*		**					+			
3	M	21.5 K	4	*								*		
4	M	23 K	4	*	**	**					+ +	** +		
5	M	23.5 D	4	**	***	***	+ +				+ +			
6	M	24 K	4		*	**	+			+		*		
7	M	16 K	5	**		**	+ +	+ C		+ +		+		
8	M	16 K	5	**	*	**	+	+ + C	+ C					
9	M	22 K	5		*	*	+		+ +	+ +			+	Bladder +
10	M	22 D	5	**		*C				+ +	+ +			Parathyroid +
11	M	22 K	5			*				+ +	+		+	
12	M	22 D	5	***	**			C						Parathyroid +
13	M	24 D	5	**	**			C						
14	M	24.5 K	5			**	+		+					
15	M	24.5 K	5	*	**	**		+ + +	+	+ +	+	+ + +		
	F	21 K	6			*			+ + +	+			+ + +	Mamma + + + Adrenal medullary + + +
17	F	27.5 D	6	*	*	**		+				*	+ +	
18	F	27.5 K	6	*	*		+				+	*	+ + +	Mamma + + +
19	M	20 K	8	***	**	**C	+ +	C		+ +	+ +	*		Bladder
20	M	21.5 K	8	***	***	**	+			+ +	+ +			
21	M	22 K	8							+ + +				
22	M	12.5 K	8 + NA			*C	+	+				+ + +		Bladder + Esophagus +
23	M	23.5 K	8 + NA	**	**					+ + +	+			
24	M	23.5 K	8 + NA							+ + +	+		+ +	
25	M	24.5 K	8 + NA		**				+	+ + +				

From Schoental, R., Joffe, A. Z., and Yagen, B., *Cancer Res.*, 39, 2279, 1979. With permission.

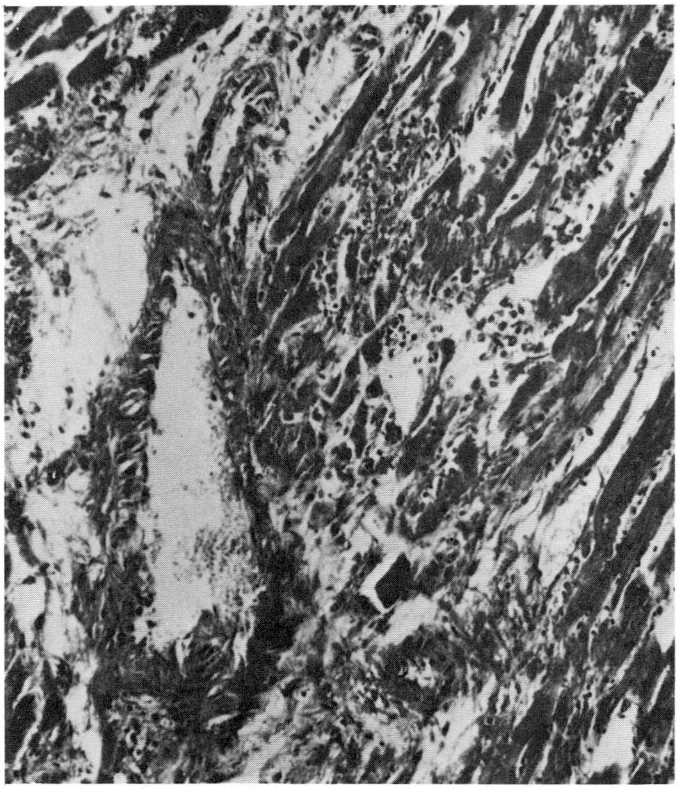

FIGURE 33. Section of heart showing degeneration of muscle fibers, cellular infiltration and fibrosis. Male rat died 24 months after first and 8 months after last of five doses of T-2 toxin. (Phosphotungstic-Mallory, magnification × 125.) (From Schoental, R., et al., *Cancer Res.*, 39, 2279, 1979. With permission.)

changes were noted in the erythrocytes, those in the treated animals exhibiting basophilic stippling, fragmentation, and increased polychromasia and anisocytes in addition to nucleate forms. These changes increased in intensity as the experiment progressed. In the bone marrow, the myeloid:erythroid ratio was greatly reduced as compared to the controls. The treated animals also showed a reduced number of lymphocytes within the bone marrow.

Rabbits

Most of the work done with rabbits has involved the toxic skin test used for detecting trichothecenes and certain other toxins (see section on Bioassay Methods). The author described an experiment on 16 rabbits which received 100 to 150 mg mycelium or 3 to 5 mℓ culture filtrate of *F. poae* #60/9 and *F. sporotrichioides* #60/10. Death occurred 8 to 12 days after p.o. feeding of the *Fusarium* cultures with decreased number of leukocytes. Autopsy showed dilated blood vessels and hemorrhages in the intestinal walls and many other organs and tissues.[112] After s.c. injection of rabbits with ethanol extract of *F. poae* #792, marked migration of leukocytes through the walls of small blood vessels was noted (Figure 49). *F. sporotrichioides* #738 also showed marked deep leukocyte infiltration of muscular layers after application of toxic extract to rabbit skin (Figure 50).

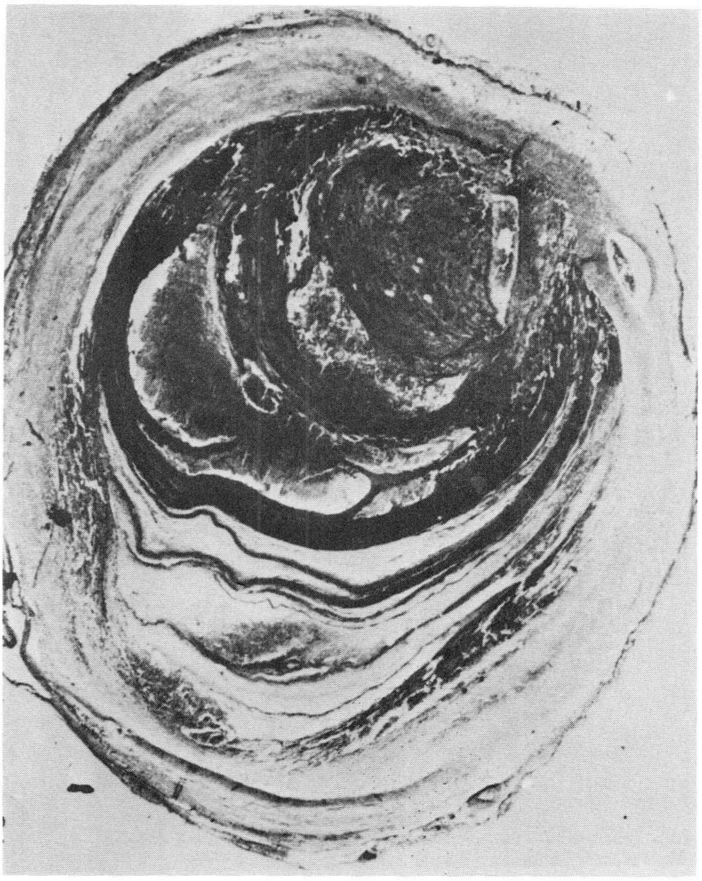

FIGURE 34. Cross-section of coronary artery greatly distended and partly occluded by fibrinoid swelling of collagen. Male rat killed 21.5 months after first and 10.5 months after last of seven doses of T-2 toxin. (Phosphotungstic-Mallory, magnification × 13.) (From Schoental, R., et al., *Cancer Res.,* 39, 2279, 1979. With permission.)

Bilai[25] fed small rabbits weighing 600 to 900 g with food infected with *F. poae* and *F. sporotrichioides* and found that the effect of the toxin depended on the weight of the animals and that rabbits were more sensitive than guinea pigs and mice. Post-mortem examinations of the rabbits revealed hyperemia of the subepithelial tissues and GI system as well as diffuse hemorrhages in the liver.

Maisuradge[168] found that when rabbits were fed germinating oats infected with *F. sporotrichioides,* they lost weight, the body temperature rose, the pulse became quicker, and the breathing heavy and slow. The leukocyte count rose to 37,000/mm^3 in the middle of the experimental period and later fell to 600 to 200/mm^3. A decrease in the red cell count was also evident.

Dogs

The effect of various quantities of agar culture of *F. poae* #396, isolated from overwintered millet grain, was studied on the motor activity of dogs fed orally and through a gastric fistula under varying conditions.[99] The fungus culture had been mixed with 200 g of meat, and gastric motility was registered by a kymograph. It was found that introduction of large doses of fungus (1 g and more) caused poisoning of the animal and cessation of stomach motor activity.

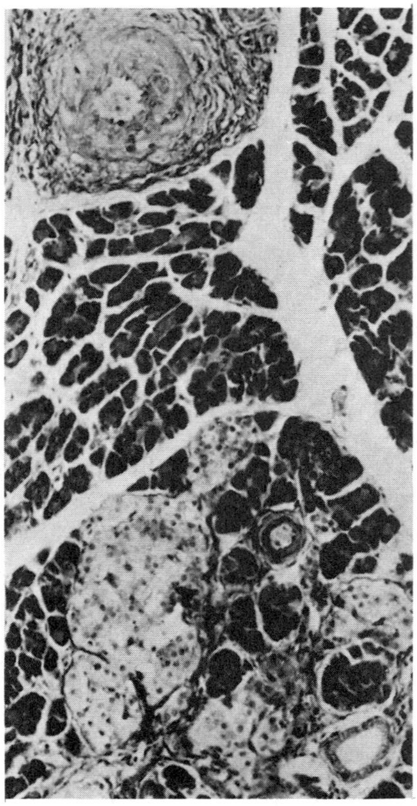

FIGURE 37. Pancreas showing hyperplasia and "budding" of small islets. A thick-walled artery is seen at top right of section. (Phsphotungstic-Mallory; magnification × 100.) (From Schoental, R., et al., *Cancer Res.*, 39, 2279, 1979. With permission.)

tensen et al.[50] carried out a study on turkey poults with *F. tricinctum*. The turkeys died after eating contaminated corn.

The 1-day- old New Hampshire chicks were obtained from a commercial hatchery and kept for further study. They were divided into 18 treatment trials and 18 corresponding control groups consisting of two replicates for each dose of treatment, and housed in electrically heated crates with feed containing all the required ingredients and water *ad libitum*. The control groups were also housed in crates under similar conditions and were given a mixture of water-ethyl alcohol-corn oil without fungi in the same amounts as the treated chicks, or uninfected autoclaved ground grains which were mixed with commercial ration in the same proportion as were the fungal cultures.

The experiments were performed on a total of 300 chicks. The treated chicks showed reduction of weight gains, clinical and gross pathological changes, and severe hematopoietic damage, which were also the principal features of ATA in man. The control groups administered the same diet without toxic material remained healthy and showed no clinical or pathological changes.

The effect of different doses of T-2 toxin isolated from the above-mentioned *Fusarium* strains on mortality and on blood components of chicks is presented in Tables 27 to 29. Chicks given the toxic ration grew slowly and the difference between treated and control groups is shown in Figures 56 and 57.

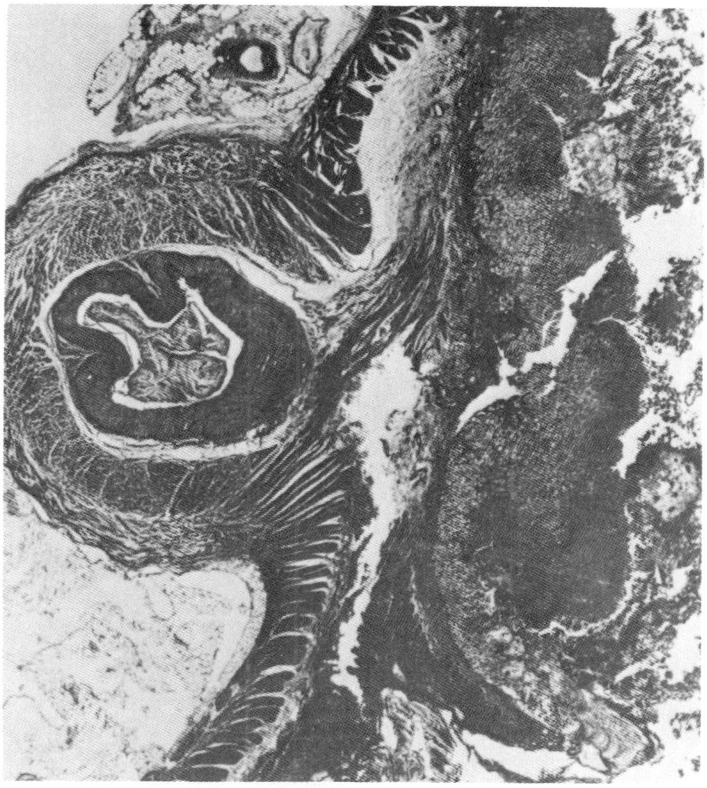

FIGURE 38. Cross-section through esophagus and part of squamous stomach showing squamous hyperplasia, edema, and extensive ulceration of stomach. Female rat died 10 months after first and 3 days after last of seven doses of T-2 toxin. (H.E.; magnification × 14.) (From Schoental, R., et al., *Cancer Res.*, 39, 2279, 1979. With permission.)

Cats

Of all laboratory animals the best results were obtained in cats, since the clinical and histopathological syndromes induced in cats by the *Fusarium* species and their toxins assumed to be responsible for ATA, closely resemble those occurring in man.

We studied the effect of *F. poae* #60/9 and *F. sporotrichioides* #60/10 in the form of agar cultures, millet cultures, dry mycelium mass, and culture liquid on 26 cats. The daily dose of agar and millet cultures was 50 to 120 mg and that of the liquid substrate 0.5 to 1.0 mℓ. In all the forms administered to cats, both *Fusarium* species were lethal after variable periods of time, depending on the daily dose of fungus and on individual properties of the respective organism. The death of cats followed a failure in blood production. A fall in hemoglobin, red cells, leukocytes, and neutrophils was observed, with a relative increase in lymphocytes. The lowest leukocyte count found was 100/mm^3. The symptoms included vomiting, a hemorrhagic diathesis, and neurological disturbances.

In the majority of cases the cats died on the 6 to 12th day. Autopsy revealed marked hyperemia of internal organs, especially of the digestive tract and kidneys, and extreme changes in the adrenal glands. Histological examination of organs of cats which died after infection by *F. poae* #60/9 and *F. sporotrichioides* #60/10 revealed changes in the blood-producing tissue which were similar to those seen in ATA. After i.p. injection of toxic extract from millet infected by *F. sporotrichioides* #60/10, an extensive

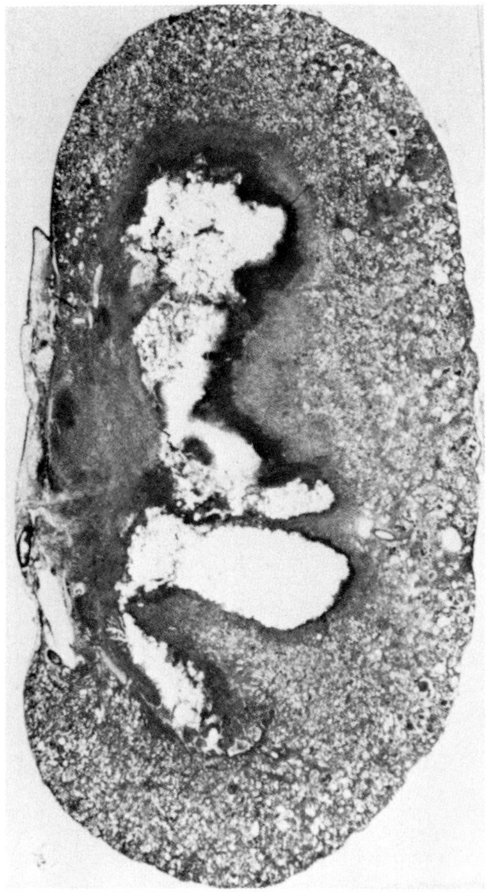

FIGURE 39. Section of kidney showing extensive degeneration and distension of pelvis with calcified walls, from which a calculus was removed. Male rat killed 12.5 months after first and 1 month after last of eight doses of T-2 toxin given in conjunction with nicotinamide treatment. (H.E.; magnification × 4.8.) (From Schoental, R., et al., *Cancer Res.*, 39, 2279, 1979. With permission.)

leukocyte migration was noted on the surface of the epiploon (ementum) of cats (Figure 58). Part of these studies were carried out in cooperation with Alisova.[7]

Sarkisov[255] fed seven cats with doses of 0.1 to 2 g of *F. sporotrichioides* infected millet. The lethal dose was 0.4 to 16.5 g, and death occurred within 2 to 34 days. The body temperature rose to 41°C and a progressive leukopenia appeared, the white cell count falling to 500 to 200 cells/mm^3.

Rubinstein and Lyass[249] fed millet also infected with *F. sporotrichioides* to cats and monkeys, which developed symptoms similar to those of ATA in man. Later, Sato et al.[259] reported that T-2 toxin caused emesis, vomiting, diarrhea, anorexia, ataxia of the hind legs, and discharge from the eyes. Both crude and pure toxin caused leukopenia after continuous administration. After each dose of T-2 toxin cats suffered temporary leukocytosis. The cats also had meningeal hemorrhage in the brain, bleeding in the lung, and vacuolic degeneration of the renal tubes.

After evaluating the specific pathogenicity of T-2 toxin on chicks,[130] we decided, in cooperation with Lutsky,[166] to carry out a study on cats using crude extract and T-2 toxin isolated from strains of *F. sporotrichioides* #921.

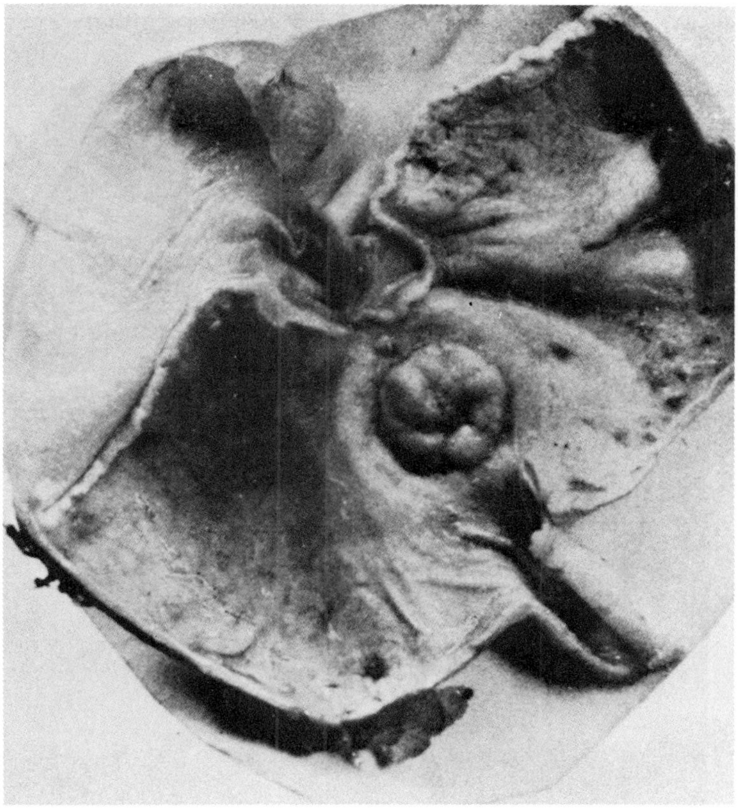

FIGURE 40. Stomach showing nodule on glandular part. Male rat killed 24.5 months after first and 8 months after last of five doses of T-2 toxin. (H.E.; magnification × 1.28.) (From Schoental R., et al., *Cancer Res.*, 39, 2279, 1979. With permission.)

Twenty-four mature male and female short-haired cats were used in the study. Before the experiments began all cats were dewormed and vaccinated against feline panleukopenia with a modified live virus vaccine. Cats were included in the study only if they appeared healthy and had normal blood values at the outset of the experiment.

The cats were crated individually and were divided into four groups according to the test material dosage. The toxin was given p.o. every 48 hr. The designated dose of T-2 toxin or crude extract was carefully measured and transferred by means of a 25-$\mu\ell$ Hamilton Microliter Syringe to a 250-$\mu\ell$ gelatin capsule. The cats were anesthetized following i.m. injection of ketamine hydrochloride. During brief general anesthesia the gelatine capsule containing the test dose was administered p.o. Concurrent with test dose administration, the cats were weighed and a blood sample for hematologic determinations was taken.

A necropsy was performed on all animals found dead in their cages, either immediately after death or after refrigeration for no more than 12 hr. Samples of organs and tissues showing gross pathological changes were fixed in 10% formalin. Histological sections were stained with hematoxylin and eosin and with other special stains as needed. Sternal bone marrow films prepared at the time of necropsy were stained with Wright's stain.

The 20 cats receiving either pure T-2 toxin or the crude extract containing T-2 toxin all showed clinical signs of toxicity, including bloody feces, vomiting, hind-leg ataxia, and conjunctivitis: loss of body weight and weakness of head and neck muscles were

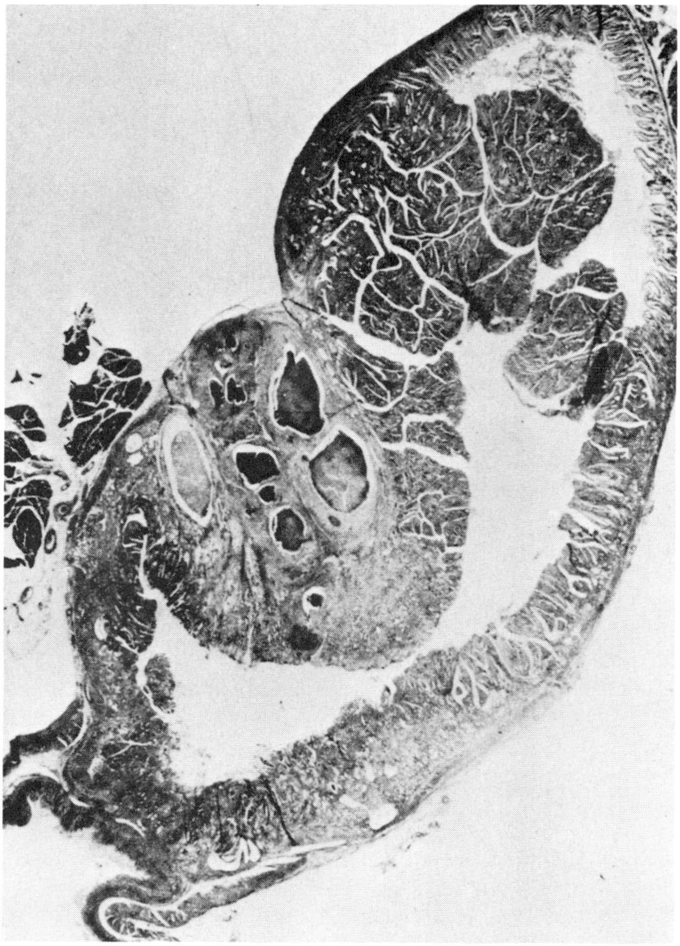

FIGURE 41. Longitudinal section of duodenum, distended by an ulcerated papillary adenocarcinoma penetrating the serosa. Female rat killed 21 months after first and 4 months after last of six doses of T-2 toxin. (H.E.; magnification × 6.) (From Schoental, R., et al., *Cancer Res.*, 39, 2279, 1979. With permission.)

frequently seen. The gross pathological findings included severe emaciation, intestinal hemorrhages (Figure 59), and enlarged lymph nodes. Histologic analyses showed necrosis of small intestinal epithelium, pulmonary edema, hemorrhagic lymph nodes, and swelling of the renal tubules of the kidneys.

Cats receiving the same extract from which T-2 toxin had been removed showed no gross or histologic changes.

Our results establish that the trichothecene T-2 toxin isolate from *F. sporotrichioides* #921 produces all the signs, symptoms, and pathological changes associated with the ATA disease in man.

Monkeys

The best models for reproducing the same clinical signs and syndromes as those occurring in ATA in man were cats and monkeys. For this reason we cooperated with Kriek (Division of Nutrition Pathology) of the National Research Institute for Nutritional Diseases of the South African Medical Research Council, in carrying out comprehensive experiments on 19 vervet monkeys of various ages, using pure T-2 toxin

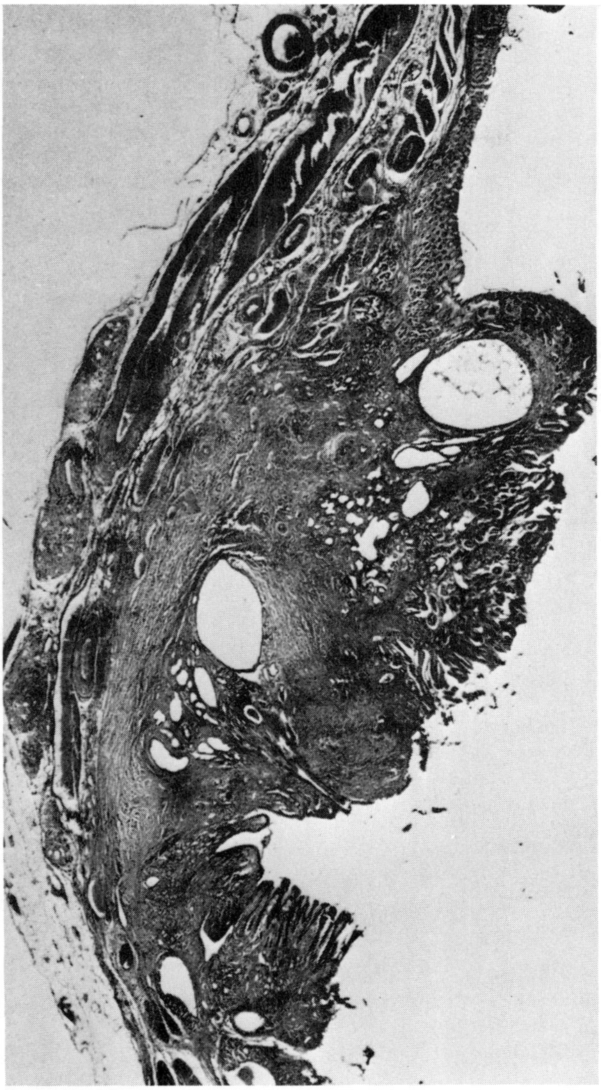

FIGURE 42. Section through nodule of stomach in Figure 40, showing adenocarcinoma penetrating through muscular layer. (H.E.; magnification × 15.) (From Schoental, R., et al., *Cancer Res.*, 39, 2279, 1979. With permission.)

isolated by Joffe and Yagen,[129] Yagen and Joffe,[331] and Yagen et al.[332] from *F. sporotrichioides* #921.

The experiments covered a wide range of toxic doses, from 10 mg/kg to 100 mg/kg administered 3 times per week in DMSO by gastric intubation, and induced syndromes ranging from peracute, with death occurring within 18 hr, to a long and very extended course with death occurring after 1 year as a consequence of an aplastic anemia. There were five controls which cover the entire period of all the experiments. Clinically, particularly at the lower dosages, a hemorrhage diathesis was induced.

The most important finding of this study is the demonstration of chronic toxicity at very low dosage rates over an extended period of time. The effect of T-2 on these primates is shown in Figures 60 to 63.

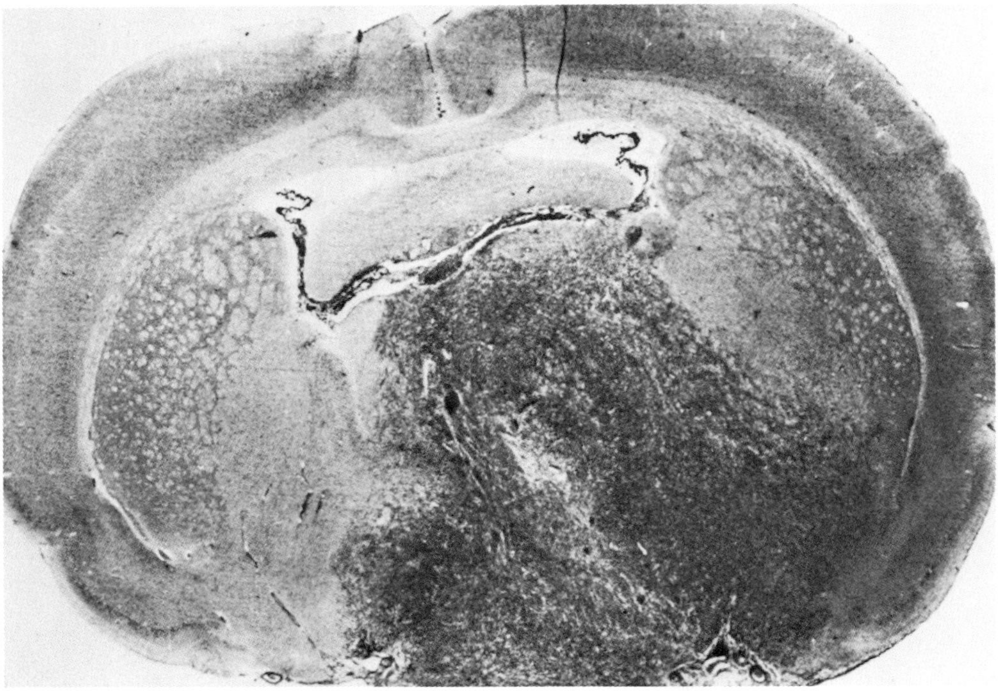

FIGURE 43. Section of brain with large tumor in frontal lobe. Male rat killed 12.5 months after first and 1 month after last of eight doses of T-2 toxin given in conjunction with nicotinamide treatment. (H.E.; magnification × 5.) (From Schoental, R., et al., *Cancer Res.*, 39, 2279, 1979. With permission.)

Horses

Detailed studies were carried out on two horses by the author in cooperation with Antonov et al.,[9] using *F. poae* #60/9. Two series of sucrose-glucose-peptone agar medium (carbohydrate-peptone medium) were used for inoculation and incubation of *F. poae* #60/9 isolated from overwintered millet grain associated with ATA in man. The first series was grown for 10 days at room temperature and then for 20 days at +1°C. The second series was cultivated at low temperatures, 2 days at room temperature and 45 days of multiple freezing and subsequent thawing. The second series proved to be more toxic than the first.

The horse in the first series received 240 g dry culture during 23 days and in the second series another horse was given only a single dose of 40 g of fungus culture by nasopharyngeal incubation. The latter produced an acute toxicosis, the horse dying 36 hr after administration of the *Fusarium* culture.

The degree of toxicity depends on the environmental conditions of incubation and growth of the *Fusarium* culture, on dosage size, and also on the sensibility of the organism. *F. poae* #60/9 of the second series had a harmful effect, causing rapid death.

Clinical symptoms showed loss of appetite, severe depression, and disturbance in movement coordination. Pathological findings revealed hemorrhagic diathesis, hemorrhages in the GI tract, liver, and kidney, hyperemia and necrosis of mucosa of small and large intestines, and a process of severe degeneration of the parenchymatous organs. Blood examination showed a slight increase in the quantity of leukocytes and lymphocytes and a drastic increase in the amount of erythrocytes and the level of hemoglobin (almost twice). The horse in the first series showed reduction in white and red cell counts.

Sarkisov[255] fed a horse 16.4 kg cereals infected by *F. sporotrichioides,* causing the development of stomatitis and gingivitis only.

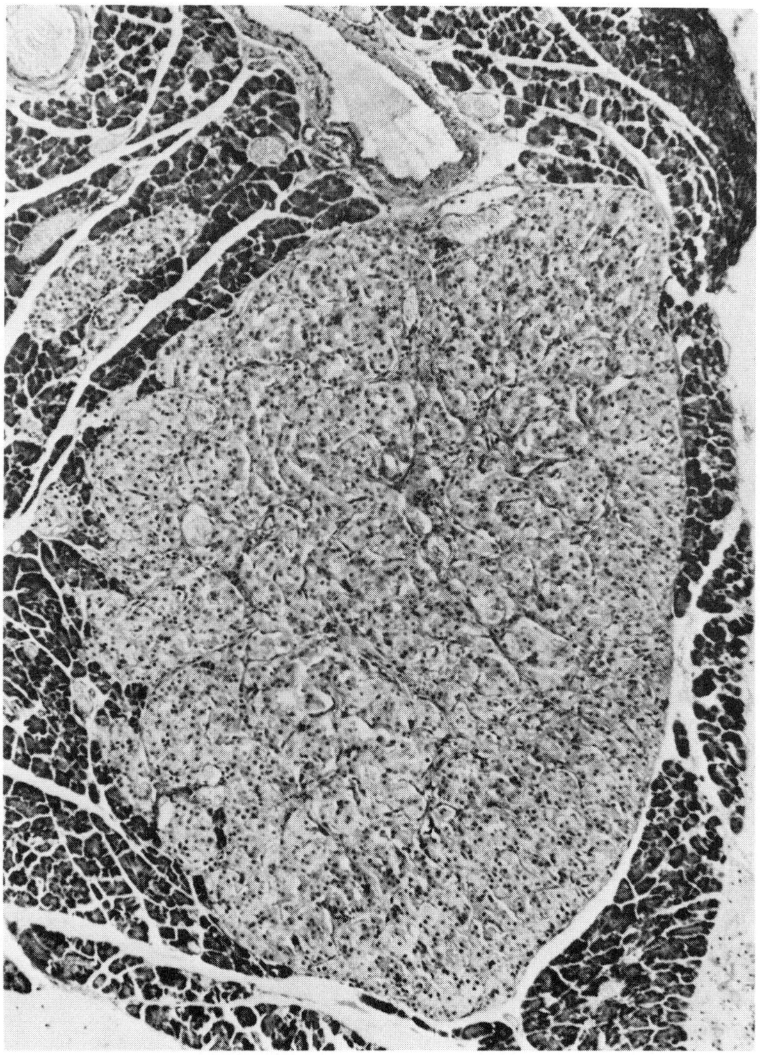

FIGURE 44. Pancreatic islet cell adenoma. Male rat killed 23.5 months after first and 13 months after last of eight doses of T-2 toxin given in conjunction with nicotinamide treatment. (H.E.; magnification × 70.) (From Schoental, R., et al., *Cancer Res.*, 39, 2279, 1979. With permission.)

Horses which had fed on herbage or whole oat grains which were toxic showed the following symptoms, according to Maisuradge:[168] 2 days after ingestion of toxic food, hyperemia was already evident in the mouth, with swelling and splitting of the lips. After 7 days, a foul-smelling slough appeared in the oral mucosa. Deformation of the horses' heads was observed and the animals lost weight because they refused to eat the toxic food after a while. Although one horse ate 23 kg of toxic food and another 18 kg, they recovered slowly and after a very long period returned to normal health.

Lukin and Berlin[163] and Lukin et al.[164,165] carried out feeding tests on horses, pigs, and sheep, and found that animals fed on large quantities of overwintered toxic cereals, even for as long as 39 days, did not develop symptoms of ATA. They concluded that climatic conditions affect toxin production differently in different geographic regions.

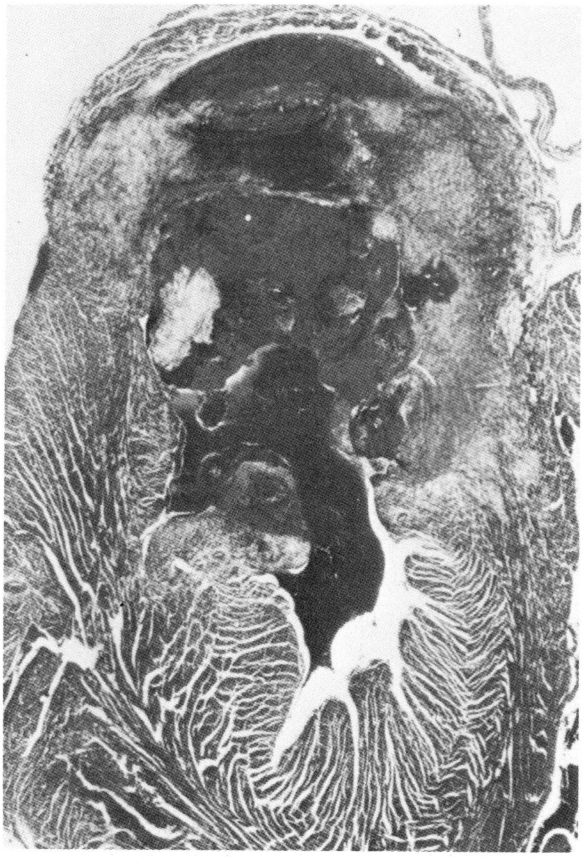

FIGURE 45. Section of heart showing areas of infection, inflammation, and cellular infiltration surrounded by fibrosis and degeneration of heart muscle. Male rat died 26 months after first and 6 months after last of 22 doses of crude alcoholic extract of cultures from *F. sporotrichioides* #921. (H.E.; magnification × 10.) (From Schoental, R., et al., *Cancer Res.*, 39, 2279, 1979. With permission.)

Pathological Changes in Animal Organs and Tissues by Toxins of *F. poae* and *F. sporotrichioides*

Some experiments have been carried out to study the effect of overwintered toxic grains and normal grains infected with *F. poae* and *F. sporotrichioides* strains on different laboratory animals by oral feeding, parenteral tests, and also by skin tests on rabbits. Skin reactions, caused by the action of toxic overwintered grains and by toxic *Fusarium* extracts, were distinguished both by their external appearance (application on skin) and histological changes (Figures 64,65). The toxins of *F. poae* and *F. sporotrichioides* had thus both a localized and generalized toxic effect. The localized effect was first apparent as an inflammatory reaction and was accompanied mainly by subsequent skin hemorrhage or necrosis at the site of the toxin application. The general effect was apparent in defective blood production, acute degenerative processes in the internal organs, and extreme hyperemia and hemorrhages, especially of the digestive tract. Pentman[219] used the skin test on rabbits, which was not used again until our laboratory [54,112,113,118,120,124,126,127,182,190-193] and the All-Union Research Laboratory of Toxic Fungi of U.S.S.R. Agriculture Ministry carried out large numbers of skin tests in 1942.[254,255,257] Today skin tests are considered a reliable method for determining the toxicity of overwintered grains and various strains of toxic *Fusarium*.

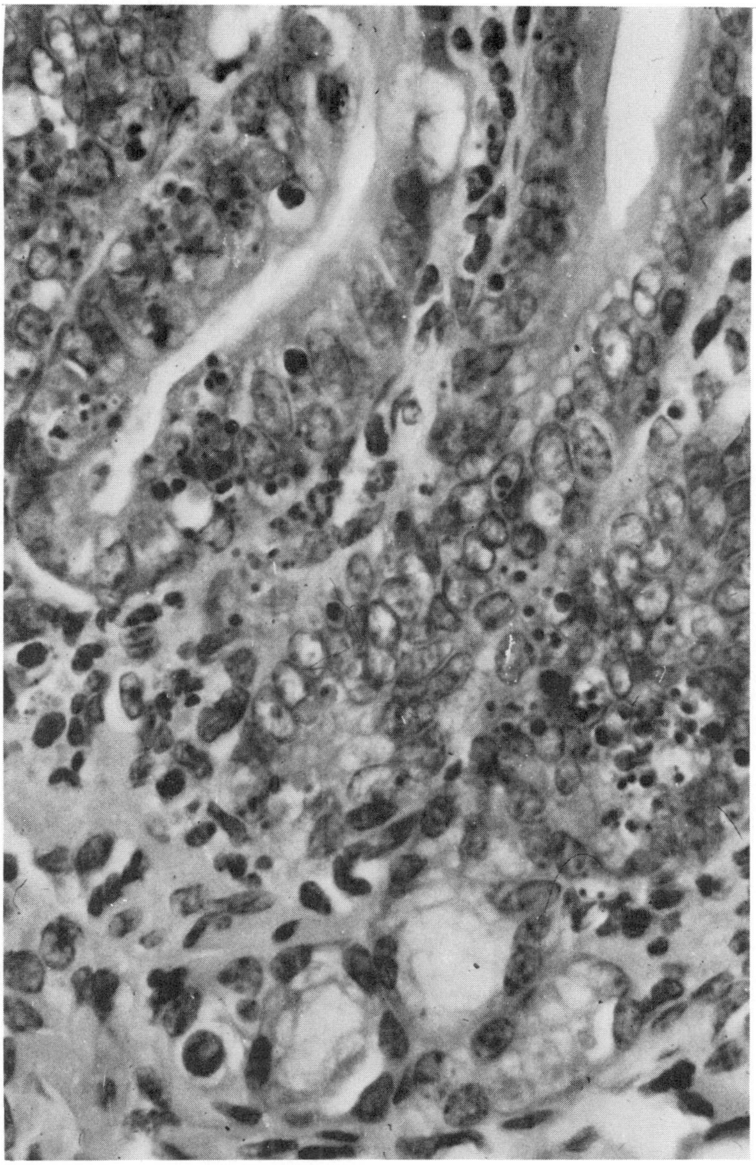

FIGURE 46. Duodenum: severe necrosis of crypt epithelium in guinea pig.

In the same way that the clinical syndrome of ATA varies from patient to patient, so do the symptoms in animals vary according to the potency and quality of the toxin, the means of administration, and the sensitivity of the animal. These variations led Bilai[25] to refer to the polymorphism of the disease. For many years the typical clinical syndrome of ATA in man could not be reproduced in laboratory animals. When toxic overwintered grains, as well as normal grains and other different solid and liquid media infected by toxic *F. poae* and *F. sporotrichioides,* were fed to mice, rats, guinea pigs, rabbits, dogs, horses, poultry, pigs, sheep, and cattle, or toxic extracts were given orally or parenterally, the animals developed clinical syndromes which differed from the typical picture of ATA in man.[8,9,25,42-45,112-116,118-120,124-141,168,181,231,238,248,252,255,295,296,298]

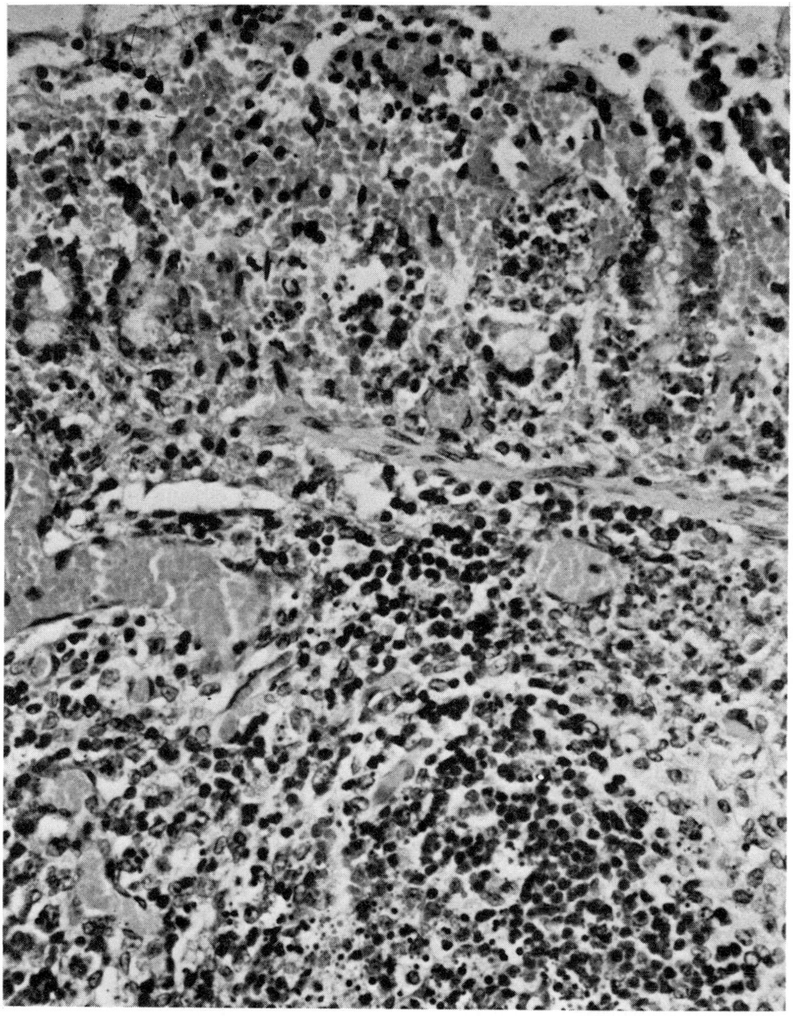

FIGURE 47. Cecum of guinea pig: hemorrhage and necrosis of mucosa extending to muscular mucosae. Capillary hyperemia. Necrosis of lymph follicle.

The clinical and pathological findings revealed stomatitis, gingivitis, glositis, esophagitis, hyperemia, hemorrhages, and necrosis of the GI tract, and even hemorrhagic diathesis and degeneration in the liver, kidneys, heart, adrenal and thyroid glands, and nervous system, which sometimes caused paralysis or convulsion. However, characteristic changes were not always apparent in the hematopoietic system. The attempts, therefore, to produce an experimental animal model of ATA in feeding studies with the above-mentioned animals have been unsuccessful. Further investigations established that cats and monkeys were the best experimental test animals for these aims.

Sarkisov,[253,258] Sarkisov et al.,[256] Alisova,[7] Alisova and Mironov,[4,6] Joffe[108,112,118,120124] and Lutskey et al.[166] studied the effect of toxic *F. poae* and *F. sporotrichioides* on cats, and Rubinstein and Lyass[249] on cats and monkeys, and all succeeded in reproducing the disease of ATA. The death of the cats followed a failure in blood production. A fall in hemoglobin, red cells, and leukocytes was observed with a relative increase in lymphocytes. The lowest leukocyte count found was 100/mm^3. The symptoms included vomiting, a hemorrhagic diathesis, and neurological disturbances.

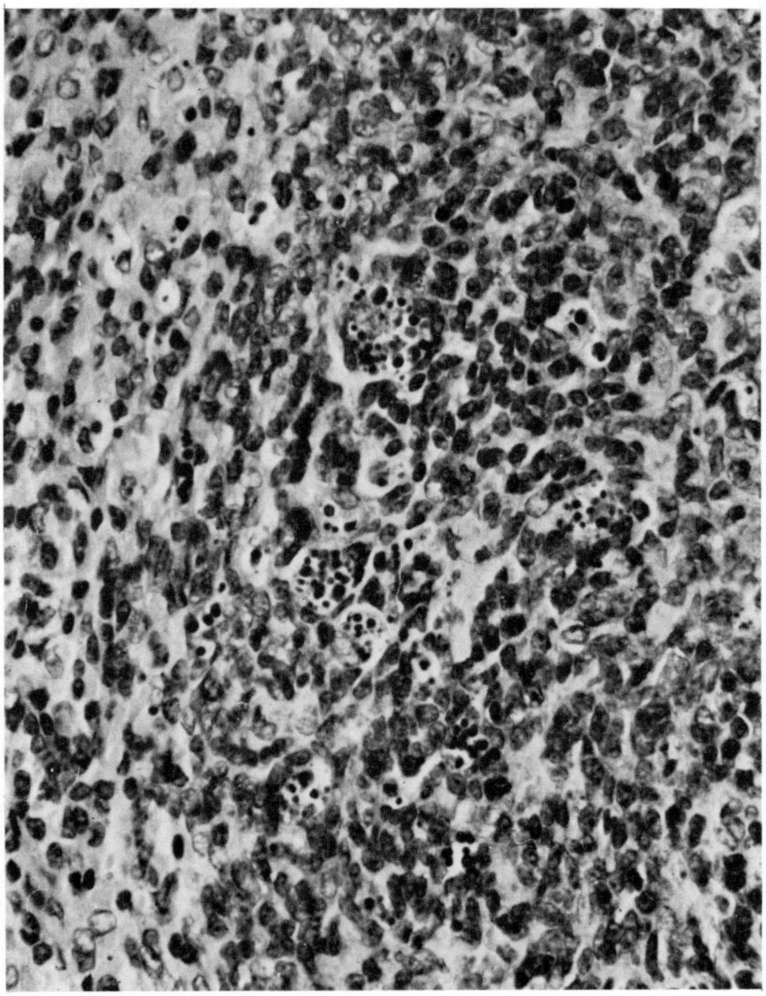

FIGURE 48. Spleen of guinea pig: necrosis in lymphoid follicles.

In the majority of cases, the cats died on the 6th to 12th day.[112] Autopsy revealed marked hyperemia and hemorrhages of internal organs, especially of the digestive tract, liver, and kidneys, and extreme changes in the adrenal glands. Histopathological examinations of organs of cats dying after infection with *F. poae* and *F. sporotrichioides* and hematological analysis revealed changes in the tissues and in the blood-component elements which were similar to those seen in ATA.[325]

However, all the scientists in the U.S.S.R. worked with overwintered cereal grains or infected grain with *Fusarium* cultures without correct identification of the toxins produced by the above *Fusarium* fungi. Only the determination of the chemical nature of the toxins isolated from *F. poae* and *F. sporotrichioides* induced us to investigate the toxicity of the most toxic strain of *F. sporotrichioides* #921 on cats. The main toxin obtained from toxic *Fusarium* strains was T-2 toxin.[332-334]

Our experiments on cats indicate that pure T-2 toxin and crude extract containing T-2 toxin caused a lethal intoxication in cats in which the severity of the symptoms was associated with hemorrhagic diathesis and drastic changes in the hematopoietic system. The characteristic leukopenia seen in our experiments is similar to that reported by Sato et al.[259] despite differences in dosage and route of administration.

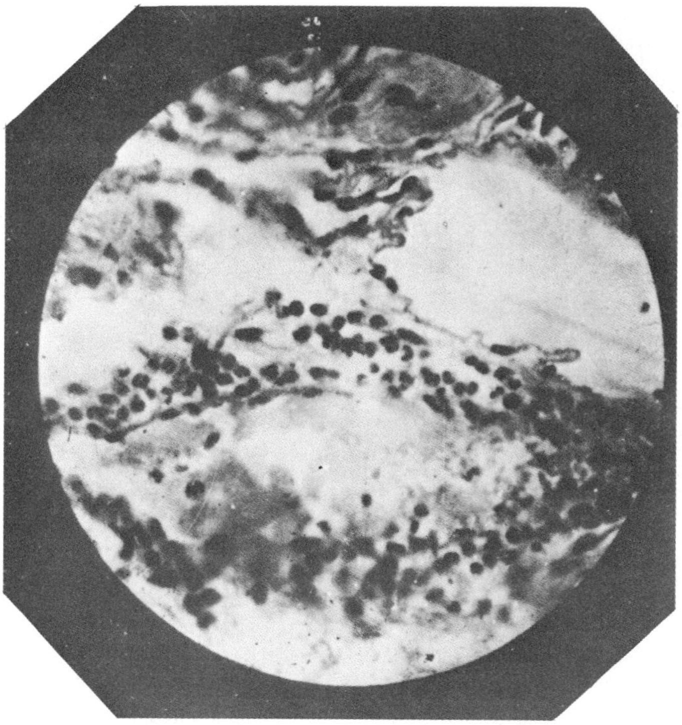

FIGURE 49. Leukocytes migrating through walls of small blood vessels after s.c. injection of *F. poae* into rabbits. (From Joffe, A. Z., in *Mycotoxins*, Elsevier, Amsterdam, 1974. With permission.)

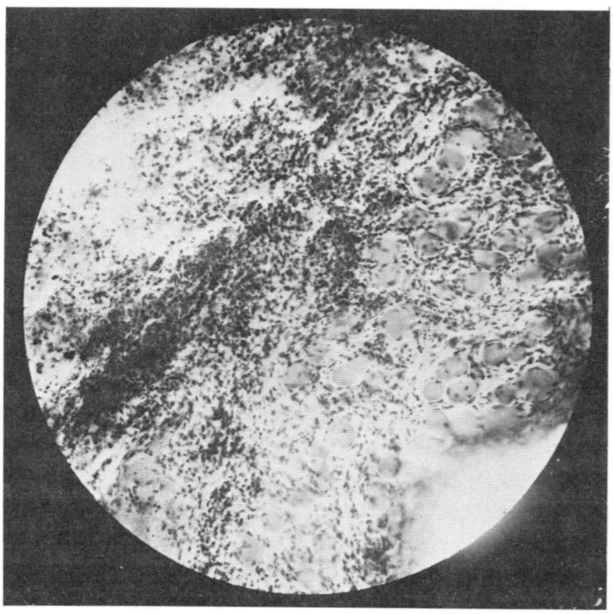

FIGURE 50. Marked leukocyte infiltration of muscle following application of toxic *F. sporotrichioides* #344 extract to rabbit skin. (From Joffe, A. Z., in *Microbial Toxins*, Vol. 7, Academic Press, New York, 1971. With permission.)

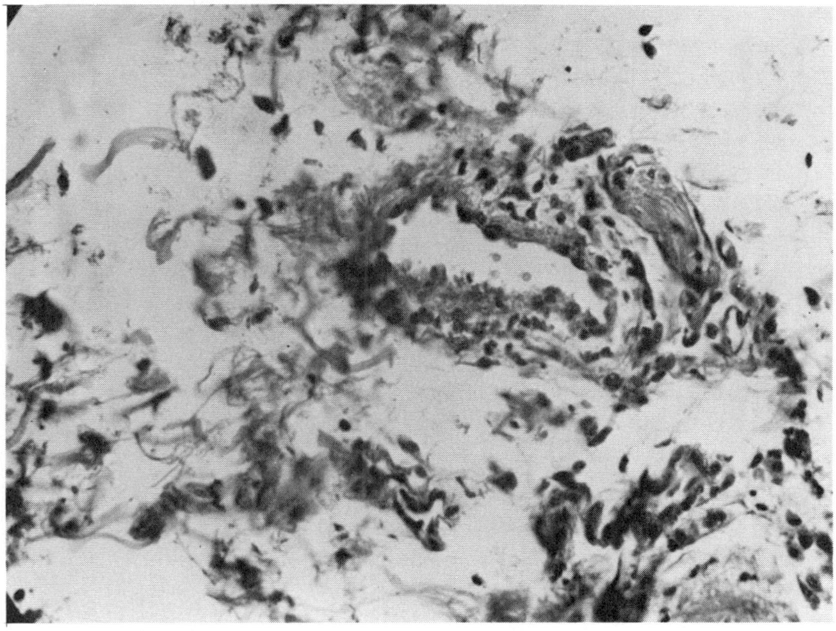

FIGURE 51. T-2 toxin from *F. poae* #958 after application of 10 μg on rabbit skin. Subcutis vasculitis, the whole of vessel infiltrated with polymorphonuclear leukocytes, most of which are eosinophils. (H.E.; magnification × 260.)

The cats that died of T-2 toxin intoxication showed gross and microscopic pathologic changes in the lymph and abdominal mesenteric nodes, contrary to Sarkisov,[255] who did not find significant changes in the lymph nodes.

Our results indicate that the trichothecene T-2 toxin isolated from *F. sporotrichiodes* #921 is that metabolite which is responsible for ATA disease in man.

Our recent studies on New Hampshire chicks with authentic strains of *F. poae* #958 and *F. sporotrichioides* #921 involved in causing ATA in the U.S.S.R. showed reduction of weight gains and drastic clinical and pathological changes, which were the principal features of ATA. The effect of the toxins isolated from the *Fusarium* species on mortality of chicks is shown in Tables 27 and 28.

F. sporotrichioides #921 produced several times more T-2 toxin than isolates from the U.S. and France, namely 4.1 g of T-2 toxin per kilogram of infected grain. Other trichothecenes obtained from the extract of this strain were neosolaniol, 0.025 g, HT-2 toxin, 0.039 g (see Figure 23), and six different steroids. None of these steroids was toxic and none of them had the structure of steroidal lactones, as described by Olifson[211] and Olifson et al.[215] These findings indicate that the trichothecene T-2 toxin is the main toxic metabolite in the extract isolated from *Fusarium* cultures. The toxicological studies on cats[166] and monkeys using crude extract containing T-2 toxin and pure T-2 toxin isolated from *F. sporotrichioides* #921 established that T-2 toxin (Figure 22) is that component in the extract which produces all the symptoms and pathologic changes associated with ATA.

T-2 toxin caused severe oral lesions and neural disturbances in chicks at levels of 4 to 16 μg/g of feed. These doses had no effect on blood components and did not cause pathological changes in the organs.[327-331] Only slight damage to hematopoietic system was seen in laying hens after they were fed on a T-2 toxin diet.

Several toxigenic studies were carried out on poultry in the U.S.S.R.[255] and revealed that chicks and ducklings, after being fed toxic overwintered prosomillet or infected prosomillet with *F. sporotrichioides* involved in causing ATA in man, remained com-

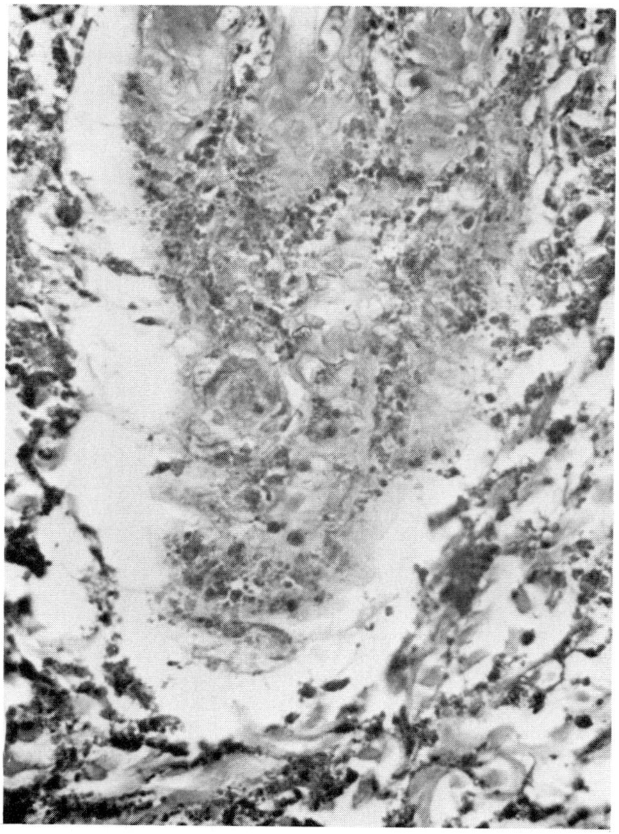

FIGURE 52. Crude extract from *F. poae* #958 after application of 25 μg on rabbit skin. Superficial necrosis of squamous epithelium infiltrated with polymorphs and eosinophils with widespread hemorrhage and cellular infiltration in necrotic hair follicles. (H.E.; magnification × 415.)

pletely healthy. Only Sergiev[280,281] reported that chickens died after feeding on overwintered grains. Yefremov[339,340] reported that geese died when fed overwintered prosomillet and according to Greenway and Puls[93] when they fed contaminated overwintered barley in the field in British Columbia. Puls and Greenway[232] associated the death of geese in the field cases and in the laboratory studies with T-2 toxin produced by *Fusarium* spp. Shalman[284] revealed the development of leukopenia in chicks fed overwintered toxic grains. Sergiev[279] found mortality of poultry in some districts of the U.S.S.R. where there were outbreaks of ATA in man. Tostanovskaja[313] and Tostanovskaja and Ratmanskaja[314] fed pigeons prosomisllet infected with toxic *Fusarium*, which caused vomiting a few hours after being fed. Specivceva found extensive hemorrhages in the kidneys of chicks fed oats mixed with cultures of *F. sporotrichiella* var. *poae*[297] and Palyusik et al. found myocardial and hepatic degenerations and severe lesions in the oral cavities and tongues of goslings.[217,218]

The marked leukopenia as well as the effect on thrombocyte and erythrocyte counts and hemoglobin levels showed in Table 29 and seen in our experiments differ from the findings of Wyatt et al.[327-330] They used doses too small to obtain the characteristic clinical and pathological changes. A similar failure and depression is caused in the hematopoietic system by ATA in man affected by progressive leukopenia.

Our experimental findings indicate that giving crude extract containing T-2 toxin at

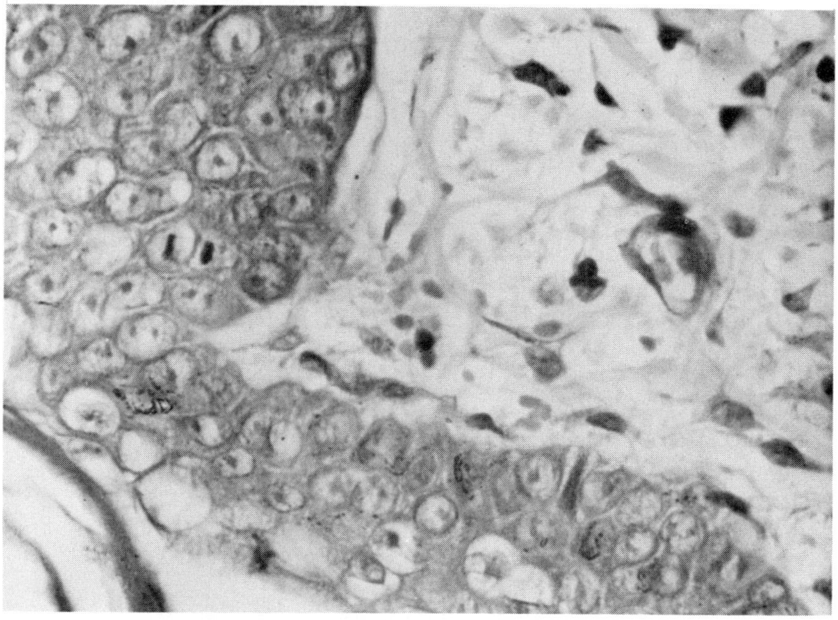

FIGURE 53. T-2 toxin from *F. sporotrichioides* #921 after application of 10 μg on rabbit skin. Heavy edema of the dermis with few inflammatory cells, some of them eosinophils and slight desquammation of epithelium. (H.E.; magnification × 415.)

the level of 0.35 to 1.2 mg/kg body weight to New Hampshire chicks produces a lethal intoxication with symptoms and pathogenic changes similar to those observed in man affected by ATA. Therefore we suggest that T-2 toxin played the essential role in ATA disease.

Schoental and Joffe[260] suggested that the crude extract obtained from toxic cultures of *F. poae* and *F. sporotrichioides* has two types of action, local and systemic, in rats and mice. We have used extracts from these *Fusarium* species to test their long-term effects on rodents when administered in various ways. Immunosuppression and induction of hyperplasia suggested that the extracts may have contained carcinogenic constituents. This appearance is clearly shown in Figures 66 to 71. The use of moldy cereals, contaminated by the inflammatory and irritant metabolites of *F. poae* and *F. sporotrichioides,* may have played a part in the development of tumors of the digestive tract. The effects of the extracts on the thymus, spleen, and lymph glands, and the development of infective foci in various tissues including the kidneys, strongly suggest immunosuppressive action.[264]

The larger doses of crude extract caused death within 1 day or longer regardless of the route of administration. The main lesions were present in the skin, lymphoid tissues, lung, heart, kidney, esophagus, and digestive tract. Figure 72 shows the cystic distension and hemorrhage in the retroperitoneal lymph node of mice after a s.c. dose of *F. sporotrichioides* extract. Some of the more acute effects induced by our crude *Fusarium* extracts resembled those described in animals treated with certain trichothecenes.[17,130,166,173,175,259,300]

Rats were also given T-2 toxin in a chronic, long-extended period, the toxin having been isolated from crude extract of *F. sporotrichioides* #921. White, weanling rats were obtained from MRC Laboratory Animals Centre, Carshalton, Surrey, England. They were given freshly prepared solutions of T-2 toxin by stomach tube (1 to 4 mg/kg body weight) and the dosing was repeated at various time intervals. Among rats which survived longer than 16 months after the first dose and 5 to 8 doses of T-2 toxin, chronic

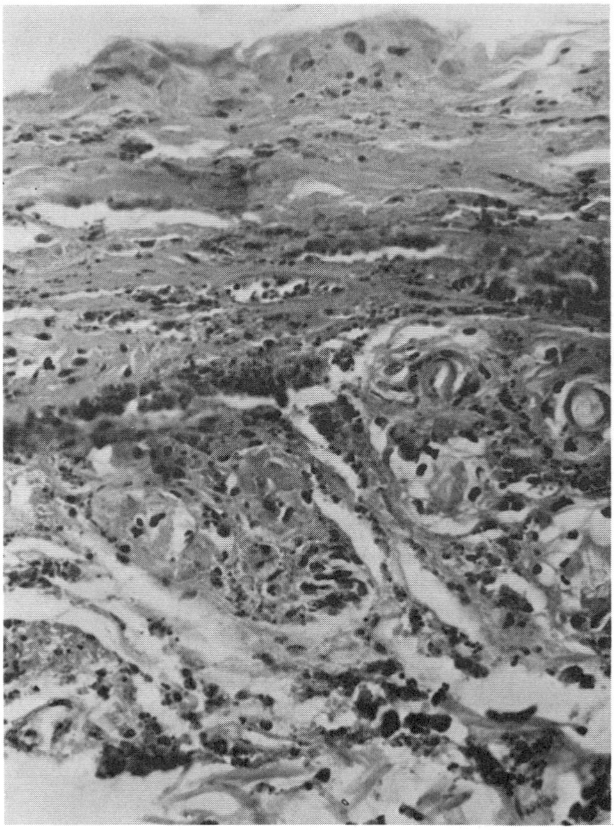

FIGURE 54. T-2 toxin from *F. sporotrichioides* #921 after application of 15 μg on rabbit skin. Superficial necrosis and severe inflammatory edema of the dermis which include infiltration with scattered leukocytes. (H.E.; magnification ×415.)

and neoplastic lesions were present, which included severe cardiovascular lesions and various tumors (benign and malignant) of the pituitary, brain, pancreas, mammary gland, as well as adenocarcinoma of the stomach. The implications of these results for human health must be seriously considered.

In recent experiments, T-2 toxin isolated from *F. sporotrichioides* #921 proved to resemble *N*-methyl-*N*-nitrosourethane in being carcinogenic for the digestive tract in rats,[264] and also in inducing pigmentation when applied to the skin of dark mice.[275] Solution of T-2 toxin, containing 0.2 to 0.3 mg/mℓ in 10% aqueous ethanol, when applied to the clipped intrascapular region of C57BL mice caused local irritation, hyperemia, edema, ulceration, and scab formation. When the scab peeled off, the healed area remained depilated and in some animals was surrounded by areas of depigmented hair.

The trichothecenes are tetracyclic sesquiterpenoids,[87] which contain an olefinic bond and an epoxy group (at 9, 10, 12, and 13, respectively) essential for biological activity and characterized as 12,13-epoxytrichothecenes.[13,14,19] The effects of esterification, oxidation, and so on, have still to be explored in detail. Only one long-term study appears to have been published.[173] These authors did not find tumors in rats and trout given T-2 toxin from *F. tricinctum* in their diet for 8 months. Our results suggest that neoplastic lesions may develop among the animals in chronic long-term experiments.[260,262,274,276]

FIGURE 55. Crude extract from *F. sporotrichioides* #921 after application of 25 μg crude extract on rabbit skin. Superficial necrosis of epidermis and heavy edema of dermis with inflammatory cells. (H.E.; magnification × 415.)

Akhmeteli et al.[1] carried out a trial on mice with extracts of barley infected by *F. sporotrichioides* and also discovered the blastogenic properties. In contradiction to these results, Lindenfelser et al.[160] carried out extensive studies on the biological effects of trichothecene compounds, particularly T-2 toxin and diacetoxyscirpenol, on mouse tissue, and revealed that these substances caused inflammation and tissue necrosis when applied to the skin of rats and mice, but did not cause papillomas.

The trichothecene compounds diacetoxyscirpenol, T-2 toxin, and HT-2 toxin caused an inflammatory skin response on rabbits, guinea pigs, and albino rats, but were not carcinogenic.[12,17,85]

Cancer of the digestive tract in man, particularly of the esophagus, has often been considered to be related to the consumption of alcoholic beverages. Pure ethanol is not known to induce tumors in experimental animals. Moldy cereals, though not usually consumed as such except during famine, may be used for making fermented foods and alcoholic drinks. For example, a condition of beer known in the brewing industry as "gushing", which has been traced to the use of barley that became contaminated with *F. sporotrichiella* during wet seasons.[86] In certain areas of the world a high incidence of digestive tract tumors has been recorded in man and in livestock.[225] The drastically irritant metabolites of the *Fusarium* species should be added to the many factors that have already been considered as possibly involved in the development of such

Table 27
EFFECTS OF METABOLITES DERIVED FROM *F. SPOROTRICHIOIDES* NO. 921 ON MORTALITY OF CHICKS[a]

No. of trial	Amount of toxin fed	Initial age of chicks (in days)	Feeding period (in days)	Duration of trial (in days)	Mortality[b]	Effect on weight: average weight ± S.E. (g) Initial	Final tested	Control
	T-2 toxin in crude extract (mg/kg BW)							
1	0.45	17	14	26	6	134 ± 2	323 ± 10	586 ± 8
2	0.9	17	9	23	9	140 ± 4	276 ± 6	498 ± 9
3	0.35	30	6	18	8	292 ± 4	334 ± 10	722 ± 10
	T-2 toxin in[c] mg/kg feed							
4	23	17	28	30	0	140 ± 7	340 ± 7	592 ± 10
5	46	17	28	30	4	145 ± 4	271 ± 7	595 ± 9
6	92	30	8	8	5	355 ± 9	262 ± 5	492 ± 9
7	230	30	8	14	10	351 ± 4	201 ± 3	497 ± 8
8	460	30	8	14	10	352 ± 7	146 ± 5	504 ± 9
	Pure T-2 toxin[d] in mg/kg feed							
9	8	17	11	26	0	133 ± 4	471 ± 12	593 ± 5
10	16	17	11	26	0	131 ± 5	338 ± 7	597 ± 8

[a] The number of treated chicks was 10 except in trial 2 which had 16 chicks. The number of control chicks in each trial was 5, except for trials 2, 9, and 10 which had 8 chicks. The values are expressed as means ± S.E. The values of treated and control groups significantly differ at $p<0.05$.
[b] Designates the number of birds that died during the experiment.
[c] The mixture obtained from the molded milled grain and commercial feed. The consumed amount varied from 30 to 8% of the feed mixture depending on the degree of toxicity.
[d] The mixture obtained from pure T-2 toxin and the commercial feed.

Adapted from Joffe, A. Z. and Yagen, B., *Toxicon*, 16, 263, 1978. With permission.

tumors. The damage which they can cause to the mucosa of the digestive tract together with suppression of immunity might facilitate carcinogenesis in this organ.

THE PHYTOTOXIC ACTION OF THE SPOROTRICHIELLA SECTION

The Phytotoxic Effects of *Fusarium* Culture Filtrates on Seed Germination

The phytotoxic effect of culture filtrates on seed germination was determined on pea, bean, wheat, and barley seed. Cultures of six isolates each of *F. poae* (#24, 60/9, 344, 396, 792, and 958) and *F. sporotrichioides* (#60/10, 347, 738, 921, and 1182), two

Table 28
EFFECTS OF METABOLITES DERIVED FROM *F. POAE* NO. 958 ON MORTALITY OF CHICKS[a]

No. of trial	Amount of toxin fed	Initial age of chicks (in days)	Feeding period (in days)	Duration of trial (in days)	Mortality[b] (in days)	Effect on weight: average weight ± S.E. (g) Initial	Final treated	Control
	T-2 toxin in crude extract (mg/kg BW)							
11	0.6	17	14	26	6	142 ± 3	371 ± 8	589 ± 9
12	1.2	17	8	23	12	139 ± 2	256 ± 4	468 ± 9
13	0.46	30	6	14	8	343 ± 7	361 ± 3	592 ± 9
	T-2 toxin[c] mg/kg feed							
14	30	17	26	30	3	143 ± 3	342 ± 6	562 ± 8
15	60	17	26	30	5	141 ± 4	234 ± 5	566 ± 10
16	120	30	10	10	6	339 ± 7	279 ± 5	434 ± 8
17	300	30	7	10	10	335 ± 6	238 ± 10	499 ± 8
18	600	30	8	10	10	337 ± 10	161 ± 10	501 ± 10

[a] The number of treated chicks was 10 except in trial 12 which had 16 chicks. The number of control chicks in each trial was 5, except for trial 12 which had 8 chicks. The values of treated and control groups are significantly different at $p<0.05$.
[b] The same as in Table 27.
[c] The same as in Table 27.

Adapted from Joffe, A. Z. and Yagen, B., *Toxicon*, 16, 263, 1978. With permission.

isolates of *F. sporotrichioides* var. *tricnictum* (M304 and 1227) and two isolates of *F. sporotrichioides* var. *chlamydosporum* (117⁴ and 4337) were inoculated on media I and II in Erlenmeyer flasks at 25°C for 7 days, at 10°C for 45 days, and at 4°C for 75 days. Cultures were then sterilized at 100°C for 15 min. Equal amounts of filtrate were applied to paper filters which were placed in large petri dishes and seeds placed for 4 to 5 days on the filters (five to ten seeds in each of ten petri dishes for every plant per *Fusarium* filtrate per temperature combination for both media). Control seeds were sown on filter paper wetted with sterile filtrates without *Fusarium* cultures.

Results showed that all filtrates substantially reduced seed germination in all four plants, reduction increasing as the temperature at which cultures were kept decreased (Table 30). Germination in control groups ranged from 89 to 91%. Beans were most, and wheat least susceptible to these effects.

The Effect of *Fusarium* Culture Filtrates and their Extracts on Leaf Blades of Young Plants

The following *Fusarium* culture filtrates and their crude ethanol extracts were studied on young pea, bean, and tomato plants: *F. sporotrichioides*, ten isolates; *F. poae*,

Table 29
EFFECTS OF METABOLITES OF *F. POAE* NO. 958 AND *F. SPOROTRICHIOIDES* NO. 921 ON SOME BLOOD COMPONENTS OF NEW HAMPSHIRE CHICKS[a]

Species and toxic material administered	No. of trial	Hemoglobin (g/100 mℓ)	Leukocytes (in thousands /1 mm³)	Erythrocytes (in millions /1 mm³)	Thrombocytes (in thousands /1 mm³)
F. poae					
T-2 toxin in crude extract 1.2 mg/kg BW	12	6.1 ± 0.2	3.0 ± 0.2	1.1 ± 0.1	9.2 ± 0.3
T-2 toxin in feed mixture 1% 60 mg/kg feed	15	7.2 ± 0.2	3.2 ± 0.2	1.4 ± 0.1	11.0 ± 0.4
F. sporotrichioides					
T-2 toxin in crude extract 0.9 mg/kg BW	2	6.4 ± 0.4	3.7 ± 0.2	1.8 ± 0.2	12.1 ± 0.4
T-2 toxin in feed mixture 1% 46 mg/kg feed	5	7.4 ± 0.4	4.2 ± 0.2	2.1 ± 0.2	13.4 ± 0.3
Pure T-2 toxin 16 mg/kg feed	10	8.2 ± 0.3	7.3 ± 0.5	2.3 ± 0.2[a]*	19.6 ± 0.5
Control		11.2 ± 0.4	19.4 ± 0.4	3.3 ± 0.2	25.5 ± 0.6

Note: The values are expressed as means ± S.E. (n = 5). The values of treated and control groups significantly differ at $p<0.05$.

[a] This result is significantly different at $p<0.1$.

Adapted from Joffe, A. Z. and Yagen, B., *Toxicon*, 16, 263, 1978.

ten isolates; *F. sporotrichioides* var. *tricinctum*, three isolates; and *F. sporotrichioides* var. *chlamydosporum*, three isolates.

All Fusaria were cultivated on medium I. *Fusarium* culture filtrates and their extracts were applied to leaf blades or leaf axils of young plants. At the site of application, various necrotized spots appeared after 6 to 38 hr. The action of crude extracts on plants was stronger than that of filtrates (see Table 31).

Effect of *Fusarium* and Pure Filtrates on Plant Branches

Four *Fusarium* species or varieties were tested by the author for their phytotoxic effect on a variety of plants. The *Fusarium* cultures were grown on PD liquid medium with pH 5.6 in Erlenmeyer flasks containing 70 mℓ of substrate. After incubation at 12°C for 21 days, the cultures were killed by 1 atm for 20 min. After cooling, the cotton corks were removed and the Erlenmeyer flasks covered with aluminum foil. Branches of test plants were introduced into each flask and the rate at which they wilted was recorded.

Each *Fusarium* strain was tested in five separate flasks on branches of the same plant. A total of 18 experiments were carried out using 180 flasks: 90 for experimental and 90 for control purposes. All branches treated with Fusaria wilted completely within 4 to 18 hr, depending on the species of *Fusarium* (Table 32 and Figures 73-75).

The Phytotoxic Effect of *Fusarium* of Sporotrichiella Section in Inoculation Test with Seedlings in Comparison with their Action on Rabbit Skin

The phytotoxic effect of eight isolates of *F. sporotrichioides*, five of *F. poae*, two of *F. sporotrichioides* var. *tricinctum*, and two of *F. sporotrichioides* var. *chlamydo-*

FIGURE 56. Changes in size of treated and control 28-day-old chicks after feeding 0.35 mg/kg T-2 toxin in crude extract from *F. sporotrichioides* #921 (trial 3).

FIGURE 57. Changes in size of treated and control 36-day-old chicks after feeding 0.46 mg/kg T-2 toxin in crude extract from *F. poae* #958 (trial 13).

sporum was tested in inoculation tests with seedlings of ten field and vegetable crops for comparison of their action on rabbit skin (by histological findings) after application of the above Fusaria ethanol crude extracts.

For determination of the phytotoxic properties of *Fusarium* fungi, 200 seedlings for each *Fusarium* isolate were used (i.e., 20 per test crop) of the following test plants: bean, cucumber, watermelon, cotton, tomato, onion, eggplant, pepper, wheat, and maize. The two last-named plants were only rarely affected by *Fusarium* isolates and

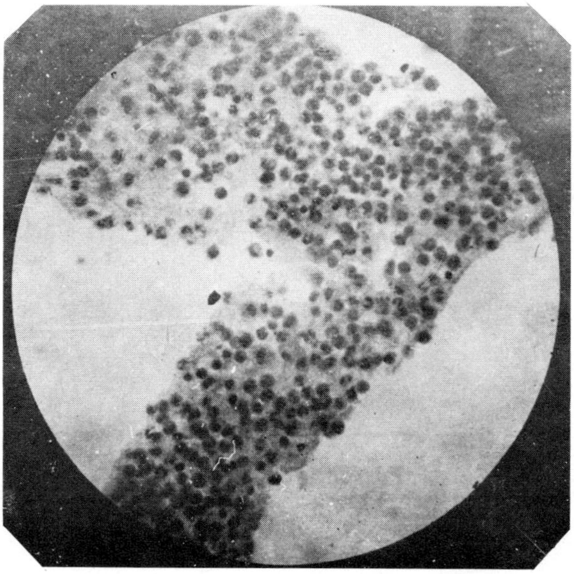

FIGURE 58. Extensive leukocyte immigration on the surface of the epiploon of a cat after i.p. injection of toxic extract of *F. sporotrichioides* #738.

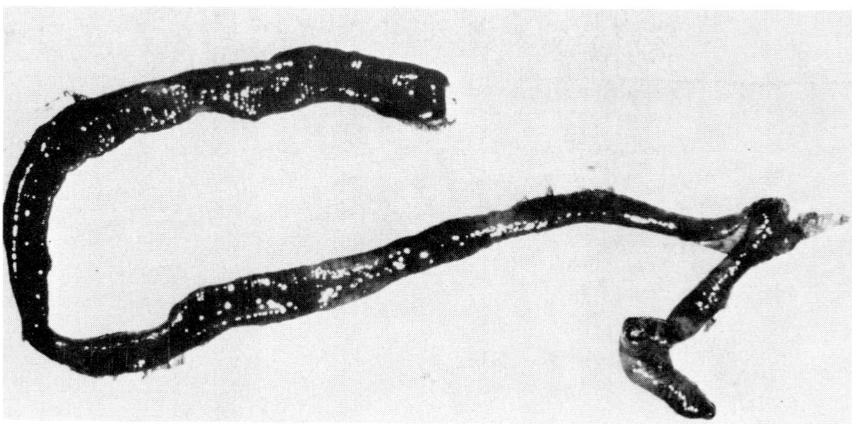

FIGURE 59. Hemorrhagic blood-filled small intestine of cat after feeding T-2 toxin.

the results relating to these crops have therefore been omitted. Bolton and Nuttall[34] found that *F. poae* readily penetrated only a few leguminous plants, but only slightly penetrated wheat and maize.

Each isolate was inoculated on two petri dishes and incubated at 6, 12, 18, 24, 30, and 35°C for 46, 21, 15, 10, and 7 days, respectively. The contents of uncontaminated dishes (separately for each temperature) were then introduced into a blender and 200 mℓ tap water added. The resulting suspension was divided over three to four beakers, into which 5- to 8-day-old seedlings prepared beforehand were dipped for 1 min. The seedlings were subsequently planted into previously prepared holes in unsterilized soil in plastic trays of 3 ℓ and the soil round the seedlings well clamped down.

The results were recorded in greenhouse after 21 days, but if any one plant of the uninoculated series of a crop wilted, the inoculated series of that crop was disregarded.

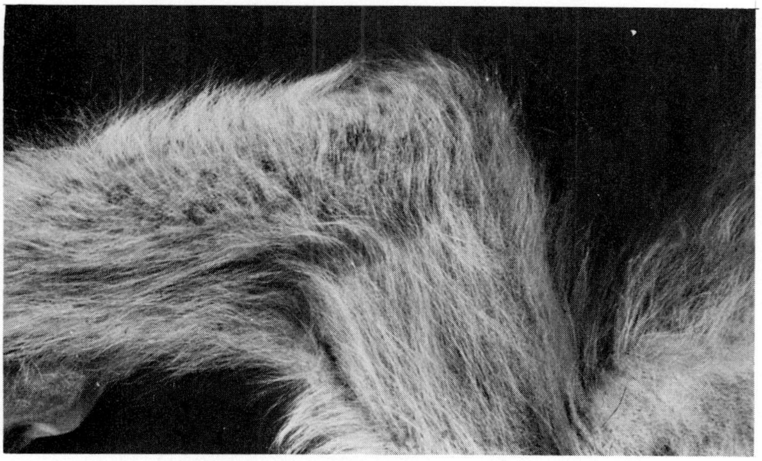

FIGURE 60. Hemorrhage in skin of arm of vervet monkey after feeding T-2 toxin. (Photo courtesy of N. P. J. Kriek).

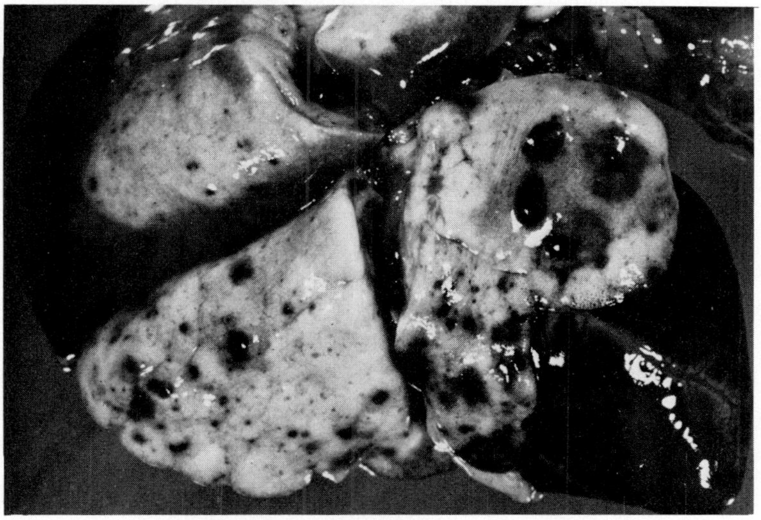

FIGURE 61. Lungs and heart of vervet monkey after feeding T-2 toxin. Extensive pulmonary hemorrhage, emphysema, and edema. (Photo courtesy of N. P. J. Kriek.)

Re-isolation of *Fusarium* isolated from wilted (dead) plants was also carried out. These re-isolates were tested as before, this time using 450 to 500 instead of 200 seedlings per *Fusarium* isolate. Results showed that isolates became more phytotoxic to seedlings upon re-isolation and exerted their strongest effect on their source plants.

For the toxicity test on rabbit skin crude ethanol extracts of *Fusarium* isolates were used. The cultivation of *Fusarium* isolates and preparation of crude extracts were carried out in the following way: natural media, wheat, or millet grain (containing 10 g grain and 20 mℓ tap water in 150-mℓ Erlenmeyer flasks) were sterilized for 30 min twice at 1 atm, then inoculated with suspensions of monospore cultures of *Fusarium* isolates and incubated also at the same temperatures, ranging from 6 to 35°C (6, 12, 18, 24, 30, and 35°C), as for experiments on seedlings. After incubation, infected

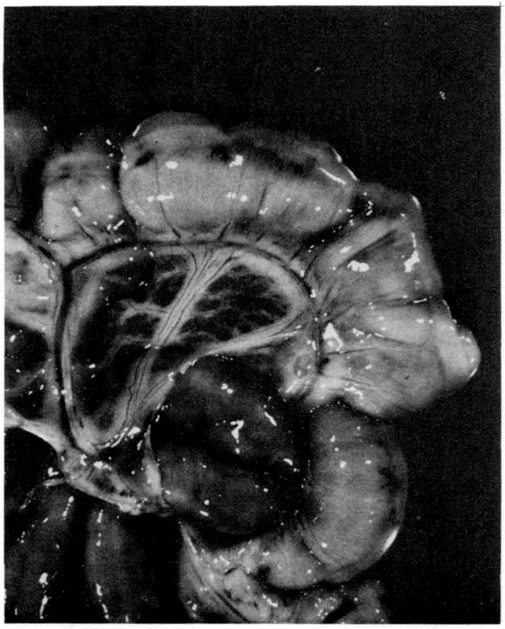

FIGURE 62. Colon of vervet monkey after feeding T-2 toxin. Scattered subserosal hemorrhages. (Photo courtesy N. P. J. Kriek.)

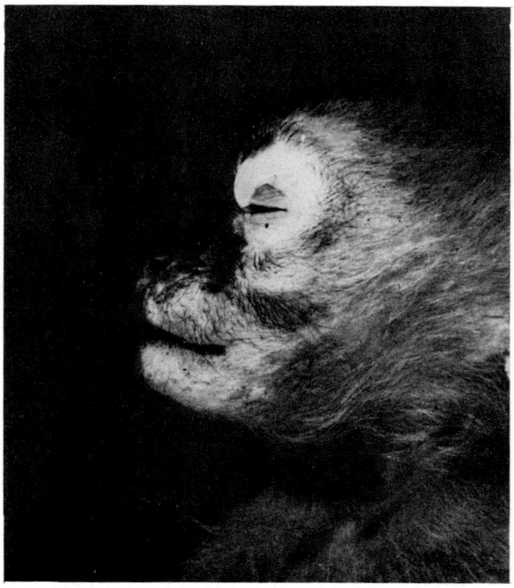

FIGURE 63. Extensive s.c. edema of head in vervet monkey after feeding T-2 toxin. (Photo courtesy of N. P. J. Kriek.)

natural media were extracted with ethyl alcohol (96%) in a vacuum rotary evaporator, yielding biologically active oil.

The bioassay was performed by applying 10 μg of *Fusarium* crude extract to the shaved rabbit skin. The intensity of the skin reaction produced by each isolate was

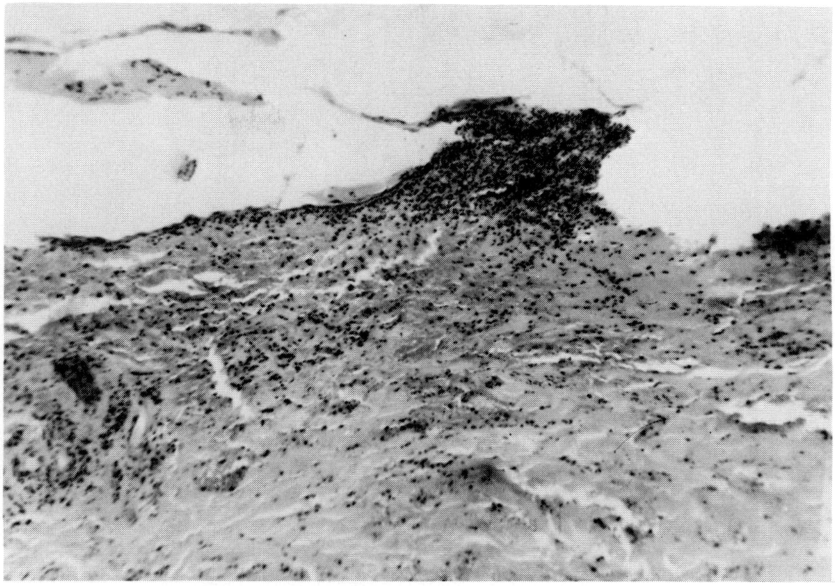

FIGURE 64. Crude extract from *F. sporotrichioides* #921 grown at + 4°C on rabbit skin. Large epidermal vesicle formation leaving only thin covering membrane of two to three layers of flattened epithelium. Floor of vesicles consists of necrotic dermis with deep massive hemorrhage and more superficial infiltration of polymorphonuclear leukocytes. Vesicles are mostly empty, but masses of granulocytes are still present. (H.E.; magnification × 145.) (From Joffe, A. Z., in *Microbial Toxins*, Vol. 7, Academic Press, New York, 1971. With permission.)

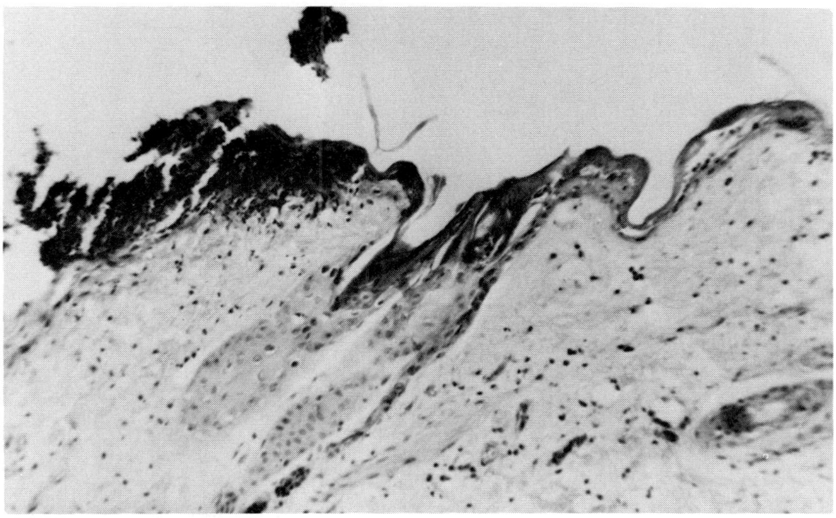

FIGURE 65. *F. poae* #958 grown at + 4°C on rabbit skin. Necrosis of entire thickness of epidermis. Only one to two layers of necrotic epidermal cells persist. Necrotic area is infiltrated with leukocytes which extend into dermis and s.c. tissue. (H.E.; magnification × 145.) (From Joffe, A. Z., in *Microbial Toxins*, Vol. 7, Academic Press, New York, 1971. With permission.)

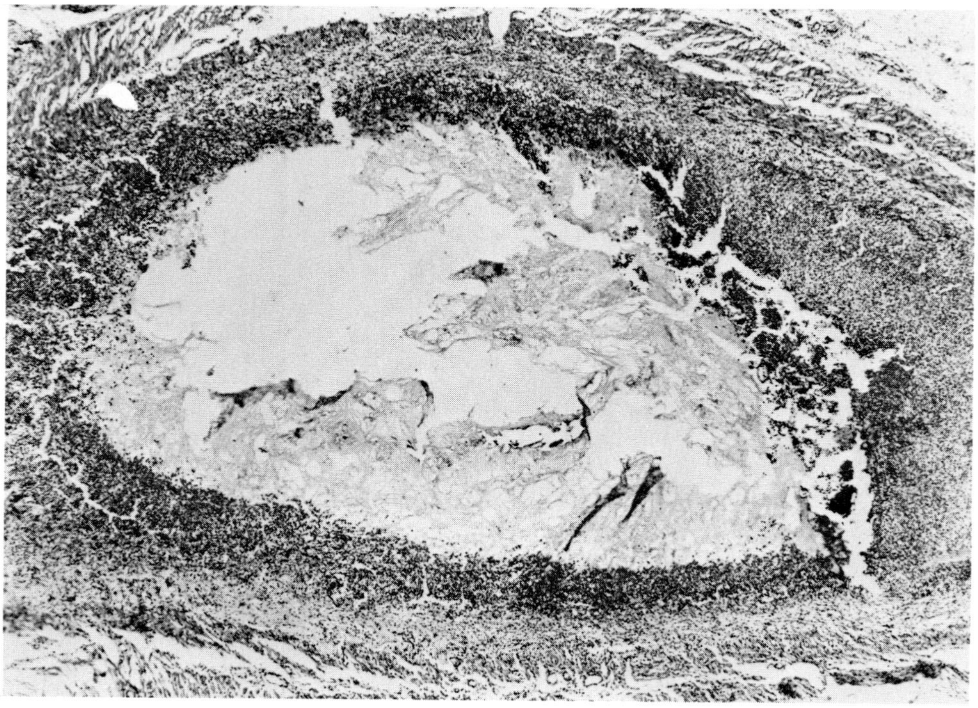

FIGURE 66. Esophagus of rat killed *in extremis* 15 days after first and 2 days after last of five doses of *F. poae* extract, given by stomach tube: local distension and blockage with cellular debris. (H.E.; magnification × 40.) (Photo courtesy of Marcel Dekker, Inc.)

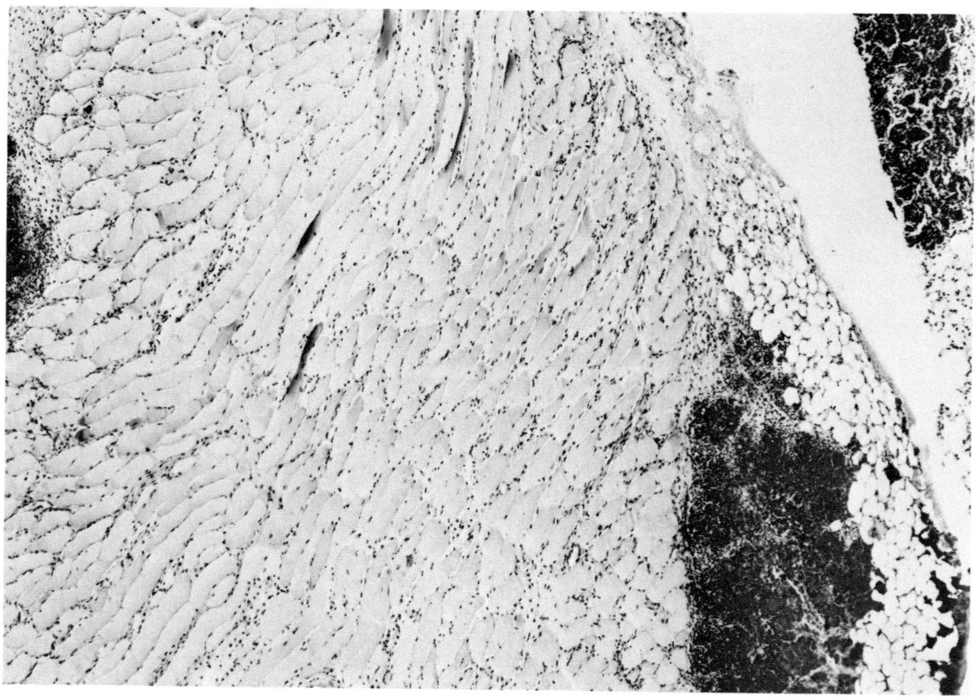

FIGURE 67. Heart and (on the right) lung of mouse dying 10 weeks after first of weekly skin applications of *F. poae* extract: foci of infection and infiltration with inflammatory cells. (H.E.; magnification × 40.) (From Schoental, R. and Joffe, A. Z., *J. Pathol.*, 112, 37, 1974. With permission.)

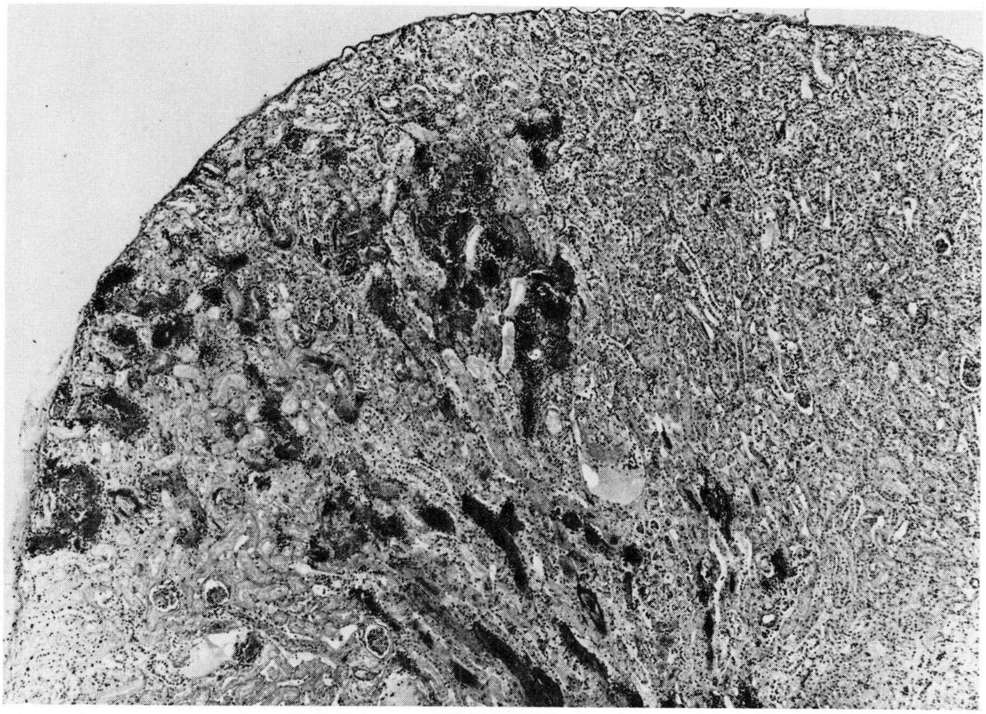

FIGURE 68. Kidney of male mouse dying 4 days after i.p. dose of *F. poae* extract: area of infection, necrosis and infiltration with inflammatory cells. (H.E.; magnification × 40.) (From Schoental, R. and Joffe, A. Z., *J. Pathol.*, 112, 37, 1974. With permission.)

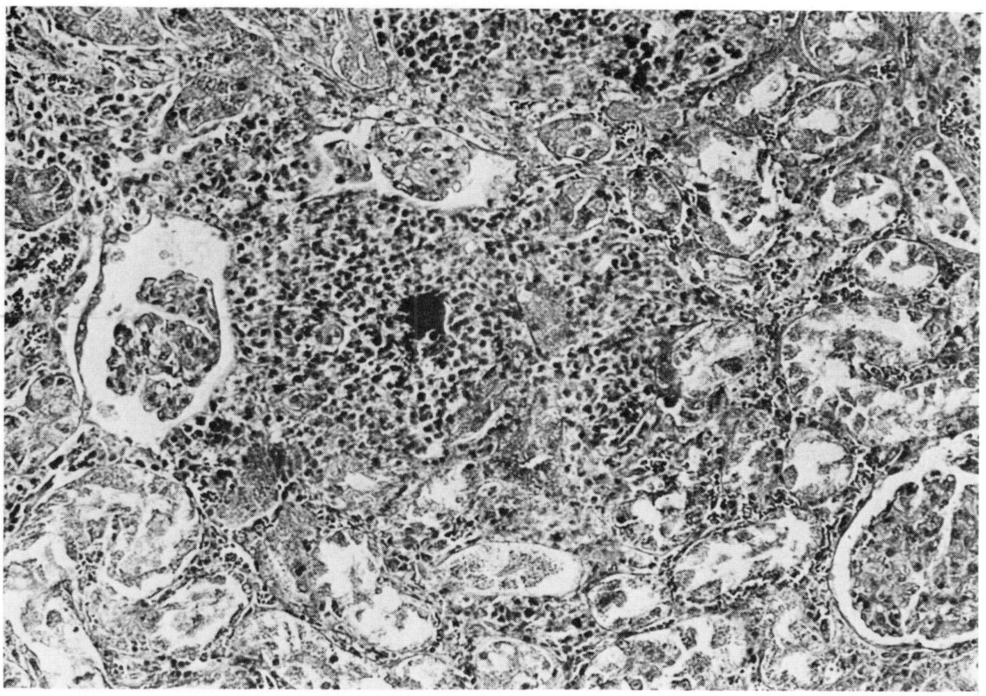

FIGURE 69. Kidney of female rat dying 23 days after first and last (doubled) of 11 doses of *F. poae* extract: basal cell hyperplasia. (H.E.; magnification × 240.) (From Schoental, R. and Joffe, A. Z., *J. Pathol.*, 112, 37, 1974. With permission.)

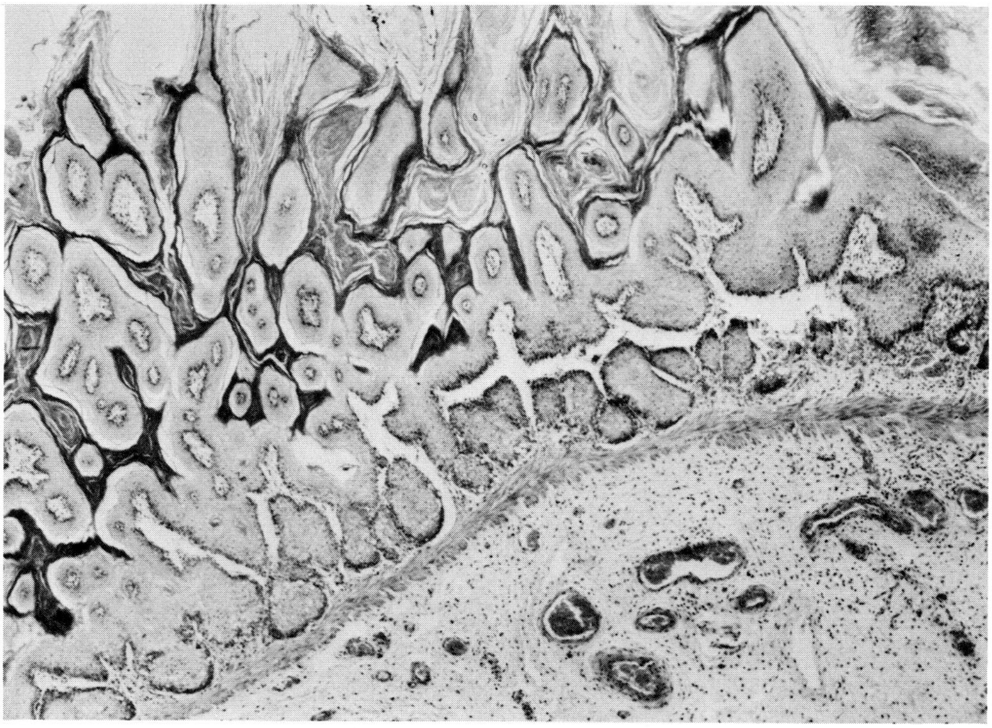

FIGURE 70. Forestomach of female rat in Figure 69: squamous cell hyperplasia, edema of stomach wall and (on right) ulceration. (H.E.; magnification × 60.) (From Schoental, R. and Joffe, A. Z., *J. Pathol.*, 112, 37, 1974. With permission.)

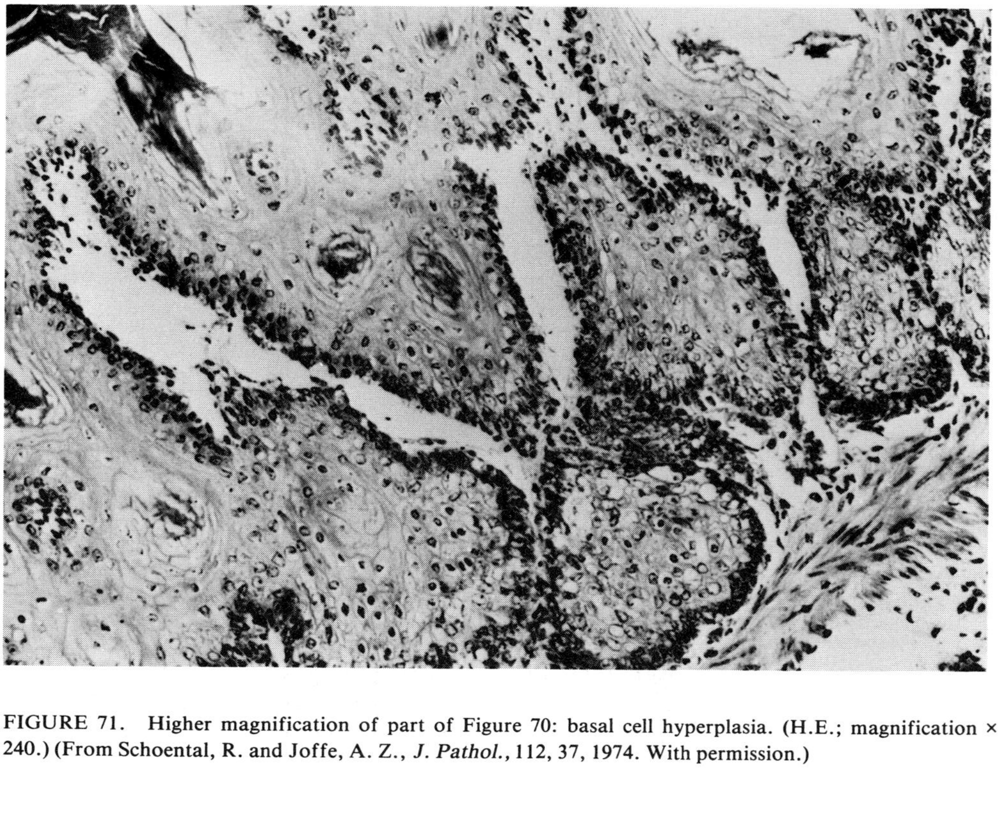

FIGURE 71. Higher magnification of part of Figure 70: basal cell hyperplasia. (H.E.; magnification × 240.) (From Schoental, R. and Joffe, A. Z., *J. Pathol.*, 112, 37, 1974. With permission.)

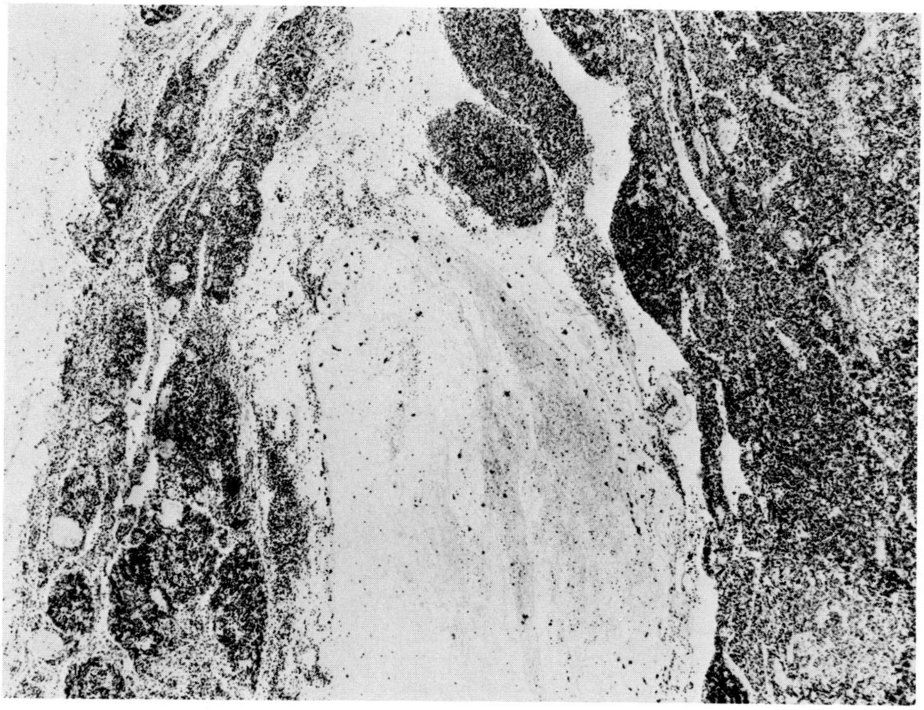

FIGURE 72. Retroperitoneal lymph node of mouse dying 1 day after s.c. dose of *F. sporotrichioides* extract: cystic distension and hemorrhage. (H.E.; magnification × 40.) (Photo courtesy of Marcel Dekker, Inc.)

assessed on the following scale: 0, no skin reaction; M, moderately toxic — slight inflammation, reddening, edema; S, toxic or strongly toxic — edematous, leukocytoxic, and severe necrotic reaction.

The effects on plants and rabbit skin (by histological findings) are presented in Table 33 and Figures 76 to 83.

Results of rabbit skin reaction are given for the optimum temperature at which phytotoxic effects are produced by each *Fusarium* species and variety. *Fusarium* cultures grown at 6 and 12°C developed the highest phytotoxic action and toxicity to rabbit skin. This action decreased with the rise in temperature.

The results indicated a strong correlation between the toxicity of *Fusarium* isolates to rabbit skin and their phytotoxic effect in seedling mortality. This is in full agreement with the results obtained by Marasas et al.[174] in connection with what these authors term *F. tricinctum,* and the question therefore arises, why species so potent as producers of toxin, and so widely distributed in soil and on cereal seeds, are not more prominent as plant pathogens, and are indeed considered by many authors[87-91, 299] to be only weak pathogens.

There are several possible answers to this question; probably the most valid is the fact that it has been proven that *F. poae* and *F. sporotrichioides* produce their toxin most actively at very low temperatures[112] at which there is in any case little plant growth. Thus, even in soils rich in these fungi the amount of toxin present during the main growing seasons is likely to be small.

UROV OR KASHIN-BECK DISEASE

It is known that *Fusarium* species, especially varieties of the Sporotrichiella section,

Table 30
EFFECT OF SPOROTRICHIELLA SECTION FUSARIA ISOLATES ON PLANT GERMINATION

Species	Conditions of incubation and range of germination (%)		
	5 days (25°C) (%)	40 days (10°C) (%)	75 days (4°C) (%)
F. sporotrichioides	28—49	11—21	Total suppression
F. poae	32—47	17—24	3—7
F. sporotrichioides var. *tricinctum*	51—57	40—51	21—33
F. sporotrichioides var. *chlamydosporum*	53—59	42—53	23—30

Note: Germination in the control groups ranged from 89—91%.

Table 31
NECROTIC EFFECT OF *FUSARIUM* ISOLATES ON YOUNG PEA, BEAN, AND TOMATO PLANTS

Fusarium	Av time required for effect to appear (hr)	Color of spots on leaf blade or axil
F. sporotrichioides		
Filtrate	18	Brown
Crude extract	6	Black
F. poae		
Filtrate	24	Red to light brown
Crude extract	9	Brown
F. sporotrichioides var. *tricinctum*		
Filtrate	31	Rose red
Crude extract	16	Purple red
F. sporotrichioides var. *chlamydosporum*		
Filtrate	38	Light carmine
Crude extract	19	Carmine red to brown

under favorable environmental conditions in the field and in storage produce toxic compounds in cereal grains, foodstuff, food products, and fodder. Depending on the degree of toxicity these fungi cause diseases with different clinical and pathological changes in the organs and tissues of men and animals.

Intoxication associated with toxic *Fusarium* isolates usually occur sporadically or spontaneously in various places and produce acute diseases as, for example, ATA. However, sometimes the fungi can only develop slight toxic substances in grains and other substrates and thus cause only a chronic course of diseases. In such cases the disease develops gradually and in its initial stage is frequently asymptomatic. Prolonged consumption of the slightly affected grain or other food products can produce harmful consequences for humans and animals. One such disease, which is connected with cereal grains in specific endemic regions and causes chronic toxicoses in man, is called Urov or Kashin-Beck disease, because it was found for the first time along the Urov River, in Transbaikal.

In the U.S.S.R. Urov disease was noted in the Transbaikal and in the Far East. In addition to edemic foci various cases of this disease are encountered near the Baikal,

Table 32
PHYTOTOXIC ACTION OF *FUSARIUM* SPECIES ON PLANT BRANCHES

Plant	*Fusarium* sp.	Wilted within (hr)
Tagetes patula	*F. poae* #958	12
Zinnia elegans	*F. poae* #958	9
Melilotus albus	*F. poae* #958	14
Xantium strumarium	*F. poae* #958	4
Allium cepa	*F. poae* #958	10
Phaseolus vulgaris	*F. poae* #958	6
Solanum lycopersicum	*F. poae* #958	18
Eschscholtzia california	*F. sporotrichioides* #921	6
Melilotus albus	*F. sporotrichioides* #921	13
Vicia faba	*F. sporotrichioides* #921	8
Solanum lycopersicum	*F. sporotrichioides* #921	16
Pisum sativum	*F. sporotrichioides* #921	12
Sorghum	*F. sporotrichioides* #921	14
Vicia faba	*F. sporotrichioides* #347	8
Solanum lycopersicum	*F. sporotrichioides* #738	18
Zea maydis	*F. sporotrichioides* NRRL5908	12
Hordeum sativum	*F. sporotrichioides* var. *tricinctum* #1227	14
Phaseolus vulgaris	*F. sporotrichioides* var. *chlamydosporum* #1174	15

in Irkutsk, Vologodsk, Pskovsk Leningrad, and Kiev districts.[246] Cases of Urov disease have been described in northern Sweden and Holland, but chiefly this disease is widespread in northern China and Korea. This disease has been known for more than 100 years and its name was dedicated to Russian scientists Kashin and Beck, who studied the disease at the end of the 19th and the beginning of the 20th century.

Kashin-Beck disease is an endemic bone-joint disease associated with an enchondrial type of ossification, which develops chiefly in children of school age. The disease is manifested chiefly as a shortening of the long bones, thickening and subsequently deformity of the joints, flexor contractures, and muscular weakness and atrophy. The disease develops gradually and has a chronic course. In the early stage, marked pains are noted in the joints until the completion of skeletal growth.[221,235,243,245]

The etiology of this disease is connected with various hypotheses and theories, for instance, water sources, deficiency of calcium, and other minerals, etc.

Recently Rubinstein[246] and Perkel[221] isolated various fungi from wheat grains in endemic regions of eastern Transbaikal, along with *F. sporotrichiella* var. *poae.* In experiments on growing white rats and puppies the authors succeeded in reproducing the characteristic symptoms of Urov disease with the isolate of *F. sporotrichiella* var. *poae.* Along with this strain, nontoxic strains which did not cause any pathological changes in animals were isolated more frequently.

The experiments showed that the disease is connected with a toxigenic fungi which had a definite, specific effect on the animals and therefore the strain was separated into a special new physiological form as *F. sporotrichioides* var. *poae* f. *osteodystrophica* Rubinst.[246] It was found capable of producing the specific substances which cause osteodystrophy in bone tissue. Therefore the Urov disease is a food fusariotoxicosis, as is ATA, with a selective effect on the development of bones showing an enchondral type of ossification.

FIGURE 73. Plant branches treated with *F. poae* #958 and control plants. (A) *Tagetes patula* wilted after 12 hr; (B) *Zinnia elegans* wilted after 9 hr; (C) *Melilotus albus* wilted after 14 hr; (D) *Solanum lycopersicum* wilted after 18 hr; (E) *Allium cepa* wilted after 10 hr.

A

B

C

D

FIGURE 74. Plant branches treated with *F. sporotrichioides* #921 and control plants. (A) *Eschscholtzia californica* wilted after 6 hr; (B) *Solanum lycopersicum* wilted after 16 hr; (C) *Vicia faba* wilted after 8 hr; (D) *Sorghum* wilted after 14 hr.

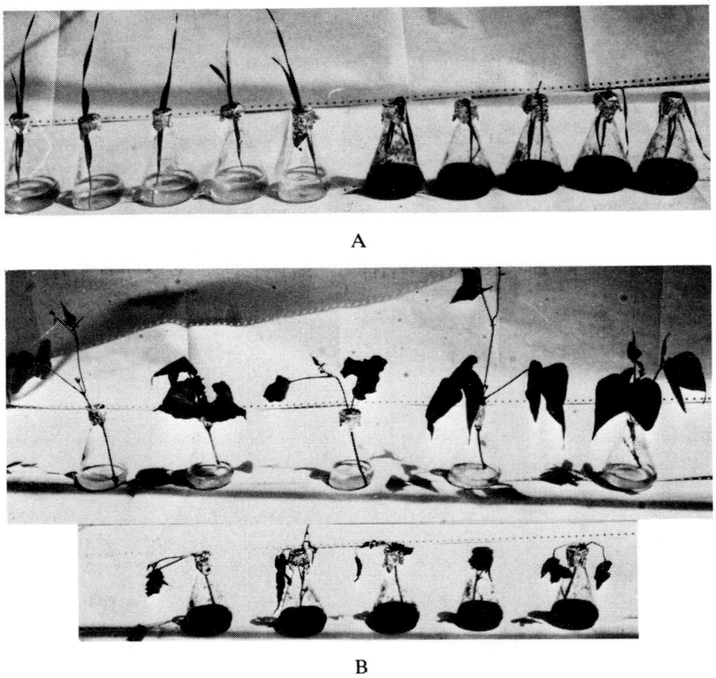

FIGURE 75. Plant branches treated with *F. sporotrichioides* var. *tricinctum* #1227 (A) and *F. sporotrichioides* var. *chlamydosporum* #1174 (B) and control plants. (A) *Hordeum sativum* wilted after 14 hr; (B) *Phaseolus vulgaris* wilted after 15 hr.

Unfortunately, not all investigators agree with this alimentary toxicosis theory which seemed to be most probable. However, this disease requires further studies together with comprehensive investigations of the ecological and environmental conditions including soil conditions of toxin formation in the endemic regions and also identification and determination of the chemical nature of the specific substances produced by the fungi.

Table 33
ISOLATES OF THE SPOROTRICHIELLA SECTION: RELATION BETWEEN INTENSITY OF TOXICITY TO RABBIT SKIN AND PHYTOTOXICITY TO SEEDLINGS

		Av % of plants killed								Rabbit Skin test			
Isolate	No. of isolates	Bean	Cucumber	Watermelon	Cotton	Tomato	Onion	Eggplant	Pepper	O	M	S	Incubation temp (°C)
F. sporotrichioides	8	52	27.5	31.5	59.5	67	53	45	26.5	0	0	8	6, 12
F. poae	5	57	41	34	23	55	81	58	38	0	0	5	6, 12
F. sporotrichioides var. *tricinctum*	2	17.5	37.5	17.5	12.5	50	30	40	10	0	1	1	6, 12
F. sporotrichioides var. *chlamydosporum*	2	65	45	20	40	55	70	20	0	0	0	2	6—18

Note: O, nontoxic; M, mildly toxic; S, strongly toxic.

FIGURE 76. Cucumber seedling wilt caused by *F. sporotrichioides* #921 isolated from rye, grown at 12°C.

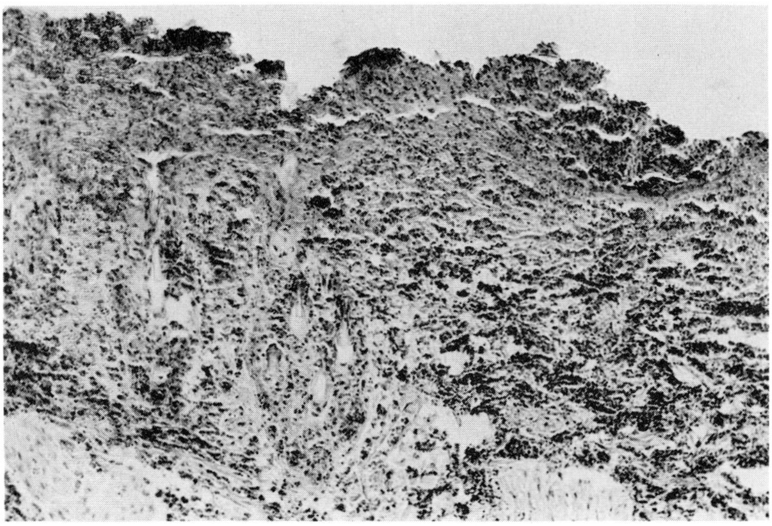

FIGURE 77. *F. sporotrichioides* #921 in rabbit skin test: acute vesicular dermatitis; superficial half of dermis is necrotic with widespread hemorrhage and cellular infiltration of necrotic hair follicles. Deeper layers of dermis are edematous and also infiltrated with leukocytes and eosinophils. (H.E.; magnification × 125.) (From Joffe, A. Z. and Yagen, B., *Mycopathologia,* 60, 93, 1977. With permission.)

FIGURE 78. Pepper seedling wilt induced by *F. poae* #958 isolated from millet, grown at 12°C.

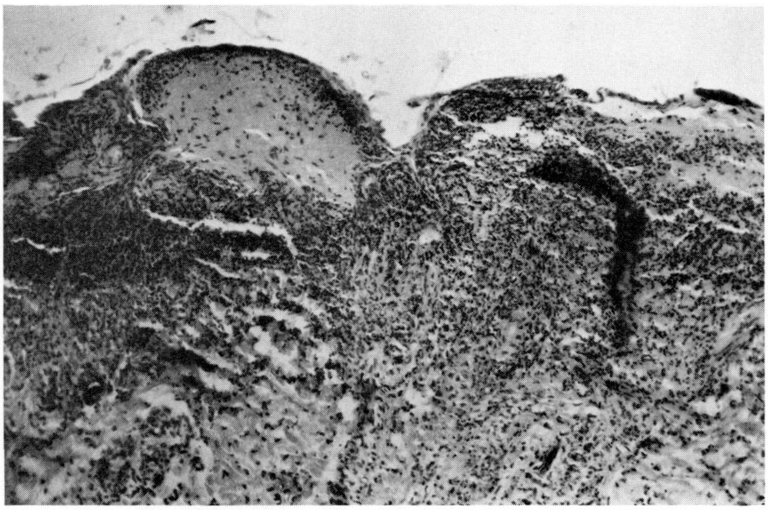

FIGURE 79. *F. poae* #958 in rabbit skin test: acute necrotizing dermatitis characterized by seropurulent exudate covering destroyed epithelium. Few dermal papillae remain imbedded in an edematous dermis, which is heavily infiltrated by inflammatory cells, among which eosinophils are seen. (H.E.; magnification × 130.) (From Joffe, A. Z. and Yagen, B., *Mycopathologia*, 60, 93, 1977. With permission.)

FIGURE 80. Eggplant seedling wilt caused by *F. sporotrichioides* var. *tricinctum* #1227 from millet, grown at 12°C.

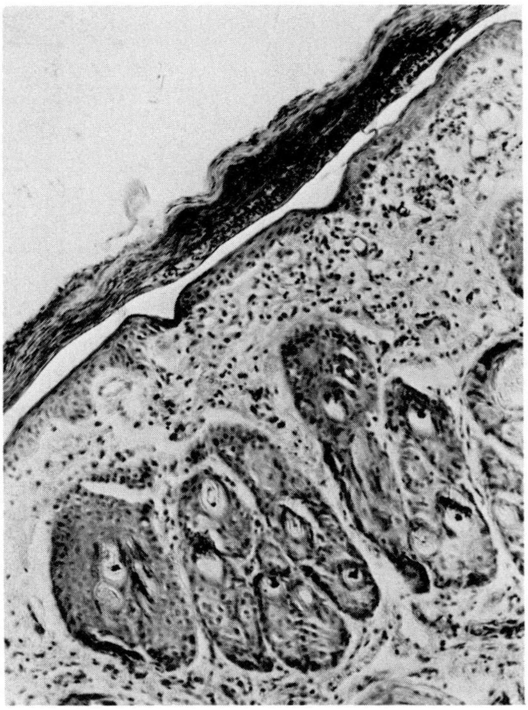

FIGURE 81. *F. sporotrichioides* var. *tricinctum* in rabbit skin test; a thick layer of parakeratotic epithelium heavily infiltrated by polymorphonuclear leukocytes is seen on the surface. Dermis shows moderate inflammatory cell infiltration. (H.E.; magnification × 132.) (Photo courtesy of Marcel Dekker, Inc.)

FIGURE 82. Cotton seedling wilt caused by *F. sporotrichioides* var. *chlamydosporum* #4337 from wheat, grown at 24°C. (From Joffe, A. Z., *Pflkrankh*, 80, 92, 1973. With permission.)

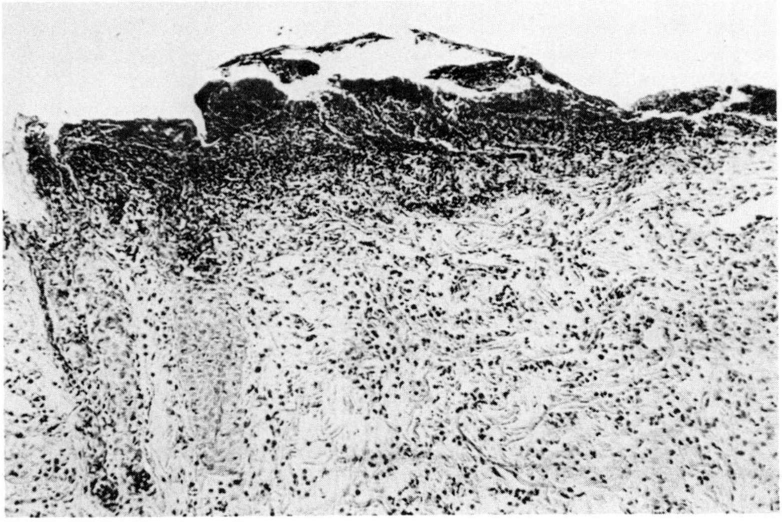

FIGURE 83. *F. sporotrichioides* var. *chlamydosporum* #4337 in rabbit skin test: necrosis of dermis with seropurulent exudate; papillae in edematous dermis are heavily infiltrated by inflammatory cells, mainly eosinophils. (H.E.; magnification × 135.)

REFERENCES

1. **Akhmeteli, M. A., Linnik, A. B., Cernov, K. S., Voronin, V. M., Hesina, A. Ja., Guseva, N. A., and Sabad, L. M.**, Study of toxins isolated from grain infected with *Fusarium sporotrichioides, Pure Appl. Chem.*, 35, 209, 1973.
2. **Aleshin, B. and Eyngorn, E.**, Changes in blood by septic angina and experimental agranulocytosis, in Data on Septic Angina, Proc. 1st Kharkov Med. Inst. a. Chkalov Inst. Epidemiol. Microbiol., Chkalov, 1, 135, 1944.
3. **Aleshin, B. V., Burshtein, Sh. A., and Chernyak, B. I.**, Hematopoietic organs and reticuloendothelial system in aleukia, in Alimentary Toxic Aleukia, Acta Chkalov Inst. Epidemiol. Microbiol., First Communic. Chkalov, 2, 125, 1947.
4. **Alisova, Z., and Mironov, S.**, On toxicity of prosomillet overwintered in the field, in Data on Septic Angina, Proc. First Kharkov Med. Inst. a. Chkalov Inst. Epidemiol. Microbiol., First Communic. Chkalov, 1, 17, 1944.
5. **Alisova, Z. I.**, General toxic action of cereal crops overwintered in the field, in Alimentary Toxic Aleukia. Acta Chkalov Inst. Epidemiol. Microbiol, Second Communic. Chkalov, 2, 104, 1947.
6. **Alisova, Z. I., and Mironov, S. G.**, General toxic action of cereal crops overwintered in the field, in: Alimentary Toxic Aleukia. Acta Chkalov Inst. Epidemiol. Microbiol., Third Communic. Chkalov, 2, 97, 1947.
7. **Alisova, Z. I.**, General action of overwintered cereal extracts and toxic fungi on laboratory animals, in Alimentary Toxic Aleukia, Acta Chkalov Inst. Epidemiol. Microbiol., Abstr., Chkalov, 2, 192, 1947.
8. **Andreyev, K. P.**, Prosomillet poisoning in pastures, *Soviet Vet.*, 9, 75, 1939.
9. **Antonov, N. A., Belkin, G. S., Joffe, A. Z., Lukin, A. Ya., and Simonov, E. N.**, Feeding experiments on horses with cultures of the toxic fungi *Fusarium poae* (Peck.) Wr. and *Cladosporium epiphyllum* (Pers.) Mart., *Acta Chkalov Agric. Inst.*, Chkalov, 4, 47, 1951.
10. **Appel, O. and Wollenweber, H. W.**, Grundlagen einer Monographic der gattung *Fusarium* (Link), *Arb. a.d. Kais. Biol. Inst. f. Land-u. Forstwirschaft*, 8, 1, 1910.
11. **Bamburg, J. R.**, Mycotoxins of the Trichothecene Family Produced by Cereal Molds, Ph.D. thesis, University of Wisconsin, Madison, 1969.
12. **Bamburg, J. R., Riggs, N. V., and Strong, F. M.**, The structures of toxins from two strains of *Fusarium tricinctum, Tetrahedron*, 24, 3329, 1968.
13. **Bamburg, J. R.**, The biological activities and detection of naturally occurring 12,13-epoxy-Δ^9-trichothecenes, *Clin. Toxicol.*, 5, 495, 1972.
14. **Bamburg, J. R.**, Chemical and biochemical studies of the trichothecene mycotoxins, in *Mycotoxins and Other Fungal Related Food Problems*, Rodricks, J. V., Ed., Adv. Chem. Ser. No. 149, American Chemical Society, Washington, D.C., 1976, 144.
15. **Bamburg, J. R., Marasas, W. F. O., Riggs, N. V., Smalley, E. B., and Strong, F. M.**, Toxic spiroepoxy compounds from *Fusarium* and other Hyphomycetes, *Biotechnol. Bioeng.*, 10, 445, 1968.
16. **Bamburg, J. R., Strong, F. M., and Smalley, E. B.**, Toxins from moldy cereals, *J. Agric. Food Chem.*, 17, 443, 1969.
17. **Bamburg, J. R. and Strong, F. M.**, Mycotoxins of the Trichothecene family produced by *Fusarium tricinctum* and *Trichoderma lignorum, Phytochemistry*, 8, 2405, 1969.
18. **Bamburg, J. R., Riggs, M. V., and Strong, F. M.**, The structures of toxins from two strains of *Fusarium tricinctum, Tetrahedron*, 24, 3329, 1968.
19. **Bamburg, J. R. and Strong, F. M.**, 12,13-Epoxytrichothecenes, in *Microbial Toxins*, Vol. 7, Kadis, S., Ciegler, A., and Ajl, S. J., Eds., Academic Press, New York, 1971, 207.
20. **Barer, G. L.**, The problem of the chemical nature of the toxic cereals, in Lecture on Republic Conference about Alimentary Toxic Aleukia, Moscow, January 1947.
21. **Bart, V. V.**, Material on the study of the toxicity of late-gathered cereals in the Latv. S.S.R., in Mycotoxicoses of Man and Agricultural Animals, Bilai, V. I., Ed., U.S. Joint Public Service, Washington, D.C., 1960, 125.
22. **Beletskij, G. N.**, Measures to prevent and fight against "septic angina" (alimentary toxic aleukia), *Hyg. Sanit. (Moscow)*, 3, 22, 1943.
23. **Bilai, V. I.**, *Fusarium* species on cereal crops and their toxic properties, *Microbiology*, 16, 11, 1947.
24. **Bilai, V. I.**, The action of extracts from toxic fungi on animal and plant tissues, *Microbiology*, 17, 142, 1948.
25. **Bilai, V. I.**, Toxic Fungi on the Grain of Cereal Crops, Academy of Sciences of Ukranian S.S.R., Kiev, 1953, 1.
26. **Bilai, V. I.**, *Fusaria*, Academy of Sciences of Ukranian S.S.R., Kiev, 1955.
27. **Bilai, V. I.**, Toxins, in *Biologically Active Substances of Microscopic Fungi*, Naukowa Dumka, Kiev, 1965, 160.

28. **Bilai, V. I.**, Experimental morphogenesis in the fungi of the genus *Fusarium* and their taxonomy, *Ann. Acad. Sci. Fenn. A. IV Biol.*, 168, 7, 1970.
29. **Bilai, V. I.**, Phytopathological and hygienic significance of representatives of the section *Sporotrichiella* in the genus *Fusarium* Link., *Ann. Acad. Sci. Fenn. A. IV Biol.*, 168, 19, 1970.
30. **Bilai, V. I.**, *Fusaria*, 2nd enlarged ed., Naukova Dumka, Kiev, 1977.
31. **Bilai, V. I.**, Principles of the taxonomy and phytopathogenic structure of species of the genus *Fusarium* Lk. ex Fr., *Mikrobiol. Zh.*, 40, 148, 1978.
32. **Bilai, V. I.**, Mycological aspects of alimentary toxicoses, *Mikrobiol. Zh.*, 40, 205, 1978.
33. **Boldyrev, T. E. and Shtenberg, A. I.**, Alimentary toxic aleukia, in *Food Hygiene*, Methodological and Reference Guide for Physician, Medgiz, Moscow, 1950, 234.
34. **Bolton, A. T. and Nuttall, V. W.**, Pathogenicity studies with *Fusarium poae*, *Can. J. Plant Sci.*, 48, 161, 1968.
35. **Booth, C.**, The genus *Fusarium*, Commonwealth Mycolotical Institute, Commonwealth Agricultural Bureau, Central Sales, Farnham Royal, Bucks, England, 1971, 1.
36. **Brown, W.**, Studies in the Genus Fusarium. VI. General description of strains, together with a discussion of the principles at present adopted in the classification of *Fusarium*, *Ann. Bot.*, 42, 285, 1928.
37. **Brown, W. and Horne, A. S.**, Studies in the Genus *Fusarium*. III. An analysis of factors which determine certain microscopic features of *Fusarium* strains, *Ann. Bot.*, 40, 203, 1926.
38. **Burmeister, H. R.**, T-2 toxin production by *Fusarium tricinctum* on solid substrate, *Appl. Microbiol.*, 21, 739, 1971.
39. **Burmeister, H. R. and Hesseltine, C. W.**, Biological assays for two mycotoxins produced by *Fusarium tricinctum*, *Appl. Microbiol.*, 20, 437, 1970.
40. **Burmeister, H. R., Ellis, J. J., and Hesseltine, C. W.**, Survey for *Fusarium* that elaborate T-2 toxin, *Appl. Microbiol.*, 23, 1165, 1972.
41. **Chernikov, E. A.**, Problems of etiopathogenesis and therapy of "septic angina", in Data of Septic Angina, *Proc. First Kharkov Med. Inst. a. Chkalov Inst. Epidemiol. Microbiol.*, 1, 173, 1944.
42. **Chi, M. S., Mirocha, C. J., Kurz, M. J., Weaver, G., Bates, F., and Shimoda, W.**, Subacute toxicity of T-2 toxin in broiler chicks, *Poultry Sci.*, 56, 306, 1977.
43. **Chi, M. S., Mirocha, C. J., Kurz, M. J., Weaver, G., Bates, F., and Shimoda, W.**, Effect of T-2 toxin on reproductive performance and health of laying hens, *Poultry Sci.*, 56, 628, 1977.
44. **Chi, M. S., Mirocha, C. J., Kurz, M. J., Weaber, G., Bates, F., and Burmeister, H. R.**, Acute toxicity of T-2 toxin in broiler chicks and laying hens, *Poultry Sci.*, 56, 103, 1977.
45. **Chi, M. S. and Mirocha, C. J.**, Necrotic oral lesions in chickens deacetoxyscirpenol, T-2 toxin, and crotocin, *Poultry Sci.*, 57, 807, 1978.
46. **Chi, S. M., Robinson, T. S., Mirocha, C. J., and Reddy, K. R.**, Acute toxicity of 12,13-Epoxytrichothecenes in one-day old broiler chicks, *Appl. Environ. Microbiol.*, 35, 636, 1978.
47. **Chilikin, V. I.**, Principal Problems of Clinical Syndromes, Pathogenesis, and Therapy of Alimentary Toxic Aleukia (Septic Angina), Reginal Publishing House, Kujbyshev, 1944, 1.
48. **Chilikin, V. I.**, Septic angina, in Clinical Aspects and Therapy of Alimentary Toxic Aleukia (Septic Angina), Nesterov, A. I., Sysin, A. H., and Karlic, L. N., Eds., State Medical Literature, Medgiz, 1945, 39.
49. **Chilikin, V. I.**, Peculiarities and problems concerning clinical aspects, pathogenesis and therapy of alimentary toxic aleukia, in Alimentary Toxic Aleukia. Acta Chkalov Inst. Epidemiol. Microbiol., Chkalov, 2, 145, 1947.
50. **Christensen, C. M., Meronuck, R. H., Nelson, G. H., and Behrens, J. C.**, Effects on turkey poults of rations containing corn invaded by *Fusarium tricinctum* (Cda.) Snyder and Hanson, *Appl. Microbiol.*, 23, 177, 1972.
51. **Christensen, C. M., Nelson, G. H., Morocha, C. J., and Bates, F.**, Toxicity to experimental animals of 943 isolates of fungi, *Cancer Res.*, 28, 2293, 1968.
52. **Chu, F. S.**, Mode of action of mycotoxins and related compounds, in *Advances in Applied Microbiology*, Vol. 22, Perlman, D., Ed., Academic Press, New York, 1977, 83.
53. **Ciegler, A.**, Trichothecenes: occurrence and toxicoses, *J. Food Prot.*, 41, 399, 1978.
54. **Davydova, V. L.**, Sensibility of human and animal skin to toxin of cereals overwintered in the field, in Alimentary Toxic Aleukia, Acta Chkalov Inst. Epidemiol. Microbiol., 2, 94, 1947.
55. **Davydovskij, E. B., Kestner, A. G.**, On so-called septic angina (morphology and pathogenesis), *Arth. Pathol. Anat.*, 1, 11, 1935.
56. **De Nicola, D. B., Rebar, A. H., and Carlton, W. W.**, T-2 toxin mycotoxicosis in the guinea-pig, *Food Cosmet. Toxicol.*, 16, 601, 1978.
57. **Dorrell, D. G.**, Fatty acid composition of buckwheat seed, *J. Am. Oil Chem. Soc.*, 48, 693, 1971.
58. **Drabkin, B. S. and Joffe, A. Z.**, The effect of extract from over-wintered cereals on *Paramaecium caudatum*, *Acta Chkalov Med. Inst.*, 2, 92, 1950.

59. **Drabkin, B. S. and Joffe, A. Z.**, The protistocide effect of certain mold, *Microbiol. Acad. Sci. U.S.S.R. (Moscow)*, 21, 700, 1952.
60. **Ellison, R. A. and Kotsonis, F. N.**, T-2 toxin on an emetic factor in moldy corn, *Appl. Microbiol.*, 25, 540, 1973.
61. **Ellison, R. A. and Kotsonis, F. N.**, In vitro metabolism of T-2 toxin, *Appl. Microbiol.*, 27, 423, 1974.
62. **Elpidina, O. K.**, The etiology of septic angina and determination of grain toxicity, in *Data on Alimentary Toxic Aleukia*, Malkin, Z. I., Ed., Kazan, 1945, 78.
63. **Elpidina, O. K.**, Phytotoxin in grain causing septic angina, *Lect. Acad. Sci. U.S.S.R. (DAN)*, 46, 2, 1945.
64. **Elpidina, O. K.**, Biological methods of determining grain toxicity causing septic angina, *Lect. Acad. Sci. U.S.S.R.*, 51, 2, 1946.
65. **Elpidina, O. K.**, Biological and antibiotic properties of poin, in Abstr. Mycotoxicoses of Men and Agric. Anim., Academy of Sciences of Ukrainian S.S.R. Kiev, 1956, 23.
66. **Elpidina, O. K.**, Antibiotic and antiblastic properties of poin, *Antibiotics*, 4, 46, 1959.
67. **Elpidina, O. K.**, Toxic and antibiotic properties of poin, in Mycotoxicoses of Man and Agricultural Animals, Bilai, V. I., Ed. Academy of Sciences of Ukrainian S.S.R. Kiev, 1960, 59.
68. **Elpidina, O. K.**, The Action of Poin Toxin from *Fusarium sporotrichioides* v. *poae* on Malignant Tumor Growth in Experiments, Abstr. Rep. Conf. Mycotoxicoses, Kiev, 1961.
69. **Elpidina, O. K.**, The antiblastic properties of poin, according to experimental data, *Kazan Med. J.*, 39, 96, 1958.
70. **Fiester, L. F. and Fiester, M.**, *Steroids*, Reinhold, New York, 1959, 727.
71. **Forgacs, J. and Carll, W. I.**, Mycotoxicoses, *Adv. Vet. Sci.*, 7, 273, 1962.
72. **Friedman, M. Yu.**, Prophylaxis of alimentary toxic aleukia (Septic Angina), in *Alimentary Toxic Aleukia (Septic Angina)*, Medical Publishing House, Moscow, 1945, 54.
73. **Friedman, M. N.**, Residual changes in the internal organs and the blood by alimentary toxicosis with overwintered wheat, in *Acta of Septic Angina*, Bashkir Publishing, Ufa, 1945, 158.
74. **Gabel, Iu, O.**, On the chemistry of the toxic substances of overwintered millet, in Alimentary Toxic Aleukia, Acta Chkalov Inst. Epidemiol. Microbiol., Chkalov, 2, 42, 1947.
75. **Gajdusek, D. C.**, Alimentary toxic aleukia, in Acute Infectious Hemorrhagie Fevers and *Mycotoxicoses* in the Union of Soviet Socialist Republic Army Med. Serv. Graduate School, Walter Reed Medical Center, Washington, D.C., 1953, 82.
76. **Gauman, E.**, Toxins and plant disease, *Endeavour*, 13, 198, 1954.
77. **Geimberg, V. G. and Babusenko, A. M.**, Dynamics of development of microflora in overwintered grain in experimental field conditions, *Hyg. Sanit.*, 5, 31, 1949.
78. **Geminov, N. B.**, Etiology of alimentary toxic aleukia, in *Septic Angina and its Treatment*, Chilikin, V. and Seminov, N., Eds., Kuybyshev, 1945, 3.
79. **Geminov, N. B.**, Epidemiology of Septic Angina in the Kuybyshev District in 1945, Lecture on Republic Conference of Alimentary Toxic Aleukia, Moscow, December 1945.
80. **Geminov, N. V.**, On the Problem of Outbreaks of Alimentary Toxic Aleukia Disease from Use of Acorns in Food of the Spring Harvest, Lecture on Republic Conference of Alimentary Toxic Aleukia, Moscow, March 1948.
81. **Genkin, A.**, Condition of the upper respiratory tract in septic angina, in Data on Septic Angina, Proc. First Kharkov, Med. Inst. a. Chakov Inst. Epidemiol. Microbiol., 1, 117, 1944.
82. **Germanov, A. I.**, Some laboratory data relating to alimentary toxic aleukia, *Sov. Med.*, 3, 9, 1945.
83. **Gerlach, W.**, Suggestion for an acceptable modern *Fusarium* system, in *Ann. Acad. Sci. Fenn. Ser. A IV Biol.*, 168, 37, 1970.
84. **Getsova, G.**, Data on experimental study of fusariotoxicosis, in Mycotoxicoses of Man and Agricultural Animals, Bilai, V., Ed., Academy of Sciences of Ukranian S.S.R., Kiev, 1960, 81.
85. **Gilgan, M. W., Smalley, E. B., and Strong, F. M.**, Isolation and partial characterization of a toxin from *Fusarium tricinctum* on moldy corn, *Arch. Biochem. Biophys.*, 114, 1, 1966.
86. **Gjertsen, P., Trolle, B., and Anderson, K.**, Studies on gushing. II. Gushing caused by microorganisms specially *Fusarium* species, in Eur. Brew. Conv. Proc. Congr., Stockholm, 1965, 428.
87. **Goldfredsen, W. O., Grove, J. F., and Tamm, C.**, On the nomenclature of a new class of sesquiterpenoids, *Helv. Chim. Acta*, 40, 1666, 1968.
88. **Gordon, W. L.**, The occurrence of *Fusarium* species in Canada. II. Prevalence and taxonomy of *Fusarium* species in cereal seed, *Can. J. Bot.*, 30, 209, 1952.
89. **Gordon, W. L.**, Taxonomy of *Fusarium* species in the seed of vegetables, forage, and miscellaneous crops, *Can. J. Bot.*, 32, 576, 1951.
90. **Gordon, W. L.**, The taxonomy and habitus of *Fusarium* species from tropical and temperate regions, *Can. J. Bot.*, 38, 643, 1960.

91. **Gordon, D. L. and Levitskij, L. U.**, Pecularities of the Clinical Course of Alimentary Toxic Aleukia and the Results of its Therapy, Lecture on Republic Conference of Alimentary Toxic Aleukia, Moscow, December 1945.
92. **Gorodijskaja, R. B.**, Blood changes induced by septic angina, in *Acta on Septic Angina*, Bashkir Publishing, Ufa, 1945, 112.
93. **Greenway, J. A. and Puls, R.**, Fusariotoxicosis from barley in British Columbia. I. Natural occurrence and diagnosis, *Can. J. Comp. Med.*, 40, 12, 1976.
94. **Grinberg, G. I.**, Clinical aspects and pathogenesis of the so-called septic angina, *Sov. Med.*, 10, 24, 1943.
95. **Gromashevskij, L. V.**, Pathogenesis of alimentary toxic aleukia, *J. Microbiol. Epidemiol. Immunol.*, 3, 65, 1945.
96. **Grove, M. D., Yates, S. G., Tallent, W. H., Ellis, J. J., Wolf, I. A., Kosuri, N. R., and Nicholas, R. E.**, Mycotoxins produced by *Fusarium tricinctum* as possible causes of cattle disease, *J. Agric. Food Chem.*, 18, 734, 1970.
97. **Gubarev, E. M. and Gubareva, N. A.**, Chemical nature and certain chemical properties of the toxic substances responsible for septic angina, *Biochemistry*, 10, 199, 1945.
98. **Gurewitch, Z. A.**, Neuropathologic peculiarities of septic angina, in Data of Septic Angina, Proc. First Kharkov Med. Inst. a. Chkalov Inst. Epidemiol. Microbiol, 1, 103, 1944.
99. **Hrootski, E. T. and Joffe, A. Z.**, The action of *Fusarium poae* on the motor activity of the stomach of dogs, *Acta Chkalov Agric. Inst.*, 6, 59, 1953.
100. **Hsu, I. C., Smalley, F. B., Strong, F. M., and Ribelin, W. E.**, Identification of T-2 toxin in moldy corn associated with a lethal toxicosis in cattle, *Appl. Microbiol.*, 24, 684, 1972.
101. **Ishii, K., Sakai, K., Ueno, Y., Tsudoda, H., and Enomoto, M.**, Solaniol, a toxic metabolite of *Fusarium solani, Appl. Microbiol.*, 22, 718, 1971.
102. **Itoh, T., Tamura, T., and Matsumoto, T.**, Sterol composition of 19 vegetable oils, *J. Am. Oil Chem. Soc.*, 50, 122, 1973.
103. **Jamalainen, E. A.**, Über die Fusarien Finlands. II, *Valt. Maatalusk. Julk. Helsinki*, 123, 1, 1943.
104. **Joffe, A. Z.**, The biological properties of fungi, isolated from overwintered cereal crops, in Alimentary Toxic Aleukia, Acta Chkalov Inst. Epidemiol. Microbiol., Abstr. Chkalov, 2, 192, 1947.
105. **Joffe, A. Z.**, The mycoflora of normal and overwintered cereals in 1943-44, in Alimentary Toxic Aleukia, Acta Chkalov Inst. Epidemiol. Microbiol., First Commun., 2, 28, 1947.
106. **Joffe, A. Z.**, The mycoflora of normal and overwintered cereals in 1944—45, in Alimentary Toxic Aleukia, Acta Chkalov Inst. Epidemiol. Microbiol., Second Commun., 2, 35, 1947.
107. **Joffe, A. Z.**, The dynamics of toxin accumulation in overwintered cereals and their mycoflora in 1945—46, in Alimentary Toxic Aleukia, Acta, Chkalov Inst. Epidemiol. Microbiol., Chkalov, Abstr. 2, 192, 1947.
108. **Joffe, A. Z.**, Toxicity of Fungi on Cereals Overwintered in the Field (on the Etiology of Alimentary Toxic Aleukia), Dissertation, Institute of Botany, Academy of Sciences of U.S.S.R., Leningrad, 1950.
109. **Joffe, A. Z.**, The etiology of alimentary toxic aleukia, in Conference on Mycotoxicosis of Man and Husbandry Animals, Academy of Sciences of Ukrainian S.S.R., Kiev, 1956, 36.
110. **Joffe, A. Z.**, The influence of overwintering on the antibiotic activity of several molds of the genus *Cladosporium, Alternaria, Fusarium, Mucor, Thamnidium* and *Aspergillus, Acta Acad. Sci. Lithuanian S.S.R., Ser. B*, 3, 85, 1956.
111. **Joffe, A. Z.**, The effect of environmental conditions on the antibiotic activity of some fungi of the genus *Penicillium, Acta Acad. Sci. Lithuanian S.S.R. Ser. B.*, 4, 101, 1956.
112. **Joffe, A. Z.**, Toxicity and antibiotic properties of some *Fusarium, Bull. Res. Counc. Isr.*, 8D, 81, 1960.
113. **Joffe, A. Z.**, The mycoflora of overwintered cereals and its toxicity, *Bull. Res. Counc. Isr.*, 9D, 101, 1960.
114. **Joffe, A. Z.**, Biological properties of some toxic fungi isolated from overwintered cereals, *Mycopathol. Mycol. Appl.*, 16, 201, 1962.
115. **Joffe, A. Z.**, Toxicity of overwintered cereals, *Plant Soil*, 18, 31, 1963.
116. **Joffe, A. Z.**, Toxin production by cereal fungi causing toxic alimentary aleukia in man, in *Mycotoxins in Foodstuffs*, Wogan, G. N., Ed., MIT Press, Cambridge, Mass., 1965, 77.
117. **Joffe, A. Z.**, Toxic properties and effects of *Fusarium poae* (Peck.) *F. sporotrichioides* Sherf., and *Aspergillus flavus* Link, *J. Stored Prod. Res.*, 5, 211, 1969.
118. **Joffe, A. Z.**, Alimentary toxic aleukia, in *Microbial Toxins*, Vol. 7, Kadis, S., Ciegler, A., and Afj, S. J., Eds., Academic Press, New York, 1971, 139.
119. **Joffe, A. Z.**, *Fusarium* species of the *Sporotrichiella* section and relations between their toxicity to plants and animals, *Pflkrankh*, 80, 92, 1973.
120. **Joffe, A. Z.**, Toxicity of *Fusarium poae* and *F. sporotrichioides* and its relation to alimentary toxic aleukia, in *Mycotoxins*, Purchase, I. F. H., Ed., Elsevier, Amsterdam, 1974, 229.

121. **Joffe, A. Z.**, A modern system of *Fusarium* taxonomy, *Mycopathol. Mycol. Appl.*, 53, 201, 1974.
122. **Joffe, A. Z.**, Growth and toxigenicity of *Fusarium* of the *Sporotrichiella* section as related to environmental factors and culture substrates, *Mycopathol. Mycol. Appl.*, 54, 35, 1974.
123. **Joffe, A. Z.**, The taxonomy of toxigenic species of *Fusarium*, in *Handbook of Mycotoxins and Mycotoxicoses*, Vol. 1, Wyllie, T. D. and Morehouse, L. G., Eds., Marcel Dekker, New York, 1977, 59.
124. **Joffe, A. Z.**, *Fusarium poae* and *Fusarium sporotrichioides* as principal causes of Alimentary Toxic Aleukia, in *Handbook of Mycotoxins and Mycotoxicoses*, Vol. 3, Wyllie, T. D. and Morehouse, L. G., Eds., Marcel Dekker, New York, 1978, 21
125. **Joffe, A.Z.**,Fusarium toxicoses of poultry, in *Handbook of Mycotoxins and Mycotoxicoses*, Vol. 2, Wyllie, T. D. and Morehouse, L. G., Eds., Marcel Dekker, New York, 1978, 309.
126. **Joffe, A. Z. and Mironov, S. G.**, Mycoflora of normal cereals and cereals overwintered in the field, Part 2, in Alimentary Toxic Aleukia, Acta Chkalov Inst. Epidemiol. Microbiol., Chkalov, 2, 35, 1947.
127. **Joffe, A. Z. and Palti, J.**, Relations between harmful effects on plants and on animals of toxins produced by species of *Fusarium*, *Mycopathol. Mycol. Appl.*, 52, 209, 1974.
128. **Joffe, A. Z. and Palti, J.**, Taxonomic study of *Fusaria Sporotrichiella* section used in recent toxicological work, *Appl. Microbiol.*, 29, 575, 1975.
129. **Joffe, A. Z. and Yagen, B.**, Comparative study of the yield of T-2 toxin produced by *Fusarium poae*, *F. sporotrichioides*, and *F. sporotrichioides* var. *tricinctum* from different sources, *Mycopathologia*, 60, 93, 1977.
130. **Joffe, A. Z. and Yagen, B.**, Intoxication produced by toxic fungi *Fusarium poae* and *F. sporotrichioides* on chicks, *Toxicon*, 16, 263, 1978.
130a. **Joffe, A. Z.**, unpublished data.
131. **Karatygin, V. M. and Rozhnova, Z. I.**, Vitamin insufficiency in alimentary toxic aleukia (septic angina), *Sov. Med.*, 5, 17, 1947.
132. **Karlik, L. N.**, The history of alimentary toxic aleukia (septic angina), in Alimentary Toxic Aleukia (Septic Angina), Nesterov, A. I., Sysin, A. N., and Karlic, L. I., Eds., State Medical Literature, Medgiz, 1945, 3.
133. **Kasirskij, I. A. and Alekseyev, G. A.**, Alimentary toxic aleukia, in *Disease of Blood and Hematopoietic System*, Medgiz, 1948, 204.
134. **Kost, E. A.**, Agranulocytosis and hemorrhagic angina, *Sov. Clin.*, 1, 5, 1935.
135. **Kavetskij, N. E. and Grinberg, B. M.**, On the problem of the therapy of septic angina (alimentary toxic aleukia), *Sov. Med.*, 3, 12, 1945.
136. **Khabibullina, G. F.**, On the method of combined therapy of septic angina of Bogomolt's antireticular cytotoxic serum (ACS) and sulfonamide compounds according to the data of the Ear clinic of Bashkir Medical Institute, in *Acta of Septic Angina*, Bashkir Publishing, Ufa, 1945, 118.
137. **Kholodenko, M. I.**, Changes in the nervous system of toxic alimentary aleukia (so called "septic angina"), *Neuropath. Psychiatr. (Moscow)*, 16, 67, 1947.
138. **Kolosova, N. I.**, Data on Chemical and Toxic Properties of some Fatty Acids Isolated from Grains Causing Alimentary Toxic Aleukia, Dissertation, Moscow, 1949, 112.
139. **Kopytkova, O. I.**, Differential diagnosis of alimentary toxic aleukia, *Health Kazakhst.*, 7, 39, 1948.
140. **Kosuri, N. R., Grove, M. D., Yates, M. S., Tallent, W. H., Ellis, J. J., Wolff, I. A., and Nichols, R. E.**, Response of cattle to mycotoxins of *Fusarium tricinctum* isolated from corn and fescue, *J. Am. Vet. Med. Assoc.*, 157, 938, 1970.
141. **Kosuri, N. R., Smalley, E. B., and Nichols, R. E.**, Toxicologic studies of *Fusarium tricinctum* (Corda), Snyder et Hanson from corn, *Am. J. Vet. Res.*, 32, 1943, 1971.
142. **Kovalev, E. N.**, The nervous system in so-called septic angina, *Neuropathol. Psychiatry*, 13, 75, 1944.
143. **Koza, M. A., Leontiev, I. A., and Yasnitskij, P. Ya.**, *Alimentary Toxic Aleukia, (Septic Angina)*, Medical Publishing House, Moscow, 1944, 3.
144. **Kozin, N. I. and Yershova, O. A.**, Cultivation method for determining toxicity of cereal grain overwintered under snow (prosomillet), in Proc. Nutr. Inst. Acad. Med. Sci. U.S.S.R., Moscow, 1945, 39.
145. **Kretovich, V. L.**, Biochemistry of Grain Causing Septic Angina, Abstr. Lect. on Republic Conference on Alimentary Toxic Aleukia, Moscow, January 1945, 9.
146. **Kretovich, V. L. and Bundel, A. A.**, Investigations of the oil of toxic overwintered prosomillet, *Biochemistry*, 10, 216, 1945.
147. **Kretovich, V. L., Mishustin, E. N., and Bundel, A. A.**, Suppression of alcohol fermentation by products of oil decomposition, *Biochemistry*, 11, 149, 1946.
148. **Kretovich, V. L. and Sosedov, N. I.**, Biochemical properties of toxic prosomillet, *Biochemistry*, 10, 279, 1946.
149. **Kretovich, V. L. and Skripkina, Z. G.**, Diagnostics of overwintered toxic grains in the field, *Lect. Acad. Sci. U.S.S.R.*, 47, 504, 1945.

150. **Kudryakov, V. T.**, Pathogenesis and treatment of alimentary toxic aleukia, *Clin. Med.*, 24, 14, 1945.
151. **Kurbatova, T. G.**, Alimentary toxic aleukia (review of the literature), *Acta of Kirousk Inst. Epidemiol. Microbiol.*, 2, 71, 1948.
152. **Kvashnina, E. S.**, Mycoflora of cereal crops overwintered in the field, in Data on Cereal Crops Wintered under Show, Ministry of Agriculture, U.S.S.R., Moscow, 1948, 86.
153. **Lacicowa, B.**, Studies on the morphology and biology of *F. poae* and its pathogenicity to wheat seedlings, *Ann. Univ. Mariae Curie-Sklodovska, Sect. C*, 18, 419, 1963.
154. **Lando, Ya. Kh.**, On the pathologic anatomy of septic angina, *Med. J. Kazakhstan*, 4—5, 88, 1935.
155. **Lando, Ya. Kh.**, Material relating to the pathologic anatomy of septic angina, *Sov. Public Health Serv. Kirghiz S.S.R.*, 6, 72, 1939.
156. **Leonian, L. H.**, Studies on the variability and dissociations in the genus *Fusarium, Phytopathology*, 19, 753, 1929.
157. **Levin, I. I.**, Treatment of alimentary toxic aleukia with sulfamid compound preparations, *Clin. Med.*, 7, 54, 1946.
158. **Levitskij, L. M.**, The Use of Penicillin in Alimentary Toxic Aleukia, Lecture on Republic Conference of Alimentary Toxic Aleukia, Moscow, March 1948.
159. **Lopatin, G. M.**, Alimentary toxic aleukia (septic angina) in children, *Pediatrics*, 1, 27, 1946.
160. **Lindenfelser, L. A., Lillehoj, E. B., and Burmeister, H. R.**, Aflatoxin trichothecene toxins: skin tumor induction synergistic acute toxicity in white mice, *J. Natl. Cancer Inst.*, 52, 113, 1974.
161. **Lovla, D. S.**, Septic angina in the Chkalov district, in Alimentary Toxic Aleukia (Septic Angina), Abstr. Lect. Publ. Lab. Septic Angina, Chkalov Inst. Epidemiol. Microbiol., Kharkov, Medical Institute and Clinic, Hospital of Orenburg Railway, Chkalov, 1944, 1.
162. **Lozanov, N. N. and Tsareva, V. Ya.**, *Septic Angina*, Tatar A.S.S.R. Publishing, Kazan, 1944, 3.
163. **Lukin, A. Ya. and Berlin, M. G.**, Toxic influences of overwintered prosomillet on the organs of horses and pigs, *Proc. Chkalov Agric. Inst.*, 3, 65, 1947.
164. **Lukin, A. Ya, Antonov, N. A., and Simonov, I. N.**, Feeding tests with toxic cereals on pigs and sheep, *Proc. Chkalov. Agric. Inst.*, 3, 78, 1947.
165. **Lukin, A. Ya., Antonov, N. A., and Simonov, I. N.**, Feeding tests with toxic cereals on horses, *Proc. Chkalov Agric. Inst.*, 3, 93, 1947.
166. **Lutsky, I., Mor, N., Yagen, B., and Joffe, A. Z.**, The role of T-2 toxin in experimental alimentary toxic aleukia: a toxicity study in cats, *Toxicol. Appl. Pharmacol.*, 43, 111, 1978.
167. **Lyass, M. A., Agranulocytosis**, *Vietbsk Med. Inst.*, 1940, 3.
168. **Maisuradge, G. I.**, The Role of Fungi in Developing Toxicosis in Horses by Feeding Them Germinating Oats, Abstr. Dissertation, Moscow, 1953, 1.
169. **Malkin, Z. I. and Odelevskaja, N. N.**, Clinical aspects and treatment of alimentary toxic aleukia (septic angina), in *Alimentary Toxic Aleukia*, Kazan, 1945, 19.
170. **Manburg, E. M.**, Clinical aspects of septic angina, in Data on septic angina, Proc. First Kharkov Med. Inst. a. Chkalov Inst. Epidemiol. Microbiol., Chkalov, 1, 85, 1944.
171. **Manburg, E. M. and Rachalskij, E. A.**, Clinical aspects and therapy of alimentary toxic aleukia, in Alimentary Toxic Aleukia, Acta Chkalov Inst. Epidemiol. Microbiol. 2, 152, 1947.
172. **Manoilova, O. S.**, Chemical Diagnosis of Toxic Overwintered Cereals, Lecture Republic Conference on Alimentary Toxic Aleukia, Moscow, January 1947.
173. **Marasas, W. F. O., Bamburg, J. R., Smalley, E. B., Strong, F. M., Ragland, W. L., and Degurse, P. E.**, Toxic effects on trout, rats and mice of T-2 toxin produced by the fungus *Fusarium tricinctum* (Cd.), Snyd. and Hans., *Toxicol. Appl. Pharmacol.*, 15, 471, 1969.
174. **Marasas, W. F. O., Smalley, E. B., Bamburg, J. B., and Strong, F. M.**, Phytotoxicity of T-2 toxin produced by *Fusarium tricinctum, Phytopathology*, 61, 1488, 1971.
175. **Marasas, W. F. O., Smalley, E. B., Degurse, P. E., Bamburg, J. R., and Nichols, R. E.**, Acute toxicity to rainbow trout *(Salmo gardineri)* of a metabolite produced by the fungus *Fusarium tricinctum, Nature (London)*, 214, 817, 1967.
176. **Matuo, T.**, Taxonomic studies of phytopathogenic *Fusarium* in Japan, *Rev. Plant Prot. Res. (Tokyo)*, 5, 34, 1977.
177. **Mayer, C. F.**, Endemic panmyelotoxicosis in the Russian grain belt. I. The clinical aspects of alimentary toxic aleukia (ATA), A comprehensive review, *Mil. Surg.*, 113, 173, 1953.
178. **Mayer, C. F.**, Endemic panmyelotoxicosis in the Russian grain belt. II. The botany, phytopathology, and toxicology of Russian cereal food, *Mil. Surg.*, 113, 295, 1953.
179. **Mikhailovskij, S. V.**, The first experience in therapeutic application of Bogomolt's antireticular cytotoxic serum (ACS) in so-called septic angina, in *Acta of Septic Angina*, Bashkir Publishing, Ufa, 1945, 94.
180. **Mirocha, C. J. and Pathre, S.**, Identification of the toxic principle in a sample of poaefusarin, *Appl. Microbiol.*, 26, 719, 1973.
181. **Mirocha, C. J. and Christensen, C. M.**, Fungus metabolites toxic to animals, *Ann. Rev. Phytopathol.*, 12, 303, 1974.

182. **Mironov, S.**, Etiology of septic angina (alimentary toxic aleukia) and measures for its prevention, *J. Microbiol. Epidemiol. Immunol.*, 6, 70, 1945.
183. **Mironov, S. and Fok, R.**, Toxicity of overwintered prosomillet in the field, Sec. Commun., in Data on Septic Angina, Proc. First Kharkov Med. Inst. a. Chkalov Inst. Epidemiol. Microbiol., 1, 37, 1944.
184. **Mironov, S., Soboleva, R., Fok, R. and Yudenich, V.**, Phytopathological analysis of overwintered prosomillet samples in the field, in Data on Septic Angina, Proc. First Kharkov Med. Inst. a. Chkalov Inst. Epidemiol. Microbiol., 1, 47, 1944.
185. **Mironov, S. G., Joffe, A. Z., Bakbardina, M. K., Fok, R., and Davydova, V. L.**, Phytopathological analysis of toxic overwintered prosomillet samples, Second communic., in Alimentary Toxic Aleukia, Acta Chkalov Inst. Epidemiol. Microbiol., 2, 11, 1947.
186. **Mironov, S. G., Myasnikov, V. A.**, Characteristics of toxins from overwintered cereals in the field, First communic., in Alimentary Toxic Aleukia, Acta Chkalov Inst. Epidemiol. Microbiol., 2, 61, 1947.
187. **Mironov, S. G. and Alisova, Z. I.**, Detection of toxic compounds of fungal derivation in overwintered cereals by means of immunity tests, in Alimentary Toxic Aleukia, Acta Chkalov Inst. Epidemiol. Microbiol., 2, 192, 1947.
188. **Mironov, S. G., and Davydova, V. L.**, Sensibility of man and animal skin to toxins of cereals overwintered in the field, in Alimentary Toxic Aleukia, Acta Chkalov Inst. Epidemiol. Microbiol, 2, 89, 1947.
189. **Mironov, S. G. and Fok, R. A.**, Skin tests of rabbits for determination of toxicity of overwintered cereals, in Alimentary Toxic Aleukia, Acta Chkalov Inst. Epidemiol. Microbiol., Second communic. Chkalov, 2, 80, 1947.
190. **Mironov, S. G. and Fok, R. A.**, Skin test of rabbits for determination of toxicity of overwintered cereals, in Alimentary Toxic Aleukia, Acta Chkalov Inst. Epidemiol. Microbiol., Third communic. Chkalov, 2, 83, 1947.
191. **Mironov, S. G., Strukov, A. I., and Fok, R. A.**, Skin test of rabbits for determination of toxicity of overwintered cereals, in Alimentary Toxic Aleukia, Acta Chkalov Inst. Epidemiol. Microbiol., First communic. Chkalov, 2, 73, 1947.
192. **Mironov, S. G. and Joffe, A. Z.**, The dynamics of toxin accumulation in overwintered cereals in the field, in Alimentary Toxic Aleukia, Acta Chkalov Inst. Epidemiol. Microbiol., First communic. Chkalov, 2, 19, 1947.
193. **Mironov, S. G. and Joffe, A. Z.**, The dynamics of toxin accumulation in overwintered cereals in the field, in Alimentary Toxic Aleukia, Acta Chkalov Inst. Epidemiol. Microbiol., Second communic. Chkalov, 2, 23, 1947.
194. **Mishustin, E. N., Kretovich, V. L., and Bundel, A. A.**, The fermentation test as a diagnostic method for toxic overwintered prosomillet, *Hyg. Sanit.*, 11, 32, 1946.
195. **Misiurenko, I. P.**, Formation of toxin by deep cultivation of *Fusarium sporotrichiella* Bilai, in Proc. Symp. Mycotoxins, Academy of Sciences of Ukrainian S.S.R., Kiev, October 3 to 9, 1972, 17.
196. **Mitiukevich, N. A.**, On the Problem of Alimentary Toxic Aleukia Foci and Methods for their Detection, Lecture on Republic Conference of Alimentary Toxic Aleukia, Moscow, January 1947.
197. **Murashkinskij, K. E.**, On the study of fusariosis of cereal crops, Species of genus *Fusarium* on cereal crops in Siberia, *Proc. Siber. Agric. Acad. (Osmk.)*, 3, 87, 1934.
198. **Myasnikov, A. L.**, Clinical aspects of alimentary hemorrhagic aleukia, in Alimentary Hemorrhagic Aleukia (Septic Angina), Western Siberia Territorial Public Health Service, Novosibirsk, 1935, 48.
199. **Myasnikov, V. A.**, Characteristics of extracts of various fungi treated by the barium method, compared to extracts of prosomillet exposed in water, *Chkalov Inst. Epidemiol. Microbiol.*, 1948.
200. **Myasnikov, V. A.**, Experiments in destruction and neutralization of toxic material from overwintered prosomillet, in Alimentary Toxic Aleukia, Acta Chkalov Inst. Epidemiol. Microbiol., Chkalov, 2, 55, 1947.
201. **Nakhapetov, M. I.**, *Septic Angina*, Medgiz, Leningrad, 1944, 14.
202. **Nesterov, V. S.**, The clinical aspects of septic angina, *Clin. Med.*, 7, 34, 1948.
203. **Okuniev, N. V.**, New Data on the Chemical Nature of the Initial Toxicity of Cereal Grain Causing Alimentary Toxic Aleukia, Abstr. Republic Conference on Alimentary Toxic Aleukia, Moscow, 1945, 15.
204. **Olifson, L. E.**, Chemical composition of millet grain overwintered in the field and spoiled in other conditions, Monitor, Orenburg Sect. of the USSR, *D. J. Mendelijev Chem. Soc.*, 4, 24, 1951.
205. **Olifson, L. E.**, The Chemical Activity of *Fusarium sporotrichiella*, Abstr. on Mycotoxicoses in Human and Agricultural Animals, Academy of Sciences of Ukranian S.S.R., Kiev, 1956, 22.
206. **Olifson, L. E.**, Toxins isolated from overwintered cereals and their chemical nature, Monitor, Orenburg Sect. of the U.S.S.R., *D. J. Mendeleyev Chem. Soc.*, 7, 21, 1957.
207. **Olifson, L. E.**, Chemical action of some fungi on overwintered cereals, Monitor, Orenburg Sect. of the U.S.S.R., *D. J. Mendeleyev Chem. Soc.*, 7, 21, 1957.

208. **Olifson, L. E.**, New Chemical Methods of Determining Toxicity of Cereal Crops, Commun. Sci. Works of Members D. J. Mendeleyev All Soviet Chem. Soc., Academy of Sciences of U.S.S.R., Moscow, 1955, 58.
209. **Olifson, L. E.**, The chemical activity of some species of fungi which affect the grain of cereals, in Mycotoxicoses of Man and Agricultural Animals, Bilai, V. I., Ed., Office of Technical Services, Washington, D.C., 1960, 58.
210. **Olifson, L. E.**, Spectrographic investigation of biologically active compounds from *Fusarium sporotrichioides*, in Annual Report of Mendeleyev, All Soviet Union Chem. Association, Moscow, 1962, 109.
211. **Olifson, L. E.**, Chemical and Biological Properties of Toxic Materials Derived from Grain Infected with the Fungus *Fusarium sporotrichiella*, Abstr. Dissertation, Moscow Technological Institute Industrial Nutrition, 1965, 1—36.
212. **Olifson, L. E.**, The problems of toxic steroids in microscopic fungi, *Fusarium sporotrichiella*, in Proc. Symp. Mycotoxins, Academy Science Ukraine S.S.R., Kiev, 1972, 12.
213. **Olifson, L. E., Drabkin, B. S., and Joffe, A. Z.**, The influence of *Fusarium* fungi on millet oil, *Acta Chkalov Med. Inst.*, 2, 103, 1950.
214. **Olifson, L. E. and Joffe, A. Z.**, On the changes in some chemical constants of millet oil under the influence of fungi developing on millet, Monitor, Orenburg Sect. of the U.S.S.R., *D. J. Mendeleyev Chem. Soc.*, 5, 61, 1954.
215. **Olifson, L. E., Kenina, S. M., and Kartashova, V. L.**, Chromatographic method to identify toxicity of grain of cereals (wheat, rye, millet and others) affected by a toxigenic strain of *Fusarium sporotrichiella*, Instruction on how to identify the toxicity of the grain, Monitor, Orenburg Sect. of the U.S.S.R., *D. J. Mendeleyev Chem. Soc.*, p. 3, 1972.
216. **Onegov, A. P. and Naumov, V. A.**, Toxicity of overwintered cereals, *Acta Kirov Zootechnol. Vet. Inst.*, 5, 110, 1943.
217. **Palyusik, M., Szep, I., and Szoke, F.**, Data on susceptibility of day-old goslings, *Acta Vet. Hung.*, 18, 363, 1968.
218. **Palyusik, M. and Karlic-Kovacs, E.**, Effect on laying geese of feeds containing the fusariotoxins T-2 and F-2, *Acta Vet. Acad. Hung. Tomas*, 25, 363, 1975.
219. **Pentman, I. S.**, Pathologic Anatomy of Alimentary Hemorrhagic Aleukia (Septic Angina), West-Siber Region, Health, Novosibirsk, 1935, 17.
220. **Peregud, G. M.**, Clinical aspects and treatment of oral cavity and upper respiratory tract in cases of alimentary toxic aleukia, in Alimentary Toxic Aleukia, Acta Chkalov Inst. Epidemiol., Microbiol., 2, 170, 1947.
221. **Perkel, N. V.**, Study of the toxicity of Transbaikal strains of *Fusarium sporotrichiella* Bilai in connection with the etiology of Novv disease (Kaschin-Beck Disease) in Mycotoxicoses of Man and Agricultural Animals, Bilai, V. I., Ed., U.S. Joint Public Research Service, Washington, D.C., 1960, 117.
222. **Perstneva, A. P.**, On some peculiarities of the clinical aspects of alimentary hemorrhagic aleukia (septic angina), *Clin. Med.*, 27, 85, 1949.
223. **Pidoplichka, N. M. and Bilai, V. I.**, Toxic Fungi on Cereal Grains, Academy of Sciences of Ukrainian S.S.R., Kiev, 1946, 1—65.
224. **Pidoplichka, N. M. and Bilai, V. I.**, Toxic fungi which develop in food products and fodder, in Mycotoxicoses of Man and Agricultural Animals, Bilai, V. I., Ed., Office of Technical Services, Washington, D.C., 1960, 3.
225. **Plowright, W., Linsell, C. A., and Peers, F. G.**, A focus of rumenal cancer in Kenya cattle, *Br. J. Cancer*, 25, 72, 1971.
226. **Poliantseva, A. I.**, Residual changes in the internal organs and blood in convalescents from alimentary toxic aleukia caused by ingestion of overwintered millet and rye, in *Acta of Septic Angina*, Bashkir Publishing, Ufa, 1945, 154.
227. **Popova, A. A.**, Alimentary toxic aleukia in some counties of Kirovsk district in 1942—1946, *Acta Kirovsk Inst. Epidemiol. Microbiol.*, 2, 77, 1948.
228. **Pouchert, C. J. and Campbell, J. R.**, The Aldrich Library of N. M. R. spectra, 1974.
229. **Pouchert, C. J.**, The Aldrich Library of Infrared Spectra, 1975.
230. **Poznanski, A. S.**, Neuropsychic disturbances in alimentary toxic aleukia, in Alimentary Toxic Aleukia, Acta Chkalov Inst. Epidemiol. Microbiol., Chkalov, 2, 176, 1947.
231. **Prentice, N. and Dickson, A. D.**, Emetic material associated with *Fusarium* species in cereal grains and artificial media, *Biotech. Bioeng.*, 10, 413, 1968.
232. **Puls, R. and Greenway, J. A.**, Fusariotoxicosis from barley in British Columbia. II. Analysis and toxicity of suspected barley, *Can. J. Comp.*, 40, 16, 1976.
233. **Radkevich, P. E.**, Poisoning with overwintered grain, in Veterinary Toxicology, State Agricultural Literature, Moscow, 1952, 119.

234. **Raillo, A. I.**, Fungi of the Genus *Fusarium*, State Agricultural Literature, Moscow, 1950.
235. **Razumov, M. I. and Rubinstein, Yu. I.**, Experimental food mycotoxic enchondral osteodystrophy (on the etiology of Kashin-Beck Disease), *Probl. Nutr.*, 1, 227, 1951.
236. **Reisler, A. V.**, *Septic Angina*, Medical House, Moscow, 1943, 19.
237. **Reisler, A. V.**, Alimentary toxic aleukia, in Hygiene of Nutrition, State Medical Publishing, Moscow, 1952, 402.
238. **Romanova, E. D.**, Clinical observation and therapy of alimentary toxic aleukia, in Alimentary Toxic Aleukia, Acta Chkalov. Inst. Epidemiol. Microbiol, Chkalov, 2, 164, 1947.
239. **Rubinstein, Yu, I.**, Microflora of Cereals Overwintered Under Snow, Proc. Nutr. Inst. Acad. Med. Sci. U.S.S.R., Moscow, 1948, 29.
240. **Rubinstein, Yu., I.**, Biochemical properties of *Fusarium* cultures (sect. *Sporotrichiella), Microbiology*, 19, 438, 1950.
241. **Rubinstein, In., I.**, On the problem of digestive mycotoxicoses, *News Med. (Moscow)*, 2, 30, 1951.
242. **Rubinstein, Yu, I.**, Some properties of the *Fusarium sporotrichioides* toxin, *Acta Acad. Med. Sci. U.S.S.R. Nutr. Probl.*, 13, 247, 1951.
243. **Rubinstein, Yu., I.**, The etiology of Urovsk disease, *Nutr. Probl.*, 12, 73, 1953.
244. **Rubinstein, Yu., I.**, Mycotoxicoses of men and the problems of their further investigation, in Conf. on Mycotoxicoses in Humans and Agric. Anim., Academy of Sciences of Ukrainian S.S.R., Kiev, 1956, 10.
245. **Rubinstein, Yu., I.**, Actual problems in study of fusariotoxicoses, *Nutr. Probl.*, 15, 8, 1956.
246. **Rubinstein, Yu., I.**, Food fusariotoxicoses, in *Mycotoxicoses of Man and Agricultural Animals*, Bilai, V. I., Ed., U.S. Joint Publ. Res. Serv., Washington, D.C., 1960b, 89.
247. **Rubinstein, Yu., I.**, The effect of the cultivation conditions on toxin formation in *Fusarium sporotrichiella* in Mycotoxicoses of Man and Agricultural Animals, Bilai, V. I., Ed., Office of Technical Services, Washington, D.C., 1960, 66.
248. **Rubinstein, Yu., I., Kykel, Yu., and Kudinova, G.**, New experimental chronic toxicosis caused by *Fusarium sporotrichiella*, Abstr. Lect. II, Conf. on Mycotoxicoses, Kiev, 1961, 16.
249. **Rubinstein, Yu., I. and Lyass, L. S.**, On the etiology of alimentary toxic aleukia (septic angina), *Hyg. Sanit.*, 7, 33, 1948.
250. **Ryazanov, B. A.**, Lecture on the Republic Conference on Alimentary Toxic Aleukia, Moscow, December 1947.
251. **Ryazanov, B. A.**, Lecture on the Republic Conference on Alimentary Toxic Aleukia, Moscow, March 1948.
252. **Sarkisov, A. Kh.**, Method of determining toxicity of cereal crops overwintered in the field, *Hyg. Sanit.*, 9, 19, 1944.
253. **Sarkisov, A. Kh.**, Data on toxicity of cereals overwintered under snow, in Cereal Crops Overwintered under Snow, Ministry of Agriculture of U.S.S.R., Moscow, 1948, 40.
254. **Sarkisov, A. Kh.**, Etiology of "septic angina" in man, *J. Microbiol. Epidemiol. Immunol.*, 1, 43, 1950.
255. **Sarkisov, A. Kh.**, Mycotoxicoses, Ministry of Agriculture of U.S.S.R., Moscow, 1954, 1—216.
256. **Sarkisov, A. Kh., Korneyev, N. E., Kvashnina, E. S., Koroleva, V. P., Gerasimova, P. A., and Akulova, N. S.**, The harm caused by overwintered cereal crops to livestock and poultry, in Cereal Crops Wintered under Snow, Public Ministry Agriculture U.S.S.R., Moscow, 1948, 10.
257. **Sarkisov, A. Kh. and Kvashnina, E. S.**, Toxico-biological properties of *Fusarium sporotrichioides*, in Cereal Crops Wintered under Snow, Ministry of Agriculture of U.S.S.R., Moscow, 1948, 86.
258. **Sarkisov, A. Kh.**, General characterization of the alimentary mycotoxicoses of agricultural animals, in Mycotoxicoses of Man and Agricultural Animals, Bilai, V. I., Ed., Office of Technical Services, Washington, D.C., 1960, 155.
259. **Sato, N., Ueno, Y., and Enomoto, M.**, Toxicological approaches to the toxic metabolites of *Fusaria*, Acute and subacute toxicities of T-2 toxin in cats, *Jpn. J. Pharmacol.*, 25, 263, 1975.
260. **Schoental, R. and Joffe, A. Z.**, Lesions induced in rodents by extracts from cultures of *Fusarium poae* and *F. sporotrichioides, J. Pathol.*, 112, 37, 1974.
261. **Schoental, R.**, Mycotoxicoses "by proxy," *Int. J. Environ. Stud.*, 8, 171, 1975.
262. **Schoental, R., Joffe, A. Z., and Yagen, B.**, Chronic lesions in rats treated with crude extracts of *Fusarium poae* and *F. sporotrichioides;* the role of mouldy food in the incidence of aesophageal, mammary and certain other abnormalities and tumors in livestock and man, *Br. J. Cancer*, 34 (Abstr.), 310, 1976.
263. **Schoental, R.**, The role of nicotinamide and certain other modifying factors in diethylnitrosamine carcinogenesis, *Fusaria* mycotoxins and "spontaneous" tumours in animals and man, *Cancer*, 401, 1833, 1977.
264. **Schoental, R., Joffe, A. Z., and Yagen, B.**, The induction of tumours of the digestive tract and of certain other organs in rats given T-2 toxin, a secondary metabolite of *Fusarium* sporotrichioides, *Br. J. Cancer*, 38 (Abstr.,) 171, 1978.

265. **Schoental, R.**, The role of *Fusarium* mycotoxins in the aetiology of tumors of digestive tract and of certain organs in man and animals, *Front. Gastrointest. Res.*, 4, 17, 1978.
266. **Schoental, R.**, Mycotoxins in food and the variations in tumour incidence among laboratory rodents, *Nutr. Cancer*, 1, 12, 1978.
267. **Schoental, R.**, The role of *Fusarium* mycotoxins in the aetiology of tumours of the digestive tract and of certain other organs in man and animals, *Front. Gastrointest. Res.*, 4, 17, 1979.
268. **Schoental, R.**, *Fusarium* mycotoxins in the aetiology of neonatal abnormalities. Cardiovascular and sexual disorders and tumours in man and animals, *Toxicon*, 17 (Suppl. 1), 164, 1979.
269. **Schoental, R.**, Possible public health significance of *Fusarium* toxins, in Proc. 3rd Meet. Mycotoxins in Anim. Dis., Agric. Dev. Advisory Ser., *Ministry of Agriculture, Fisheries and Food*, Weybridge, 1978, 67.
270. **Schoental, R.**, Retationships of *Fusarium* mycotoxins to disorders and tumours associated with alcoholic drinks, *Nutr. Cancer*, 2, 88, 1980.
271. **Schoental, R.**, Mouldy grain and the aetiology of pellagra. The role of toxic metabolites of *Fusarium*, *Biochem. Soc. Trans.*, 8, 147, 1980.
272. **Schoental, R.**, Mycotoxins and fetal abnormalities, *Int. J. Environ. Stud.*, 17, 25, 1981.
273. **Schoental, R.**, Relationships of *Fusarium* toxins to tumours and other disorders in livestock, *J. Vet. Pharmacol. Ther.*, 4, 1, 1981.
274. **Schoental, R., Joffe, A. Z., and Yagen, B.**, Comparison of the effects of neosolaniol, a trichothecene metabolite of *Fusarium* species, with those observed in rodents given T-2 toxin, *Br. J. Cancer*, 40 (Abstr.), 301, 1979.
275. **Schoental, R., Joffe, A. Z., and Yagen, B.**, Irreversible depigmentation of dark hair by T-2 toxin (a metabolite of *Fusarium sporotrichioides*) and by calcium pantothenate, *Experientia*, 34, 763, 1978.
276. **Schoental, R., Joffe, A. Z., and Yagen, B.**, Cardiovascular lesions and various tumours found in rats given T-2 toxin, a trichothecene metabolite of *Fusarium* species, *Cancer Res.*, 39, 2279, 1979.
277. **Scott, P. M. and Somers, E.**, Biologically active compounds from field fungi, *J. Agric. Food Chem.*, 17, 430, 1969.
278. **Seemüller, E.**, Untersuchungen über die morphologische und biologische Differenzierung in der *Fusarium*, Section *Sporotrichiella. Mitt. Biol. Bundesanst. Berlin-Dahlem*, 127, 1, 1968.
279. **Sergiev, P. G.**, Epidemiology of alimentary toxic aleukia, in *Alimentary Toxic Aleukia (Septic Angina)*, Nesterov, A. I., Sysin, A. N., and Karlic, L. I., Eds., State Medical Literature, Medgiz, 1945, 7—11.
280. **Sergiev, P. G.**, Harvesting losses and overwintered grain in the field-source of disease, *Kolkhozn. Proizvod.*, 7, 16, 1946.
281. **Sergiev, P. G.**, Septic angina, in Artic. Lect. Conversat. on Sanit., Instructive Subjects, Barnaul, 1948, 62.
282. **Serafimov, B. M.**, Mental disorder, caused by alimentary toxic aleukia, *Neuropathol. Psychiatry (Moscow)*, 5, 25, 1945.
283. **Serafimov, B. N.**, Mental symptomatology of alimentary toxic aleukia, *Neuropathol. Psychiatry (Moscow)*, 6, 50, 1946.
284. **Shalman, L. B.**, Observation of the clinical aspects and epidemiology by "Septic Angina", in *Alimentary Toxic Aleukia (Septic Angina)*, Malkin, Z. I., Ed., Tatar State Publishers, Kazan, 1945, 61.
285. **Shimshelevitch, S. B. and Dubniakova, A. M.**, Clinico-Hematological Characteristics of Alimentary Toxic Aleukia, Abstr. on Republic Conference of Alimentary Toxic Aleukia, Moscow, December 24 to 26, 1945, 13—14.
286. **Shklovskaja, R. S. and Brodskaja, F. P.**, The so-called "septic angina", in *Date Scient.-Pract. Works of Town Sterlitamak Physicians*, Lurie, G. S., Ed., Sterlitamak, Bashkir, S.S.R., 1944, 5.
287. **Sirotinina, O. N.**, Toxicity of Cereals Overwintered in the Field, Dissertation Saratov, 1945, 232.
288. **Slonevski, S. I.**, Alimentary toxic aleukia (Septic Angina) in Udmurt A.S.S.R., *Hyg. Sanit.*, 6, 23, 1946.
289. **Smalley, E. B.**, T-2 toxin, *J. Am. Vet. Med. Assoc.*, 163, 1278, 1973.
290. **Smalley, E. B., Marasas, W. F. O., Strong, F. M., Bamburg, J. R., Nichols, R. E., and Kosuri, N. R.**, Mycotoxicoses associated with moldy corn, in Toxic Microorganisms, Herzberg, M., Ed., U.S. Department of the Interior, Washington, D.C., 1970, 163.
291. **Smalley, E. B. and Strong, F. M.**, Toxic trichothecenes, in *Mycotoxins*, Purchase, I. F. H., Ed., Elsevier, Amsterdam, 1974, 435.
292. **Smirnova, V. A.**, Long-term consequences of septic angine in laryngeal organs, in Data on *Septic Angina*, Bashkin Publishing, Ufa, 1945, 145.
293. **Snyder, W. C. and Hansen, H. N.**, The species concept in *Fusarium* with reference to Discolor and other sections, *Am. J. Bot.*, 32, 657, 1945.
294. **Snyder, W. C. and Toussoun, T. A.**, Current status of taxonomy in *Fusarium* species and their perfect stages, *Phytopathology*, 55, 833, 1965.

295. **Speers, G. M., Mirocha, C. J., and Christensen, C. M.**, Effect of feeding *F. tricinctum* and *F. roseum* isolate "oxyrose" invaded corn and the purified T-2 mycotoxin on S. C. W. L. hens, *Poultry Sci.*, 52, 2088, 1973.
296. **Speers, G. M., Mirocha, C. J., Christensen, C. M., and Behrem, J. C.**, Effect on laying hens of feeding corn invaded by two species of *Fusarium* and pure T-2 mycotoxin, *Poultry Sci.*, 56, 92, 1977.
297. **Specivceva, N. H.**, Fusariotoxicoses of poultry, in *Mycoses and Mycotoxicoses*, Kolos, Moscow, 1964, 435.
298. **Specivceva, N. H.**, Fusariotoxicosis in horses, *Proc. All Union Inst. Vet. Sanit.*, 28, 11, 1967.
299. **Sprague, R.**, *Disease of Cereals and Grasses in North America*, Ronald Press, New York, 1950.
300. **Stähelin, H., Kalberer, M. E., Kalberer-Rush, Y., Signer, E., and Lazary, S.**, Über einige biologische Wirkungen des Cytostaticum Diacetoxyscirpenol, *Arzneim. Forsch.*, 18, 989, 1968.
301. **Strukov, A. and Tishchenko, M.**, Pathomorphology of septic angina, in *Data on Septic Angina, Proc. First Kharkov Med. Inst. a. Chkalov Inst. Epidemiol. Microbiol.*, Chkalov, 1, 53, 1944.
302. **Strukov, A. I.**, Pathological changes in animal tissues caused by toxin from overwintered cereals, in *Alimentary Toxic Aleukia, Acta Chkalov Inst. Epidemiol. Microbiol.*, Sec. communic. Chkalov, 2, 117, 1947.
303. **Strukov, A. I. and Mironov, S. G.**, Pathological changes in animal tissues caused by toxin from overwintered cereals, in *Alimentary Toxic Aleukia, Acta Chkalov Inst. Epidemiol. Microbiol.*, First communic. Chkalov, 2, 109, 1947.
304. **Strukov, A. I. and Tishchenko, M. A.**, Some suppositions on pathomorphology and pathogenesis of alimentary toxic aleukia, in *Alimentary Toxic Aleukia, Acta Chkalov Inst. Epidemiol. Microbiol.*, Chkalov, 2, 120, 1947.
305. **Svojskaja, E. D.**, Experiments in isolating toxic substances from overwintered prosomillet, in *Alimentary Toxic Aleukia, Acta Chkalov Inst.*, Epidemiol. Microbiol. Chkalov, 2, 45, 1947.
306. **Szathmary, Cs. I., Mirocha, C. J., Palyusik, M., and Pathre, S. V.**, Identification of mycotoxins produced by species of *Fusarium* and *Stachybotrys* obtained from Eastern Europe, *Appl. Environ. Microbiol.*, 32, 579, 1976.
307. **Talayev, B. T., Mogunov, B. I., and Sharbe, E. N.**, Alimentary hemorrhagic aleukia, *Nutr. Probl.*, 5, 27, 1936.
308. **Tatarinov, D. I.**, Problems to be discussed in the clinical aspects and therapy of alimentary toxic aleukia, in *Acta of Septic Angina*, Bashkir Publishing, Ufa, 1945, 12.
309. **Teregulov, G. N.**, Clinical aspects and treatment of alimentary toxic aleukia by propaeduetic-therapy materials of Bashkir Med. Inst., in *Data on Septic Angina*, Bashkir Publishing, Ufa, 1945, 30.
310. **Tkachev, T. Ya.**, The fight against septic angina, *Hyg. Sanit.*, 6, 24, 1945.
311. **Tomina, M. V.**, Distribution of Fungal Toxin from Overwintered Cereals in Animal Organs, Abstr. Republic Conference on Alimentary Toxic Aleukia, Moscow, 1948, 10.
312. **Tookey, H., Yates, S. G., Ellis, J. J., Grove, M. D., and Nochols, R. E.**, Toxic effects of a butenolide mycotoxin and of *Fusarium tricinctum* cultures in cattle, *J. Am. Vet. Med. Assoc.*, 60, 1522, 1972.
313. **Tostanovskaja, A. A.**, Toxicological characteristics of grain infected with *Fusarium sporotrichiella* var. *poae* and other fungi species from cereal crops in the U.S.S.R., in *Conference on Mycotoxicoses of Man and Husbandry Animals*, Academy of Sciences of Ukranian S.S.R., Kiev, 1958, 32.
314. **Tostanovskaja, A. A. and Ratmanskaja, U. M.**, A study in toxic grains infected with some fungi of the genus *Fusarium*, *Microbiol. J. Acad. Sci. Ukr. S.S.R.*, 13, 1, 1951.
315. **Toussoun, T. A., and Nelson, D. E.**, A Pictoral Guide to the Identification of *Fusarium* Species According to the Taxonomic System of Snyder and Hansen, Pennsylvania State University Press, 1968.
316. **Tsvetkov, I. I.**, On the incidence of alimentary toxic aleukia in urban settlements, in the Jubillee Proc., dedicated to 25 years of Astrakhansk Lunacharsk State Med. Inst., Astra Khan, 1946, 142.
317. **Ueno, Y., Ishii, K., Sakai, K., Kanaeda, S., Tsunoda, H., Tanaka, T., and Enomoto, M.**, Toxicological approaches to the metabolites of *Fusaria*. IV. Microbial survey on "Bean Hulls Poisoning of Horses" with isolation of toxic trichothecenes neosolaniol and T-2 toxin of *Fusarium solani* M-1-1, *Jpn. J. Exp. Med.*, 42, 187, 1972.
318. **Ueno, Y., Sato, N., Ishii, K., and Enomoto, M.**, Toxicological approaches to the metabolites of *Fusaria*. V. Neosolaniol, T-2 toxin and butenolide, toxic metabolites of *Fusarium sporotrichoides* NRRL 3510 and *Fusarium poae* 3287, *Jpn. J. Exp. Med.*, 42, 461, 1972.
319. **Ueno, Y., Sato, N., Ishii, K., Sakai, K., Tsunoda, H., and Enomoto, M.**, Biological and chemical detection of trichothecene mycotoxins of *Fusarium* species, *Appl. Microbiol.*, 25, 699, 1973.
320. **Ueno, Y., Sawano, M., and Ishii, K.**, Production of trichothecene mycotoxins by *Fusarium* species in shake culture, *Appl. Microbiol.*, 36, 4, 1975.
321. **Veindrach, G. M. and Fadeyeva, S. V.**, The blood-picture in septic granulocytic angina, *Kazan Med. J.*, 9, 1065, 1937.

322. **Vertinskij, K. I. and Adutskevich, V. A.**, Pathomorphologic studies of cats who died by experimental alimentary toxic aleukia, in Cereal Crops Wintered under Snow, Ministry of Agriculture of U.S.S.R., Moscow, 1948, 80.
323. **Wollenweber, H. W. and Reinking, O. A.**, *Die Fusarien, ihre Beschreibung, Schadwirkung und Bekampfung,* P. Parey, Berlin, 1935, 1—355.
324. **Wollenweber, H. W., Sherbakoff, C. D., Reinking, O. K., Johann, H., and Bailey, A. D.**, Fundamentals for taxonomic studies of *Fusarium, J. Agric. Res.,* 30, 833, 1925.
325. **Wollenweber, H. W.**, *Fusaria Autographice Delineata,* 1200 Fafeln. Selbstverlag., Berlin, 1916—1935.
326. **Wyatt, R. D., Harris, J. R., Hamilton, P. B., and Burmeister, H. R.**, Possible outbreaks of fusariotoxicosis in avians, *Avian Dis.,* 16, 1123, 1972.
327. **Wyatt, R. D., Weeks, B. A., Hamilton, P. B., and Burmeister, H. R.**, Several oral lesions in chickens caused by ingestion of dietary fusariotoxin T-2, *Appl. Microbiol.,* 24, 251, 1972.
328. **Wyatt, R. D., Colwell, W. M., Hamilton, P. B., and Burmeister, H. R.**, Neural disturbances in chickens caused by dietary T-2, *Appl. Microbiol.,* 26, 757, 1973.
329. **Wyatt, R. D., Hamilton, P. B., and Burmeister, H. R.**, The effect of T-2 toxin in broiler chickens, *Poultry Sci.,* 53, 1853, 1973.
330. **Wyatt, R. D., Doerr, J. A., Hamilton, P. B., and Burmeister, H. R.**, Egg production, shell thickness and other physiological parameters of laying hens affected by T-2 toxin, *Appl. Microbiol.,* 29, 641, 1975.
331. **Yagen, B. and Joffe, A. Z.**, Screening of toxic isolates of *Fusarium* poae and *f. sporotrichioides* involved in causing alimentary toxic aleukia, *Appl. Environ. Microbiol.,* 32, 423, 1976.
332. **Yagen, B., Joffe, A. Z., Horn, P., Mor, N., and Lutsky, I. I.**, Toxins from a strain involved in ATA, in *Mycotoxins in Human and Animal Health,* Rodricks, J. V., Hesseltine, C. W., and Mehlman, M. A., Pathotox Publishers, Park Forest South, Ill., 1977, 327.
333. **Yagen, B., Horn, P., Joffe, A. Z., and Cox, R. H.**, Isolation and structural elucidation of a novel sterol metabolite of *Fusarium sporotrichioides* 921, *J. Chem. Soc. Perkin I,* 2914, 1980.
334. **Yamazaki, M.**, Chemistry of mycotoxins, in *Toxicology, Biochemistry and Pathology of Mycotoxins,* Uraguchi, K. and Yamazaki, M., Eds., Kodansha, Tokyo, 1978, 65.
335. **Yates, S. G., Tookey, H. L., Eliss, J. J., Tallent, W. H., and Wolff, I. A.**, Mycotoxins as a possible cause of fescue toxicity, *J. Agric. Food Chem.,* 17, 437, 1969.
336. **Yates, S. G., Tookey, H. L., and Ellis, J. J.**, Survey of tall-fescue pasture: correlation of toxicity of *Fusarium* isolates to known toxins, *Appl. Microbiol.,* 19, 103, 1970.
337. **Yates, S. G.**, Toxin-production fungi from fescue posture, in *Microbial Toxins,* Vol. 7, Kadis, S., Ciegler, C., and Ajl, S. J., Eds., Academic Press, New York, 1971, 191.
338. **Yefremov, V. V.**, On the so-called alimentary toxic aleukia ("septic angina"), *Sov. Med.,* 1, 19, 1944.
339. **Yefremov, V. V.**, Alimentary toxic aleukia (septic angina), *Hyg. Sanit.,* 7, 18, 1944.
340. **Yefremov, V. V.**, *Alimentary Toxic Aleukia.* Medical Literature, Moscow, 1948, 120.
341. **Yudenich, V., Mironov, C., Soboleva, P., and Fok, R.**, The septic angina in the Chkalov district in 1944, in *Data on Septic Angina, Proc. First Kharkov Med. Inst. a. Chkalov Epidemiol. Microbiol.,* 1, 5, 1944.
342. **Zavyalova, A. P.**, Chemical and Toxicological Characteristics of Lipoproteins Isolated from Wheat, Causing Alimentary Toxic Aleukia, Dissertation, Kuybyshev, 1946, 164.
343. **Zhodzishkij, B. Ya.**, Data on the Study of Clinical Aspects, Pathogenesis and Alimentary Panhematopathy Treatment in So-Called Septic Angina (Agranulocytosis, Hemorrhagic Aleukia, Panmyelophthisis), Dissertation, Novosibirsk, 1933, 137.
344. **Zhukhin, V. A.**, Data on pathogenic anatomy and pathogenesis in so-called septic angina (alimentary toxic aleukia), in *Acta of Septic Angina,* Bashkir Publishing, Ufa, 1945, 82.

PARALYTIC SHELLFISH POISONING

Edward J. Schantz

Paralytic shellfish poisoning, in some areas called mussel poisoning, is a disease caused by eating shellfish such as clams or mussels that have been feeding on a poisonous marine dinoflagellate. Symptoms of the disease usually are apparent shortly after eating poisonous shellfish and begin with a tingling sensation and numbness in the lips, tongue, and finger tips followed by numbness in the legs, arms, and neck with general muscular incoordination. A feeling of lightness as though floating on air is often described by the afflicted persons. Other associated symptoms are dizziness, weakness, drowsiness, incoherence, and headache. As the illness progresses, respiratory distress and muscular paralysis become more and more severe. These poisons cause sickness and death by blocking specifically the sodium channels in nerve and muscle cell membranes.[2,4] The passage of sodium ions through these membranes into the cells is essential for the passage of an impulse in a normal manner along a nerve and muscle fiber. The diaphragm apparently is most sensitive to the poisons and death results from respiratory paralysis within 2 to 24 hr depending upon the magnitude of the dose. If one survives 24 hr, the prognosis is good and normal functions are regained within a few days. There is no effective antidote for shellfish poisoning and patients should be given medical treatment without delay. Artificial respiration is the recommended method of treatment if respiratory distress is apparent. The amount of poison to cause illness from eating poisonous shellfish may be as little as 1 mg and death may result from as little as 1 to 4 mg. This amount of poison could be contained in one small mussel.

The poisons that cause the disease are produced by certain species of marine dinoflagellates that sporadically occur in waters usually located in areas greater than 30° north or south latitude where the temperature ranges around 15 to 17°C. The observation that shellfish became poisonous by feeding on a certain dinoflagellate was first made by Sommer and associates at the University of California.[12] The two main species that produce the paralytic poisons are *Gonyaulax catenella*,[13] which occur most commonly along the north Pacific coast of North America from central California northward to Oregon, Washington, British Columbia, Alaska, and on to Japan and *G. tamarensis* var. *excavata*,[5] which occurs along the east coast of the New England states, the Maritime Provinces of Canada, and along the coasts of countries bordering the North Sea. When conditions such as temperature, pH, salinity, and food requirements become favorable for the growth of these particular organisms, they multiply and produce blooms containing anywhere from a few hundred cells (40 to 50 μm in diameter) to several thousand per milliliter. Blooms that contain 20,000 or more cells per milliliter produce what is called a "red tide". Another dinoflagellate, *G. acatenella*,[6] found in some areas along the coast of British Columbia produces several poisons. These dinoflagellates bloom to a maximum within 10 or 15 days, remain for a week or two and gradually disappear as other organisms bloom in their place. Shellfish feed on these organisms the same as other organisms but store the poison in the body organs such as the hepatopancreas in the case of mussels and some clams and in the siphon in the case of the Alaska butter clam. Poison stored in the hepatopancreas is usually excreted or destroyed within a week or two after the bloom of poisonous dinoflagellates has disappeared, but poison stored in the siphon is usually retained for many months after the dinoflagellate has disappeared.[8] Mussels and soft shell clams are usually safe to eat again within a week or two after the poisonous bloom has receded.

The paralytic poison in California sea mussels and Alaska butter clams was first

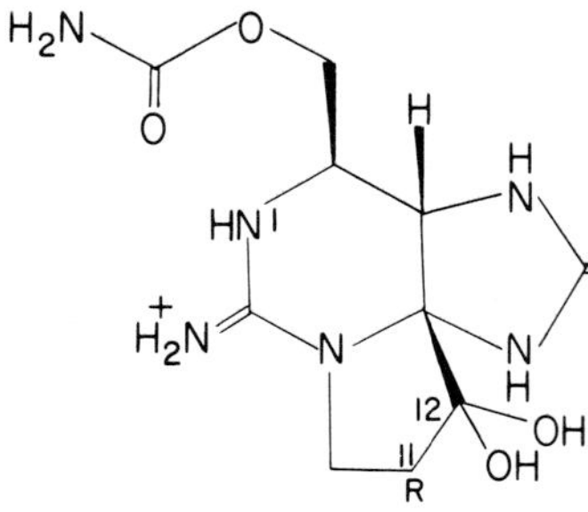

FIGURE 1. Structure of saxitoxin and related poisons. Saxitoxin, R = H.[10] 11-Hydroxysaxitoxin sulfate, R = OSO^-_3.[1] Neosaxitoxin, OH group at 1 replacing H.[11]

purified by chromatography on cation-exchange resins and finally by chromatography on acid washed alumina.[7] This poison, now called saxitoxin, is a white hygroscopic solid, very soluble in water, slightly soluble in methanol and ethanol, but insoluble in lipid solvents such as ethyl and petroleum ethers. The poison was also purified from cultures of *G. catenella* and found to be identical to the poison from the mussels and clams.[9] The specific toxicity of the purified poison was found to be 5.5×10^6 mouse lethal doses per gram making the mouse unit (MU) equivalent to 0.18 μg. Saxitoxin is the major poison produced by *G. catenella* and is a substituted dibasic tetrahydropurine (pK_a 8.2 and 11.5) with a molecular formula of $C_{10}H_{17}N_7O_4$ (mol wt 299). The structure of saxitoxin, shown in Figure 1, was determined by researchers at the University of Wisconsin in cooperation with researchers at Iowa State University by X-ray analyses of a crystalline derivative of saxitoxin formed with *p*-bromobenzene sulfonic acid.[10] The dihydroxy or hydrated ketone group on the five-membered ring (position 12, Figure 1) is essential for its poisonous activity of blocking the sodium channels in nerve and muscle cell membranes. Catalytic reduction of this group with hydrogen at atmospheric pressure and 20°C or reduction with sodium borohydride to a monohydroxy group eliminates the activity. Removal of the carbamyl group side chain on the six-membered ring with 7 *N* hydrochloric acid at 100°C, leaving a hydroxy group in its place, produces a molecule with about 60% of the original activity.[3]

The major poison produced by *G. tamarensis* var. *excavata* is a sulfonic acid ester of saxitoxin and causes paralytic poisoning in the same manner as saxitoxin. In this case the sulfonic acid group is on the five-membered ring at position 11 (designated by the letter R) as indicated in Figure 1.[1] The presence of the sulfonic acid group neutralizes the strongly basic guanidinum group of the molecule making the 11-hydroxysaxitoxin sulfate a neutral to slightly basic substance in contrast to saxitoxin.

Small amounts of other poisons, amounting to about 5% of the total toxicity, are found in extracts of the poisonous dinoflagellates and in shellfish that have consumed the poisonous dinoflagellates. One of these produced by *G. tamarensis* var. *excavata* is neosaxitoxin,[11] which has a hydroxyl group on the nitrogen at position 1 (see Figure 1). Some epimers of these poisons occur naturally and in most cases they are at least partially biologically active.[15]

An important factor that makes these poisons so dangerous in the seafood industry is that they are not destroyed to any extent by ordinary cooking, and poisonous shellfish are not distinguishable from nonpoisonous ones unless tested for toxicity with mice. For this reason they have caused many deaths and much sickness throughout areas where shellfish commonly are collected for food in home and commercial establishments. Interstate and intrastate commercial shipment of shellfish are checked for poison content by local and federal food and drug authorities. The assay for poison is carried out with mice. Serial dilutions of an extract of the shellfish are injected i.p. into white mice for the test. The dilution that will kill a 20-g mouse in 15 min is said to contain 1 MU of poison, equivalent to 0.18 μg. If no deaths occur with the undiluted extract within 15 min, the shellfish are safe for human consumption. The extract as described by the Food and Drug Administration is one part shellfish meats and one part acidified water at pH 2. Any shellfish meats containing more than 400 MU (80 μg) or more per 100 g are not marketable as designated by the U.S. Food and Drug Administration.[14]

REFERENCES

1. Boyer, G. L., Schantz, E. J., and Schnoes, H. K., *J. Chem. Soc. Chem. Commun.*, p. 888, 1978.
2. Evans, M. H., *Br. J. Pharmacol.*, 22, 478, 1964.
3. Ghazarossian, V. E., Ph.D. thesis, University of Wisconsin, Madison, 1977.
4. Kao, C. Y. and Nishiyama, A., *J. Physiol. (London)*, 180, 50, 1965.
5. Prakash, A., *J. Fish. Res. Board Can.*, 20, 983, 1963.
6. Prakash, A. and Taylor, J. J. R., *J. Fish. Res. Board Can.*, 23, 1265, 1966.
7. Schantz, E. J., Mold, J. D., Stanger, D. W., Shavel, J., Reil, F. J., Bowden, J. P., Lynch, J. M., Wyler, R. S., Riegel, B., and Sommer, H., *J. Am. Chem. Soc.*, 79, 5230, 1957.
8. Schantz, E. J. and Magnusson, H. W., *J. Protozool.*, 11, 239, 1964.
9. Schantz, E. J., Lynch, J. M., Vayada, G., Matsumoto, K., and Rapoport, H., *Biochemistry*, 5, 1191, 1966.
10. Schantz, E. J., Ghazarossian, V. E., Schnoes, H. K., Strong, F. M., Springer, J. P., Pezzanite, J. O., and Clardy, J., *J. Am. Chem. Soc.*, 97, 1238, 1975.
11. Shimizu, Y., Hsu, C., Fallon, W. E., Oshima, Y., Miura, I., and Nakanishi, K., *J. Am. Chem. Soc.*, 100, 6791, 1978.
12. Sommer, H. and Myer, K. F., *Am. Med. Assoc. Arch. Pathol.*, 24, 560, 1937.
13. Sommer, H., Whedon, W. F., Kofoid, C. A., and Stohler, R., *Am. Med. Assoc. Arch. Pathol.*, 24, 537, 1937.
14. U.S. Public Health Service, Manual of Recommended Practice for Sanitary Control of the Shellfish Industry, U.S. Department of Health, Education and Welfare, Washington, D.C., 1959.
15. Wichmann, C. F., Boyer, G. L., Divan, C. L., Schantz, E. J., and Schnoes, H. K., *Tetrahedron Lett.*, 22, 1941, 1981.

Index

INDEX

A

B

C

G

H

I

O

P

Q

R

S

T

U

V

W

X

Y

Z